Algorithms and Combinatorics 24

Editorial Board

R.L. Graham, La Jolla   B. Korte, Bonn
L. Lovász, Budapest   A. Wigderson, Princeton
G.M. Ziegler, Berlin

Springer
*Berlin
Heidelberg
New York
Hong Kong
London
Milan
Paris
Tokyo*

Alexander Schrijver

Combinatorial Optimization

Polyhedra and Efficiency

Volume B
Matroids, Trees, Stable Sets
Chapters 39 – 69

 Springer

Alexander Schrijver
CWI
Kruislaan 413
1098 SJ Amsterdam
The Netherlands
e-mail: lex@cwi.nl

Library of Congress Cataloging-in-Publication Data applied for

A catalog record for this book is available from the Library of Congress.

Bibliographic information published by Die Deutsche Bibliothek
Die Deutsche Bibliothek lists this publication in the Deutsche Nationalbibliografie;
detailed bibliographic data is available in the Internet at http://dnb.ddb.de

Mathematics Subject Classification (2000): 90C27, 05Cxx, 05C85, 68R05, 68Q25

ISSN 0937-5511
ISBN 3-540-44389-4 Springer-Verlag Berlin Heidelberg New York

This work is subject to copyright. All rights are reserved, whether the whole or part of the material is concerned, specifically the rights of translation, reprinting, reuse of illustrations, recitation, broadcasting, reproduction on microfilm or in any other way, and storage in data banks. Duplication of this publication or parts thereof is permitted only under the provisions of the German Copyright Law of September 9, 1965, in its current version, and permission for use must always be obtained from Springer-Verlag. Violations are liable for prosecution under the German Copyright Law.

Springer-Verlag Berlin Heidelberg New York
a member of BertelsmannSpringer Science+Business Media GmbH

http://www.springer.de

© Springer-Verlag Berlin Heidelberg 2003
Printed in Germany

The use of general descriptive names, registered names, trademarks etc. in this publication does not imply, even in the absence of a specific statement, that such names are exempt from the relevant protective laws and regulations and therefore free for general use.

Cover design: *design & production*, Heidelberg
Typesetting: by the author using a Springer LaTeX macro package
Printed on acid-free paper   46/3111ck-5 4 3 2 1

Table of Contents

Volume B

Part IV: Matroids and Submodular Functions	649

39	Matroids		651
	39.1	Matroids	651
	39.2	The dual matroid	652
	39.3	Deletion, contraction, and truncation	653
	39.4	Examples of matroids	654
		39.4a Relations between transversal matroids and gammoids	659
	39.5	Characterizing matroids by bases	662
	39.6	Characterizing matroids by circuits	662
		39.6a A characterization of Lehman	663
	39.7	Characterizing matroids by rank functions	664
	39.8	The span function and flats	666
		39.8a Characterizing matroids by span functions	666
		39.8b Characterizing matroids by flats	667
		39.8c Characterizing matroids in terms of lattices	668
	39.9	Further exchange properties	669
		39.9a Further properties of bases	671
	39.10	Further results and notes	671
		39.10a Further notes	671
		39.10b Historical notes on matroids	672

40	The greedy algorithm and the independent set polytope		688
	40.1	The greedy algorithm	688
	40.2	The independent set polytope	690
	40.3	The most violated inequality	693
		40.3a Facets and adjacency on the independent set polytope	698
		40.3b Further notes	699

41 Matroid intersection 700
- 41.1 Matroid intersection theorem 700
  - 41.1a Applications of the matroid intersection theorem 702
  - 41.1b Woodall's proof of the matroid intersection theorem .. 704
- 41.2 Cardinality matroid intersection algorithm 705
- 41.3 Weighted matroid intersection algorithm 707
  - 41.3a Speeding up the weighted matroid intersection algorithm 710
- 41.4 Intersection of the independent set polytopes 712
  - 41.4a Facets of the common independent set polytope 717
  - 41.4b Up and down hull of the common base polytope 719
- 41.5 Further results and notes 720
  - 41.5a Menger's theorem for matroids 720
  - 41.5b Exchange properties 721
  - 41.5c Jump systems 722
  - 41.5d Further notes 723

42 Matroid union 725
- 42.1 Matroid union theorem 725
  - 42.1a Applications of the matroid union theorem 727
  - 42.1b Horn's proof 729
- 42.2 Polyhedral applications 730
- 42.3 Matroid union algorithm 731
- 42.4 The capacitated case: fractional packing and covering of bases 732
- 42.5 The capacitated case: integer packing and covering of bases .. 734
- 42.6 Further results and notes 736
  - 42.6a Induction of matroids 736
  - 42.6b List-colouring 737
  - 42.6c Strongly base orderable matroids 738
  - 42.6d Blocking and antiblocking polyhedra 741
  - 42.6e Further notes 743
  - 42.6f Historical notes on matroid union 743

43 Matroid matching 745
- 43.1 Infinite matroids 745
- 43.2 Matroid matchings 746
- 43.3 Circuits 747
- 43.4 A special class of matroids 747
- 43.5 A min-max formula for maximum-size matroid matching 751
- 43.6 Applications of the matroid matching theorem 753
- 43.7 A Gallai theorem for matroid matching and covering 756
- 43.8 Linear matroid matching algorithm 757
- 43.9 Matroid matching is not polynomial-time solvable in general 762

		43.10	Further results and notes 763
			43.10a Optimal path-matching 763
			43.10b Further notes 764

44 Submodular functions and polymatroids 766
    44.1   Submodular functions and polymatroids 766
            44.1a Examples... 768
    44.2   Optimization over polymatroids by the greedy method 771
    44.3   Total dual integrality................................... 773
    44.4 f is determined by EP_f 773
    44.5   Supermodular functions and contrapolymatroids........... 774
    44.6   Further results and notes 775
            44.6a Submodular functions and matroids................. 775
            44.6b Reducing integer polymatroids to matroids 776
            44.6c The structure of polymatroids..................... 776
            44.6d Characterization of polymatroids 779
            44.6e Operations on submodular functions and polymatroids....................................... 781
            44.6f Duals of polymatroids............................. 782
            44.6g Induction of polymatroids 782
            44.6h Lovász's generalization of Kőnig's matching theorem.. 783
            44.6i Further notes 784

45 Submodular function minimization 786
    45.1   Submodular function minimization....................... 786
    45.2   Orders and base vectors 787
    45.3   A subroutine.. 787
    45.4   Minimizing a submodular function 789
    45.5   Running time of the algorithm 790
    45.6   Minimizing a symmetric submodular function 792
    45.7   Minimizing a submodular function over the odd sets 793

46 Polymatroid intersection................................... 795
    46.1   Box-total dual integrality of polymatroid intersection....... 795
    46.2   Consequences 796
    46.3   Contrapolymatroid intersection 797
    46.4   Intersecting a polymatroid and a contrapolymatroid........ 798
    46.5   Frank's discrete sandwich theorem 799
    46.6   Integer decomposition 800
    46.7   Further results and notes 801
            46.7a Up and down hull of the common base vectors 801
            46.7b Further notes 804

47 Polymatroid intersection algorithmically ... 805
- 47.1 A maximum-size common vector in two polymatroids ... 805
- 47.2 Maximizing a coordinate of a common base vector ... 807
- 47.3 Weighted polymatroid intersection in polynomial time ... 809
- 47.4 Weighted polymatroid intersection in strongly polynomial time ... 811
- 47.5 Contrapolymatroids ... 818
- 47.6 Intersecting a polymatroid and a contrapolymatroid ... 818
  - 47.6a Further notes ... 819

48 Dilworth truncation ... 820
- 48.1 If $f(\emptyset) < 0$ ... 820
- 48.2 Dilworth truncation ... 821
  - 48.2a Applications and interpretations ... 823
- 48.3 Intersection ... 825

49 Submodularity more generally ... 826
- 49.1 Submodular functions on a lattice family ... 826
- 49.2 Intersection ... 828
- 49.3 Complexity ... 829
- 49.4 Submodular functions on an intersecting family ... 832
- 49.5 Intersection ... 833
- 49.6 From an intersecting family to a lattice family ... 834
- 49.7 Complexity ... 835
- 49.8 Intersecting a polymatroid and a contrapolymatroid ... 837
- 49.9 Submodular functions on a crossing family ... 838
- 49.10 Complexity ... 840
  - 49.10a Nonemptiness of the base polyhedron ... 841
- 49.11 Further results and notes ... 842
  - 49.11a Minimizing a submodular function over a subcollection of a lattice family ... 842
  - 49.11b Generalized polymatroids ... 845
  - 49.11c Supermodular colourings ... 849
  - 49.11d Further notes ... 851

Part V: Trees, Branchings, and Connectors   853

50 Shortest spanning trees ... 855
- 50.1 Shortest spanning trees ... 855
- 50.2 Implementing Prim's method ... 857
- 50.3 Implementing Kruskal's method ... 858
  - 50.3a Parallel forest-merging ... 859
  - 50.3b A dual greedy algorithm ... 859
- 50.4 The longest forest and the forest polytope ... 860

	50.5	The shortest connector and the connector polytope 862
	50.6	Further results and notes 864
		50.6a Complexity survey for shortest spanning tree 864
		50.6b Characterization of shortest spanning trees 865
		50.6c The maximum reliability problem................... 866
		50.6d Exchange properties of forests..................... 867
		50.6e Uniqueness of shortest spanning tree 868
		50.6f Forest covers....................................... 869
		50.6g Further notes 870
		50.6h Historical notes on shortest spanning trees 871

51 Packing and covering of trees 877
    51.1 Unions of forests... 877
    51.2 Disjoint spanning trees 877
    51.3 Covering by forests 878
    51.4 Complexity 879
    51.5 Further results and notes 889
        51.5a Complexity survey for tree packing and covering..... 889
        51.5b Further notes 892

52 Longest branchings and shortest arborescences 893
    52.1 Finding a shortest r-arborescence 893
        52.1a r-arborescences as common bases of two matroids.... 895
    52.2 Related problems 895
    52.3 A min-max relation for shortest r-arborescences 896
    52.4 The r-arborescence polytope 897
        52.4a Uncrossing cuts 899
    52.5 A min-max relation for longest branchings 900
    52.6 The branching polytope 901
    52.7 The arborescence polytope................................ 901
    52.8 Further results and notes 902
        52.8a Complexity survey for shortest r-arborescence 902
        52.8b Concise LP-formulation for shortest r-arborescence .. 902
        52.8c Further notes 903

53 Packing and covering of branchings and arborescences 904
    53.1 Disjoint branchings 904
    53.2 Disjoint r-arborescences 905
    53.3 The capacitated case 907
    53.4 Disjoint arborescences................................... 908
    53.5 Covering by branchings.................................. 908
    53.6 An exchange property of branchings...................... 909
    53.7 Covering by r-arborescences............................. 911
    53.8 Minimum-length unions of k r-arborescences 913
    53.9 The complexity of finding disjoint arborescences 918

X   Table of Contents

    53.10 Further results and notes 921
        53.10a Complexity survey for disjoint arborescences 921
        53.10b Arborescences with roots in given subsets........... 923
        53.10c Disclaimers 925
        53.10d Further notes 926

54 Biconnectors and bibranchings............................ 928
    54.1 Shortest $R - S$ biconnectors 928
    54.2 Longest $R - S$ biforests 930
    54.3 Disjoint $R - S$ biconnectors............................. 931
    54.4 Covering by $R - S$ biforests............................. 934
    54.5 Minimum-size bibranchings 934
    54.6 Shortest bibranchings 935
        54.6a Longest bifurcations 937
    54.7 Disjoint bibranchings................................... 940
        54.7a Proof using supermodular colourings 943
        54.7b Covering by bifurcations........................... 943
        54.7c Disjoint $R - S$ biconnectors and $R - S$ bibranchings.. 944
        54.7d Covering by $R - S$ biforests and by $R - S$
             bifurcations....................................... 944

55 Minimum directed cut covers and packing directed cuts .. 946
    55.1 Minimum directed cut covers and packing directed cuts..... 946
    55.2 The Lucchesi-Younger theorem 947
    55.3 Directed cut k-covers................................... 949
    55.4 Feedback arc sets 951
    55.5 Complexity 953
        55.5a Finding a dual solution............................ 954
    55.6 Further results and notes 956
        55.6a Complexity survey for minimum-size directed cut
             cover... 956
        55.6b Feedback arc sets in linklessly embeddable digraphs .. 956
        55.6c Feedback vertex sets 958
        55.6d The bipartite case 959
        55.6e Further notes 960

56 Minimum directed cuts and packing directed cut covers .. 962
    56.1 Minimum directed cuts and packing directed cut covers..... 962
    56.2 Source-sink connected digraphs.......................... 964
    56.3 Other cases where Woodall's conjecture is true 967
        56.3a Further notes 968

57 Strong connectors 969
- 57.1 Making a directed graph strongly connected 969
- 57.2 Shortest strong connectors 970
- 57.3 Polyhedrally 973
- 57.4 Disjoint strong connectors 973
- 57.5 Complexity 975
  - 57.5a Crossing families 976

58 The traveling salesman problem 981
- 58.1 The traveling salesman problem 981
- 58.2 NP-completeness of the TSP 982
- 58.3 Branch-and-bound techniques 982
- 58.4 The symmetric traveling salesman polytope 983
- 58.5 The subtour elimination constraints 984
- 58.6 1-trees and Lagrangean relaxation 985
- 58.7 The 2-factor constraints 986
- 58.8 The clique tree inequalities 987
  - 58.8a Christofides' heuristic for the TSP 989
  - 58.8b Further notes on the symmetric traveling salesman problem 990
- 58.9 The asymmetric traveling salesman problem 992
- 58.10 Directed 1-trees 993
  - 58.10a An integer programming formulation 993
  - 58.10b Further notes on the asymmetric traveling salesman problem 994
- 58.11 Further notes on the traveling salesman problem 995
  - 58.11a Further notes 995
  - 58.11b Historical notes on the traveling salesman problem ... 996

59 Matching forests 1005
- 59.1 Introduction 1005
- 59.2 The maximum size of a matching forest 1006
- 59.3 Perfect matching forests 1007
- 59.4 An exchange property of matching forests 1008
- 59.5 The matching forest polytope 1011
- 59.6 Further results and notes 1015
  - 59.6a Matching forests in partitionable mixed graphs 1015
  - 59.6b Further notes 1017

60 Submodular functions on directed graphs 1018
- 60.1 The Edmonds-Giles theorem 1018
  - 60.1a Applications 1020
  - 60.1b Generalized polymatroids and the Edmonds-Giles theorem 1020
- 60.2 A variant 1021

	60.2a Applications 1023
	60.3 Further results and notes 1025
	60.3a Lattice polyhedra............................. 1025
	60.3b Polymatroidal network flows 1028
	60.3c A general model.............................. 1029
	60.3d Packing cuts and Győri's theorem 1030
	60.3e Further notes 1034

61 Graph orientation 1035
61.1 Orientations with bounds on in- and outdegrees 1035
61.2 2-edge-connectivity and strongly connected orientations ... 1037
   61.2a Strongly connected orientations with bounds on degrees.. 1038
61.3 Nash-Williams' orientation theorem 1040
61.4 k-arc-connected orientations of $2k$-edge-connected graphs .. 1044
   61.4a Complexity 1045
   61.4b k-arc-connected orientations with bounds on degrees.. 1045
   61.4c Orientations of graphs with lower bounds on indegrees of sets 1046
   61.4d Further notes 1047

62 Network synthesis 1049
62.1 Minimal k-(edge-)connected graphs 1049
62.2 The network synthesis problem 1051
62.3 Minimum-capacity network design 1052
62.4 Integer realizations and r-edge-connected graphs.......... 1055

63 Connectivity augmentation 1058
63.1 Making a directed graph k-arc-connected 1058
   63.1a k-arc-connectors with bounds on degrees 1061
63.2 Making an undirected graph 2-edge-connected............ 1062
63.3 Making an undirected graph k-edge-connected............ 1063
   63.3a k-edge-connectors with bounds on degrees 1066
63.4 r-edge-connectivity and r-edge-connectors 1067
63.5 Making a directed graph k-vertex-connected 1074
63.6 Making an undirected graph k-vertex-connected 1077
   63.6a Further notes 1078

Part VI: Cliques, Stable Sets, and Colouring   1081

64 Cliques, stable sets, and colouring 1083
- 64.1 Terminology and notation 1083
- 64.2 NP-completeness 1084
- 64.3 Bounds on the colouring number 1085
  - 64.3a Brooks' upper bound on the colouring number 1086
  - 64.3b Hadwiger's conjecture 1086
- 64.4 The stable set, clique, and vertex cover polytope.......... 1088
  - 64.4a Facets and adjacency on the stable set polytope 1088
- 64.5 Fractional stable sets 1090
  - 64.5a Further on the fractional stable set polytope 1091
- 64.6 Fractional vertex covers 1093
  - 64.6a A bound of Lorentzen 1095
- 64.7 The clique inequalities................................. 1095
- 64.8 Fractional and weighted colouring numbers 1096
  - 64.8a The ratio of $\chi(G)$ and $\chi^*(G)$.................... 1098
  - 64.8b The Chvátal rank 1098
- 64.9 Further results and notes 1099
  - 64.9a Graphs with polynomial-time stable set algorithm .. 1099
  - 64.9b Colourings and orientations 1101
  - 64.9c Algebraic methods............................... 1102
  - 64.9d Approximation algorithms 1103
  - 64.9e Further notes 1104

65 Perfect graphs: general theory 1106
- 65.1 Introduction to perfect graphs......................... 1106
- 65.2 The perfect graph theorem............................. 1108
- 65.3 Replication 1109
- 65.4 Perfect graphs and polyhedra 1110
  - 65.4a Lovász's proof of the replication lemma............ 1111
- 65.5 Decomposition of Berge graphs 1112
  - 65.5a 0- and 1-joins 1112
  - 65.5b The 2-join 1113
- 65.6 Pre-proof work on the strong perfect graph conjecture..... 1115
  - 65.6a Partitionable graphs 1116
  - 65.6b More characterizations of perfect graphs........... 1118
  - 65.6c The stable set polytope of minimally imperfect graphs 1118
  - 65.6d Graph classes 1120
  - 65.6e The P_4-structure of a graph and a semi-strong perfect graph theorem........................... 1122
  - 65.6f Further notes on the strong perfect graph conjecture 1123

	65.7	Further results and notes 1125

       65.7 Further results and notes 1125
             65.7a Perz and Rolewicz's proof of the perfect graph
                    theorem 1125
             65.7b Kernel solvability 1126
             65.7c The amalgam 1130
             65.7d Diperfect graphs................................... 1131
             65.7e Further notes 1133

66 Classes of perfect graphs 1135
   66.1 Bipartite graphs and their line graphs 1135
   66.2 Comparability graphs 1137
   66.3 Chordal graphs.. 1138
        66.3a Chordal graphs as intersection graphs of subtrees of
              a tree 1142
   66.4 Meyniel graphs 1143
   66.5 Further results and notes 1145
        66.5a Strongly perfect graphs............................. 1145
        66.5b Perfectly orderable graphs 1146
        66.5c Unimodular graphs 1147
        66.5d Further classes of perfect graphs................... 1148
        66.5e Further notes 1149

67 Perfect graphs: polynomial-time solvability 1152
   67.1 Optimum clique and colouring in perfect graphs
        algorithmically 1152
   67.2 Weighted clique and colouring algorithmically 1155
   67.3 Strong polynomial-time solvability 1159
   67.4 Further results and notes 1159
        67.4a Further on $\vartheta(G)$ 1159
        67.4b The Shannon capacity $\Theta(G)$ 1167
        67.4c Clique cover numbers of products of graphs 1172
        67.4d A sharper upper bound $\vartheta'(G)$ on $\alpha(G)$ 1173
        67.4e An operator strengthening convex bodies 1173
        67.4f Further notes 1175
        67.4g Historical notes on perfect graphs.................. 1176

68 T-perfect graphs... 1186
   68.1 T-perfect graphs.. 1186
   68.2 Strongly t-perfect graphs 1187
   68.3 Strong t-perfection of odd-K_4-free graphs 1188
   68.4 On characterizing t-perfection 1194
   68.5 A combinatorial min-max relation 1196
   68.6 Further results and notes 1200
        68.6a The w-stable set polyhedron 1200
        68.6b Bidirected graphs................................... 1201

	68.6c	Characterizing odd-K_4-free graphs by mixing stable sets and vertex covers 1203
	68.6d	Orientations of discrepancy 1 1204
	68.6e	Colourings and odd K_4-subdivisions 1206
	68.6f	Homomorphisms 1207
	68.6g	Further notes 1207

69 Claw-free graphs... 1208
    69.1 Introduction 1208
    69.2 Maximum-size stable set in a claw-free graph............. 1208
    69.3 Maximum-weight stable set in a claw-free graph 1213
    69.4 Further results and notes 1216
        69.4a On the stable set polytope of a claw-free graph 1216
        69.4b Further notes 1217

Volume A

1 **Introduction** 1
  1.1 Introduction 1
  1.2 Matchings 2
  1.3 But what about nonbipartite graphs? 4
  1.4 Hamiltonian circuits and the traveling salesman problem 5
  1.5 Historical and further notes 6
      1.5a Historical sketch on polyhedral combinatorics 6
      1.5b Further notes 8

2 **General preliminaries** 9
  2.1 Sets 9
  2.2 Orders 11
  2.3 Numbers 11
  2.4 Vectors, matrices, and functions 11
  2.5 Maxima, minima, and infinity 14
  2.6 Fekete's lemma 14

3 **Preliminaries on graphs** 16
  3.1 Undirected graphs 16
  3.2 Directed graphs 28
  3.3 Hypergraphs 36
      3.3a Background references on graph theory 37

4 **Preliminaries on algorithms and complexity** 38
  4.1 Introduction 38
  4.2 The random access machine 39
  4.3 Polynomial-time solvability 39
  4.4 P 40
  4.5 NP 40
  4.6 co-NP and good characterizations 42
  4.7 Optimization problems 42
  4.8 NP-complete problems 43
  4.9 The satisfiability problem 44
  4.10 NP-completeness of the satisfiability problem 44
  4.11 NP-completeness of some other problems 46
  4.12 Strongly polynomial-time 47
  4.13 Lists and pointers 48
  4.14 Further notes 49
      4.14a Background literature on algorithms and complexity .. 49
      4.14b Efficiency and complexity historically 49

5 Preliminaries on polyhedra and linear and integer programming ... 59
- 5.1 Convexity and halfspaces ... 59
- 5.2 Cones ... 60
- 5.3 Polyhedra and polytopes ... 60
- 5.4 Farkas' lemma ... 61
- 5.5 Linear programming ... 61
- 5.6 Faces, facets, and vertices ... 63
- 5.7 Polarity ... 65
- 5.8 Blocking polyhedra ... 65
- 5.9 Antiblocking polyhedra ... 67
- 5.10 Methods for linear programming ... 67
- 5.11 The ellipsoid method ... 68
- 5.12 Polyhedra and NP and co-NP ... 71
- 5.13 Primal-dual methods ... 72
- 5.14 Integer linear programming ... 73
- 5.15 Integer polyhedra ... 74
- 5.16 Totally unimodular matrices ... 75
- 5.17 Total dual integrality ... 76
- 5.18 Hilbert bases and minimal TDI systems ... 81
- 5.19 The integer rounding and decomposition properties ... 82
- 5.20 Box-total dual integrality ... 83
- 5.21 The integer hull and cutting planes ... 83
  - 5.21a Background literature ... 84

Part I: Paths and Flows   85

6 Shortest paths: unit lengths ... 87
- 6.1 Shortest paths with unit lengths ... 87
- 6.2 Shortest paths with unit lengths algorithmically: breadth-first search ... 88
- 6.3 Depth-first search ... 89
- 6.4 Finding an Eulerian orientation ... 91
- 6.5 Further results and notes ... 91
  - 6.5a All-pairs shortest paths in undirected graphs ... 91
  - 6.5b Complexity survey ... 93
  - 6.5c Ear-decomposition of strongly connected digraphs ... 93
  - 6.5d Transitive closure ... 94
  - 6.5e Further notes ... 94

7 Shortest paths: nonnegative lengths ... 96
- 7.1 Shortest paths with nonnegative lengths ... 96
- 7.2 Dijkstra's method ... 97
- 7.3 Speeding up Dijkstra's algorithm with k-heaps ... 98

	7.4	Speeding up Dijkstra's algorithm with Fibonacci heaps	99
	7.5	Further results and notes	101
		7.5a Weakly polynomial-time algorithms	101
		7.5b Complexity survey for shortest paths with nonnegative lengths	103
		7.5c Further notes	105
8	**Shortest paths: arbitrary lengths**		**107**
	8.1	Shortest paths with arbitrary lengths but no negative circuits	107
	8.2	Potentials	107
	8.3	The Bellman-Ford method	109
	8.4	All-pairs shortest paths	110
	8.5	Finding a minimum-mean length directed circuit	111
	8.6	Further results and notes	112
		8.6a Complexity survey for shortest path without negative-length circuits	112
		8.6b NP-completeness of the shortest path problem	114
		8.6c Nonpolynomiality of Ford's method	115
		8.6d Shortest and longest paths in acyclic graphs	116
		8.6e Bottleneck shortest path	117
		8.6f Further notes	118
		8.6g Historical notes on shortest paths	119
9	**Disjoint paths**		**131**
	9.1	Menger's theorem	131
		9.1a Other proofs of Menger's theorem	133
	9.2	Path packing algorithmically	134
	9.3	Speeding up by blocking path packings	135
	9.4	A sometimes better bound	136
	9.5	Complexity of the vertex-disjoint case	137
	9.6	Further results and notes	138
		9.6a Complexity survey for the disjoint $s-t$ paths problem	138
		9.6b Partially disjoint paths	140
		9.6c Exchange properties of disjoint paths	140
		9.6d Further notes	141
		9.6e Historical notes on Menger's theorem	142
10	**Maximum flow**		**148**
	10.1	Flows: concepts	148
	10.2	The max-flow min-cut theorem	150
	10.3	Paths and flows	151
	10.4	Finding a maximum flow	151
		10.4a Nontermination for irrational capacities	152

	10.5	A strongly polynomial bound on the number of iterations ... 153
	10.6	Dinits' $O(n^2m)$ algorithm................................. 154
		10.6a Karzanov's $O(n^3)$ algorithm 155
	10.7	Goldberg's push-relabel method 156
	10.8	Further results and notes 159
		10.8a A weakly polynomial bound 159
		10.8b Complexity survey for the maximum flow problem ... 160
		10.8c An exchange property 162
		10.8d Further notes 162
		10.8e Historical notes on maximum flow 164

11 Circulations and transshipments 170
11.1 A useful fact on arc functions 170
11.2 Circulations.. 171
11.3 Flows with upper and lower bounds 172
11.4 b-transshipments 173
11.5 Upper and lower bounds on excess_f 174
11.6 Finding circulations and transshipments algorithmically 175
    11.6a Further notes 176

12 Minimum-cost flows and circulations..................... 177
12.1 Minimum-cost flows and circulations 177
12.2 Minimum-cost circulations and the residual graph D_f 178
12.3 Strongly polynomial-time algorithm 179
12.4 Related problems 182
    12.4a A dual approach................................. 183
    12.4b A strongly polynomial-time algorithm using
            capacity-scaling 186
12.5 Further results and notes 190
    12.5a Complexity survey for minimum-cost circulation 190
    12.5b Min-max relations for minimum-cost flows and
            circulations 191
    12.5c Dynamic flows 192
    12.5d Further notes 195

13 Path and flow polyhedra and total unimodularity 198
13.1 Path polyhedra....................................... 198
    13.1a Vertices, adjacency, and facets..................... 202
    13.1b The $s-t$ connector polytope 203
13.2 Total unimodularity................................... 204
    13.2a Consequences for flows 205
    13.2b Consequences for circulations 207
    13.2c Consequences for transshipments 207
    13.2d Unions of disjoint paths and cuts 210
13.3 Network matrices 213

	13.4 Cross-free and laminar families	214
14	**Partially ordered sets and path coverings**	**217**
	14.1 Partially ordered sets	217
	14.2 Dilworth's decomposition theorem	218
	14.3 Path coverings	219
	14.4 The weighted case	220
	14.5 The chain and antichain polytopes	221
	14.5a Path coverings algorithmically	222
	14.6 Unions of directed cuts and antichains	224
	14.6a Common saturating collections of chains	226
	14.7 Unions of directed paths and chains	227
	14.7a Common saturating collections of antichains	229
	14.7b Conjugacy of partitions	230
	14.8 Further results and notes	232
	14.8a The Gallai-Milgram theorem	232
	14.8b Partially ordered sets and distributive lattices	233
	14.8c Maximal chains	235
	14.8d Further notes	236
15	**Connectivity and Gomory-Hu trees**	**237**
	15.1 Vertex-, edge-, and arc-connectivity	237
	15.2 Vertex-connectivity algorithmically	239
	15.2a Complexity survey for vertex-connectivity	241
	15.2b Finding the 2-connected components	242
	15.3 Arc- and edge-connectivity algorithmically	243
	15.3a Complexity survey for arc- and edge-connectivity	246
	15.3b Finding the 2-edge-connected components	247
	15.4 Gomory-Hu trees	248
	15.4a Minimum-requirement spanning tree	251
	15.5 Further results and notes	252
	15.5a Ear-decomposition of undirected graphs	252
	15.5b Further notes	253

Part II: Bipartite Matching and Covering   257

16	**Cardinality bipartite matching and vertex cover**	**259**
	16.1 M-augmenting paths	259
	16.2 Frobenius' and Kőnig's theorems	260
	16.2a Frobenius' proof of his theorem	262
	16.2b Linear-algebraic proof of Frobenius' theorem	262
	16.2c Rizzi's proof of Kőnig's matching theorem	263
	16.3 Maximum-size bipartite matching algorithm	263
	16.4 An $O(n^{1/2}m)$ algorithm	264

16.5 Finding a minimum-size vertex cover 265
16.6 Matchings covering given vertices 265
16.7 Further results and notes 267
    16.7a Complexity survey for cardinality bipartite matching 267
    16.7b Finding perfect matchings in regular bipartite graphs 267
    16.7c The equivalence of Menger's theorem and Kőnig's theorem 275
    16.7d Equivalent formulations in terms of matrices 276
    16.7e Equivalent formulations in terms of partitions 276
    16.7f On the complexity of bipartite matching and vertex cover... 277
    16.7g Further notes 277
    16.7h Historical notes on bipartite matching............. 278

17 Weighted bipartite matching and the assignment problem 285
17.1 Weighted bipartite matching 285
17.2 The Hungarian method................................. 286
17.3 Perfect matching and assignment problems................ 288
17.4 Finding a minimum-size w-vertex cover.................. 289
17.5 Further results and notes 290
    17.5a Complexity survey for maximum-weight bipartite matching 290
    17.5b Further notes 290
    17.5c Historical notes on weighted bipartite matching and optimum assignment 292

18 Linear programming methods and the bipartite matching polytope.. 301
18.1 The matching and the perfect matching polytope 301
18.2 Totally unimodular matrices from bipartite graphs 303
18.3 Consequences of total unimodularity 304
18.4 The vertex cover polytope 305
18.5 Further results and notes 305
    18.5a Derivation of Kőnig's matching theorem from the matching polytope................................ 305
    18.5b Dual, primal-dual, primal?......................... 305
    18.5c Adjacency and diameter of the matching polytope ... 307
    18.5d The perfect matching space of a bipartite graph 308
    18.5e Up and down hull of the perfect matching polytope .. 309
    18.5f Matchings of given size 310
    18.5g Stable matchings 311
    18.5h Further notes 314

19 Bipartite edge cover and stable set ... 315
- 19.1 Matchings, edge covers, and Gallai's theorem ... 315
- 19.2 The Kőnig-Rado edge cover theorem ... 317
- 19.3 Finding a minimum-weight edge cover ... 317
- 19.4 Bipartite edge covers and total unimodularity ... 318
- 19.5 The edge cover and stable set polytope ... 318
  - 19.5a Some historical notes on bipartite edge covers ... 319

20 Bipartite edge-colouring ... 321
- 20.1 Edge-colourings of bipartite graphs ... 321
  - 20.1a Edge-colouring regular bipartite graphs ... 322
- 20.2 The capacitated case ... 322
- 20.3 Edge-colouring polyhedrally ... 323
- 20.4 Packing edge covers ... 324
- 20.5 Balanced colours ... 325
- 20.6 Packing perfect matchings ... 326
  - 20.6a Polyhedral interpretation ... 327
  - 20.6b Extensions ... 328
- 20.7 Covering by perfect matchings ... 329
  - 20.7a Polyhedral interpretation ... 330
- 20.8 The perfect matching lattice of a bipartite graph ... 331
- 20.9 Further results and notes ... 333
  - 20.9a Some further edge-colouring algorithms ... 333
  - 20.9b Complexity survey for bipartite edge-colouring ... 334
  - 20.9c List-edge-colouring ... 335
  - 20.9d Further notes ... 336

21 Bipartite b-matchings and transportation ... 337
- 21.1 b-matchings and w-vertex covers ... 337
- 21.2 The b-matching polytope and the w-vertex cover polyhedron ... 338
- 21.3 Simple b-matchings and b-factors ... 339
- 21.4 Capacitated b-matchings ... 341
- 21.5 Bipartite b-matching and w-vertex cover algorithmically ... 342
- 21.6 Transportation ... 343
  - 21.6a Reduction of transshipment to transportation ... 345
  - 21.6b The transportation polytope ... 346
- 21.7 b-edge covers and w-stable sets ... 347
- 21.8 The b-edge cover and the w-stable set polyhedron ... 348
- 21.9 Simple b-edge covers ... 349
- 21.10 Capacitated b-edge covers ... 350
- 21.11 Relations between b-matchings and b-edge covers ... 351
- 21.12 Upper and lower bounds ... 353
- 21.13 Further results and notes ... 355

           21.13a Complexity survey on weighted bipartite b-matching
                  and transportation................................. 355
           21.13b The matchable set polytope...................... 359
           21.13c Existence of matrices 359
           21.13d Further notes 361
           21.13e Historical notes on the transportation and
                  transshipment problems 362

22 **Transversals** 378
   22.1 Transversals 378
        22.1a Alternative proofs of Hall's marriage theorem 379
   22.2 Partial transversals 380
   22.3 Weighted transversals 382
   22.4 Min-max relations for weighted transversals 382
   22.5 The transversal polytope 383
   22.6 Packing and covering of transversals 385
   22.7 Further results and notes 387
        22.7a The capacitated case 387
        22.7b A theorem of Rado 389
        22.7c Further notes 389
        22.7d Historical notes on transversals.................. 390

23 **Common transversals**...................................... 393
   23.1 Common transversals 393
   23.2 Weighted common transversals 395
   23.3 Weighted common partial transversals.................... 397
   23.4 The common partial transversal polytope 399
   23.5 The common transversal polytope 401
   23.6 Packing and covering of common transversals 402
   23.7 Further results and notes 407
        23.7a Capacitated common transversals................... 407
        23.7b Exchange properties 407
        23.7c Common transversals of three families............. 408
        23.7d Further notes 409

Part III: Nonbipartite Matching and Covering   411

---

24 **Cardinality nonbipartite matching** 413
   24.1 Tutte's 1-factor theorem and the Tutte-Berge formula...... 413
        24.1a Tutte's proof of his 1-factor theorem 415
        24.1b Petersen's theorem................................. 415
   24.2 Cardinality matching algorithm 415
        24.2a An $O(n^3)$ algorithm 418
   24.3 Matchings covering given vertices 421

24.4 Further results and notes 422
    24.4a Complexity survey for cardinality nonbipartite matching 422
    24.4b The Edmonds-Gallai decomposition of a graph 423
    24.4c Strengthening of Tutte's 1-factor theorem........... 425
    24.4d Ear-decomposition of factor-critical graphs.......... 425
    24.4e Ear-decomposition of matching-covered graphs 426
    24.4f Barriers in matching-covered graphs 427
    24.4g Two-processor scheduling 428
    24.4h The Tutte matrix and an algebraic matching algorithm.. 429
    24.4i Further notes 430
    24.4j Historical notes on nonbipartite matching........... 431

25 The matching polytope 438
25.1 The perfect matching polytope 438
25.2 The matching polytope 439
25.3 Total dual integrality: the Cunningham-Marsh formula 440
    25.3a Direct proof of the Cunningham-Marsh formula 442
25.4 On the total dual integrality of the perfect matching constraints.. 443
25.5 Further results and notes 444
    25.5a Adjacency and diameter of the matching polytope ... 444
    25.5b Facets of the matching polytope 446
    25.5c Polynomial-time solvability with the ellipsoid method 448
    25.5d The matchable set polytope 450
    25.5e Further notes 452

26 Weighted nonbipartite matching algorithmically 453
26.1 Introduction and preliminaries 453
26.2 Weighted matching algorithm 454
    26.2a An $O(n^3)$ algorithm 456
26.3 Further results and notes 458
    26.3a Complexity survey for weighted nonbipartite matching 458
    26.3b Derivation of the matching polytope characterization from the algorithm 459
    26.3c Further notes 459

27 Nonbipartite edge cover 461
27.1 Minimum-size edge cover 461
27.2 The edge cover polytope and total dual integrality 462
27.3 Further notes on edge covers 464
    27.3a Further notes 464

		27.3b Historical notes on edge covers 464

28 Edge-colouring 465
28.1 Vizing's theorem for simple graphs 465
28.2 Vizing's theorem for general graphs 467
28.3 NP-completeness of edge-colouring 468
28.4 Nowhere-zero flows and edge-colouring 470
28.5 Fractional edge-colouring 474
28.6 Conjectures 475
28.7 Edge-colouring polyhedrally 477
28.8 Packing edge covers 478
28.9 Further results and notes 480
     28.9a Shannon's theorem 480
     28.9b Further notes 480
     28.9c Historical notes on edge-colouring 482

29 T-joins, undirected shortest paths, and the Chinese postman 485
29.1 T-joins 485
29.2 The shortest path problem for undirected graphs 487
29.3 The Chinese postman problem 487
29.4 T-joins and T-cuts 488
29.5 The up hull of the T-join polytope 490
29.6 The T-join polytope 491
29.7 Sums of circuits 493
29.8 Integer sums of circuits 494
29.9 The T-cut polytope 498
29.10 Finding a minimum-capacity T-cut 499
29.11 Further results and notes 500
     29.11a Minimum-mean length circuit 500
     29.11b Packing T-cuts 501
     29.11c Packing T-joins 507
     29.11d Maximum joins................................... 510
     29.11e Odd paths 515
     29.11f Further notes 517
     29.11g On the history of the Chinese postman problem 519

30 2-matchings, 2-covers, and 2-factors 520
30.1 2-matchings and 2-vertex covers 520
30.2 Fractional matchings and vertex covers 521
30.3 The fractional matching polytope 522
30.4 The 2-matching polytope 522
30.5 The weighted 2-matching problem 523
     30.5a Maximum-size 2-matchings and maximum-size
           matchings 524

30.6	Simple 2-matchings and 2-factors	526
30.7	The simple 2-matching polytope and the 2-factor polytope	528
30.8	Total dual integrality	531
30.9	2-edge covers and 2-stable sets	531
30.10	Fractional edge covers and stable sets	532
30.11	The fractional edge cover polyhedron	533
30.12	The 2-edge cover polyhedron	533
30.13	Total dual integrality of the 2-edge cover constraints	534
30.14	Simple 2-edge covers	535
30.15	Graphs with $\nu(G) = \tau(G)$ and $\alpha(G) = \rho(G)$	536
30.16	Excluding triangles	539
	30.16a Excluding higher polygons	544
	30.16b Packing edges and factor-critical subgraphs	544
	30.16c 2-factors without short circuits	545

31 b-matchings 546
- 31.1 b-matchings 546
- 31.2 The b-matching polytope 547
- 31.3 Total dual integrality 550
- 31.4 The weighted b-matching problem 554
- 31.5 If b is even 556
- 31.6 If b is constant 558
- 31.7 Further results and notes 559
  - 31.7a Complexity survey for the b-matching problem 559
  - 31.7b Facets and minimal systems for the b-matching polytope 559
  - 31.7c Regularizable graphs 560
  - 31.7d Further notes 561

32 Capacitated b-matchings 562
- 32.1 Capacitated b-matchings 562
- 32.2 The capacitated b-matching polytope 564
- 32.3 Total dual integrality 566
- 32.4 The weighted capacitated b-matching problem 567
  - 32.4a Further notes 567

33 Simple b-matchings and b-factors 569
- 33.1 Simple b-matchings and b-factors 569
- 33.2 The simple b-matching polytope and the b-factor polytope .. 570
- 33.3 Total dual integrality 570
- 33.4 The weighted simple b-matching and b-factor problem 571
- 33.5 If b is constant 572
- 33.6 Further results and notes 573
  - 33.6a Complexity results 573
  - 33.6b Degree-sequences 573

33.6c Further notes 574

34 b-edge covers 575
  34.1 b-edge covers 575
  34.2 The b-edge cover polyhedron 576
  34.3 Total dual integrality.................................... 576
  34.4 The weighted b-edge cover problem 577
  34.5 If b is even 578
  34.6 If b is constant 578
  34.7 Capacitated b-edge covers............................... 579
  34.8 Simple b-edge covers 581
        34.8a Simple b-edge covers and b-matchings 582
        34.8b Capacitated b-edge covers and b-matchings......... 583

35 Upper and lower bounds................................... 584
  35.1 Upper and lower bounds................................ 584
  35.2 Convex hull 586
  35.3 Total dual integrality................................. 589
  35.4 Further results and notes 591
        35.4a Further results on subgraphs with prescribed
                degrees.. 591
        35.4b Odd walks 593

36 Bidirected graphs... 594
  36.1 Bidirected graphs....................................... 594
  36.2 Convex hull 597
  36.3 Total dual integrality................................. 598
  36.4 Including parity conditions............................ 600
  36.5 Convex hull 604
        36.5a Convex hull of vertex-disjoint circuits 605
  36.6 Total dual integrality................................. 605
  36.7 Further results and notes 607
        36.7a The Chvátal rank 607
        36.7b Further notes 608

37 The dimension of the perfect matching polytope........... 609
  37.1 The dimension of the perfect matching polytope 609
  37.2 The perfect matching space 611
  37.3 The brick decomposition................................ 612
  37.4 The brick decomposition of a bipartite graph............. 613
  37.5 Braces 614
  37.6 Bricks 614
  37.7 Matching-covered graphs without nontrivial tight cuts...... 617

38 The perfect matching lattice 619
- 38.1 The perfect matching lattice 619
- 38.2 The perfect matching lattice of the Petersen graph 620
- 38.3 A further fact on the Petersen graph 621
- 38.4 Various useful observations 622
- 38.5 Simple barriers 624
- 38.6 The perfect matching lattice of a brick 630
- 38.7 Synthesis and further consequences of the previous results .. 643
- 38.8 What further might (not) be true 644
- 38.9 Further results and notes 646
  - 38.9a The perfect 2-matching space and lattice 646
  - 38.9b Further notes 647

Volume C

Part VII: Multiflows and Disjoint Paths 1219

70 Multiflows and disjoint paths 1221
- 70.1 Directed multiflow problems 1221
- 70.2 Undirected multiflow problems 1222
- 70.3 Disjoint paths problems 1223
- 70.4 Reductions 1223
- 70.5 Complexity of the disjoint paths problem 1224
- 70.6 Complexity of the fractional multiflow problem 1225
- 70.7 The cut condition for directed graphs 1227
- 70.8 The cut condition for undirected graphs 1228
- 70.9 Relations between fractional, half-integer, and integer solutions 1230
- 70.10 The Euler condition 1233
- 70.11 Survey of cases where a good characterization has been found 1234
- 70.12 Relation between the cut condition and fractional cut packing 1236
  - 70.12a Sufficiency of the cut condition sometimes implies an integer multiflow 1238
  - 70.12b The cut condition and integer multiflows in directed graphs 1241
- 70.13 Further results and notes 1242
  - 70.13a Fixing the number of commodities in undirected graphs 1242
  - 70.13b Fixing the number of commodities in directed graphs 1243
  - 70.13c Disjoint paths in acyclic digraphs 1244
  - 70.13d A column generation technique for multiflows 1245
  - 70.13e Approximate max-flow min-cut theorems for multiflows 1247
  - 70.13f Further notes 1248
  - 70.13g Historical notes on multicommodity flows 1249

71 Two commodities 1251
- 71.1 The Rothschild-Whinston theorem and Hu's 2-commodity flow theorem 1251
  - 71.1a Nash-Williams' proof of the Rothschild-Whinston theorem 1254
- 71.2 Consequences 1255
- 71.3 2-commodity cut packing 1257
- 71.4 Further results and notes 1261

71.4a　Two disjoint paths in undirected graphs 1261
　　　71.4b　A directed 2-commodity flow theorem 1262
　　　71.4c　Kleitman, Martin-Löf, Rothschild, and Whinston's
　　　　　　theorem 1263
　　　71.4d　Further notes 1265

72　Three or more commodities 1266
　72.1　Demand graphs for which the cut condition is sufficient. . . . 1266
　72.2　Three commodities 1271
　　　72.2a　The $K_{2,3}$-metric condition 1273
　　　72.2b　Six terminals. 1275
　72.3　Cut packing. 1276

73　T-paths. 1279
　73.1　Disjoint T-paths 1279
　　　73.1a　Disjoint T-paths with the matroid matching
　　　　　　algorithm. 1283
　　　73.1b　Polynomial-time findability of edge-disjoint
　　　　　　T-paths 1285
　　　73.1c　A feasibility characterization for integer K_3-flows . . . 1286
　73.2　Fractional packing of T-paths 1287
　　　73.2a　Direct proof of Corollary 73.2d 1288
　73.3　Further results and notes 1289
　　　73.3a　Further notes on Mader's theorem 1289
　　　73.3b　A generalization of fractionally packing T-paths 1290
　　　73.3c　Lockable collections 1291
　　　73.3d　Mader matroids 1292
　　　73.3e　Minimum-cost maximum-value multiflows 1294
　　　73.3f　Further notes 1295

74　Planar graphs 1296
　74.1　All nets spanned by one face: the Okamura-Seymour
　　　theorem 1296
　　　74.1a　Complexity survey. 1299
　　　74.1b　Graphs on the projective plane 1299
　　　74.1c　If only inner vertices satisfy the Euler condition 1302
　　　74.1d　Distances and cut packing 1304
　　　74.1e　Linear algebra and distance realizability 1305
　　　74.1f　Directed planar graphs with all terminals on the
　　　　　　outer boundary. 1307
　74.2　$G + H$ planar 1307
　　　74.2a　Distances and cut packing 1308
　　　74.2b　Deleting the Euler condition if $G + H$ is planar. 1309
　74.3　Okamura's theorem 1311
　　　74.3a　Distances and cut packing 1313

		74.3b	The Klein bottle 1314

        74.3c Commodities spanned by three or more faces....... 1316
  74.4 Further results and notes 1318
        74.4a Another theorem of Okamura 1318
        74.4b Some other planar cases where the cut condition is
               sufficient 1320
        74.4c Vertex-disjoint paths in planar graphs............. 1320
        74.4d Grid graphs..................................... 1323
        74.4e Further notes 1325

75 Cuts, odd circuits, and multiflows 1326
  75.1 Weakly and strongly bipartite graphs 1326
        75.1a NP-completeness of maximum cut 1328
        75.1b Planar graphs................................... 1328
  75.2 Signed graphs 1329
  75.3 Weakly, evenly, and strongly bipartite signed graphs 1330
  75.4 Characterizing strongly bipartite signed graphs 1331
  75.5 Characterizing weakly and evenly bipartite signed graphs .. 1334
  75.6 Applications to multiflows 1341
  75.7 The cut cone and the cut polytope....................... 1342
  75.8 The maximum cut problem and semidefinite programming.. 1345
  75.9 Further results and notes 1348
        75.9a Cuts and stable sets 1348
        75.9b Further notes 1350

76 Homotopy and graphs on surfaces 1352
  76.1 Graphs, curves, and their intersections: terminology and
       notation 1352
  76.2 Making curves minimally crossing by Reidemeister moves .. 1353
  76.3 Decomposing the edges of an Eulerian graph on a surface .. 1354
  76.4 A corollary on lengths of closed curves 1356
  76.5 A homotopic circulation theorem 1357
  76.6 Homotopic paths in planar graphs with holes............. 1361
  76.7 Vertex-disjoint paths and circuits of prescribed homotopies.. 1367
        76.7a Vertex-disjoint circuits of prescribed homotopies.... 1367
        76.7b Vertex-disjoint homotopic paths in planar graphs
               with holes 1368
        76.7c Disjoint trees 1371

Part VIII: Hypergraphs — 1373

77 Packing and blocking in hypergraphs: elementary notions 1375
- 77.1 Elementary hypergraph terminology and notation 1375
- 77.2 Deletion, restriction, and contraction 1376
- 77.3 Duplication and parallelization 1376
- 77.4 Clutters 1376
- 77.5 Packing and blocking 1377
- 77.6 The blocker 1377
- 77.7 Fractional matchings and vertex covers 1378
- 77.8 k-matchings and k-vertex covers 1378
- 77.9 Further results and notes 1379
  - 77.9a Bottleneck extrema 1379
  - 77.9b The ratio of τ and τ^* 1380
  - 77.9c Further notes 1381

78 Ideal hypergraphs 1383
- 78.1 Ideal hypergraphs 1383
- 78.2 Characterizations of ideal hypergraphs 1384
- 78.3 Minimally nonideal hypergraphs 1386
- 78.4 Properties of minimally nonideal hypergraphs: Lehman's theorem 1387
  - 78.4a Application of Lehman's theorem: Guenin's theorem 1392
  - 78.4b Ideality is in co-NP 1394
- 78.5 Further results and notes 1395
  - 78.5a Composition of clutters 1395
  - 78.5b Further notes 1395

79 Mengerian hypergraphs 1397
- 79.1 Mengerian hypergraphs 1397
  - 79.1a Examples of Mengerian hypergraphs 1399
- 79.2 Minimally non-Mengerian hypergraphs 1400
- 79.3 Further results and notes 1401
  - 79.3a Packing hypergraphs 1401
  - 79.3b Restrictions instead of parallelizations 1402
  - 79.3c Equivalences for k-matchings and k-vertex covers ... 1402
  - 79.3d A general technique 1403
  - 79.3e Further notes 1404

80 Binary hypergraphs 1406
- 80.1 Binary hypergraphs 1406
- 80.2 Binary hypergraphs and binary matroids 1406
- 80.3 The blocker of a binary hypergraph 1407
  - 80.3a Further characterizations of binary clutters 1408
- 80.4 On characterizing binary ideal hypergraphs 1408
- 80.5 Seymour's characterization of binary Mengerian hypergraphs 1409
  - 80.5a Applications of Seymour's theorem 1413
- 80.6 Mengerian matroids.................................... 1415
  - 80.6a Oriented matroids 1415
- 80.7 Further results and notes 1416
  - 80.7a $\tau_2(H) = 2\tau(H)$ for binary hypergraphs H 1416
  - 80.7b Application: T-joins and T-cuts 1417
  - 80.7c Box-integrality of $k \cdot P_H$ 1418

81 Matroids and multiflows 1419
- 81.1 Multiflows in matroids 1419
- 81.2 Integer k-flowing 1420
- 81.3 1-flowing and 1-cycling 1421
- 81.4 2-flowing and 2-cycling 1421
- 81.5 3-flowing and 3-cycling 1422
- 81.6 4-flowing, 4-cycling, ∞-flowing, and ∞-cycling............ 1423
- 81.7 The circuit cone and cycle polytope of a matroid 1424
- 81.8 The circuit space and circuit lattice of a matroid 1425
- 81.9 Nonnegative integer sums of circuits 1425
- 81.10 Nowhere-zero flows and circuit double covers in matroids .. 1426

82 Covering and antiblocking in hypergraphs 1428
- 82.1 Elementary concepts 1428
- 82.2 Fractional edge covers and stable sets 1429
- 82.3 k-edge covers and k-stable sets 1429
- 82.4 The antiblocker and conformality 1430
  - 82.4a Gilmore's characterization of conformality 1431
- 82.5 Perfect hypergraphs................................... 1431
- 82.6 Further notes 1434
  - 82.6a Some equivalences for the k-parameters 1434
  - 82.6b Further notes 1437

83 Balanced and unimodular hypergraphs 1439
- 83.1 Balanced hypergraphs................................. 1439
- 83.2 Characterizations of balanced hypergraphs............... 1440
  - 83.2a Totally balanced matrices....................... 1444
  - 83.2b Examples of balanced hypergraphs............... 1447
  - 83.2c Balanced $0, \pm 1$ matrices 1447

    83.3   Unimodular hypergraphs 1448
          83.3a Further notes 1450

Survey of Problems, Questions, and Conjectures 1453

References 1463

Name Index 1767

Subject Index 1807

Greek graph and hypergraph functions 1880

Part IV

Matroids and Submodular Functions

Part IV: Matroids and Submodular Functions

Matroids form an important tool in combinatorial optimization. Among other, they apply to shortest and disjoint trees in undirected graphs, to bipartite matching, and to directed cut covering.

Matroids were introduced by Whitney in 1935, and equivalent axiom systems were considered in the 1930s by Nakasawa, Birkhoff, and van der Waerden. They were motivated by questions from algebra, geometry, and graph theory. The importance of matroids for combinatorial optimization was revealed by J. Edmonds in the 1960s, who found efficient algorithms and min-max relations for optimization problems involving matroids.

Matroids are exactly those structures where the greedy algorithm yields an optimum solution. Edmonds discovered that matroids have an even stronger algorithmic property: also optimization over intersections of two different matroids can be done efficiently. It is closely related to matroid union. Among the consequences of matroid intersection and union methods and results are min-max relations, polyhedral characterizations, and algorithms for bipartite matching, common transversals, and tree packing and covering. (In fact, tree packing and covering are best investigated within the structures offered by matroids. This insight was obtained already in the original paper of Nash-Williams on tree packing. That is why we discuss matroids before Part V on trees and forests.)

While bipartite matching is generalized by matroid intersection, nonbipartite matching is generalized by *matroid matching*. We prove in Chapter 43 Lovász's matroid matching theorem for linear matroids. For general matroids the problem is intractable.

The rank function of a matroid is a special case of a submodular function. Submodular functions give rise to a polyhedral generalization of matroids, the *polymatroids*. Most of matroid theory can be lifted to the level of submodular functions and polymatroids. Next to having applications by its own, it will also be used in Part V where we consider submodular functions defined on digraphs (Chapter 60). This applies to directed variants of tree and cut packing and covering, and to graph orientation and connectivity augmentation.

Chapters:

39. Matroids 651
40. The greedy algorithm and the independent set polytope 688
41. Matroid intersection 700
42. Matroid union 725
43. Matroid matching 745
44. Submodular functions and polymatroids 766
45. Submodular function minimization 786
46. Polymatroid intersection 795
47. Polymatroid intersection algorithmically 805
48. Dilworth truncation 820
49. Submodularity more generally 826

Chapter 39

Matroids

This chapter gives the basic definitions, examples, and properties of matroids. We use the shorthand notation

$$X + y := X \cup \{y\} \text{ and } X - y := X \setminus \{y\}.$$

39.1. Matroids

A pair $(S, \mathcal{I})$ is called a *matroid* if S is a finite set and $\mathcal{I}$ is a nonempty collection of subsets of S satisfying:

(39.1) (i) if $I \in \mathcal{I}$ and $J \subseteq I$, then $J \in \mathcal{I}$,
 (ii) if $I, J \in \mathcal{I}$ and $|I| < |J|$, then $I + z \in \mathcal{I}$ for some $z \in J \setminus I$.

(These axioms are given by Whitney [1935].)

Given a matroid $M = (S, \mathcal{I})$, a subset I of S is called *independent* if I belongs to $\mathcal{I}$, and *dependent* otherwise. For $U \subseteq S$, a subset B of U is called a *base* of U if B is an inclusionwise maximal independent subset of U. That is, $B \in \mathcal{I}$ and there is no $Z \in \mathcal{I}$ with $B \subset Z \subseteq U$.

It is not difficult to see that, under condition (39.1)(i), condition (39.1)(ii) is equivalent to:

(39.2) for any subset U of S, any two bases of U have the same size.

The common size of the bases of a subset U of S is called the *rank* of U, denoted by $r_M(U)$. If the matroid is clear from the context, we write $r(U)$ for $r_M(U)$.

A set is called simply a *base* if it is a base of S. The common size of all bases is called the *rank* of the matroid. A subset of S is called *spanning* if it contains a base as a subset. So bases are just the inclusionwise minimal spanning sets, and also just the independent spanning sets. A *circuit* of a matroid is an inclusionwise minimal dependent set. A *loop* is an element s such that $\{s\}$ is a circuit. Two elements s, t of S are called *parallel* if $\{s, t\}$ is a circuit.

Nakasawa [1935] showed the equivalence of axiom system (39.1) with an ostensibly weaker system, which will be useful in proofs:

Theorem 39.1. *Let S be a finite set and let $\mathcal{I}$ be a nonempty collection of subsets satisfying (39.1)(i). Then (39.1)(ii) is equivalent to:*

(39.3) $\quad$ *if $I, J \in \mathcal{I}$ and $|I \setminus J| = 1$, $|J \setminus I| = 2$, then $I + z \in \mathcal{I}$ for some $z \in J \setminus I$.*

Proof. Obviously, (39.1)(ii) implies (39.3). Conversely, (39.1)(ii) follows from (39.3) by induction on $|I \setminus J|$, the case $|I \setminus J| = 0$ being trivial. If $|I \setminus J| \geq 1$, choose $i \in I \setminus J$. We apply the induction hypothesis twice: first to $I - i$ and J to find $j \in J \setminus I$ with $I - i + j \in \mathcal{I}$, and then to $I - i + j$ and J to find $j' \in J \setminus (I + j)$ with $I - i + j + j' \in \mathcal{I}$. Then by (39.3) applied to I and $I - i + j + j'$, we have that $I + j \in \mathcal{I}$ or $I + j' \in \mathcal{I}$. ∎

39.2. The dual matroid

With each matroid M, a dual matroid M^* can be associated, in such a way that $(M^*)^* = M$. Let $M = (S, \mathcal{I})$ be a matroid, and define

(39.4) $\quad \mathcal{I}^* := \{I \subseteq S \mid S \setminus I \text{ is a spanning set of } M\}$.

Then (Whitney [1935]):

Theorem 39.2. $M^* = (S, \mathcal{I}^*)$ *is a matroid.*

Proof. Condition (39.1)(i) trivially holds for $\mathcal{I}^*$. To see (39.1)(ii), consider $I, J \in \mathcal{I}^*$ with $|I| < |J|$. By definition of $\mathcal{I}^*$, $S \setminus J$ contains some base B of M. As also $S \setminus I$ contains some base of M, and as $B \setminus I \subseteq S \setminus I$, there exists a base B' of M with $B \setminus I \subseteq B' \subseteq S \setminus I$. Then $J \setminus I \not\subseteq B'$, since otherwise (as $B \cap I \subseteq I \setminus J$, and as $B \setminus I$ and $J \setminus I$ are disjoint, since $B \cap J = \emptyset$)

(39.5) $\quad |B| = |B \cap I| + |B \setminus I| \leq |I \setminus J| + |B \setminus I| < |J \setminus I| + |B \setminus I| \leq |B'|$,

which is a contradiction. As $J \setminus I \not\subseteq B'$, there is a $z \in J \setminus I$ with $z \notin B'$. So B' is disjoint from $I + z$. Hence $I + z \in \mathcal{I}^*$. ∎

The matroid M^* is called the *dual matroid* of M. The bases of M^* are precisely the complements of the bases of M. This implies $(M^*)^* = M$, which justifies the name dual.

Theorem 39.3. *The rank function r_{M^*} of the dual matroid M^* satisfies, for $U \subseteq S$:*

(39.6) $\quad r_{M^*}(U) = |U| + r_M(S \setminus U) - r_M(S)$.

Proof. Let $\mathcal{B}$ and $\mathcal{B}^*$ denote the collections of bases of M and of M^*, respectively. Then

(39.7) $\quad r_{M^*}(U) = \max\{|U \cap A| \mid A \in \mathcal{B}^*\} = \max\{|U \setminus B| \mid B \in \mathcal{B}\}$
$= |U| - \min\{|B \cap U| \mid B \in \mathcal{B}\}$
$= |U| - r_M(S) + \max\{|B \setminus U| \mid B \in \mathcal{B}\}$
$= |U| - r_M(S) + r_M(S \setminus U).$ ∎

The circuits of M^* are called the *cocircuits* of M. They are the inclusionwise minimal sets intersecting each base of M (as they are the inclusionwise minimal sets contained in no base of M^*, that is, not contained in the complement of any base of M). The loops of M^* are the *coloops* or *bridges* of M, and parallel elements of M^* are called *coparallel* or *in series* in M.

Let $M = (S, \mathcal{I})$ be a matroid, and suppose that we can test in polynomial time if any subset of S is independent in M (or we have an oracle for that). Then we can calculate, for any subset U of S, the rank $r_M(U)$ of U in polynomial time (by growing an independent set (starting from $\emptyset$) to an inclusionwise maximal independent subset of U). It follows that we can test in polynomial time if any subset U of S in independent in M^*, just by testing if $r_M(S \setminus U) = r_M(S)$.

A matroid $M = (S, \mathcal{I})$ is called *connected* if $r_M(U) + r_M(S \setminus U) > r_M(S)$ for each nonempty proper subset U of S. This is equivalent to: for any two elements $s, t \in S$ there exists a circuit containing both s and t. One may derive from (39.6) that a matroid M is connected if and only if M^* is connected.

39.3. Deletion, contraction, and truncation

We can derive matroids from matroids by 'deletion' and 'contraction'. Let $M = (S, \mathcal{I})$ be a matroid and let $Y \subseteq S$. Define

(39.8) $\quad \mathcal{I}' := \{Z \mid Z \subseteq Y, Z \in \mathcal{I}\}.$

Then $M' = (Y, \mathcal{I}')$ is a matroid again, as directly follows from the matroid axioms (39.1). M' is called the *restriction* of M to Y, denoted by $M|Y$. If $Y = S \setminus Z$ with $Z \subseteq S$, we say that M' arises by *deleting* Z, and denote M' by $M \setminus Z$. Clearly, the rank function of $M|Y$ is the restriction of the rank function of M to subsets of Y.

Contraction is the operation dual to deletion. *Contracting* Z means replacing M by $(M^* \setminus Z)^*$. This matroid is denoted by M/Z. If $Y = S \setminus Z$, then we denote $M \cdot Y := M/Z$. Theorem 39.3 implies that the rank function r' of M/Z satisfies

(39.9) $\quad r_{M/Z}(X) = r(X \cup Z) - r(Z)$

for $X \subseteq S \setminus Z$.

We can describe contraction as follows. Let $Z \subseteq S$ and let X be a base of Z. Then

(39.10) $\quad$ a subset I of $S \setminus Z$ is independent in M/Z if and only if $I \cup X$ is independent in M.

Note that for disjoint subsets Y, Z of S one has $(M \setminus Y) \setminus Z = M \setminus (Y \cup Z)$ and hence $(M/Y)/Z = M/(Y \cup Z)$. Moreover, deletion and contraction commute, as for any two distinct $x, y \in S$ and any $Z \subseteq S \setminus \{x, y\}$ one has (using (39.9)):

(39.11) $\quad r_{M \setminus x/y}(Z) = r_{M \setminus x}(Z \cup \{y\}) - r_{M \setminus x}(\{y\}) = r_M(Z \cup \{y\}) - r_M(\{y\})$
$= r_{M/y}(Z) = r_{M/y \setminus x}(Z).$

If matroid M' arises from M by a series of deletions and contractions, M' is called a *minor* of M.

The circuits of $M|Y$ are exactly the circuits of M contained in Y, and the circuits of $M \cdot Y$ are exactly the minimal nonempty sets $C \cap Y$, where C is a circuit of M.

Another operation is that of 'truncation'. Let $M = (S, \mathcal{I})$ be a matroid and let k be a natural number. Define $\mathcal{I}' := \{I \in \mathcal{I} \mid |I| \leq k\}$. Then $(S, \mathcal{I}')$ is again a matroid, called the *k-truncation* of M.

39.4. Examples of matroids

We describe some basic classes of matroids.

Uniform matroids. An easy class of matroids is given by the *uniform matroids*. They are determined by a set S and a number k: the independent sets are the subsets I of S with $|I| \leq k$. This trivially gives a matroid, called a *k-uniform matroid* and denoted by U_n^k, where $n := |S|$.

Linear matroids (Grassmann [1862], Steinitz [1913]). Let A be an $m \times n$ matrix. Let $S := \{1, \ldots, n\}$ and let $\mathcal{I}$ be the collection of all those subsets I of S such that the columns of A with index in I are linearly independent. That is, such that the submatrix of A consisting of the columns with index in I has rank $|I|$.

Then $(S, \mathcal{I})$ is a matroid (property (39.1)(ii) was proved by Grassmann [1862] and by Steinitz [1913], and is called *Steinitz' exchange property*). Condition (39.1)(i) is trivial. To see condition (39.1)(ii), let $I, J \in \mathcal{I}$ with $|I| < |J|$. Then I spans an $|I|$-dimensional space $\overline{I}$. So $J \not\subseteq \overline{I}$. Take $j \in J \setminus \overline{I}$. Then $I + j \in \mathcal{I}$ and $j \in J \setminus I$.

Any matroid obtained in this way, or isomorphic to such a matroid, is called a *linear matroid*. If A has entries in a field $\mathbb{F}$, then M is called *representable over* $\mathbb{F}$. We will also say that M is *represented by* (the columns of) A, and A is called a *representation* of M.

Note that the rank $r_M(U)$ of any subset U of S is equal to the rank of the matrix formed by the columns indexed by U.

The dual matroid of a matroid representable over a field $\mathbb{F}$ is again representable over $\mathbb{F}$. Indeed, we can assume that the matrix A is of the form $[I_m \ B]$, where I_m is the $m \times m$ identity matrix, and B is an $m \times (n - m)$

matrix. Then the dual matroid can be represented by the matrix $[B^\mathsf{T}\ I_{n-m}]$, as follows directly from elementary linear algebra. This implies that the class of matroids representable over $\mathbb{F}$ is closed under taking minors.

MacLane [1936] (and also Lazarson [1958]) showed that nonlinear matroids exist.

Binary matroids. A matroid representable over GF(2) — the field with two elements — is called a *binary matroid*. For later purposes, we give some characterizations of binary matroids. The following is direct (Whitney [1935]):

(39.12)   a matroid M is binary if and only if for each choice of circuits $C_1, \ldots, C_t$, the set $C_1 \triangle \cdots \triangle C_t$ can be partitioned into circuits.

In a binary matroid M, disjoint unions of circuits are called the *cycles* of M. Of special interest is the *Fano matroid* F_7, represented by the nonzero vectors in $\text{GF}(2)^3$.

Tutte [1958a,1958b] showed that the unique minor-minimal nonbinary matroid is U_4^2, the 2-uniform matroid on 4 elements. (We follow the proof suggested by A.M.H. Gerards.)

Theorem 39.4. *A matroid is binary if and only if it has no U_4^2 minor.*

Proof. Necessity follows from the facts that the class of binary matroids is closed under taking minors and that U_4^2 is not binary.

To see sufficiency, we first show the following. Let M and N be matroids on the same set S. Call a set *wrong* if it is a base of precisely one of M and N. A *far base* is a common base B of M and N such that there is no wrong set X with $|B \triangle X| = 2$. We first show:

(39.13)   if M and N are different and have a far base, then M or N has a U_4^2 minor.

Let M, N form a counterexample with S as small as possible. Let B be a far base and X be a wrong set with $|B \triangle X|$ minimal. Then $B \cup X = S$, since we can delete $S \setminus (B \cup X)$. Similarly (by considering M^* and N^*), $B \cap X = \emptyset$. Then, by the minimality of $|B \triangle X|$, X is the only wrong set. By symmetry, we may assume that X is a base of M. Then M has a base B' with $|B \triangle B'| = 2$. By the uniqueness of X, B' is also a base of N. By the minimality of $|B \triangle X|$, B' is not far. Hence, by the uniqueness of X, $|B' \triangle X| = 2$. So $|S| = 4$.

Let $S = \{a,b,c,d\}$, $B = \{a,b\}$, $X = \{c,d\}$. Since $M \neq U_4^2$ by assumption, we may assume that $\{a,c\}$ is not a base of M. Hence, since $\{a\}$ and $\{c,d\}$ are independent in M, $\{a,d\}$ is a base of M. Similarly, since $\{c\}$ and $\{a,b\}$ are independent in M, $\{b,c\}$ is a base of M.

Since B is far, $\{a,d\}$ and $\{b,c\}$ are bases also of N, and $\{a,c\}$ is not a base of N. So $\{c\}$ is independent in N, implying that $\{c,a\}$ or $\{c,d\}$ is a base of N, a contradiction. This proves (39.13).

Now let M be a nonbinary matroid on a set S. Choose a base B of M. Let $\{x_b \mid b \in B\}$ be a collection of linearly independent vectors over GF(2). For each $s \in S \setminus B$, let C_s be the circuit contained in $B \cup \{s\}$, and define

$$(39.14) \qquad x_s := \sum_{b \in C_s \setminus \{s\}} x_b.$$

Let N be the binary matroid represented by $\{x_s \mid s \in S\}$. Now for each $b \in B$ and each $s \in S \setminus B$ one has that $(B \setminus \{b\}) \cup \{s\}$ is a base of M if and only if it is a base of N. So B is a far base. Since N is binary, we know that $N \neq M$ and that N has no U_4^2 minor. Hence, by (39.13), M has a U_4^2 minor. ∎

Regular matroids. A matroid is called *regular* if it is representable over each field. It is equivalent to requiring that it can be represented over $\mathbb{R}$ by the columns of a totally unimodular matrix.

Regular matroids are characterized by Tutte [1958a,1958b] as those binary matroids not having an F_7 or F_7^* minor. (Gerards [1989b] gave a short proof.)

A basic decomposition theorem of Seymour [1980a] states that each regular matroid can be obtained by taking 1-, 2-, and 3-sums from graphic and cographic matroids and from copies of a 10-element matroid called R_{10}. (We do not use this theorem in this book. Background can be found in the book of Truemper [1992].)

Algebraic matroids (Steinitz [1910]). Let L be a field extension of a field K and let S be a finite subset of L. Let $\mathcal{I}$ be the collection of all subsets $\{s_1, \ldots, s_n\}$ of S that consist of algebraically independent elements over K. That is, there is no nonzero polynomial $p(x_1, \ldots, x_n) \in K[x_1, \ldots, x_n]$ with $p(s_1, \ldots, s_n) = 0$. Then $(S, \mathcal{I})$ is a matroid, and matroids arising in this way are called *algebraic (over K)*. (Steinitz [1910] showed that $(S, \mathcal{I})$ satisfies the matroid axioms, although the term matroid was not yet introduced.)

To see that $(S, \mathcal{I})$ is a matroid, we check (39.3). It suffices to show that for all $s_1, \ldots, s_n \in S$ one has:

(39.15) 	if $\{s_1, s_2, s_3, \ldots, s_{n-1}\} \in \mathcal{I}$ and $\{s_3, \ldots, s_{n-1}, s_n\} \in \mathcal{I}$, then $\{s_1, s_3, \ldots, s_n\} \in \mathcal{I}$ or $\{s_2, s_3, \ldots, s_n\} \in \mathcal{I}$.

Suppose not. Then there exist nonzero polynomials $p(x_1, x_3, \ldots, x_n)$ and $q(x_2, x_3, \ldots, x_n)$ over K with $p(s_1, s_3, \ldots, s_n) = 0$ and $q(s_2, s_3, \ldots, s_n) = 0$. We may assume that p and q are irreducible. Moreover, since $\{s_3, \ldots, s_n\} \in \mathcal{I}$, p and q are relatively prime. Define $F := K(x_1, x_2, \ldots, x_{n-1})$. So p and q belong to the Euclidean ring $F[x_n]$. Let r be the g.c.d. of p and q in $F[x_n]$. As p and q are relatively prime, we know $r \in F$, and hence we may assume $r \in K[x_1, \ldots, x_{n-1}]$. Now $r = \alpha p + \beta q$ for some $\alpha, \beta \in F[x_n]$. So $r(s_1, \ldots, s_{n-1}) = 0$, contradicting the fact that $\{s_1, \ldots, s_{n-1}\} \in \mathcal{I}$. This proves (39.15).

Each linear matroid is algebraic (as we can consider the linear relations between the elements as polynomials of rank 1), while Ingleton [1971] gave an example of a nonlinear algebraic matroid. Examples of nonalgebraic matroids were given by Ingleton and Main [1975] and Lindström [1984,1986]. The class of algebraic matroids can be easily seen to be closed under taking minors (deletion is direct, while contraction of an element t corresponds to replacing K by $K(t)$), but it is unknown if it is closed under duality.

In fact, for any field K, the class of matroids that are algebraic over K is closed under taking minors, since Lindström [1989] showed that any matroid algebraic over $K(t)$ (for any t), is also algebraic over K.

For an in-depth survey on algebraic matroids, see Oxley [1992].

Graphic matroids (Birkhoff [1935c], Whitney [1935]). Let $G = (V, E)$ be a graph and let $\mathcal{I}$ be the collection of all subsets of E that form a forest. Then $M = (E, \mathcal{I})$ is a matroid. Condition (39.1)(i) is trivial. To see that condition (39.2) holds, let $F \subseteq E$. Then, by definition, each base U of F is an inclusionwise maximal forest contained in F. Hence U forms a spanning tree in each component of the graph (V, F). So U has $|V| - k$ elements, where k is the number of components of (V, F). So each base of F has $|V| - k$ elements, proving (39.2).

The matroid M is called the *cycle matroid* of G, denoted by $M(G)$. Any matroid obtained in this way, or isomorphic to such a matroid, is called a *graphic matroid*.

Trivially, the circuits of $M(G)$, in the matroid sense, are exactly the circuits of G, in the graph sense. The bases of $M(G)$ are exactly the inclusionwise maximal forests F of G. So if G is connected, the bases are the spanning trees.

The rank function of $M(G)$ can be described as follows. For each subset F of E, let $\kappa(V, F)$ denote the number of components of the graph (V, F). Then for each $F \subseteq E$:

(39.16) $r_{M(G)}(F) = |V| - \kappa(V, F)$.

Note that deletion and contraction in the matroid correspond to deletion and contraction of edges in the graph.

Graphic matroids are regular, that is, representable over any field: orient the edges of G arbitrarily, and consider the $V \times E$ matrix L given by: $L_{v,e} = +1$ if v is the head of e, $L_{v,e} := -1$ if v is the tail of e, and $L_{v,e} := 0$ otherwise (for $v \in V$, $e \in E$). Then a subset F of E is a forest if and only if the set of columns with index in F is linearly independent.

By a theorem of Tutte [1959], the graphic matroids are precisely those regular matroids containing no $M(K_5)^*$ and $M(K_{3,3})^*$ minor. (Alternative proofs were given by Ghouila-Houri [1964] (Chapitre III), Seymour [1980d], Truemper [1985], Wagner [1985], and Gerards [1995b].)

Cographic matroids (Whitney [1935]). The dual of the cycle matroid $M(G)$ of a graph $G = (V, E)$ is called the *cocycle matroid* of G, and denoted by $M^*(G)$. Any matroid obtained in this way, or isomorphic to such a matroid, is called a *cographic matroid*.

So the bases of $M^*(G)$ are the complements of maximal forests of G. (So if G is connected, these are exactly the complements of the spanning trees in G.)

Hence the independent sets are those edge sets F for which $E \setminus F$ contains a maximal forest of G; that is, $(V, E \setminus F)$ has the same number of components as G.

A subset C of E is a circuit of $M^*(G)$ if and only if C is an inclusionwise minimal set with the property that $(V, E \setminus C)$ has more components than G. Hence C is a circuit of $M^*(G)$ if and only if C is an inclusionwise minimal nonempty cut in G.

The rank function of $M^*(G)$ can be described as follows. Again, for each subset F of E, let $\kappa(V, F)$ denote the number of components of the graph (V, F). Then (39.6) and (39.16) give that for each $F \subseteq E$:

(39.17) $\quad r_{M^*(G)}(F) = |F| - \kappa(V, E \setminus F) + \kappa(V, E).$

Let G be an (embedded) planar graph, and let G^* be the dual planar graph of G. Then the cycle matroid $M(G^*)$ of G^* is isomorphic to the cocycle matroid $M^*(G)$ of G.

A theorem of Whitney [1933] implies that a matroid is both graphic and cographic if and only if it is isomorphic to the cycle matroid of a planar graph.

Transversal matroids (Edmonds and Fulkerson [1965], Mirsky and Perfect [1967]). Let $\mathcal{X} = (X_1, \ldots, X_n)$ be a family of subsets of a finite set S and let $\mathcal{I}$ be the collection of all partial transversals of $\mathcal{X}$. Then $M = (S, \mathcal{I})$ is a matroid, as follows directly from Corollary 22.4a. Any matroid obtained in this way, or isomorphic to such a matroid, is called a *transversal matroid* (*induced by* $\mathcal{X}$).

The bases of this matroid are the inclusionwise maximal partial transversals. If $\mathcal{X}$ has a transversal, the bases of M are the transversals of $\mathcal{X}$. In fact, Theorem 22.5 implies that we can assume the latter situation:

(39.18) Let M be the transversal matroid induced by the family $\mathcal{X}$. Then $\mathcal{X}$ has a subfamily $\mathcal{Y}$ such that M is equal to the transversal matroid induced by $\mathcal{Y}$ and such that $\mathcal{Y}$ has a transversal.

So we can assume that any transversal matroid has the transversals of a family of sets as bases.

It follows from Kőnig's matching theorem that the rank function r of the transversal matroid induced by $\mathcal{X}$ is given by

(39.19) $\quad r(U) = \min_{T \subseteq U}(|U \setminus T| + |\{i \mid X_i \cap T \neq \emptyset\}|)$

$$= \min_{I \subseteq \{1,\ldots,n\}}(n - |I| + |\bigcup_{i \in I}(X_i \cap U)|)$$

for $U \subseteq S$. This follows directly from Theorem 22.2 and Corollary 22.2a, applied to the family $(X_1 \cap U, \ldots, X_n \cap U)$.

Piff and Welsh [1970] (cf. Atkin [1972]) showed that

(39.20)   any transversal matroid is representable over all fields, except for finitely many finite fields.

If the sets $X_1, \ldots, X_m$ form a partition of S, one speaks of a *partition matroid*. Trivially, each partition matroid is graphic and cographic (by considering a graph consisting of vertex-disjoint parallel classes of edges). Also uniform matroids are special cases of transversal matroids.

Gammoids (Perfect [1968]). An extension of transversal matroids is obtained by taking a directed graph $D = (V, A)$ and subsets U and S of V. For $X, Y \subseteq V$, call X *linked to* Y if $|X| = |Y|$ and D has $|X|$ vertex-disjoint $X - Y$ paths. (So X is the set of starting vertices of these paths, and Y the set of end vertices.)

Let $\mathcal{I}$ be the collection of subsets I of S such that some subset of U is linked to I. Then $M = (S, \mathcal{I})$ is a matroid. This follows from Theorem 9.11: let $I, J \in \mathcal{I}$ with $|I| < |J|$. Let $T := I \cup J$. Let k be the maximum number of disjoint $U - T$ paths. So $k \geq |J| > |I|$. By Theorem 9.11, there exist k disjoint $U - T$ paths covering I. Hence $I + j \in \mathcal{I}$ for some $j \in J \setminus I$. So M is a matroid.

Matroids obtained in this way are called *gammoids*. If $S = V$, the gammoid is called a *strict gammoid (induced by D, U)*. Hence:

(39.21)   gammoids are exactly the restrictions of strict gammoids.

The bases of the strict gammoid induced by D, U are the subsets B of V such that U is linked to B. In particular, U is a base.

From Menger's theorem (Corollary 9.1a) one easily derives the following formula for the rank function r_M of M:

(39.22) $r_M(X) = \min\{|Y| \mid Y \text{ intersects each } U - X \text{ path}\}$

for $X \subseteq S$. (One may prove easily that the right-hand side of (39.22) satisfies Theorem 39.8 below, thus proving again that M is a matroid.)

39.4a. Relations between transversal matroids and gammoids

Ingleton and Piff [1973] showed the following theorem (based on a duality of bipartite graphs and directed graphs similar to that described in Section 16.7c). The proof provides an alternative proof that gammoids are indeed matroids.

Theorem 39.5. *Strict gammoids are exactly the duals of the transversal matroids.*

Proof. Let M be the strict gammoid induced by the directed graph $D = (V, A)$ and $U \subseteq V$. We can assume that $(v, v) \in A$ for each $v \in V$. For each $v \in V$, let

(39.23) $X_v := \{u \in V \mid (u, v) \in A\}$.

Let L be the transversal matroid induced by the family $\mathcal{X} := (X_v \mid v \in V \setminus U)$. We show that $L = M^*$.

As $v \in X_v$ for each $v \in V \setminus U$, the set $V \setminus U$ is a transversal of $\mathcal{X}$. Hence the bases of L are the transversals of $\mathcal{X}$. As U is a base of the strict gammoid induced by D, U, it suffices to show, for each $B \subseteq V$:

(39.24) U is linked to B in D if and only if $V \setminus B$ is a transversal of $\mathcal{X}$.

To see necessity in (39.24), let U be linked to B in D and let $\mathcal{P}$ be a set of $|U|$ disjoint $U - B$ paths. Then for each $v \in V \setminus U$, let $x_v := u$ if v is entered by an arc (u, v) in a path P in $\mathcal{P}$ and let $x_v := v$ otherwise. Then:

(39.25)   (i) $x_v \in X_v$, (ii) $x_v \neq x_{v'}$ for $v \neq v' \in V \setminus U$, and (iii) $\{x_v \mid v \in V \setminus U\} = V \setminus B$.

So $V \setminus B$ is a transversal of $\mathcal{X}$.

To see sufficiency in (39.24), let $V \setminus B$ be a transversal of $\mathcal{X}$. Hence there exist x_v for $v \in V \setminus U$ satisfying (39.25). Let A' be the set of arcs (x_v, v) of D with $v \in V \setminus U$. Then $V \setminus U$ is the set of vertices entered by an arc in A', and $V \setminus B$ is the set of vertices left by an arc in A'. Hence U is linked to B in D.

This shows (39.24), and hence that $M^* = L$. So the dual of a strict gammoid is a transversal matroid.

To see that each transversal matroid is the dual of a strict gammoid, we show that the construction described above can be reversed. Let L be the transversal matroid induced by the family $\mathcal{X} = (X_i \mid i = 1, \ldots, m)$ of sets. By (39.18) we can assume that $\mathcal{X}$ has a transversal. Hence we can assume that $i \in X_i$ for $i = 1, \ldots, m$ (by renaming). Let $V := X_1 \cup \cdots \cup X_m$ and let

(39.26) $A := \{(u, v) \mid v \in \{1, \ldots, m\}, u \in X_v\}$.

Let $D = (V, A)$ and define $U := V \setminus \{1, \ldots, m\}$. Since D, U and $\mathcal{X}$ are related as in (39.23), we again have (39.25). So L is equal to the dual of the strict gammoid induced by D, U. ∎

This theorem has a number of implications for the interrelations of the classes of transversal matroids and of gammoids. Consider the following class of matroids, introduced by Ingleton and Piff [1973]. Let $G = (V, E)$ be a bipartite graph, with colour classes U and W. Let $M = (V, \mathcal{I})$ be the transversal matroid induced by the family $(\{v\} \cup N(v) \mid v \in U)$ (where $N(v)$ is the set of neighbours of v). So $B \subseteq V$ is a base of M if and only if $(U \setminus B) \cup (W \cap B)$ is matchable in G (that is, it induces a subgraph of G having a perfect matching).

Any such matroid M is called a *deltoid* (*induced by* G, U, W). Then M^* is the deltoid induced by G, W, U. So

(39.27)   the dual of a deltoid is a deltoid again.

Now

(39.28)   transversal matroids are exactly those matroids that are the restriction of a deltoid.

Indeed, each deltoid is a transversal matroid, and hence the restriction of any deltoid is a transversal matroid (as the class of transversal matroids is closed under taking restrictions). Conversely, any transversal matroid, induced by (say) $X_1, \ldots, X_m$ is the restriction to W of the deltoid induced by the bipartite graph G with colour classes $U := \{1, \ldots, m\}$ and $W := X_1 \cup \cdots \cup X_m$, with $i \in U$ and $x \in W$ adjacent if and only if $x \in X_i$. (Assuming without loss of generality that $U \cap W = \emptyset$.) This shows (39.28).

Then (39.27) and (39.28) give with Theorem 39.5:

(39.29)   the strict gammoids are exactly the contractions of the deltoids.

Indeed, the strict gammoids are the duals of transversal matroids, hence the duals of restrictions of deltoids, and therefore the contractions of (the duals of) deltoids.

This gives:

Corollary 39.5a. *The gammoids are exactly the contractions of the transversal matroids.*

Proof. Gammoids are the restrictions of strict gammoids, hence the restrictions of contractions of deltoids, hence the contractions of restrictions of deltoids, therefore the contractions of transversal matroids. ∎

Similarly:

(39.30)   the gammoids are exactly the minors of deltoids,

which implies (with (39.27)) a result of Mason [1972]:

(39.31)   the class of gammoids is closed under taking minors and duals.

Theorem 39.5 also implies, with (39.20), that gammoids are representable over all fields, except for a finite number of finite fields (Mason [1972]). In fact, Lindström [1973] showed that any gammoid $(S, \mathcal{I})$ is representable over each field with at least $2^{|S|}$ elements.

Edmonds and Fulkerson [1965] showed that one gets a transversal matroid as follows. Let $G = (V, E)$ be an undirected graph and let $S \subseteq V$. Let $\mathcal{I}$ be the collection of subsets of S which are covered by some matching in G. Then $M = (S, \mathcal{I})$ is a matroid (which is easy to show), called the *matching matroid* of G. In fact, any matching matroid is a transversal matroid. To prove this, we may assume $S = V$. Let $D(G)$, $A(G)$, $C(G)$ form the Edmonds-Gallai decomposition of G (Section 24.4b). Let $\mathcal{K}$ be the collection of components of $G[D(G)]$. Let $\mathcal{X}$ be the family of sets

(39.32) $\{v\}$ for each $v \in A(G) \cup C(G)$,
             $N(v) \cap D(G)$ for each $v \in A(G)$,
             K, repeated $|K| - 1$ times,   for each $K \in \mathcal{K}$.

Then M is equal to the transversal matroid induced by $\mathcal{X}$, as is easy to derive from the properties of the Edmonds-Gallai decomposition. A min-max relation for the rank function is given by Theorem 24.6.

It is straightforward to see that, conversely, each transversal matroid is a matching matroid, by taking G bipartite.

39.5. Characterizing matroids by bases

In Section 39.1, the notion of matroid is defined by 'axioms' in terms of the independent sets. There are several other axiom systems that characterize matroids. In this and the next sections we give a number of them.

Clearly, a matroid is determined by the collection of its bases, since a set is independent if and only if it is contained in a base. Conditions characterizing a collection of bases of a matroid are given in the following theorem (Whitney [1935]).

Theorem 39.6. *Let S be a set and let $\mathcal{B}$ be a nonempty collection of subsets of S. Then the following are equivalent:*

(39.33) (i) *$\mathcal{B}$ is the collection of bases of a matroid;*
(ii) *if $B, B' \in \mathcal{B}$ and $x \in B' \setminus B$, then $B' - x + y \in \mathcal{B}$ for some $y \in B \setminus B'$;*
(iii) *if $B, B' \in \mathcal{B}$ and $x \in B' \setminus B$, then $B - y + x \in \mathcal{B}$ for some $y \in B \setminus B'$.*

Proof. (i)$\Rightarrow$(ii): Let $\mathcal{B}$ be the collection of bases of a matroid $(S, \mathcal{I})$. Then all sets in $\mathcal{B}$ have the same size. Now let $B, B' \in \mathcal{B}$ and $x \in B' \setminus B$. Since $B' - x \in \mathcal{I}$, there exists a $y \in B \setminus B'$ with $B'' := B' - x + y \in \mathcal{I}$. Since $|B''| = |B'|$, we know $B'' \in \mathcal{B}$.

(iii)$\Rightarrow$(i): (iii) directly implies that no set in $\mathcal{B}$ is contained in another. Let $\mathcal{I}$ be the collection of sets I with $I \subseteq B$ for some $B \in \mathcal{B}$. We check (39.3). Let $I, J \in \mathcal{I}$ with $|I \setminus J| = 1$ and $|J \setminus I| = 2$. Let $I \setminus J = \{x\}$.

Consider sets $B, B' \in \mathcal{B}$ with $I \subseteq B$, $J \subseteq B'$. If $x \in B'$, we are done. So assume $x \notin B'$. Then by (iii), $B' - y + x \in \mathcal{B}$ for some $y \in B' \setminus B$. As $|J \setminus I| = 2$, there is a $z \in J \setminus I$ with $z \neq y$. Then $I + z \subseteq B' - y + x$, and so $I + z \in \mathcal{I}$.

(ii)$\Rightarrow$(iii): By the foregoing we know that (iii) implies (ii). Now axioms (ii) and (iii) interchange if we replace $\mathcal{B}$ by the collection of complements of sets in $\mathcal{B}$. Hence also the implication (ii)$\Rightarrow$(iii) holds. ∎

The equivalence of (ii) and (iii) also follows from the fact that the collection of complements of bases of a matroid is the collection of bases of the dual matroid. Conversely, Theorem 39.6 implies that the dual indeed is a matroid.

39.6. Characterizing matroids by circuits

A matroid is determined by the collection of its circuits, since a set is independent if and only if it contains no circuit. Conditions characterizing a collection of circuits of a matroid are given in the following theorem (Whitney

[1935] proved (i)⇔(iii), and Robertson and Weston [1958] (and also Lehman [1964] and Asche [1966]) proved (i)⇔(ii)).

Theorem 39.7. *Let S be a set and let $\mathcal{C}$ be a collection of nonempty subsets of S, such that no two sets in $\mathcal{C}$ are contained in each other. Then the following are equivalent:*

(39.34) (i) $\mathcal{C}$ *is the collection of circuits of a matroid;*
(ii) *if $C, C' \in \mathcal{C}$ with $C \neq C'$ and $x \in C \cap C'$, then $(C \cup C') \setminus \{x\}$ contains a set in $\mathcal{C}$;*
(iii) *if $C, C' \in \mathcal{C}$, $x \in C \cap C'$, and $y \in C \setminus C'$, then $(C \cup C') \setminus \{x\}$ contains a set in $\mathcal{C}$ containing y.*

Proof. (i)⇒(iii): Let $\mathcal{C}$ be the collection of circuits of a matroid $(S, \mathcal{I})$ and let $\mathcal{B}$ be its collection of bases. Let $C, C' \in \mathcal{C}$, $x \in C \cap C'$, and $y \in C \setminus C'$. We can assume that $S = C \cup C'$. Let $B, B' \in \mathcal{B}$ with $B \supseteq C - y$ and $B' \supseteq C' - x$. Then $y \notin B$ and $x \notin B'$ (since $C \not\subseteq B$ and $C' \not\subseteq B'$).

We can assume that $y \notin B'$. Otherwise, $y \in B' \setminus B$, and hence by (ii) of Theorem 39.6, there exists a $z \in B \setminus B'$ with $B'' := B' - y + z \in \mathcal{B}$. Then $z \neq x$, since otherwise $C' \subseteq B''$. Hence, replacing B' by B'' gives $y \notin B'$.

As $y \notin B'$, we know $B' \cup \{y\} \notin \mathcal{I}$, and hence there exists a $C''' \in \mathcal{C}$ contained in $B' \cup \{y\}$. As $C''' \not\subseteq B'$, we know $y \in C'''$. Moreover, as $x \notin B'$ we know $x \notin C'''$.

(iii)⇒(ii): is trivial.

(ii)⇒(i): Let $\mathcal{I}$ be the collection of sets containing no set in $\mathcal{C}$ as a subset. We check (39.3). Let $I, J \in \mathcal{I}$ with $|I \setminus J| = 1$ and $|J \setminus I| = 2$. Assume that $I + z \notin \mathcal{I}$ for each $z \in J \setminus I$. Let y be the element of $I \setminus J$. If $J + y \in \mathcal{I}$, then $I \cup J \in \mathcal{I}$, contradicting our assumption. So $J + y$ contains a set $C \in \mathcal{C}$. Then C is the unique set in $\mathcal{C}$ contained in $J + y$. For suppose that there is another, C' say. Again, $y \in C'$, and hence by (39.34)(ii) there exists a $C'' \in \mathcal{C}$ contained in $(C \cup C') \setminus \{y\}$. But then $C'' \subseteq J$, a contradiction.

As $C \not\subseteq I$, C intersects $J \setminus I$. Choose $x \in C \cap (J \setminus I)$. Then $X := J + y - x$ contains no set in $\mathcal{C}$ (as C is the only set in $\mathcal{C}$ contained in $J + y$). So $X \in \mathcal{I}$, implying that $I + z \in \mathcal{I}$ for the $z \in J \setminus I$ with $z \neq x$. ∎

This theorem implies the following important property for a matroid $M = (S, \mathcal{I})$:

(39.35) for any independent set I and any $s \in S \setminus I$ there is at most one circuit contained in $I \cup \{s\}$.

39.6a. A characterization of Lehman

Lehman [1964] showed that the cocircuits of a matroid M are exactly the inclusionwise minimal nonempty subsets D of S with $|D \cap C| \neq 1$ for each circuit C of M.

To show this, it suffices to show that

(39.36)   (i) $|D \cap C| \neq 1$ for each cocircuit D and circuit C,
         (ii) for each nonempty $D \subseteq S$, if $|D \cap C| \neq 1$ for each circuit C, then D contains a cocircuit; that is, then D is dependent in M^*.

To see (i), suppose that $D \cap C = \{s\}$ for some circuit C and cocircuit D. As $D - s$ is independent in M^*, M has a base B disjoint from $D - s$. Since $C - s$ is disjoint from $D - s$ and since $C - s \in \mathcal{I}$, we can assume that $C - s \subseteq B$. Then $s \notin B$, and so B is disjoint from D. This implies that D is independent in M^*, contradicting the fact that D is a circuit in M^*. This shows (i).

To see (ii), let $\emptyset \neq D \subseteq S$ with $|D \cap C| \neq 1$ for each circuit C. We show that D is dependent in M^*. Suppose not. Then M has a base B disjoint from D. Choose $s \in D$. Then $B + s$ contains a circuit C with $s \in C$. Hence $D \cap C = \{s\}$, contradicting our assumption, thus showing (ii).

39.7. Characterizing matroids by rank functions

The *rank function* of a matroid $M = (S, \mathcal{I})$ is the function $r_M : \mathcal{P}(S) \to \mathbb{Z}_+$ given by:

(39.37) $r_M(U) := \max\{|Z| \mid Z \in \mathcal{I}, Z \subseteq U\}$

for $U \subseteq S$. Again, a matroid is determined by its rank function, as a set U is independent if and only if $r(U) = |U|$. Conditions characterizing a rank function are given by the following theorem (Whitney [1935]; necessity was also shown (in a different terminology) by Bergmann [1929] and Nakasawa [1935]):

Theorem 39.8. *Let S be a set and let $r : \mathcal{P}(S) \to \mathbb{Z}_+$. Then r is the rank function of a matroid if and only if for all $T, U \subseteq S$:*

(39.38)   (i) $r(T) \leq r(U) \leq |U|$ if $T \subseteq U$,
         (ii) $r(T \cap U) + r(T \cup U) \leq r(T) + r(U)$.

Proof. *Necessity.* Let r be the rank function of a matroid $(S, \mathcal{I})$. Choose $T, U \subseteq S$. Clearly (39.38)(i) holds. To see (ii), let I be an inclusionwise maximal set in $\mathcal{I}$ with $I \subseteq T \cap U$ and let J be an inclusionwise maximal set in $\mathcal{I}$ with $I \subseteq J \subseteq T \cup U$. Since $(S, \mathcal{I})$ is a matroid, we know that $r(T \cap U) = |I|$ and $r(T \cup U) = |J|$. Then

(39.39) $r(T) + r(U) \geq |J \cap T| + |J \cap U| = |J \cap (T \cap U)| + |J \cap (T \cup U)|$
          $\geq |I| + |J| = r(T \cap U) + r(T \cup U);$

that is, we have (39.38)(ii).

Sufficiency. Let $\mathcal{I}$ be the collection of subsets I of S with $r(I) = |I|$. We show that $(S, \mathcal{I})$ is a matroid, with rank function r.

Trivially, $\emptyset \in \mathcal{I}$. Moreover, if $I \in \mathcal{I}$ and $J \subseteq I$, then

Section 39.7. Characterizing matroids by rank functions   665

(39.40) $r(J) \geq r(I) - r(I \setminus J) \geq |I| - |I \setminus J| = |J|$.

So $J \in \mathcal{I}$.

In order to check (39.3), let $I, J \in \mathcal{I}$ with $|I \setminus J| = 1$ and $|J \setminus I| = 2$. Let $J \setminus I = \{z_1, z_2\}$. If $I + z_1, I + z_2 \notin \mathcal{I}$, we have $r(I + z_1) = r(I + z_2) = |I|$. Then by (39.38)(ii),

(39.41) $r(J) \leq r(I + z_1 + z_2) \leq r(I + z_1) + r(I + z_2) - r(I) = |I| < |J|$,

contradicting the fact that $J \in \mathcal{I}$.

So $(S, \mathcal{I})$ is a matroid. Its rank function is r, since $r(U) = \max\{|I| \mid I \subseteq U, I \in \mathcal{I}\}$ for each $U \subseteq S$. Here $\geq$ follows from (39.38)(i), since if $I \subseteq U$ and $I \in \mathcal{I}$, then $r(U) \geq r(I) = |I|$. Equality can be shown by induction on $|U|$, the case $U = \emptyset$ being trivial. If $U \neq \emptyset$, choose $y \in U$. By induction, there is an $I \subseteq U - y$ with $I \in \mathcal{I}$ and $|I| = r(U - y)$. If $r(U) = r(U - y)$ we are done, so assume $r(U) > r(U - y)$. Then $I + y \in \mathcal{I}$, since $r(I + y) \geq r(I) + r(U) - r(U - y) \geq |I| + 1$. Moreover, $r(U) \leq r(U - y) + r(\{y\}) \leq |I| + 1$. This proves equality for U.   ∎

Set functions satisfying condition (39.38)(ii) are called *submodular*, and will be studied in Chapter 44.

Whitney [1935] also showed that (39.38) is equivalent to:

(39.42)   (i) $r(\emptyset) = 0$,
           (ii) $r(U) \leq r(U + s) \leq r(U) + 1$ for $U \subseteq S$, $s \in S \setminus U$,
           (iii) for all $U \subseteq S$, $s, t \in S \setminus U$, if $r(U + s) = r(U + t) = r(U)$, then $r(U + s + t) = r(U)$.

The proof above in fact uses only these properties of r.

The following equivalent form of Theorem 39.8 will be useful.

Corollary 39.8a. *Let S be a finite set and let $\mathcal{I}$ be a nonempty collection of subsets of S, closed under taking subsets. For $U \subseteq S$, let $r(U)$ be the maximum size of a subset of U that belongs to $\mathcal{I}$. Then $(S, \mathcal{I})$ is a matroid if and only if r satisfies (39.38)(ii) for all $T, U \subseteq S$.*

Proof. Necessity follows directly from Theorem 39.8. To see sufficiency, it is easy to see that r satisfies (39.38)(i). So by Theorem 39.8, r is the rank function of some matroid $M = (S, \mathcal{J})$. Now: $I \in \mathcal{J} \iff r(I) = |I| \iff I \in \mathcal{I}$. Hence $\mathcal{I} = \mathcal{J}$, and so $(S, \mathcal{I})$ is a matroid.   ∎

Note that if we can test in polynomial time if a given set is independent, we can also test in polynomial time if a given set is a base, or a circuit, and we can determine the rank of a given set in polynomial time.

39.8. The span function and flats

With any matroid $M = (S, \mathcal{I})$ we can define the *span function* $\operatorname{span}_M : \mathcal{P}(S) \to \mathcal{P}(S)$ as follows:

(39.43) $\quad \operatorname{span}_M(T) := \{s \in S \mid r_M(T \cup \{s\}) = r_M(T)\}$

for $T \subseteq S$. If the matroid M is clear from the context, we write $\operatorname{span}(T)$ for $\operatorname{span}_M(T)$. Note that $T \subseteq \operatorname{span}_M(T)$ and that

(39.44) $\quad r_M(\operatorname{span}_M(T)) = r_M(T)$.

This follows directly from the fact that if $r_M(Y) > r_M(T)$, then $r_M(T \cup \{s\}) > r_M(T)$ for some $s \in Y$.

Note also that

(39.45) $\quad T$ is spanning $\iff \operatorname{span}_M(T) = S$

for any $T \subseteq S$. To see $\Longrightarrow$, let T be spanning. Then for each $s \in T$: $r_M(T + s) \leq r_M(S) = r_M(T)$. To see $\Longleftarrow$, suppose $\operatorname{span}_M(T) = S$. Then $r_M(T) = r_M(\operatorname{span}_M(T)) = r_M(S)$.

A *flat* in a matroid $M = (S, \mathcal{I})$ is a subset F of S with $\operatorname{span}_M(F) = F$. A matroid is determined by its collection of flats, as is shown by:

(39.46) $\quad$ a subset I of S is independent if and only if for each $y \in I$ there is a flat F with $I - y \subseteq F$ and $y \notin F$.

Indeed, if I is independent and $y \in I$, let $F := \operatorname{span}_M(I - y)$. Then F is a flat containing $I - y$, but not y, since $r_M(F + y) \geq r_M(I) > r_M(I - y) = r_M(F)$. Conversely, if I is not independent, then $y \in \operatorname{span}_M(I - y)$ for some $y \in I$, and hence each flat containing $I - y$ also contains y.

39.8a. Characterizing matroids by span functions

It was observed by Mac Lane [1938] that the following characterizes span functions of matroids (sufficiency was shown by van der Waerden [1937]).

Theorem 39.9. *Let S be a finite set. A function* $\operatorname{span} : \mathcal{P}(S) \to \mathcal{P}(S)$ *is the span function of a matroid if and only if:*

(39.47) $\quad$ (i) *if* $T \subseteq S$, *then* $T \subseteq \operatorname{span}(T)$;
$\qquad$ (ii) *if* $T, U \subseteq S$ *and* $U \subseteq \operatorname{span}(T)$, *then* $\operatorname{span}(U) \subseteq \operatorname{span}(T)$;
$\qquad$ (iii) *if* $T \subseteq S$, $t \in S \setminus T$, *and* $s \in \operatorname{span}(T + t) \setminus \operatorname{span}(T)$, *then* $t \in \operatorname{span}(T + s)$.

Proof. *Necessity.* Let span be the span function of a matroid $M = (S, \mathcal{I})$ with rank function r. Clearly, (39.47)(i) is satisfied. To see (39.47)(ii), let $U \subseteq \operatorname{span}(T)$ and $s \in \operatorname{span}(U)$. We show $s \in \operatorname{span}(T)$. We can assume $s \notin T$. Then, by the submodularity of r,

(39.48) $\quad r(T \cup \{s\}) \leq r(T \cup U \cup \{s\}) \leq r(T \cup U) + r(U \cup \{s\}) - r(U)$
$\qquad \quad = r(T \cup U) = r(T)$.

(The last equality follows from (39.44).) This shows that $s \in \mathrm{span}(T)$.

To see (39.47)(iii), note that $s \in \mathrm{span}(T+t) \setminus \mathrm{span}(T)$ is equivalent to: $r(T+t+s) = r(T+t)$ and $r(T+s) > r(T)$. Hence

(39.49) $\qquad r(T+t+s) = r(T+t) \leq r(T) + 1 \leq r(T+s),$

that is, $t \in \mathrm{span}(T+s)$. This shows necessity of the conditions (39.47).

Sufficiency. Let a function span satisfy (39.47), and define

(39.50) $\qquad \mathcal{I} := \{I \subseteq S \mid s \notin \mathrm{span}(I-s) \text{ for each } s \in I\}.$

We first show the following:

(39.51) $\qquad$ if $I \in \mathcal{I}$, then $\mathrm{span}(I) = I \cup \{t \mid I + t \notin \mathcal{I}\}.$

Indeed, if $t \in \mathrm{span}(I) \setminus I$, then $I + t \notin \mathcal{I}$, by definition of $\mathcal{I}$. Conversely, $I \subseteq \mathrm{span}(I)$ by (39.47)(i). Moreover, if $I + t \notin \mathcal{I}$, then by definition of $\mathcal{I}$, $s \in \mathrm{span}(I+t-s)$ for some $s \in I + t$. If $s = t$, then $t \in \mathrm{span}(I)$ and we are done. So assume $s \neq t$; that is, $s \in I$. As $I \in \mathcal{I}$, we know that $s \notin \mathrm{span}(I-s)$. So by (39.47)(iii) (for $T := I - s$), $t \in \mathrm{span}(I)$, proving (39.51).

We now show that $M = (S, \mathcal{I})$ is a matroid. Trivially, $\emptyset \in \mathcal{I}$. To see that $\mathcal{I}$ is closed under taking subsets, let $I \in \mathcal{I}$ and $J \subseteq I$. We show that $J \in \mathcal{I}$. Suppose to the contrary that $s \in \mathrm{span}(J-s)$ for some $s \in J$. By (39.47)(ii), $\mathrm{span}(J-s) \subseteq \mathrm{span}(I-s)$. Hence $s \in \mathrm{span}(I-s)$, contradicting the fact that $I \in \mathcal{I}$.

In order to check (39.3), let $I, J \in \mathcal{I}$ with $|I \setminus J| = 1$ and $|J \setminus I| = 2$. Let $I \setminus J = \{i\}$ and $J \setminus I = \{j_1, j_2\}$. Assume that $I + j_1 \notin \mathcal{I}$. That is, $J + i - j_2 \notin \mathcal{I}$, and so, by (39.51) applied to $J - j_2$, $i \in \mathrm{span}(J - j_2)$. Therefore, $I \subseteq \mathrm{span}(J - j_2)$, and so $\mathrm{span}(I) \subseteq \mathrm{span}(J - j_2)$. So $j_2 \notin \mathrm{span}(I)$ (as $J \in \mathcal{I}$), and therefore, by (39.51) applied to I, $I + j_2 \in \mathcal{I}$.

So M is a matroid. We finally show that $\mathrm{span} = \mathrm{span}_M$. Choose $T \subseteq S$. To see that $\mathrm{span}(T) = \mathrm{span}_M(T)$, let I be a base of T (in M). Then (using (39.51)),

(39.52) $\qquad \mathrm{span}_M(T) = I \cup \{x \mid I + x \notin \mathcal{I}\} = \mathrm{span}(I) \subseteq \mathrm{span}(T).$

So we are done by showing $\mathrm{span}(T) \subseteq \mathrm{span}(I)$; that is, by (39.47)(ii), $T \subseteq \mathrm{span}(I)$. Choose $t \in T \setminus I$. By the maximality of I, we know $I + t \notin \mathcal{I}$, and hence, by (39.51), $t \in \mathrm{span}(I)$. ∎

39.8b. Characterizing matroids by flats

Conditions characterizing collections of flats of a matroid are given in the following theorem (Bergmann [1929]):

Theorem 39.10. *Let S be a set and let $\mathcal{F}$ be a collection of subsets of S. Then $\mathcal{F}$ is the collection of flats of a matroid if and only if:*

(39.53) $\qquad$ (i) $S \in \mathcal{F}$;
$\qquad\qquad$ (ii) *if $F_1, F_2 \in \mathcal{F}$, then $F_1 \cap F_2 \in \mathcal{F}$;*
$\qquad\qquad$ (iii) *if $F \in \mathcal{F}$ and $t \in S \setminus F$, and F' is the smallest flat containing $F + t$, then there is no flat F'' with $F \subset F'' \subset F'$.*

Proof. *Necessity.* Let $\mathcal{F}$ be the collection of flats of a matroid $M = (S, \mathcal{I})$. Condition (39.53)(i) is trivial, and condition (39.53)(ii) follows from $\text{span}_M(F_1 \cap F_2) \subseteq \text{span}_M(F_1) \cap \text{span}_M(F_2) = F_1 \cap F_2$. To see (39.53)(iii), suppose that such an F'' exists. Choose $s \in F'' \setminus F$. So $s \notin \text{span}_M(F)$. As $F' \not\subseteq F''$, we have $t \notin \text{span}_M(F+s)$. Therefore, by (39.47)(iii) for $T := F$, $s \notin \text{span}_M(F) = F'$, a contradiction.

Sufficiency. Let $\mathcal{F}$ satisfy (39.53). For $Y \subseteq S$, let $\text{span}(Y)$ be the smallest set in $\mathcal{F}$ containing Y. Since $F \in \mathcal{F} \iff \text{span}(F) = F$, it suffices to show that span satisfies the conditions (39.47). Here (39.47)(i) and (ii) are trivial. To see (39.47)(iii), let $T \subseteq S$, $t \in S \setminus T$, and $s \in \text{span}(T+t) \setminus \text{span}(T)$. Then $\text{span}(T) \subset \text{span}(T+s) \subseteq \text{span}(T+t)$. Hence, by (39.53)(iii), $\text{span}(T+s) = \text{span}(T+t)$, and hence $t \in \text{span}(T+s)$. ∎

39.8c. Characterizing matroids in terms of lattices

Bergmann [1929] and Birkhoff [1935a] characterized matroids in terms of lattices. A partially ordered set $(L, \leq)$ is called a *lattice* if

(39.54) (i) for all $A, B \in L$ there is a unique element, called $A \wedge B$, satisfying $A \wedge B \leq A, B$ and $C \leq A \wedge B$ for all $C \leq A, B$;
(ii) for all $A, B \in L$ there is a unique element, called $A \vee B$, satisfying $A \vee B \geq A, B$ and $C \geq A \vee B$ for all $C \geq A, B$.

$A \wedge B$ and $A \vee B$ are called the *meet* and *join* respectively of A and B. Here we assume lattices to be finite. Then a lattice has a unique minimal element, denoted by 0. The *rank* of an element A is the maximum number n of elements $x_1, \ldots, x_n$ with $0 < x_1 < \cdots < x_n = A$. An element of rank 1 is called a *point* or *atom*.

Call a lattice a *point lattice* if each element is a join of points, and a *matroid lattice* (or a *geometric lattice*) if it is isomorphic to the lattice of flats of a matroid. Trivially, each matroid lattice is a point lattice. Moreover, a matroid without loops and parallel elements is completely determined by the lattice of flats.

In the following theorem, the equivalence of (i) and (ii), and the implication (ii)⇒(iv) are due (in a different terminology) to Bergmann [1929]; the equivalence of (iii) and (iv) was shown by Birkhoff [1933], and the implication (iii)⇒(i) was shown by Birkhoff [1935a].

In a partially ordered set $(L, \leq)$ an element y is said to *cover* an element x if $x < y$ and there is no z with $x < z < y$.

Theorem 39.11. *For any finite point lattice $(L, \leq)$, with rank function r, the following are equivalent:*

(39.55) (i) *L is a matroid lattice;*
(ii) *for each $a \in L$ and each point p, if $p \not\leq a$, then $a \vee p$ covers a;*
(iii) *for each $a, b \in L$, if a and b cover $a \wedge b$, then $a \vee b$ covers a and b;*
(iv) *$r(a) + r(b) \geq r(a \vee b) + r(a \wedge b)$ for all $a, b \in L$.*

Proof. (i)⇒(iv): Let L be the lattice of flats of a matroid $M = (S, \mathcal{I})$, with rank function r_M. We can assume that M has no loops and no parallel elements. Then for any flat F we have $r(F) = r_M(F)$, since $r_M(F)$ is equal to the maximum number k of nonempty flats $F_1 \subset \cdots \subset F_k$ with $F_k = F$. So (iv) follows from Theorem 39.8.

(iv)⇒(iii): We first show that (iv) implies that if b covers a, then $r(b) = r(a)+1$. As b is a join of points, and as b covers a, we know that $b = a \vee p$ for some point p with $p \not\leq a$. Hence $r(b) = r(a \vee p) \leq r(a) + r(p) - r(a \wedge p) = r(a) + r(p) - r(0) = r(a) + 1$. As $r(b) > r(a)$, we have $r(b) = r(a) + 1$.

To derive (iii) from (iv), let a and b cover $a \wedge b$. Then $r(a) = r(b) = r(a \wedge b) + 1$. Hence $r(a \vee b) \leq r(a) + r(b) - r(a \wedge b) = r(a) + 1$. Hence $a \vee b$ covers a. Similarly, $a \vee b$ covers b.

(iii)⇒(ii): We derive (ii) from (iii) by induction on $r(a)$. If $a = 0$, the statement is trivial. If $a > 0$, let a' be an element covered by a. Then, by induction, $a' \vee p$ covers a'. So $a' = a \wedge (a' \vee p)$. Hence by (iii), $a \vee (a' \vee p) = a \vee p$ covers a.

(ii)⇒(i): Let S be the set of points of L, and for $f \in L$ define $F_f := \{s \in S \mid s \leq f\}$. Let $\mathcal{F} := \{F_f \mid f \in L\}$. Then for all $f_1, f_2 \in L$ we have:

(39.56) $\quad f_1 \leq f_2 \iff F_{f_1} \subseteq F_{f_2}$.

Here ⇒ is trivial, while ⇐ follows from the fact that for each $f \in L$ we have $f = \bigvee F_f$, as L is a point lattice.

By (39.56), $(L, \leq)$ is isomorphic to $(\mathcal{F}, \subseteq)$. Moreover, by (39.54)(i), $F_{f_1 \wedge f_2} = F_{f_1} \cap F_{f_2}$. So $\mathcal{F}$ is closed under intersections, implying (39.53)(ii), while (39.53)(i) is trivial. Finally, (39.53)(iii) follows from (39.55)(ii). ∎

Lattices satisfying (39.55)(iii) are called *upper semimodular*.

39.9. Further exchange properties

In this section we prove a number of exchange properties of bases, as a preparation to the forthcoming sections on matroid intersection algorithms.

An exchange property of bases, stronger than given in Theorem 39.6, is (Brualdi [1969c]):

Theorem 39.12. *Let $M = (S, \mathcal{I})$ be a matroid. Let B_1 and B_2 be bases and let $x \in B_1 \setminus B_2$. Then there exists a $y \in B_2 \setminus B_1$ such that both $B_1 - x + y$ and $B_2 - y + x$ are bases.*

Proof. Let C be the unique circuit in $B_2 + x$ (cf. (39.35)). Then $(B_1 \cup C) - x$ is spanning, since $x \in \text{span}_M(C - x) \subseteq \text{span}_M((B_1 \cup C) - x)$, implying $\text{span}((B_1 \cup C) - x) = \text{span}(B_1 \cup C) = S$.

Hence there is a base B_3 with $B_1 - x \subseteq B_3 \subseteq (B_1 \cup C) - x$. So $B_3 = B_1 - x + y$ for some y in $C - x$. Therefore, $B_2 - y + x$ is a base, as it contains no circuit (since C is the only circuit in $B_2 + x$). ∎

Let $M = (S, \mathcal{I})$ be a matroid. For any $I \in \mathcal{I}$ define the (bipartite) directed graph $D_M(I) = (S, A_M(I))$, or briefly $(S, A(I))$, by:

(39.57) $\quad A(I) := \{(y, z) \mid y \in I, z \in S \setminus I, I - y + z \in \mathcal{I}\}$.

Repeated application of the exchange property described in Theorem 39.12 gives (Brualdi [1969c]):

Corollary 39.12a. *Let $M = (S, \mathcal{I})$ be a matroid and let $I, J \in \mathcal{I}$ with $|I| = |J|$. Then $A(I)$ contains a perfect matching on $I \triangle J$.*[1]

Proof. By truncating M, we can assume that I and J are bases of M. We prove the lemma by induction on $|I \setminus J|$. We can assume $|I \setminus J| \geq 1$. Choose $y \in I \setminus J$. By Theorem 39.12, $I - y + z \in \mathcal{I}$ and $J - z + y \in \mathcal{I}$ for some $z \in J \setminus I$. By induction, applied to I and $J' := J - z + y$, $A(I)$ has a perfect matching N on $I \triangle J'$. Then $N \cup \{(y, z)\}$ is a perfect matching on $I \triangle J$. ∎

Corollary 39.12a implies the following characterization of maximum-weight bases:

Corollary 39.12b. *Let $M = (S, \mathcal{I})$ be a matroid, let B be a base of M, and let $w : S \to \mathbb{R}$ be a weight function. Then B is a base of maximum weight $\iff w(B') \leq w(B)$ for every base B' with $|B' \setminus B| = 1$.*

Proof. Necessity being trivial, we show sufficiency. Suppose to the contrary that there is a base B' with $w(B') > w(B)$. Let N be a perfect matching in $A(B)$ covering $B \triangle B'$. As $w(B') > w(B)$, there is an edge (y, z) in N with $w(z) > w(y)$, where $y \in B \setminus B'$ and $z \in B' \setminus B$. Hence $w(B - y + z) > w(B)$, contradicting the condition. ∎

The following forms a counterpart to Corollary 39.12a (Krogdahl [1974, 1976,1977]):

Theorem 39.13. *Let $M = (S, \mathcal{I})$ be a matroid and let $I \in \mathcal{I}$. Let $J \subseteq S$ be such that $|I| = |J|$ and such that $A(I)$ contains a unique perfect matching N on $I \triangle J$. Then J belongs to $\mathcal{I}$.*

Proof. Since N is unique, we can order N as $(y_1, z_1), \ldots, (y_t, z_t)$ such that $(y_i, z_j) \notin A(I)$ if $1 \leq i < j \leq t$. Suppose that $J \notin \mathcal{I}$, and let C be a circuit contained in J. Choose the smallest i with $z_i \in C$. Then $(y_i, z) \notin A(I)$ for all $z \in C - z_i$ (since $z = z_j$ for some $j > i$). Therefore, $z \in \text{span}(I - y_i)$ for all $z \in C - z_i$. So $C - z_i \subseteq \text{span}(I - y_i)$, and therefore $z_i \in C \subseteq \text{span}(C - z_i) \subseteq \text{span}(I - y_i)$, contradicting the fact that $I - y_i + z_i$ is independent. ∎

This implies:

Corollary 39.13a. *Let $M = (S, \mathcal{I})$ be a matroid and let $I \in \mathcal{I}$. Let $J \subseteq S$ be such that $|I| = |J|$ and $r_M(I \cup J) = |I|$, and such that $A(I)$ contains a unique perfect matching N on $I \triangle J$. Let $s \notin I \cup J$ with $I + s \in \mathcal{I}$. Then $J + s \in \mathcal{I}$.*

Proof. Let t be a new element and let $M' = (S \cup \{t\}, \mathcal{I}')$ be the matroid with $F \in \mathcal{I}'$ if and only if $F \setminus \{t\} \in \mathcal{I}$. Then $N' := N \cup \{(t, s)\}$ forms a

---

[1] A *perfect matching on* a vertex set U in a digraph is a set of vertex-disjoint arcs such that U is the set of tails and heads of these arcs.

unique perfect matching on $(I \triangle J) \cup \{s,t\}$ in $D_{M'}(I \cup \{t\})$ (since there is no arc from t to $J \setminus I$, as $I + j \notin \mathcal{I}$ for all $j \in J \setminus I$, since $r_M(I \cup J) = |I|$). So by Theorem 39.13, $J \cup \{s\}$ is independent in M', and hence in M. ∎

39.9a. Further properties of bases

Bases satisfy the following exchange property, stronger than that described in Theorem 39.12 (conjectured by G.-C. Rota, and proved by Brylawski [1973], Greene [1973], Woodall [1974a]):

(39.58)   if B_1 and B_2 are bases and B_1 is partitioned into X_1 and Y_1, then B_2 can be partitioned into X_2 and Y_2 such that $X_1 \cup Y_2$ and $Y_1 \cup X_2$ are bases.

This will be proved in Section 42.1a (using the matroid union theorem).

Other exchange properties of bases were given by Greene [1974a] and Kung [1978a]. Decomposing exchanges was studied by Gabow [1976b].

In Schrijver [1979c] it was shown that the exchange property described in Corollary 16.8b for bipartite graphs and, more generally, in Theorem 9.12 for directed graphs, in fact characterizes systems that correspond to matroids.

To this end, let U and W be disjoint sets and let Λ be a collection of pairs (X, Y) with $X \subseteq U$ and $Y \subseteq W$. Call (U, W, Λ) a *bimatroid* (or *linking system*) if:

(39.59)   (i) $(\emptyset, \emptyset) \in \Lambda$;
        (ii) if $(X, Y) \in \Lambda$ and $x \in X$, then $(X - x, Y - y) \in \Lambda$ for some $y \in Y$;
        (iii) if $(X, Y) \in \Lambda$ and $y \in Y$, then $(X - x, Y - y) \in \Lambda$ for some $x \in X$;
        (iv) if $(X_1, Y_1), (X_2, Y_2) \in \Lambda$, then there is an $(X, Y) \in \Lambda$ with $X_1 \subseteq X \subseteq X_1 \cup X_2$ and $Y_2 \subseteq Y \subseteq Y_1 \cup Y_2$.

Note that (ii) and (iii) imply that $|X| = |Y|$ for each $(X, Y) \in \Lambda$.

To describe the relation with matroids, define:

(39.60) $\mathcal{B} := \{(U \setminus X) \cup Y \mid (X, Y) \in \Lambda\}$.

So $\mathcal{B}$ determines Λ. Then (Schrijver [1979c]):

(39.61) (U, W, Λ) is a bimatroid if and only if $\mathcal{B}$ is the collection of bases of a matroid on $U \cup W$, with $U \in \mathcal{B}$.

So bimatroids are in one-to-one correspondence with pairs (M, B) of a matroid M and a base B of M, and the conditions (39.59) yield a characterization of matroids. An equivalent axiom system characterizing matroids was given by Kung [1978b].

(Bapat [1994] gave an extension of König's matching theorem to bimatroids.)

39.10. Further results and notes

39.10a. Further notes

Dilworth [1944] showed that if $r : \mathcal{P}(S) \to \mathbb{Z}$ satisfies (39.38) and $r(U) \geq 0$ if $U \neq \emptyset$, then

(39.62) $\mathcal{I} := \{I \subseteq S \mid \forall \text{ nonempty } U \subseteq I : |U| \leq r(U)\}$

is the collection of independent sets of a matroid M. Its rank function satisfies:

(39.63) $r_M(U) = \min(r(U_1) + \cdots + r(U_t))$,

where the minimum ranges over partitions of U into nonempty subsets $U_1, \ldots, U_t$ ($t \geq 0$). If $G = (V, E)$ is a graph, and we define $r(F) := |\bigcup F| - 1$ for $F \subseteq E$, we obtain the cycle matroid of G (this also was shown by Dilworth [1944]).[2]

Conforti and Laurent [1988] showed the following sharpening of Corollary 39.8a. Let $\mathcal{C}$ be a collection of subsets of a set S and let $f : \mathcal{C} \to \mathbb{Z}_+$. Let $\mathcal{I}$ be the collection of subsets T of S with $|T \cap U| \leq f(U)$ for each $U \in \mathcal{C}$. For $T \subseteq S$, let $r(T)$ be the maximum size of a subset of T that belongs to $\mathcal{I}$. Then $(S, \mathcal{I})$ is a matroid if and only if r satisfies the submodular inequality (39.38)(ii) for all $Y, Z \in \mathcal{C}$ with $Y \cap Z \neq \emptyset$. In fact, in the right-hand side of this inequality, r may be replaced by f.

Jensen and Korte [1982] showed that there is no polynomial-time algorithm to find the minimum size of a circuit of a matroid, if the matroid is given by an oracle for testing independence. For binary matroids (represented by binary vectors), the problem of finding a minimum-size circuit was shown by Vardy [1997] to be NP-complete (solving a problem of Berlekamp, McEliece, and van Tilborg [1978], who showed the NP-completeness of finding the minimum size of a circuit containing a given element of the matroid, and of finding a circuit of given size). If we know that a matroid is binary, a vector representation can be derived by a polynomially bounded number of calls from an independence testing oracle.

For further studies of the complexity of matroid properties, see Hausmann and Korte [1978], Robinson and Welsh [1980], and Jensen and Korte [1982].

Extensions of matroid theory to infinite structures were considered by Rado [1949a], Bleicher and Preston [1961], Johnson [1961], and Dlab [1962,1965].

Standard references on matroid theory are Welsh [1976] and Oxley [1992]. The book by Truemper [1992] focuses on decomposition of matroids. Earlier texts were given by Tutte [1965a,1971]. Elementary introductions to matroids were given by Wilson [1972b,1973], and a survey with applications to electrical networks and statics by Recski [1989]. Bixby [1982], Faigle [1987], Lee and Ryan [1992], and Bixby and Cunningham [1995] survey matroid optimization and algorithms. White [1986, 1987,1992] offers a collection of surveys on matroids, and Kung [1986] is a source book on matroids. Stern [1999] focuses on semimodular lattices. Books discussing matroid optimization include Lawler [1976b], Papadimitriou and Steiglitz [1982], Gondran and Minoux [1984], Nemhauser and Wolsey [1988], Parker and Rardin [1988], Cook, Cunningham, Pulleyblank, and Schrijver [1998], and Korte and Vygen [2000].

39.10b. Historical notes on matroids

The idea of a matroid, that is, of abstract dependence, seems to have been developed historically along a number of independent lines during the period 1900-1935. Independently, different axiom systems were given, each of which is equivalent to

---
[2] $\bigcup F$ denotes the union of the edges (as sets) in F.

Section 39.10b. Historical notes on matroids   673

that of a matroid. It indicates the naturalness of the concept. Only at the end of the 1930s a synthesis of the different streams was obtained.

There is a line, starting with the *Dualgruppen* (dual groups = lattices) of Dedekind [1897,1900], introduced in order to study modules (= additive subgroups) of numbers. They give rise to lattices satisfying what Dedekind called the *Modulgesetz* (module law). Later, independently, Birkhoff [1933] studied such lattices, calling them initially *B*-lattices, and later (after he had learned about Dedekind's earlier work), *modular lattices*. Both Dedekind and Birkhoff considered, in their studies of modular lattices, an auxiliary property that characterizes so-called *semimodular lattices*. If the lattice is a point lattice (that is, each element of the lattice is a join of atoms (points)), then such semimodular lattices are exactly the lattices of flats of a matroid. This connection was pointed out by Birkhoff [1935a] directly after Whitney's introduction of matroids.

A second line concerns exchange properties of bases. It starts with the new edition of the *Ausdehnungslehre* of Grassmann [1862], where he showed that each linearly independent set can be extended to a bases, using elements from a given base. Next Steinitz [1910], in his fundamental paper *Algebraische Theorie der Körper* (Algebraic Theory of Fields), showed that algebraic dependence has a number of basic properties, which makes it into a matroid (like the equicardinality of bases), and he derived some other properties from these basic properties (thus deriving essentially properties of matroids). In a subsequent paper, Steinitz [1913] gave, as an auxiliary result, the property that is now called *Steinitz' exchange property* for linearly independent sets of vectors. Steinitz did not mention the similarities to his earlier results on algebraic dependence. These similarities were observed by Haupt [1929a] and van der Waerden [1930] in their books on 'modern' algebra. They formulated properties shared by linear and algebraic dependence that are equivalent to matroids. In the second edition of his book, van der Waerden [1937] condensed these properties to three properties, and gave a unified treatment of linear and algebraic dependence. Mac Lane [1938] observed the relation of this work to the work on lattices and matroids.

A third line pursued the axiomatization of geometry, which clearly can be rooted back to as early as Euclid. At the beginning of the 20th century this was considered by, among others, Hilbert and Veblen. Bergmann [1929] aimed at giving a lattice-theoretical basis for affine geometry, and from lattice-theoretical conditions equivalent to matroids (cf. Theorem 39.11 above) he derived a number of properties, like the equicardinality of bases and the submodularity of the rank function. In their book *Grundlagen der Mathematik I* (Foundations of Mathematics I), Hilbert and Bernays [1934] gave axioms for the collinearity of triples of points, amounting to the fact that any two distinct points belong to exactly one line. A direct extension of these axioms to general dimensions gives the axioms described by Nakasawa [1935], that are again equivalent to the matroid axioms. He introduced the concept of a $\mathcal{B}_1$-space, equivalent to a matroid. In fact, the only reference in Nakasawa [1935] is to the book *Grundlagen der Elementargeometrie* (Foundations of Elementary Geometry) of Thomsen [1933], in which a different axiom system, the *Zyklenkalkül* (cycle calculus), was given (not equivalent to matroids). Nakasawa only gave subsets of linear spaces as an example. In a sequel to his paper, Nakasawa [1936b] observed that his axioms are equivalent to those of Whitney. The same axiom system as Nakasawa's, added with a continuity axiom, was given by Pauc [1937]. In Haupt,

Nöbeling, and Pauc [1940] the concept of an *Abhängigkeitsraum* (dependence space) based on these axioms was investigated.

The fourth 'line' was that of Whitney [1935], who introduced the notion of a matroid as a concept by itself. He was motivated by generalizing certain separability and duality phenomena in graphs, studied by him before. This led him to show that each matroid has a dual. While Whitney showed the equivalence of several axiom systems for matroids, he did not consider an axiom system based on a closure operation or on flats. Whitney gave linear dependence as an example, but not algebraic dependence. In a paper in the same year and journal, Birkhoff [1935a] showed the relation of Whitney's work with lattices.

We now discuss some historical papers more extensively, in a more or less chronological order.

1894-1900: Dedekind: lattices

In the supplements to the fourth edition of *Vorlesungen über Zahlentheorie* (Lectures on Number Theory) by Lejeune Dirichlet [1894], R. Dedekind introduced the notion of a *module* as any nonempty set of (real or complex) numbers closed under addition and subtraction, and he studied the lattice of all modules ordered by inclusion. He called A *divisible by* B if $A \subseteq B$. Trivially, the lattice operations are given by $A \wedge B = A \cap B$ and $A \vee B = A + B$. In fact, Dedekind denoted $A \cap B$ by $A - B$.

He gave the following 'charakteristischen Satz' (characteristic theorem):

Ist m theilbar durch d, und a ein beliebiger Modul, so ist
$$m + (a - d) = (m + a) - d.\ ^3$$

In modern notation, for all a, b, c:

(39.64)   if $a \leq c$, then $a \vee (b \wedge c) = (a \vee b) \wedge c$,

which is now known as the *modular law*, and lattices obeying it are called *modular lattices*.

Next, Dedekind [1897] introduced the notion of a lattice under the name *Dualgruppe* (dual group), motivated by similarities observed by him between operations on modules and those for logical statements as given in the book *Algebra der Logik* (Algebra of Logic) by Schröder [1890]. Dedekind mentioned, as examples, subsets of a set, modules, ideals in a finite field, subgroups of a group, and all fields, and he introduced the name *module law* for property (39.64):

> ich will es daher das Modulgesetz nennen, und jede Dualgruppe, in welcher es herrscht, mag eine Dualgruppe vom Modultypus heißen.[4]

---

[3] *If m is divisible by d, and a is an arbitrary module, then*
$$m + (a - d) = (m + a) - d.$$

[4] I will therefore call it the module law, and every dual group in which it holds, may be called a dual group of module type.

Dedekind [1900] continued the study of modular lattices, and showed that each modular lattice allows a rank function $r : M \to \mathbb{Z}_+$ with the property that for all a, b:

(39.65) (i) $r(0) = 0$;
(ii) $r(b) = r(a) + 1$ if b covers a;
(iii) $r(a \wedge b) + r(a \vee b) = r(a) + r(b)$.

In fact, this characterizes modular lattices.

In proving (39.65), Dedekind showed that each modular lattice satisfies

(39.66) if a and b cover c, and $a \neq b$, then $a \vee b$ covers a and b,

which is the property characterizing *upper semimodular lattices*, a structure equivalent to matroids.

1862-1913: Grassmann, Steinitz: linear and algebraic dependence

The basic exchange property of linear independence was formulated by Grassmann [1862], in his book *Die Ausdehnungslehre*, as follows (in his terminology, vectors are quantities):

> 20. *Wenn m Grössen $a_1, \ldots a_m$, die in keiner Zahlbeziehung zu einander stehen, aus n Grössen $b_1, \ldots b_n$ numerisch ableitbar sind, so kann man stets zu den m Grössen $a_1, \ldots a_m$ noch $(n - m)$ Grössen $a_{m+1}, \ldots a_n$ von der Art hinzufügen, dass sich die Grössen $b_1, \ldots b_n$ auch aus $a_1, \ldots a_n$ numerisch ableiten lassen, und also das Gebiet der Grössen $a_1, \ldots a_n$ identisch ist dem Gebiete der Grössen $b_1, \ldots b_n$; auch kann man jene $(n - m)$ Grössen aus den Grössen $b_1, \ldots b_n$ selbst entnehmen.*[5]

This property was also given by Steinitz [1913] (see below), but before that, Steinitz proved it for algebraic independence. In his fundamental paper *Algebraische Theorie der Körper* (Algebraic Theory of Fields), Steinitz [1910] studied, in § 22, algebraic dependence in field extensions. The statements proved are as follows, where L is a field extension of field K. Throughout, a is *algebraically dependent* on S if a is algebraic with respect to the field extension $K(S)$; in other words, if there is a nonzero polynomial $p(x) \in K(S)[x]$ with $p(a) = 0$.

Calling a set a *system*, he first observed:

> 1. *Hängt das Element a vom System S algebraisch ab, so gibt es ein endliches Teilsystem S' von S, von welchem a algebraisch abhängt.*[6]

and next he showed:

> 2. *Hängt S_3 von S_2, S_2 von S_1 algebraisch ab, so ist S_3 algebraisch abhängig von S_1.*[7]

---

[5] 20. *If m quantities $a_1, \ldots a_m$, that stand in no number relation to each other, are numerically derivable from n quantities $b_1, \ldots b_n$, then one can always add to the m quantities $a_1, \ldots a_m$ another $(n - m)$ quantities $a_{m+1}, \ldots a_n$ such that the quantities $b_1, \ldots b_n$ can also be derived numerically from $a_1, \ldots a_n$, and that hence the domain of the quantities $a_1, \ldots a_n$ is identical to the domain of the quantities $b_1, \ldots b_n$; one also can take those $(n - m)$ quantities from the quantities $b_1, \ldots b_n$ themselves.*

[6] 1. *If element a depends algebraically on the system S, then there is a finite subsystem S' of S on which a depends algebraically.*

[7] 2. *If S_3 depends algebraically on S_2, and S_2 on S_1, then S_3 is algebraically dependent on S_1.*

He called two sets S_1 and S_2 *equivalent* if S_1 depends algebraically on S_2, and conversely. A set is *reducible* if it has a proper subset equivalent to it. He showed:

> 3. *Jedes Teilsystem eines irreduziblen Systems ist irreduzibel.*
> 4. *Jedes reduzible System enthält ein endliches reduzibles Teilsystem.*[8]

and (after statement 5, saying that any two field extensions by equicardinal irreducible systems are isomorphic):

> 6. *Wird ein irreduzibles System S durch Hinzufügung eines Elementes a reduzibel, so ist a von S algebraisch abhängig.*[9]

From these properties, Steinitz derived:

> 7. *Ist S ein (in bezug auf K) irreduzibles System, das Element a in bezug auf K transzendent, aber von S algebraisch abhängig, so enthält S ein bestimmtes endliches Teilsystem T von folgender Beschaffenheit: a ist von T algebraisch abhängig; jedes Teilsystem von S, von welchem a algebraisch abhängt, enthält das System T; wird irgendein Element aus T durch a ersetzt, so geht S in ein äquivalentes irreduzibles System über; keinem der übrigen Elemente von S kommt diese Eigenschaft zu.*[10]

Steinitz proved this using only the properties given above (together with the fact that any $s \in S$ is algebraically dependent on S). Moreover, he derived from 7, (what is now called) *Steinitz' exchange property* for algebraic dependence:

> 8. *Es seien U und B endliche irreduzible Systeme von m bzw. n Elementen; es sei $n \leq m$ und B algebraisch abhängig von U. Dann sind im Falle $m = n$ die Systeme U und B äquivalent, im Falle $n < m$ aber ist U einem irreduziblen System äquivalent, welches aus B und $m - n$ Elementen aus U besteht.*[11]

This in particular implies that any two equivalent irreducible systems have the same size, and that the properties are equivalent to that determining a matroid.

In a subsequent paper, Steinitz [1913] proved a number of auxiliary statements on linear equations. Among other things, he showed (in his terminology, vectors are numbers, and a vector space is a module):

> *Besitzt der Modul M eine Basis von p Zahlen, und enthält er r linear unabhängige Zahlen $\beta_1, \ldots, \beta_r$, so besitzt er auch eine Basis von p Zahlen, unter denen die Zahlen $\beta_1, \ldots, \beta_r$ sämtlich vorkommen.*[12]

---

[8] 3. *Every subsystem of an irreducible system is irreducible.*
4. *Every reducible system contains a finite reducible subsystem.*

[9] 6. *If an irreducible system S becomes reducible by adding an element a, then a is algebraically dependent on S.*

[10] 7. *If S is an irreducible system (with respect to K), [and] the element a transcendent with respect to K, but algebraically dependent on S, then S contains a certain finite subsystem T with the following quality: a is algebraically dependent on T; every subsystem of S on which a depends algebraically, contains the system T; if any element from T is replaced by a, then S passes into an equivalent irreducible system; this property belongs to none of the other elements of S.*

[11] 8. *Let U and B be finite irreducible systems of m and n elements respectively; let $n \leq m$ and let B be algebraically dependent on U. Then, in case $m = n$, the systems U and B are equivalent, but in case $n < m$, U is equivalent to an irreducible system which consists of B and $m - n$ elements from U.*

[12] *If a module M possesses a base of p numbers, and it contains r linearly independent numbers $\beta_1, \ldots, \beta_r$, then it possesses also a base of p numbers, among which the numbers $\beta_1, \ldots, \beta_r$ all occur.*

Steinitz' proof of this in fact gives a stronger result, known as *Steinitz' exchange property*: the new base is obtained by extending $\beta_1, \ldots, \beta_r$ with vectors from the given base. So Steinitz came to the same result as Grassmann [1862] quoted above. In his paper, Steinitz [1913] did not make a link with similar earlier results in Steinitz [1910] on algebraic dependence.

1929: Bergmann

Inspired by Menger [1928a], who aimed at giving an axiomatic foundation for projective geometry on a lattice-theoretical basis, Bergmann [1929] gave an axiomatic foundation of affine geometry, again on the basis of lattices. Bergmann's article contains a number of proofs that in fact concern matroids, while he assumed, but not used, a complementation axiom (since he aimed at characterizing full affine spaces, not subsets of it): for each pair of elements $A \leq B$ there exist C_1 and C_2 with $A \vee C_1 = B$, $A \wedge C_1 = 0$, $B \wedge C_2 = A$, and $B \vee C_2 = 1$. This obviously implies (in the finite case) that

(39.67)   each element of the lattice is a join of points.

(A *point* is a minimal nonzero element.) It is property (39.67) that Bergmann uses in a number of subsequent arguments (and not the complementation axiom). His further axiom is:

(39.68)   for any element A and any point P of the lattice, there is no element B with $A < B < A \vee P$.

He called an ordered sequence $(P_1, \ldots, P_n)$ of points a *chain* (Kette) (*of* an element A), if $P_i \not\leq P_1 \vee \cdots \vee P_{i-1}$ for $i = 1, \ldots, n$ (and $A = P_1 \vee \cdots \vee P_n$). He derived from (39.67) and (39.68) that being a chain is independent of the order of the elements in the chain, and that any two chains of an element A have the same length:

Satz: Alle Ketten eines Elementes A haben dieselbe Gliederzahl.[13]

He remarked that under condition (39.67), this in turn implies (39.68).

Denoting the length of any chain of A by $|A|$, Bergmann showed that it is equal to the rank of A in the lattice, and he derived the submodular inequality:

$$|A| + |B| \geq |A + B| + |A \cdot B|.$$

(Bergmann denoted $\vee$ and $\wedge$ by $+$ and $\cdot$.) Thus he proved the submodularity of the rank function of a matroid. These results were also given by Alt [1936] in Menger's *mathematischen Kolloquium* in Vienna on 1 March 1935 (cf. Menger [1936a,1936b]).

1929-1937: Haupt, van der Waerden

Inspired by the work of Steinitz, in the books *Einführung in die Algebra* (Introduction to Algebra) by Haupt [1929a,1929b] and *Moderne Algebra* (Modern Algebra) by van der Waerden [1930], the analogies between proof methods for linear and algebraic dependence were observed.

Haupt mentioned in his preface (after saying that his book will contain the modern developments of algebra):

---

[13] *Theorem: All chains of an element A have the same number of members.*

Demgemäß ist das vorliegende Buch durchweg beeinflußt von der bahnbrechenden „Algebraischen Theorie der Körper" von Herrn E. Steinitz, was hier ein für allemal hervorgehoben sei. Ferner stützt sich die Behandlung der linearen Gleichungen (vgl. *9,1* bis *9,4*), einer Anregung von Frl. E. Noether folgend, auf die von Herrn E. Steinitz gegebene Darstellung (vgl. das Zitat in *9,0*).[14]

(The quotation in Haupt's '*9,0*' is to Steinitz [1910,1913].)

A number of theorems on algebraic dependence were proved in Chapter 23 of Haupt [1929b] by referring to the proofs of the corresponding results on linear dependence in Chapter 9 of Haupt [1929a]. In the introduction of his Chapter 9, Haupt wrote:

Die Behandlung der linearen Gleichungen ist (soweit es geht) so angelegt, daß sich ein Teil der dabei gewonnenen Sätze auf *Systeme von algebraisch abhängigen Elementen überträgt*, was später (*23,6*) dargelegt wird.[15]

In the first edition of his book, van der Waerden [1930] listed the properties of algebraic dependence:

Die Relation der algebraischen Abhängigkeit hat demnach die folgenden Eigenschaften:
1. a ist abhängig von sich selbst, d.h. von der Menge $\{a\}$.
2. Ist a abhängig von M, so hängt es auch von jeder Obermenge von M ab.
3. Ist a abhängig von M, so ist a schon von einer endlichen Untermenge $\{m_1, \ldots, m_n\}$ von M (die auch leer sein kann) abhängig.
4. Wählt man diese Untermenge minimal, so ist jedes m_i von a und den übrigen m_j abhängig.
Weiter gilt:
5. Ist a abhängig von M und jedes Element von M abhängig von N, so ist a abhängig von N.[16]

Following Steinitz, van der Waerden called two sets *equivalent* if each element of the one set depends algebraically on the other set, and vice versa, while a set is *irreducible* if no element of it depends algebraically on the remaining.

Using only the properties 1-5, van der Waerden derived that each set contains an irreducible set equivalent to it, and that if $M \subseteq N$, then each irreducible subset of M equivalent to M can be extended to an irreducible subset of N equivalent to N — in other words, inclusionwise minimal subsets of M equivalent to M are

---

[14] Accordingly, the present book is invariably influenced by the pioneering 'Algebraic Theory of Fields' by Mr E. Steinitz, which be emphasized here once and for all. Further, following a suggestion by Miss E. Noether, the treatment of linear equations (cf. *9,1* to *9,4*) leans on the presentation by Mr E. Steinitz (cf. the quotation in *9,0*).

[15] The treatment of linear equations is (as far as it goes) made such that a part of the theorems obtained therewith *transfers to systems of algebraically dependent elements*, which will be discussed later (*23,6*).

[16] The relation of algebraic dependence has therefore the following properties:
1. a is dependent on itself, that is, on the set $\{a\}$.
2. If a is dependent on M, then it also depends on every superset of M.
3. If a is dependent on M, then a is dependent already on a finite subset $\{m_1, \ldots, m_n\}$ of M (that can also be empty).
4. If one chooses this subset minimal, then every m_i is dependent on a and the remaining m_j.
Further it holds:
5. If a is dependent on M and every element of M is dependent on N, then a is dependent on N.

Section 39.10b. Historical notes on matroids   679

independent, and inclusionwise maximal independent subsets of M are equivalent to M.

Van der Waerden [1930] also showed that two equivalent irreducible systems have the same size, but in the proof he uses polynomials. This is not necessary, since the properties 1-5 determine a matroid.

Van der Waerden noticed the analogy with linear dependence, treated in his § 28, where he uses specific facts on linear equations:

> Tatsächlich gelten für die dort betrachtete lineare Abhängigkeit dieselben Regeln 1 bis 5, die für die algebraische Abhängigkeit in § 61 aufgestellt wurden; man kann also alle Beweise wörtlich übertragen.[17]

In the second edition of his book, van der Waerden [1937] gave a unified treatment of linear and algebraic dependence, slightly different from the first edition. As for linear dependence he stated in § 33:

> Drei Grundsätze genügen. Der erste ist ganz selbstverständlich.
> Grundsatz 1. *Jedes u_i ($i = 1, \ldots, n$) ist von $u_1, \ldots, u_n$ linear abhängig.*
> Grundsatz 2. *Ist v linear abhängig von $u_1, \ldots, u_n$, aber nicht von $u_1, \ldots, u_{n-1}$, so ist u_n linear abhängig von $u_1, \ldots, u_{n-1}, v$.*
> [$\cdots$]
> Grundsatz 3. *Ist w linear abhängig von $v_1, \ldots, v_s$ und ist jedes v_j ($j = 1, \ldots, s$) linear abhängig von $u_1, \ldots, u_n$, so ist w linear abhängig von $u_1, \ldots, u_n$.*[18]

The same axioms are given in § 64 of van der Waerden [1937], with 'linear' replaced by 'algebraisch'.

Next, van der Waerden called elements $u_1, \ldots, u_n$ (linearly or algebraically) *independent* if none of them depend on the rest of them. Among the consequences of these principles, he mentioned that if $u_1, \ldots, u_{n-1}$ are independent but $u_1, \ldots, u_{n-1}, u_n$ are not, then u_n is dependent on $u_1, \ldots, u_{n-1}$, and that each finite system of elements $u_1, \ldots, u_n$ contains a (possibly empty) independent subsystem on which each u_i is dependent. He called two systems $u_1, \ldots, u_n$ and $v_1, \ldots, v_s$ *equivalent* if each v_k depends on $u_1, \ldots, u_n$ and each u_i depends on $v_1, \ldots, v_s$, and he now derived from the three principles that two equivalent independent systems have the same size.

Mac Lane [1938] observed that the axioms introduced by Whitney [1935] and those by van der Waerden [1937] determine equivalent structures.

1934: Hilbert, Bernays: collinearity axioms

Axiom systems for points and lines in a plane were given by Hilbert [1899] in his book *Grundlagen der Geometrie* (Foundations of Geometry), and by Veblen [1904].

---
[17] In fact, the same rules 1 to 5, that were formulated for algebraic dependence in § 61, hold for the linear dependence considered there; one can transfer therefore all proofs word for word.
[18] Three principles suffice. The first one is fully self-evident.
   Principle 1. *Every u_i ($i = 1, \ldots, n$) is linearly dependent on $u_1, \ldots, u_n$.*
   Principle 2. *If v is linearly dependent on $u_1, \ldots, u_n$, but not on $u_1, \ldots, u_{n-1}$, then u_n is linearly dependent on $u_1, \ldots, u_{n-1}, v$.*
   [$\cdots$]
   Principle 3. *If w is linearly dependent on $v_1, \ldots, v_s$ and every v_j ($j = 1, \ldots, s$) is linearly dependent on $u_1, \ldots, u_n$, then w is linearly dependent on $u_1, \ldots, u_n$.*

Basis is the axiom that any two distinct points are in exactly one line. Note that this axiom determines precisely all matroids of rank at most 3 with no parallel elements (by taking the lines as maximal flats).

One of the axioms of Veblen is:

> Axiom VI. If points C and D ($C \neq D$) lie on the line AB, then A lies on the line CD.

This axiom corresponds to axiom 3) in the book *Grundlagen der Mathematik* (Foundations of Mathematics) of Hilbert and Bernays [1934], who aim to make an axiom system based on points only:

> Dabei empfiehlt es sich für unseren Zweck, von dem HILBERTschen Axiomensystem darin abzuweichen, daß wir nicht die Punkte und die Geraden als zwei Systeme von Dingen zugrunde legen, sondern *nur die Punkte als Individuen nehmen*.[19]

The axiom system of Hilbert and Bernays is in terms of a relation Gr to describe collinearity of triples of points (where (x) stands for $\forall x$, (Ex) for $\exists x$, and $\overline{P}$ for the negation of P):

> I. Axiome der Verknüpfung.
>
> 1) $(x)(y)Gr(x,x,y)$.
> „x,x,y liegen stets auf einer Geraden."
> 2) $(x)(y)(z)(Gr(x,y,z) \to Gr(y,x,z)\&Gr(x,z,y))$.
> „Wenn x,y,z auf einer Geraden liegen, so liegen stets auch y,x,z sowie auch x,z,y auf einer Geraden."
> 3) $(x)(y)(z)(u)(Gr(x,y,z)\&Gr(x,y,u)\&x \neq y \to Gr(x,z,u))$.
> „Wenn x,y, verschiedene Punkte sind und wenn x,y,z sowie x,y,u auf einer Geraden liegen, so liegen stets auch x,z,u auf einer Geraden."
> 4) $(Ex)(Ey)(Ez)\overline{Gr}(x,y,z)$.
> „Es gibt Punkte x,y,z, die nicht auf einer Geraden liegen."[20]

The axioms 1) and 2) in fact tell that the relation Gr is determined by unordered triples of distinct points. The exchange axiom 3) is a special case of the matroid axiom for circuits in a matroid.

Hilbert and Bernays extended the system by axioms for a betweenness relation Zw for ordered triples of points, and a parallelism relation Par for ordered quadruples of points.

---

[19] At that it is advisable for our purpose to deviate from HILBERT's axiom system in that we do not lay the points and the lines as two systems of things as base, but *take only the points as individuals*.

[20] I. Axioms of connection.

1) $(x)(y)Gr(x,x,y)$.
'x,x,y always lie on a line.'
2) $(x)(y)(z)(Gr(x,y,z) \to Gr(y,x,z)\&Gr(x,z,y))$.
'If x,y,z lie on a line, then also y,x,z as well as x,z,y always lie on a line.'
3) $(x)(y)(z)(u)(Gr(x,y,z)\&Gr(x,y,u)\&x \neq y \to Gr(x,z,u))$.
'If x,y, are different points and if x,y,z as well as x,y,u lie on a line, then also x,z,u always lie on a line.'
4) $(Ex)(Ey)(Ez)\overline{Gr}(x,y,z)$.
'There are points x,y,z, that do not lie on a line.'

1933–1935: Birkhoff: Lattices

In his paper 'On the combination of subalgebras', Birkhoff [1933] ('Received 15 May 1933') wrote:

> The purpose of this paper is to provide a point of vantage from which to attack combinatorial problems in what may be termed modern, synthetic, or abstract algebra. In this spirit, a research has been made into the consequences and applications of seven or eight axioms, only one [V] of which itself is new.

The axioms are those for a lattice, added with axiom V, that amounts to (39.64) above. Any lattice satisfying this condition is called by Birkhoff in this paper a 'B-lattice'. In an addendum, Birkhoff [1934b] mentioned that O. Ore had informed him that part of his results had been obtained before by Dedekind [1900]. Therefore, Birkhoff [1935b] renamed it to *modular lattice*.

Birkhoff [1933] mentioned, as examples, the classes of normal subgroups and of characteristic subgroups of a group. Other examples mentioned are the ideals of a ring, and the linear subspaces of Euclidean space. (Both examples actually give sublattices of the lattice of all normal subgroups of the corresponding groups.)

Like Dedekind, Birkhoff [1933] showed that (39.64) implies (39.66). Lattices satisfying (39.66) are called (upper) *semimodular*. Birkhoff showed that any upper semimodular lattice has a rank function satisfying (39.65)(i) and (ii) and satisfying the submodular law:

$$(39.69) \qquad r(a \cap b) + r(a \cup b) \leq r(a) + r(b).$$

This characterizes upper semimodular lattices.

Birkhoff noticed that this implies that the modular lattices are exactly those lattices satisfying both (39.66) and its symmetric form:

$$(39.70) \qquad \text{if } c \text{ covers } a \text{ and } b \text{ and } a \neq b, \text{ then } a \text{ and } b \text{ cover } a \wedge b.$$

Birkhoff [1935c] showed that the partition lattice is upper semimodular, that is, satisfies (39.66), and hence has a rank function satisfying the submodular inequality[21]. Thus the complete graph, and hence any graph, gives a geometric lattice (and hence a matroid — however, Whitney's work seems not to have been known yet to Birkhoff at the time of writing this paper).

In a number of other papers, Birkhoff [1934a,1934c,1935b] made a further study of modular lattices, and gave relations to projective geometries (in which the collection of all flats gives a modular lattice). Klein-Barmen [1937] further investigated semimodular lattices (called by him *Birkhoffsche Verbände* (Birkhoff lattices)), of which he found several lattice-theoretical characterizations.

1935: Whitney: Matroids

Whitney [1935] (presented to the American Mathematical Society, September 1934) introduces the notion of matroid as follows:

---

[21] In fact, Birkhoff [1935c] claimed the modular equality for the rank function of a partition lattice (page 448), but this must be a typo, witness the formulation of, and the reference in, the first footnote on that page.

Let $C_1, C_2, \cdots, C_n$ be the columns of a matrix M. Any subset of these columns is either linearly independent or linearly dependent; the subsets thus fall into two classes. These classes are not arbitrary; for instance, the two following theorems must hold:
    (a) Any subset of an independent set is independent.
    (b) If N_p and N_{p+1} are independent sets of p and $p+1$ columns respectively, then N_p together with some column of N_{p+1} forms an independent set of $p+1$ columns.
There are other theorems not deducible from this; for in § 16 we give an example of a system satisfying these two theorems but not representing any matrix. Further theorems seem, however, to be quite difficult to find. Let us call a system obeying (a) and (b) a "matroid." The present paper is devoted to a study of the elementary properties of matroids. The fundamental question of completely characterizing systems which represent matrices is left unsolved. In place of the columns of a matrix we may equally well consider points or vectors in a Euclidean space, or polynomials, etc.

In the paper, Whitney observed that forests in a graph form the independent sets of a matroid, for which reason he carried over various terms from graphs to matroids.

Whitney described several equivalent axiom systems for the notion of matroid. First, he showed that the rank function is characterized by (39.42), and he derived that it is submodular. Next, he showed that the collection of bases is characterized by (39.33)(ii), and the collection of circuits by (39.34)(iii). Moreover, he showed that complementing all bases gives again a matroid, the dual matroid, and that the dual of a linear matroid is again a linear matroid. In the paper, he also studied separability and representability of matroids. The example given in Whitney's § 16 (mentioned in the above quotation), is in fact the well-known Fano matroid — he apparently did not consider matrices over GF(2). However, in an appendix of the paper, he characterized the matroids representable by a matrix 'of integers mod 2': a matroid is representable over GF(2) if and only if any sum (mod 2) of circuits can be partitioned into circuits.

In a subsequent paper 'Abstract linear independence and lattices', Birkhoff [1935a] pointed out the relations of Whitney's work with Birkhoff's earlier work on semimodular lattices. He stated:

> In a preceding paper, Hassler Whitney has shown that it is difficult to distinguish theoretically between the properties of linear dependence of ordinary vectors, and those of elements of a considerably wider class of systems, which he has called "matroids."
> Now it is obviously impossible to incorporate all of the heterogeneous abstract systems which are constantly being invented, into a body of systematic theory, until they have been classified into two or three main species. The purpose of this note is to correlate matroids with abstract systems of a very common type, which I have called "lattices."

Birkhoff showed that a lattice is isomorphic to the lattice of flats of a matroid if and only if the lattice is semimodular, that is, satisfies (39.66), and each element is a join of atoms.

In the paper 'Some interpretations of abstract linear dependence in terms of projective geometry', MacLane [1936] gave a geometric interpretation of matroids. He introduced the notion of a 'schematic n-dimensional figure', consisting of 'k-dimensional planes' for $k = 1, 2, \ldots$. Each such plane is a subset of an (abstract) set of 'points', with the following axioms (for any appropriate k):

(39.71) (i) any k points belonging to no $k-1$-dimensional plane, belong to a unique k-dimensional plane; moreover, this plane is contained in any plane containing these k points;

(ii) every k-dimensional plane contains k points that belong to no $k-1$-dimensional plane.

MacLane mentioned that there is a 1-1 correspondence between schematic figures and the collections of flats of matroids. As a consequence he mentioned that a schematic n-dimensional figure is completely determined by its collection of $n-1$-dimensional planes (as a matroid is determined by its hyperplanes = complements of cocircuits).

1935: Nakasawa: Abhängigkeitsräume

In the paper *Zur Axiomatik der linearen Abhängigkeit. I* (On the axiomatics of linear dependence. I) in *Science Reports of the Tokyo Bunrika Daigaku* (Tokyo University of Literature and Science), Nakasawa [1935] introduced an axiom system for dependence, that he proved to be equivalent to matroids (in a different terminology).

He was motivated by an axiom system described by Thomsen [1933] in his book *Grundlagen der Elementargeometrie* (Foundations of Elementary Geometry). Thomsen's 'cycle calculus' is an attempt to axiomatize relations (like coincidence, orthogonality, parallelism) between geometric objects (points, lines, etc.). Thomsen emphasized that existence questions often are inessential in elementary geometry:

> In der Tat erscheinen uns ja auch die Existenzaussagen als ein verhältnismäßig unwesentliches Beiwerk der Elementargeometrie. Ohne Zweifel empfinden wir als die eigentlich inhaltsvollsten und die wichtigsten Einzelaussagen der Elementargeometrie die von der folgenden reinen Form: „Wenn eine Reihe von geometrischen Gebilden, d.h. eine Anzahl von Punkten, Geraden, usw., gegeben vorliegt, und zwar derart, daß zwischen den gegebenen Punkten, Geraden usw. die und die geometrischen Lagebeziehungen bestehen (Koinzidenz, Senkrechtstehen, Parallellaufen, „Mittelpunkt sein" und anderes mehr), dann ist eine notwendige Folge dieser Annahme, daß auch noch diese bestimmte weitere geometrische Lagebeziehung gleichzeitig besteht." In Sätzen dieser Form kommt nichts von Existenzaussagen vor. Was das Wichtigste ist, nicht in den Folgerungen. Dann aber auch nicht in den Annahmen. Wir nehmen an: *Wenn* die und die Dinge in den und den Beziehungen gegeben vorliegen..., usw. Wir machen aber keinerlei Voraussetzungen darüber, ob eine solche Konfiguration in unserer Geometrie existieren kann. Der Schluß ist nur: Wenn sie existieren, dann .... Falls die Konfiguration gar nicht existiert, der Satz also gegenstandslos wird, betrachten wir ihn nach der üblichen Konvention „gegenstandslos, also richtig" als richtig.[22]

---

[22] Indeed, also the existence statements seem to us a relatively inessential side issue of elementary geometry. Undoubtedly, we find as the really most substantial and most important special statements of elementary geometry those of the following pure form: 'If a sequence of geometric creations, that is, a number of points, lines etc., are given to us, and that in such a way, that those and those geometric position relations exist between the given points, lines etc. (coincidence, orthogonality, parallelism, "being a centre", and other), then a necessary consequence of this assumption is that also this certain further geometric position relation exists at the same time.' In theorems of this form, no existence statements occur. What is most important: not in the consequences. But then neither in the assumptions. We assume: *If* those and those things are given

Thomsen aimed at founding axiomatically 'the partial geometry of all elementary geometric theorems without existence statements'. To that end, he introduced the concept of a *cycle*, which is an ordered finite sequence of abstract objects, which can be thought of as points, lines, etc. Certain cycles are 'correct' and the other 'incorrect' (essentially they represent a system of relations defining any binary group):

> A) *Axiom der Grundzyklen:* Der Zyklus $\alpha\alpha$ ist für jedes α richtig, der Zyklus α für kein α.
> B) *Axiom des Löschens:* $\beta_1\beta_2\ldots\beta_n\alpha\alpha \to \beta_1\beta_2\ldots\beta_n$; in Worten: Aus der Richtigkeit des Zyklus $\beta_1\beta_2\ldots\beta_n\alpha\alpha$ folgt auch die des Zyklus $\beta_1\beta_2\ldots\beta_n$.
> C) *Axiom des Umstellens:* $\beta_1\beta_2\ldots\beta_n \to \beta_2\beta_3\ldots\beta_n\beta_1$.
> D) *Axiom des Umkehrens:* $\beta_1\beta_2\ldots\beta_{n-1}\beta_n \to \beta_n\beta_{n-1}\ldots\beta_2\beta_1$.
> E) *Axiom des Anfügens:* $\beta_1\beta_2\ldots\beta_n$ und $\gamma_1\gamma_2\ldots\gamma_r \to \beta_1\beta_2\ldots\beta_n\gamma_1\gamma_2\ldots\gamma_r$.[23]

Axiom B) can be considered as a variant of Steinitz' exchange property. With the other axioms it implies that if $\beta_1\cdots\beta_n\alpha$ and $\gamma_1\cdots\gamma_r\alpha$ are cycles, then $\beta_1\cdots\beta_n\gamma_1\cdots\gamma_r$ is a cycle. Therefore, the set of all inclusionwise minimal nonempty sets containing a cycle form the circuits of a matroid.

The purpose of Nakasawa [1935] is to generalize Thomsen's axiom system:

> In der vorliegenden Untersuchung soll ein Axiomensystem für eine neue Formulierung der linearen Abhängigkeit des n-dimensionalen projektiven Raumes angegeben werden, indem wir hauptsächlich den *Zyklenkalkül*, den Herr G. Thomsen bei seiner Grundlegung der elementaren Geometrie hergestellt hat, hier in einem noch abstrakteren Sinne verwenden.[24]

While Thomsen's cycles relate to unions of circuits in a matroid, those of Nakasawa form the dependent sets of a matroid. His axiom system can be considered as a direct extension to higher dimensions of the collinearity axioms of Hilbert and Bernays given above.

He called the structure *der erste Verknüpfungsraum* (the first connection space), or a $\mathcal{B}_1$-*Raum* ($\mathcal{B}_1$-space), writing $a_1\cdots a_s$ for $a_1\cdots a_s = 0$:

> **Grundannahme:** Wir denken uns eine gewisse Menge der Elementen; $\mathcal{B}_1 \ni a_1, a_2,\cdots, a_s,\cdots$. Für gewisse Reihen der Elementen, die wir *Zyklen* nennen wollen, denken wir dazu die Relationen "*gelten*" oder "*gültig sein*", in Zeichen $a_1\cdots a_s = 0$, bzw. "*nicht gelten*" oder "*nicht gültig sein*", in Zeichen $a_1\cdots a_s \neq 0$. Diese Relationen sollen nun folgenden Axiomen genügen;

---

to us in those and those relations..., etc. We do not make any assumption on the fact if such a configuration can exist in our geometry. The conclusion is only: If they exist, then .... In case the configuration does not exist at all, and the theorem thus becomes meaningless, we consider it by the usual convention 'meaningless, hence correct' as correct.

[23]
A) *Axiom of ground cycles:* The cycle $\alpha\alpha$ is correct for each α, the cycle α for no α.
B) *Axiom of solving:* $\beta_1\beta_2\ldots\beta_n\alpha\alpha \to \beta_1\beta_2\ldots\beta_n$; in words: From the correctness of the cycle $\beta_1\beta_2\ldots\beta_n\alpha\alpha$ follows that of the cycle $\beta_1\beta_2\ldots\beta_n$.
C) *Axiom of transposition:* $\beta_1\beta_2\ldots\beta_n \to \beta_2\beta_3\ldots\beta_n\beta_1$.
D) *Axiom of inversion:* $\beta_1\beta_2\ldots\beta_{n-1}\beta_n \to \beta_n\beta_{n-1}\ldots\beta_2\beta_1$.
E) *Axiom of addition:* $\beta_1\beta_2\ldots\beta_n$ and $\gamma_1\gamma_2\ldots\gamma_r \to \beta_1\beta_2\ldots\beta_n\gamma_1\gamma_2\ldots\gamma_r$.

[24] In the present research, an axiom system for a new formulation of linear dependence of the n-dimensional projective space should be indicated, while we use here mainly the *cycle calculus*, which Mr G. Thomsen has constructed in his foundation of elementary geometry, in a still more abstract sense.

Axiom 1. (Reflexivität)   : aa.
Axiom 2. (Folgerung)   : $a_1 \cdots a_s \to a_1 \cdots a_s x, (s = 1, 2, \cdots)$.
Axiom 3. (Vertauschung)   : $a_1 \cdots a_i \cdots a_s \to a_i \cdots a_1 \cdots a_s$,
                                $(s = 2, 3, \cdots; i = 2, \cdots, s)$.
Axiom 4. (Transitivität)   : $a_1 \cdots a_s \neq 0, xa_1 \cdots a_s, a_1 \cdots a_s y$
                                $\to xa_1 \cdots a_{s-1} y, (s = 1, 2, \cdots)$.

Definition I. Eine solche Menge B_1 heisst der erste Verknüpfungsraum, in kurzen Worten, B_1-**Raum**.[25]

Axiom 3 corresponds to condition (39.3).

Nakasawa introduced the concept of span, and he derived that any two independent sets having the same span, have the same size. It implies that B_1-spaces are the same structures as matroids. Moreover, he gave a submodular law for a rank concept.

In a second paper, Nakasawa [1936a] added a further axiom on intersections of subspaces, yielding a 'B_2-space', which corresponds to a projective space (in which the rank is modular), and in a third paper, Nakasawa [1936b] observed that his B_1-spaces form the same structure as the matroids of Whitney.

1937-1940: Pauc, Haupt, Nöbeling

The axioms presented by Nakasawa were also given by Pauc [1937], added with an axiom describing the limit behaviour of dependence, if the underlying set is endowed with a topology:

> INTRODUCTION AXIOMATIQUE D'UNE NOTION DE DÉPENDANCE SUR UNE CLASSE LIMITE. — Soit D un prédicat relatif aux systèmes finis non ordonnés de points d'une classe limite $\mathcal{L}$, assujetti aux axiomes (notation d'Hilbert-Bernays)
>
> (A$_1$) $\qquad (x_1)(x_2)(D[x_1, x_2] \sim (x_1 = x_2))$,
> (A$_2$) $\qquad (x_1)(x_2)\ldots(x_p)(y)(D[x_1, x_2, \ldots, x_p] \to D[x_1, x_2, \ldots, x_p, y])$,
> (A$_3$) $\qquad (x_1)(x_2)\ldots(x_p)(y)(z)(\overline{D[x_1, \ldots, x_p]}\,\&\,D[x_1, \ldots, x_p, y]\,\&$
> $\qquad D[x_1, \ldots, x_p, z] \to D[x_2, \ldots, x_p, y, z])$,
> (A$_4$) $\begin{cases} \text{Quels que soient les points } x_1, x_2, \ldots, x_p \text{ et la suite } y_1, y_2, \ldots, y_q, \\ \ldots \text{ de } \mathcal{L} \\ (\lim_{q \to \infty} y_q = y)\,\&\,(q)D[x_1, x_2, \ldots, x_p, y_q] \to D[x_1, x_2, \ldots, x_p, y]. \end{cases}$ [26]

In a subsequent paper, Haupt, Nöbeling, and Pauc [1940] studied systems, called *A-Mannigfaltigkeit*, (*A*-manifolds) that satisfy the axioms A$_1$-A$_3$. They mentioned

---

[25] **Basic assumption:** We imagine ourselves a certain set of elements; $B_1 \ni a_1, a_2, \cdots, a_s, \cdots$. For certain sequences of the elements, which we want to call *cycles*, we think the relations on them '*to hold*' or '*to be valid*', in notation $a_1 \cdots a_s = 0$, and '*not to hold*' or '*not to be valid*', in notation $a_1 \cdots a_s \neq 0$, respectively. These relations now should satisfy the following axioms;
Axiom 1. (reflexivity)   : aa.
Axiom 2. (deduction)   : $a_1 \cdots a_s \to a_1 \cdots a_s x, (s = 1, 2, \cdots)$.
Axiom 3. (exchange)   : $a_1 \cdots a_i \cdots a_s \to a_i \cdots a_1 \cdots a_s$,
                              $(s = 2, 3, \cdots; i = 2, \cdots, s)$.
Axiom 4. (transitivity)   : $a_1 \cdots a_s \neq 0, xa_1 \cdots a_s, a_1 \cdots a_s y$
                              $\to xa_1 \cdots a_{s-1} y, (s = 1, 2, \cdots)$.
Definition I. Such a set B_1 is called the first connection space, in short, B_1-**space**.

[26] AXIOMATIC INTRODUCTION OF A NOTION OF DEPENDENCE ON A LIMIT CLASS. — Let D be a predicate relative to the finite unordered systems of points from a limit class $\mathcal{L}$, subject to the axioms (notation of Hilbert-Bernays)

that this axiom system was indeed inspired by those for collinearity of Hilbert-Bernays quoted above. They commented that its relation with Birkhoff's lattices, is analogous to the relation of the Hilbert-Bernays collinearity axioms with those of Hilbert for points and lines.

Haupt, Nöbeling, and Pauc [1940] gave, as examples, linear and algebraic dependence, and derived several basic facts (all bases have the same size, each independent set is contained in a base, for each pair of bases B, B' and $x \in B \setminus B'$ there is a $y \in B' \setminus B$ such that $B - x + y$ is a base, and the rank is submodular).

The authors mentioned that they were informed by G. Köthe about the relations of their work with the lattice formulation of algebraic dependence of Mac Lane [1938], but no connection is made with Whitney's matroid.

Among the further papers related to matroids are Menger [1936b], giving axioms for (full) affine spaces, and Wilcox [1939,1941,1942,1944] and Dilworth [1941a, 1941b,1944] on matroid lattices. The notion of M-symmetric lattice introduced by Wilcox [1942] was shown in Wilcox [1944] to be equivalent to upper semimodular lattice.

Rado

Rado was one of the first to take the independence structure as a source for further theorems, and to connect it with matching type theorems and combinatorial optimization. He had been interested in Kőnig-Hall type theorems (Rado [1933,1938]), and in his paper Rado [1942], he extended Hall's marriage theorem to transversals that are independent in a given matroid — a precursor of matroid intersection. In fact, with an elementary construction, Rado's theorem implies the matroid union theorem, and hence also the matroid intersection theorem (to be discussed in Chapters 41 and 42).

Rado [1942] did not refer to any earlier literature when introducing the concept of an *independence relation*, but the axioms are similar to those of Whitney for the independent sets in a matroid. Rado mentioned only linear independence as a special case.

He proved that a family of subsets of a matroid has an independent transversal if and only if the union of any k of the subsets contains an independent set of size k, for all k. Rado also showed that this theorem characterizes matroids.

Rado [1949a] extended the concept of matroid to infinite matroids, where he says that he extends the axioms of Whitney [1935].

Rado [1957] showed that if the elements of a matroid are linearly ordered by $\leq$, there is a unique minimal base $\{b_1, \ldots, b_r\}$ with $b_1 < b_2 < \cdots < b_r$ such that for each $i = 1, \ldots, r$ all elements $s < b_i$ belong to $\text{span}(\{b_1, \ldots, b_{i-1}\})$. Rado derived that for any independent set $\{a_1, \ldots, a_k\}$ with $a_1 < \cdots < a_k$ one has $b_i \leq a_i$ for $i = 1, \ldots, k$. Therefore, the greedy method gives an optimum solution when

(A_1) $\qquad (x_1)(x_2)(D[x_1, x_2] \sim (x_1 = x_2))$,

(A_2) $\qquad (x_1)(x_2) \ldots (x_p)(y)(D[x_1, x_2, \ldots, x_p] \to D[x_1, x_2, \ldots, x_p, y])$,

(A_3) $\qquad (x_1)(x_2) \ldots (x_p)(y)(z)(\overline{D[x_1, \ldots, x_p]} \& D[x_1, \ldots, x_p, y] \&$
$\qquad\qquad D[x_1, \ldots, x_p, z] \to D[x_2, \ldots, x_p, y, z])$,

(A_4) $\begin{cases} \text{Whatever are the points } x_1, x_2, \ldots, x_p \text{ and the sequence } y_1, y_2, \ldots, y_q, \\ \ldots \text{ from } \mathcal{L} \\ (\lim_{q \to \infty} y_q = y) \& (q) D[x_1, x_2, \ldots, x_p, y_q] \to D[x_1, x_2, \ldots, x_p, y]. \end{cases}$

applied to find a minimum-weight base. Rado mentioned that it extends the work of Borůvka and Kruskal on finding a shortest spanning tree in a graph.

For notes on the history of matroid union, see Section 42.6f. For an excellent survey of early literature on matroids, with reprints of basic articles, see Kung [1986].

Chapter 40

The greedy algorithm and the independent set polytope

We now pass to algorithmic and polyhedral aspects of matroids. We show that the greedy algorithm characterizes matroids and that it implies a characterization of the independent set polytope (the convex hull of the incidence vectors of the independent sets).

Algorithmic and polyhedral aspects of the *intersection* of two matroids will be studied in Chapter 41.

40.1. The greedy algorithm

Let $\mathcal{I}$ be a nonempty collection of subsets of a finite set S closed under taking subsets. For any weight function $w : S \to \mathbb{R}$ we want to find a set I in $\mathcal{I}$ maximizing $w(I)$. The *greedy algorithm* consists of setting $I := \emptyset$, and next repeatedly choosing $y \in S \setminus I$ with $I \cup \{y\} \in \mathcal{I}$ and with $w(y)$ as large as possible. We stop if no such y exists.

For general collections $\mathcal{I}$ of this kind this need not lead to an optimum solution. Indeed, matroids are precisely the structures where it always works, as the following theorem shows (Rado [1957] (necessity) and Gale [1968] and Edmonds [1971] (sufficiency)):

Theorem 40.1. *Let $\mathcal{I}$ be a nonempty collection of subsets of a set S, closed under taking subsets. Then the pair $(S, \mathcal{I})$ is a matroid if and only if for each weight function $w : S \to \mathbb{R}_+$, the greedy algorithm leads to a set I in $\mathcal{I}$ of maximum weight $w(I)$.*

Proof. *Necessity.* Let $(S, \mathcal{I})$ be a matroid and let $w : S \to \mathbb{R}_+$ be any weight function on S. Call an independent set I *good* if it is contained in a maximum-weight base. It suffices to show that if I is good, and y is an element in $S \setminus I$ with $I + y \in \mathcal{I}$ and with $w(y)$ as large as possible, then $I + y$ is good.

As I is good, there exists a maximum-weight base $B \supseteq I$. If $y \in B$, then $I + y$ is good again. If $y \notin B$, then there exists a base B' containing $I + y$ and contained in $B + y$. So $B' = B - z + y$ for some $z \in B \setminus I$. As $w(y)$ is chosen maximum and as $I + z \in \mathcal{I}$ since $I + z \subseteq B$, we know $w(y) \geq w(z)$.

Hence $w(B') \geq w(B)$, and therefore B' is a maximum-weight base. So $I + y$ is good.

Sufficiency. Suppose that the greedy algorithm leads to an independent set of maximum weight for each weight function $w : S \to \mathbb{R}_+$. We show that $(S, \mathcal{I})$ is a matroid.

Condition (39.1)(i) is satisfied by assumption. To see condition (39.1)(ii), let $I, J \in \mathcal{I}$ with $|I| < |J|$. Suppose that $I + z \notin \mathcal{I}$ for each $z \in J \setminus I$.

Let $k := |I|$. Consider the following weight function w on S:

(40.1) $\qquad w(s) := \begin{cases} k+2 & \text{if } s \in I, \\ k+1 & \text{if } s \in J \setminus I, \\ 0 & \text{if } s \in S \setminus (I \cup J). \end{cases}$

Now in the first k iterations of the greedy algorithm we find the k elements in I. By assumption, at any further iteration, we cannot choose any element in $J \setminus I$. Hence any further element chosen, has weight 0. So the greedy algorithm yields an independent set of weight $k(k+2)$.

However, J has weight at least $|J|(k+1) \geq (k+1)(k+1) > k(k+2)$. Hence the greedy algorithm does not give a maximum-weight independent set, contradicting our assumption. ∎

The theorem restricts w to nonnegative weight functions. However, it is shown similarly that for matroids $M = (S, \mathcal{I})$ and arbitrary weight functions $w : S \to \mathbb{R}$, the greedy algorithm finds a maximum-weight base. By replacing 'as large as possible' in the greedy algorithm by 'as small as possible', one obtains an algorithm finding a *minimum*-weight base in a matroid. Moreover, by deleting elements of negative weight, the algorithm can be adapted to yield an independent set of maximum weight, for any weight function $w : S \to \mathbb{R}$.

Throughout we assume that the matroid $M = (S, \mathcal{I})$ is given by an algorithm testing if a given subset of S belongs to $\mathcal{I}$. We call this an *independence testing oracle*. So the full list of all independent sets is not given explicitly (such a list would increase the size of the input exponentially, making most complexity issues meaningless).

In explicit applications, the matroid usually can be described by such a polynomial-time algorithm (polynomial in $|S|$). For instance, we can test if a given set of edges of a graph $G = (V, E)$ is a forest in time polynomially bounded by $|V| + |E|$. So the matroid $(E, \mathcal{F})$ can be described by such an algorithm.

Under these assumptions we have:

Corollary 40.1a. *A maximum-weight independent set in a matroid can be found in strongly polynomial time.*

Proof. See above. ∎

Similarly, for minimum-weight bases:

Corollary 40.1b. *A minimum-weight base in a matroid can be found in strongly polynomial time.*

Proof. See above. ∎

40.2. The independent set polytope

The algorithmic results obtained in the previous section have interesting consequences for polyhedra associated with matroids, as was shown by Edmonds [1970b,1971,1979].

The *independent set polytope* $P_{\text{independent set}}(M)$ of a matroid $M = (S, \mathcal{I})$ is, by definition, the convex hull of the incidence vectors of the independent sets of M. So $P_{\text{independent set}}(M)$ is a polytope in $\mathbb{R}^S$.

Each vector x in $P_{\text{independent set}}(M)$ satisfies the following linear inequalities:

(40.2) $\quad x_s \geq 0 \qquad$ for $s \in S$,
$\quad\quad\quad\;\; x(U) \leq r_M(U) \quad$ for $U \subseteq S$,

because the incidence vector χ^I of any independent set I of M satisfies (40.2). Note that x is an integer vector satisfying (40.2) if and only if x is the incidence vector of some independent set of M.

Edmonds showed that system (40.2) fully determines the independent set polytope, by deriving it from the following formula (yielding a good characterization):

Theorem 40.2. *Let $M = (S, \mathcal{I})$ be a matroid, with rank function r. Then for any weight function $w : S \to \mathbb{R}_+$:*

(40.3) $\quad \max\{w(I) \mid I \in \mathcal{I}\} = \sum_{i=1}^{n} \lambda_i r(U_i),$

where $U_1 \subset \cdots \subset U_n \subseteq S$ and where $\lambda_i \geq 0$ satisfy

(40.4) $\quad w = \sum_{i=1}^{n} \lambda_i \chi^{U_i}.$

Proof. Order the elements of S as $s_1, \ldots, s_n$ such that $w(s_1) \geq w(s_2) \geq \cdots \geq w(s_n)$. Define

(40.5) $\quad U_i := \{s_1, \ldots, s_i\}$

for $i = 0, \ldots, n$, and

(40.6) $\quad I := \{s_i \mid r(U_i) > r(U_{i-1})\}.$

So I is the output of the greedy algorithm. Hence I is a maximum-weight independent set.

Next let:

(40.7) $\quad \lambda_i := w(s_i) - w(s_{i+1})$ for $i = 1, \ldots, n-1$,
$\quad\quad\quad \lambda_n := w(s_n)$.

This implies (40.3):

(40.8) $\quad w(I) = \sum_{s \in I} w(s) = \sum_{i=1}^{n} w(s_i)(r(U_i) - r(U_{i-1}))$

$= w(s_n)r(U_n) + \sum_{i=1}^{n-1}(w(s_i) - w(s_{i+1}))r(U_i) = \sum_{i=1}^{n} \lambda_i r(U_i).$

By taking any ordering of S for which w is nonincreasing, (40.5) gives any chain of subsets U_i satisfying (40.4) for some $\lambda_i \geq 0$. Hence we have the theorem. ∎

This can be interpreted in terms of LP-duality. For any weight function $w : S \to \mathbb{R}$, consider the linear programming problem

(40.9) $\quad$ maximize $\quad w^\top x$,
$\quad\quad\quad$ subject to $\quad x_s \geq 0 \quad\quad\quad (s \in S)$,
$\quad\quad\quad\quad\quad\quad\quad\quad x(U) \leq r_M(U) \quad (U \subseteq S)$,

and its dual:

(40.10) $\quad$ minimize $\quad \sum_{U \subseteq S} y_U r_M(U)$,
$\quad\quad\quad$ subject to $\quad y_U \geq 0 \quad\quad\quad (U \subseteq S)$,
$\quad\quad\quad\quad\quad\quad\quad\quad \sum_{U \subseteq S} y_U \chi^U \geq w.$

Corollary 40.2a. *If $w : S \to \mathbb{Z}$, then (40.9) and (40.10) have integer optimum solutions.*

Proof. We can assume that $w(s) \geq 0$ for each $s \in S$ (as neither the maximum nor the minimum changes by resetting $w(s)$ to 0 if negative). Then (40.4) implies that the λ_i are integer. This gives integer optimum solutions of (40.9) and (40.10). ∎

In polyhedral terms, Theorem 40.2 implies:

Corollary 40.2b. *The independent set polytope is determined by (40.2).*

Proof. Immediately from Theorem 40.2 (with (40.10)). ∎

Moreover, in TDI terms:

692   Chapter 40. The greedy algorithm and the independent set polytope

Corollary 40.2c. *System (40.2) is totally dual integral.*

Proof. Immediately from Corollary 40.2a. ∎

Similar results hold for the base polytope. For any matroid M, let $P_{\text{base}}(M)$ be the *base polytope* of M, defined as the convex hull of the incidence vectors of bases of M. Then:

Corollary 40.2d. *The base polytope of a matroid $M = (S, \mathcal{I})$ is determined by*

(40.11) $\quad x_s \geq 0 \qquad$ for $s \in S$,
$\qquad\qquad x(U) \leq r_M(U) \quad$ for $U \subseteq S$,
$\qquad\qquad x(S) = r_M(S)$.

Proof. This follows directly from Corollary 40.2b, since the base polytope is the intersection of the independent set polytope with the hyperplane $\{x \mid x(S) = r_M(S)\}$, as an independent set I is a base if and only if $|I| \geq r_M(S)$. ∎

The corresponding TDI result reads:

Corollary 40.2e. *System (40.11) is totally dual integral.*

Proof. By Theorem 5.25 from Corollary 40.2c. ∎

One can similarly describe the *spanning set polytope* $P_{\text{spanning set}}(M)$ of M, which is, by definition, the convex hull of the incidence vectors of the spanning sets of M. It is determined by the system:

(40.12) $\quad 0 \leq x_s \leq 1 \qquad$ for $s \in S$,
$\qquad\qquad x(U) \geq r_M(S) - r_M(S \setminus U) \quad$ for $U \subseteq S$.

Corollary 40.2f. *The spanning set polytope is determined by (40.12).*

Proof. A subset U of S is spanning in M if and only if $S \setminus U$ is independent in M^*. Hence for any $x \in \mathbb{R}^S$ we have:

(40.13) $\quad x \in P_{\text{spanning set}}(M) \iff 1 - x \in P_{\text{independent set}}(M^*)$.

By Corollary 40.2b, $1 - x$ belongs to $P_{\text{independent set}}(M^*)$ if and only if x satisfies:

(40.14) $\quad 1 - x_s \geq 0 \qquad$ for $s \in S$,
$\qquad\qquad |U| - x(U) \leq r_{M^*}(U) \quad$ for $U \subseteq S$.

Since $r_{M^*}(U) = |U| + r_M(S \setminus U) - r_M(S)$, the present corollary follows. ∎

Corollary 40.2c gives similarly the TDI result:

Corollary 40.2g. *System* (40.12) *is totally dual integral.*

Proof. By reduction to Corollary 40.2c, by a similar reduction as in the proof of the previous corollary. ∎

Note that

(40.15) $\quad P_{\text{base}}(M) = P_{\text{independent set}}(M) \cap P_{\text{spanning set}}(M),$
$\quad\quad\quad\; P_{\text{independent set}}(M) = P_{\text{base}}^{\downarrow}(M) \cap [0,1]^S,$
$\quad\quad\quad\; P_{\text{spanning set}}(M) = P_{\text{base}}^{\uparrow}(M) \cap [0,1]^S.$

The following consequence on the intersection of the base polytope with a box was observed by Hell and Speer [1984]:

Corollary 40.2h. *Let $M = (S, \mathcal{I})$ be a matroid and let $l, u \in \mathbb{R}^S$ with $l \leq u$. Then there is an $x \in P_{\text{base}}(M)$ with $l \leq x \leq u$ if and only if $l \in P_{\text{base}}^{\downarrow}(M)$ and $u \in P_{\text{base}}^{\uparrow}(M)$.*

Proof. Necessity being trivial, we show sufficiency. We may assume that $l, u \in [0,1]^S$. So $l \in P_{\text{independent set}}(M)$ and $u \in P_{\text{spanning set}}(M)$. Choose l', u' such that $l \leq l' \leq u' \leq u$, $l' \in P_{\text{independent set}}(M)$, $u' \in P_{\text{spanning set}}(M)$, and $\|u' - l'\|_1$ minimal.

If $l' = u'$ we are done, so assume that there is an $s \in S$ with $l'(s) < u'(s)$. As we cannot increase $l'(s)$, there is a $T \subseteq S$ with $s \in T$ and $l'(T) = r(T)$. Similarly, as we cannot decrease $u'(s)$, there is a $U \subseteq S$ with $s \notin U$ and $u'(S \setminus U) = r(S) - r(U)$. Then we have the contradiction

(40.16) $\quad l'(T \cap U) + u'(T \cup U) \leq r(T \cap U) + u'(S) + r(T \cup U) - r(S)$
$\quad\quad\quad\; \leq r(T) + r(U) + u'(S) - r(S) = l'(T) + u'(U)$
$\quad\quad\quad\; < l'(T \cap U) + u'(T \cup U).$

The last inequality follows from

(40.17) $\quad u'(T \cup U) - u'(U) = u'(T \setminus U) > l'(T \setminus U) = l'(T) - l'(T \cap U),$

since $s \in T \setminus U$ and $u'(s) > l'(s)$. ∎

40.3. The most violated inequality

We now consider the problem to find, for any matroid $M = (S, \mathcal{I})$ and any $x \in \mathbb{R}_+^S$ not in the independent set polytope of M, an inequality among (40.2) most violated by x. That is, to find $U \subseteq S$ maximizing $x(U) - r_M(U)$.

The following theorem implies a min-max relation for this (Edmonds [1970b]):

Theorem 40.3. *Let $M = (S, \mathcal{I})$ be a matroid and let $x \in \mathbb{R}_+^S$. Then*

694   Chapter 40. The greedy algorithm and the independent set polytope

(40.18) $\max\{z(S) \mid z \in P_{\text{independent set}}(M), z \leq x\}$
$= \min\{r_M(U) + x(S \setminus U) \mid U \subseteq S\}.$

Proof. The inequality $\leq$ in (40.18) follows from

(40.19) $z(S) = z(U) + z(S \setminus U) \leq r_M(U) + x(S \setminus U).$

To see equality, let z attain the maximum. Then for each $s \in S$ with $z_s < x_s$ there exists a $U \subseteq S$ with $s \in U$ and $z(U) = r_M(U)$ (otherwise we can increase z_s). Now the collection of sets $U \subseteq S$ satisfying $z(U) = r_M(U)$ is closed under taking unions (and intersections), since if $z(T) = r_M(T)$ and $z(U) = r_M(U)$, then

(40.20) $z(T \cup U) = z(T) + z(U) - z(T \cap U) \geq r_M(T) + r_M(U) - r_M(T \cap U)$
$\geq r_M(T \cup U).$

Hence there exists a $U \subseteq S$ such that $z(U) = r_M(U)$ and such that U contains each $s \in S$ with $z_s < x_s$. Hence:

(40.21) $z(S) = z(U) + z(S \setminus U) = r_M(U) + x(S \setminus U),$

giving (40.18). ∎

Cunningham [1984] showed that from an independence testing oracle for a matroid one can derive a strongly polynomial time algorithm to find for any given vector x, a maximum violated inequality for the independent set polytope.

More strongly, Cunningham showed that one can solve the following problem in strongly polynomial time:

(40.22) given: a matroid $M = (S, \mathcal{I})$, by an independence testing oracle, and an $x \in \mathbb{Q}_+^S$;
find: a $z \in P_{\text{independent set}}(M)$ with $z \leq x$ maximizing $z(S)$, with a decomposition of z as convex combination of incidence vectors of independent sets, and a subset U of S satisfying $z(S) = r_M(U) + x(S \setminus U).$

By (40.18), the set U certifies that z maximizes $z(S)$. In the algorithm for (40.22), Cunningham utilized the 'consistent breadth-first search' based on lexicographic order, given by Schönsleben [1980] and Lawler and Martel [1982a].

To prove Cunningham's result, we first show two lemmas. The first lemma is used only to prove the second lemma. As in Section 39.9, we define for any independent set I of a matroid $M = (S, \mathcal{I})$:

(40.23) $A(I) := \{(y, z) \mid y \in I, z \in S \setminus I, I - y + z \in \mathcal{I}\}.$

Lemma 40.4α. *Let $M = (S, \mathcal{I})$ be a matroid and let $I \in \mathcal{I}$. Let $(s, t) \in A(I)$, define $I' := I - s + t$, and let $(u, v) \in A(I') \setminus A(I)$. Then $t = u$ or $(u, t) \in A(I)$, and $s = v$ or $(s, v) \in A(I)$.*

Proof. By symmetry, it suffices to show that $t = u$ or $(u, t) \in A(I)$ (as we may assume that I is a base, and hence the second part follows by duality). We can assume that $t \neq u$. Then $t \neq v$, since $v \notin I' = I - s + t$, as $(u, v) \in A(I')$.

If $v = s$, then $I - u + t = I - u - s + t + v = I' - u + v \in \mathcal{I}$ and hence $(u, t) \in A(I)$. If $v \neq s$, then $I - u \in \mathcal{I}$ and $I - u - s + t + v \in \mathcal{I}$, and therefore $I - u + t \in \mathcal{I}$ or $I - u + v \in \mathcal{I}$; that is, $(u, v) \in A(I)$ or $(u, t) \in A(I)$. ∎

Lemma 40.4β. *Let $M = (S, \mathcal{I})$ be a matroid and let q be a new element. For any $I \in \mathcal{I}$, define*

(40.24) $\quad \widetilde{A}(I) := \{(u, v) \mid u \in I + q, v \in S \setminus I, I - u + v \in \mathcal{I}\}.$

Let $(s, t) \in A(I)$, define $I' := I - s + t$, and let $(u, v) \in \widetilde{A}(I') \setminus \widetilde{A}(I)$. Then $t = u$ or $(u, t) \in \widetilde{A}(I)$, and $s = v$ or $(s, v) \in \widetilde{A}(I)$.

Proof. Let $\widetilde{\mathcal{I}} := \{J \subseteq S + q \mid J - q \in \mathcal{I}\}$. Then the present lemma follows from Lemma 40.4α applied to the matroid $(S + q, \widetilde{\mathcal{I}})$. ∎

Now we can derive Cunningham's result:

Theorem 40.4. *Problem (40.22) is solvable in strongly polynomial time.*

Proof. We keep a vector $z \leq x$ in the independent set polytope of M and a decomposition

(40.25) $\quad z = \sum_{i=1}^{k} \lambda_i \chi^{I_i},$

with $I_1, \ldots, I_k \in \mathcal{I}$, $\lambda_1, \ldots, \lambda_k > 0$, and $\sum_i \lambda_i = 1$. Initially $z := \mathbf{0}$, $k := 1$, $I_1 := \emptyset$, $\lambda_1 := 1$.

Let

(40.26) $\quad T := \{s \in S \mid z_s < x_s\}.$

Let q be a new element. For each i, define $\widetilde{A}(I_i)$ as in (40.24), and let $D = (S + q, A)$ be the directed graph with

(40.27) $\quad A := \widetilde{A}(I_1) \cup \cdots \cup \widetilde{A}(I_k).$

Fix an arbitrary linear order of the elements of $S + q$, by setting $S + q = \{1, \ldots, n\}$.

Case 1: D has no $q - T$ path. Let U be the set of $s \in S$ for which D has an $s - T$ path. As $T \subseteq U$, we know $z(S \setminus U) = x(S \setminus U)$. Also, as no arc of D enters U, we have $|U \cap I_i| = r_M(U)$ for all i, implying

(40.28) $\quad z(U) = \sum_{i=1}^{k} \lambda_i |U \cap I_i| = \sum_{i=1}^{k} \lambda_i r_M(U) = r_M(U).$

Hence $z(S) = r_M(U) + x(S \setminus U)$ as required.

Case 2: D has a $q-T$ path. For each $v \in S+q$, let $d(v)$ denote the distance in D from q to v (set to ∞ if no $q-v$ path exists). Choose a $t \in T$ with $d(t)$ finite and maximal, and among these t we choose the largest t. Let $(s,t) \in A$, with $d(s) = d(t) - 1$, and s largest. We can assume that $(s,t) \in \widetilde{A}(I_1)$. Let

(40.29) $\qquad \alpha := \min\{x_t - z_t, \lambda_1\}$

and define z' by

(40.30) $\qquad z' := z + \alpha(\chi^t - \chi^s)$ if $s \neq q$, and $z' := z + \alpha\chi^t$ if $s = q$.

Let $I_1' := I_1 - s + t$ (so $I_1' = I_1 + t$ if $s = q$).
Then

(40.31) $\qquad z' = \alpha\chi^{I_1'} + (\lambda_1 - \alpha)\chi^{I_1} + \sum_{i=2}^{k} \lambda_i \chi^{I_i}.$

If $\alpha = \lambda_1$, we delete the second term. We obtain a decomposition of z' as a convex combination of at most $k+1$ independent sets, and we can iterate.

Running time. We show that the number of iterations is at most $|S|^9$. Consider any iteration. Let d' and A' be the objects d and A of the next iteration. We first show:

(40.32) $\qquad$ for each $v \in S + q$: $d'(v) \geq d(v)$.

To show this, we can assume that $d'(v) < \infty$. We show (40.32) by induction on $d'(v)$, the case $d'(v) = 0$ being trivial (as it means $v = q$). Assume $d'(v) > 0$. Let u be such that $(u,v) \in A'$ and $d'(u) = d'(v) - 1$. By induction we know $d'(u) \geq d(u)$.

If $(u,v) \in A$, then $d(v) \leq d(u) + 1 \leq d'(u) + 1 = d'(v)$, as required. If $(u,v) \notin A$, then $(u,v) \in \widetilde{A}(I_1')$ and $(u,v) \notin \widetilde{A}(I_1)$. By Lemma 40.4$\beta$, $t = u$ or $(u,t) \in \widetilde{A}(I_1)$, and $s = v$ or $(s,v) \in \widetilde{A}(I_1)$. Hence

(40.33) $\qquad d(v) \leq d(s) + 1 = d(t) \leq d(u) + 1 \leq d'(u) + 1 = d'(v).$

So $d(v) \leq d'(v)$. This shows (40.32).

Let β be the number of $j = 1, \ldots, k$ with $(s,t) \in \widetilde{A}(I_j)$. Let T', t', s', and β' be the objects T, t, s, β in the next iteration. We show:

(40.34) $\qquad$ if $d'(v) = d(v)$ for each $v \in S+q$, then $(d'(t'), t', s', \beta')$ is lexicographically less than $(d(t), t, s, \beta)$.

Indeed, if $\alpha = x_t - z_t$, then $T' = T - t + s$ or $T' = T - t$. So $d'(t') < d(t)$, or $d'(t') = d(t)$ and $t' < t$. If $\alpha < x_t - z_t$, then $T' = T + s$ or $T' = T$. Moreover, $\alpha = \lambda_1$, so I_1 has been omitted from the convex combination. So, as $t \in T'$ and $d(s) < d(t)$, we know that $t' = t$ and $d'(t') = d(t)$. As $t \in I_1'$, we know $(s',t) \notin \widetilde{A}(I_1')$. Hence, as $(s',t) \in A'$, we have $(s',t) \in \widetilde{A}(I_j)$ for some $j = 2, \ldots, k$. Hence $(s',t) \in A$. By the choice of s, we know $s' \leq s$. If $s' < s$, we have (40.34), so assume $s' = s$. Then $\beta' = \beta - 1$, as $(s,t) \notin \widetilde{A}(I_1')$. This proves (40.34).

The number k of independent sets in the decomposition grows by 1 if $\alpha = x_t - z_t < \lambda_1$. In that case, $d'(v) = d(v)$ for each $v \in S + q$ (by (40.32), as $A' \supseteq A$). Moreover, $d'(t') < d(t)$ or $t' < t$ (since $T' \subseteq T - t + s$). So k does not exceed $|S|^4$, and hence β is at most $|S|^4$. Concluding, the number of iterations is at most $|S|^9$. ∎

With Gaussian elimination, we can reduce the number k in each iteration to at most $|S|$ (by Carathéodory's theorem). Incorporating this reduces the number of iterations to $|S|^6$.

Theorem 40.4 immediately implies that one can test if a given vector belongs to the independent set polytope of a matroid:

Corollary 40.4a. *Given a matroid $M = (S, \mathcal{I})$ by an independence testing oracle and an $x \in \mathbb{Q}^S$, one can test in strongly polynomial time if x belongs to $P_{\text{independent set}}(M)$, and if so, decompose x as a convex combination of incidence vectors of independent sets.*

Proof. Directly from Theorem 40.4. ∎

One can derive a similar result for the spanning set polytope:

Corollary 40.4b. *Given a matroid $M = (S, \mathcal{I})$ by an independence testing oracle and an $x \in \mathbb{Q}^S$, one can test in strongly polynomial time if x belongs to $P_{\text{spanning set}}(M)$, and if so, decompose x as a convex combination of incidence vectors of spanning sets.*

Proof. x belongs to the spanning set polytope of M if and only if $\mathbf{1} - x$ belongs to the independent set polytope of the dual matroid M^*. Also convex combinations of spanning sets of M and independent sets of M^* transfer to each other by this operation. Since $r_{M^*}(U) = |U| + r_M(S \setminus U) - r_M(S)$ for each $U \subseteq S$, also an independence testing oracle for M^* is easily obtained from one for M. ∎

The theorem also implies that the following *most violated inequality problem* can be solved in strongly polynomial time:

(40.35)   given: a matroid $M = (S, \mathcal{I})$ by an independence testing oracle, and a vector $x \in \mathbb{Q}^S$;
find: a subset U of S minimizing $r_M(U) - x(U)$.

Corollary 40.4c. *The most violated inequality problem can be solved in strongly polynomial time.*

Proof. Any negative component of x can be reset to 0, as this does not change the problem. So we can assume that $x \geq \mathbf{0}$. Then by Theorem 40.4 we can find a $U \subseteq S$ minimizing $r_M(U) + x(S \setminus U)$ in strongly polynomial time. This U is as required. ∎

40.3a. Facets and adjacency on the independent set polytope

Let $M = (S, \mathcal{I})$ be a matroid, with rank function r. Trivially, the independent set polytope P of M is full-dimensional if and only if M has no loops. If P is full-dimensional there is a unique minimal collection of linear inequalities defining P (up to scalar multiplication), which corresponds to the facets of P. Edmonds [1970b] found that this collection is given by the following theorem. Recall that a subset F of S is called a *flat* if for all s in $S \setminus F$ one has $r(F + s) > r(F)$. A subset F is called *inseparable* if there is no partition of F into nonempty sets F_1 and F_2 with $r(F) = r(F_1) + r(F_2)$. Then:

Theorem 40.5. *If M is loopless, the following is a minimal system for the independent set polytope of M:*

(40.36)   (i) $x_s \geq 0$ $(s \in S)$,
                (ii) $x(F) \leq r(F)$ (F is a nonempty inseparable flat).

Proof. As M is loopless, the independent set polytope of M is full-dimensional. It is easy to see that (40.36) determines the independent set polytope, as any other inequality $x(U) \leq r(U)$ is implied by the inequalities $x(F_i) \leq r(F_i)$, where $F_1, \ldots, F_t$ is a maximal partition of $F := \mathrm{span}_M(U)$ such that $r(F_1) + \cdots + r(F_t) = r(F)$.

The irredundancy of collection (40.36) can be seen as follows. Each inequality $x_s \geq 0$ is irredundant, since the vector $-\chi^s$ satisfies all other inequalities.

We show that also the inequalities (40.36)(ii) are irredundant, by showing that for any two nonempty nonseparable flats T, U there exists a base I of T with $|I \cap U| < r(U)$ (implying that the face determined by T is contained in no (other) facet).

To show this, let I be a base of T with $|I \cap (T \setminus U)| = r(T \setminus U)$. Suppose $|I \cap U| = r(U)$. Then

(40.37) $r(U) \geq r(T \cap U) \geq r(T) - r(T \setminus U) = |I \cap U| = r(U)$.

Hence we have equality throughout. This implies (as T is inseparable) that $T \setminus U = \emptyset$ or $T \cap U = \emptyset$, and that $r(U) = r(T \cap U)$. If $T \setminus U = \emptyset$, then $T \subset U$, and hence (as T is a flat) $r(U) > r(T) \geq r(T \cap U)$, a contradiction. If $T \cap U = \emptyset$, then $r(U) = r(T \cap U) = 0$, implying that $U = \emptyset$ (as M has no loops), again a contradiction. ∎

It follows that the base polytope, which is the face $\{x \in P \mid x(S) = r(S)\}$ of P, has dimension $|S| - 1$ if and only if S is inseparable (that is, the matroid is *connected*).

As for adjacency of vertices of the independent set polytope, we have:

Theorem 40.6. *Let $M = (S, \mathcal{I})$ be a loopless matroid and let I and J be distinct independent sets. Then χ^I and χ^J are adjacent vertices of the independent set*

polytope of M if and only if $|I \triangle J| = 1$, or $|I \setminus J| = |J \setminus I| = 1$ and $r_M(I \cup J) = |I| = |J|$.

Proof. To see sufficiency, note that the condition implies that I and J are the only two independent sets with incidence vector x satisfying $x(I \cap J) = r_M(I \cap J)$, $x_s = 0$ for $s \notin I \cup J$, and (if $|I \triangle J| = 2$) $x(I \cup J) = r_M(I \cup J)$. Hence I and J are adjacent.

To see necessity, assume that χ^I and χ^J are adjacent. If I is not a base of $I \cup J$, then $I + j$ is independent for some $j \in J \setminus I$. Hence

(40.38) $\quad \frac{1}{2}(\chi^I + \chi^J) = \frac{1}{2}(\chi^{I+j} + \chi^{J-j})$,

implying (as χ^I and χ^J are adjacent) that $I+j = J$ and $J-j = I$, that is $|I \triangle J| = 1$.

So we can assume that I and J are bases of $I \cup J$. Choose $i \in I \setminus J$. By Theorem 39.12, there is a $j \in J \setminus I$ such that $I - i + j$ and $J - j + i$ are bases of $I \cup J$. Then

(40.39) $\quad \frac{1}{2}(\chi^I + \chi^J) = \frac{1}{2}(\chi^{I-i+j} + \chi^{J-j+i})$,

implying (as χ^I and χ^J are adjacent) that $I - i + j = J$ and $J - j + i = I$, that is we have the second alternative in the condition. ∎

More on the combinatorial structure of the independent set polytope can be found in Naddef and Pulleyblank [1981a].

40.3b. Further notes

Prodon [1984] showed that the separation problem for the independent set polytope of a matching matroid can be solved by finding a minimum-capacity cut in an auxiliary directed graph.

Frederickson and Solis-Oba [1997,1998] gave strongly polynomial-time algorithm for measuring the sensitivity of the minimum weight of a base under perturbing the weight. (Related analysis was given by Libura [1991].)

Narayanan [1995] described a rounding technique for the independent set polytope membership problem, leading to an $O(n^3 r^2)$-time algorithm, where n is the size of the underlying set of the matroid and r is the rank of the matroid.

A strongly polynomial-time algorithm maximizing certain convex objective functions over the bases was given by Hassin and Tamir [1989].

For studies of structures where the greedy algorithm applies if condition (39.1)(i) is deleted, see Faigle [1979,1984b], Hausmann, Korte, and Jenkyns [1980], Korte and Lovász [1983,1984a,1984b,1984c,1985a,1985b,1989], Bouchet [1987a], Goecke [1988], Dress and Wenzel [1990], Korte, Lovász, and Schrader [1991], Helman, Moret, and Shapiro [1993], and Faigle and Kern [1996].

Chapter 41

Matroid intersection

Edmonds discovered that matroids have even more algorithmic power than just that of the greedy method. He showed that there exist efficient algorithms also for *intersections* of matroids. That is, a maximum-weight common independent set in *two* matroids can be found in strongly polynomial time. Edmonds also found good min-max characterizations for matroid intersection.

Matroid intersection yields a motivation for studying matroids: we may apply it to two matroids from different classes of examples of matroids, and thus we obtain methods that exceed the bounds of any particular class.

We should note here that if $M_1 = (S, \mathcal{I}_1)$ and $M_2 = (S, \mathcal{I}_2)$ are matroids, then $(S, \mathcal{I}_1 \cap \mathcal{I}_2)$ need not be a matroid. (An example with $|S| = 3$ is easy to construct.)

Moreover, the problem of finding a maximum-size common independent set in *three* matroids is NP-complete (as finding a Hamiltonian circuit in a directed graph is a special case; also, finding a common transversal of three partitions is a special case).

41.1. Matroid intersection theorem

Let $M_1 = (S, \mathcal{I}_1)$ and $M_2 = (S, \mathcal{I}_2)$ be two matroids, on the same set S. Consider the collection $\mathcal{I}_1 \cap \mathcal{I}_2$ of *common independent sets*. The pair $(S, \mathcal{I}_1 \cap \mathcal{I}_2)$ is generally *not* a matroid again.

Edmonds [1970b] showed the following formula, for which he gave two proofs — one based on linear programming duality and total unimodularity (see the proof of Theorem 41.12 below), and one reducing it to the matroid union theorem (see Corollary 42.1a and the remark thereafter). We give the direct proof implicit in Brualdi [1971e].

Theorem 41.1 (matroid intersection theorem). *Let $M_1 = (S, \mathcal{I}_1)$ and $M_2 = (S, \mathcal{I}_2)$ be matroids, with rank functions r_1 and r_2, respectively. Then the maximum size of a set in $\mathcal{I}_1 \cap \mathcal{I}_2$ is equal to*

(41.1) $$\min_{U \subseteq S}(r_1(U) + r_2(S \setminus U)).$$

Proof. Let k be equal to (41.1). It is easy to see that the maximum is not more than k, since for any common independent set I and any $U \subseteq S$:

(41.2) $\quad |I| = |I \cap U| + |I \setminus U| \leq r_1(U) + r_2(S \setminus U).$

We prove equality by induction on $|S|$, the case $|S| \leq 1$ being trivial. So assume that $|S| \geq 2$.

If minimum (41.1) is attained only by $U = S$ or $U = \emptyset$, choose $s \in S$. Then $r_1(U) + r_2(S \setminus (U \cup \{s\})) \geq k$ for each $U \subseteq S \setminus \{s\}$, since otherwise both U and $U \cup \{s\}$ would attain (41.1), whence $\{U, U \cup \{s\}\} = \{\emptyset, S\}$, contradicting the fact that $|S| \geq 2$. Hence, by induction, $M_1 \setminus s$ and $M_2 \setminus s$ have a common independent set of size k, implying the theorem.

So we can assume that (41.1) is attained by some U with $\emptyset \neq U \neq S$. Then $M_1|U$ and $M_2 \cdot U$ have a common independent set I of size $r_1(U)$. Otherwise, by induction, there exists a subset T of U with

(41.3) $\quad r_1(U) > r_{M_1|U}(T) + r_{M_2 \cdot U}(U \setminus T) = r_1(T) + r_2(S \setminus T) - r_2(S \setminus U),$

contradicting the fact that U attains (41.1). Similarly, $M_1 \cdot (S \setminus U)$ and $M_2|(S \setminus U)$ have a common independent set J of size $r_2(S \setminus U)$.

Now $I \cup J$ is a common independent set of M_1 and M_2. Indeed, $I \cup J$ is independent in M_1, as I is independent in $M_1|U$ and J is independent in $M_1 \cdot (S \setminus U) = M_1/U$ (cf. (39.10)). Similarly, $I \cup J$ is independent in M_2. As $|I \cup J| = r_1(U) + r_2(S \setminus U)$, this proves the theorem. ∎

This implies a characterization of the existence of a common base in two matroids:

Corollary 41.1a. *Let $M_1 = (S, \mathcal{I}_1)$ and $M_2 = (S, \mathcal{I}_2)$ be matroids, with rank functions r_1 and r_2, respectively, such that $r_1(S) = r_2(S)$. Then M_1 and M_2 have a common base if and only if $r_1(U) + r_2(S \setminus U) \geq r_1(S)$ for each $U \subseteq S$.*

Proof. Directly from Theorem 41.1. ∎

It is easy to derive from the matroid intersection theorem a similar min-max relation for the minimum size of a common spanning set:

Corollary 41.1b. *Let $M_1 = (S, \mathcal{I}_1)$ and $M_2 = (S, \mathcal{I}_2)$ be matroids, with rank functions r_1 and r_2, respectively. Then the minimum size of a common spanning set of M_1 and M_2 is equal to*

(41.4) $\quad \max_{U \subseteq S}(r_1(S) - r_1(U) + r_2(S) - r_2(S \setminus U)).$

Proof. The minimum is equal to the minimum of $|B_1 \cup B_2|$ where B_1 and B_2 are bases of M_1 and M_2 respectively. Hence the minimum is equal to $r_1(S) + r_2(S)$ minus the maximum of $|B_1 \cap B_2|$ over such B_1, B_2. This last maximum is characterized in the matroid intersection theorem, yielding the present corollary. ∎

The following result of Rado [1942] (a generalization of Hall's marriage theorem (Theorem 22.1), and therefore sometimes called the Rado-Hall theorem) may be derived from the matroid intersection theorem, applied to M and the transversal matroid M_2 induced by $\mathcal{X}$.

Corollary 41.1c (Rado's theorem). *Let $M = (S, \mathcal{I})$ be a matroid, with rank function r, and let $\mathcal{X} = (X_1, \ldots, X_n)$ be a family of subsets of S. Then $\mathcal{X}$ has a transversal which is independent in M if and only if*

(41.5) $\quad r(\bigcup_{i \in I} X_i) \geq |I|$

for each $I \subseteq \{1, \ldots, n\}$.

Proof. Let r_2 be the rank function of the transversal matroid M_2 induced by $\mathcal{X}$. By the matroid intersection theorem, M and M_2 have a common independent set of size n if and only if

(41.6) $\quad r(U) + r_2(S \setminus U) \geq n$ for each $U \subseteq S$.

Now for each $T \subseteq S$ one has (by König's matching theorem (cf. Corollary 22.2a)):

(41.7) $\quad r_2(T) = \min_{I \subseteq \{1,\ldots,n\}} (|\bigcup_{i \in I} X_i \cap T| + n - |I|)$.

So (41.6) is equivalent to:

(41.8) $\quad r(U) + |\bigcup_{i \in I} X_i \setminus U| + n - |I| \geq n$

for all $U \subseteq S$ and $I \subseteq \{1, \ldots, n\}$. We can assume that $U = \bigcup_{i \in I} X_i$, since replacing U by $\bigcup_{i \in I} X_i$ does not increase the left-hand side in (41.8). So the condition is equivalent to (41.5), proving the corollary. ∎

Notes. Mirsky [1971a] gave an alternative proof of Rado's theorem. Welsh [1970] showed that, in turn, Rado's theorem implies the matroid intersection theorem. Las Vergnas [1970] gave an extension of Rado's theorem. Rado [1942] (and also Welsh [1971]) showed that Rado's theorem in fact characterizes matroids. Perfect [1969a] generalized Rado's theorem to characterizing the maximum size of an independent partial transversal. Related results are in Perfect [1971].

41.1a. Applications of the matroid intersection theorem

In this section we mention a number of applications of the matroid intersection theorem. Further applications will be given in the next chapter on matroid union.

König's theorems. Let $G = (V, E)$ be a bipartite graph, with colour classes U_1 and U_2. For $i = 1, 2$, let $M_i = (E, \mathcal{I}_i)$ be the matroid with $F \subseteq E$ independent if and only if each vertex in U_i is covered by at most one edge in F.

Section 41.1a. Applications of the matroid intersection theorem

So M_1 and M_2 are partition matroids. The common independent sets in M_1 and M_2 are the matchings in G, and the common spanning sets are the edge covers in G. For $i = 1, 2$ and $F \subseteq E$, the rank $r_i(F)$ of F in M_i is equal to the number of vertices in U_i covered by F.

By the matroid intersection theorem, the maximum size of a matching in G is equal to the minimum of $r_1(F) + r_2(E \setminus F)$ taken over $F \subseteq E$. This last is equal to the minimum size of a vertex cover in G. So we have König's matching theorem (Theorem 16.2).

Similarly, by Corollary 41.1b, the minimum size of an edge cover in G (assuming G has no isolated vertices), is equal to the maximum of $|V| - r_1(F) - r_2(E \setminus F)$ taken over $F \subseteq E$. This last is equal to the maximum size of a stable set in G. So we have the König-Rado edge cover theorem (Theorem 19.4).

Common transversals. Let $\mathcal{X} = (X_1, \ldots, X_m)$ and $\mathcal{Y} = (Y_1, \ldots, Y_m)$ be families of subsets of a finite set S. Then the matroid intersection theorem implies Theorem 23.1 of Ford and Fulkerson [1958c]: $\mathcal{X}$ and $\mathcal{Y}$ have a common transversal if and only if

(41.9) $\quad |X_I \cap Y_J| \geq |I| + |J| - m$

for all subsets I and J of $\{1, \ldots, m\}$, where $X_I := \bigcup_{i \in I} X_i$ and $Y_J := \bigcup_{j \in J} Y_j$.

To see this, let M_1 and M_2 be the transversal matroids induced by $\mathcal{X}$ and $\mathcal{Y}$ respectively, with rank functions r_1 and r_2 say. So $\mathcal{X}$ and $\mathcal{Y}$ have a common transversal if and only if M_1 and M_2 have a common independent set of size m. By Theorem 41.1, this last holds if and only if $r_1(Z) + r_2(S \setminus Z) \geq m$ for each $Z \subseteq S$. Using König's matching theorem, this is equivalent to:

(41.10) $\quad \min_{I \subseteq \{1,\ldots,m\}} (m - |I| + |X_I \cap Z|) + \min_{J \subseteq \{1,\ldots,m\}} (m - |J| + |Y_J \setminus Z|) \geq m$

for each $Z \subseteq S$. Equivalently, for all $I, J \subseteq \{1, \ldots, m\}$:

(41.11) $\quad \min_{Z \subseteq S}(m - |I| + |X_I \cap Z| + m - |J| + |Y_J \setminus Z|) \geq m.$

As this minimum is attained by $Z := Y_J$, this is equivalent to (41.9).

Coloured trees. Let $G = (V, E)$ be a graph and let the edges of G be coloured with k colours. That is, we have partitioned E into sets $E_1, \ldots, E_k$, called *colours*. Then there exists a spanning tree with all edges coloured differently if and only if $G - F$ has at most $t + 1$ components, for any union F of t colours, for any $t \geq 0$. This follows from the matroid intersection theorem applied to the cycle matroid $M(G)$ of G and the partition matroid N induced by $E_1, \ldots, E_k$.

Indeed, $M(G)$ and N have a common independent set of size $|V| - 1$ if and only if $r_{M(G)}(E \setminus F) + r_N(F) \geq |V| - 1$ for each $F \subseteq E$. Now $r_N(F)$ is equal to the number of E_i intersecting F. So we can assume that F is equal to the union of t of the E_i, with $t := r_N(F)$. Moreover, $r_{M(G)}(E \setminus F)$ is equal to $|V| - \kappa(G - F)$, where $\kappa(G - F)$ is the number of components of $G - F$. So the requirement is that $|V| - \kappa(G - F) + t \geq |V| - 1$. In other words, $\kappa(G - F) \leq t + 1$.

Detachments. The following is a special case of a theorem of Nash-Williams [1985], which he derived from the matroid intersection theorem — in fact it is a consequence of the result on coloured trees given above.

Let $G = (V, E)$ be a graph and let $b : V \longrightarrow \mathbb{Z}_+$. Call a graph $\widetilde{G} = (\widetilde{V}, \widetilde{E})$ a *b-detachment* of G if there is a function $\phi : \widetilde{V} \longrightarrow V$ such that $|\phi^{-1}(v)| = b(v)$ for each $v \in V$, and such that there is a one-to-one function $\psi : \widetilde{E} \longrightarrow E$ with $\psi(e) = \{\phi(u), \phi(v)\}$ for each edge $e = uv$ of $\widetilde{G}$.

Then there exists a connected b-detachment if and only if

(41.12) $\quad b(U) + \kappa(G - U) \leq |E_U| + 1$ for each $U \subseteq V$,

where $\kappa(G')$ denotes the number of components of graph G' and where E_U denotes the set of edges intersecting U.

To see this, let $H = (\widetilde{V}, E')$ be the graph obtained from G by replacing each vertex v by $b(v)$ new vertices, and by connecting for each edge $e = uv$ of G, the $b(u)$ new vertices associated with u with the $b(v)$ new vertices associated with v. We assign to these $b(u)b(v)$ edges the 'colour' e.

Then there exists a connected b-detachment if and only if H has a spanning tree in which all edges have a different colour. By the previous example, such a spanning tree exists if and only if for each $F \subseteq E$, deleting from H the edges with colour in F gives a graph H' with at most $|F| + 1$ components.

Now the number of components of H' is equal to the $\kappa(G - F) + b(I_F) - |I_F|$, where I_F denotes the set of isolated (hence loopless) vertices of $G - F$. So the condition is equivalent to: $\kappa(G - F) - |F| + b(I_F) - |I_F| \leq 1$. As $\kappa(G - F) - |F|$ does not decrease by removing edges from F, we can assume that F is equal to the set of edges incident with I_F. So F is determined by $U := I_F$, namely $F = E_U$. Then $\kappa(G - F) - |I_F| = \kappa(G - U)$. So the condition is equivalent to (41.12).

41.1b. Woodall's proof of the matroid intersection theorem

P.D. Seymour attributed the following proof of the matroid intersection theorem to D.R. Woodall (cf. Seymour [1976a]):

Let k be the value of (41.1). Let $x \in S$ be such that $r_1(\{x\}) = r_2(\{x\}) = 1$. (If no such x exists the theorem is trivial, as in that case the minimum is 0.) Let $Y := S \setminus \{x\}$. Now we may assume that the restrictions $M_1 \setminus x$ and $M_2 \setminus x$ have no common independent set of size k. So, by induction,

(41.13) $\quad r_1(A_1) + r_2(A_2) \leq k - 1$,

for some partition A_1, A_2 of Y. Moreover, the contractions M_1/x and M_2/x have no common independent set of size $k - 1$ (otherwise we can add x to obtain a common independent set of size k for M_1 and M_2). So, by induction,

(41.14) $\quad r_1(B_1 \cup \{x\}) - 1 + r_2(B_2 \cup \{x\}) - 1 \leq k - 2$

(cf. (39.9) above), for some partition B_1, B_2 of Y. However,

(41.15) $\quad r_1(A_1 \cap B_1) + r_1(A_1 \cup B_1 \cup \{x\}) \leq r_1(A_1) + r_1(B_1 \cup \{x\})$,
$\qquad r_2(A_2 \cap B_2) + r_2(A_2 \cup B_2 \cup \{x\}) \leq r_2(A_2) + r_2(B_2 \cup \{x\})$,

by the submodularity (cf. (39.38)(ii)) of the rank functions. Moreover, by the definition of k,

(41.16) $\quad k \leq r_1(A_1 \cap B_1) + r_2(A_2 \cup B_2 \cup \{x\})$,
$\qquad k \leq r_1(A_1 \cup B_1 \cup \{x\}) + r_2(A_2 \cap B_2)$,

as $A_1 \cap B_1, A_2 \cup B_2 \cup \{x\}$ and $A_1 \cup B_1 \cup \{x\}, A_2 \cap B_2$ form partitions of S. Adding the inequalities in (41.13), (41.14), (41.15), and (41.16) gives a contradiction.

41.2. Cardinality matroid intersection algorithm

A maximum-size common independent set can be found in polynomial time. This result follows from the matroid union algorithm of Edmonds [1968], since (as Edmonds [1970b] and Lawler [1970] observed) cardinality matroid intersection can be reduced to matroid union.

We describe below the direct algorithm given by Aigner and Dowling [1971] and Lawler [1975], based on finding paths in auxiliary graphs. A different algorithm was given by Edmonds [1979].

Note that the examples given in Section 41.1a provide applications for the matroid intersection algorithm. We should note that in the algorithm we require that in any matroid $M = (S, \mathcal{I})$, we can test in polynomial time if any subset of S belongs to $\mathcal{I}$ — no explicit list of all sets in $\mathcal{I}$ is required. Thus complexity results are all relative to the complexity of testing independence. As such a membership testing algorithm exists in each example mentioned, we obtain polynomial-time algorithms for these special cases.

For any two matroids $M_1 = (S, \mathcal{I}_1)$ and $M_2 = (S, \mathcal{I}_2)$ and any $I \in \mathcal{I}_1 \cap \mathcal{I}_2$, we define a directed graph $D_{M_1, M_2}(I)$, with vertex set S, as follows. For any $y \in I, x \in S \setminus I$,

(41.17) (y, x) is an arc of $D_{M_1, M_2}(I)$ if and only if $I - y + x \in \mathcal{I}_1$,
(x, y) is an arc of $D_{M_1, M_2}(I)$ if and only if $I - y + x \in \mathcal{I}_2$.

These are all arcs of $D_{M_1, M_2}(I)$. So this graph is the union of the graphs $D_{M_1}(I)$ and the *reverse* of $D_{M_2}(I)$ defined in Section 39.9.

The following is the base for finding a maximum-size common independent set in two matroids.

Cardinality common independent set augmenting algorithm

input: matroids $M_1 = (S, \mathcal{I}_1)$ and $M_2 = (S, \mathcal{I}_2)$ and a set $I \in \mathcal{I}_1 \cap \mathcal{I}_2$;
output: a set $I' \in \mathcal{I}_1 \cap \mathcal{I}_2$ with $|I'| > |I|$ (if any).
description of the algorithm: Consider the sets

(41.18) $X_1 := \{x \in S \setminus I \mid I \cup \{x\} \in \mathcal{I}_1\}$,
$X_2 := \{x \in S \setminus I \mid I \cup \{x\} \in \mathcal{I}_2\}$.

Moreover, consider the directed graph $D_{M_1, M_2}(I)$ defined above. There are two cases.

Case 1: $D_{M_1, M_2}(I)$ **has an** $X_1 - X_2$ **path** P**.** (Possibly of length 0 if $X_1 \cap X_2 \neq \emptyset$.) We take a shortest such path P (that is, with a minimum number of arcs). Now output $I' := I \triangle V P$.

Case 2: $D_{M_1, M_2}(I)$ **has no** $X_1 - X_2$ **path.** Then I is a maximum-size common independent set. ∎

706   Chapter 41. Matroid intersection

This finishes the description of the algorithm. The correctness of the algorithm is given by the following two theorems.

Theorem 41.2. *If Case 1 applies, then $I' \in \mathcal{I}_1 \cap \mathcal{I}_2$.*

Proof. Assume that Case 1 applies. By symmetry it suffices to show that I' belongs to $\mathcal{I}_1$.

Let P start at $z_0 \in X_1$. The arcs in P leaving I form the only matching in $D_{M_1}(I)$ with union equal to $VP - z_0$, since otherwise P would have a shortcut. Moreover, for each $z \in VP \setminus I$ with $z \neq z_0$, one has $I + z \notin \mathcal{I}_1$, since otherwise $z \in X_1$, and hence P would have a shortcut. So by Corollary 39.13a, I' belongs to $\mathcal{I}_1$. ∎

Theorem 41.3. *If Case 2 applies, then I is a maximum-size common independent set.*

Proof. As Case 2 applies, there is no $X_1 - X_2$ path in $D_{M_1, M_2}(I)$. Hence there is a subset U of S with $X_1 \cap U = \emptyset$ and $X_2 \subseteq U$, and such that no arc enters U. We show

(41.19) $r_{M_1}(U) + r_{M_2}(S \setminus U) \leq |I|$.

To this end, we first show

(41.20) $r_{M_1}(U) \leq |I \cap U|$.

Suppose that $r_{M_1}(U) > |I \cap U|$. Then there exists an x in $U \setminus I$ such that $(I \cap U) \cup \{x\} \in \mathcal{I}_1$. Since $I \cup \{x\} \notin \mathcal{I}_1$ (as $x \notin X_1$), there is a $y \in I \setminus U$ with $I - y + x \in \mathcal{I}_1$. But then $D_{M_1}(I)$ has an arc from y to x, contradicting the facts that $x \in U$ and $y \notin U$ and that no arc enters U.

This shows (41.20). Similarly, $r_{M_2}(S \setminus U) \leq |I \setminus U|$. Hence we have (41.19). So by the matroid intersection theorem, I is a maximum-size common independent set. ∎

Clearly, the running time of the algorithm is polynomially bounded, since we can construct the auxiliary directed graph $D_{M_1, M_2}(I)$ and find the path P (if any), in polynomial time. Therefore:

Theorem 41.4. *A maximum-size common independent set in two matroids can be found in polynomial time.*

Proof. Directly from the above, as we can find a maximum-size common independent set after applying at most $|S|$ times the common independent set augmenting algorithm. ∎

The algorithm also yields a proof of the matroid intersection theorem (Theorem 41.1 above): if the algorithm stops with set I, we obtain a set U for which (41.19) holds.

Notes. The above algorithm can be shown to take $O(n^2m(n+Q))$ time, where n is the maximum size of a common independent set, m is the size of the underlying set, and Q is the time needed to test if a given set is independent (in either matroid). Cunningham [1986] showed that if one chooses a shortest path as augmenting path, the sum of the lengths of all augmenting paths chosen is $O(n \log n)$, which gives an $O(n^{3/2}mQ)$-time algorithm. This algorithm extends several of the ideas behind the $O(n^{1/2}m)$ algorithm of Hopcroft, Karp, and Karzanov for cardinality bipartite matching (see Section 16.4). For more efficient algorithms, see Gabow and Tarjan [1984], Gusfield [1984], Gabow and Stallmann [1985], Frederickson and Srinivas [1989], Gabow and Xu [1989,1996], and Fujishige and Zhang [1995].

The problem of finding a maximum-size common independent set in *three* matroids is NP-complete, as finding a Hamiltonian circuit in a directed graph is a special case (as was observed by Held and Karp [1970]). Another special case is finding a common transversal of three collections of sets, which is also NP-complete (Theorem 23.16). In particular, the k-intersection problem can be reduced to the 3-intersection problem (cf. Lawler [1976b]).

Barvinok [1995] gave an algorithm for finding a maximum-size common independent set in k linear matroids, represented by given vectors over the rationals. The running time is linear in the cardinality of the underlying set and singly polynomial in the maximum rank of the matroids.

41.3. Weighted matroid intersection algorithm

Also a maximum-*weight* common independent set can be found in strongly polynomial time. This result was announced by Edmonds [1970b], who published an algorithm in Edmonds [1979]. An alternative algorithm (which we describe below) was announced by Lawler [1970] and described in Lawler [1975,1976b] — the correctness of this algorithm was proved by Krogdahl [1974,1976], using the results described in Section 39.9. A similar algorithm was described by Iri and Tomizawa [1976].

This algorithm is an extension of the cardinality matroid intersection algorithm given in Section 41.2. In each iteration, instead of finding a path P with a minimum number of arcs in $D_{M_1,M_2}(I)$, we will now require P to have minimum length with respect to some length function defined on $D_{M_1,M_2}(I)$.

To describe the algorithm, if matroids $M_1 = (S, \mathcal{I}_1)$ and $M_2 = (S, \mathcal{I}_2)$ and a weight function $w : S \to \mathbb{R}$ are given, call a set $I \in \mathcal{I}_1 \cap \mathcal{I}_2$ *extreme* if $w(J) \leq w(I)$ for each $J \in \mathcal{I}_1 \cap \mathcal{I}_2$ satisfying $|J| = |I|$.

Weighted common independent set augmenting algorithm

input: matroids $M_1 = (S, \mathcal{I}_1)$ and $M_2 = (S, \mathcal{I}_2)$, a weight function $w : S \to \mathbb{Q}$, and an extreme common independent set I;
output: an extreme common independent set I' with $|I'| = |I| + 1$ (if any).
description of the algorithm: Consider again the sets X_1 and X_2 and the directed graph $D_{M_1,M_2}(I)$ on S, as in the cardinality case.

For any $x \in S$ define the 'length' $l(x)$ of x by:

708   Chapter 41. Matroid intersection

(41.21) $l(x) := \begin{cases} w(x) & \text{if } x \in I, \\ -w(x) & \text{if } x \notin I. \end{cases}$

The *length* of a path P, denoted by $l(P)$, is equal to the sum of the lengths of the vertices traversed by P.

Case 1: $D_{M_1,M_2}(I)$ has an $X_1 - X_2$ path P. We choose P such that $l(P)$ is minimal and such that (secondly) P has a minimum number of arcs among all minimum-length $X_1 - X_2$ paths. Set $I' := I \triangle VP$.

Case 2: $D_{M_1,M_2}(I)$ has no $X_1 - X_2$ path. Then there is no common independent set larger than I. ∎

This finishes the description of the algorithm. The correctness of the algorithm if Case 2 applies follows directly from Theorem 41.3. In order to show the correctness if Case 1 applies, we first prove the following basic property of the length function l.

Lemma 41.5α. *Let C be a directed circuit in $D_{M_1,M_2}(I)$ and let $t \in VC$. Define $J := I \triangle VC$. If $J \notin \mathcal{I}_1 \cap \mathcal{I}_2$, then there exists a directed circuit C' with $VC' \subset VC$ such that $l(VC') < 0$, or $l(VC') \le l(VC)$ and $t \in VC'$.*

Proof. By symmetry we can assume that $J \notin \mathcal{I}_1$. Let N_1 and N_2 be the sets of arcs in C belonging to $D_{M_1}(I)$ and $D_{M_2}(I)$ respectively. As $J \notin \mathcal{I}_1$, there exists, by Theorem 39.13, a matching N_1' in $D_{M_1}(I)$ with union VC and with $N_1' \ne N_1$. Consider the directed graph $D = (VC, A)$ formed by the arcs in N_1, N_1' (taking arcs in $N_1 \cap N_1'$ parallel), and by the arcs in N_2 taking each of them twice (parallel). Then each vertex in VC is entered and left by exactly two arcs of D. Moreover, since $N_1' \ne N_1$, D contains a directed circuit C_1 with $VC_1 \subset VC$ (as N_1' contains a chord of C). As D is Eulerian, we can extend this to a decomposition of A into directed circuits $C_1, \ldots, C_k$. Then

(41.22) $\chi^{VC_1} + \cdots + \chi^{VC_k} = 2 \cdot \chi^{VC}.$

Since $VC_1 \ne VC$ we know that $VC_j = VC$ for at most one j. If, say $VC_k = VC$, then (41.22) implies that either $l(VC_j) < 0$ for some $j < k$ or $l(VC_j) \le l(VC)$ for all $j < k$, implying the proposition.

Suppose next that $VC_j \ne VC$ for all j. If $l(VC_j) < 0$ for some $j \le k$ we are done. So assume $l(VC_j) \ge 0$ for each $j \le k$. We can assume that C_1 and C_2 traverse t. Then

(41.23) $l(VC_1) + l(VC_2) \le l(VC_1) + \cdots + l(VC_k) = 2l(VC).$

Hence $l(VC_1) \le l(VC)$ or $l(VC_2) \le l(VC)$, and again we are done. ∎

This implies (Krogdahl [1976], Fujishige [1977a]):

Theorem 41.5. *Let $I \in \mathcal{I}_1 \cap \mathcal{I}_2$. Then I is extreme if and only if $D_{M_1,M_2}(I)$ has no directed circuit of negative length.*

Section 41.3. Weighted matroid intersection algorithm

Proof. To see necessity, suppose that $D_{M_1,M_2}(I)$ has a directed circuit C of negative length. Choose C with $|VC|$ minimal. Consider $J := I \triangle VC$. Since $w(J) = w(I) - l(C) > w(I)$, while $|J| = |I|$, we know that $J \notin \mathcal{I}_1 \cap \mathcal{I}_2$. Hence by Lemma 41.5α, $D_{M_1,M_2}(I)$ has a negative-length directed circuit covering fewer than $|VC|$ vertices, contradicting our assumption.

To see sufficiency, consider a $J \in \mathcal{I}_1 \cap \mathcal{I}_2$ with $|J| = |I|$. By Corollary 39.12a, both $D_{M_1}(I)$ and $D_{M_2}(I)$ have a perfect matching on $I \triangle J$. These two matchings together form a vertex-disjoint union of a number of directed circuits $C_1, \ldots, C_t$. Then

$$(41.24) \qquad w(I) - w(J) = \sum_{j=1}^{t} l(VC_j) \geq 0,$$

implying $w(J) \leq w(I)$. So I is extreme. ∎

This theorem implies that we can find a shortest path P, in Case 1 of the algorithm, in strongly polynomial time (with the Bellman-Ford method). It also gives:

Theorem 41.6. *If Case 1 applies, I' is an extreme common independent set.*

Proof. We first show that $I' \in \mathcal{I}_1 \cap \mathcal{I}_2$. To this end, let t be a new element, and extend (for each $i = 1, 2$), M_i to a matroid $M'_i = (S + t, \mathcal{I}'_i)$, where for each $T \subseteq S + t$:

$$(41.25) \qquad T \in \mathcal{I}'_i \text{ if and only if } T - t \in \mathcal{I}_i.$$

Note that $D_{M'_1, M'_2}(I + t)$ arises from $D_{M_1, M_2}(I)$ by extending it with a new vertex t and adding arcs from t to each vertex in X_1, and from each vertex in X_2 to t.

Let P be the path found in the algorithm. Define

$$(41.26) \qquad w(t) := l(t) := -l(P).$$

As P is a shortest $X_1 - X_2$ path, this makes that $D_{M'_1, M'_2}(I + t)$ has no negative-length directed circuit. Hence, by Theorem 41.5, $I + t$ is an extreme common independent set of M'_1 and M'_2.

Let P run from $z_1 \in X_1$ to $z_2 \in X_2$. Extend P by the arcs (t, z_1) and (z_2, t) to a directed circuit C. So $J = (I + t) \triangle VC$. As P has a minimum number of arcs among all shortest $X_1 - X_2$ paths, and as $D_{M'_1, M'_2}(I+t)$ has no negative-length directed circuits, by Lemma 41.5α we know that $J \in \mathcal{I}_1 \cap \mathcal{I}_2$.

Moreover, J is extreme, since $I + t$ is extreme and $w(J) = w(I + t)$. ∎

So the weighted common independent set augmenting algorithm is correct. It obviously has strongly polynomially bounded running time. Therefore:

710   Chapter 41. Matroid intersection

Theorem 41.7. *A maximum-weight common independent set in two matroids can be found in strongly polynomial time.*

Proof. Starting with the extreme common independent set $I_0 := \emptyset$ we can find iteratively extreme common independent sets $I_0, I_1, \ldots, I_k$, where $|I_i| = i$ for $i = 0, \ldots, k$ and where I_k is a maximum-size common independent set. Taking one among $I_0, \ldots, I_k$ of maximum weight, we have a maximum-weight common independent set. ∎

The above algorithm gives a maximum-weight common independent set of size k, for each k. In particular, a maximum-weight common base can be found with the algorithm. Similarly for minimum-weight:

Theorem 41.8. *A minimum-weight common base in two matroids can be found in strongly polynomial time.*

Proof. The last extreme common independent set in the above algorithm is a maximum-weight common base. By flipping the signs of the weights, this can be turned into a minimum-weight common base algorithm. ∎

Notes. Frank [1981a] gave an $O(\tau n^3)$-time implementation of this algorithm, where τ is the time needed to test for any $I \in \mathcal{I}_i$ and any $s \in S$ whether or not $I \cup \{s\} \in \mathcal{I}_i$, and if not, to find a circuit of M_i contained in $I \cup \{s\}$.

Clearly, a maximum-weight common independent set need not be a common base, even if common bases exist and all weights are positive: Let $S = \{1,2,3\}$ and let M_i be the matroid on S with unique circuit $S \setminus \{i\}$ (for $i = 1, 2$). Define $w(1) := w(2) := 1$ and $w(3) := 3$. Then $\{3\}$ is the unique maximum-weight common independent set, while $\{1, 2\}$ is the unique common base.

41.3a. Speeding up the weighted matroid intersection algorithm

The algorithm described in Section 41.3 is strongly polynomial-time, since we can find a shortest path P in strongly polynomial time, as in each iteration the graph $D_{M_1,M_2}(I)$ has no negative-length directed circuit. Hence we can apply the Bellman-Ford method. To bound the running time, suppose that we can construct, for any $I \in \mathcal{I}_1 \cap \mathcal{I}_2$ the graph $D_{M_1,M_2}(I)$ in time T. Then any iteration can be done in time $O(T + n^3)$, where $n := |S|$.

We can improve this to $O(T + n \log n)$ as follows (Frank [1981a], Brezovec, Cornuéjols, and Glover [1986]). The idea is that, in each iteration, with the extreme common independent set I, we give a 'certificate' of extremity, by specifying a potential for the length function; that is, a function $p \in \mathbb{Q}^S$ satisfying

(41.27) $l(v) \geq p(v) - p(u)$

for each arc (u,v) of $D_{M_1,M_2}(I)$. By Theorem 41.5, such a potential certifies extremity of I. We call such a p a potential *for* I.

Having the potential, we can apply Dijkstra's method instead of the Bellman-Ford method, as with the potential we can transform the length function (if defined on arcs) to a nonnegative length function.

Section 41.3a. Speeding up the weighted matroid intersection algorithm

It is convenient to associate the following functions $w_1, w_2 : S \to \mathbb{R}$ to $p, w : S \to \mathbb{R}$:

(41.28) $w_1(v) = p(v)$ and $w_2(v) = w(v) - p(v)$ if $v \in I$,
$w_1(v) = w(v) + p(v)$ and $w_2(v) = -p(v)$ if $v \in S \setminus I$.

So $w = w_1 + w_2$. Then:

Theorem 41.9. *Let $I \in \mathcal{I}_1 \cap \mathcal{I}_2$ and let $p, w, w_1, w_2 : S \to \mathbb{R}$ satisfy (41.28). Then p is a potential for $D_{M_1, M_2}(I)$ if and only if for $i = 1, 2$ one has*

(41.29) *I maximizes $w_i(X)$ over all $J \in \mathcal{I}_i$ satisfying $|J| = |I|$.*

Proof. The theorem follows easily with Corollary 39.12b. Indeed, there is an arc (u, v) leaving I if and only if $I - u + v \in \mathcal{I}_1$. Then

(41.30) $w_1(v) \leq w_1(u) \iff l(v) \geq p(v) - p(u)$,

since $l(v) = -w(v) = -w_2(v) - w_1(v)$ and $-w_2(v) - w_1(u) = p(v) - p(u)$.

Similarly, there is an arc (u, v) entering I if and only if $I - v + u \in \mathcal{I}_2$. Then

(41.31) $w_2(v) \geq w_2(u) \iff l(v) \geq p(v) - p(u)$,

since $l(v) = w(v) = w_2(v) + w_1(v)$ and $w_2(u) + w_1(v) = p(v) - p(u)$. ∎

We trivially have a potential for $I := \emptyset$. Consider next an arbitrary iteration, with as input a common independent set I and a potential p for I. Construct $D_{M_1, M_2}(I)$ and l as before. Let P be an $X_1 - X_2$ path with $l(P)$ minimum, and, under this condition, with $|VP|$ minimum. (Using the potential described above, we can find P with Dijkstra's algorithm.) Let $I' := I \triangle VP$.

We now reset the potential p such that for any $v \in S$ with v reachable from X_1, $p(v)$ is equal to the distance from X_1 to v (= the minimum of $l(VQ)$ over all $X_1 - v$ paths Q in $D_{M_1, M_2}(I)$).

Let w_1 and w_2 satisfy (41.28) with respect to I, (the new) p, and w. Then:

Theorem 41.10. *w_1, w_2 satisfy (41.29) with respect to I'.*

Proof. Extend M_1 and M_2 to matroids $M_1' = (S + t, \mathcal{I}_1')$ and $M_2' = (S + t, \mathcal{I}_2')$ as in (41.25). Let P run from $z_1 \in X_1$ to $z_2 \in X_2$. Define $w(t) := l(t) := -l(P)$, $p(t) := 0$, $w_1(t) := 0$, and $w_2(t) := w(t)$. Now it suffices to show:

(41.32) (i) $w_i(I + t) = w_i(I')$ for $i = 1, 2$;
(ii) w_1, w_2 satisfy (41.29) with respect to M_1', M_2', and $I + t$.

Let C be the directed circuit obtained by extending P by the arcs (t, z_1) and (z_2, t). Now, since $I' = (I + t) \triangle VC$, to show (41.32), it suffices to show, for each arc (u, v):

(41.33) if (u, v) leaves $I + t$, then $w_1(v) \leq w_1(u)$, with equality if (u, v) is on C;
if (u, v) enters $I + t$, then $w_2(u) \leq w_2(v)$, with equality if (u, v) is on C.

Note that for each arc (u, v) of $D_{M_1', M_2'}(I + t)$ one has $p(v) \leq p(u) + l(v)$, with equality if (u, v) is on C. Hence, if (u, v) leaves $I + t$, then:

(41.34) $\qquad w_1(v) = p(v) + w(v) = p(v) - l(v) \leq p(u) = w_1(u),$

with equality if (u,v) is on C.

Similarly, if (u,v) enters $I + t$, then:

(41.35) $\qquad w_2(v) = w(v) - p(v) = l(v) - p(v) \geq -p(u) = w_2(u),$

with equality if (u,v) is on C. This proves (41.33). ∎

Using (41.28) and Theorem 41.9, we can obtain from w_1, w_2 a potential for I'. This implies:

Corollary 41.10a. *A maximum-weight common independent set can be found in time $O(k(T+n\log n))$, where $n := |S|$, k is the maximum size of a common independent set, and T is the time needed to find $D_{M_1,M_2}(I)$ for any common independent set I.*

Proof. Each iteration can be done in time $O(T + n\log n)$, since constructing the graph $D_{M_1,M_2}(I)$ takes T time, implying that there are $O(T)$ arcs. Hence, by Corollary 7.7a, a shortest $X_1 - X_2$ path P can be found in $O(T + n\log n)$ time. Hence I', and a potential for I' can be found in time $O(T + n\log n)$.

Since there are k iterations, we have the time bound given. ∎

In applications where the matroids are specifically given, one can often derive a better time bound, by obtaining $D_{M_1,M_2}(I')$ not from scratch, but by adapting $D_{M_1,M_2}(I)$. See also Brezovec, Cornuéjols, and Glover [1986] and Gabow and Xu [1989,1996].

41.4. Intersection of the independent set polytopes

It turns out that the intersection of the independent set polytopes of two matroids gives exactly the convex hull of the common independent sets, as was shown by Edmonds [1970b][27].

We first prove a very useful theorem, due to Edmonds [1970b], which we often will apply in this part. (A more general statement and interpretation in terms of network matrices will be given in Section 13.4.)

A family $\mathcal{C}$ of sets is called *laminar* if

(41.36) $\qquad Y \subseteq Z \text{ or } Z \subseteq Y \text{ or } Y \cap Z = \emptyset$

for all $Y, Z \in \mathcal{C}$.

Theorem 41.11. *Let $\mathcal{C}$ be the union of two laminar families of subsets of a set X. Let A be the $\mathcal{C} \times X$ incidence matrix of $\mathcal{C}$. Then A is totally unimodular.*

---

[27] Lawler [1976b] wrote that this result was announced by Edmonds 'at least as long ago as 1964'.

Proof. Let A be a counterexample with $|\mathcal{C}| + |X|$ minimal, and (secondly) with a minimal number of 1's. Then A is nonsingular and has determinant $\neq \pm 1$. Let $\mathcal{C}_1$ and $\mathcal{C}_2$ be laminar families, with union $\mathcal{C}$.

If each $\mathcal{C}_i$ consists of pairwise disjoint sets, then A is the incidence matrix of a bipartite graph, added with some unit base vectors. Hence A is totally unimodular, a contradiction.

If say $\mathcal{C}_1$ does not consist of pairwise disjoint sets, $\mathcal{C}_1$ contains a smallest nonempty set Y that is contained in some other set Z in $\mathcal{C}_1$. Choose Z smallest. Replacing Z by $Z \setminus Y$, maintains laminarity of $\mathcal{C}_1$. As this does not change the determinant of the corresponding matrix (as it amounts to subtracting row indexed Y from row indexed Z), we would have a counterexample with a smaller number of 1's, a contradiction. ∎

Let $M_1 = (S, \mathcal{I}_1)$ and $M_2 = (S, \mathcal{I}_2)$ be matroids, with rank functions r_1 and r_2. By Corollary 40.2a, the intersection $P_{\text{independent set}}(M_1) \cap P_{\text{independent set}}(M_2)$ of the independent set polytopes associated with the matroids $M_1 = (S, \mathcal{I}_1)$ and $M_2 = (S, \mathcal{I}_2)$ is determined by:

(41.37)   (i) $x_s \geq 0$ for $s \in S$,
           (ii) $x(U) \leq r_i(U)$ for $i = 1, 2$ and $U \subseteq S$.

Trivially, this intersection contains the convex hull of the incidence vectors of common independent sets of M_1 and M_2. We shall see that these two polytopes are equal.

Basis is the following result of Edmonds [1970b], whose proof we follow (it constitutes the base of a fundamental technique developed further in several other results).

Theorem 41.12. *System (41.37) is box-totally dual integral.*

Proof. Choose $w \in \mathbb{Z}^S$. Consider the linear programming problem dual to maximizing $w^\mathsf{T} x$ over the constraints (41.37)(ii):

(41.38)   minimize $\displaystyle\sum_{U \subseteq S} (y_1(U) r_1(U) + y_2(U) r_2(U))$

        where $y_1, y_2 \in \mathbb{R}_+^{\mathcal{P}(S)}$,

                  $\displaystyle\sum_{U \subseteq S} (y_1(U) + y_2(U)) \chi^U = w$.

Let y_1, y_2 attain this minimum, such that

(41.39) $\displaystyle\sum_{U \subseteq S} (y_1(U) + y_2(U)) |U| |S \setminus U|$

is minimized. Define

(41.40) $\mathcal{F}_i := \{U \subseteq S \mid y_i(U) > 0\}$,

for $i = 1, 2$. We show that for $i = 1, 2$, the collection $\mathcal{F}_i$ is a chain; that is,

714   Chapter 41. Matroid intersection

(41.41)   if $T, U \in \mathcal{F}_i$, then $T \subseteq U$ or $U \subseteq T$.

Suppose not. Choose $\alpha := \min\{y_i(T), y_i(U)\}$, and decrease $y_i(T)$ and $y_i(U)$ by α, and increase $y_i(T \cap U)$ and $y_i(T \cup U)$ by α. Since

(41.42) $\chi^T + \chi^U = \chi^{T \cap U} + \chi^{T \cup U}$,

y_1, y_2 remains a feasible solution of (41.38); and since

(41.43) $r_i(T) + r_i(U) \geq r_i(T \cap U) + r_i(T \cup U)$,

it remains optimum. However, sum (41.39) decreases (by Theorem 2.1), contradicting the minimality assumption. So $\mathcal{F}_1$ and $\mathcal{F}_2$ are chains.

As the constraints in (41.37)(ii) corresponding to $\mathcal{F}_1$ and $\mathcal{F}_2$ form a totally unimodular matrix (by Theorem 41.11), by Theorem 5.35 system (41.37)(ii) is box-TDI, and hence (41.37) is box-TDI.   ∎

(The fact that the $\mathcal{F}_i$ can be taken to be chains also follows directly from the proof method of Theorem 40.2.)

This implies a characterization of the *common independent set polytope*

(41.44) $P_{\text{common independent set}}(M_1, M_2)$

of two matroids $M_1 = (S, \mathcal{I}_1)$ and $M_2 = (S, \mathcal{I}_2)$, being the convex hull of the incidence vectors of the common independent sets of M_1 and M_2:

Corollary 41.12a. $P_{\text{common independent set}}(M_1, M_2)$ *is determined by* (41.37).

Proof. Directly from Theorem 41.12, since it implies that the vertices of the polytope determined by (41.37) are integer, and hence are the incidence vectors of common independent sets.   ∎

Another way of stating this is:

Corollary 41.12b.

(41.45) $P_{\text{common independent set}}(M_1, M_2)$
           $= P_{\text{independent set}}(M_1) \cap P_{\text{independent set}}(M_2)$.

Proof. From Corollary 41.12a, using the fact that (41.37) is the union of the constraints for the independent set polytopes of M_1 and M_2, by Corollary 40.2b.   ∎

The total dual integrality of (41.37) gives the following extension of the matroid intersection theorem:

Corollary 41.12c. *Let* $M_1 = (S, \mathcal{I}_1)$ *and* $M_2 = (S, \mathcal{I}_2)$ *be matroids, with rank functions* r_1 *and* r_2, *respectively, and let* $w \in \mathbb{Z}_+^S$. *Then the maximum value of* $w(I)$ *over* $I \in \mathcal{I}_1 \cap \mathcal{I}_2$ *is equal to the minimum value of*

(41.46) $\quad r_1(U_1) + \cdots + r_1(U_k) + r_2(T_1) + \cdots + r_2(T_l),$

where $U_1 \subseteq \cdots \subseteq U_k \subseteq S$ and $T_1 \subseteq \cdots \subseteq T_l \subseteq S$ such that each element s of S occurs in precisely $w(s)$ sets among $U_1, \ldots, U_k, T_1, \ldots, T_l$.

Proof. Directly from Theorem 41.12 and its proof. ∎

(Edmonds [1979] gave an algorithmic proof of this result.)

These corollaries cannot be extended to the intersection of the independent set polytopes of three matroids. Let $S = \{1, 2, 3\}$, and for $i = 1, 2, 3$, let M_i be the matroid on S with $S \setminus \{i\}$ as unique circuit. Then $P_{\text{independent set}}(M_1) \cap P_{\text{independent set}}(M_2) \cap P_{\text{independent set}}(M_3)$ contains the all-$\frac{1}{2}$ vector, while each integer vector in this intersection contains at most one 1. So the intersection is *not* the convex hull of the common independent sets.

Similar results hold for the common base polytope. For matroids M_1 and M_2, let the *common base polytope* $P_{\text{common base}}(M_1, M_2)$ be the convex hull of the incidence vectors of common bases of M_1 and M_2. Then:

Corollary 41.12d. $P_{\text{common base}}(M_1, M_2) = P_{\text{base}}(M_1) \cap P_{\text{base}}(M_2)$.

Proof. Directly from the foregoing. ∎

So the common base polytope is determined by:

(41.47) $\quad \begin{aligned} & x_s \geq 0 && \text{for } s \in S, \\ & x(U) \leq r_i(U) && \text{for } i = 1, 2 \text{ and } U \subseteq S, \\ & x(S) = r_i(S) && \text{for } i = 1, 2. \end{aligned}$

Corollary 41.12e. *System* (41.47) *is box-TDI.*

Proof. From Theorem 41.12, with Theorem 5.25. ∎

Moreover, similar results hold for the common spanning set polytope. For matroids M_1 and M_2, let the *common spanning set polytope*, in notation $P_{\text{common spanning set}}(M_1, M_2)$, be the convex hull of the incidence vectors of common spanning sets of M_1 and M_2. Then:

Corollary 41.12f.

(41.48) $\quad \begin{aligned} & P_{\text{common spanning set}}(M_1, M_2) \\ & = P_{\text{spanning set}}(M_1) \cap P_{\text{spanning set}}(M_2). \end{aligned}$

Proof. This can be reduced to Corollary 41.12b on the common independent set polytope, by duality: x belongs to the spanning set polytope of M_i if and only if $\mathbf{1} - x$ belongs to the independent set polytope of M_i^*.

Similarly, x belongs to the common spanning set polytope of M_1 and M_2 if and only if $\mathbf{1} - x$ belongs to the common independent set polytope of M_1^* and M_2^*. ∎

So the common spanning set polytope is determined by:

(41.49) $\quad 0 \leq x_s \leq 1 \qquad\qquad\qquad\qquad$ for $s \in S$,
$\qquad\quad\;\; x(U) \leq r_i(S) - r_i(S \setminus U) \quad$ for $i = 1, 2$ and $U \subseteq S$.

Corollary 41.12g. *System* (41.49) *is box-TDI.*

Proof. Again, this can be derived from Theorem 41.12, by replacing x by $\mathbf{1} - x$. ∎

Another consequence of Theorem 41.12 is:

Corollary 41.12h. *Let $M_1 = (S, \mathcal{I}_1)$ and $M_2 = (S, \mathcal{I}_2)$ be matroids and let $x \in \mathbb{R}_+^S$. Then*

(41.50) $\quad \max\{z(S) \mid z \leq x, z \in P_{\text{common independent set}}(M_1, M_2)\}$
$\qquad\quad = \min\{r(U) + x(S \setminus U) \mid U \subseteq S\},$

where $r(U)$ denotes the maximum size of a common independent set contained in U.

Proof. This follows from the box-total dual integrality of (41.37), using the fact that $r(U_1 \cup U_2) \leq r_1(U_1) + r_2(U_2)$ for disjoint U_1, U_2. ∎

Cunningham [1984] showed that, if matroids $M_1 = (S, \mathcal{I}_1)$ and $M_2 = (S, \mathcal{I}_2)$ are given by independence testing oracles, one can find in strongly polynomial time for any $x \in \mathbb{Q}^S$, optimum solutions of (41.50). This will follow from the results in Section 47.4.

The result of Cunningham [1984] also implies:

Theorem 41.13. *Given matroids $M_1 = (S, \mathcal{I}_1)$ and $M_2 = (S, \mathcal{I}_2)$ by independence testing oracles, and given $x \in \mathbb{Q}^S$, one can test in strongly polynomial time if x belongs to the common independent set polytope, and if so, decompose x as a convex combination of incidence vectors of common independent sets.*

Proof. Let r_i be the rank function of M_i ($i = 1, 2$) and let $r(U) := \min\{r_1(U), r_2(U)\}$ for $i = 1, 2$. Let P be the common independent set polytope. Corollaries 40.4a and 41.12b imply that one can test in strongly polynomial time if x belongs to P.

So we can assume that x belongs to P. We decompose x as a convex combination of incidence vectors of common independent sets. Iteratively resetting x, we keep a collection $\mathcal{U}$ of subsets of S with $x(U) = r(U)$ for each $U \in \mathcal{U}$. Initially, $\mathcal{U} := \emptyset$. We describe the iteration.

Define

(41.51) $\quad F := \{y \in P \mid \forall s \in S : x_s = 0 \Rightarrow y_s = 0; \forall U \in \mathcal{U} : y(U) = r(U)\}.$

So F is a face of P containing x.

Find a common independent set I with $\chi^I \in F$. This can be done by finding a common independent set $I \subseteq \mathrm{supp}(x)$ maximizing $w^\mathsf{T} x$, where $w := \sum_{U \in \mathcal{U}} \chi^U$. (Here $\mathrm{supp}(x)$ is the support of x; so $\mathrm{supp}(x) = \{s \in S \mid x_s > 0\}$.)

If $x = \chi^I$ we stop. Otherwise, define $u := x - \chi^I$. Let λ be the largest rational such that

(41.52) $\quad \chi^I + \lambda u$

belongs to P.

We describe an inner iteration to find λ. We consider vectors z along the halfline $L = \{\chi^I + \lambda u \mid \lambda \geq 0\}$. First we let λ be the largest rational with $\chi^I + \lambda u \geq \mathbf{0}$, and set $z := \chi^I + \lambda u$.

We iteratively reset z. We check if z belongs to the common independent set polytope, and if not, we find a $U \subseteq S$ minimizing $r(U) - z(U)$ (with Corollary 40.4c). Let z' be the (unique) vector on L achieving $x(U) \leq r(U)$ with equality; that is, satisfying $z'(U) = r(U)$.

Consider any inequality $x(U') \leq r(U')$ violated by z'. Then

(41.53) $\quad r(U') - |U' \cap I| < r(U) - |U \cap I|.$

This can be seen by considering the function

(41.54) $\quad d(y) := (r(U) - y(U)) - (r(U') - y(U')).$

We have $d(z) \leq 0$ (since U minimizes $r(U) - z(U)$) and $d(z') > 0$ (since $z'(U) = r(U)$ and $z'(U') > r(U')$). Hence, as d is linear, $d(\chi^I) > 0$; that is, we have (41.53). This implies that resetting $z := z'$, there are at most $r(S)$ inner iterations.

Let x' be the final z found. If we apply no inner iteration, then $x'_s = 0$ for some $s \in I \subseteq \mathrm{supp}(x)$ (since we chose λ largest with $\chi^I + \lambda u \geq \mathbf{0}$). If we do at least one inner iteration, we find a U such that x' satisfies $x'(U) = r(U)$ while $|U \cap I| < r(U)$ (since x' is the unique vector on L satisfying $x'(U) = r(U)$ and since $x' \neq \chi^I$).

In the latter case, set $\mathcal{U}' := \mathcal{U} \cup \{U\}$; otherwise set $\mathcal{U}' := \mathcal{U}$. Then resetting x to x' and $\mathcal{U}$ to $\mathcal{U}'$, the dimension of F decreases (as χ^I does not belong to the new F). So the number of iterations is at most $|S|$. This shows that the method is strongly polynomial-time. ∎

41.4a. Facets of the common independent set polytope

Since the common independent set polytope of two matroids is the intersection of their independent set polytopes, each facet-inducing inequality for the intersection is facet-inducing for (at least) one of the independent set polytopes, but not necessarily conversely. Giles [1975] characterized which inequalities are facet-inducing

718   Chapter 41. Matroid intersection

for the common independent set polytope. If this polytope is full-dimensional, then each inequality $x_s \geq 0$ is facet-inducing. As for the other inequalities, Giles proved:

Theorem 41.14. *Let $M_1 = (S, \mathcal{I}_1)$ and $M_2 = (S, \mathcal{I}_2)$ be loopless matroids, with rank functions r_1 and r_2. For $U \subseteq S$, define $r(U) := \min\{r_1(U), r_2(U)\}$. Then, for $U \subseteq S$, the inequality*

(41.55) $\qquad x(U) \leq r(U)$

is facet-inducing for $P_{\text{common independent set}}(M_1, M_2)$ if and only if there is no partition of U into nonempty proper subsets U_1, U_2 with

(41.56) $\qquad r(U) \geq r(U_1) + r(U_2)$

and there is no proper superset U' of U with $r(U') \leq r(U)$.

Proof. By symmetry, we can assume that $r(U) = r_1(U)$.

Necessity is easy: Assume that $x(U) \leq r_1(U)$ is facet-inducing. If (41.56) would hold, then each common independent set I with $|I \cap U| = r_1(U)$ satisfies $|I \cap U_1| = r(U_1)$ (since $|I \cap U_1| = |I \cap U| - |I \cap U_2| \geq r(U) - r(U_2) \geq r(U_1)$). Hence each x in the facet determined by $x(U) \leq r_1(U)$ satisfies $x(U_1) = r(U_1)$, a contradiction. Similarly, if $r(U') \leq r_1(U)$ for some proper superset U' of U, then each common independent set I with $|I \cap U| = r_1(U)$ satisfies $|I \cap U'| = r(U')$, implying that each x in the facet determined by $x(U) \leq r_1(U)$ satisfies $x(U') = r(U')$, again a contradiction.

To see sufficiency, suppose that (41.55) satisfies the conditions, but is not facet-inducing for the common independent set polytope. This implies that the inequality $x(U) \leq r_1(U)$ is implied by other inequalities in (41.37). So there exist $\lambda_i : \mathcal{P}(S) \to \mathbb{Q}_+$ (for $i = 1, 2$) such that

(41.57) $\qquad \sum_{T \in \mathcal{P}(S)} (\lambda_1(T) + \lambda_2(T))\chi^T \geq \chi^U$ and

$\qquad \sum_{T \in \mathcal{P}(S)} (\lambda_1(T)r_1(T) + \lambda_2(T)r_2(T)) \leq r_1(U)$,

and such that $\lambda_i(U) = 0$ for $i = 1, 2$. Let D be the least common denominator of the values of the λ_i. Choose the λ_i such that D is as small as possible and (secondly) such that

(41.58) $\qquad D \cdot \sum_{T \subseteq S} (\lambda_1(T) + \lambda_2(T))|T|(|S \setminus T| + 1)$

is as small as possible. For $i = 1, 2$, define

(41.59) $\qquad \mathcal{F}_i := \{T \subseteq S \mid \lambda_i(T) > 0\}$.

We claim that for $i = 1, 2$:

(41.60) $\qquad \mathcal{F}_i$ is a chain.

Suppose to the contrary that $T_1, T_2 \in \mathcal{F}_i$ satisfy $T_1 \not\subseteq T_2 \not\subseteq T_1$. Then decreasing $\lambda_i(T_1)$ and $\lambda_i(T_2)$ by $1/D$ and increasing $\lambda_i(T_1 \cap T_2)$ and $\lambda_i(T_1 \cup T_2)$ by $1/D$ maintains (41.57) but decreases (41.58). This would be a contradiction, except if $T_1 \cap T_2$ or $T_1 \cup T_2$ equals U. If one of these sets equals U and $D \geq 2$, we can reset $\lambda_i(U) := 0$, and multiply all values of λ_1 and λ_2 by $D/(D-1)$. This again maintains (41.57) but decreases the least common divisor of the denominators. So the contradiction would remain, except if $D = 1$. Then (41.57) implies $r_i(T_1) + r_i(T_2) \le r_1(U)$. Now if $T_1 \cap T_2 = U$, then $U \subset T_1$ and

(41.61) $\qquad r(T_1) \le r_i(T_1) \le r_i(T_1) + r_i(T_2) \le r_1(U),$

contradicting the condition. If $T_1 \cup T_2 = U$, then

(41.62) $\qquad r(T_1) + r(U \setminus T_1) \le r_i(T_1) + r_i(U \setminus T_1) \le r_i(T_1) + r_i(T_2) \le r_1(U),$

again contradicting the condition.

This proves (41.60). As each $\mathcal{F}_i$ is a chain, the incidence matrix of $\mathcal{F}_1 \cup \mathcal{F}_2$ is totally unimodular (by Theorem 41.11). Therefore, there are integer-valued λ_i satisfying (41.57), with $\lambda_i(T) = 0$ for $T \notin \mathcal{F}_i$. Then we can assume that $|\mathcal{F}_i| \le 1$ for $i = 1, 2$, since if $T, T' \in \mathcal{F}_i$ and $T \subset T'$, we can decrease $\lambda_i(T)$ by 1 without violating (41.57). If $U' \in \mathcal{F}_i$ with $U' \supset U$, then $r(U') \le r_i(U') \le r(U)$, contradicting the condition. So each $\mathcal{F}_i$ contains a set $U_i \not\supseteq U$, implying $r(U_1) + r(U \setminus U_1) \le r(U_1) + r(U_2) \le r_1(U_1) + r_2(U_2) \le r(U)$, again contradicting the condition. ∎

This theorem can be seen to imply a variant of it, in which, instead of $r(U) := \min\{r_1(U), r_2(U)\}$, we define

(41.63) $\qquad r(U) := \max\{|I| \mid I \in \mathcal{I}_1 \cap \mathcal{I}_2\} = \min_{T \subseteq U}(r_1(T) + r_2(U \setminus T)).$

Fonlupt and Zemirline [1983] characterized the dimension of the common base polytope of two matroids.

41.4b. Up and down hull of the common base polytope

We saw in Corollary 41.12d a characterization of the common base polytope $P_{\text{common base}}(M_1, M_2)$ of two matroids $M_1 = (S, \mathcal{I}_1)$ and $M_2 = (S, \mathcal{I}_2)$. The up hull of this polytope:

(41.64) $\qquad P^\uparrow_{\text{common base}}(M_1, M_2) := P_{\text{common base}}(M_1, M_2) + \mathbb{R}^S_+$

was characterized by Cunningham [1977] and McDiarmid [1978] as follows (proving a conjecture of Fulkerson [1971a]).

Let $M_1 = (S, \mathcal{I}_1)$ and $M_2 = (S, \mathcal{I}_2)$ be matroids having a common base. Then $P^\uparrow_{\text{common base}}(M_1, M_2)$ is determined by:

(41.65) $\qquad x(U) \ge r(S) - r(S \setminus U)$ for $U \subseteq S$,

where $r(Z) :=$ the maximum size of a common independent set contained in Z. (A weaker version of this was proved by Edmonds and Giles [1977].)

For a proof we refer to Section 46.7a, where it is also shown that (41.65) is TDI (Gröflin and Hoffman [1981]). (Frank and Tardos [1984a] derived this, with a direct algorithmic construction, from the total dual integrality of (41.47).)

Note that by the matroid intersection theorem, the inequalities (41.65) are equivalent to:

(41.66) $\qquad x(U) \ge k - r_1(A) - r_2(B)$ for each partition U, A, B of S, where r_1 and r_2 are the rank functions of M_1 and M_2 respectively, and where k is the size of a common base. This implies that if we add $x \leq \mathbf{1}$ to (41.66) we obtain the convex hull of the subsets of S that contain a common base.

Similarly, the down hull of the common base polytope:

(41.67) $P^{\downarrow}_{\text{common base}}(M_1, M_2) := P_{\text{common base}}(M_1, M_2) - \mathbb{R}^S_+,$

is determined by

(41.68) $x(U) \leq r_1(S \setminus A) + r_2(S \setminus B) - k$ for each partition U, A, B of S.

This can be derived from the description of the up hull of the common base polytope, since

(41.69) $P^{\downarrow}_{\text{common base}}(M_1, M_2) = \mathbf{1} - P^{\uparrow}_{\text{common base}}(M_1^*, M_2^*)$

(where $\mathbf{1}$ stands for the all-one vector in $\mathbb{R}^S$).

This implies that the convex hull of the incidence vectors of the subsets of common bases is determined by $x \geq \mathbf{0}$ and (41.68).

Cunningham [1984] gave a strongly polynomial-time algorithm to test if a vector belongs to $P^{\uparrow}_{\text{common base}}(M_1, M_2)$, or to $P^{\downarrow}_{\text{common base}}(M_1, M_2)$, using only independence testing oracles for M_1 and M_2.

41.5. Further results and notes

41.5a. Menger's theorem for matroids

Tutte [1965b] showed a special case of the matroid intersection theorem, namely when both M_1 and M_2 are minors of one matroid. Specialized to graphic matroids, it gives the vertex-disjoint, undirected version of Menger's theorem.

Let $M = (E, \mathcal{I})$ be a matroid, with rank function r, and let U and W be disjoint subsets of E. Then the maximum size of a common independent set in $M/U \setminus W$ and $M/W \setminus U$ is equal to the minimum value of

(41.70) $r(X) - r(U) + r(E \setminus X) - r(W)$

taken over sets X with $U \subseteq X \subseteq E \setminus W$. This is the special case of the matroid intersection theorem for the matroids $M/U \setminus W$ and $M/W \setminus U$, since for $Y \subseteq E \setminus (U \cup W)$ one has

(41.71) $r_{M/U \setminus W}(Y) = r(Y \cup U) - r(U),$

and similarly for $M/W \setminus U$.

To see that this implies the vertex-disjoint, undirected version of Menger's theorem, let $G = (V, E)$ be a graph and let S and T be disjoint nonempty subsets of V. We show that the above theorem implies that the maximum number of disjoint $S - T$ paths in G is equal to the minimum number of vertices intersecting each $S - T$ path.

To this end, we can assume that G is connected, and that E contains subsets U and W such that (S, U) and (T, W) are trees. (Adding appropriate edges does not modify the result to be proved.)

Let $M := M(G)$ be the cycle matroid of G. Define $R := V \setminus (S \cup T)$. Then

(41.72) the maximum number of disjoint $S-T$ paths is at least the maximum size of a common independent set I of $M/U \setminus W$ and $M/W \setminus U$, minus $|R|$.

(In fact, there is equality.)

To prove (41.72), let I be a maximum-size common independent in $M/U \setminus W$ and $M/W \setminus U$. So I is a forest. Consider any component K of I. Since I is independent in M/U, K intersects S in at most one vertex. Similarly, K intersects T in at most one vertex. Let p be the number of components K intersecting both S and T. By deleting p edges we obtain a forest I' such that no component of I' intersects both S and T. So $|I'| \leq |R|$ (since I' remains a forest after contracting (in the graphical sense) $S \cup T$ to one vertex). Hence $p = |I| - |I'| \geq |I| - |R|$. So we have (41.72).

On the other hand,

(41.73) the minimum size of a set of vertices intersecting each $S-T$ path is at most the minimum value of (41.70), minus $|R|$.

(Again, we have in fact equality.)

To prove (41.73), let X attain the minimum value of (41.70). So $U \subseteq X \subseteq E \setminus W$. Let K be the component of (V, X) containing S and let L be the component of $(V, E \setminus X)$ containing T. We choose X with $|K \cup L|$ maximized.

Then $K \cup L = V$. For suppose not. Then, as G is connected, there is an edge e of G leaving $K \cup L$. By symmetry, we can assume that $e \in X$. Let K' be the component of (V, X) containing e. So $K' \neq K$ and $E[K'] \cap U = \emptyset$. Resetting X by $X \setminus E[K']$, $r(X)$ decreases by $|K'| - 1$, while $r(E \setminus X)$ increases by at most $|K'| - 1$. So the new X again attains the minimum in (41.70), while $K \cup L$ increases. This contradicts our maximality assumption.

So $K \cup L = V$. Hence $K \cap L$ intersects each $S-T$ path (since $S \subseteq K$ and $T \subseteq L$, and there is no edge connecting $K \setminus L$ and $L \setminus K$). Moreover

(41.74) $\quad |K \cap L| = |K| + |L| - |V| \leq (r(X) + 1) + (r(E \setminus X) + 1) - |V|$
$= r(X) + r(E \setminus X) - |V| + 2 = r(X) + r(E \setminus X) - r(U) - r(W) - |R|$.

So we have (41.73).

Since the maximum number of disjoint $S-T$ paths is trivially not more than the minimum number of vertices intersecting all $S-T$ paths, we thus obtain Menger's theorem (and also equality in (41.72) and (41.73)).

(Tomizawa [1976a] gave an algorithm for Menger's theorem for matroids.)

41.5b. Exchange properties

Kundu and Lawler [1973] showed the following extension of the exchange property of bipartite graphs given in Theorem 16.8. Let $M_1 = (S, \mathcal{I}_1)$ and $M_2 = (S, \mathcal{I}_2)$ be matroids, with span functions span_1 and span_2. Then

(41.75) For any $I_1, I_2 \in \mathcal{I}_1 \cap \mathcal{I}_2$ there exists an $I \in \mathcal{I}_1 \cap \mathcal{I}_2$ with $I_1 \subseteq \text{span}_1(I)$ and $I_2 \subseteq \text{span}_2(I)$.

(Theorem 16.8 is equivalent to the case where the M_i are partition matroids.)

To prove (41.75), choose $I \in \mathcal{I}_1 \cap \mathcal{I}_2$ with $I_1 \subseteq \text{span}_1(I)$ and $|I \cap I_2|$ maximized. Suppose that $I_2 \not\subseteq \text{span}_2(I)$. Choose $s \in I_2 \setminus \text{span}_2(I)$ with $I \cup \{s\} \in \mathcal{I}_2$. By the maximality of $|I \cap I_2|$ we know that $I \cup \{s\} \notin \mathcal{I}_1$. So M_1 has a circuit C

contained in $I \cup \{s\}$. Since $I_2 \in \mathcal{I}_1$ we know that $C \not\subseteq I_2$. Choose $t \in C \setminus I_2$. Then for $I' := I - t + s$ we have $I' \in \mathcal{I}_1 \cap \mathcal{I}_2$, while $\mathrm{span}_1(I') = \mathrm{span}_1(I)$. Since $|I' \cap I_2| > |I \cap I_2|$ this contradicts the maximality assumption.

A second exchange property was shown by Davies [1976]:

(41.76) Two matroids M_1 and M_2 have bases B_1 and B_2 (respectively) with $|B_1 \cap B_2| = k$ if and only if M_1 has bases X_1 and Y_1 and M_2 has bases X_2 and Y_2 with $|X_1 \cap X_2| \leq k$ and $|Y_1 \cap Y_2| \geq k$.

To see this, we may assume that $X_2 = Y_2$, since if $|X_1 \cap Y_2| \leq k$ we can reset $X_2 := Y_2$, and if $|X_1 \cap Y_2| > k$ we can reset $Y_1 := X_1$ and exchange indices.

By (39.33)(ii), there exists a series of bases $Z_0, \ldots, Z_t$ of M_1 such that $Z_0 = X_1$, $Z_t = Y_1$, and $|Z_{i-1} \triangle Z_i| = 2$ for $i = 1, \ldots, t$. Hence

(41.77) $\big||Z_{i-1} \cap X_2| - |Z_i \cap X_2|\big| \leq 1$

for $i = 1, \ldots, t$. Since $|Z_0 \cap X_2| \leq k$ and $|Z_t \cap X_2| \geq k$, we know $|Z_i \cap X_2| = k$ for some i. This proves (41.76).

41.5c. Jump systems

A framework that includes both matroid intersection and maximum-size matching was introduced by Bouchet and Cunningham [1995]. For $x, y \in \mathbb{Z}^n$, let $[x, y]$ be the set of vectors $z \in \mathbb{Z}^n$ with $\|x - y\|_1 = \|x - z\|_1 + \|z - y\|_1$. So $[x, y]$ consists of all integer vectors z in the box $x \wedge y \leq z \leq x \vee y$.

Call a vector z a *step from x to y* if $z \in [x, y]$ and $\|z - x\|_1 = 1$. A *jump system* is a finite subset J of $\mathbb{Z}^n$ satisfying the following axiom:

(41.78) if $x, y \in J$ and z is a step from x to y, then $z \in J$ or J contains a step from z to y.

Trivially, for any jump system J and any $x, y \in \mathbb{Z}^n$, the intersection $J \cap [x, y]$ is again a jump system. Moreover, being a jump system is maintained under translations by an integer vector and by reflections in a coordinate hyperplane. Bouchet and Cunningham [1995] showed that the sum of jump systems is again a jump system (attributing the proof below to A. Sebő):

Theorem 41.15. *If J_1 and J_2 are jump systems in $\mathbb{Z}^n$, then $J_1 + J_2$ is a jump system.*

Proof. For $x, y \in J_1 + J_2$ we prove (41.78) by induction on the minimum value of

(41.79) $\|y' - x'\|_1 + \|y'' - x''\|_1$,

where $x', y' \in J_1$, $x'', y'' \in J_2$, $x' + x'' = x$, and $y' + y'' = y$.

Let z be a step from x to y. By reflection and permutation of coordinates, we can assume that $z = x + \chi^1$. So $x_1 < y_1$. Hence, by symmetry of J_1 and J_2, we can assume that $x'_1 < y'_1$. Next, by reflection, we can assume that $x' \leq y'$.

Now $x' + \chi^1$ is a step from x' to y'. If $x' + \chi^1 \in J_1$, then $z = x' + \chi^1 + x'' \in J_1 + J_2$, and we have (41.78). So we can assume that $x' + \chi^1 \notin J_1$. Hence, by (41.78) applied to J_1, there is an $i \in \{1, \ldots, n\}$ with $\tilde{x}' := x' + \chi^1 + \chi^i \in J_1$ and $\tilde{x}' \leq y'$.

So $z + \chi^i = \tilde{x}' + x'' \in J_1 + J_2$. If $z + \chi^i \in [x, y]$, we have (41.78). If $z + \chi^i \notin [x, y]$, then as $z \in [x, y]$, we have $z_i = y_i$. So z is a step from $z + \chi^i$ to y. Also,

$\|y'-\tilde{x}'\|_1 = \|y'-x'\|_1 - 2$. Hence, by our induction hypothesis applied to $z + \chi^i$ and y, we have (41.78). ∎

As Bouchet and Cunningham [1995] observed, this theorem implies that the following two constructions give jump systems $J \subseteq \mathbb{Z}^V$.

For any matroid $M = (S, \mathcal{I})$, the set $\{\chi^B \mid B \text{ base of } M\}$ is a jump system in $\mathbb{Z}^S$, as follows directly from the axioms (39.33). With Theorem 41.15, this implies that for matroids $M_1 = (S, \mathcal{I}_1)$ and $M_2 = (S, \mathcal{I}_2)$, the set

(41.80) $\quad J := \{\chi^{B_1} - \chi^{B_2} \mid B_i \text{ base of } M_i \ (i=1,2)\}$

is a jump system.

Let $G = (V, E)$ be an undirected graph and let

(41.81) $\quad J := \{\deg_F \mid F \subseteq E\} \subseteq \mathbb{Z}^V$;

that is, J is the collection of degree sequences of spanning subgraphs of G. Again, J is a jump system. This follows from Theorem 41.15, since for each edge $e = uv$ the set $\{\mathbf{0}, \chi^{\{u,v\}}\}$ is trivially a jump system in $\mathbb{Z}^V$ and since J is the sum of these jump systems.

Bouchet and Cunningham [1995] showed that the following greedy approach finds, for any $w \in \mathbb{R}^n$, a vector $x \in J$ maximizing $w^\mathsf{T} x$. By reflecting, we can assume that $w \geq \mathbf{0}$. We can also assume that $w_1 \geq w_2 \geq \cdots \geq w_n$. Let $J_0 := J$, and for $i = 1, \ldots, n$, let J_i be the set of vectors x in J_{i-1} maximizing x_i over J_{i-1}. Trivially, J_n consists of one vector, y say. Then:

Theorem 41.16. y *maximizes* $w^\mathsf{T} x$ *over* J.

Proof. It suffices to show that the maximum value of $w^\mathsf{T} x$ over J_1 is the same as over J (since applying this to the jump systems $J_1, \ldots, J_n$ gives the theorem). Let the maximum over J be attained by x and over J_1 by y. Suppose $w^\mathsf{T} y < w^\mathsf{T} x$. So $x \notin J_1$, and hence $x_1 < y_1$. We choose x, y such that $y_1 - x_1$ is minimal. Let $z := x + \chi^1$. So z is a step from x to y.

Then $w^\mathsf{T} z = w^\mathsf{T} x + w_1 \geq w^\mathsf{T} x$. Hence $z \notin J$, since otherwise we can replace x by z, contradicting the minimality of $y_1 - x_1$. So, by (41.78), J contains a step u from z to y. So $u = z \pm \chi^i$ for some $i \in \{1, \ldots, n\}$. Then

(41.82) $\quad w^\mathsf{T} u = w^\mathsf{T} z \pm w_i \geq w^\mathsf{T} z - w_i = w^\mathsf{T} x + w_1 - w_i \geq w^\mathsf{T} x$.

So we can replace x by u, again contradicting the minimality of $y_1 - x_1$ (as $u_1 > x_1$). ∎

Lovász [1997] gave a min-max relation for the minimum l_1-distance of an integer vector to a jump system of special type. It can be considered as a common generalization of the matroid intersection theorem (Theorem 41.1) and the Tutte-Berge formula (Theorem 24.1).

For a survey, see Cunningham [2002].

41.5d. Further notes

A special case of the weighted matroid intersection algorithm (where one matroid is a partition matroid) was studied by Brezovec, Cornuéjols, and Glover [1988].

Data structures for on-line updating of matroid intersection solutions were given by Frederickson and Srinivas [1984,1987], and a randomized parallel algorithm for linear matroid intersection by Narayanan, Saran, and Vazirani [1992,1994].

An extension of matroid intersection to 'supermatroid' intersection was given by Tardos [1990]. Fujishige [1977a] gave a primal approach to weighted matroid intersection, and Shigeno and Iwata [1995] a dual approximation approach. Camerini and Maffioli [1975,1978] studied 3-matroid intersection problems.

Chapter 42

Matroid union

Matroid union is closely related to matroid intersection, and most of the basic matroid union results follow from basic matroid intersection results, and vice versa. But matroid union also gives a shift in focus and offers a number of specific algorithmic questions.

42.1. Matroid union theorem

The matroid union theorem will be derived from the following basic result given by Nash-Williams [1967], suggested by earlier unpublished work of J. Edmonds[28]:

Theorem 42.1. *Let $M' = (S', \mathcal{I}')$ be a matroid, with rank function r', and let $f : S' \to S$. Define*

(42.1) $\quad \mathcal{I} := \{f(I') \mid I' \in \mathcal{I}'\}$

(where $f(I') := \{f(s) \mid s \in I'\}$). Then $M = (S, \mathcal{I})$ is a matroid, with rank function r given by

(42.2) $\quad r(U) = \min_{T \subseteq U}(|U \setminus T| + r'(f^{-1}(T)))$

for $U \subseteq S$.

Proof. Trivially, $\mathcal{I}$ is nonempty and closed under taking subsets. To see condition (39.1)(ii), let $I, J \in \mathcal{I}$ with $|I| < |J|$. Choose $I', J' \in \mathcal{I}'$ with $f(I') = I$, $f(J') = J$, $|I'| = |I|$, $|J'| = |J|$, and $|I' \cap J'|$ as large as possible. As M' is a matroid, $I' + j \in \mathcal{I}'$ for some $j \in J' \setminus I'$. If $f(j) \in f(I')$, say $f(j) = f(i)$ for $i \in I'$, replacing I' by $I' - i + j$ would increase $|I' \cap J'|$, contradicting our assumption. So $f(j) \in J \setminus I$ and $f(I') + f(j) = f(I' + j) \in \mathcal{I}$. This proves (39.1)(ii), and hence M is a matroid.

The rank $r(U)$ of a subset U of S is equal to the maximum size of a common independent set in M' and the partition matroid $N = (S', \mathcal{J})$ induced by the family $(f^{-1}(s) \mid s \in U)$. By the matroid intersection theorem (Theorem 41.1), this is equal to the right-hand side of (42.2). ∎

---

[28] as mentioned in the footnote on page 20 of Pym and Perfect [1970] (quoted in Section 42.6f below).

(In his paper, Nash-Williams suggested a direct proof, by decomposing f as a product of 'elementary' functions in which only two elements are merged. Welsh [1970] observed that the rank formula (42.2) also follows directly from Rado's theorem (Corollary 41.1c) of Rado [1942].)

Theorem 42.1 implies the following result, formulated explicitly by Edmonds [1968] (and for all M_i equal by Nash-Williams [1967]).

Let $M_1 = (S_1, \mathcal{I}_1), \ldots, M_k = (S_k, \mathcal{I}_k)$ be matroids. Define the *union* of these matroids as $M_1 \vee \cdots \vee M_k = (S_1 \cup \cdots \cup S_k, \mathcal{I}_1 \vee \cdots \vee \mathcal{I}_k)$, where

(42.3) $\quad \mathcal{I}_1 \vee \cdots \vee \mathcal{I}_k := \{I_1 \cup \ldots \cup I_k \mid I_1 \in \mathcal{I}_1, \ldots, I_k \in \mathcal{I}_k\}$.

Corollary 42.1a (matroid union theorem). *Let $M_1 = (S_1, \mathcal{I}_1), \ldots, M_k = (S_k, \mathcal{I}_k)$ be matroids, with rank functions $r_1, \ldots, r_k$, respectively. Then $M_1 \vee \cdots \vee M_k$ is a matroid again, with rank function r given by:*

(42.4) $\quad r(U) = \min_{T \subseteq U}(|U \setminus T| + r_1(T \cap S_1) + \cdots + r_k(T \cap S_k))$.

for $U \subseteq S_1 \cup \cdots \cup S_k$.

Proof. To see that $M_1 \vee \cdots \vee M_k$ is a matroid, let for each i, $M_i' = (S_i', \mathcal{I}_i')$ be a copy of M_i with $S_1', \ldots, S_k'$ disjoint. Then trivially $M_1' \vee \cdots \vee M_k'$ is a matroid. Now define $f : S_1' \cup \cdots \cup S_k' \to S_1 \cup \cdots \cup S_k$ by, for $i = 1, \ldots, k$ and $s \in S_i'$: $f(s)$ is the original of s in S_i. Then the matroid obtained in Theorem 42.1 is equal to $M_1 \vee \cdots \vee M_k$, proving that the latter indeed is a matroid, and (42.4) follows from (42.2). ∎

Conversely, the matroid intersection theorem may be derived from the matroid union theorem (as was shown by Edmonds [1970b]): the maximum size of a common independent set in two matroids M_1 and M_2, is equal to the maximum size of an independent set in the union $M_1 \vee M_2^*$, minus the rank of M_2^*.

Application of the matroid union theorem to a number of copies of the same matroid gives the following results. First:

Corollary 42.1b. *Let $M = (S, \mathcal{I})$ be a matroid, with rank function r, and let $k \in \mathbb{Z}_+$. Then the maximum size of the union of k independent sets is equal to*

(42.5) $\quad \min_{U \subseteq S}(|S \setminus U| + k \cdot r(U))$.

Proof. This follows by applying Corollary 42.1a to $M_1 = \cdots = M_k = M$. ∎

This implies that the minimum number of independent sets (or bases) needed to cover the underlying set is described by the following result of Edmonds [1965c][29]:

---

[29] This result was also given, without proof, by Rado [1966], saying that the argument of Horn [1955] for linear matroids can be extended to arbitrary matroids. The result con-

Corollary 42.1c (matroid base covering theorem). *Let $M = (S, \mathcal{I})$ be a matroid, with rank function r, and let $k \in \mathbb{Z}_+$. Then S can be covered by k independent sets if and only if*

(42.6) $\qquad k \cdot r(U) \geq |U|$

for each $U \subseteq S$.

Proof. M can be covered by k independent sets if and only if there is a union of k independent sets of size $|S|$. By Corollary 42.1b, this is the case if and only if

(42.7) $\qquad \min_{U \subseteq S}(|S \setminus U| + k \cdot r(U)) \geq |S|$,

that is, if and only if $k \cdot r(U) \geq |U|$ for each subset U of S. ∎

One similarly has for the maximum number of disjoint bases in a matroid (Edmonds [1965a]):

Corollary 42.1d (matroid base packing theorem). *Let $M = (S, \mathcal{I})$ be a matroid, with rank function r, and let $k \in \mathbb{Z}_+$. Then there exist k disjoint bases if and only if*

(42.8) $\qquad k \cdot (r(S) - r(U)) \leq |S \setminus U|$

for each $U \subseteq S$.

Proof. M has k disjoint bases if and only if the maximum size of the union of k independent sets is equal to $k \cdot r(S)$. By Corollary 42.1b, this is the case if and only if

(42.9) $\qquad \min_{U \subseteq S}(|S \setminus U| + k \cdot r(U)) \geq k \cdot r(S)$,

that is, if and only if $|S \setminus U| \geq k \cdot (r(S) - r(U))$ for each subset U of S. ∎

The more general forms of Corollaries 42.1c and 42.1d, with different matroids, were shown by Edmonds and Fulkerson [1965].

42.1a. Applications of the matroid union theorem

We describe a number of applications of the matroid union theorem. Further applications will follow in Chapter 51 on packing and covering of trees and forests.

Transversal matroids. Let $\mathcal{X} = (X_1, \ldots, X_n)$ be a family of subsets of a finite set S, and define for each $i = 1, \ldots, n$ a matroid M on S by: Y is independent in M_i if and only if $Y \subseteq X_i$ and $|Y| \leq 1$. Now the union $M_1 \vee \cdots \vee M_n$ is the same

firms a question of Rado [1962a,1962b] (in fact, the result also follows by an elementary construction from Rado's theorem (Corollary 41.1c) given in Rado [1942]).

Disjoint transversals. Let $\mathcal{X} = (X_1, \ldots, X_n)$ be a family of subsets of a finite set S. Then $\mathcal{X}$ has k disjoint transversals if and only if

(42.10) $$\left|\bigcup_{i \in I} X_i\right| \geq k \cdot |I|$$

for each $I \subseteq \{1, \ldots, n\}$. This easy consequence of Hall's marriage theorem (cf. Theorem 22.10) can also be derived by applying the matroid base packing theorem to the transversal matroid induced by $\mathcal{X}$, using (39.19).

Similarly, it can be derived from the matroid base covering theorem that S can be partitioned into k partial transversals of $\mathcal{X}$ if and only if

(42.11) $$k(n - |I|) \geq \left|S \setminus \bigcup_{i \in I} X_i\right|$$

for each $I \subseteq \{1, \ldots, n\}$ (cf. Theorem 22.12).

Vector spaces. A finite subset S of a vector space can be covered by k linearly independent sets if and only if

(42.12) $$|U| \leq k \cdot \mathrm{rank}(U) \text{ for each } U \subseteq S.$$

This conjecture of K.F. Roth and R. Rado was shown by Horn [1955][30]. It is the special case of the matroid base covering theorem for linear matroids (see also Section 42.1b below).

As a similar consequence of the matroid base packing theorem one has that the n-dimensional vector space S over the field $GF(q)$ contains $k := \lfloor (q^n - 1)/n \rfloor$ disjoint bases. Indeed, for each $U \subseteq S$ one has $k(n - r(U)) \leq q^n - |U|$, as $|U| \leq q^{r(U)}$.

An exchange property of bases. The matroid union theorem also implies the following stronger exchange property of bases of a matroid (stronger than given in the 'axioms' in Theorem 39.6). In any matroid $M = (S, \mathcal{I})$,

(42.13)   for any two bases B_1 and B_2 and for any partition of B_1 into X_1 and Y_1, there is a partition of B_2 into X_2 and Y_2 such that both $X_1 \cup Y_2$ and $X_2 \cup Y_1$ are bases.

This property was conjectured by G.-C. Rota, and proved by Brylawski [1973], Greene [1973], and Woodall [1974a] — we follow the proof of McDiarmid [1975a].

Consider the matroids $M_1 := M/Y_1$ and $M_2 := M/X_1$. Note that M_1 has rank $|X_1|$ and that M_2 has rank $|Y_1|$. We must show that B_2 is the union of an independent set X_2 of M_1 and an independent set Y_2 of M_2. By the submodularity of the rank functions ((39.38)(ii)) we have for each $T \subseteq B_2$:

---

[30] Horn [1955] thanked Rado 'for improvements in the setting out of the argument'. The result was also published, in the same journal, by Rado [1962a]. This paper does not mention Horn's paper. The proof by Rado [1962a] is the same as that of Horn [1955] and uses the same notation. But Rado [1966] said that the theorem was first proved by Horn [1955].

(42.14) $|B_2 \setminus T| + r_{M_1}(T \setminus Y_1) + r_{M_2}(T \setminus X_1)$
$= |B_2 \setminus T| + r(T \cup Y_1) - |Y_1| + r(T \cup X_1) - |X_1|$
$\geq r(T) + r(T \cup Y_1 \cup X_1) - |T| = |B_2|.$

Hence, by the matroid union theorem (Corollary 42.1a), we have the required result.

Repeated application of this exchange phenomenon implies the following stronger property, given by Greene and Magnanti [1975]:

(42.15) for any two bases B_1 and B_2 and any partition of B_1 into $X_1, \ldots, X_k$, there is a partition of B_2 into $Y_1, \ldots, Y_k$ such that $(B_1 \setminus X_i) \cup Y_i$ is a base, for each $i = 1, \ldots, k$.

This extends Corollary 39.12a, which is the special case where each X_i is a singleton.

42.1b. Horn's proof

The proof of Horn [1955] of the matroid base covering theorem for linear matroids directly extends to general matroids (as was observed by Rado [1966]):

Consider a counterexample to the matroid base covering theorem (Corollary 42.1c) with smallest $|S|$. For subsets $S_1, \ldots, S_n$ of S, define inductively:

(42.16) $[S_1, \ldots, S_n] := \begin{cases} S & \text{if } n = 0, \\ \text{span}([S_1, \ldots, S_{n-1}] \cap S_n) & \text{if } n \geq 1. \end{cases}$

By the minimality of $|S|$, we know that for each $s \in S$, $S \setminus \{s\}$ can be partitioned into k independent sets $I_1, \ldots, I_k$. We first show:

(42.17) for each $s \in S$ and $I_1, \ldots, I_k$ partitioning $S \setminus \{s\}$, there exist $j_1, \ldots, j_n \in \{1, \ldots, k\}$ with $s \notin [I_{j_1}, \ldots, I_{j_n}]$.

Indeed, choose $j_1, \ldots, j_n \in \{1, \ldots, k\}$ with the rank of $[I_{j_1}, \ldots, I_{j_n}]$ as small as possible. Define $A := [I_{j_1}, \ldots, I_{j_n}]$. By the minimality of the rank of A, we have $r(A \cap I_j) = r(A)$ for each $j = 1, \ldots, k$. Hence, by (42.6),

(42.18) $|A| \leq k \cdot r(A) = \sum_{j=1}^{k} r(A \cap I_j) \leq \sum_{j=1}^{k} |A \cap I_j| = |A \setminus \{s\}|.$

So $s \notin A$, proving (42.17).

Now choose $s, I_1, \ldots, I_k$, and $j_1, \ldots, j_n$ as in (42.17) with n as small as possible. For $t = 0, \ldots, n$, define

(42.19) $B_t := [I_{j_1}, \ldots, I_{j_t}].$

As we have a counterexample, we know that $s \in \text{span}(I_{j_n})$ (otherwise we can add s to I_{j_n}). Let C be the circuit in $I_{j_n} \cup \{s\}$. As $s \notin B_n = \text{span}(B_{n-1} \cap I_{j_n})$, we know that $C \setminus \{s\}$ is not contained in B_{n-1} (otherwise $C \setminus \{s\} \subseteq B_{n-1} \cap I_{j_n}$, and hence $s \in \text{span}(B_{n-1} \cap I_{j_n})$). So we can choose $z \in C \setminus \{s\}$ with $z \notin B_{n-1}$.

Define $I'_{j_n} := I_{j_n} - z + s$ and $I'_j := I_j$ for $j \neq j_n$. Then $I'_1, \ldots, I'_k$ are independent sets partitioning $S \setminus \{z\}$. Define, for $t = 0, \ldots, n$:

(42.20) $B'_t := [I'_{j_1}, \ldots, I'_{j_t}].$

By the minimality of n we know that $z \in B'_{n-1}$. Since $z \notin B_{n-1}$, we have $B'_{n-1} \not\subseteq B_{n-1}$. Choose the smallest $q \leq n-1$ with $B'_q \not\subseteq B_q$. Then $q \geq 1$ and $B'_{q-1} \subseteq B_{q-1}$. By the minimality of n we know that $s \in B_q$ (as $q < n$). So

(42.21) $B'_q = \text{span}(B'_{q-1} \cap I'_{j_q}) \subseteq \text{span}((B_{q-1} \cap I_{j_q}) \cup \{s\}) \subseteq \text{span}(B_q) = B_q,$

a contradiction.

42.2. Polyhedral applications

The matroid base packing and covering theorems imply (in fact, are equivalent to) the following polyhedral result:

Corollary 42.1e. *For any matroid, the independent set polytope, the base polytope, and the spanning set polytope have the integer decomposition property.*

Proof. Let $M = (S, \mathcal{I})$ be a matroid. Choose $k \in \mathbb{Z}_+$ and an integer vector $x \in k \cdot P_{\text{independent set}}(M)$. Replace each element s of S by x_s parallel elements, thus obtaining the matroid $N = (T, \mathcal{J})$ say. Now for each $U \subseteq T$, one has $k \cdot r_N(U) \geq |U|$, since if W denotes the set of elements s in S such that U intersects the parallel class of s, then

(42.22) $\quad r_N(U) = r_M(W) \geq x(W)/k \geq |U|/k,$

since x/k belongs to $P_{\text{independent set}}(M)$. So by the matroid base covering theorem (Corollary 42.1c), T can be partitioned into k independent sets of N. Hence x is the sum of k incidence vectors of independent sets of M.

To see that the base polytope has the integer decomposition property, let $x \in k \cdot P_{\text{base}}(M)$. By the above, x is the sum of the incidence vectors of k independent sets. As $x(S) = k \cdot r(S)$, each of these independent sets is a base.

One similarly derives from the matroid base packing theorem (Corollary 42.1d) that the spanning set polytope has the integer decomposition property. ∎

The matroid base packing and covering theorems imply generalizations to the capacitated case, by splitting elements into parallel elements. For the matroid base covering theorem this gives:

Theorem 42.2. *Let $M = (S, \mathcal{I})$ be a matroid, with rank function r, and let $c : S \to \mathbb{Z}_+$. Then the minimum value of $\sum_{I \in \mathcal{I}} \lambda_I$, where $\lambda : \mathcal{I} \to \mathbb{Z}_+$ satisfies*

(42.23) $\quad \displaystyle\sum_{I \in \mathcal{I}} \lambda_I \chi^I = c,$

is equal to the maximum value of

(42.24) $\quad \lceil \dfrac{c(U)}{r(U)} \rceil$

taken over $U \subseteq S$ with $r(U) \geq 1$.

Proof. Directly from the matroid base covering theorem (Corollary 42.1c), by splitting each $s \in S$ into $c(s)$ parallel elements. ∎

In other words, the system defining the antiblocking polyhedron of the independent set polytope:

(42.25) $\quad x_s \geq 0 \quad$ for $s \in S$,
$\quad\quad\quad\quad x(I) \leq 1 \quad$ for $I \in \mathcal{I}$,

has the integer rounding property (the optimum integer solution to the dual of maximizing $c^\mathsf{T} x$ over (42.25) has value equal to the upper integer part of the value of the optimum (fractional) solution, for any integer objective function c).

Similarly, the matroid base packing theorem gives:

Theorem 42.3. *Let* $M = (S, \mathcal{I})$ *be a matroid, with rank function* r, *and let* $c : S \to \mathbb{Z}_+$. *Let* $\mathcal{B}$ *be the collection of bases of* M. *Then the maximum value of* $\sum_{B \in \mathcal{B}} \lambda_B$, *where* $\lambda : \mathcal{B} \to \mathbb{Z}_+$ *satisfies*

(42.26) $\quad \displaystyle\sum_{B \in \mathcal{B}} \lambda_B \chi^B \leq c,$

is equal to the minimum value of

(42.27) $\quad \lfloor \dfrac{c(S \setminus U)}{r(S) - r(U)} \rfloor$

taken over $U \subseteq S$ *with* $r(S) - r(U) \geq 1$.

Proof. Directly from the matroid base packing theorem (Corollary 42.1d), by splitting each $s \in S$ into $c(s)$ parallel elements. ∎

In other words, the system defining the blocking polyhedron of the base polytope:

(42.28) $\quad x_s \geq 0 \quad$ for $s \in S$,
$\quad\quad\quad\quad x(B) \geq 1 \quad$ for $B \in \mathcal{B}$,

has the integer rounding property.

De Pina and Soares [2000] showed that, in Theorem 42.3, the number of bases B with $\lambda_B > 0$ can be restricted to at most $|S| + r$, where r is the rank of M. This strengthens a result of Cook, Fonlupt, and Schrijver [1986].

42.3. Matroid union algorithm

A polynomial-time algorithm for partitioning a matroid in as few independent sets as possible may be derived from the matroid intersection algorithm, with the construction given in the proof of Theorem 42.1. A direct algorithm was given by Edmonds [1968]. We give the algorithm described by Knuth [1973] and Greene and Magnanti [1975], which is similar to the algorithm described in Section 41.2 for cardinality matroid intersection.

Let $M_1 = (S, \mathcal{I}_1), \ldots, (S, \mathcal{I}_k)$ be matroids. Let $I_i \in \mathcal{I}_i$, for $i = 1, \ldots, k$, with $I_i \cap I_j = \emptyset$ if $i \neq j$. Let D be the union of the graphs $D_{M_i}(I_i)$ as defined in Section 39.9.

For each i, let F_i be the set of elements $s \notin I_i$ with $I_i \cup \{s\} \in \mathcal{I}_i$. Define $I := I_1 \cup \cdots \cup I_k$, $F := F_1 \cup \cdots \cup F_k$, and $\mathcal{I} := \mathcal{I}_1 \vee \cdots \vee \mathcal{I}_k$.

Theorem 42.4. *For any $s \in S \setminus I$ one has: $I \cup \{s\} \in \mathcal{I} \iff D$ has an $F - s$ path.*

Proof. To see necessity, suppose that D has no $F - s$ path. Let T be the set of elements of S that can reach s in D. So $s \in T$, $T \cap F = \emptyset$, and no arc of D enters T. Then $r_i(T) = |I_i \cap T|$ for each $i = 1, \ldots, k$. Otherwise, there exists a $t \in T \setminus I_i$ with $(I_i \cap T) \cup \{t\} \in \mathcal{I}_i$. Since $t \notin F$, $I_i \cup \{t\} \notin \mathcal{I}_i$. So there is a $u \in I_i \setminus T$ with $I_i - u + t \in \mathcal{I}_i$. But then (u, t) is an arc of D entering T, a contradiction.

So $r_i(T) = |I_i \cap T|$ for each i. Hence $r_1(T) + \cdots + r_k(T) = |I \cap T|$. As $s \in T \setminus I$, this implies $(I \cap T) \cup \{s\} \notin \mathcal{I}$, and so $I \cup \{s\} \notin \mathcal{I}$.

To see sufficiency, let $P = (s_0, s_1, \ldots, s_p)$ be a shortest $F - s$ path in D. We can assume by symmetry that $s_0 \in F_1$; so $s_0 \notin I_1$ and $I_1 \cup \{s_0\} \in \mathcal{I}_1$. Since P is a shortest path, for each $i = 1, \ldots, k$, the set N_i of edges (s_{j-1}, s_j) with $j = 1, \ldots, p$ and $s_{j-1} \in I_i$, forms a unique perfect matching in $D_{M_i}(I_i)$ on the set S_i covered by N_i. So by Theorem 39.13, $I_i \triangle S_i$ belongs to $\mathcal{I}_i$ for each i. Moreover, by Corollary 39.13a, $(I_1 \triangle S_1) \cup \{s_0\} \in \mathcal{I}_1$. So $I \cup \{s\} \in \mathcal{I}$. ∎

This implies that a maximum-size set in $\mathcal{I}_1 \vee \cdots \vee \mathcal{I}_k$ can be found in polynomial time (by greedily growing an independent set in $M_1 \vee \cdots \vee M_k$). Similarly, we can find with the greedy algorithm a maximum-weight set in $\mathcal{I}_1 \vee \cdots \vee \mathcal{I}_k$.

In particular, we can test if a given set is independent in $M_1 \vee \cdots \vee M_k$. Cunningham [1986] gave an $O((n^{3/2} + k)mQ + n^{1/2}km)$ algorithm to find a maximum-size set in $\mathcal{I}_1 \vee \cdots \vee \mathcal{I}_k$, where n is the maximum size of a set in $\mathcal{I}_1 \vee \cdots \vee \mathcal{I}_k$, m is the size of the underlying set, and Q is the time needed to test if a given set belongs to $\mathcal{I}_j$ for any given j.

These methods (including the reduction to matroid intersection) also imply:

Theorem 42.5. *Given a matroid $M = (S, \mathcal{I})$ by an independence testing oracle, we can find a maximum number of disjoint bases, and a minimum number of independent sets covering S, in polynomial time.*

Proof. See above. ∎

42.4. The capacitated case: fractional packing and covering of bases

The complexity of the capacitated and fractional cases of the above packing and covering problems can be studied with the help of the strong polynomial-

Section 42.4. The capacitated case: fractional packing and covering of bases 733

time solvability of the *most violated inequality problem* for a matroid $M = (S, \mathcal{I})$, with rank function r:

(42.29) given: a vector $x \in \mathbb{Q}_+^S$;
find: a subset U of S minimizing $r(U) - x(U)$.

The strong polynomial-time solvability of this problem was shown in Corollary 40.4c, and is a result of Cunningham [1984].

If x belongs to $P_{\text{independent set}}(M)$, we can decompose x as a convex combination of incidence vectors of independent sets. This decomposition can be found in strongly polynomial time, by Corollary 40.4a.

We now consider the problem of finding a maximum fractional packing of bases subject to a given capacity function, and its dual, finding a minimum fractional covering by independent sets of a demand function.

With a method given by Picard and Queyranne [1982a] and Padberg and Wolsey [1984] one finds:

Theorem 42.6. *Given a matroid $M = (S, \mathcal{I})$ by an independence testing oracle and given $y \in \mathbb{Q}_+^S$, we can find the minimum value of λ such that $y \in \lambda \cdot P_{\text{independent set}}(M)$ in strongly polynomial time.*

Proof. Let r be the rank function of M. We can assume that y does not belong to the independent set polytope. Let L be the line through 0 and y. We iteratively reset y as follows. By Corollary 40.4c, we can find a subset U of S minimizing $r(U) - y(U)$. Let y' be the vector on L with $y'(U) = r(U)$.

Now, for any $U' \subseteq S$, if y' violates $x(U') \leq r(U')$, then $r(U') < r(U)$, since the function $d(x) := (r(U) - x(U)) - (r(U') - x(U'))$ is nonpositive at y and positive at y', implying that it is positive at 0 (as d is linear in x).

We reset $y := y'$ and iterate, until y belongs to $P_{\text{independent set}}(M)$. So after at most $r(S)$ iterations the process terminates, with a y on the boundary of $P_{\text{independent set}}(M)$. Comparing the final y with the original y gives the required λ. ∎

Theorem 42.6 implies an algorithm for capacitated fractional covering by independent sets:

Corollary 42.6a. *Given a matroid $M = (S, \mathcal{I})$ by an independence testing oracle and given $y \in \mathbb{Q}_+^S$, we can find independent sets $I_1, \ldots, I_k$ and rationals $\lambda_1, \ldots, \lambda_k \geq 0$ such that*

(42.30) $y = \lambda_1 \chi^{I_1} + \cdots + \lambda_k \chi^{I_k}$

with $\lambda_1 + \cdots + \lambda_k$ minimal, in strongly polynomial time.

Proof. Without loss of generality, $y \neq 0$. By Theorem 51.7, we can find the minimum value of λ such that y belongs to $\lambda \cdot P_{\text{independent set}}(M)$. By Corollary 40.4a, we can decompose $\frac{1}{\lambda} \cdot y$ as a convex combination of incidence vectors of independent sets. This gives the required decomposition of y. ∎

One similarly shows for the spanning set polytope:

Theorem 42.7. *Given a matroid $M = (S, \mathcal{I})$ by an independence testing oracle and given $y \in \mathbb{Q}_+^S$, we can find the maximum value of λ such that $y \in \lambda \cdot P_{\text{spanning set}}(M)$, in strongly polynomial time.*

Proof. Let r be the rank function of M. By Corollary 40.2f, the spanning set polytope of M is determined by the constraints $0 \leq x \leq 1$ and

(42.31) $\quad r(U) - x(U) \geq r(S) - x(S)$ for $U \subseteq S$.

We can assume that $y \notin P_{\text{spanning set}}(M)$ and that the support of y is a spanning set. Let L be the line through 0 and y. We iteratively reset y as follows.

Find a $U \subseteq S$ minimizing $r(U) - y(U)$ (this can be done in strongly polynomial time, by Corollary 40.4c). If y does not belong to the spanning set polytope, we know that y violates the constraint $r(U) - x(U) \geq r(S) - x(S)$. Let y' be the vector on L satisfying $r(U) - y'(U) = r(S) - y'(S)$.

Now for any $U' \subseteq S$, if y' violates $r(U') - x(U') \geq r(S) - x(S)$, then $r(U') > r(U)$, since the function $d(x) := (r(U) - x(U)) - (r(U') - x(U'))$ is nonpositive at y and positive at y', implying that it is negative at 0 (as d is linear in x).

We reset $y := y'$ and iterate, until y belongs to $P_{\text{spanning set}}(M)$. So after at most $r(S)$ iterations the process terminates, in which case y is on the boundary of $P_{\text{spanning set}}(M)$. Comparing the final y with the original y gives the required λ. ∎

In turn, this gives an algorithm for capacitated fractional base packing:

Corollary 42.7a. *Given a matroid $M = (S, \mathcal{I})$ by an independence testing oracle and given $y \in \mathbb{Q}_+^S$, we can find bases $B_1, \ldots, B_k$ and rationals $\lambda_1, \ldots, \lambda_k \geq 0$ such that*

(42.32) $\quad y \geq \lambda_1 \chi^{B_1} + \cdots + \lambda_k \chi^{B_k}$

with $\lambda_1 + \cdots + \lambda_k$ maximal, in strongly polynomial time.

Proof. By Theorem 42.7, we can find the maximum value of λ such that y belongs to $\lambda \cdot P_{\text{spanning set}}(M)$. If $\lambda = 0$, we take $k = 0$. If $\lambda > 0$, by Corollary 40.4b we can decompose $\frac{1}{\lambda} \cdot y$ as a convex combination of incidence vectors of spanning sets. This gives the required decomposition of y. ∎

42.5. The capacitated case: integer packing and covering of bases

It is not difficult to derive integer versions of the above algorithms, but they are not strongly polynomial-time, as we round numbers in it. In fact, an

Section 42.5. The capacitated case: integer packing and covering of bases

integer packing or covering cannot be found in strongly polynomial time, as it would imply a strongly polynomial-time algorithm for testing if an integer k is even (which algorithm does not exist[31]): Let M be the 2-uniform matroid on 3 elements and let $k \in \mathbb{Z}_+$. Then k is even if and only if M has $\frac{3}{2}k$ bases containing each element of M at most k times.

Polynomial-time algorithms follow directly from the fractional versions with the help of the matroid base packing and covering theorems.

Theorem 42.8. *Given a matroid $M = (S, \mathcal{I})$ by an independence testing oracle and given $y \in \mathbb{Z}_+^S$, we can find independent sets $I_1, \ldots, I_t$ and integers $\lambda_1, \ldots, \lambda_t \geq 0$ such that*

$$(42.33) \qquad y = \lambda_1 \chi^{I_1} + \cdots + \lambda_t \chi^{I_t}$$

with $\lambda_1 + \cdots + \lambda_t$ minimal, in polynomial time.

Proof. First find $I_1, \ldots, I_k$ and $\lambda_1, \ldots, \lambda_k$ as in Corollary 42.6a. We can assume that $k \leq |S|$ (by Carathéodory's theorem, applying Gaussian elimination). Let

$$(42.34) \qquad y' := \sum_{i=1}^{k} (\lambda_i - \lfloor \lambda_i \rfloor) \chi^{I_i} = y - \sum_{i=1}^{k} \lfloor \lambda_i \rfloor \chi^{I_i}.$$

So y' is integer.

Replace each $s \in S$ by $y'(s)$ parallel elements, making matroid $M' = (S', \mathcal{I}')$. By Theorem 42.5, we can find a minimum number of independent sets partitioning S', in polynomial time (as $y'(s) \leq |S|$ for each $s \in S$). This gives independent sets $I_{k+1}, \ldots, I_t$ of M.

Setting $\lambda_i := 1$ for $i = k+1, \ldots, t$, we show that this gives a solution of our problem. Trivially, (42.33) is satisfied (with λ_i replaced by $\lfloor \lambda_i \rfloor$). By the matroid base covering theorem applied to M' (as (42.34) gives a fractional decomposition of S' into independent sets),

$$(42.35) \qquad t - k \leq \left\lceil \sum_{i=1}^{k} (\lambda_i - \lfloor \lambda_i \rfloor) \right\rceil.$$

Therefore,

$$(42.36) \qquad \sum_{i=1}^{t} \lfloor \lambda_i \rfloor = (t-k) + \sum_{i=1}^{k} \lfloor \lambda_i \rfloor \leq \left\lceil \sum_{i=1}^{k} \lambda_i \right\rceil,$$

---

[31] For any strongly polynomial-time algorithm with one integer k as input, there is a number L and a rational function $q : \mathbb{Z} \to \mathbb{Q}$ such that if $k > L$, then the output equals $q(k)$. (This can be proved by induction on the number of steps of the algorithm, which is a fixed number as the input consists of only one number.) However, there do not exist a rational function q and number L such that for $k > L$, $q(k) = 0$ if k is even, and $q(k) = 1$ if k is odd.

proving that the decomposition is optimum (cf. Theorem 42.2). ∎

One similarly shows for packing bases:

Theorem 42.9. *Given a matroid $M = (S, \mathcal{I})$ by an independence testing oracle and given $y \in \mathbb{Z}_+^S$, we can find bases $B_1, \ldots, B_t$ and integers $\lambda_1, \ldots, \lambda_t \geq 0$ such that*

$$(42.37) \qquad y \geq \lambda_1 \chi^{B_1} + \cdots + \lambda_t \chi^{B_t}$$

with $\lambda_1 + \cdots + \lambda_t$ maximal, in polynomial time.

Proof. First find bases $B_1, \ldots, B_k$ and $\lambda_1, \ldots, \lambda_k$ as in Corollary 42.7a. Again we can assume that $k \leq |S|$. Let

$$(42.38) \qquad y' := \lceil \sum_{i=1}^{k} (\lambda_i - \lfloor \lambda_i \rfloor) \chi^{B_i} \rceil.$$

Replace each $s \in S$ by $y'(s)$ parallel elements, making matroid M'. By Theorem 42.5, we can find a maximum number of disjoint bases in M' in polynomial time (as $y'(s) \leq |S|$ for each $s \in S$). This gives bases $B_{k+1}, \ldots, B_t$ in M.

Setting $\lambda_i := 1$ for $i = k+1, \ldots, t$, we show that this gives a solution of our problem. Trivially, (42.37) is satisfied (with λ_i replaced by $\lfloor \lambda_i \rfloor$). Again, now by the matroid base packing theorem applied to M', using (42.38),

$$(42.39) \qquad t - k \geq \lfloor \sum_{i=1}^{k} (\lambda_i - \lfloor \lambda_i \rfloor) \rfloor.$$

Therefore,

$$(42.40) \qquad \sum_{i=1}^{t} \lfloor \lambda_i \rfloor = (t - k) + \sum_{i=1}^{k} \lfloor \lambda_i \rfloor \geq \lfloor \sum_{i=1}^{k} \lambda_i \rfloor,$$

proving that the decomposition is optimum (cf. Theorem 42.3). ∎

De Pina and Soares [2000] showed that, in this theorem we can make the additional condition that $t \leq |S| + r$, where r is the rank of M.

42.6. Further results and notes

42.6a. Induction of matroids

An application of matroid intersection and union is the following 'induction of a matroid through a directed graph', discovered by Perfect [1969b] (for bipartite graphs) and Brualdi [1971c]. In fact, it forms a generalization of the basic Theorem 42.1.

Let $D = (V, A)$ be a directed graph, let $U, W \subseteq V$, and let $M = (U, \mathcal{I})$ be a matroid. Let $\mathcal{J}$ be the collection of subsets Y of W such that there exists an $X \in \mathcal{I}$ with X linked to Y. (Set X is *linked to* Y if $|X| = |Y|$ and D has $|X|$ disjoint $X - Y$ paths.)

Then:

(42.41) $\quad N = (W, \mathcal{J})$ is a matroid.

To show that N is a matroid, we can assume that U and W are disjoint. (Otherwise, add a new vertex w' and new arc (w, w') for each $w \in W$.) Let L be the gammoid induced by $D, U, U \cup W$. Then $N = (M \vee L)/U$. Indeed, since U is independent in L and hence in $M \vee L$, a subset Y of W is independent in $(M \vee L)/U$ if and only if $Y \cup U$ is independent in $M \vee L$. This is easily seen to be equivalent to: $Y \in \mathcal{J}$. So N is a matroid.

The rank function r_N of N can be described by (for $Y \subseteq W$):

(42.42) $\quad r_N(Y) = \min\{r_M(X) + |Z| \mid X \subseteq U, Z \subseteq V, Z$ intersects each $U \setminus X - Y$ path$\}$.

This can be derived from the matroid union theorem, but also (and simpler) from the matroid intersection theorem, as follows. Let K be the gammoid induced by D^{-1}, Y, U, where D^{-1} arises from D by reversing the orientations of all arcs. Then $r_N(Y)$ is equal to the maximum size of a common independent set in M and K. So, by the matroid intersection theorem (Theorem 41.1),

(42.43) $\quad r_N(Y) = \min_{X \subseteq U} (r_M(X) + r_K(U \setminus X))$,

which by Menger's theorem is equal to the right-hand side of (42.42).

Applying the matroid intersection theorem again gives the following result of Brualdi [1971e] (generalizing Brualdi [1970a]).

Let $D = (V, A)$ be a directed graph, let $U, W \subseteq V$, and let $M = (U, \mathcal{I})$ and $M' = (W, \mathcal{I}')$ be matroids. Then the maximum size of an independent set in M that is linked to an independent set in M' is equal to the minimum value of

(42.44) $\quad r_M(X) + |Z| + r_{M'}(Y)$,

where $X \subseteq U$, $Y \subseteq W$, and $Z \subseteq V$, such that Z intersects each $U \setminus X - W \setminus Y$ path. (This follows directly by considering the maximum size of a common independent set in M' and N as defined above.)

Related results are given by McDiarmid [1975b] and Woodall [1975]. These results are generalized in Schrijver [1979c]. For an algorithm, see Fujishige [1977b].

42.6b. List-colouring

Seymour [1998] showed the following matroid list-colouring theorem (cf. Section 20.9c):

Theorem 42.10. *Let $M = (S, \mathcal{I})$ be a matroid such that S can be partitioned into k independent sets, and let $m \in \mathbb{Z}_+$. For each $s \in S$, let $L_s \subseteq \{1, \ldots, m\}$ be a set of size k. Then S can be partitioned into independent sets $I_1, \ldots, I_m$ such that for each $j = 1, \ldots, m$: if $s \in I_j$, then $j \in L_s$.*

738   Chapter 42. Matroid union

Proof. For each $j = 1, \ldots, m$, let $U_j := \{s \in S \mid j \in L_s\}$. We need to prove that for all j, there exists an independent set $I_j \subseteq U_j$ such that $S = I_1 \cup \cdots \cup I_n$.

Since S can be partitioned into k independent sets, we know that $|X| \leq k \cdot r_M(X)$ for each $X \subseteq S$. Hence, for each $T \subseteq S$,

$$(42.45) \qquad \sum_{j=1}^{m} r_M(U_j \cap T) \geq \sum_{j=1}^{m} \frac{1}{k} |U_j \cap T| = |T|,$$

since each $s \in T$ belongs to k of the U_j. So by the matroid union theorem (Corollary 42.1a), applied to the matroids $M|U_j$, the independent sets I_j as required exist. ∎

42.6c. Strongly base orderable matroids

In general it is not true that given two matroids $M_1 = (S, \mathcal{I}_1)$ and $M_2 = (S, \mathcal{I}_2)$ such that S can be partitioned into k independent sets of M_1, and also into k independent sets of M_2, then S can be partitioned into k common independent sets of M_1 and M_2. This could yield a 'matroid union intersection theorem'. However, taking for M_1 is the cycle matroid of K_4 and for M_2 the matroid with independent sets all sets of pairwise intersecting edges of K_4 (which is a partition matroid), shows that the statement is false for $k = 2$.

But the assertion is true if both M_1 and M_2 belong to the class of so-called strongly base orderable matroids, introduced by Brualdi [1970b]. A matroid $M = (S, \mathcal{I})$ is called *strongly base orderable* if for each two bases B_1, B_2 of M there exists a bijection $\pi : B_1 \to B_2$ such that for each subset X of B_1 the set $\pi(X) \cup (B_1 \setminus X)$ is a base again.

One easily checks that for such π, the function $\pi|B_1 \cap B_2$ is the identity map. It is also straightforward to check that if M is strongly base orderable, then also the dual of M and any contraction of M is strongly base orderable, and hence also any restriction, and therefore any minor is strongly base orderable. Moreover, Brualdi [1970b] showed:

Theorem 42.11. *Any truncation of a strongly base orderable matroid is strongly base orderable again.*

Proof. Let $M = (S, \mathcal{I})$ be a strongly base orderable matroid, with rank function r, and let $k := r(S) - 1$. It suffices to show that the k-truncation of M is strongly base orderable. Let I and J be independent sets of size k, and restrict M to $I \cup J$. If $r(I \cup J) = k$, we are done, since then I and J are bases of the strongly base orderable matroid $M|I \cup J$. So suppose $r(I \cup J) = r(S) = k + 1$, and let $i \in I \setminus J$ and $j \in J \setminus I$ be such that $I \cup \{j\}$ and $J \cup \{i\}$ are bases of M. As M is strongly base orderable, there exists a bijection $\pi : I \cup \{j\} \to J \cup \{i\}$ with the prescribed exchange property. So $\pi(j) = j$ and $\pi(i) = i$. Define $\pi' : I \to J$ by $\pi'(s) := \pi(s)$ if $s \neq i$, and $\pi'(i) = j$. We show that this bijection is as required. To prove this, choose $X \subseteq I$. We must show that $\pi'(X) \cup (I \setminus X)$ is independent.

If $i \notin X$, then $\pi'(X) = \pi(X)$, hence $\pi'(X) \cup (I \setminus X)$ is independent, since

$$(42.46) \qquad \pi'(X) \cup (I \setminus X) = \pi(X) \cup (I \setminus X) \subseteq \pi(X) \cup ((I \cup \{j\}) \setminus X)$$

and the last set is independent.

If $i \in X$, then $\pi'(X) = \pi(X \setminus \{i\}) \cup \{j\}$, hence $\pi'(X) \cup (I \setminus X)$ is independent, since

(42.47) $\pi'(X) \cup (I \setminus X) = \pi(X \setminus \{i\}) \cup \{j\} \cup (I \setminus X) = \pi(X \setminus \{i\}) \cup ((I \cup \{j\}) \setminus X)$
$\subseteq \pi(X \setminus \{i\}) \cup ((I \cup \{j\}) \setminus (X \setminus \{i\}))$

and the last set is independent. ∎

One also easily checks that strong base orderability is closed under making parallel extensions. (Given a matroid $M = (S, \mathcal{I})$ a *parallel extension* in $s \in S$ is obtained by extending S with some new element s', and $\mathcal{I}$ with $\{(I \setminus \{s\}) \cup \{s'\} \mid s \in I \in \mathcal{I}\}$.)

Since transversal matroids are strongly base orderable, also gammoids are strongly base orderable (Brualdi [1971c]):

Theorem 42.12. *Each gammoid is strongly base orderable.*

Proof. Since strong base orderability is closed under taking contractions and since each gammoid is a contraction of a transversal matroid (by Corollary 39.5a), it suffices to show that any transversal matroid is strongly base orderable.

Let M be the transversal matroid induced by a family $\mathcal{X} = (X_1, \ldots, X_m)$ of subsets of a set S. We may assume that $\mathcal{X}$ has a transversal (cf. (39.18)). Consider two transversals $T_1 = \{x_1, \ldots, x_m\}$ and $T_2 = \{y_1, \ldots, y_m\}$ of $\mathcal{X}$, where $x_i, y_i \in X_i$ for $i = 1, \ldots, m$.

Consider the bipartite graph on $\{1, \ldots, m\} \cup S$ with edges all pairs $\{i, s\}$ with $i \in \{1, \ldots, m\}$ and $s \in X_i$ (assuming without loss of generality that $\{1, \ldots, m\} \cap S = \emptyset$). Then $M_1 := \{\{i, x_i\} \mid i = 1, \ldots, m\}$ and $M_2 := \{\{i, y_i\} \mid i = 1, \ldots, m\}$ are matchings in G. Define $\pi : T_1 \to T_2$ as follows. If $s \in T_1 \cap T_2$, define $\pi(s) := s$. If $s \in T_1 \setminus T_2$, let $\pi(s)$ be the (other) end of the path in $M_1 \cup M_2$ starting at s. This defines a bijection as required. ∎

Brualdi [1971c] showed more generally that strong base orderability is maintained under induction of matroids through a directed graph, as described in Section 42.6a. However, not every strongly base orderable matroid is a gammoid (cf. Oxley [1992] p. 411).

Davies and McDiarmid [1976] (cf. McDiarmid [1976]) showed the following.

Theorem 42.13. *Let $M_1 = (S, \mathcal{I}_1)$ and $M_2 = (S, \mathcal{I}_2)$ be strongly base orderable matroids, let $k \in \mathbb{Z}_+$, and suppose that S can be split into k independent sets of M_1, and also into k independent sets of M_2. Then S can be split into k common independent sets of M_1 and M_2.*

Proof. In order to prove this, let $\mathcal{X} = (X_1, \ldots, X_k)$ and $\mathcal{Y} = (Y_1, \ldots, Y_k)$ be partitions of S into independent sets of M_1 and M_2, respectively, with

(42.48) $\sum_{i=1}^{k} |X_i \cap Y_i|$

as large as possible. If this sum is equal to $|S|$ we are done, so suppose that this sum is less than $|S|$. Hence there are i and j with $X_i \cap Y_j \neq \emptyset$ and $i \neq j$. Extend X_i and X_j to bases C_i and C_j of M_1. Similarly, extend Y_i and Y_j to bases D_i and D_j of M_2. Since M_1 and M_2 are strongly base orderable, there exist bijections

$\pi_1 : C_i \to C_j$ and $\pi_2 : D_i \to D_j$ with the exchange property. So $p_1(s) = s$ for each $s \in C_i \cap C_j$ and $p_2(s) = s$ for each $s \in D_i \cap D_j$.

Let G be the bipartite graph with vertex set $C_i \cup C_j \cup D_i \cup D_j$, and edges the pairs $\{s, \pi_1(s)\}$ with s in $C_i \setminus C_j$ and the pairs $\{s, \pi_2(s)\}$ with s in $D_i \setminus D_j$. Split the vertex set into colour classes S and T, say. Define

(42.49) $\quad X_i' := S \cap (X_i \cup X_j),\ X_j' := T \cap (X_i \cup X_j),$
$\qquad Y_i' := S \cap (Y_i \cup Y_j),\ Y_j' := T \cap (Y_i \cup Y_j).$

So $X_i' \cap Y_j' = \emptyset$ and $X_j' \cap Y_i' = \emptyset$. Moreover, X_i' and X_j' are independent in M_1, since, by the exchange property of π, $S \cap (C_i \cup C_j)$ and $T \cup (C_i \cup C_j)$ are independent in M_1. Similarly, Y_i' and Y_j' are independent in M_2.

So replacing the classes X_i and X_j of $\mathcal{X}$ by X_i' and X_j', and the classes Y_i and Y_j of $\mathcal{Y}$ by Y_i' and Y_j' yields partitions as required. However, since $X_i' \cap Y_j' = \emptyset$ and $X_j' \cap Y_i' = \emptyset$, we have

(42.50) $\quad |X_i' \cap Y_i'| + |X_j' \cap Y_j'| > |X_i \cap Y_i| + |X_j \cap Y_j|,$

contradicting the maximality of (42.48). ∎

The proof also shows that the required partition can be found in polynomial time, provided that there is a polynomial-time algorithm to find the exchange bijection π. (This is the case for transversal matroids induced by a given family of sets.)

By the matroid base covering theorem (Corollary 42.1c), Theorem 42.13 is equivalent to:

Corollary 42.13a. *Let $M_1 = (S, \mathcal{I}_1)$ and $M_2 = (S, \mathcal{I}_2)$ be loopless, strongly base orderable matroids, with rank functions r_1 and r_2. Then the minimum number of common independent sets needed to cover S, is equal to*

(42.51) $\quad \max\{\lceil \frac{|U|}{r_i(U)} \rceil \mid \emptyset \neq U \subseteq S, i = 1, 2\}.$

Proof. Directly from Theorem 42.13 with the matroid base covering theorem. ∎

Applying Corollary 42.13a to transversal matroids gives Corollary 23.9a. Similarly, it follows from Theorem 42.13 that:

Corollary 42.13b. *Let $M_1 = (S, \mathcal{I}_1)$ and $M_2 = (S, \mathcal{I}_2)$ be strongly base orderable matroids, with rank functions r_1 and r_2, satisfying $r_1(S) = r_2(S)$. Then M_1 and M_2 have k disjoint common bases if and only if*

(42.52) $\quad |S \setminus (T \cup U)| \geq k(r_1(S) - r_1(T) - r_2(U))$

for all $T, U \subseteq S$.

Proof. Indeed, from Theorem 42.13 we have that M_1 and M_2 have k disjoint common bases if and only if the matroids $M_1 \vee \cdots \vee M_1$ and $M_2 \vee \cdots \vee M_2$ (k-fold unions) have a common independent set of size $k \cdot r_1(S)$. By the matroid union and intersection theorems, this last is equivalent to the condition stated in the present corollary. ∎

By truncating M_1 and M_2 one has similar results if we replace 'common bases' by 'common independent sets of size t'. Application to transversal matroids yields Corollary 23.9d.

Another consequence of Theorem 42.13 is:

Corollary 42.13c. *Let $M_1 = (S, \mathcal{I}_1)$ and $M_2 = (S, \mathcal{I}_2)$ be strongly base orderable matroids. Then M_1 and M_2 have k disjoint common spanning sets if and only if both M_1 and M_2 have k disjoint bases.*

Proof. This can be deduced as follows. Let N_i arise from the dual matroid of M_i by replacing each element s of S by $k - 1$ parallel elements (for $i = 1, 2$). So N_1 and N_2 are strongly base orderable again, with an underlying ground set of size $(k - 1)|S|$. Now M_1 and M_2 have k disjoint (common) spanning sets, if and only if N_1 and N_2 have k (common) independent sets covering the underlying set. This directly implies the present corollary. ∎

Applying Corollary 42.13c to transversal matroids gives Theorem 23.11.

Corollary 42.13d. *Let $M_1 = (S, \mathcal{I}_1)$ and $M_2 = (S, \mathcal{I}_2)$ be strongly base orderable matroids, with rank functions r_1 and r_2, satisfying $r_1(S) = r_2(S)$. Then S can be covered by k common bases of M_1 and M_2 if and only if*

(42.53) $\qquad k(r_1(T) + r_2(U) - r_1(S)) \geq |T \cap U|$

for all $T, U \subseteq S$.

Proof. Condition (42.53) is equivalent to:

(42.54) $\qquad (k-1)|S \setminus (T \cup U)| \geq k(r_1^*(S) - r_1^*(T) - r_2^*(U))$

for all $T, U \subseteq S$. Let N_1 and N_2 be the matroids defined in the proof of Corollary 42.13c. By Corollary 42.13b, condition (42.54) implies that N_1 and N_2 contain k disjoint common bases. So M_1^* and M_2^* have k common bases covering each element at most $k - 1$ times. Hence M_1 and M_2 have k common bases covering S. ∎

Applying Corollary 42.13d to transversal matroids gives Theorem 23.12.

42.6d. Blocking and antiblocking polyhedra

We next investigate the blocking and antiblocking polyhedra corresponding to intersections of independent set polytopes of two matroids. Let $M_1 = (S, \mathcal{I}_1)$ and $M_2 = (S, \mathcal{I}_2)$ be loopless matroids, with rank functions r_1 and r_2 respectively, and independent set polytopes P_1 and P_2 respectively. So $P_1 \cap P_2$ is the convex hull of the incidence vectors of common independent sets. Hence its antiblocking polyhedron $A(P_1 \cap P_2)$ is determined by the linear inequalities

(42.55) $\qquad \begin{array}{ll} x_s \geq 0 & (s \in S), \\ x(I) \leq 1 & (I \in \mathcal{I}_1 \cap \mathcal{I}_2). \end{array}$

Since $P_1 \cap P_2$ is determined by the linear inequalities (41.37), $A(P_1 \cap P_2)$ consists of all vectors $x \geq \mathbf{0}$ for which there exists a $y \geq x$ which is a convex combination of vectors

(42.56) $$\frac{1}{r_i(U)}\chi^U$$

where U is a nonempty subset of S and $i = 1, 2$. Then $A(P_1 \cap P_2)$ gives rise to the following linear programming duality equation, for $c : S \to \mathbb{R}_+$:

(42.57) $$\max\{c^T x \mid x \in A(P_1 \cap P_2)\} = \max\{\frac{c(U)}{r_i(U)} \mid \emptyset \neq U \subseteq S; i = 1, 2\}$$
$$= \min\{\sum_{I \in \mathcal{I}_1 \cap \mathcal{I}_2} y(I) \mid y \in \mathbb{R}_+^{\mathcal{I}_1 \cap \mathcal{I}_2}, \sum_{I \in \mathcal{I}_1 \cap \mathcal{I}_2} y(I)\chi^I \geq c\}.$$

For integer c, an integer optimum solution y need not exist (for instance, if $|S| = 3$, $r_i(U) := \min\{|U|, 2\}$, and $c = 1$). That is, system (42.55) need not be totally dual integral. In fact, it generally does not have the integer rounding property. That is, it is not true, for each pair of matroids, that the minimum in (42.57) with y restricted to be integer:

(42.58) $$\min\{\sum_{I \in \mathcal{I}_1 \cap \mathcal{I}_2} y(I) \mid y \in \mathbb{Z}_+^{\mathcal{I}_1 \cap \mathcal{I}_2}, \sum_{I \in \mathcal{I}_1 \cap \mathcal{I}_2} y(I)\chi^I \geq c\},$$

is equal to the upper integer part of the common value of (42.57). For instance, take for M_1 the cycle matroid of K_4, and for M_2 the matroid with independent sets all sets of pairwise intersecting edges in K_4, and let $c = 1$; then the common value in (42.57) is 2, while (42.58) is equal to 3. However, Corollary 42.13a implies that if M_1 and M_2 are strongly base orderable matroids, then (42.58) is equal to the upper integer part of (42.57). That is, for strongly base orderable matroids, system (42.57) has the integer rounding property.

Similar results hold if we consider the blocker $B(Q_1 \cap Q_2)$ of the intersection of the spanning set polytopes Q_1 and Q_2 of M_1 and M_2. In particular, Corollary 42.13c implies that the system

(42.59) $\quad x_s \geq 0 \quad (s \in S),$
$\quad\quad\quad x(U) \geq 1 \quad (U \text{ common spanning set of } M_1 \text{ and } M_2)$

has the integer rounding property, if M_1 and M_2 are strongly base orderable.

Moreover, Corollaries 42.13b and 42.13d imply that the systems

(42.60) $\quad x_s \geq 0 \quad (s \in S),$
$\quad\quad\quad x(B) \geq 1 \quad (B \text{ common base of } M_1 \text{ and } M_2)$

and

(42.61) $\quad x_s \geq 0 \quad (s \in S),$
$\quad\quad\quad x(B) \leq 1 \quad (B \text{ common base of } M_1 \text{ and } M_2)$

have the integer rounding property, if M_1 and M_2 are strongly base orderable. Here the results of Section 41.4b are used: to prove that (42.60) has the integer rounding property, let $w \in \mathbb{Z}_+^S$. Let Q be the polytope determined by (42.60), let $r(U)$ be the maximum size of a common independent set contained in U, and let $\mathcal{B}$ denote the collection of common bases. Then

(42.62) $$\lceil \min\{w^T x \mid x \in Q\}\rceil$$
$$= \min\{\lceil \frac{w(U)}{r(S) - r(S \setminus U)} \rceil \mid U \subseteq S, r(S) > r(S \setminus U)\}$$
$$= \max\{\sum_{B \in \mathcal{B}} y_B \mid y \in \mathbb{Z}_+^{\mathcal{B}}, \sum_{B \in \mathcal{B}} y_B \chi^B \leq w\}.$$

The first equality holds as the vertices of Q are given by the vectors

(42.63) $$\frac{1}{r(S) - r(S \setminus U)} \chi^U,$$

since Q is the blocking polyhedron of the common base polytope (cf. Section 41.4b). The second equality follows from Corollary 42.13b, using the fact that strong base orderability is maintained under adding parallel elements.

Related results on integer decomposition of the intersection of the independent set polytopes of two strongly base orderable matroids can be found in McDiarmid [1983].

42.6e. Further notes

Krogdahl [1976] observed that the following, general problem is solvable in polynomial time, by reduction to matroid intersection: given matroids $(S, \mathcal{I}_1), \ldots, (S, \mathcal{I}_k)$, weight functions $w_1, \ldots, w_k \in \mathbb{R}^S$, and $l \leq k$, find the maximum value of $w_1(I_1) + \cdots + w_k(I_k)$, where $I_1 \in \mathcal{I}_1, \ldots, I_k \in \mathcal{I}_k$, with $I_1, \ldots, I_l$ disjoint and $I_{l+1}, \ldots, I_k$ disjoint, and with $I_1 \cup \ldots \cup I_l = I_{l+1} \cup \ldots \cup I_k$.

With matroid union, several new classes of matroids can be constructed. One of them is formed by the *bicircular matroids*, which are the union of the cycle matroid $M(G)$ of a graph $G = (V, E)$ and the matroid on E in which $F \subseteq E$ is independent if and only if $|F| \leq 1$. The independent sets of this matroid are the edge sets containing at most one circuit.

A randomized parallel algorithm for linear matroid union was given by Narayanan, Saran, and Vazirani [1992,1994]. For matroid base packing algorithms, see Knuth [1973] and Karger [1993,1998].

42.6f. Historical notes on matroid union

As the matroid base covering theorem can be derived by an elementary construction from Rado's theorem (proved by Rado [1942]), it is surprising that, for a long time, it had remained an open question, posed by Rado himself.

In fact, it was Horn [1955] who showed that a set X of vectors is the union of k linearly independent sets of vectors if and only if each finite subset Y of X has rank at least $|Y|/k$. He mentioned that this was conjectured by K.F. Roth and R. Rado, and he did not refer to matroids. Horn also acknowledged the help of Rado.

Surprisingly, the same theorem was also published by Rado [1962a] (in the same journal). The proof method (including notation) is the same as that of Horn, but no reference to Horn's paper is given. Rado wondered if the theorem can be generalized to matroids:

> It can be seen that some steps of the argument can be adapted to the more general situation of abstract independence functions but there does not appear to be an obvious way of making the whole argument apply to the more general case.

Rado [1962b] presented the vector theorem at the International Congress of Mathematicians in Stockholm in 1962, where he mentioned again that its proof has not yet been extended to 'abstract independence relations' (matroids). He wondered if the property in fact would *characterize* linear matroids.

Finally, two years later, at the Conference on General Algebra in Warsaw, 7–11 September 1964, Rado announced the base covering theorem. Simultaneously, there was the Seminar on Matroids at the National Bureau of Standards in Washington, D.C., 31 August–11 September 1964, where Edmonds [1965c] presented the base covering theorem.

In the paper based on his lecture in Warsaw, Rado [1966] did not give a proof of the matroid base cover theorem, but just said that the argument of Horn [1955] can be adapted so as to yield the more general version (as we did in Section 42.1b).

The matroid base covering theorem generalizes also the min-max relation of Nash-Williams [1964] for the minimum number of forests needed to cover the edges of a graph. (As each graphic matroid is linear, this follows also from the result of Horn [1955] described above.)

The basic unifying result (Theorem 42.1) on matroid union was given in Nash-Williams [1967], which has as special case the matroid union theorem given by Edmonds [1968]. In a footnote on page 20 of Pym and Perfect [1970], it is remarked that:

> Professor Nash-Williams has written to inform us that these results were suggested by earlier unpublished work of Professor J. Edmonds on the relation between independence structures and submodular functions.

It seems in fact much easier to prove the matroid union theorem in general, than just its special case for graphic matroids (for instance, the covering forests theorem). It also generalizes theorems of Higgins [1959] on disjoint transversals (Theorem 22.11), and of Tutte [1961a] and Nash-Williams [1961b] on disjoint spanning trees in a graph (Corollary 51.1a). (These papers mention no possible generalization to matroids.)

Welsh [1976] mentioned on these results:

> They illustrate perfectly the principle that mathematical generalization often lays bare the important bits of information about the problem at hand.

Chapter 43

Matroid matching

We saw two generalizations of Kőnig's matching theorem for bipartite graphs: the Tutte-Berge formula on matchings in arbitrary graphs and the matroid intersection theorem. This raises the demand for a common generalization of these last two theorems. A solution to the following *matroid matching problem*, posed by Lawler [1971b,1976b], could yield such a generalization: given an undirected graph $G = (S, E)$ and a matroid $M = (S, \mathcal{I})$, what is the maximum number of disjoint edges of G whose union is independent in M?

By taking M trivial, the matroid matching problem reduces to the matching problem, and by taking G regular of degree one, and M to be the disjoint sum of two matroids defined on the two colour classes of the bipartite graph G, we obtain the matroid intersection problem.

However, the general matroid matching problem has been shown to be NP-complete in the regular NP-framework, and unsolvable in polynomial time in an oracle framework.

On the other hand, Lovász [1980b] gave a strongly polynomial-time algorithm in case the matroid M is linear. Moreover, Lovász [1980a] gave a min-max relation, which was extended by Dress and Lovász [1987] to algebraic matroids.

No extension to the weighted case has been discovered, even not for the linear case: no polyhedral characterization or polynomial-time algorithm for finding a maximum-weight matroid matching has been found.

43.1. Infinite matroids

In this chapter, we need an extension of the notion of matroids to infinite matroids. An *infinite matroid* is defined as a pair $M = (S, \mathcal{I})$, where S is an infinite set and $\mathcal{I}$ is a nonempty collection of subsets of S satisfying:

(43.1) (i) if $I \in \mathcal{I}$ and $J \subseteq I$, then $J \in \mathcal{I}$,
 (ii) if $I \subseteq S$ and each finite subset of I belongs to $\mathcal{I}$, then I belongs to $\mathcal{I}$;
 (iii) if I, J are finite sets in $\mathcal{I}$ and $|I| < |J|$, then $I \cup \{j\} \in \mathcal{I}$ for some $j \in J \setminus I$.

746   Chapter 43. Matroid matching

Standard matroid terminology transfers to infinite matroids. The sets in $\mathcal{I}$ are called *independent* and those subsets of S not in $\mathcal{I}$ *dependent*. An inclusionwise minimal dependent set is a *circuit*. By (43.1)(ii), each *circuit* of M is finite. We will restrict ourselves to infinite matroids of *finite rank*. That is, there is a finite upper bound on the size of the sets in $\mathcal{I}$.

Examples of infinite matroids are linear spaces, where $\mathcal{I}$ is the collection of linearly independent subsets, and field extensions L of a field K, where $\mathcal{I}$ is the collection of subsets of L that are algebraically independent over K. In fact, these are the only two classes of infinite matroids that we will consider.

We call a matroid $M = (S, \mathcal{I})$ with S finite also a *finite matroid*.

43.2. Matroid matchings

Let $(S, \mathcal{I})$ be a (finite or infinite) matroid, with rank function r and span function span. Let E be a finite collection of unordered pairs from S, such that each pair is an independent set of $(S, \mathcal{I})$. For $F \subseteq E$ define

(43.2) $\text{span}(F) := \text{span}(\bigcup F)$

(where $\bigcup F$ denotes the union of the pairs in F), and

(43.3) $r(F) := r(\text{span}(F))$.

Then for $X, Y \subseteq E$ one has

(43.4) $r(X) + r(Y) \geq r(X \cap Y) + r(X \cup Y)$,

since

(43.5) $r(X) + r(Y) = r(\text{span}(X)) + r(\text{span}(Y))$
$\geq r(\text{span}(X) \cap \text{span}(Y)) + r(\text{span}(X) \cup \text{span}(Y))$
$\geq r(\text{span}(X \cap Y)) + r(\text{span}(X \cup Y)) = r(X \cap Y) + r(X \cup Y)$.

Call a subset M of E a *matroid matching*, or just a *matching*, if

(43.6) $r(M) = 2|M|$.

So M is a matroid matching if and only if M consists of disjoint pairs and the union of the pairs in M belongs to $\mathcal{I}$. Hence each subset of a matching is a matching again. The maximum size of a matching in E is denoted by $\nu(E)$, or just by ν. A matching of size $\nu(E)$ is called a *base* of E. (We should be aware of the difference between a matching in a graph and a matroid matching, and between a base of a matroid and a base of a collection of pairs in a matroid. Below we will see moreover the notion of a circuit in a set of pairs in a matroid. We will be careful to avoid confusion.[32])

Consider the function s defined on subsets F of E by

(43.7) $s(F) := 2|F| - r(F)$.

---
[32] As we denote a matching by M, we denote a matroid, for the time being, just by $(S, \mathcal{I})$.

So a subset M of E is a matching if and only if $s(M) = 0$.
Then for all collections X and Y:

(43.8) (i) $s(X) \leq s(Y)$ if $X \subseteq Y$,
(ii) $s(X) + s(Y) \leq s(X \cap Y) + s(X \cup Y)$.

Here (i) follows from

(43.9) $\quad r(Y) \leq r(X) + r(Y \setminus X) \leq r(X) + 2|Y| - 2|X|.$

(43.8)(ii) follows from (43.4).
(43.8) implies:

(43.10) each $F \subseteq E$ contains a unique inclusionwise minimal subset X with $s(X) = s(F)$.

For let F contain subsets X and Y with $s(X) = s(Y) = s(F)$. Then by (43.8)(i), $s(X \cap Y) \leq s(F)$ and $s(X \cup Y) = s(F)$, and by (43.8)(ii), $s(X \cap Y) \geq s(X) + s(Y) - s(X \cup Y) = s(F)$. So $s(X \cap Y) = s(F)$.

43.3. Circuits

A subset C of E is called a *circuit* if it is an inclusionwise minimal set satisfying $r(C) = 2|C| - 1$. By (43.10):

(43.11) each $F \subseteq E$ with $r(F) = 2|F| - 1$ contains a unique circuit.

It implies that for each matching M and each $e \in E$ with $r(M + e) = r(M) + 1$, there is a unique circuit contained in $M + e$. This circuit is denoted by $C(M, e)$, and is called a *fundamental circuit (of M)*. (Here and below, $M + e := M \cup \{e\}$ and $M - e := M \setminus \{e\}$.)
Such circuits have a useful exchange property:

(43.12) for each $f \in C(M, e)$, $M + e - f$ is a matching again.

Indeed, if $M + e - f$ is not a matching, then $s(M + e - f) \geq 1$. In fact, $s(M + e - f) = 1$, since $s(M + e - f) \leq s(M + e) = 1$. So $M + e - f$ contains a circuit C. As $f \notin C$, we know $C \neq C(M, e)$, contradicting (43.11).

43.4. A special class of matroids

The min-max equality for matroid matching to be proved, holds for (finite or infinite) matroids $(S, \mathcal{I})$ satisfying the following condition:

(43.13) for each pair of circuits C_1, C_2 of $(S, \mathcal{I})$ with $C_1 \cap C_2 \neq \emptyset$ and $r(C_1 \cup C_2) = |C_1 \cup C_2| - 2$, the intersection of span(C) taken over all circuits $C \subseteq C_1 \cup C_2$ has positive rank.

Examples of such matroids will be seen in Section 43.6.

In (43.13), 'circuits' are meant in the original meaning: as subsets of S. But the property transfers to subsets of E, as follows:

Lemma 43.1α. *Let $(S, \mathcal{I})$ be a matroid satisfying (43.13) and let E be a collection of pairs from S. Then for each pair of circuits $C_1, C_2 \subseteq E$ with $C_1 \cap C_2 \neq \emptyset$ and $s(C_1 \cup C_2) = 2$, the intersection of $\mathrm{span}(C)$ taken over all circuits $C \subseteq C_1 \cup C_2$ has positive rank.*

Proof. Let $F := C_1 \cup C_2$. By assumption, $s(F) = 2$. Each proper subcollection F' of F satisfies $s(F') \leq 1$, since if $e \in C_i$, then $s(F - e) \leq s(F) + s(C_i - e) - s(C_i) = 2 + 0 - 1 = 1$.

Let $C_1, \ldots, C_k$ be the circuits contained in F. We can assume that $k \geq 3$ (otherwise the lemma trivially holds, since $C_1 \cap C_2 \neq \emptyset$ by assumption). Then

(43.14) $\quad C_i \cup C_j = F$ for all distinct $i, j = 1, \ldots, k$,

since for any $e \in F \setminus (C_i \cup C_j)$ we would have that $s(F - e) = 1$ and that $F - e$ contains two distinct circuits, which contradicts (43.11).

An equivalent way of stating (43.14) is:

(43.15) $\quad F \setminus C_1, \ldots, F \setminus C_k$ are pairwise disjoint.

Now first assume that there exist distinct $e, f \in F$ with $e \cap f \neq \emptyset$. Then $|e \cup f| = 3$, so $\{e, f\}$ is a circuit, and therefore by (43.15), each C_i intersects $\{e, f\}$ (as $k \geq 3$). So each $\mathrm{span}(C_i)$ contains $e \cap f$, and therefore the intersection of the $\mathrm{span}(C_i)$ is nonempty, as required.

So we can assume that the pairs in F are disjoint. Consider any i. Then $\bigcup C_i$ is a subset of S, containing a unique circuit C_i' (as subset of S). This follows from:

(43.16) $\quad r(\bigcup C_i) = |\bigcup C_i| - 1$

(as C_i is a circuit in E), because (43.16) implies that $\bigcup C_i$ contains an independent set of size $|\bigcup C_i| - 1$.

Then

(43.17) $\quad C_i' \neq C_j'$ if $i \neq j$.

Indeed, C_i' intersects each pair in C_i, since for each $e \in C_i$ the union of the $f \in C_i - e$ has rank $2|C_i - e|$, hence is independent. As the pairs in F are disjoint, this shows (43.17).

Moreover, if $i \neq j$ and $h \in \{1, \ldots, k\}$, then

(43.18) $\quad C_h' \subseteq C_i' \cup C_j'$.

Otherwise, choose $x \in C_i'$, $y \in C_j' \setminus C_i'$, and $z \in C_h' \setminus (C_i' \cup C_j')$. So $x, y, z \in \mathrm{span}((C_i' \cup C_j' \cup C_h') \setminus \{x, y, z\})$. Hence $r(C_i' \cup C_j' \cup C_h') \leq |C_i' \cup C_j' \cup C_h'| - 3$, and so

(43.19) $\quad r(F) \leq r(C'_i \cup C'_j \cup C'_h) + |\bigcup F| - |C'_i \cup C'_j \cup C'_h| \leq |\bigcup F| - 3,$

a contradiction, since $s(F) = 2$.

This proves (43.18), which implies that $C'_1 \cap C'_2 \neq \emptyset$ (since $C'_1 \subseteq C'_2 \cup C'_3$ and $C'_1 \not\subseteq C'_3$). Then by (43.13), the intersection of $\mathrm{span}(C'_i)$ over all i has positive rank. Hence the intersection of $\mathrm{span}(C_i)$ over all i has positive rank. ∎

For any collection E of pairs from S, let H_E be the hypergraph with vertex set E and edges all fundamental circuits. The following theorem will be used in deriving a general min-max relation.

Theorem 43.1. *Let $(S, \mathcal{I})$ be a matroid satisfying (43.13) and let E be a collection of pairs from S such that the intersection of $\mathrm{span}(B)$ over all bases B of E has rank 0. Then*

(43.20) $\quad |B \cap F| = \lfloor \frac{1}{2} r(F) \rfloor$

for each base B and each component F of H_E.

Proof. I. Call two fundamental circuits C, D *far* if there exist a base B and $e, g \in E$ with $r(B + e + g) = 2\nu + 2$ and with $C = C(B, e)$ and $D = C(B, g)$. We first show:

(43.21) $\quad$ far fundamental circuits are disjoint.

Suppose to the contrary that there exist a base B and $e, g \in E$ with $r(B + e + g) = 2\nu + 2$ and $C(B, e) \cap C(B, g) \neq \emptyset$. Let $D := C(B, e) \cup C(B, g)$. Then

(43.22) $\quad s(D) \geq s(C(B, e)) + s(C(B, g)) - s(C(B, e) \cap C(B, g)) = 2$

and

(43.23) $\quad s(D) \leq s(B + e + g) = 2.$

So $s(D) = 2$. If C is any circuit contained in $B + e + g$, then $C \subseteq D$, since otherwise $s(C \cap D) = 0$, and hence

(43.24) $\quad 2 = 0 + s(B + e + g) \geq s(C \cap D) + s(C \cup D) \geq s(C) + s(D) = 3,$

a contradiction.

By Lemma 43.1α, there is a nonloop p that is contained in $\mathrm{span}(C)$ for each circuit $C \subseteq D$. By assumption, there is a base B' with $p \notin \mathrm{span}(B')$. Choose B' with $|B' \cap (B+e+g)|$ maximal. Then $r(B'+p) = 2\nu+1 < r(B+e+g)$, and hence $f \not\subseteq \mathrm{span}(B'+p)$ for some $f \in B+e+g$. Then $p \notin \mathrm{span}(B'+f)$ (since $r(B'+f) \leq 2\nu+1$), and therefore $p \notin \mathrm{span}(C(B', f))$. So $C(B', f)$ is not one of the circuits contained in $B+e+g$. Choose $h \in C(B', f) \setminus (B+e+g)$. Hence, resetting B' to $B'-h+f$ would give a larger intersection with $B+e+g$, a contradiction. This shows (43.21).

II. We next show the theorem assuming that H_E is connected. Suppose to the contrary that $r(E) \geq 2\nu(E) + 2$. Then far fundamental circuits exist, since for any base B, there exist $e,g \in E$ with $r(B+e+g) = 2\nu + 2$, since $r(E) \geq r(B) + 2$. Then (43.21) implies, as H_E is connected, that there exist fundamental circuits C, C', D with C and D far, $C \cap C' \neq \emptyset$, and C' and D not far.

Choose $e \in C \cap C'$ and $f \in D$. As C and D are far fundamental circuits, there is a base B with $r(B+e+f) = 2\nu+2$ and $C = C(B,e)$, $D = C(B,f)$. Also, as C' is a fundamental circuit, there is a base B' with $r(B'+e) = 2\nu+1$ and $C' = C(B',e)$. Choose such a B' with $|B' \cap (B+f)|$ maximal.

As $r(B+e+f) > r(B'+e)$, there exists a $g \in B+f$ with $r(B'+e+g) = 2\nu + 2$. As C' and D are not far, $C(B',g) \neq D = C(B,f)$. So $C(B',g) \not\subseteq B+f$, and hence there exists an $h \in C(B',g) \setminus (B+f)$. Set $B'' := B' - h + g$. Then $r(B''+h+e) = r(B'+g+e) = 2\nu+2$, and hence $r(B''+e) = 2\nu+1$. As, by (43.21), $C(B',g)$ and $C(B',e)$ are disjoint, we know $h \notin C(B',e)$, so $C(B',e) \subseteq B''+e$, and hence $C(B'',e) = C(B',e) = C'$. As $|B'' \cap (B+f)| > |B' \cap (B+f)|$ this contradicts the maximality of $|B' \cap (B+f)|$.

III. We finally prove the theorem in general. Let F be a component of H_E. Suppose that there is a base B of E with $|B \cap F| < \lfloor \frac{1}{2}r(F) \rfloor$. Then

(43.25)   there is a base B of E and a base M of F with $|M| > |B \cap F|$.

Otherwise, for each base B of E, $B \cap F$ is a base of F. Then H_F consists of one component (as each fundamental circuit of E contained in F is a fundamental circuit of F). Hence, by part II of this proof, $|B \cap F| = \nu(F) = \lfloor \frac{1}{2}r(F) \rfloor$, contradicting our assumption.

So (43.25) holds. Choose B and M as in (43.25) with $|M \cap B|$ maximal. Then

(43.26) $\mathrm{span}(M) \subseteq \mathrm{span}(B)$,

since otherwise there is an $e \in M$ with $e \not\subseteq \mathrm{span}(B)$, and we can choose $f \in C(B,e) \setminus M$ and replace B by $B - f + e$, thereby increasing $|M \cap B|$, contradicting the maximality of $|M \cap B|$.

Moreover,

(43.27)   for each $e \in F$ with $e \not\subseteq \mathrm{span}(B)$, we have $C(M,e) = C(B,e)$.

Otherwise, choose $f \in C(B,e) \setminus (M+e)$ and $g \in C(M,e) \setminus (B+e)$. Replacing B and M by $B - f + e$ and $M - g + e$ respectively, increases $|M \cap B|$, a contradiction.

As $M \setminus B \neq \emptyset$, there is an $h \in M \setminus B$. Then there is a base B' of E with $h \not\subseteq \mathrm{span}(B')$ (as by the condition in the theorem, there is no nonloop that is contained in the span of each base). We assume that we have chosen M, B, and B' with $|B \cap B'|$ maximal (under the primary condition that $|M \cap B|$ is maximum).

Since $h \not\subseteq \mathrm{span}(B')$, we know by (43.26) that $\mathrm{span}(B) \neq \mathrm{span}(B')$. Hence there exists an $e \in B'$ with $e \not\subseteq \mathrm{span}(B)$.

If $e \notin F$, then $C(B,e)$ is disjoint from F (as F is a component of H_E). Choose $f \in C(B,e) \setminus B'$. Then replacing B by $B - f + e$ maintains M, $B \cap F$, and $M \cap B$, but increases $|B \cap B'|$, contradicting our assumption.

So $e \in F$. By (43.27), $C(B,e) = C(M,e)$. Choose $f \in C(B,e) \setminus B'$. Then replacing M and B by $M - f + e$ and $B - f + e$ respectively, maintains $|M|$, $|B \cap F|$, and $|M \cap B|$, but increases $|B \cap B'|$, contradicting our assumption. ∎

43.5. A min-max formula for maximum-size matroid matching

We can now derive a min-max formula for the maximum size of a matching in matroids satisfying (43.13) in an hereditary way, due to Lovász [1980a]:

Theorem 43.2 (matroid matching theorem). *Let $M = (S, \mathcal{I})$ be a (finite or infinite) matroid (with rank function r) such that each contraction of M satisfies (43.13). Let E be a finite set of pairs from S. Then the maximum size $\nu(E)$ of a matching in E satisfies*

$$(43.28) \qquad \nu(E) = \min(r(F) + \sum_{i=1}^{k} \lfloor \tfrac{1}{2}(r(F_i) - r(F)) \rfloor),$$

where $F, F_1, \ldots, F_k$ are flats such that $F \subseteq F_i$ for $i = 1, \ldots, k$, and such that each $e \in E$ is contained in some F_i.

Proof. We first show that $\leq$ holds in (43.28). Let B be a base of E, and partition B into $B_1, \ldots, B_k$ such that $\text{span}(B_i) \subseteq F_i$ for $i = 1, \ldots, k$. Define $F'_i := \text{span}(B_i)$.

By induction on l we show that for each $l = 0, \ldots, k$:

$$(43.29) \qquad r(F \cup F'_1 \cup \cdots \cup F'_l) \leq r(F) + \sum_{i=1}^{l}(|B_i| + \lfloor \tfrac{1}{2}(r(F \cup F'_i) - r(F)) \rfloor).$$

For $l = 0$ this is trivial. For $l \geq 1$ we have (by induction and submodularity):

$$(43.30) \qquad r(F \cup F'_1 \cup \cdots \cup F'_l) \leq r(F \cup F'_1 \cup \cdots \cup F'_{l-1}) + r(F \cup F'_l) - r(F)$$
$$\leq r(F \cup F'_l) + \sum_{i=1}^{l-1}(|B_i| + \lfloor \tfrac{1}{2}(r(F \cup F'_i) - r(F)) \rfloor)$$
$$\leq r(F) + \sum_{i=1}^{l}(|B_i| + \lfloor \tfrac{1}{2}(r(F \cup F'_i) - r(F)) \rfloor),$$

since

$$(43.31) \qquad r(F \cup F'_l) \leq r(F) + |B_l| + \tfrac{1}{2}(r(F \cup F'_l) - r(F)),$$

as $|B_l| = \frac{1}{2}r(F'_l)$. This shows (43.29), which for $l = k$ implies that $\nu(E)$ is at most (43.28), since

(43.32)
$$2\nu(E) \leq r(F \cup F'_1 \cup \cdots \cup F'_k)$$
$$\leq r(F) + \sum_{i=1}^{k}\left(|B_i| + \lfloor\tfrac{1}{2}(r(F \cup F'_i) - r(F))\rfloor\right)$$
$$= \nu(E) + r(F) + \sum_{i=1}^{l}\lfloor\tfrac{1}{2}(r(F \cup F_i) - r(F))\rfloor.$$

Equality is shown by induction on $r(M)$. First assume that there is a nonloop p that is contained in span(B) for each base B of E. Let M' be the matroid M/p obtained by contracting p. Let E' be the set of pairs $\{s,t\}$ in E such that $s, t \neq p$ and such that s and t are not parallel in M'. Let ν' be the maximum size of a base $B' \subseteq E'$ with respect to M'.

Then $\nu' < \nu(E)$. For suppose that $\nu' \geq \nu(E)$. Let B' be a base of E' with respect to M'. As $|B'| \geq \nu(E)$, B' is also a base of E with respect to M. As $r_{M'}(B') = 2|B'| = r_M(B')$, we have $p \notin \text{span}_M(B)$. This contradicts our assumption.

So $\nu' < \nu(E)$. By induction, M' has flats $F', F'_1, \ldots, F'_{k'}$ with $F' \subseteq F'_i$ for $i = 1, \ldots, k'$, such that each $e \in E'$ is contained in some F'_i and such that

(43.33)
$$\nu' = r_{M'}(F') + \sum_{i=1}^{k'}\lfloor\tfrac{1}{2}(r_{M'}(F'_i) - r_{M'}(F'))\rfloor.$$

Define $F := \text{span}_M(F' + p)$ and $F_i := \text{span}_M(F'_i + p)$ for $i = 1, \ldots, k'$. Moreover, for each $e \in E$ not occurring in E', introduce a new F_i with $F_i := \text{span}_M(F + e)$. As $p \in F$, we have $r_M(F_i) \leq r_M(F) + 1$ for each of these F_i.

This gives $F, F_1, \ldots, F_k$ such that $F \subseteq F_i$ for $i = 1, \ldots, k$, such that each $e \in E$ is contained in some F_i and such that

(43.34)
$$\nu(E) \geq \nu' + 1 = r_{M'}(F') + 1 + \sum_{i=1}^{k'}\lfloor\tfrac{1}{2}(r_{M'}(F'_i) - r_{M'}(F'))\rfloor$$
$$= r(F) + \sum_{i=1}^{k}\lfloor\tfrac{1}{2}(r(F_i) - r(F))\rfloor.$$

So we can assume that there is no nonloop p contained in span(B) for all bases B of E. Let $E_1, \ldots, E_k$ be the components of H_E and let $F_i := \text{span}(E_i)$ for $i = 1, \ldots, k$. Let B be a base of E. Then by (43.20),

(43.35)
$$\nu(E) = |B| = \sum_{i=1}^{k}|B \cap E_i| = \sum_{i=1}^{k}\lfloor\tfrac{1}{2}r(F_i)\rfloor.$$

So taking $F := \emptyset$ gives (43.28). ∎

43.6. Applications of the matroid matching theorem

We now consider specific classes of matroids satisfying (43.13), such that we know that the min-max equality holds. First, the linear matroids (Lovász [1980b]):

Corollary 43.2a. *If E is a finite set of pairs from a linear space S, then (43.28) holds, where flats are linear subspaces of S.*

Proof. Let $\mathcal{I}$ be the collection of sets of linearly independent vectors in S. We must show that each contraction of the infinite matroid $M = (S, \mathcal{I})$ satisfies (43.13). It suffices to show that M satisfies (43.13), since each contraction of M is again coming from a linear space, up to loops and parallel elements.

Let C_1 and C_2 be intersecting circuits in M with $r(C_1 \cup C_2) = |C_1 \cup C_2| - 2$. As C_1 is a circuit, there is a nonzero vector p in $\text{span}(C_1 \setminus C_2) \cap \text{span}(C_1 \cap C_2)$, since $r(C_1 \setminus C_2) + r(C_1 \cap C_2) > r(C_1)$. Consider any circuit C contained in $C_1 \cup C_2$.

Suppose $p \notin \text{span}(C)$. As $p \in \text{span}(C_1 \setminus C_2) \cap \text{span}(C_1 \cap C_2)$, C misses an element $s \in C_1 \setminus C_2$ and an element $t \in C_1 \cap C_2$. Now $t \in \text{span}(C_2 - t)$ and $s \in \text{span}(C_1 - s)$. Hence $(C_1 \cup C_2) - s - t$ spans $C_1 \cup C_2$, and hence, as $C_1 \cup C_2$ has rank $|C_1 \cup C_2| - 2$, we have that $(C_1 \cup C_2) - s - t$ is independent. This contradicts the fact that C is contained in $(C_1 \cup C_2) - s - t$. ∎

Dress and Lovász [1987] proved that a similar result holds for algebraic dependence in field extensions (where $\text{tr}_K(E)$ denotes the transcendence degree of $\bigcup E$ over K):

Corollary 43.2b. *Let E be a finite set of pairs from a field extension L of a field K. Then the maximum number of disjoint pairs from E such that the union is algebraically independent over K is equal to the minimum value of*

$$(43.36) \qquad \text{tr}_K(F) + \sum_{i=1}^{k} \lfloor \tfrac{1}{2} \text{tr}_F(E_i) \rfloor,$$

where F ranges over all field extensions of K in L and where $E_1, \ldots, E_k$ ranges over all partitions of E.

Proof. Let $M = (L, \mathcal{I})$ be the infinite matroid with $\mathcal{I}$ consisting of all subsets of L that are algebraically independent over K.

Similarly as for the previous corollary, it suffices to show that for any two intersecting circuits C_1 and C_2 of M with $r(C_1 \cup C_2) = |C_1 \cup C_2| - 2$ there is an $\alpha \in L \setminus \text{span}(K)$ such that α belongs to $\text{span}(C)$ for each circuit C contained in $C_1 \cup C_2$.

Let $I := C_1 \setminus C_2$. Then

(43.37) I is a circuit in M/C_2.

To see this, trivially I is dependent in M/C_2. Consider any circuit $C \subseteq C_1 \cup C_2$ intersecting I. We must show that $I \subseteq C$. Suppose that there is an $s \in I \setminus C$. As C intersects I, C misses at least one element of C_2, say t. So $C \subseteq (C_1 \cup C_2) - s - t$. Now $(C_1 \cup C_2) - s - t$ spans $C_1 \cup C_2$ (since $t \in \text{span}(C-t)$ and $s \in \text{span}(C_1 \cup C_2 - s)$). This implies that $(C_1 \cup C_2) - s - t$ is independent (as $r(C_1 \cup C_2) = |C_1 \cup C_2| - 2$), contradicting the fact that it contains a circuit. This proves (43.37).

Let $I = \{s_1, \ldots, s_n\}$. Since I is a circuit in M/C_2, there exists an irreducible polynomial p in $\text{span}(C_2)[x_1, \ldots, x_n]$ with $p(s_1, \ldots, s_n) = 0$. We can choose p such that at least one coefficient of p equals 1. Note that p has at least one coefficient, α say, that is not in $\text{span}(K)$, since I is independent over K. It therefore is enough to show that all coefficients of p belong to $\text{span}(C)$ for each circuit C contained in $C_1 \cup C_2$, since then α belongs to each $\text{span}(C)$.

Choose a circuit $C \neq C_2$ with $C \subseteq C_1 \cup C_2$. As I is a circuit in M/C_2, we have $C \setminus C_2 = I$. So I is a circuit in $M/(C \cap C_2)$. Hence there exists an irreducible polynomial q in $\text{span}(C \cap C_2)[x_1, \ldots, x_n]$ with $q(s_1, \ldots, s_n) = 0$. As $\text{span}(C \cap C_2)$ is algebraically closed in $\text{span}(C_2)$, q is also irreducible in $\text{span}(C_2)[x_1, \ldots, x_n]$[33]. Then p and q are also irreducible in $\text{span}(C_2)(x_1, \ldots, x_{n-1})[x_n]$ (cf., for instance, Section IV:6 of Jacobson [1951]). Therefore, p is a multiple of q in $\text{span}(C_2)(x_1, \ldots, x_{n-1})$; that is, there are nonzero $r, s \in \text{span}(C_2)[x_1, \ldots, x_{n-1}]$ with $rp = sq$. Hence by the unique factorization theorem (cf., for instance, Section IV:6 of Jacobson [1951]), $p = \lambda q$ for some $\lambda \in \text{span}(C_2)$. As some coefficient of p equals 1, $\lambda \in \text{span}(C \cap C_2)$. Hence $p \in \text{span}(C \cap C_2)[x_1, \ldots, x_n]$. ∎

(The property of algebraic matroids shown in this proof generalizes a property shown by Ingleton and Main [1975].)

We also formulate the special case of graphic matroids:

Corollary 43.2c. *Let $G = (V, E)$ be a graph and let $\mathcal{P}$ be a partition of E into pairs. Then the maximum size of a forest $F \subseteq E$ that is the union of classes of $\mathcal{P}$ is equal to the minimum value of*

$$(43.38) \quad 2|V| - 2|\mathcal{Q}| + 2\sum_{i=1}^{k} \lfloor \tfrac{1}{2} \delta_{\mathcal{Q}}(E_i) \rfloor,$$

---

[33] This can be seen as follows. Let L be a field extension of field K, such that K is algebraically closed in L. Then if p is an irreducible polynomial in $K[x_1, \ldots, x_n]$, then p is irreducible also in $L[x_1, \ldots, x_n]$. For suppose to the contrary that $p = p_1 p_2$ for nonconstant polynomials p_1, p_2 in $L[x_1, \ldots, x_n]$. We can assume that p_1 has at least one coefficient in K. Hence, as p is irreducible in $K[x_1, \ldots, x_n]$, p_1 has at least one coefficient not in K. Choose a large enough natural number k such that substituting x_i by x^{k^i} for $i = 1, \ldots, n$, transforms p_1 to a polynomial $\tilde{p}_1$ in $L[x] \setminus K[x]$. Let $\tilde{p} \in K[x]$ be obtained similarly from p. Now the algebraic closure of K contains all roots of $\tilde{p}$, hence all roots of $\tilde{p}_1$, and hence all coefficients of $\tilde{p}_1$. As each element in $L \setminus K$ is transcendental over K, we have a contradiction.

Section 43.6. Applications of the matroid matching theorem   755

where $\mathcal{Q}$ ranges over partitions of V into nonempty classes and where $E_1,\ldots,E_k$ ranges over partitions of E such that each E_i is a union of pairs in $\mathcal{P}$. In (43.38), $\delta_\mathcal{Q}(E_i)$ denotes the size of a largest forest in the graph obtained from (V,E_i) by contracting each class in $\mathcal{Q}$ to one vertex.

Proof. We apply Theorem 43.2 to the cycle matroid M of the graph H obtained from the complete graph on V by adding a parallel edge for each edge in E. Then (43.13) is satisfied for each contraction of M.

Now for each flat F of M there is a partition $\mathcal{Q}$ of V such that F is the set of edges of H contained in a class of $\mathcal{Q}$. The rank $r(F)$ of F (in M) is equal to $|V| - |\mathcal{Q}|$. For any $E' \subseteq E$, the smallest flat F' containing $F \cup E'$ has rank $r(F') = \delta_\mathcal{Q}(E') + r(F)$. Hence the corollary follows from Theorem 43.2. ∎

This corollary implies the following result on 3-uniform hypergraphs. A *hypergraph* is a pair $H = (V, \mathcal{E})$, where V is a finite set and $\mathcal{E}$ is a family of subsets of V. The hypergraph is called *k-uniform* if $|U| = k$ for each $U \in \mathcal{E}$.

A subfamily $\mathcal{F}$ of $\mathcal{E}$ is called a *forest* if there do not exist distinct $v_1,\ldots,v_t \in V$ and distinct $U_1,\ldots,U_t \in \mathcal{F}$ such that $t \geq 2$ and $v_{i-1}, v_i \in U_i$ for $i = 1,\ldots,t$, setting $v_0 := v_t$.

Corollary 43.2c implies a min-max relation for the maximum size of a forest in a given 3-uniform hypergraph (Lovász [1980a]):

Corollary 43.2d. *Let $H = (V, \mathcal{E})$ be a 3-uniform hypergraph. Then the maximum size of a forest $\mathcal{F} \subseteq \mathcal{E}$ is equal to the minimum value of*

(43.39) $$|V| - |\mathcal{Q}| + \sum_{\mathcal{S} \in \Sigma} \lfloor \tfrac{1}{2}(\phi_\mathcal{Q}(\mathcal{S}) - 1) \rfloor,$$

where $\mathcal{Q}$ and Σ range over partitions of V and $\mathcal{E}$, respectively. Here $\phi_\mathcal{Q}(\mathcal{S})$ denotes the number of classes of $\mathcal{Q}$ intersected by $\bigcup \mathcal{S}$.

Proof. For each $U \in \mathcal{E}$, choose two different pairs $e_U, f_U \subseteq U$, and let $G = (V, E)$ be the graph with edges all e_U and f_U. Let $\mathcal{P}$ be the partition of E into the pairs e_U, f_U. Then the maximum size of a forest $\mathcal{F} \subseteq \mathcal{E}$ is equal to half of the maximum size of a forest in E that is the union of pairs in $\mathcal{P}$. So to see that Corollary 43.2c implies the present corollary, it suffices to show that minimum (43.39) is equal to half of minimum (43.38).

First, let $\mathcal{Q}$ and Σ attain minimum (43.39). The partition Σ of $\mathcal{E}$ induces a partition of E into classes $\{e_U, f_U \mid U \in \mathcal{S}\}$ for $\mathcal{S} \in \Sigma$. One easily checks that for each $\mathcal{S} \in \Sigma$:

(43.40) $$\delta_\mathcal{Q}(\{e_U, f_U \mid U \in \mathcal{S}\}) \leq \phi_\mathcal{Q}(\mathcal{S}) - 1,$$

which implies that the minimum (43.39) is not less than half of minimum (43.38).

Second, to see the reverse inequality, let $\mathcal{Q}, E_1,\ldots,E_k$ attain minimum (43.38). Consider any $i = 1,\ldots,k$. Let $\mathcal{Q}'$ be the set of those classes in $\mathcal{Q}$

intersected by E_i and let t be the number of components of the hypergraph $(V, \mathcal{Q}' \cup E_i)$. Then $\delta_\mathcal{Q}(E_i) = |\mathcal{Q}'| - t$. The components partition E_i into $E_{i,1}, \ldots, E_{i,t}$. Then

$$(43.41) \qquad \delta_\mathcal{Q}(E_i) = |\mathcal{Q}'| - t = \sum_{j=1}^{t}(\phi_\mathcal{Q}(E_{i,j}) - 1).$$

So letting Σ to be the partition of $\mathcal{E}$ into classes $\mathcal{S}_{i,j} := \{U \mid e_U, f_U \in E_{i,j}\}$ (for all i, j), we have that minimum (43.38) is not less than twice minimum (43.39). ∎

(Szigeti [1998a] gave a direct proof of this theorem for the case where the hypergraph consists of all triangles of a given graph.)

Other applications of matroid matching are a derivation of Mader's theorem on maximum packings of T-paths (cf. Chapter 73), to rigidity (see Lovász [1980a]), and to matching forests (an easy application, see Section 59.6b).

43.7. A Gallai theorem for matroid matching and covering

We prove a Gallai-type theorem that relates the maximum size of a matroid matching to the minimum number of pairs spanning the matroid.

Let E be a collection of pairs of elements from a matroid $(S, \mathcal{I})$ such that each pair is an independent set and such that $\text{span}(E) = S$. Call $F \subseteq E$ a *matroid cover* if $\text{span}(F) = S$. Let $\rho(E)$ be the minimum size of a matroid cover. The following relation between $\nu(E)$ and $\rho(E)$ was observed by Lovász and extends Gallai's theorem (Theorem 19.1):

Theorem 43.3. *Let $(S, \mathcal{I})$ be a matroid, with rank function r, and let E be a collection of pairs from S spanning S. Then $\nu(E) + \rho(E) = r(S)$.*

Proof. To see $\leq$, let M be matching of size $\nu(E)$. Then by adding at most $r(S) - r(M)$ pairs from E to M we obtain a matroid cover F. So $\rho(E) \leq |F| \leq |M| + (r(S) - r(M)) = \nu(E) + r(S) - 2\nu(E) = r(S) - \nu(E)$.

To see $\geq$, let F be a matroid cover of size $\rho(E)$. Let $M := F$. As long as M contains an element e with $r(M - e) \geq r(M) - 1$, delete e from M. We end up with a matching M. For suppose not. Let M' be a maximum-size matching contained in M, and choose $e \in M \setminus M'$. Then $r(M-e) \leq r(M) - 2$ (otherwise we would delete e from M). Hence:

$$(43.42) \qquad r(M'+e) \geq r(M') + r(M) - r(M-e) \geq r(M') + 2 = 2|M'| + 2.$$

So $M' + e$ is a matching, contradicting the maximality of M'.

So M is a matching. Each time we have deleted an edge from M, its rank drops by at most 1. Hence $r(M) \geq r(S) - (|F| - |M|)$. Therefore $\nu(E) \geq |M| = r(M) - |M| \geq r(S) - |F| = r(S) - \rho(E)$. ∎

This theorem implies that formula (43.28) for the maximum size of a matching yields a formula for the minimum number of lines spanning all space.

43.8. Linear matroid matching algorithm

Jensen and Korte [1982] and Lovász [1981] showed that no polynomial-time algorithm exists for the matroid matching problem in general (see Section 43.9). On the other hand, Lovász [1981] gave a strongly polynomial-time algorithm for the matroid matching problem for linear matroids (an explicit representation over a field is required). This extends, e.g., Edmonds' polynomial-time algorithm finding a maximum matching in an undirected graph (cf. Section 24.2). It does not extend Edmonds' algorithm for a maximum-size common independent set in two matroids, as this algorithm also works for nonlinear matroids.

Theorem 43.4. *Given a set E of pairs of vectors in a linear space L, a maximum-size matching can be found in strongly polynomial time.*

Proof. The algorithm is a 'brute-force' polynomial-time algorithm, based on collecting many matchings and utilizing standard linear-algebraic operations, which can be performed in strongly polynomial time. Since we deal with subsets of a vector space, we can use $X + Y := \{x + y \mid x \in X, y \in Y\}$. For each $X \subseteq L$, span(X) is a subspace of L.

Throughout this proof, $\mathcal{B}$ will be a collection of matchings, all of the same size ν (say). Define:

(43.43) $\quad K_\mathcal{B} := \bigcap\{\text{span}(B) \mid B \in \mathcal{B}\}$ and $H_\mathcal{B} :=$ the hypergraph with vertex set E and edges all fundamental circuits of all $B \in \mathcal{B}$.

We say that we *improve* $\mathcal{B}$ if we find, in strongly polynomial time, either a matching B of size $\nu + 1$, or of size ν such that $K_\mathcal{B} \not\subseteq \text{span}(B)$, or of size ν such that $H_{\mathcal{B} \cup \{B\}}$ has fewer components than $H_\mathcal{B}$. So replacing $\mathcal{B}$ by $\{B\}$ if $|B| = \nu + 1$, and by $\mathcal{B} \cup \{B\}$ if $|B| = \nu$, we can have at most $2|E|$ improvements.

I. We first show (where a component is called *nontrivial* if it has more than one element):

(43.44) $\quad$ We can improve $\mathcal{B}$ if we have a union F of nontrivial components of $H_\mathcal{B}$, a matching $M \subseteq F$, and a $B \in \mathcal{B}$ such that $r(M \cup A) > |B \cap F| + |M|$, where $A := \text{span}(B \cap F) \cap K_\mathcal{B}$.

Here and below, $r(X \cup Y) := r(\bigcup X \cup Y)$ for $X \subseteq E$ and $Y \subseteq S$.

To see (43.44), apply the first applicable case of the following five cases, and then iterate. If Case 1 applies, we improve $\mathcal{B}$. In any of the other cases, we reset B or M or both, add the reset B to $\mathcal{B}$, and iterate with the reset B and M. The input condition given in (43.44) is maintained, as will be shown after describing the five cases.

Case 1: There is a $B' \in \mathcal{B}$ and an $e \in E$ such that $B' + e$ is a matching, or such that $C(B', e)$ intersects both F and $E \setminus F$, or such that $K_{\mathcal{B}} \not\subseteq \text{span}(B' - f + e)$ for some $f \in C(B', e)$. Output $B' + e$, B', or $B' - f + e$ (thus we improve $\mathcal{B}$).

Note that if Case 1 does not apply, then

(43.45) $\quad f \not\subseteq K_{\mathcal{B}}$ for each $f \in F$.

Indeed, as f is in a nontrivial component of $H_{\mathcal{B}}$, f is contained in some fundamental circuit $C(B', e)$ for some $B' \in \mathcal{B}$. As Case 1 does not apply, we know $K_{\mathcal{B}} \subseteq \text{span}(B' - f + e)$. Hence, if $f \subseteq K_{\mathcal{B}}$, then $f \subseteq \text{span}(B' - f + e)$, hence $2\nu + 1 = r(B' + e) = r(B' - f + e) = 2\nu$, a contradiction.

Case 2: There is an $e \in F$ such that $M + e$ is a matching and $r((M \cup A) + e) \geq r(M \cup A) + 1$. Reset $M := M + e$.

Case 3: $\text{span}(M) \not\subseteq \text{span}(B)$. Choose $e \in M$ with $e \not\subseteq \text{span}(B)$, choose $f \in C(B, e) \setminus M$, and reset $B := B - f + e$.

Case 4: There is an $e \in F$ such that $e \not\subseteq \text{span}(B)$ and $C(B, e) \neq C(M, e)$. (Note: $e \not\subseteq \text{span}(M) + A$, since $\text{span}(M) + A \subseteq \text{span}(B)$ (as Case 3 does not apply). So, as Case 2 does not apply, $M + e$ is not a matching. Hence $C(M, E)$ is defined.)

Choose $f \in C(B, e) \setminus (M + e)$ and $g \in C(M, e) \setminus (B + e)$, and reset $B := B - f + e$ and $M := M - g + e$.

Case 5. Choose $B' \in \mathcal{B}$ with $\text{span}(M \triangle (B \cap F)) \not\subseteq \text{span}(B')$ and with $|B \cap B'|$ maximal. (This is possible, since $M \neq B \cap F$, since $r((B \cap F) \cup A) = r(B \cap F) = 2|B \cap F|$ and $r(M \cup A) > |B \cap F| + |M|$ by assumption. As $M \triangle (B \cap F) \subseteq F$, such a B' exists, by (43.45).)

Choose $e \in B'$ with $e \not\subseteq \text{span}(B)$. (This is possible since $\text{span}(M \triangle (B \cap F)) \subseteq \text{span}(B)$, so $\text{span}(B) \neq \text{span}(B')$.)

Choose $f \in C(B, e) \setminus B'$. If $e \notin F$, reset $B := B - f + e$. If $e \in F$, reset $B := B - f + e$ and $M := M - f + e$. (Note that if $e \in F$, then $C(B, e) = C(M, e)$ as Case 4 does not apply.)

Running time. The number of iterations is polynomially bounded, since in each iteration (except the last, where Case 1 applies), the vector $(|M|, |M \cap B|, |B \cap B'|)$ increases lexicographically. Here it is important to note that Case 5 does not modify the set $M \triangle (B \cap F)$, and increases the intersection of this set with B'.

We finally prove that the resettings in Cases 2-5 indeed maintain the condition given in (43.44). Let $\widetilde{B}$, $\widetilde{M}$, and $\widetilde{A}$ denote B, M, and A after resetting (taking $\widetilde{B}$ or $\widetilde{M}$ equal to B or M if they are not reset). We must show

(43.46) $\quad r(\widetilde{M} \cup \widetilde{A}) > |\widetilde{B} \cap F| + |\widetilde{M}|$.

Note that, as Case 1 does not apply, $|\widetilde{B} \cap F| = |B \cap F|$.

We first show:

(43.47) $A \subseteq \widetilde{A}$.

This is equivalent to (since K_B does not change, as Case 1 does not apply):

(43.48) $A \subseteq \text{span}(\widetilde{B} \cap F)$.

This is trivial if $\widetilde{B} \cap F = B \cap F$. So we can assume that $\widetilde{B} \cap F \neq B \cap F$. Hence $\widetilde{B} = B - f + e$ for some $e, f \in F$. Then (43.48) follows from

(43.49) $r((\widetilde{B} \cap F) \cup A) \leq r((\widetilde{B} \cap F) \cup A + f) - 1 = r((B \cap F) \cup A + e) - 1$
$= r((B \cap F) + e) - 1 \leq r(B \cap F) = 2|B \cap F| = 2|\widetilde{B} \cap F|$
$= r(\widetilde{B} \cap F)$.

Here the first inequality holds as $f \not\subseteq \text{span}(\widetilde{B} \cap F) + A$, since $f \not\subseteq \text{span}(\widetilde{B})$ and $\text{span}(\widetilde{B} \cap F) + A \subseteq \text{span}(\widetilde{B})$. (We use that $A \subseteq K_B \subseteq \text{span}(\widetilde{B})$, as Case 1 does not apply.) The last inequality holds as $(B \cap F) + e$ is not a matching, since it contains $C(B, e)$ (as Case 1 does not apply). This shows (43.48), and hence (43.47).

We finally show (43.46). In Case 2, we have $\widetilde{B} = B$, $\widetilde{M} = M + e$, and $\widetilde{A} = A$, and hence

(43.50) $r(\widetilde{M} \cup \widetilde{A}) = r((M \cup A) + e) \geq r(M \cup A) + 1 > |B \cap F| + |M| + 1$
$= |\widetilde{B} \cap F| + |\widetilde{M}|$,

as required.

In Case 3, (43.47) implies (as $\widetilde{M} = M$) that $r(\widetilde{M} \cup \widetilde{A}) \geq r(M \cup A) > |B \cap F| + |M| = |\widetilde{B} \cap F| + |\widetilde{M}|$.

In Cases 4 and 5 we have $\widetilde{M} = M - g + e$ (possible $g = f$). Then

(43.51) $r(\widetilde{M} \cup \widetilde{A}) \geq r(\widetilde{M} \cup A) \geq r((\widetilde{M} \cup A) + g) - 1 = r((M \cup A) + e) - 1$
$\geq r(M \cup A) > |B \cap F| + |M| = |\widetilde{B} \cap F| + |\widetilde{M}|$.

The first inequality follows from (43.47). Next, $e \not\subseteq \text{span}(M \cup A)$ (as $e \not\subseteq \text{span}(B)$ and as $A \subseteq \text{span}(B)$ and $\text{span}(M) \subseteq \text{span}(B)$, since Case 3 does not apply). This gives the third inequality. To see the second inequality, suppose it does not hold. Then $\widetilde{M} + g$ is a matching, hence $M + e$ is a matching. Therefore, as Case 2 does not apply, $r(M \cup A + e) = r(M \cup A)$, contradicting the fact that $e \not\subseteq \text{span}(M \cup A)$.

II. Secondly,

(43.52)   we can improve $\mathcal{B}$ if $K_B = \{0\}$, H_B is connected, and $\nu < \lfloor \frac{1}{2} r(E) \rfloor$.

(In this case, $\mathcal{B}$ can only be improved by finding a matching larger than B.)

The algorithm follows the framework of parts I and II in the proof of Theorem 43.1. Again, the algorithm iteratively applies the first applicable case. Call two circuits C_1, C_2 *far* if there exist $B \in \mathcal{B}$ and $e, g \in E$ with $r(B + e + g) = 2\nu + 2$ and $C_1 = C(B, e)$ and $C_2 = C(B, g)$.

Case 1: There exists a $B \in \mathcal{B}$ and $e \in E$ such that $B + e$ is a matching of size $\nu + 1$. Output $B + e$.

Case 2: There exist far circuits C_1 and C_2 with $C_1 \cap C_2 \neq \emptyset$. We will create a matching of size $\nu + 1$.

Let $C_1 = C(B, e)$ and $C_2 = C(B, g)$ for some $B \in \mathcal{B}$ with $e, g \in E$ and $r(B+e+g) = 2\nu+2$. Define $D := C_1 \cup C_2$. As is shown in the proof of Corollary 43.2a, there is a $p \neq 0$ contained in span(C) for each circuit $C \subseteq D$. Since $K_B = \{0\}$, there is a $B' \in \mathcal{B}$ with $p \notin \text{span}(B')$. Now $r(B' + p) = 2\nu + 1 < r(B + e + g)$, and hence $f \not\subseteq \text{span}(B' + p)$ for some $f \in B + e + g$. Then (as $B' + f$ is not a matching, since Case 1 does not apply) $p \notin \text{span}(B' + f)$, and therefore $p \notin \text{span}(C(B', f))$. So $C(B', f)$ is not contained in $B + e + g$. Choose $h \in C(B', f) \setminus (B + e + g)$. Hence, resetting B' to $B' - h + f$ increases $|B' \cap (B + e + g)|$. So iterating this, we finally obtain a matching larger than ν.

Case 3. We show that we can create a matching of size $\nu + 1$, or make that Case 1 or 2 applies.

Far circuits exist, since for any base B, there exist $e, g \in E$ with $r(B+e+g) = 2\nu+2$, since $r(E) \geq r(B) + 2$. Choose far circuits C, D that are closest[34] together in the hypergraph H_B. Assuming that Case 2 does not apply, we know $C \cap D = \emptyset$. Hence there is an intermediate set C' on a shortest path from C to D. Let $C = C(B, e)$, $D = C(B, g)$, and $C' = C(B', f)$ for $B, B' \in \mathcal{B}$ and $e, f, g \in E$ with $r(B+e+g) = 2\nu+2$. We choose B' such that $|B' \cap (B+e+g)|$ is maximal. Choose $h \in B + e + g$ with $h \not\subseteq \text{span}(B' + f)$.

$C(B', h)$ and $C(B', f)$ are disjoint, since otherwise we can apply Case 2. Moreover,

(43.53) $\quad C(B', h) \not\subseteq B + e + g$,

Otherwise, $C(B', h) = C(B, e)$ or $C(B', h) = C(B, g)$. Hence C' and C or D are far, contradicting the minimality of the distance of C and D.

Hence we have (43.53). Choose $i \in C(B', h) \setminus (B + e + g)$ and add $B'' := B' - i + h$ to $\mathcal{B}$. Iterate Case 3 with B' replaced by B'' (note that $C' = C(B'', f)$). As $|B'' \cap B| > |B' \cap B|$, the number of iterations of Case 3 is at most ν.

III. Combination of the previous two algorithms implies:

(43.54) $\quad$ we can improve $\mathcal{B}$ if $K_B = \{0\}$ and $\nu < \nu(E)$.

As $\nu < \nu(E)$, there is a component F of H_B with $|B \cap F| < \nu(F) \leq \lfloor \frac{1}{2} r(F) \rfloor$ for at least one $B \in \mathcal{B}$. If there exist $B, B' \in \mathcal{B}$ with $|B \cap F| < |B' \cap F|$, set

---

[34] Here the distance of fundamental circuits C, D is the minimum length of a path connecting C and D. A *path* connecting C and D is a sequence $C = C_0, \ldots, C_k = D$ of fundamental circuits such that $C_{i-1} \cap C_i \neq \emptyset$ for $i = 1, \ldots, k$. Its length is k.

$M := B' \cap F$. Otherwise (that is, if $|B \cap F| = |B' \cap F|$ for all $B' \in \mathcal{B}$), apply (43.52) to $\mathcal{B}' := \{B \cap F \mid B \in \mathcal{B}\}$ and $B \cap F$ for any $B \in \mathcal{B}$, to obtain a matching $M \subseteq F$ with $|M| = |B \cap F| + 1$.

Now applying (43.44) to $\mathcal{B}$, F, B, and M improves $\mathcal{B}$. (Since $A \subseteq K_\mathcal{B}$, we have $A = \{\mathbf{0}\}$, and hence $r(M \cup A) = r(M) = 2|M| > |M| + |B \cap F|$.)

IV. Finally:

(43.55)   We can improve $\mathcal{B}$ if $\mathcal{B} \neq \emptyset$ and $\nu < \nu(E)$.

Define F to be the union of all fundamental circuits of the $B \in \mathcal{B}$. This implies

(43.56) $\mathrm{span}(E \setminus F) \subseteq K_\mathcal{B}$.

If there exist $B, B' \in \mathcal{B}$ with $|B \cap F| < |B' \cap F|$, then applying (43.44) to B and $M := B' \cap F$ improves $\mathcal{B}$. So we can assume that $|B \cap F| = \beta$ for all $B \in \mathcal{B}$. Choose $B_0 \in \mathcal{B}$ with $r(\mathrm{span}(B_0 \cap F) \cap K_\mathcal{B})$ maximal. Define

(43.57) $A := \mathrm{span}(B_0 \cap F) \cap K_\mathcal{B}$, $E' := F/A$, and $\nu' := \beta - r(A)$.

For each $B \in \mathcal{B}$ there is a matching M_B in $(B \cap F)/A$ of size ν', since

(43.58) $\beta - r(\mathrm{span}(B \cap F) \cap A) \geq \beta - r(\mathrm{span}(B \cap F) \cap K_\mathcal{B})$
$\geq \beta - r(\mathrm{span}(B_0 \cap F) \cap K_\mathcal{B}) = \beta - r(A) = \nu'$.

Let $\mathcal{B}' := \{M_B \mid B \in \mathcal{B}\}$. Then $K_{\mathcal{B}'} = \{\mathbf{0}\}$, since

(43.59) $\bigcap_{B \in \mathcal{B}} \mathrm{span}(B \cap F) \subseteq \mathrm{span}(B_0 \cap F) \cap \bigcap_{B \in \mathcal{B}} \mathrm{span}(B)$
$= \mathrm{span}(B_0 \cap F) \cap K_\mathcal{B} = A$.

Since $\nu(E) > \nu$ we have $\nu(E') > \nu'$. Indeed, let B' be a matching in E of size $\nu + 1$. Then B'/A contains a matching of size $|B' \cap F| - r(\mathrm{span}(B' \cap F) \cap A)$. Hence

(43.60) $\nu(E') \geq |B' \cap F| - r(\mathrm{span}(B' \cap F) \cap A)$
$= |B'| - |B' \setminus F| - r(\mathrm{span}(B' \cap F) \cap A)$
$= |B'| - \tfrac{1}{2}(r(B' \setminus F) + r(\mathrm{span}(B' \cap F) \cap A)) - \tfrac{1}{2}r(\mathrm{span}(B' \cap F) \cap A)$
$\geq |B'| - \tfrac{1}{2}r(K_\mathcal{B}) - \tfrac{1}{2}r(A) > |B_0| - \tfrac{1}{2}r(K_\mathcal{B}) - \tfrac{1}{2}r(A)$
$\geq |B_0| - \tfrac{1}{2}(r(B_0 \setminus F) + r(A)) - \tfrac{1}{2}r(A) = |B_0 \cap F| - r(A) = \nu'$.

The second inequality holds as $\mathrm{span}(B' \setminus F)$ and $\mathrm{span}(B' \cap F) \cap A$ are subspaces of $K_\mathcal{B}$ having intersection $\{\mathbf{0}\}$ (since B is a matching and as (43.56) holds). The last inequality follows from

(43.61) $r(K_\mathcal{B}) = r(\mathrm{span}(B_0) \cap K_\mathcal{B})$
$= r((\mathrm{span}(B_0 \setminus F) + \mathrm{span}(B_0 \cap F)) \cap K_\mathcal{B})$
$= r(\mathrm{span}(B_0 \setminus F) + (\mathrm{span}(B_0 \cap F) \cap K_\mathcal{B}))$
$= r(\mathrm{span}(B_0 \setminus F) + r(\mathrm{span}(B_0 \cap F) \cap K_\mathcal{B}))$.

Here we use that

(43.62) $\quad(\text{span}(B_0 \setminus F) + \text{span}(B_0 \cap F)) \cap K_B$
$= \text{span}(B_0 \setminus F) + (\text{span}(B_0 \cap F) \cap K_B),$

which holds since if $x \in \text{span}(B_0 \setminus F)$ and $y \in \text{span}(B_0 \cap F)$ with $x+y \in K_B$, then $y \in K_B$ (since $x \in \text{span}(B_0 \setminus F) \subseteq K_B$ by (43.56)).

Now applying (43.54) repeatedly to $\mathcal{B}'$, we finally find a matching M' in E' with $|M'| = \nu' + 1$. It corresponds to a matching M in F with

(43.63) $\quad r(M \cup A) = 2|M| + r(A) = |M| + \nu' + 1 + r(A) = |B_0 \cap F| + |M| + 1.$

Then applying (43.44) improves $\mathcal{B}$. ∎

The proof also yields an alternative proof of Theorem 43.2.

While most of the matroids we meet in daily life are linear, it might yet be interesting to extend the algorithm to the class of algebraic matroids. As Dress and Lovász [1987] remark, this requires the development of algorithmic techniques for algebraic matroids, for instance, for testing algebraic independence, and for finding a point p in the intersection of certain flats. If such techniques are available, pursuing the layout of the above algorithm for linear matroids might yield a polynomial-time algorithm for algebraic matroids.

An augmenting path algorithm for linear matroid matching, of complexity $O(n^3 m)$ (where $n := \text{rank}$, $m := |S|$) was given by Stallmann and Gabow [1984] and Gabow and Stallmann [1986] and an $O(n^4 m)$-time algorithm (by solving a sequence of matroid intersection algorithms) by Orlin and Vande Vate [1990] (these bounds can be improved to $O(n^{2.376} m)$ and $O(n^{3.376} m)$, respectively, with fast matrix multiplication).

43.9. Matroid matching is not polynomial-time solvable in general

Theorem 43.2 characterizes the matroid matching problem for algebraic matroids, and one is challenged to extend this to general matroids. A main objection to do this in a direct way is that in Theorem 43.2 a line of E may intersect the flat F in a point not contained in the original matroid. So we need to extend the matroid in some way, which is quite natural for linear matroids, but, as Lovász remarks, 'in general, there seems to be no hope to extend the original matroid so as to achieve the validity of [Theorem 43.2]. The possibility of "simulating" the flat F inside the matroid seems to be a difficult, and probably not only technical, question.'

Jensen and Korte [1982] and Lovász [1981] showed that, for matroids in general, the matroid matching problem is not solvable in polynomial time, if the matroid is given by an independence testing oracle (an oracle telling if a given set is independent or not). The construction in both papers is as follows.

Let $\nu \in \mathbb{Z}$, let S be a set, and let E be a partition of S into pairs. Let M be the matroid on S of rank 2ν, where $T \subseteq S$ is independent if and only if $|T| \leq 2\nu - 1$, or $|T| = 2\nu$ and T is not the union of ν pairs in E.

For each subset F of E of size ν, let M_F be the matroid on S obtained from M by adding $\bigcup F$ as independent subset.

It is easy to check that M and each of the M_F are matroids, and that E has no matroid matching of size ν with respect to M, while F is the unique matroid matching of size ν in M_F.

Suppose now that we want to find the maximum size of a matroid matching in a matroid, and that we know that the matroid is equal to M or to M_F for some ν-element $F \subseteq E$. Then we must ask the oracle for the independence of $\bigcup F$ for each ν-element subset F of E, in order to know if there exists a matroid matching of size ν. This takes exponential time.

This example shows that the matroid matching problem even does not belong to (oracle) co-NP, since any certificate that the matching number is at most $\nu - 1$, needs the oracle output that $\bigcup F$ is dependent, for all ν-element subsets F of E.

The example can be easily adapted to remove the oracle, and to obtain a proper problem in NP that is NP-complete. Let G be an undirected graph with vertex set V and let $\nu \in \mathbb{Z}_+$. For each vertex v of G, let p_v be a pair of elements, such that $p_u \cap p_v = \emptyset$ if $u \neq v$. Let $S := \bigcup_{v \in V} p_v$ and $E := \{p_v | v \in V\}$. So E is a partition of S into pairs. Define a matroid on S by extending the matroid M above by an independent set

(43.64) $\quad I := \bigcup_{v \in C} p_v$

for each clique C of G with $|C| = \nu$. Then E contains a matroid matching of size ν if and only if G has a clique of size ν. As the maximum-size clique problem is NP-complete, also the matroid matching problem for such matroids is NP-complete.

43.10. Further results and notes

43.10a. Optimal path-matching

Cunningham and Geelen [1996,1997] gave the following generalization of nonbipartite matching and matroid intersection.

Let $G = (V, E)$ be an undirected graph, let S_1 and S_2 be two disjoint stable subsets of V, and let $M_1 = (S_1, \mathcal{I}_1)$ and $M_2 = (S_2, \mathcal{I}_2)$ be matroids, with rank functions r_1 and r_2, such that $r_1(S_2) = r_2(S_2) =: \rho$. Define $R := V \setminus (S_1 \cup S_2)$. A *basic path-matching* is a collection of ρ vertex-disjoint $B_1 - B_2$ paths, each having all internal vertices in R, where B_1 and B_2 are bases of M_1 and M_2 respectively, together with a perfect matching on the vertices of R not covered by these paths.

If $R = V$, a basic path-matching is just a perfect matching. If $R = \emptyset$ and E consists of disjoint edges linking S_1 and S_2, then a basic path-matching corresponds to a common base.

Geelen and Cunningham showed that a basic path-matching exists if and only if for each $U_1 \subseteq S_1 \cup R$ and $U_2 \subseteq S_2 \cup R$ such that there is no edge connecting two sets among $U_1 \cap U_2$, $U_1 \setminus U_2$, $U_2 \setminus U_1$, one has

(43.65) $\quad r_1(S_1 \setminus U_1) + r_2(S_2 \setminus U_2) + |R \setminus (U_1 \cup U_2)| \geq \rho + o(G[U_1 \cap U_2])$,

where $o(H)$ is the number of odd components of a graph H. Moreover, they gave a polynomial-time algorithm to decide whether there exists a basic path-matching.

More generally, they introduced the concept of an *independent path-matching*, which is a set F of edges such that each nonsingleton component of the graph (V, F) is an $S_1 \cup R - S_2 \cup R$ path all of whose internal vertices are in R, and such that the vertices in S_i covered by the paths is independent in M_i ($i = 1, 2$). The corresponding *independent path-matching vector* is the vector $x \in \mathbb{Z}_+^E$ with $x(e) = 0$ if $e \notin F$, $x(e) = 2$ if $e \in F$ forms a component of (V, F) with both ends of e in R, and $x(e) = 1$ otherwise.

Geelen and Cunningham showed that the convex hull of the independent path-matching vectors is determined by:

(43.66) $\quad \begin{array}{ll} x_e \geq 0 & \text{for } e \in E, \\ x(\delta(v)) \leq 2 & \text{for } v \in R, \\ x(E[U]) \leq |U \cap R| & \text{for } U \subseteq V \text{ with } U \cap S_1 = \emptyset \text{ or } U \cap S_2 = \emptyset, \\ x(E[U]) \leq |U| - 1 & \text{for } U \subseteq R, \\ x(\delta(U)) \leq r_i(U) & \text{for } U \subseteq S_i \text{ and } i = 1, 2, \end{array}$

and that this system is TDI. It implies that the maximum of $\mathbf{1}^\mathsf{T} x$ over independent path-matching vectors is equal to the minimum of

(43.67) $\quad r_1(S_1 \setminus U_1) + r_2(S_2 \setminus U_2) + |R \setminus (U_1 \cup U_2)| + |R| - o(G[U_1 \cap U_2])$

over all $U_i \subseteq S_i \cup R$ ($i = 1, 2$) such that there is no edge connecting two sets among $U_1 \cap U_2$, $U_1 \setminus U_2$, $U_2 \setminus U_1$. (A simplified proof of this was given by Frank and Szegő [2002].)

Cunningham and Geelen argue that the set of inequalities (43.66) can be checked in polynomial time, implying (with the ellipsoid method) that, for any weight function w, an independent path-matching vector x maximizing $w^\mathsf{T} x$ can be found in strongly polynomial time. A combinatorial algorithm for the unweighted version was given by Spille and Weismantel [2002a,2002b].

For a survey, see Cunningham [2002].

43.10b. Further notes

Hochstättler and Kern [1989] showed that condition (43.13) is implied by the following:

(43.68) for any three flats A, B, C with

$$r(A \cup C) - r(A) = r(B \cup C) - r(B) = r(A \cup B \cup C) - r(A \cup B),$$

one has

$$r(\mathrm{span}(A \cup C) \cap \mathrm{span}(B \cup C)) - r(A \cap B) = r(A \cup C) - r(A).$$

Matroids with this property are called *pseudomodular* by Björner and Lovász [1987], who proved that full linear matroids (infinite matroids determined by linear independence of a linear space), full algebraic matroids (infinite matroids determined by algebraic independence of a field extension of a field), and full graphic matroids (cycle matroids of a complete graph) are pseudomodular. See also Lindström [1988], Dress, Hochstättler, and Kern [1994], and Tan [1997].

A randomized parallel algorithm for linear matroid matching was given by Narayanan, Saran, and Vazirani [1992,1994]. Stallmann and Gabow [1984] gave an algorithm for graphic matroid matching with running time $O(n^2 m)$, which was improved by Gabow and Stallmann [1985] to $O(nm \log^6 n)$. Tong, Lawler, and Vazirani [1984] found a polynomial-time algorithm for *weighted* matroid matching for gammoids (by reduction to weighted matching). Structural properties of matroid matching, including an Edmonds-Gallai type decomposition, were given by Vande Vate [1992], which paper also studied the matroid matching polytope and a fractional relaxation of it.

The matroid matching problem generalizes the *matchoid problem* of J. Edmonds (cf. Jenkyns [1974]): given a graph $G = (V, E)$ and a matroid $M_v = (\delta(v), \mathcal{I}_v)$ for each v in V, what is the maximum number of edges such that the restriction to $\delta(v)$ forms an independent set in M_v, for each v in V?

Chapter 44

Submodular functions and polymatroids

In this chapter we describe some of the basic properties of a second main object of the present part, the submodular function. Each submodular function gives a polymatroid, which is a generalization of the independent set polytope of a matroid. We prove as a main result the theorem of Edmonds [1970b] that the vertices of a polymatroid are integer if and only if the associated submodular function is integer.

44.1. Submodular functions and polymatroids

Let f be a *set function* on a set S, that is, a function defined on the collection $\mathcal{P}(S)$ of all subsets of S. The function f is called *submodular* if

(44.1) $\quad f(T) + f(U) \geq f(T \cap U) + f(T \cup U)$

for all subsets T, U of S. Similarly, f is called *supermodular* if $-f$ is submodular, i.e., if f satisfies (44.1) with the opposite inequality sign. f is *modular* if f is both submodular and supermodular, i.e., if f satisfies (44.1) with equality.

A set function f on S is called *nondecreasing* if $f(T) \leq f(U)$ whenever $T \subseteq U \subseteq S$, and *nonincreasing* if $f(T) \geq f(U)$ whenever $T \subseteq U \subseteq S$.

As usual, denote for each function $w : S \to \mathbb{R}$ and for each subset U of S,

(44.2) $\quad w(U) := \sum_{s \in U} w(s).$

So w may be considered also as a set function on S, and one easily sees that w is modular, and that each modular set function f on S with $f(\emptyset) = 0$ may be obtained in this way. (More generally, each modular set function f on S satisfies $f(U) = w(U) + \gamma$ (for $U \subseteq S$), for some unique function $w : S \to \mathbb{R}$ and some unique real number γ.)

In a sense, submodularity is the discrete analogue of convexity. If we define, for any $f : \mathcal{P}(S) \to \mathbb{R}$ and any $x \in S$, a function $\delta f_x : \mathcal{P}(S) \to \mathbb{R}$ by: $\delta f_x(T) := f(T \cup \{x\}) - f(T)$, then f is submodular if and only if δf_x is nonincreasing for each $x \in S$.

In other words:

Section 44.1. Submodular functions and polymatroids 767

Theorem 44.1. *A set function f on S is submodular if and only if*

(44.3) $\qquad f(U \cup \{s\}) + f(U \cup \{t\}) \geq f(U) + f(U \cup \{s,t\})$

for each $U \subseteq S$ and distinct $s, t \in S \setminus U$.

Proof. Necessity being trivial, we show sufficiency. We prove (44.1) by induction on $|T \triangle U|$, the case $|T \triangle U| \leq 2$ being trivial (if $T \subseteq U$ or $U \subseteq T$) or being implied by (44.3). If $|T \triangle U| \geq 3$, we may assume by symmetry that $|T \setminus U| \geq 2$. Choose $t \in T \setminus U$. Then, by induction,

(44.4) $\qquad f(T \cup U) - f(T) \leq f((T \setminus \{t\}) \cup U) - f(T \setminus \{t\}) \leq f(U) - f(T \cap U),$

(as $|T \triangle ((T \setminus \{t\}) \cup U)| < |T \triangle U|$ and $|(T \setminus \{t\}) \triangle U| < |T \triangle U|$). This shows (44.1). ∎

Define two polyhedra associated with a set function f on S:

(44.5) $\qquad P_f := \{x \in \mathbb{R}^S \mid x \geq \mathbf{0}, x(U) \leq f(U) \text{ for each } U \subseteq S\},$
$\qquad\quad EP_f := \{x \in \mathbb{R}^S \mid x(U) \leq f(U) \text{ for each } U \subseteq S\}.$

Note that P_f is nonempty if and only if $f \geq \mathbf{0}$, and that EP_f is nonempty if and only if $f(\emptyset) \geq 0$.

If f is a submodular function, then P_f is called the *polymatroid associated with f*, and EP_f the *extended polymatroid associated with f*. A polyhedron is called an (extended) polymatroid if it is the (extended) polymatroid associated with some submodular function. A polymatroid is bounded (since $0 \leq x_s \leq f(\{s\})$ for each $s \in S$), and hence is a polytope.

The following observation presents a basic technique in proofs for submodular functions, which we often use without further reference:

Theorem 44.2. *Let f be a submodular set function on S and let $x \in EP_f$. Then the collection of sets $U \subseteq S$ satisfying $x(U) = f(U)$ is closed under taking unions and intersections.*

Proof. Suppose $x(T) = f(T)$ and $x(U) = f(U)$. Then

(44.6) $\qquad f(T) + f(U) \geq f(T \cap U) + f(T \cup U) \geq x(T \cap U) + x(T \cup U)$
$\qquad\qquad = x(T) + x(U) = f(T) + f(U),$

implying that equality holds throughout. So $x(T \cap U) = f(T \cap U)$ and $x(T \cup U) = f(T \cup U)$. ∎

A vector x in EP_f (or in P_f) is called a *base vector* of EP_f (or of P_f) if $x(S) = f(S)$. A *base vector* of f is a base vector of EP_f. The set of all base vectors of f is called the *base polytope* of EP_f or of f. It is a face of EP_f, and denoted by B_f. So

(44.7) $\qquad B_f = \{x \in \mathbb{R}^S \mid x(U) \leq f(U) \text{ for all } U \subseteq S, x(S) = f(S)\}.$

(It is a polytope, since $x_s = x(S) - x(S \setminus \{s\}) \geq f(S) - f(S \setminus \{s\})$ for each $s \in S$.)

Let f be a submodular set function on S and let $a \in \mathbb{R}^S$. Define the set function $f|a$ on S by

(44.8) $\quad (f|a)(U) := \min_{T \subseteq U}(f(T) + a(U \setminus T))$

for $U \subseteq S$. It is easy to check that $f|a$ again is submodular and that

(44.9) $\quad EP_{f|a} = \{x \in EP_f \mid x \leq a\}$ and $P_{f|a} = \{x \in P_f \mid x \leq a\}$.

It follows that if P is an (extended) polymatroid, then also the set $P \cap \{x \mid x \leq a\}$ is an (extended) polymatroid, for any vector a. In fact, as Lovász [1983c] observed, if $f(\emptyset) = 0$, then $f|a$ is the unique largest submodular function f' satisfying $f'(\emptyset) = 0$, $f' \leq f$, and $f'(U) \leq a(U)$ for each $U \subseteq V$.

44.1a. Examples

Matroids. Let $M = (S, \mathcal{I})$ be a matroid. Then the rank function r of M is submodular and nondecreasing. In Theorem 39.8 we saw that a set function r on S is the rank function of a matroid if and only if r is nonnegative, integer, nondecreasing and submodular with $r(U) \leq |U|$ for all $U \subseteq S$. (This last condition may be replaced by: $r(\emptyset) = 0$ and $r(\{s\}) \leq 1$ for each s in S.) Then the polymatroid P_r associated with r is equal to the independent set polytope of M (by Corollary 40.2b).

A generalization is obtained by partitioning S into sets $S_1, \ldots, S_k$, and defining

(44.10) $\quad f(J) := r(\bigcup_{i \in J} S_i)$

for $J \subseteq \{1, \ldots, k\}$. It is not difficult to show that each integer nondecreasing submodular function f with $f(\emptyset) = 0$ can be constructed in this way (see Section 44.6b).

As another generalization, if $w : S \to \mathbb{R}_+$, define $f(U)$ to be the maximum of $w(I)$ over $I \in \mathcal{I}$ with $I \subseteq U$. Then f is submodular. (To see this, write $w = \lambda_1 \chi^{T_1} + \cdots + \lambda_n \chi^{T_n}$, with $\emptyset \neq T_1 \subset T_2 \subset \cdots \subset T_n \subseteq S$. Then by (40.3), $f(U) = \sum_{i=1}^n \lambda_i r(U \cap T_i)$, implying that f is submodular.)

For more on the relation between submodular functions and matroids, see Sections 44.6a and 44.6b.

Matroid intersection. Let $M_1 = (S, \mathcal{I}_1)$ and $M_2 = (S, \mathcal{I}_2)$ be matroids, with rank functions r_1 and r_2 respectively. Then the function f given by

(44.11) $\quad f(U) := r_1(U) + r_2(S \setminus U)$

for $U \subseteq S$, is submodular. By the matroid intersection theorem (Theorem 41.1), the minimum value of f is equal to the maximum size of a common independent set.

Set unions. Let $T_1, \ldots, T_n$ be subsets of a finite set T and let $S = \{1, \ldots, n\}$. Define

(44.12) $$f(U) := |\bigcup_{i \in U} T_i|$$

for $U \subseteq S$. Then f is nondecreasing and submodular. More generally, for $w : T \to \mathbb{R}_+$, the function f defined by

(44.13) $$f(U) := w(\bigcup_{i \in U} T_i)$$

for $U \subseteq S$, is nondecreasing and submodular.

More generally, for any nondecreasing submodular set function g on T, the function f defined by

(44.14) $$f(U) := g(\bigcup_{i \in U} T_i)$$

for $U \subseteq S$, again is nondecreasing and submodular.

Let $G = (V, E)$ be the bipartite graph corresponding to $T_1, \ldots, T_n$. That is, G has colour classes S and T, and $s \in S$ and $t \in T$ are adjacent if and only if $t \in T_s$. Then we have: $x \in P_f$ if and only if there exist $z \in P_g$ and $y : E \to \mathbb{Z}_+$ such that

(44.15) $\quad y(\delta(v)) = x(v) \quad$ for all $v \in S$,
$\quad\quad\quad\ \ y(\delta(v)) = z(v) \quad$ for all $v \in T$.

So y may be considered as an 'assignment' of a 'supply' z to a 'demand' x. If g and x are integer we can take also y and z integer.

Directed graph cut functions. Let $D = (V, A)$ be a directed graph and let $c : A \to \mathbb{R}_+$ be a 'capacity' function on A. Define

(44.16) $$f(U) := c(\delta^{\text{out}}(U))$$

for $U \subseteq V$ (where $\delta^{\text{out}}(U)$ denotes the set of arcs leaving U). Then f is submodular (but in general not nondecreasing). A function f arising in this way is called a *cut function*.

Hypergraph cut functions. Let $(V, \mathcal{E})$ be a hypergraph. For $U \subseteq V$, let $f(U)$ be the number of edges $E \in \mathcal{E}$ split by U (that is, with both $E \cap U$ and $E \setminus U$ nonempty). Then f is submodular.

Directed hypergraph cut functions. Let V be a finite set and let $(E_1, F_1), \ldots, (E_m, F_m)$ be pairs of subsets of V. For $U \subseteq V$, let $f(U)$ be the number of indices i with $U \cap E_i \neq \emptyset$ and $F_i \not\subseteq U$. Then f is submodular. (In proving this, we can assume $m = 1$, since any sum of submodular functions is submodular again.)

More generally, we can choose $c_1, \ldots, c_m \in \mathbb{R}_+$ and define

(44.17) $$f(U) = \sum(c_i \mid U \cap E_i \neq \emptyset, F_i \not\subseteq U)$$

for $U \subseteq V$. Again, f is submodular. This generalizes the previous two examples (where $E_i = F_i$ for each i or $|E_i| = |F_i| = 1$ for each i).

Maximal element. Let V be a finite set and let $h : V \to \mathbb{R}$. For nonempty $U \subseteq V$, define

(44.18) $$f(U) := \max\{h(u) \mid u \in U\},$$

and define $f(\emptyset)$ to be the minimum of $h(v)$ over $v \in V$. Then f is submodular.

Subtree diameter. Let $G = (V, E)$ be a forest (a graph without circuits), and for each $X \subseteq E$ define

(44.19) $\qquad f(X) := \sum_{K} \text{diameter}(K),$

where K ranges over the components of the graph (V, X). Here diameter(K) is the length of a longest path in K. Then f is submodular (Tamir [1993]); that is:

(44.20) $\qquad f(X) + f(Y) \geq f(X \cap Y) + f(X \cup Y)$

for $X, Y \subseteq E$.

To see this, denote, for any $X \subseteq E$, the set of vertices covered by X by VX. We first show (44.20) for $X, Y \subseteq E$ with (VX, X) and (VY, Y) connected and $VX \cap VY \neq \emptyset$. Note that in this case $X \cap Y$ and $X \cup Y$ give connected subgraphs again.

The proof of (44.20) is based on the fact that for all $s, t, u, v \in V$ one has:

(44.21) $\qquad \text{dist}(s, u) + \text{dist}(t, v) \geq \text{dist}(s, t) + \text{dist}(u, v)$
$\qquad$ or $\text{dist}(t, u) + \text{dist}(s, v) \geq \text{dist}(s, t) + \text{dist}(u, v),$

where dist denotes the distance in G.

To prove (44.20), let P and Q be longest paths in $X \cap Y$ and $X \cup Y$ respectively. If EQ is contained in X or in Y, then (44.20) follows, since P is contained in X and in Y. So we can assume that EQ is contained neither in X nor in Y. Let Q have ends u, v, with $u \in VX$ and $v \in VY$. Let P have ends s, t. So $s, t, u \in VX$ and $s, t, v \in VY$. Hence (44.21) implies (44.20).

We now derive (44.20) for all $X, Y \subseteq E$. Let $\mathcal{X}$ and $\mathcal{Y}$ be the collections of edge sets of the components of (V, X) and of (V, Y) respectively. Let $\mathcal{F}$ be the family made by the union of $\mathcal{X}$ and $\mathcal{Y}$, taking the sets in $\mathcal{X} \cap \mathcal{Y}$ twice. Then

(44.22) $\qquad f(X) + f(Y) \geq \sum_{Z \in \mathcal{F}} f(Z).$

We now modify $\mathcal{F}$ iteratively as follows. If $Z, Z' \in \mathcal{F}$, $Z \not\subseteq Z' \not\subseteq Z$, and $VZ \cap VZ' \neq \emptyset$, we replace Z, Z' by $Z \cap Z'$ and $Z \cup Z'$. By (44.20), (44.22) is maintained. By Theorem 2.1, these iterations stop. We delete the empty sets in the final $\mathcal{F}$.

Then the inclusionwise maximal sets in $\mathcal{F}$ have union equal to $X \cup Y$ and form the nonempty edge sets of the components of $(V, X \cup Y)$. Similarly, the inclusionwise minimal sets in $\mathcal{F}$ form the nonempty edge sets of the components of $(V, X \cap Y)$. So

(44.23) $\qquad \sum_{Z \in \mathcal{F}} f(Z) = f(X \cap Y) + f(X \cup Y),$

and we have (44.20).

Further examples. Choquet [1951,1955] showed that the classical Newtonian capacity in $\mathbb{R}^3$ is submodular. Examples of submodular functions based on probability are given by Fujishige [1978b] and Han [1979], and other examples by Lovász [1983c].

44.2. Optimization over polymatroids by the greedy method

Edmonds [1970b] showed that one can optimize a linear function $w^\mathsf{T} x$ over an (extended) polymatroid by an extension of the greedy algorithm. The submodular set function f on S is given by a *value giving oracle*, that is, by an oracle that returns $f(U)$ for any $U \subseteq S$.

Let f be a submodular set function on S, and suppose that we want to maximize $w^\mathsf{T} x$ over EP_f, for some $w : S \to \mathbb{R}$. We can assume that $EP_f \neq \emptyset$, that is $f(\emptyset) \geq 0$, and hence that $f(\emptyset) = 0$ (since decreasing $f(\emptyset)$ maintains submodularity). We can also assume that $w \geq \mathbf{0}$, since if some component of w is negative, the maximum value is unbounded.

Now order the elements in S as $s_1, \ldots, s_n$ such that $w(s_1) \geq \cdots \geq w(s_n)$. Define

(44.24) $\quad U_i := \{s_1, \ldots, s_i\}$ for $i = 0, \ldots, n$,

and define $x \in \mathbb{R}^S$ by

(44.25) $\quad x(s_i) := f(U_i) - f(U_{i-1})$ for $i = 1, \ldots, n$.

Then x maximizes $w^\mathsf{T} x$ over EP_f, as will be shown in the following theorem.

To prove it, consider the following linear programming duality equation:

(44.26) $\quad \max\{w^\mathsf{T} x \mid x \in EP_f\}$
$= \min\{\sum_{T \subseteq S} y(T)f(T) | y \in \mathbb{R}_+^{\mathcal{P}(S)}, \sum_{T \in \mathcal{P}(S)} y(T)\chi^T = w\}.$

Define:

(44.27) $\quad y(U_i) := w(s_i) - w(s_{i+1}) \quad (i = 1, \ldots, n-1),$
$ y(S) := w(s_n),$
$ y(T) := 0 \qquad\qquad\qquad\qquad (T \neq U_i \text{ for each } i).$

Theorem 44.3. *Let f be a submodular set function on S with $f(\emptyset) = 0$ and let $w : S \to \mathbb{R}_+$. Then x and y given by (44.25) and (44.27) are optimum solutions of (44.26).*

Proof. We first show that x belongs to EP_f; that is, $x(T) \leq f(T)$ for each $T \subseteq S$. This is shown by induction on $|T|$, the case $T = \emptyset$ being trivial. Let $T \neq \emptyset$ and let k be the largest index with $s_k \in T$. Then by induction,

(44.28) $\quad x(T \setminus \{s_k\}) \leq f(T \setminus \{s_k\}).$

Hence

(44.29) $\quad x(T) \leq f(T\setminus\{s_k\}) + x(s_k) = f(T\setminus\{s_k\}) + f(U_k) - f(U_{k-1}) \leq f(T)$

(the last inequality follows from the submodularity of f). So $x \in EP_f$.

Also, y is feasible for (44.26). Trivially, $y \geq \mathbf{0}$. Moreover, for any i we have by (44.27):

(44.30) $$\sum_{T \ni s_i} y(T) = \sum_{j \geq i} y(U_j) = w(s_i).$$

So y is a feasible solution of (44.26).

Optimality of x and y follows from:

(44.31) $$w^\mathsf{T} x = \sum_{s \in S} w(s) x_s = \sum_{i=1}^{n} w(s_i)(f(U_i) - f(U_{i-1}))$$
$$= \sum_{i=1}^{n-1} f(U_i)(w(s_i) - w(s_{i+1})) + f(S)w(s_n) = \sum_{T \subseteq S} y(T) f(T).$$

The third equality follows from a straightforward reordering of the terms, using that $f(\emptyset) = 0$. ∎

Note that if f is integer, then x is integer, and that if w is integer, then y is integer. Moreover, if f is nondecreasing, then x is nonnegative. Hence, in that case, x and y are optimum solutions of

(44.32) $$\max\{w^\mathsf{T} x \mid x \in P_f\}$$
$$= \min\{\sum_{T \subseteq S} y(T) f(T) \mid y \in \mathbb{R}_+^{\mathcal{P}(S)}, \sum_{T \in \mathcal{P}(S)} y(T)\chi^T \geq w\}.$$

Therefore:

Corollary 44.3a. *Let f be a nondecreasing submodular set function on S with $f(\emptyset) = 0$ and let $w : S \to \mathbb{R}_+$. Then x and y given by (44.25) and (44.27) are optimum solutions for (44.32).*

Proof. Directly from Theorem 44.3, using the fact that $x \geq 0$ if f is nondecreasing. ∎

As for complexity we have:

Corollary 44.3b. *Given a submodular set function f on a set S (by a value giving oracle) and a function $w \in \mathbb{Q}^S$, we can find an $x \in EP_f$ maximizing $w^\mathsf{T} x$ in strongly polynomial time. If f is moreover nondecreasing, then $x \in P_f$ (and hence x maximizes $w^\mathsf{T} x$ over P_f).*

Proof. By the extension of the greedy method given above. ∎

The greedy algorithm can be interpreted geometrically as follows. Let w be some linear objective function on S, with $w(s_1) \geq \ldots \geq w(s_n)$. Travel via the vertices of P_f along the edges of P_f, by starting at the origin, as follows: first go from the origin as far as possible (in P_f) in the positive s_1-direction, say to vertex x_1; next go from x_1 as far as possible in the positive s_2-direction, say to x_2, and so on. After n steps one reaches a vertex x_n maximizing $w^\mathsf{T} x$

over P_f. In fact, the effectiveness of this algorithm characterizes polymatroids (Dunstan and Welsh [1973]).

44.3. Total dual integrality

Theorem 44.3 implies the box-total dual integrality of the following system:
(44.33) $x(U) \leq f(U)$ for $U \subseteq S$.

Corollary 44.3c. *If f is submodular, then (44.33) is box-totally dual integral.*

Proof. Consider the dual of maximizing $w^T x$ over (44.33), for some $w \in \mathbb{Z}_+^S$. By Theorem 44.3, it has an optimum solution $y : \mathcal{P}(S) \to \mathbb{R}_+$ with the sets $U \subseteq S$ having $y(U) > 0$ forming a chain. So these constraints give a totally unimodular submatrix of the constraint matrix (by Theorem 41.11). Therefore, by Theorem 5.35, (44.33) is box-TDI. ∎

This gives the integrality of polyhedra:

Corollary 44.3d. *For any integer submodular set function f, the polymatroid P_f and the extended polymatroid EP_f are integer.*

Proof. Directly from Corollary 44.3c. (In fact, integer optimum solutions are explicitly given by Theorem 44.3 and Corollary 44.3a.) ∎

44.4. f is determined by EP_f

Theorem 44.3 implies that for any extended polymatroid P there is a unique submodular function f satisfying $f(\emptyset) = 0$ and $EP_f = P$, since:

Corollary 44.3e. *Let f be a submodular set function on S with $f(\emptyset) = 0$. Then*
(44.34) $f(U) = \max\{x(U) \mid x \in EP_f\}$
for each $U \subseteq S$.

Proof. Directly from Theorem 44.3 by taking $w := \chi^U$. ∎

So there is a one-to-one correspondence between nonempty extended polymatroids and submodular set functions f with $f(\emptyset) = 0$. The correspondence relates integer extended polymatroids with integer submodular functions.

There is a similar correspondence between nonempty polymatroids and *nondecreasing* submodular set functions f with $f(\emptyset) = 0$. For any (not necessarily nondecreasing) nonnegative submodular set function f, define $\bar{f}$ by:

(44.35) $\bar{f}(\emptyset) = 0,$
$\bar{f}(U) = \min_{T \supseteq U} f(T)$ for nonempty $U \subseteq S.$

It is easy to see that $\bar{f}$ is nondecreasing and submodular and that $P_{\bar{f}} = P_f$ (Dunstan [1973]). In fact, $\bar{f}$ is the unique nondecreasing submodular set function associated with P_f, with $\bar{f}(\emptyset) = 0$, as (Kelley [1959]):

Corollary 44.3f. *If f is a nondecreasing submodular function with $f(\emptyset) = 0$, then*

(44.36) $f(U) = \max\{x(U) \mid x \in P_f\}$

for each $U \subseteq S$.

Proof. This follows from Corollary 44.3a by taking $w := \chi^T$. ∎

This one-to-one correspondence between polymatroids and nondecreasing submodular set functions f with $f(\emptyset) = 0$ relates integer polymatroids to integer such functions:

Corollary 44.3g. *For each integer polymatroid P there exists a unique integer nondecreasing submodular function f with $f(\emptyset) = 0$ and $P = P_f$.*

Proof. By Corollary 44.3d and (44.36). ∎

By (44.36) we have for any nonnegative submodular set function f that $\bar{f}(U) = \max\{x(U) \mid x \in P_f\}$. Since we can optimize over EP_f in polynomial time (with the greedy algorithm described above), with the ellipsoid method we can optimize over $P_f = EP_f \cap \mathbb{R}^S_+$ in polynomial time. Hence we can calculate $\bar{f}(U)$ in polynomial time. Alternatively, calculating $\bar{f}(U)$ amounts to minimizing the submodular function $f'(T) := f(T \cup U)$.

In fact $\bar{f}$ is the largest among all nondecreasing submodular set functions g on S with $g(\emptyset) = 0$ and $g \leq f$, as can be checked straightforwardly.

44.5. Supermodular functions and contrapolymatroids

Similar results hold for supermodular functions and the associated contrapolymatroids. Associate the following polyhedra with a set function g on S:

(44.37) $Q_g := \{x \in \mathbb{R}^S \mid x \geq 0, x(U) \geq g(U) \text{ for each } U \subseteq S\},$
$EQ_g := \{x \in \mathbb{R}^S \mid x(U) \geq g(U) \text{ for each } U \subseteq S\}.$

If g is supermodular, then Q_g and EQ_g are called the *contrapolymatroid* and the *extended contrapolymatroid associated with g*, respectively. A vector $x \in EQ_g$ (or Q_g) is called a *base vector* of EQ_g (or Q_g) if $x(S) = g(S)$. A base vector of g is a base vector of EQ_g.

Since $EQ_g = -EP_{-g}$, we can reduce most problems on (extended) contrapolymatroids to (extended) polymatroids. Again we can minimize a linear function $w^\mathsf{T} x$ over EQ_g with the greedy algorithm, as described in Section 44.2. (In fact, we can apply the same formulas (44.25) and (44.27) for g instead of f.) If g is nondecreasing, it yields a nonnegative optimum solution, and hence a vector x minimizing $w^\mathsf{T} x$ over Q_g.

Similarly, the system

(44.38) $\qquad x(U) \geq g(U)$ for $U \subseteq S$

is box-TDI, as follows directly from the box-total dual integrality of

(44.39) $\qquad x(U) \leq -g(U)$ for $U \subseteq S$.

Let EP_f be the extended polymatroid associated with the submodular function f with $f(\emptyset) = 0$. Let B_f be the face of base vectors of EP_f, i.e.,

(44.40) $\qquad B_f = \{x \in EP_f \mid x(S) = f(S)\}$.

A vector $y \in \mathbb{R}^S$ is called *spanning* if there exists an x in B_f with $x \leq y$. Let Q be the set of spanning vectors.

A vector y belongs to Q if and only if $(f|y)(S) = f(S)$, that is (by (44.8) and (44.9)) if and only if

(44.41) $\qquad y(U) \geq f(S) - f(S \setminus U)$

for each $U \subseteq S$. So Q is equal to the contrapolymatroid EQ_g associated with the submodular function g defined by $g(U) := f(S) - f(S \setminus U)$ for $U \subseteq S$. Then B_f is equal to the face of minimal elements of EQ_g.

There is a one-to-one correspondence between submodular set functions f on S with $f(\emptyset) = 0$ and supermodular set functions g on S with $g(\emptyset) = 0$, given by the relations

(44.42) $\qquad g(U) = f(S) - f(S \setminus U)$ and $f(U) = g(S) - g(S \setminus U)$

for $U \subseteq S$.

Then the pair $(-g, -Q)$ is related to the pair (f, P) by a relation similar to the duality relation of matroids (cf. Section 44.6f).

44.6. Further results and notes

44.6a. Submodular functions and matroids

Let P be the polymatroid associated with the nondecreasing integer submodular set function f on S, with $f(\emptyset) = 0$. Then the collection

(44.43) $\qquad \mathcal{I} := \{I \subseteq S \mid \chi^I \in P\}$

forms the collection of independent sets of a matroid $M = (S, \mathcal{I})$ (this result was announced by Edmonds and Rota [1966] and proved by Pym and Perfect [1970]). By Corollary 40.2b, the subpolymatroid (cf. Section 44.6c)

(44.44) $$P|1 = \{x \in P \mid x \leq 1\}$$

is the convex hull of the incidence vectors of the independent sets of M. By (44.8), the rank function r of M satisfies

(44.45) $$r(U) = \min_{T \subseteq U}(|U \setminus T| + f(T))$$

for $U \subseteq S$.

As an example, if f is the submodular function given in the set union example in Section 44.1a, we obtain the transversal matroid on $\{1, \ldots, n\}$ with $I \subseteq \{1, \ldots, n\}$ independent if and only if the family $(T_i \mid i \in I)$ has a transversal (Edmonds [1970b]).

44.6b. Reducing integer polymatroids to matroids

In fact, each integer polymatroid can be derived from a matroid as follows (Helgason [1974]). Let f be a nondecreasing submodular set function on S with $f(\emptyset) = 0$. Choose for each s in S, a set X_s of size $f(\{s\})$, such that the sets X_s ($s \in S$) are disjoint. Let $X := \bigcup_{s \in S} X_s$, and define a set function r on X by

(44.46) $$r(U) := \min_{T \subseteq S}(|U \setminus \bigcup_{s \in T} X_s| + f(T))$$

for $U \subseteq X$. One easily checks that r is the rank function of a matroid M (by checking the axioms (39.38)), and that for each subset T of S

(44.47) $$f(T) = r(\bigcup_{s \in T} X_s).$$

Therefore, f arises from the rank function of M, as in the Matroids example in Section 44.1a. The polymatroid P_f associated with f is just the convex hull of all vectors x for which there exists an independent set I in M with $x_s = |I \cap X_s|$ for all s in S.

Given a nondecreasing submodular set function f on S with $f(\emptyset) = 0$, Lovász [1980a] called a subset $U \subseteq S$ a *matching* if

(44.48) $$f(U) = \sum_{s \in U} f(\{s\}).$$

If $f(\{s\}) = 1$ for each s in S, f is the rank function of a matroid, and U is a matching if and only if U is independent in this matroid. If $f(\{s\}) = 2$ for each s in S, the elements of S correspond to certain flats of rank 2 in a matroid. Now determining the maximum size of a matching is just the matroid matching problem (cf. Chapter 43).

44.6c. The structure of polymatroids

Vertices of polymatroids (Edmonds [1970b], Shapley [1965,1971]). Let f be a submodular set function on a set $S = \{s_1, \ldots, s_n\}$ with $f(\emptyset) = 0$. Let P_f be the polymatroid associated with f. It follows immediately from the greedy algorithm, as in the proof of Corollary 44.3a, that the vertices of P_f are given by (for $i = 1, \ldots, n$):

(44.49) $\quad x(s_{\pi(i)}) = \begin{cases} f(\{s_{\pi(1)}, \ldots, s_{\pi(i)}\}) - f(\{s_{\pi(1)}, \ldots, s_{\pi(i-1)}\}) & \text{if } i \leq k, \\ 0 & \text{if } i > k, \end{cases}$

where π ranges over all permutations of $\{1, \ldots, n\}$ and where k ranges over $0, \ldots, n$.

Similarly, for any submodular set function f on S with $f(\emptyset) = 0$, the vertices of the extended polymatroid EP_f are given by

(44.50) $\quad x(s_{\pi(i)}) = f(\{s_{\pi(1)}, \ldots, s_{\pi(i)}\}) - f(\{s_{\pi(1)}, \ldots, s_{\pi(i-1)}\})$

for $i = 1, \ldots, n$, where π ranges over all permutations of $\{1, \ldots, n\}$.

Topkis [1984] characterized adjacency of the vertices of a polymatroid, while Bixby, Cunningham, and Topkis [1985] and Topkis [1992] gave further results on vertices of and paths on a polymatroid and on related partial orders of S.

Facets of polymatroids. Let f be a nondecreasing submodular set function on S with $f(\emptyset) = 0$. One easily checks that P_f is full-dimensional if and only if $f(\{s\}) > 0$ for all s in S. If P_f is full-dimensional there is a unique minimal collection of linear inequalities defining P_f (clearly, up to scalar multiplication). They correspond to the facets of P_f. Edmonds [1970b] found that this collection is given by the following theorem. A subset $U \subseteq S$ is called an f-*flat* if $f(U \cup \{s\}) > f(U)$ for all $s \in S \setminus U$, and U is called f-*inseparable* if there is no partition of U into nonempty sets U_1 and U_2 with $f(U) = f(U_1) + f(U_2)$. Then:

Theorem 44.4. *Let f be a nondecreasing submodular set function on S with $f(\emptyset) = 0$ and $f(\{s\}) > 0$ for each $s \in S$. The following is a minimal system determining the polymatroid P_f:*

(44.51) $\quad x_s \geq 0 \quad\quad (s \in S),$
$\quad\quad\quad\quad x(U) \leq f(U) \quad (U \text{ is a nonempty } f\text{-inseparable } f\text{-flat}).$

Proof. It is easy to see that (44.51) determines P_f, as any other inequality $x(U) \leq f(U)$ follows from (44.51). The irredundancy of collection (44.51) can be seen as follows.

Clearly, each inequality $x_s \geq 0$ determines a facet. Next consider a nonempty f-inseparable f-flat U. Suppose that the face determined by U is not a facet. Then it is contained in another face, say determined by T. Let x be a vertex of P_f with $x(U \setminus T) = f(U \setminus T)$, $x(U) = f(U)$, and $x(S \setminus U) = 0$. Such a vertex exists by the greedy algorithm (cf. (44.49)).

Since x is on the face determined by U, it is also on the face determined by T. So $x(T) = f(T)$. Hence $f(T) = x(T) = x(T \cap U) = f(U) - f(U \setminus T)$. So we have equality throughout in:

(44.52) $\quad f(U \setminus T) + f(T) \geq f(U \setminus T) + f(T \cap U) \geq f(U).$

This implies that $U \setminus T = \emptyset$ or $T \cap U = \emptyset$ (as U is f-inseparable), and that $f(T) = f(T \cap U)$. If $U \setminus T = \emptyset$, then $U \subset T$, and hence (as U is an f-flat) $f(T) > f(U) \geq f(T \cap U)$, a contradiction. If $T \cap U = \emptyset$, then $f(T) = f(T \cap U) = 0$, implying that $T = \emptyset$, again a contradiction. ∎

It follows that the face $\{x \in P_f \mid x(S) = f(S)\}$ of maximal vectors in P_f is a facet if and only if $f(U) + f(S \setminus U) > f(S)$ for each proper nonempty subset U of S. More generally, its codimension is equal to the number of inclusionwise minimal nonempty sets U with $f(U) + f(S \setminus U) = f(S)$ (cf. Fujishige [1984a]).

Faces of polymatroids (Giles [1975]). We now extend the characterizations of vertices and facets of polymatroids given above to arbitrary faces. Let P be the polymatroid associated with the nondecreasing submodular set function f on S with $f(\emptyset) = 0$. Suppose that P is full-dimensional. If $\emptyset \neq S_1 \subset \cdots \subset S_k \subseteq T \subseteq S$, then

(44.53) $\qquad F = \{x \in P \mid x(S_1) = f(S_1), \ldots, x(S_k) = f(S_k), x(S \setminus T) = 0\}$

is a face of P of dimension at most $|T| - k$. (Indeed, F is nonempty by the characterization (44.49) of vertices, while $\dim(F) \leq |T| - k$, as the incidence vectors of $S_1, \ldots, S_k$ are linearly independent.)

In fact, each face has a representation (44.53). Indeed, let F be a face of P. Define $T = \{s \in S \mid x_s > 0 \text{ for some } x \text{ in } F\}$, and let $S_1 \subset \cdots \subset S_k$ be any maximal chain of nonempty subsets of T with the property that

(44.54) $\qquad F \subseteq \{x \in P \mid x(S_1) = f(S_1), \ldots, x(S_k) = f(S_k), x(S \setminus T) = 0\}$.

Then we have equality in (44.54), and $\dim(F) = |T| - k$. (Here a maximal chain is a chain which is contained in no larger chain satisfying (44.54) — since the empty chain satisfies (44.54), there exist maximal chains.)

In order to prove this assertion, suppose that F has dimension d. As the right-hand side of (44.54) is a face of P of dimension at most $|T| - k$, it suffices to show that $d = |T| - k$. Therefore, suppose $d < |T| - k$. Then there exists a subset U of S such that $x(U) = f(U)$ for all x in F, and such that the incidence vector of $U \cap T$ is linearly independent of the incidence vectors of $S_1, \ldots, S_k$. That is, $U \cap T$ is not the union of some of the sets $S_i \setminus S_{i-1}$ $(i = 1, \ldots, k)$. Since $x(U \cap T) = x(U) = f(U) \geq f(U \cap T)$ for all x in F, we may assume that $U \subseteq T$. Since the collection of subsets U of S with $x(U) = f(U)$ is closed under taking unions and intersections, we may assume moreover that U is comparable with each of the sets in the chain $S_1 \subset \cdots \subset S_k$. Hence U could be added to the chain to obtain a larger chain, contradicting our assumption. So $d = |T| - k$.

Note that a chain $S_1 \subset \cdots \subset S_k$ of nonempty subsets of T is a maximal chain satisfying (44.54) if and only if there is equality in (44.54) and (setting $S_0 := \emptyset$):

(44.55) $\qquad f(S_k \cup \{s\}) > f(S_k)$ for all s in $T \setminus S_k$, and each of the sets $S_i \setminus S_{i-1}$ is f_i-inseparable, where f_i is the submodular set function on $S_i \setminus S_{i-1}$ given by $f_i(U) := f(U \cup S_{i-1}) - f(S_{i-1})$ for $U \subseteq S_i \setminus S_{i-1}$.

This may be derived straightforwardly from the existence, by (44.49), of appropriate vertices of F.

It is not difficult to show that if F has a representation (44.53), then F is the direct sum of $F_1, \ldots, F_k$ and Q, where F_i is the face of maximal vectors in the polymatroid associated with f_i $(i = 1, \ldots, k)$, and Q is the polymatroid associated with the submodular set function g on $T \setminus S_k$ given by $g(U) := f(U \cup S_k) - f(S_k)$ for $U \subseteq T \setminus S_k$. Since $\dim(F_i) \leq |S_i \setminus S_{i-1}| - 1$ and $\dim(Q) \leq |T \setminus S_k|$, this yields that $\dim(F) = |T| - k$ if and only if $\dim(F_i) = |S_i \setminus S_{i-1}| - 1$ $(i = 1, \ldots, k)$ and $\dim(Q) = |T \setminus S_k|$. From this, characterization (44.55) can be derived again. It also yields that if F, represented by (44.53), has dimension $|T| - k$, then the unordered partition $\{S_1, S_2 \setminus S_1, \ldots, S_k \setminus S_{k-1}, T \setminus S_k\}$ is the same for all maximal chains $S_1 \subset \cdots \subset S_k$.

For a characterization of the faces of a polymatroid, see Fujishige [1984a].

44.6d. Characterization of polymatroids

Let P be the polymatroid associated with the nondecreasing submodular set function f on S with $f(\emptyset) = 0$. The following three observations are easily derived from the representation (44.49) of vertices of P. (a) If x_0 is a vertex of P, there exists a vertex x_1 of P such that $x_1 \geq x_0$ and x_1 has the form (44.49) with $k = n$. (b) A vertex x_1 of P can be represented as (44.49) with $k = n$ if and only if $x_1(S) = f(S)$. (c) The convex hull of the vertices x_1 of P with $x_1(S) = f(S)$ is the face $\{x \in P \mid x(S) = f(S)\}$ of P. It follows directly from (a), (b) and (c) that $x \in P$ is a maximal element of P (with respect to $\leq$) if and only if $x(S) = f(S)$. So for each vector y in P there is a vector x in P with $y \leq x$ and $x(S) = f(S)$.

Applying this to the subpolymatroids $P|a = P \cap \{x \mid x \leq a\}$ (cf. Section 44.1), one finds the following property of polymatroids:

(44.56)   for each $a \in \mathbb{R}_+^S$ there exists a number $r(a)$ such that each maximal vector x of $P \cap \{x \mid x \leq a\}$ satisfies $x(S) = r(a)$.

Here *maximal* is maximal in the partial order $\leq$ on vectors. The number $r(a)$ is called the *rank* of a, and any x with the properties mentioned in (44.56) is called a *base* of a.

Edmonds [1970b] (cf. Dunstan [1973], Woodall [1974b]) noticed the following (we follow the proof of Welsh [1976]):

Theorem 44.5. *Let $P \subseteq \mathbb{R}_+^S$. Then P is a polymatroid if and only if P is compact, and satisfies* (44.56) *and*

(44.57)   *if $0 \leq y \leq x \in P$, then $y \in P$.*

Proof. Necessity was observed above. To see sufficiency, let f be the set function on S defined by

(44.58) $f(U) := \max\{x(U) \mid x \in P\}$

for $U \subseteq S$. Then f is nonnegative and nondecreasing. Moreover, f is submodular. To see this, consider $T, U \subseteq S$. Let x be a maximal vector in P satisfying $x_s = 0$ if $s \notin T \cup U$, and let y be a maximal vector in P satisfying $y(s) = 0$ if $s \notin T \cap U$ and $x \leq y$. Note that (44.56) and (44.58) imply that $x(T \cap U) = f(T \cap U)$ and $y(T \cup U) = f(T \cup U)$. Hence

(44.59) $f(T) + f(U) \geq y(T) + y(U) = y(T \cap U) + y(T \cup U) \geq x(T \cap U) + y(T \cup U)$
                   $= f(T \cap U) + f(T \cup U)$,

that is, f is submodular.

We finally show that P is equal to the polymatroid P_f associated to f. Clearly, $P \subseteq P_f$, since if $x \in P$ then $x(U) \leq f(U)$ for each $U \subseteq S$, by definition (44.58) of f.

To see that $P_f = P$, suppose $v \in P_f \setminus P$. Let u be a base of v (that is, a maximal vector $u \in P$ satisfying $u \leq v$). Choose u such that the set

(44.60) $U := \{s \in S \mid u_s < v_s\}$

is as large as possible. Since $v \notin P$, we have $u \neq v$, and hence $U \neq \emptyset$. As $v \in P_f$, we know

(44.61) $\quad u(U) < v(U) \leq f(U)$.

Define

(44.62) $\quad w := \frac{1}{2}(u+v)$.

So $u \leq w \leq v$. Hence u is a base of w, and each base of w is a base of v.

For any $z \in \mathbb{R}^S$, define z' as the projection of z on the subspace $L := \{x \in \mathbb{R}^S \mid x_s = 0 \text{ if } s \in S \setminus U\}$. That is:

(44.63) $\quad z'(s) := z(s)$ if $s \in U$, and $z'(s) := 0$ if $s \in S \setminus U$.

By definition of f, there is an $x \in P$ with $x(U) = f(U)$. We may assume that $x \in L$. Choose $y \in L$ with $x \leq y$ and $u' \leq y$. Then

(44.64) $\quad x(S) = x(U) = f(U) > u(U) = u'(U) = u'(S)$.

So $r(y) > u'(S)$. Hence, by (44.56), there exists a base z of y with $u' \leq z$ and $z(S) > u'(S)$. So $u'_s < z_s$ for at least one $s \in U$. This implies, since $u'_s < w'_s$ for each $s \in S$, that there is an $a \in P$ with $u' \leq a \leq w'$ and $a \neq u'$, hence $a(U) > u'(U)$.

Since $a \leq w' \leq w$, there is a base b of w with $a \leq b$. Then $b(S) = u(S)$ (since also u is a base of w) and $b(U) \geq a(U) > u'(U) = u(U)$. Hence $b_s < u_s = v_s$ for some $s \in S \setminus U$. Moreover, $b_s \leq w_s < v_s$ for each $s \in U$. So U is properly contained in $\{s \in S \mid b_s < v_s\}$, contradicting the maximality of U. ∎

(For an alternative characterization, see Welsh [1976].)

By (44.8) and (44.9) the rank of a is given by

(44.65) $\quad r(a) = \min_{U \subseteq S}(a(S \setminus U) + f(U))$

(from this one may derive a 'submodular law' for r: $r(a \wedge b) + r(a \vee b) \leq r(a) + r(b)$, where $\wedge$ and $\vee$ are the meet and join in the lattice $(\mathbb{R}^S, \leq)$ (Edmonds [1970b])).

Since if P has integer vertices and a is integer, the intersection $P|a = \{x \in P \mid x \leq a\}$ is integer again, we know that for integer polymatroids (44.56) also holds if we restrict a and x to integer vectors. So if a is integer, then there exists an integer vector $x \leq a$ in P with $x(S) = r(a)$.

Theorem 44.5 yields an analogous characterization of extended polymatroids. Let f be a submodular set function on S with $f(\emptyset) = 0$. Choose $c \in \mathbb{R}_+^S$ such that

(44.66) $\quad g(U) := f(U) + c(U)$

is nonnegative for all $U \subseteq S$. Clearly, g again is submodular, and $g(\emptyset) = 0$. Then the extended polymatroid EP_f associated with f and the polymatroid P_g associated with g are related by:

(44.67) $\quad P_g = \{x \mid x \geq 0, x - c \in EP_f\} = (c + EP_f) \cap \mathbb{R}_+^S$.

Since P_g is a polymatroid, by (44.56) we know that EP_f satisfies:

(44.68) $\quad$ for each a in $\mathbb{R}^S$ there exists a number $r(a)$ such that each maximal vector x in $EP_f \cap \{x \in \mathbb{R}^S \mid x \leq a\}$ satisfies $x(S) = r(a)$.

One easily derives from Theorem 44.5 that (44.68) together with

(44.69) $\quad$ if $y \leq x \in EP_f$, then $y \in EP_f$,

characterizes the class of all extended polymatroids among the closed subsets of $\mathbb{R}^S$.

44.6e. Operations on submodular functions and polymatroids

The class of submodular set functions on a given set is closed under certain operations. Obviously, the sum of two submodular functions is submodular again. In particular, adding a constant t to all values of a submodular function maintains submodularity. Also the multiplication of a submodular function by a nonnegative scalar maintains submodularity. Moreover, if f is a nondecreasing submodular set function on S, and q is a real number, then the function f' given by $f'(U) := \min\{q, f(U)\}$ for $U \subseteq S$, is submodular again. (Monotonicity cannot be deleted, as is shown by taking $S := \{a, b\}$, $f(\emptyset) = f(S) = 1$, $f(\{a\}) = 0$, $f(\{b\}) = 2$, and $q = 1$.)

It follows that the class of all submodular set functions on S forms a convex cone C in $\mathbb{R}^{\mathcal{P}(S)}$. This cone is polyhedral as the constraints (44.1) form a finite set of linear inequalities defining C. Edmonds [1970b] raised the problem of determining the extreme rays of the cone of all nonnegative nondecreasing submodular set functions f on S with $f(\emptyset) = 0$. It is not difficult to show that the rank function r of a matroid M determines an extreme ray of this cone if and only if r is not the sum of the rank functions of two other matroids, i.e., if and only if M is the sum of a connected matroid and a number of loops. But these do not represent all extreme rays: if $S = \{1,\ldots,5\}$ and $w(1) = 2, w(s) = 1$ for $s \in S \setminus \{1\}$, let $f(U) := \min\{3, w(U)\}$ for $U \subseteq S$; then f is on an extreme ray, but cannot be decomposed as the sum of rank functions of matroids (L. Lovász's example; cf. also Murty and Simon [1978] and Nguyen [1978]).

Lovász [1983c] observed that if f_1 and f_2 are submodular and $f_1 - f_2$ is nondecreasing, then $\min\{f_1, f_2\}$ is submodular.

Let f be a nonnegative submodular set function on S. Clearly, for any $\lambda \geq 0$ we have $P_{\lambda f} = \lambda P_f$ (where $\lambda P_f = \{\lambda x \mid x \in P_f\}$). If $q \geq 0$, and f' is given by $f'(U) = \min\{q, f(U)\}$ for $U \subseteq S$, then f' is submodular and

(44.70) $\qquad P_{f'} = \{x \in P_f \mid x(S) \leq q\}$,

as can be checked easily. So the class of polymatroids is closed under intersections with affine halfspaces of the form $\{x \in \mathbb{R}^S \mid x(S) \leq q\}$, for $q \geq 0$.

Let f_1 and f_2 be nondecreasing submodular set functions on S, with $f_1(\emptyset) = f_2(\emptyset) = 0$, and associated polymatroids P_1 and P_2 respectively. Let P be the polymatroid associated with $f := f_1 + f_2$. Then (McDiarmid [1975c]):

Theorem 44.6. $P_{f_1+f_2} = P_{f_1} + P_{f_2}$.

Proof. It is easy to see that $P_{f_1+f_2} \supseteq P_{f_1} + P_{f_2}$. To prove the reverse inclusion, let x be a vertex of $P_{f_1+f_2}$. Then x has the form (44.49). Hence, by taking the same permutation π and the same k, $x = x_1 + x_2$ for certain vertices x_1 of P_{f_1} and x_2 of P_{f_2}. Since $P_{f_1} + P_{f_2}$ is convex it follows that $P_{f_1+f_2} = P_{f_1} + P_{f_2}$. ∎

In fact, if f_1 and f_2 are integer, each *integer* vector in $P_{f_1} + P_{f_2}$ is the sum of *integer* vectors in P_{f_1} and P_{f_2} — see Corollary 46.2c. Similarly, if f_1 and f_2 are integer, each integer vector in $EP_{f_1} + EP_{f_2}$ is the sum of integer vectors in EP_{f_1} and EP_{f_2}.

Faigle [1984a] derived from Theorem 44.6 that, for any submodular function f, if $x, y \in P_f$ and $x = x_1 + x_2$ with $x_1, x_2 \in P_f$, then there exist $y_1, y_2 \in P_f$ with

782   Chapter 44. Submodular functions and polymatroids

$y = y_1 + y_2$ and $x_1 + y_1, x_2 + y_2 \in P_f$. (Proof: $y \in P_f \subseteq P_{2f-x} = P_{f-x_1} + P_{f-x_2}$.) An integer version of this can be derived from Corollary 46.2c and generalizes (42.13).

If $M_1 = (S, \mathcal{I}_1)$ and $M_2 = (S, \mathcal{I}_2)$ are matroids, with rank functions r_1 and r_2 and corresponding independent set polytopes P_1 and P_2, respectively, then by Section 44.6c above, $P_1 + P_2$ is the convex hull of sums of incidence vectors of independent sets in M_1 and M_2. Hence the 0,1 vectors in $P_1 + P_2$ are just the incidence vectors of the sets $I_1 \cup I_2$, for $I_1 \in \mathcal{I}_1$ and $I_2 \in \mathcal{I}_2$. Therefore, the polyhedron

(44.71) $(P_1 + P_2)|\mathbf{1} = \{x \in P_1 + P_2 \mid x \leq \mathbf{1}\}$

is the convex hull of the independent sets of $M_1 \vee M_2$. By Theorem 44.6 and (44.45), it follows that the rank function r of $M_1 \vee M_2$ satisfies

(44.72) $r(U) = \min_{T \subseteq U}(|U \setminus T| + r_1(T) + r_2(T))$

for $U \subseteq S$. Thus we have derived the matroid union theorem (Corollary 42.1a).

44.6f. Duals of polymatroids

McDiarmid [1975c] described the following duality of polymatroids. Let P be the polymatroid associated with the nondecreasing submodular set function f on S with $f(\emptyset) = 0$ and let a be a vector in $\mathbb{R}^S$ with $a \geq x$ for all x in P (i.e., $a(s) \geq f(\{s\})$ for all s in S). Define

(44.73) $f^*(U) := a(U) + f(S \setminus U) - f(S)$

for $U \subseteq S$. One easily checks that f^* again is nondecreasing and submodular, and that $f^*(\emptyset) = 0$. We call f^* the *dual* of f (with respect to a). Then $f^{**} = f$ taking the second dual with respect to the same a, as follows immediately from (44.73).

Let P^* be the polymatroid associated with f^*, and call P^* the *dual* polymatroid of P (with respect to a). Now the maximal vertices of P and P^* are given by (44.49) by choosing $k = n$. It follows that x is a maximal vertex of P if and only if $a - x$ is a maximal vertex of P^*. Since the maximal vectors of a polymatroid form just the convex hull of the maximal vertices, we may replace in the previous sentence the word 'vertex' by 'vector'. So the set of maximal vectors of P^* arises from the set of maximal vectors of P by reflection in the point $\frac{1}{2}a$.

Clearly, duals of matroids correspond in the obvious way to duals of the related polymatroids (with respect to the vector $\mathbf{1}$).

44.6g. Induction of polymatroids

Let $G = (V, E)$ be a bipartite graph, with colour classes S and T. Let f be a nondecreasing submodular set function on S with $f(\emptyset) = 0$, and define

(44.74) $g(U) := f(N(U))$

for $U \subseteq T$ (cf. Section 44.1a). (As usual, $N(U)$ denotes the set of vertices not in U adjacent to at least one vertex in U.)

The function g again is nondecreasing and submodular. Similarly to Rado's theorem (Corollary 41.1c), one may prove that a vector x belongs to P_g if and only if there exist $y \in \mathbb{R}^E_+$ and $z \in P_f$ such that

(44.75) $\quad y(\delta(t)) = x_t \quad (t \in T),$
$\quad\quad\quad\ y(\delta(s)) = z_s \quad (s \in S).$

Moreover, if f and g are integer, we can take y and z to be integer. This procedure gives an 'induction' of polymatroids through bipartite graphs, and yields 'Rado's theorem for polymatroids' (cf. McDiarmid [1975c]).

In case f is the rank function of a matroid on S, a 0,1 vector x belongs to P_g if and only if there exists a matching in G whose end vertices in S form an independent set of the matroid, and the end vertices in T have x as incidence vector. So these 0,1 vectors determine a matroid on T, with rank function r given by

(44.76) $\quad r(U) = \min_{W \subseteq U} (|U \setminus W| + f(N(W)))$

for $U \subseteq T$ (cf. (44.45) and (44.74)).

Another extension is the following. Let $D = (V, A)$ be a directed graph and let V be partitioned into classes S and T. Let furthermore a 'capacity' function $c : A \to \mathbb{R}_+$ be given. Define the set function g on T by

(44.77) $\quad g(U) := c(\delta^{\mathrm{out}}(U))$

for $U \subseteq T$, where $\delta^{\mathrm{out}}(U)$ denotes the set of arcs leaving U. Then g is nonnegative and submodular, and it may be derived straightforwardly from the max-flow min-cut theorem (Theorem 10.3) that a vector x in $\mathbb{R}_+^T$ belongs to P_g if and only if there exist $T - S$ paths $Q_1, \ldots, Q_k$ and nonnegative numbers $\lambda_1, \ldots, \lambda_k$ (for some k), such that

(44.78) $\quad \sum_{i=1}^{k} \lambda_i \chi^{AQ_i} \leq c \text{ and } \sum_{i=1}^{k} \lambda_i \chi^{b(Q_i)} = x,$

where $b(Q_i)$ is the beginning vertex of Q_i. If the c and x are integer, we can take also the λ_i integer.

Here the function g in general is not nondecreasing, but the value

(44.79) $\quad \bar{g}(U) = \min\{g(W) \mid U \subseteq W \subseteq T\}$

of the associated nondecreasing submodular function (cf. (44.35)) is equal to the minimum capacity of a cut separating U and S, which is equal to the maximum amount of flow from U to S, subject to the capacity function c (by the max-flow min-cut theorem).

In an analogous way, one can construct polymatroids by taking vertex-capacities instead of arc-capacities. Moreover, the notion of induction of polymatroids through bipartite graphs can be extended in a natural way to the induction of polymatroids through directed graphs (cf. McDiarmid [1975c], Schrijver [1978]).

44.6h. Lovász's generalization of Kőnig's matching theorem

Lovász [1970a] gave the following generalization of Kőnig's matching theorem (Theorem 16.2).

For a graph $G = (V, E)$, $U \subseteq V$, and $F \subseteq E$, let $N_F(U)$ denote the set of vertices not in U that are adjacent in (V, F) to at least one vertex in U. Kőnig's matching theorem follows by taking $g(X) := |X|$ in the following theorem.

Theorem 44.7. *Let $G = (V, E)$ be a simple bipartite graph, with colour classes S and T. Let g be a supermodular set function on S, such that $g(\{v\}) \geq 0$ for each $v \in S$ and such that*

(44.80) $g(U \cup \{v\}) \leq g(U) + g(\{v\})$ *for nonempty $U \subseteq S$ and $v \in S \setminus U$.*

Then E has a subset F with $\deg_F(v) = g(\{v\})$ for each $v \in V$ and $|N_F(U)| \geq g(U)$ for each nonempty $U \subseteq S$ if and only if $|N_E(U)| \geq g(U)$ for each nonempty $U \subseteq S$.

Proof. Necessity being trivial, we show sufficiency. Choose $F \subseteq E$ such that

(44.81) $|N_F(U)| \geq g(U)$

for each nonempty $U \subseteq S$, with $|F|$ as small as possible. We show that F is as required.

Suppose to the contrary that $\deg_F(v) > g(\{v\})$ for some $v \in S$. By the minimality of F, for each edge $e = vw \in F$, there is a subset U_e of S with $v \in U_e$, $|N_F(U_e)| = g(U_e)$, and $w \notin N_F(U_e \setminus \{v\})$. Since the function $|N_F(U)|$ is submodular, the intersection U of the U_e over $e \in \delta(v)$ satisfies $|N_F(U)| = g(U)$ (using (44.81)). Then no neighbour w of v is adjacent to U. Hence $N_F(v)$ and $N_F(U \setminus \{v\})$ are disjoint. Moreover, $U \neq \{v\}$, since $N_F(U) = g(U)$ and $N_F(\{v\}) > g(v)$. This gives the contradiction

(44.82) $g(U) \leq g(U \setminus \{v\}) + g(\{v\}) < |N_F(U \setminus \{v\})| + |N_F(v)| = |N_F(U)|.$ ∎

For a derivation of this theorem with the Edmonds-Giles method, see Frank and Tardos [1989].

44.6i. Further notes

Edmonds [1970b] and D.A. Higgs (as mentioned in Edmonds [1970b]) observed that if f is a set function on a set S, we can define recursively a submodular function $\bar{f}$ as follows:

(44.83) $\bar{f}(T) := \min\{f(T), \min(\bar{f}(S_1) + \bar{f}(S_2) - \bar{f}(S_1 \cap S_2))\},$

where the second minimum ranges over all pairs S_1, S_2 of proper subsets of T with $S_1 \cup S_2 = T$.

Lovász [1983c] gave the following characterization of submodularity in terms of convexity. Let f be a set function on S and define for each $c \in \mathbb{R}_+^S$

(44.84) $\hat{f}(c) := \sum_{i=1}^{k} \lambda_i f(U_i),$

where $\emptyset \neq U_1 \subset U_2 \subset \cdots \subset U_k \subseteq S$ and $\lambda_1, \ldots, \lambda_k > 0$ are such that $c = \sum_{i=1}^{k} \lambda_i \chi^{U_i}$. Then f is submodular if and only if $\hat{f}$ is convex. Similarly, f is supermodular if and only if $\hat{f}$ is concave. Related is the 'subdifferential' of a submodular function, investigated by Fujishige [1984d].

Korte and Lovász [1985c] and Nakamura [1988a] studied polyhedral structures where the greedy algorithm applies. Federgruen and Groenevelt [1986] extended the greedy method for polymatroids to 'weakly concave' objective functions (instead of linear functions). (Related work was reported by Bhattacharya, Georgiadis, and

Tsoucas [1992].) Nakamura [1993] extended polymatroids and submodular functions to Δ-polymatroids and Δ-submodular functions.

Gröflin and Liebling [1981] studied the following example of 'transversal polymatroids'. Let $G = (V, E)$ be an undirected graph, and define the submodular set function f on E by $f(F) := |\bigcup F|$ for $F \subseteq E$. Then the vertices of the associated polymatroid are all $\{0, 1, 2\}$ vectors x in $\mathbb{R}^E$ with the property that the set $F := \{e \in E \mid x_e \geq 1\}$ forms a forest each component of which contains at most one edge e with $x_e = 2$. If x is a maximal vertex, then each component contains exactly one edge e with $x_e = 2$.

Narayanan [1991] studied, for a given submodular function f on S, the lattice of all partitions $\mathcal{P}$ of S into nonempty sets such that there exists a $\lambda \in \mathbb{R}$ for which $\mathcal{P}$ attains $\min \sum_{U \in \mathcal{P}} (f(U) - \lambda)$ (taken over all partitions $\mathcal{P}$). Fujishige [1980b] studied minimum values of submodular functions.

For results on the (NP-hard) problems of *maximizing* a submodular function and of submodular set cover, see Fisher, Nemhauser, and Wolsey [1978], Nemhauser and Wolsey [1978,1981], Nemhauser, Wolsey, and Fisher [1978], Wolsey [1982a,1982b], Conforti and Cornuéjols [1984], and Fujito [1999].

Cunningham [1983], Fujishige [1983], and Nakamura [1988c] presented decomposition theories for submodular functions. Benczúr and Frank [1999] considered covering symmetric supermodular functions by graphs.

For surveys and books on polymatroids and submodular functions, see McDiarmid [1975c], Welsh [1976], Lovász [1983c], Lawler [1985], Nemhauser and Wolsey [1988], Fujishige [1991], Narayanan [1997], and Murota [2002]. For a survey on applications of submodular functions, see Frank [1993a].

Historically, submodular functions arose in lattice theory (Bergmann [1929], Birkhoff [1933]), while submodularity of the rank function of a matroid was shown by Bergmann [1929] and Whitney [1935]. Choquet [1951,1955] and Kelley [1959] studied submodular functions in relation to the Newton capacity and to measures in Boolean algebras. The relevance of submodularity for optimization was revealed by Edmonds [1970b].

Several alternative names have been proposed for submodular functions, like sub-valuation (Choquet [1955]), β-function (Edmonds [1970b]), and ground set rank function (McDiarmid [1975c]). The set of integer vectors in an integer polymatroid was called a hypermatroid by Helgason [1974] and Lovász [1977c]. A generalization of polymatroids (called supermatroids) was studied by Dunstan, Ingleton, and Welsh [1972].

Chapter 45

Submodular function minimization

This chapter describes a strongly polynomial-time algorithm to find the minimum value of a submodular function. It suffices that the submodular function is given by a value giving oracle.

One application of submodular function minimization is optimizing over the intersection of two polymatroids. This will be discussed in Chapter 47.

45.1. Submodular function minimization

It was shown by Grötschel, Lovász, and Schrijver [1981] that the minimum value of a rational-valued submodular set function f on S can be found in polynomial time, if f is given by a value giving oracle and an upper bound B is given on the numerators and denominators of the values of f. The running time is bounded by a polynomial in $|S|$ and $\log B$. This algorithm is based on the ellipsoid method: we can assume that $f(\emptyset) = 0$ (by resetting $f(U) := f(U) - f(\emptyset)$ for all $U \subseteq S$); then with the greedy algorithm, we can optimize over EP_f in polynomial time (Corollary 44.3b), hence the separation problem for EP_f is solvable in polynomial time, hence also the separation problem for

(45.1) $P := EP_f \cap \{x \mid x \leq \mathbf{0}\}$,

and therefore also the optimization problem for P. Now the maximum value of $x(S)$ over P is equal to the minimum value of f (by (44.8), (44.9), and (44.34)).

Having a polynomial-time method to find the minimum value of a submodular function, we can turn it into a polynomial-time method to find a subset T of S minimizing $f(T)$: For each $s \in S$, we can determine if the minimum value of f over all subsets of S is equal to the minimum value of f over subsets of $S \setminus \{s\}$. If so, we reset $S := S \setminus \{s\}$. Doing this for all elements of S, we are left with a set T minimizing f over all subsets of (the original) S.

Grötschel, Lovász, and Schrijver [1988] showed that this algorithm can be turned into a strongly polynomial-time method. Cunningham [1985b] gave a combinatorial, pseudo-polynomial-time algorithm for minimizing a submodular function f (polynomial in the size of the underlying set and the maximum absolute value of f (assuming f to be integer)). Inspired by Cunningham's method, combinatorial strongly polynomial-time algorithms were found by Iwata, Fleischer, and Fujishige [2000,2001] and Schrijver [2000a]. We will describe the latter algorithm.

45.2. Orders and base vectors

Let f be a submodular set function on a set S. In finding the minimum value of f, we can assume $f(\emptyset) = 0$, as resetting $f(U) := f(U) - f(\emptyset)$ for all $U \subseteq S$ does not change the problem. So throughout we assume that $f(\emptyset) = 0$.

Moreover, we assume that f is given by a *value giving oracle*, that is, an oracle that returns $f(U)$ for any given subset U of S. We also assume that the numbers returned by the oracle are rational (or belong to any ordered field in which we can perform the elementary arithmetic operations algorithmically).

Recall that the *base polytope* B_f of f is defined as the set of base vectors of f:

(45.2) $\quad B_f := \{x \in \mathbb{R}^S \mid x(U) \leq f(U) \text{ for all } U \subseteq S, x(S) = f(S)\}$.

Consider any total order $\prec$ on S.[35] For any $v \in S$, denote

(45.3) $\quad v_\prec := \{u \in S \mid u \prec v\}$.

Define a vector $b^\prec$ in $\mathbb{R}^S$ by:

(45.4) $\quad b^\prec(v) := f(v_\prec \cup \{v\}) - f(v_\prec)$

for $v \in S$. Theorem 44.3 implies that $b^\prec$ belongs to B_f.

Note that $b^\prec(U) = f(U)$ for each lower ideal U of $\prec$ (where a *lower ideal* of $\prec$ is a subset U of S such that if $v \in U$ and $u \prec v$, then $u \in U$).

45.3. A subroutine

In this section we describe a subroutine that is important in the algorithm. It replaces a total order $\prec$ by other total orders, thereby reducing some interval $(s,t]_\prec$, where

(45.5) $\quad (s,t]_\prec := \{v \mid s \prec v \preceq t\}$

for $s, t \in S$.

Let $\prec$ be a total order on S. For any $s, u \in S$ with $s \prec u$, let $\prec^{s,u}$ be the total order on S obtained from $\prec$ by resetting $v \prec u$ to $u \prec v$ for each

---

[35] As usual, we use $\prec$ for strict inequality and $\preceq$ for nonstrict inequality. We refer to the order by the strict inequality sign $\prec$.

v satisfying $s \preceq v \prec u$. Thus in the ordering, we move u to the position just before s. Hence $(s,t]_{\prec^{s,u}} = (s,t]_\prec \setminus \{u\}$ if $u \in (s,t]_\prec$.

We show that there is a strongly polynomial-time subroutine that

(45.6)   for any $s, t \in S$ with $s \prec t$, finds a $\delta \geq 0$ and describes $b^\prec +$ $\delta(\chi^t - \chi^s)$ as a convex combination of the $b^{\prec^{s,u}}$ for $u \in (s,t]_\prec$.

To describe the subroutine, we can assume that $b^\prec = \mathbf{0}$, by replacing (temporarily) $f(U)$ by $f(U) - b^\prec(U)$ for each $U \subseteq S$.

We investigate the signs of the vector $b^{\prec^{s,u}}$. We show that for each $v \in S$:

(45.7) $b^{\prec^{s,u}}(v) \leq 0$ if $s \preceq v \prec u$,
$b^{\prec^{s,u}}(v) \geq 0$ if $v = u$,
$b^{\prec^{s,u}}(v) = 0$ otherwise.

To prove this, observe that if $T \subseteq U \subseteq S$, then for any $v \in S \setminus U$ we have by the submodularity of f:

(45.8) $f(U \cup \{v\}) - f(U) \leq f(T \cup \{v\}) - f(T)$.

To see (45.7), if $s \preceq v \prec u$, then by (45.8),

(45.9) $b^{\prec^{s,u}}(v) = f(v_{\prec^{s,u}} \cup \{v\}) - f(v_{\prec^{s,u}}) \leq f(v_\prec \cup \{v\}) - f(v_\prec)$
$= b^\prec(v) = 0$,

since $v_{\prec^{s,u}} = v_\prec \cup \{u\} \supset v_\prec$.

Similarly,

(45.10) $b^{\prec^{s,u}}(u) = f(u_{\prec^{s,u}} \cup \{u\}) - f(u_{\prec^{s,u}}) \geq f(u_\prec \cup \{u\}) - f(u_\prec)$
$= b^\prec(u) = 0$,

since $u_{\prec^{s,u}} = s_\prec \subset u_\prec$.

Finally, if $v \prec s$ or $u \prec v$, then $v_{\prec^{s,u}} = v_\prec$, and hence $b^{\prec^{s,u}}(v) = b^\prec(v) = 0$. This shows (45.7).

By (45.7), the matrix $M = (b^{\prec^{s,u}}(v))_{u,v}$ with rows indexed by $u \in (s,t]_\prec$ and columns indexed by $v \in S$, in the order given by $\prec$, has the following, partially triangular, shape, where $+$ means that the entry is ≥ 0, and $-$ that the entry is ≤ 0:

			s						t			
	0	⋯	0	−	+	0	⋯	⋯	⋯	0	0	0	⋯	0
	⋮		⋮	−	−	+	⋱			⋮	⋮	⋮		⋮
	⋮		⋮	−	−	−	⋱	⋱		⋮	⋮	⋮		⋮
	⋮		⋮	⋮	⋮	⋮		⋱	⋱	⋮	⋮	⋮		⋮
	⋮		⋮	⋮	⋮	⋮			⋱	0	0	⋮		⋮
	⋮		⋮	⋮	⋮	⋮				⋱	+	0	⋮		⋮
t	0	⋯	0	−	−	−	⋯	⋯	⋯	−	+	0	⋯	0

As each row of M represents a vector $b^{\prec^{s,u}}$, to obtain (45.6) we must describe $\delta(\chi^t - \chi^s)$ as a convex combination of the rows of M, for some $\delta \geq 0$.

We call the $+$ entries in the matrix the 'diagonal' elements. Now for each row of M, the sum of its entries is 0, as $b^{\prec^{s,u}}(S) = f(S) = b^{\prec}(S) = 0$. Hence, if a 'diagonal' element $b^{\prec^{s,u}}(u)$ is equal to 0 for some $u \in (s,t]_{\prec}$, then the corresponding row of M is all-zero. So in this case we can take $\delta = 0$ in (45.6).

If $b^{\prec^{s,u}}(u) > 0$ for each $u \in (s,t]_{\prec}$ (that is, if each 'diagonal' element is strictly positive), then $\chi^t - \chi^s$ can be described as a nonnegative combination of the rows of M (by the sign pattern of M and since the entries in each row of M add up to 0). Hence $\delta(\chi^t - \chi^s)$ is a convex combination of the rows of M for some $\delta > 0$, yielding again (45.6).

45.4. Minimizing a submodular function

We now describe the algorithm to find the minimum value of a submodular set function f on S. We assume $f(\emptyset) = 0$ and $S = \{1, \ldots, n\}$.

We iteratively update a vector $x \in B_f$, given as a convex combination

(45.11) $\qquad x = \lambda_1 b^{\prec_1} + \cdots + \lambda_k b^{\prec_k}$,

where the $\prec_i$ are total orders of S, and where the λ_i are positive and sum to 1. Initially, we choose an arbitrary total order $\prec$ and set $x := b^{\prec}$ (so $k = 1$ and $\prec_1 = \prec$).

We describe the iteration. Consider the directed graph $D = (S, A)$, with

(45.12) $\qquad A := \{(u, v) \mid \exists i = 1, \ldots, k : u \prec_i v\}$.

Define

(45.13) $\qquad P := \{v \in S \mid x(v) > 0\}$ and $N := \{v \in S \mid x(v) < 0\}$.

Case 1: D has no path from P to N. Then let U be the set of vertices of D that can reach N by a directed path. So $N \subseteq U$ and $U \cap P = \emptyset$; that is, U contains all negative components of x and no positive components. Hence $x(W) \geq x(U)$ for each $W \subseteq S$. As no arcs of D enter U, U is a lower ideal of $\prec_i$, and hence $b^{\prec_i}(U) = f(U)$, for each $i = 1, \ldots, k$. Therefore, for each $W \subseteq S$:

(45.14) $\qquad f(U) = \sum_{i=1}^{k} \lambda_i b^{\prec_i}(U) = x(U) \leq x(W) \leq f(W)$.

So U minimizes f.

Case 2: D has a path from P to N. Let $d(v)$ denote the distance in D from P to v ($=$ minimum number of arcs in a directed path from P to v). Set $d(v) := \infty$ if v is not reachable from P. Choose $s, t \in S$ as follows.

Let t be the element in N reachable from P with $d(t)$ maximum, such that t is largest. Let s be the element with $(s,t) \in A$, $d(s) = d(t) - 1$, and s largest. Let α be the maximum of $|(s,t]_{\prec_i}|$ over $i = 1, \ldots, k$. Reorder indices such that $|(s,t]_{\prec_1}| = \alpha$.

By (45.6), we can find $\delta \geq 0$ and describe

(45.15) $\qquad b^{\prec_1} + \delta(\chi^t - \chi^s)$

as a convex combination of the $b^{\prec_1^{s,u}}$ for $u \in (s,t]_{\prec_1}$. Then with (45.11) we obtain

(45.16) $\qquad y := x + \lambda_1 \delta(\chi^t - \chi^s)$

as a convex combination of $b^{\prec_i}$ ($i = 2, \ldots, k$) and $b^{\prec_1^{s,u}}$ ($u \in (s,t]_{\prec_1}$).

Let x' be the point on the line segment $\overline{xy}$ closest to y satisfying $x'(t) \leq 0$. (So $x'(t) = 0$ or $x' = y$.) We can describe x' as a convex combination of $b^{\prec_i}$ ($i = 1, \ldots, k$) and $b^{\prec_1^{s,u}}$ ($u \in (s,t]_{\prec_1}$). Moreover, if $x'(t) < 0$, then we can do without $b^{\prec_1}$.

We reduce the number of terms in the convex decomposition of x' to at most $|S|$ by linear algebra: any affine dependence of the vectors in the decomposition yields a reduction of the number of terms in the decomposition, as in the standard proof of Carathéodory's theorem (subtract an appropriate multiple of the linear expression giving the affine dependence, from the linear expression giving the convex combination, such that all coefficients remain nonnegative, and at least one becomes 0). As all $b^{\prec}$ belong to a hyperplane, this reduces the number of terms to at most $|S|$.

Then reset $x := x'$ and iterate. This finishes the description of the algorithm.

45.5. Running time of the algorithm

We show that the number of iterations is at most $|S|^6$. Consider any iteration. Let

(45.17) $\qquad \beta := $ number of $i \in \{1, \ldots, k\}$ with $|(s,t]_{\prec_i}| = \alpha$.

Let x', d', A', P', N', t', s', α', β' be the objects x, d, A, P, N, t, s, α, β in the next iteration (if any). Then

(45.18) $\qquad$ for all $v \in S$, $d'(v) \geq d(v)$,

and

(45.19) $\qquad$ if $d'(v) = d(v)$ for all $v \in S$, then $(d'(t'), t', s', \alpha', \beta')$ is lexicographically less than $(d(t), t, s, \alpha, \beta)$.

Since each of $d(t), t, s, \alpha, \beta$ is at most $|S|$, and since (if $d(v)$ is unchanged for all v) there are at most $|S|$ pairs $(d(t), t)$, (45.19) implies that in at most $|S|^4$

iterations $d(v)$ increases for at least one v. Any fixed v can have at most $|S|$ such increases, and hence the number of iterations is at most $|S|^6$.

In order to prove (45.18) and (45.19), notice that

(45.20)   for each arc $(v,w) \in A' \setminus A$ we have $s \preceq_1 w \prec_1 v \preceq_1 t$.

Indeed, as $(v,w) \notin A$ we have $w \prec_1 v$. As $(v,w) \in A'$, we have $v \prec_1^{s,u} w$ for some $u \in (s,t]_{\prec_1}$. Hence the definition of $\prec_1^{s,u}$ gives $v = u$ and $s \preceq_1 w \prec_1 u$. This shows (45.20).

If (45.18) does not hold, then $A' \setminus A$ contains an arc (v,w) with $d(w) \geq d(v) + 2$ (using that $P' \subseteq P$). By (45.20), $s \preceq_1 w \prec_1 v \preceq_1 t$, and so $d(w) \leq d(s) + 1 = d(t) \leq d(v) + 1$, a contradiction. This shows (45.18).

To prove (45.19), assume that $d'(v) = d(v)$ for all $v \in S$. As $x'(t') < 0$, we have $x(t') < 0$ or $t' = s$. So by our criterion for choosing t (maximizing $(d(t),t)$ lexicographically), and since $d(s) < d(t)$, we know that $d(t') \leq d(t)$, and that if $d(t') = d(t)$, then $t' \leq t$.

Next assume that moreover $d(t') = d(t)$ and $t' = t$. As $(s',t) \in A'$, and as (by (45.20)) $A' \setminus A$ contains no arc entering t, we have $(s',t) \in A$, and so $s' \leq s$, by the maximality of s.

Finally assume that moreover $s' = s$. As $(s,t]_{\prec_1^{s,u}}$ is a proper subset of $(s,t]_{\prec_1}$ for each $u \in (s,t]_{\prec_1}$, we know that $\alpha' \leq \alpha$. Moreover, if $\alpha' = \alpha$, then $\beta' < \beta$, since $\prec_1$ does not occur anymore among the linear orders making the convex combination, as $x'(t) < 0$. This proves (45.19).

We therefore have proved:

Theorem 45.1. *Given a submodular function f by a value giving oracle, a set U minimizing $f(U)$ can be found in strongly polynomial time.*

Proof. See above.   ∎

This algorithm performs the elementary arithmetic operations on function values, including multiplication and division (in order to solve certain systems of linear equations). One would wish to have a 'fully combinatorial' algorithm, in which the function values are only compared, added, and subtracted. The existence of such an algorithm was shown by Iwata [2002a, 2002c], by extending the algorithm of Iwata, Fleischer, and Fujishige [2000, 2001].

Notes. In the algorithm, we have chosen t and s largest possible, in some fixed order of S. To obtain the above running time bound it only suffices to choose t and s in a consistent way. That is, if the set of choices for t is the same as in the previous iteration, then we should choose the same t — and similarly for s. This roots in the idea of 'consistent breadth-first search' of Schönsleben [1980].

The observation that the number of iterations in the algorithm of Section 45.4 is $O(|S|^6)$ instead of $O(|S|^7)$ is due to L.K. Fleischer. Vygen [2002] showed that the number of iterations can in fact be bounded by $O(|S|^5)$.

The algorithm described above has been speeded up by Fleischer and Iwata [2000,2002], by incorporating a push-relabel type of iteration based on approximate distances instead of precise distances (like Goldberg's method for maximum flow, given in Section 10.7). Iwata [2002b] combined the approaches of Iwata, Fleischer, and Fujishige [2000,2001] and Schrijver [2000a] to obtain a faster algorithm. A descent method for submodular function minimization based on an oracle for membership of the base polytope was given by Fujishige and Iwata [2002].

Surveys and background on submodular function minimization are given by Fleischer [2000b] and McCormick [2001].

45.6. Minimizing a symmetric submodular function

A set function f on S is called *symmetric* if $f(U) = f(S \setminus U)$ for each $U \subseteq S$. The minimum of a symmetric submodular function f is attained by $\emptyset$, since for each $U \subseteq S$ one has

(45.21) $\quad 2f(U) = f(U) + f(S \setminus U) \geq f(\emptyset) + f(S) = 2f(\emptyset)$.

By extending a method of Nagamochi and Ibaraki [1992b] for finding the minimum nonempty cut in an undirected graph, Queyranne [1995,1998] gave an easy combinatorial algorithm to find a nonempty proper subset U of S minimizing $f(U)$, where f is given by a value giving oracle. We may assume that $f(\emptyset) = f(S) = 0$, by resetting $f(U) := f(U) - f(\emptyset)$ for all $U \subseteq S$.

Call an ordering $s_1, \ldots, s_n$ of the elements of S a *legal order* of S for f, if, for each $i = 1, \ldots, n$,

(45.22) $\quad f(\{s_1, \ldots, s_{i-1}, x\}) - f(\{x\})$

is minimized over $x \in S \setminus \{s_1, \ldots, s_{i-1}\}$ by $x = s_i$. One easily finds a legal order, by $O(|S|^2)$ oracle calls (for the value of f).

Now the algorithm is (where a set U *splits* a set X if both $X \cap U$ and $X \setminus U$ are nonempty):

(45.23) $\quad$ Find a legal order $(s_1, \ldots, s_n)$ of S for f.
Determine (recursively) a nonempty proper subset T of S not splitting $\{s_{n-1}, s_n\}$, minimizing $f(T)$. (This can be done by identifying s_{n-1} and s_n.)
Then the minimum value of $f(U)$ over nonempty proper subsets U of S is equal to $\min\{f(T), f(\{s_n\})\}$.

The correctness of the algorithm follows from, for $n \geq 2$:

(45.24) $\quad f(U) \geq f(\{s_n\})$ for each $U \subseteq S$ splitting $\{s_{n-1}, s_n\}$.

This can be proved as follows. Define $t_0 := s_1$. For $i = 1, \ldots, n-1$, define $t_i := s_j$, where j is the smallest index such that $j > i$ and such that U splits $\{s_i, s_j\}$. For $i = 0, \ldots, n$, let $U_i := \{s_1, \ldots, s_i\}$. Note that for each $i = 1, \ldots, n-1$ one has

(45.25) $\quad f(U_{i-1} \cup \{t_i\}) - f(\{t_i\}) \geq f(U_{i-1} \cup \{t_{i-1}\}) - f(\{t_{i-1}\}),$

since if $t_{i-1} = t_i$ this is trivial, and if $t_{i-1} \neq t_i$, then $t_{i-1} = s_i$, in which case (45.25) follows from the legality of the order.

Moreover, for each $i = 1, \ldots, n-1$ (setting $\overline{U} := S \setminus U$):

(45.26) $\quad f(U_i \cup U) - f(U_{i-1} \cup U) + f(U_i \cup \overline{U}) - f(U_{i-1} \cup \overline{U})$
$\leq f(U_i \cup \{t_i\}) - f(U_{i-1} \cup \{t_i\}).$

In proving this, we may assume (by symmetry of U and $\overline{U}$) that $s_i \in \overline{U}$. Then $U_i \cup \overline{U} = U_{i-1} \cup \overline{U}$ and $t_i \in U$. So $f(U_i \cup \{t_i\}) + f(U_{i-1} \cup U) \geq f(U_{i-1} \cup \{t_i\}) + f(U_i \cup U)$, by submodularity. This gives (45.26).

Then we have:

(45.27)
$$\begin{aligned}
f(s_n) &- 2f(U) \\
&= f(U_{n-1} \cup U) + f(U_{n-1} \cup \overline{U}) - f(U_0 \cup U) - f(U_0 \cup \overline{U}) \\
&= \sum_{i=1}^{n-1}(f(U_i \cup U) - f(U_{i-1} \cup U) + f(U_i \cup \overline{U}) - f(U_{i-1} \cup \overline{U})) \\
&\leq \sum_{i=1}^{n-1}(f(U_i \cup \{t_i\}) - f(U_{i-1} \cup \{t_i\})) \\
&\leq \sum_{i=1}^{n-1}(f(U_i \cup \{t_i\}) - f(U_{i-1} \cup \{t_{i-1}\}) + f(\{t_{i-1}\}) - f(\{t_i\})) \\
&= f(U_{n-1} \cup \{t_{n-1}\}) - f(\{t_{n-1}\}) - f(\{t_0\}) + f(\{t_0\}) = -f(s_n)
\end{aligned}$$

(where the first inequality follows from (45.26), and the second inequality from (45.25)). This shows (45.24).

Notes. Fujishige [1998] gave an alternative correctness proof. Nagamochi and Ibaraki [1998] extended the algorithm to minimizing submodular functions f satisfying

(45.28) $\quad f(T) + f(U) \geq f(T \setminus U) + f(U \setminus T)$

for all $T, U \subseteq S$. Rizzi [2000b] gave an extension.

45.7. Minimizing a submodular function over the odd sets

From the strong polynomial-time solvability of submodular function minimization, one can derive that also a set of odd cardinality minimizing f (over the odd sets) is solvable in strongly polynomial time (Grötschel, Lovász, and Schrijver [1981,1984a,1988] (the second paper corrects a wrong argument given in the first paper)).

Theorem 45.2. *Given a submodular set function f on S (by a value giving oracle) and a nonempty subset T of S, one can find in strongly polynomial time a set $W \subseteq S$ minimizing $f(W)$ over W with $|W \cap T|$ odd.*

Proof. The case T odd can be reduced to the case T even as follows. Find for each $t \in T$ a subset W_t of $S - t$ with $W_t \cap (T - t)$ odd, and minimizing $f(W_t)$. Moreover, find a subset U of S minimizing $f(U)$ over $U \supseteq T$. Then a set that attains the minimum among $f(U)$ and the $f(W_t)$, is an output as required.

So we can assume that T is even. We describe a recursive algorithm. Say that a set U *splits* T if both $T \cap U$ and $T \setminus U$ are nonempty. First find a set U minimizing $f(U)$ over all subsets U of S splitting T. This can be done by finding for all $s, t \in T$ a set $U_{s,t}$ minimizing $f(U_{s,t})$ over all subsets of S containing s and not containing t (this amounts to submodular function minimization), and taking for U a set that minimizes $f(U_{s,t})$ over all such s, t.

If $U \cap T$ is odd, we output $W := U$. If $U \cap T$ is even, then recursively we find a set X minimizing $f(X)$ over all X with $X \cap (T \cap U)$ odd, and not splitting $T \setminus U$. This can be done by shrinking $T \setminus U$ to one element. Also, recursively we can find a set Y minimizing $f(Y)$ over all Y with $Y \cap (T \setminus U)$ odd, and not splitting $T \cap U$. Output an X or Y attaining the minimum of $f(X)$ and $f(Y)$.

This gives a strongly polynomial-time algorithm as the total number of recursive calls is at most $|T| - 2$ (since $2 + (|T \cap U| - 2) + (|T \setminus U| - 2) = |T| - 2$).

To see the correctness, let W minimize $f(W)$ over those W with $|W \cap T|$ odd. Suppose that $f(W) < f(X)$ and $f(W) < f(Y)$. As $f(W) < f(X)$, W splits $T \setminus U$, and hence $W \cup U$ splits T. Similarly, $f(W) < f(Y)$ implies that $W \cap U$ splits T.

Since $W \cap T$ is odd and $U \cap T$ is even, either $(W \cap U) \cap T$ or $(W \cup U) \cap T$ is odd.

If $(W \cap U) \cap T$ is odd, then $f(W \cap U) \geq f(W)$ (as W minimizes $f(W)$ over W with $W \cap T$ odd) and $f(W \cup U) \geq f(U)$ (as $W \cup U$ splits T and as U minimizes $f(U)$ over U splitting T). Hence, by the submodularity of f, $f(W \cap U) = f(W)$. Since $(W \cap U) \cap (T \cap U) = (W \cap U) \cap T$ is odd and since $W \cap U$ does not split $T \setminus U$, we have $f(W) = f(W \cap U) \geq f(X)$, contradicting our assumption.

If $(W \cup U) \cap T$ is odd, a symmetric argument gives a contradiction. ∎

This generalizes the strong polynomial-time solvability of finding a minimum-capacity odd cut in a graph, proved by Padberg and Rao [1982] (Corollary 25.6a). For a further generalization, see Section 49.11a.

Chapter 46

Polymatroid intersection

The intersection of two polymatroids behaves as nice as the intersection of two matroids, as was shown by Edmonds again. We study in this chapter min-max relations, polyhedral characterizations, and total dual integrality results. In the next chapter we go over to the algorithmic questions.

46.1. Box-total dual integrality of polymatroid intersection

We saw in Section 44.2 that the greedy algorithm yields a proof that an integer-valued submodular function gives an integer polymatroid. The interest of polymatroids for combinatorial optimization is enlarged by the fundamental result of Edmonds [1970b] that also the intersection of two integer polymatroids is integer, thus generalizing the matroid intersection theorem. In order to obtain this result, we first show a more general theorem (also due to Edmonds [1970b]).

Consider, for submodular set functions f_1, f_2 on S, the system:

(46.1) $\quad x(U) \leq f_1(U) \quad \text{for } U \subseteq S,$
$\qquad x(U) \leq f_2(U) \quad \text{for } U \subseteq S.$

Then:

Theorem 46.1. *If f_1 and f_2 are submodular, then (46.1) is box-TDI.*

Proof. Choose $w : S \to \mathbb{R}$. Let y_1, y_2 attain

(46.2) $\quad \min\{\sum_{U \subseteq S}(y_1(U)f_1(U) + y_2(U)f_2(U)) \mid\mid$
$\qquad y_1, y_2 \in \mathbb{R}_+^{\mathcal{P}(S)}, \sum_{U \subseteq S}(y_1(U) + y_2(U))\chi^U = w\}.$

For $i = 1, 2$, define $w_i : S \to \mathbb{R}$ by

(46.3) $\quad w_i := \sum_{U \subseteq S} y_i(U)\chi^U.$

Then y_i attains

(46.4) $$\min\{\sum_{U \subseteq S} y_i(U) f_i(U) \mid y_i \in \mathbb{R}_+^{\mathcal{P}(S)}, \sum_{U \subseteq S} y_i(U) \chi^U = w_i\}.$$

So by Theorem 44.3, we can assume that the collections

(46.5) $$\mathcal{F}_i := \{U \subseteq S \mid y_i(U) > 0\}$$

are chains. Hence, by Theorem 41.11, (46.2) has an optimum solution such that the inequalities with positive coefficients have a totally unimodular constraint matrix. Therefore, by Theorem 5.35, (46.1) is box-TDI. ∎

(This proof method is due to Edmonds [1970b].)

46.2. Consequences

Theorem 46.1 has the following consequences. First, the integrality of the intersection of two polymatroids:

Corollary 46.1a (polymatroid intersection theorem). *The intersection of two integer (extended) polymatroids is box-integer.*

Proof. If P_{f_1} and P_{f_2} are integer polymatroids, f_1 and f_2 can be taken to be integer-valued, by Corollary 44.3g. Hence (46.1) determines a box-integer polyhedron. ∎

Next, a min-max relation:

Corollary 46.1b. *Let f_1 and f_2 be submodular set functions on S with $f_1(\emptyset) = f_2(\emptyset) = 0$. Then*

(46.6) $$\max\{x(U) \mid x \in EP_{f_1} \cap EP_{f_2}\} = \min_{T \subseteq U}(f_1(T) + f_2(U \setminus T))$$

for each $U \subseteq S$.

Proof. This follows by maximizing $w^T x$ over (46.1) for $w := \chi^U$, and applying Theorem 46.1. ∎

Similarly, for (nonextended) polymatroids:

Corollary 46.1c. *Let f_1 and f_2 be nondecreasing submodular set functions on S with $f_1(\emptyset) = f_2(\emptyset) = 0$. Then*

(46.7) $$\max\{x(U) \mid x \in P_{f_1} \cap P_{f_2}\} = \min_{T \subseteq U}(f_1(T) + f_2(U \setminus T))$$

for each $U \subseteq S$.

Proof. As the previous corollary. ∎

Let f_1 and f_2 be submodular set functions on S with $f_1(\emptyset) = f_2(\emptyset) = 0$. Define

(46.8) $\quad f(U) := \min_{T \subseteq U}(f_1(T) + f_2(U \setminus T))$

for $U \subseteq S$. It is easy to see that a vector x belongs to $P_{f_1} \cap P_{f_2}$ if and only if

(46.9) $\quad\quad x_s \geq 0 \quad\quad (s \in S),$
$\quad\quad\quad\quad x(U) \leq f(U) \quad (U \subseteq S).$

Moreover, system (46.9) is box-totally dual integral, since $f(U) \leq f_i(U)$ for each $U \subseteq S$ and $i = 1, 2$.

A consequence is that $P_{f_1} \cap P_{f_2}$ is integer if and only if f is integer. It may occur that P_{f_1} and P_{f_2} are not integer (i.e., f_1 and f_2 are not integer), while $P_{f_1} \cap P_{f_2}$ is integer (i.e., f is integer). For instance, take $P_{f_1} = \{(x_1, x_2) \mid 0 \leq x_1 \leq 1, 0 \leq x_2 \leq \frac{3}{2}\}$ and $P_{f_2} = \{(x_1, x_2) \mid (x_2, x_1) \in P_{f_1}\}$.

Many other results on polymatroid intersection may be deduced from Theorem 46.1, by considering derived polymatroids (cf. McDiarmid [1978]). For instance, if P_{f_1} and P_{f_2} are integer polymatroids in $\mathbb{R}^S$, v and w are integer vectors, and k and ℓ are integers, then the polytope

(46.10) $\quad \{x \in P_{f_1} \cap P_{f_2} \mid v \leq x \leq w, k \leq x(S) \leq \ell\}$

is integer again. To see this, it suffices to show that the polytope $P_{f_1} \cap P_{f_2} \cap \{x \mid x(S) = k\}$ is integer for any integer k. We can reset $f_1(S) := \min\{f_1(S), k\}$. Then the polytope is a face of $P_{f_1} \cap P_{f_2}$, and hence is integer. In fact, the system determining (46.10) is box-TDI — see Corollary 49.12d.

The intersection of *three* integer polymatroids can have noninteger vertices, as the following example shows. Let $S = \{1, 2, 3, 4\}$ and let P_1, P_2 and P_3 be the following polymatroids:

(46.11) $\quad P_1 := \{x \in \mathbb{R}^S \mid x \geq 0, x(\{1,2\}) \leq 1, x(\{3,4\}) \leq 1\},$
$\quad\quad\quad\quad P_2 := \{x \in \mathbb{R}^S \mid x \geq 0, x(\{1,3\}) \leq 1, x(\{2,4\}) \leq 1\},$
$\quad\quad\quad\quad P_3 := \{x \in \mathbb{R}^S \mid x \geq 0, x(\{1,4\}) \leq 1, x(\{2,3\}) \leq 1\}.$

(So each P_i is the independent set polytope of a partition matroid.) Now the vector $(\frac{1}{2}, \frac{1}{2}, \frac{1}{2}, \frac{1}{2})$ is in $P_1 \cap P_2 \cap P_3$, but the only integer vectors in $P_1 \cap P_2 \cap P_3$ are the 0,1 vectors with at most one 1.

46.3. Contrapolymatroid intersection

Similar results as in the previous sections can be shown for the intersection of two contrapolymatroids. Such results can be proved similarly, or can be derived from the corresponding results for polymatroids.

Consider the system, for set functions g_1, g_2 on S:

(46.12) $\quad x(U) \geq g_1(U) \quad$ for $U \subseteq S,$
$\quad\quad\quad\quad x(U) \geq g_2(U) \quad$ for $U \subseteq S.$

Then Theorem 46.1 gives:

Corollary 46.1d. *If g_1 and g_2 are supermodular, then (46.12) is box-TDI.*

Proof. This follows from the box-total dual integrality of (46.1) taking $f_i := -g_i$ for $i = 1, 2$. ∎

46.4. Intersecting a polymatroid and a contrapolymatroid

Let S be a finite set. For set functions f and g on S consider the system

(46.13) $\quad x(U) \leq f(U) \quad \text{for } U \subseteq S,$
$\qquad\quad x(U) \geq g(U) \quad \text{for } U \subseteq S.$

Theorem 46.2. *If f is submodular and g is supermodular, then system (46.13) is box-TDI.*

Proof. We can assume that $f(\emptyset) \geq 0$ and $g(\emptyset) \leq 0$. Choose $w \in \mathbb{R}^S$, and consider the dual problem of maximizing $w^\mathsf{T} x$ over (46.13):

(46.14) $\quad \min\{\sum_{U \subseteq S} y(U)f(U) - \sum_{U \subseteq S} z(U)g(U) \mid$
$\qquad\qquad y, z \in \mathbb{R}_+^{\mathcal{P}(S)}, \sum_{U \subseteq S} y(U)\chi^U - \sum_{U \subseteq S} z(U)\chi^U = w\}.$

Let y, z attain this minimum. Define

(46.15) $\quad u := \sum_{U \subseteq S} y(U)\chi^U \text{ and } v := \sum_{U \subseteq S} z(U)\chi^U.$

So y attains

(46.16) $\quad \min\{\sum_{U \subseteq S} y(U)f(U) \mid y \in \mathbb{R}_+^{\mathcal{P}(S)}, \sum_{U \subseteq S} y(U)\chi^U = u\}$

and z attains

(46.17) $\quad \max\{\sum_{U \subseteq S} z(U)g(U) \mid z \in \mathbb{R}_+^{\mathcal{P}(S)}, \sum_{U \subseteq S} z(U)\chi^U = v\}.$

By Theorem 44.3, (46.16) has an optimum solution y with $\mathcal{F} := \{U \mid y(U) > 0\}$ is a chain. Similarly, (46.17) has an optimum solution z with $\mathcal{G} := \{U \mid z(U) > 0\}$ is a chain. Hence by Theorem 41.11, minimum (46.14) has an optimum solution such that the inequalities corresponding to positive coefficients have a totally unimodular constraint matrix. Hence by Theorem 5.35, (46.13) is box-TDI. ∎

So for the intersection of a polymatroid and a contrapolymatroid one gets:

Corollary 46.2a. *The intersection of an integer extended polymatroid and an integer extended contrapolymatroid is integer.*

Proof. Directly from the fact that an integer extended polymatroid is the solution set of $x(U) \leq f(U)$ ($U \subseteq S$) for some integer submodular set function on S, and an integer extended contrapolymatroid is the solution set of $x(U) \geq g(U)$ ($U \subseteq S$) for some integer submodular set function on S. Hence, by Theorem 46.2, the intersection is determined by a TDI system with integer right-hand sides. So the intersection is integer. ∎

46.5. Frank's discrete sandwich theorem

Frank [1982b] showed the following 'discrete sandwich theorem' (analogous to the 'continuous sandwich theorem', stating the existence of a linear function between a convex and a concave function):

Corollary 46.2b (Frank's discrete sandwich theorem). *Let f be a submodular and g a supermodular set function on S, with $g \leq f$. Then there exists a modular set function h on S with $g \leq h \leq f$. If f and g are integer, h can be taken integer.*

Proof. We can assume that $g(\emptyset) = 0 = f(\emptyset)$, by resetting $f(U) := f(U) - f(\emptyset)$ and $g(U) := g(U) - f(\emptyset)$, for each $U \subseteq S$, and $g(\emptyset) := 0$.

Define $f'(U) := f(S) - g(S \setminus U)$ for each $U \subseteq S$. Then f' is submodular. Now by Corollary 46.1b:

(46.18) $\quad \max\{x(S) \mid x(U) \leq f(U), x(U) \leq f'(U) \text{ for each } U \subseteq S\}$
$\quad = \min\{f(T) + f'(S \setminus T) \mid T \subseteq S\}.$

The minimum is at least $f(S)$, since $f(T) + f'(S \setminus T) = f(T) + f(S) - g(T) \geq f(S)$. Hence there exists an $x \in \mathbb{R}^S$ with $x(U) \leq f(U)$ and $x(U) \leq f'(U)$ for each $U \subseteq S$ and with $x(S) = f(S)$. Defining $h(U) := x(U)$, gives the modular function as required, since for each $U \subseteq S$:

(46.19) $\quad g(U) = f(S) - f'(S \setminus U) \leq x(S) - x(S \setminus U) = x(U) \leq f(U).$

If f and g are integer, we can choose x integer, implying that h is integer. ∎

As Lovász [1983c] observed, the first part of this result can be derived from the continuous sandwich theorem: define $\tilde{f} : \mathbb{R}_+^S \to \mathbb{R}$ by

(46.20) $\quad \tilde{f}(x) := \sum_{i=1}^{k} \lambda_i f(U_i),$

where $\emptyset \neq U_1 \subset U_2 \subset \cdots \subset U_n \subseteq S$ and $\lambda_1, \ldots, \lambda_k > 0$ are such that $x = \sum_{i=1}^{k} \lambda_i \chi^{U_i}$. Define $\tilde{g}$ similarly. Then $\tilde{f}$ is convex and $\tilde{g}$ is concave, and $\tilde{g} \leq \tilde{f}$. Hence there is a linear function $\tilde{h}$ satisfying $\tilde{g} \leq \tilde{h} \leq \tilde{f}$. This gives the function h as required.

46.6. Integer decomposition

Integer polymatroids have the integer decomposition property. More generally:

Corollary 46.2c. *Let $P_1, \ldots, P_k$ be integer polymatroids. Then each integer vector in $P_1 + \cdots + P_k$ is a sum $x_1 + \cdots + x_k$ of integer vectors $x_1 \in P_1, \ldots, x_k \in P_k$.*

Proof. It suffices to show this for $k = 2$; the general case follows by induction (as the sum of integer polymatroids is again an integer polymatroid, by Theorem 44.6). Choose an integer vector $x \in P_1 + P_2$. Let Q be the contrapolymatroid given by

(46.21) $Q := x - P_2$.

Then $P_1 \cap Q \neq \emptyset$, since $x = x_1 + x_2$ for some $x_1 \in P_1$ and $x_2 \in P_2$, implying $x_1 \in P_1 \cap Q$. Now Q is integer as x and P_2 are integer. Hence by Corollary 46.2a, $P_1 \cap Q$ contains an integer vector y. Then $x - y \in P_2$, and so x is the sum of $y \in P_1$ and $x - y \in P_2$. ∎

This implies the integer decomposition property for integer polymatroids, proved by Giles [1975] (also by Baum and Trotter [1981]):

Corollary 46.2d. *An integer polymatroid has the integer decomposition property.*

Proof. Directly from Corollary 46.2c, by taking all P_i equal. ∎

This gives the following integer rounding properties (Baum and Trotter [1981]). Let P_f be the integer polymatroid determined by some integer submodular set function f on S. Let $\mathcal{B}$ be the collection of integer base vectors of P_f. Let B be the $\mathcal{B} \times S$ incidence matrix. Then for each $c \in \mathbb{Z}_+^S$, one has

(46.22) $\min\{y^\mathsf{T} \mathbf{1} \mid y \in \mathbb{Z}_+^\mathcal{B}, y^\mathsf{T} B \geq c^\mathsf{T}\}$
$= \lceil \min\{y^\mathsf{T} \mathbf{1} \mid y \in \mathbb{R}_+^\mathcal{B}, y^\mathsf{T} B \geq c^\mathsf{T}\} \rceil$.

Indeed, $\geq$ is trivial. To see equality, let k be equal to the right-hand side. Then $c \in k \cdot P_f$, and hence, by Corollary 46.2d, $c \leq b_1 + \cdots + b_k$ for rows $b_1, \ldots, b_k$ of B. This shows equality.

Note that the right-hand side in (46.22) is equal to $\lceil \max\{c^\mathsf{T} x \mid x \in A(P_f)\} \rceil$, where $A(P_f)$ is the antiblocking polyhedron of P_f.

Similarly, one has:

(46.23) $\max\{y^\mathsf{T} \mathbf{1} \mid y \in \mathbb{Z}_+^\mathcal{B}, y^\mathsf{T} B \leq c^\mathsf{T}\}$
$= \lfloor \max\{y^\mathsf{T} \mathbf{1} \mid y \in \mathbb{R}_+^\mathcal{B}, y^\mathsf{T} B \leq c^\mathsf{T}\} \rfloor$.

Now the right-hand side is equal to $\lfloor\min\{c^\mathsf{T} x \mid x \in B(Q)\}\rfloor$, where $B(Q)$ is the blocking polyhedron of $Q := \{x \in \mathbb{R}^S \mid x(U) \geq f(S) - f(S \setminus U)$ for $U \subseteq S\}$.

If f is the rank function of a matroid, then (46.22) describes the minimum number of bases covering S, while (46.23) describes the maximum number of disjoint bases.

46.7. Further results and notes

46.7a. Up and down hull of the common base vectors

Let f_1 and f_2 be nondecreasing submodular set functions on S, with $f_1(\emptyset) = f_2(\emptyset) = 0$ and $f_1(S) = f_2(S)$, and let P_1 and P_2 be the associated polymatroids. Let F_1 and F_2 be the faces of base vectors of P_1 and of P_2, respectively. Suppose that $F_1 \cap F_2 \neq \emptyset$, equivalently that

(46.24) $\quad f_1(S) = f_2(S) = \max\{x(S) \mid x \in P_1 \cap P_2\} = \min_{U \subseteq S} f_1(U) + f_2(S \setminus U)$.

Consider the polyhedra P and Q defined by

(46.25) $\quad P := \{x \in \mathbb{R}^S_+ \mid x \leq y \text{ for some } y \text{ in } F_1 \cap F_2\}$,
$\quad\quad\quad\quad Q := \{x \in \mathbb{R}^S_+ \mid x \geq y \text{ for some } y \text{ in } F_1 \cap F_2\}$.

So if f_1 and f_2 are the rank functions of matroids on S, then P is just the convex hull of incidence vectors of subsets of S which are contained in a common base.

Note that F_1 and F_2 are the faces of minimal vectors in the contrapolymatroids Q_1 and Q_2 associated with the supermodular set functions g_1 and g_2 on S given by

(46.26) $\quad g_i(U) := f_i(S) - f_i(S \setminus U)$

for $U \subseteq S$ and $i = 1, 2$ (cf. Section 44.5). So $P \subseteq P_1 \cap P_2$ and $Q \subseteq Q_1 \cap Q_2$.

Let the set functions f and g on S be defined by

(46.27) $\quad f(U) := \max\{x(U) \mid x \in P_1 \cap P_2\} = \min_{T \subseteq U}(f_1(T) + f_2(U \setminus T))$,
$\quad\quad\quad\quad g(U) := \min\{x(U) \mid x \in Q_1 \cap Q_2\} = \max_{T \subseteq U}(g_1(T) + g_2(U \setminus T))$,

for $U \subseteq S$ (cf. Corollary 46.1c). Then $f(S) = g(S) = f_1(S) = f_2(S) = g_1(S) = g_2(S)$.

It is easy to see that if x belongs to Q, then

(46.28) $\quad x(U) \geq f(S) - f(S \setminus U)$ for each $U \subseteq S$

(note that $x \geq \mathbf{0}$ follows from (46.28) by taking $U = \{s\}$). Indeed, if $x \geq z$ with $z \in F_1 \cap F_2$, then $x(U) \geq z(U) = f(S) - z(S \setminus U) \geq f(S) - f(S \setminus U)$, as $z \in P_1 \cap P_2$.

Similarly, if x belongs to P, then

(46.29) $\quad\quad x_s \geq 0 \quad\quad\quad\quad\quad\quad\quad\quad (s \in S)$,
$\quad\quad\quad\quad x(U) \leq g(S) - g(S \setminus U) \quad (U \subseteq S)$.

In fact, the systems (46.28) and (46.29) determine Q and P respectively. This was shown by Cunningham [1977] and McDiarmid [1978], thus proving a conjecture of Fulkerson [1971a] (cf. Weinberger [1976]).

Theorem 46.3. *Polyhedron Q is determined by (46.28). Polyhedron P is determined by (46.29).*

Proof. Consider $x \in \mathbb{R}_+^S$ and let P_i' be the polymatroid $P_i' := \{y \in P_i \mid y \leq x\}$ for $i = 1, 2$ (cf. Section 44.1). By (44.8), the submodular function f_i' associated with P_i' is given by

(46.30) $\qquad f_i'(U) = \min_{T \subseteq U}(f_i(T) + x(U \setminus T))$

for $U \subseteq S$ and $i = 1, 2$. Now x is in Q if and only if there is a vector z in $P_1 \cap P_2$ with $z \leq x$ and $z(S) = f(S)$, i.e., if and only if there is a vector z in $P_1' \cap P_2'$ with $z(S) = f(S)$. By (46.7) such a vector exists if and only if

(46.31) $\qquad \min_{U \subseteq S}(f_1'(U) + f_2'(S \setminus U)) \geq f(S).$

Substituting (46.30) one finds that (46.31) is equivalent to (46.28).

The second statement of Theorem 46.3 is proved similarly. ∎

This theorem has a self-refining character. If k is a rational number with $k \leq f(S)$ and if $w \in Q$, then

(46.32) $\qquad \{x \in \mathbb{R}_+^S \mid x \geq z \text{ for some } z \text{ in } P_1 \cap P_2 \text{ with } z(S) = k\}$
$\qquad = \{x \in \mathbb{R}_+^S \mid x(U) \geq k - f(S \setminus U) \text{ for all } U \subseteq S\}$

and

(46.33) $\qquad \{x \in \mathbb{R}_+^S \mid x \geq z \text{ for some } z \text{ in } F_1 \cap F_2 \text{ with } z \leq w\}$
$\qquad = \{x \in \mathbb{R}_+^S \mid x(S \setminus (T \cup U)) \geq f(S) - w(U) - f(T) \text{ for disjoint } T, U \subseteq S\},$

as can be seen by taking appropriate subpolymatroids of P_1 and P_2 (cf. also McDiarmid [1976,1978]).

This has the following applications. Let $G = (V, E)$ be a bipartite graph, let $x \in \mathbb{R}_+^E$, and let k be a natural number. Then there exists a vector $z \leq x$ such that z is a convex combination of incidence vectors of matchings in G of size k if and only if

(46.34) $\qquad x(E[U]) \geq k - |V| + |U|$

for all $U \subseteq V$ (where $E[U]$ denotes the set of edges spanned by U). This can be derived as follows. Let V_1 and V_2 be the colour classes of G. For $F \subseteq E$, let $f_i(F)$ be the number of vertices in V_i covered by F (for $i = 1, 2$). Then $f(F)$ equals the maximum size of a matching in F, which is equal to the minimum number of vertices covering F. Hence the inequalities $x(F) \geq k - f(E \setminus F)$ (for $F \subseteq E$) follow from $x(E[U]) \geq k - |V \setminus U|$ (for $U \subseteq V$).

Another application is Corollary 52.3a on the up hull of the r-arborescence polytope (cf. Section 52.1a).

Gröflin and Hoffman [1981] gave a method to show the following:

Theorem 46.4. (46.28) *and* (46.29) *are box-TDI.*

Proof. We prove that (46.28) is box-TDI. The box-total dual integrality of (46.29) is proved similarly.

Let $\mathcal{R}$ be the collection of all pairs (T,U) of subsets of S with $T \cap U = \emptyset$. Then the system

(46.35) $\qquad x(S \setminus (T \cup U)) \geq f(S) - f_1(T) - f_2(U) \quad ((T,U) \in \mathcal{R})$

is equivalent to (46.28), in the following sense: by (46.27), (46.35) determines Q, and (46.35) contains all inequalities occurring in (46.28); moreover, all inequalities in (46.35) satisfied with equality by some $x \in Q$, also occur in (46.28). Hence, if (46.35) is box-totally dual integral, also (46.28) is box-totally dual integral. So it suffices to show the box-total dual integrality of (46.35). To this end, let $w \in \mathbb{Z}_+^S$, and consider the dual of minimizing $w^\mathsf{T} x$ over (46.35):

(46.36) $\qquad \max\{ \sum_{(T,U) \in \mathcal{R}} y(T,U)(f(S) - f_1(T) - f_2(U)) \mid$
$\qquad\qquad y \in \mathbb{R}_+^\mathcal{R}, \sum_{(T,U) \in \mathcal{R}} y(T,U) \chi^{S \setminus (T \cup U)} = w \}.$

We show that it is attained by an integer vector y if w is integer.

To this end, let y attain the maximum (46.36) such that

(46.37) $\qquad \sum_{(T,U) \in \mathcal{R}} y(T,U)(|T| + |S \setminus U|)(|U| + |S \setminus T|)$

is as small as possible. Then:

(46.38) $\qquad$ if $y(A,B) > 0$ and $y(C,D) > 0$, then either $A \subseteq C$ and $B \supseteq D$, or $A \supseteq C$ and $B \subseteq D$.

Suppose not. Define $\alpha := \min\{y(A,B), y(C,D)\}$. Define $y' : \mathcal{R} \to \mathbb{R}_+$ by

(46.39) $\qquad y'(A,B) := y(A,B) - \alpha,$
$\qquad\qquad y'(C,D) := y(C,D) - \alpha,$
$\qquad\qquad y'(A \cap C, B \cup D) := y(A \cap C, B \cup D) + \alpha,$
$\qquad\qquad y'(A \cup C, B \cap D) := y(A \cup C, B \cap D) + \alpha,$

and let y' coincide with y in the other components. One easily checks that

(46.40) $\qquad \sum_{(T,U) \in \mathcal{R}} y'(T,U) \chi^{S \setminus (T \cup U)} = \sum_{(T,U) \in \mathcal{R}} y(T,U) \chi^{S \setminus (T \cup U)}$ and
$\qquad\qquad \sum_{(T,U) \in \mathcal{R}} y'(T,U)(f(S) - f_1(T) - f_2(U))$
$\qquad\qquad \geq \sum_{(T,U) \in \mathcal{R}} y(T,U)(f(S) - f_1(T) - f_2(U)),$

by the submodularity of f_1 and f_2. So y' also attains the maximum (46.36). Moreover, one straightforwardly checks that replacing y by y' decreases (46.37).[36] This contradicts our assumption, and therefore proves (46.38).

---

[36] This can be seen with Theorem 2.1: Make a copy $\widetilde{S}$ of S, and, for any $U \subseteq S$, let $\widetilde{U}$ be the set of copies of elements of U. Define $X_{T,U} := T \cup (\widetilde{S} \setminus \widetilde{U})$. Then $|T| + |S \setminus U| = |X_{T,U}|$ and $|U| + |S \setminus T| = |(S \cup \widetilde{S}) \setminus X_{T,U}|$. Moreover, for (A,B) and (C,D) in $\mathcal{R}$ we have $X_{A \cap C, B \cup D} = X_{A,B} \cap X_{C,D}$ and $X_{A \cup C, B \cap D} = X_{A,B} \cup X_{C,D}$. So the replacements decrease (46.37) by Theorem 2.1, since $X_{A,B} \not\subseteq X_{C,D} \not\subseteq X_{A,B}$.

Let $\mathcal{R}_0 := \{(T,U) \in \mathcal{R} \mid y(T,U) > 0\} = \{(T_1, U_1), \ldots, (T_n, U_n)\}$ with $T_1 \subseteq \cdots \subseteq T_n$ and $U_n \subseteq \cdots \subseteq U_1$ (this is possible by (46.38)). Let M be the $\{0,1\}$ matrix with rows indexed by $\mathcal{R}_0$ and columns indexed by S, such that $M_{(T,U),s} = 1$ if and only if $s \notin T \cup U$. Then for each s in S, the indices i for which $M_{(T_i, U_i),s} = 1$ form a contiguous interval of $\{1, \ldots, n\}$. Hence M is totally unimodular (as it is a network matrix with directed tree being a directed path). So we have the box-total dual integrality of (46.35) by Theorem 5.35. ∎

Frank and Tardos [1984a] indicated a direct derivation of this theorem from the total dual integrality of (46.1).

There are a number of straightforward corollaries. As for the integrality of polyhedra:

Corollary 46.4a. *If f (or, equivalently, g) is integer, then the polyhedra P, Q, and $F_1 \cap F_2$ are integer.*

Proof. This follows directly from Theorem 46.4. Note that $F_1 \cap F_2$ is integer if and only if P is integer. ∎

Also a min-max relation follows:

Corollary 46.4b. *Let $M_1 = (S, \mathcal{I}_1)$ and $M_2 = (S, \mathcal{I}_2)$ be matroids, with rank functions r_1 and r_2 and let k be the maximum size of a common independent set. Then for any subset U of S,*

$$(46.41) \qquad \min_I |U \cap I| = \max_{S_1, \ldots, S_t} \sum_{i=1}^{t} (k - r(S \setminus S_i)),$$

where the minimum ranges over all common independent sets I with $|I| = k$, and where the maximum ranges over all partitions of U into sets $S_1, \ldots, S_t$ ($t \geq 0$), and where $r(T)$ denotes the maximum size of a common independent set contained in T.

Proof. Apply Theorem 46.4, taking $c := \chi^U$, $f_i := r_i$, and $f := r$. ∎

It is not necessarily true that if $F_1 \cap F_2$ is integer, then also $P_1 \cap P_2$ (or $Q_1 \cap Q_2$) is integer — i.e., that the converse implication of Corollary 46.4a holds. For instance, if

$$(46.42) \quad \begin{aligned} P_1 &:= \{(x,y,z)^\mathsf{T} \mid 0 \leq x \leq 1, 0 \leq y \leq 1, 0 \leq z \leq \tfrac{3}{2}, x+z \leq 2\}, \\ P_2 &:= \{(x,y,z)^\mathsf{T} \mid 0 \leq x \leq 1, 0 \leq y \leq 1, 0 \leq z \leq \tfrac{3}{2}, y+z \leq 2\}, \end{aligned}$$

then $F_1 \cap F_2 = \{(1,1,1)^\mathsf{T}\}$, but $(\tfrac{1}{2}, \tfrac{1}{2}, \tfrac{3}{2})^\mathsf{T}$ is a vertex of $P_1 \cap P_2$.

Related results on integer decomposition of integer polymatroids in McDiarmid [1983].

46.7b. Further notes

Giles [1975] characterized the facets of the intersection of two polymatroids. Ageev [1988] studied the problem of maximizing a concave function over the intersection of polymatroids.

Chapter 47

Polymatroid intersection algorithmically

In this chapter we consider the problem of finding a vector of maximum weight in the intersection of two (extended) polymatroids algorithmically. We describe a strongly polynomial-time algorithm for this problem in four stages (where f_1 and f_2 are submodular set functions on S):
- a strongly polynomial-time algorithm finding a maximum-size vector in $EP_{f_1} \cap EP_{f_2}$ (Section 47.1),
- a strongly polynomial-time algorithm finding a common base vector of f_1 and f_2 maximizing $x(s)$ for some prescribed $s \in S$ (Section 47.2),
- a polynomial-time algorithm finding a maximum-weight common base vector of f_1 and f_2 (Section 47.3),
- a strongly polynomial-time algorithm finding a maximum-weight common base vector of f_1 and f_2 (Section 47.4).

At the base of the algorithms is submodular function minimization, which leads back to the 'consistent breadth-first search' technique proposed in a pioneering paper of Schönsleben [1980] on polymatroid intersection.

47.1. A maximum-size common vector in two polymatroids

We first consider the problem:

(47.1)   given: submodular set functions f_1 and f_2 on a set S (by value giving oracles),
find: an $x \in EP_{f_1} \cap EP_{f_2}$ maximizing $x(S)$, and a $T \subseteq S$ with $x(S) = f_1(T) + f_2(S \setminus T)$.

So T certifies that x indeed maximizes $x(S)$ over $EP_{f_1} \cap EP_{f_2}$, since for any $x' \in EP_{f_1} \cap EP_{f_2}$ we have:

(47.2) $x'(S) = x'(T) + x'(S \setminus T) \leq f_1(T) + f_2(S \setminus T) = x(S)$.

On the other hand, x certifies that T minimizes $f_1(T) + f_2(S \setminus T)$.

Then (Lawler and Martel [1982a], extending a weakly polynomial bound of Schönsleben [1980]):

806   Chapter 47. Polymatroid intersection algorithmically

Theorem 47.1. *Problem* (47.1) *is solvable in strongly polynomial-time.*

Proof. We can assume that $f_1(\emptyset) = 0$ and $f_2(\emptyset) = 0$. Define the submodular set function f on S by

(47.3) $\qquad f(U) := f_1(U) + f_2(S \setminus U) - f_2(S)$

for $U \subseteq S$. With the submodular function minimization algorithm described in Section 45.4 we find a subset T of S minimizing f. So, by Corollary 46.1b, $f(T) + f_2(S)$ is equal to the maximum of $x(S)$ over $EP_{f_1} \cap EP_{f_2}$.

The submodular function minimization algorithm of Section 45.4 also gives vertices $b_1, \ldots, b_k$ of EP_f and $\lambda_1, \ldots, \lambda_k \geq 0$ with $\lambda_1 + \cdots + \lambda_k = 1$ such that for

(47.4) $\qquad y := \lambda_1 b_1 + \cdots + \lambda_k b_k$

we have $y(T) = f(T)$, $\text{supp}^-(y) \subseteq T$, and $\text{supp}^+(y) \subseteq S \setminus T$. (Here, as usual, $\text{supp}^+(x) := \{s \in S \mid x(s) > 0\}$ and $\text{supp}^-(x) := \{s \in S \mid x(s) < 0\}$.)

Now for each $i = 1, \ldots, k$, we can find $b'_i \in EP_{f_1}$ and $b''_i \in EP_{f_2}$ with $b_i = b'_i - b''_i$. Indeed, let $u_1, \ldots, u_n$ be a total order of S generating b_i. (That is, $b_i(u_j) = f(\{u_1, \ldots, u_j\}) - f(\{u_1, \ldots, u_{j-1}\})$ for $j = 1, \ldots, n$. These orderings are also implied by the submodular function minimization algorithm.) Let b'_i be the vertex of EP_{f_1} generated by the order $u_1, \ldots, u_n$ (that is, $b'_i(u_j) = f_1(\{u_1, \ldots, u_j\}) - f_1(\{u_1, \ldots, u_{j-1}\})$ for $j = 1, \ldots, n$). Let b''_i be the vertex of EP_{f_2} generated by the order $u_n, u_{n-1}, \ldots, u_1$ (that is, $b''_i(u_j) = f_2(\{u_j, \ldots, u_n\}) - f_2(\{u_{j+1}, \ldots, u_n\})$ for $j = 1, \ldots, n$). Then by definition of f we have $b_i = b'_i - b''_i$, since for each j:

(47.5) $\qquad b_i(u_j) = f(\{u_1, \ldots, u_j\}) - f(\{u_1, \ldots, u_{j-1}\})$
$\qquad\qquad = f_1(\{u_1, \ldots, u_j\}) + f_2(\{u_{j+1}, \ldots, u_n\}) - f_1(\{u_1, \ldots, u_{j-1}\})$
$\qquad\qquad - f_2(\{u_j, \ldots, u_n\}) = b'_i(u_j) - b''_i(u_j).$

Define

(47.6) $\qquad x' := \sum_{i=1}^{k} \lambda_i b'_i, \; x'' := \sum_{i=1}^{k} \lambda_i b''_i, \text{ and } x := x' \wedge x'',$

where $\wedge$ stands for taking coordinatewise minimum. As $x' \in EP_{f_1}$ and $x'' \in EP_{f_2}$, we know $x \in EP_{f_1} \cap EP_{f_2}$. Also, as $y = x' - x''$, we know that if $u \in T$, then $y(u) \leq 0$, hence $x''(u) \geq x'(u)$, and therefore $x(u) = x'(u)$. Similarly, if $u \in S \setminus T$, then $x(u) = x''(u)$. Hence

(47.7) $\qquad x(S) = x(T) + x(S \setminus T) = x'(T) + x''(S \setminus T) = (x' - x'')(T) + x''(S)$
$\qquad\qquad = y(T) + x''(S) = f(T) + f_2(S) = f_1(T) + f_2(S \setminus T),$

as required. ∎

47.2. Maximizing a coordinate of a common base vector

Theorem 47.1 implies the strong polynomial-time solvability of:

(47.8) given: submodular set functions f_1 and f_2 on a set S (by value giving oracles) and an element $s \in S$,
find: a common base vector x of f_1 and f_2 maximizing $x(s)$, and subsets S_1 and S_2 of S with $S_1 \cap S_2 = \{s\}$, $S_1 \cup S_2 = S$, and $x(S_i) = f_i(S_i)$ for $i = 1, 2$.

This is a result of Frank [1984a]:

Theorem 47.2. *Problem* (47.8) *is solvable in strongly polynomial time.*

Proof. We can assume that $f_1(S) = f_2(S)$ and that f_1 and f_2 have a common base vector (this can be checked by Theorem 47.1). Hence

(47.9) $f_1(S) \leq f_1(U) + f_2(S \setminus U)$

for each $U \subseteq S$. Define $S' := S \setminus \{s\}$.

First determine S_1, S_2 with $S_1 \cap S_2 = \{s\}$ and $S_1 \cup S_2 = S$ and minimizing $f_1(S_1) + f_2(S_2)$. This can be done by minimizing the submodular function $f_1(U + s) + f_2(S \setminus U)$ over $U \subseteq S'$.

Define

(47.10) $\alpha := f_1(S_1) + f_2(S_2) - f_1(S)$.

For $i = 1, 2$ and $U \subseteq S'$, define

(47.11) $g_i(U) := \min\{f_i(U), f_i(U + s) - \alpha\}$.

Then g_1 and g_2 are submodular set functions on S', as is easy to check. Moreover,

(47.12) $g_i(S') = f_i(S) - \alpha$.

To show this, we may assume that $i = 1$. Then we must show:

(47.13) $f_1(S') \geq f_1(S) - \alpha = 2f_1(S) - f_1(S_1) - f_2(S_2)$.

Now $f_1(S_1 \setminus \{s\}) + f_2(S_2) \geq f_1(S)$ (since f_1 and f_2 have a common base vector) and $f_1(S_1) - f_1(S_1 \setminus \{s\}) \geq f_1(S) - f_1(S')$ (by the submodularity of f_1). These two inequalities imply (47.13).

Then

(47.14) g_1 and g_2 have a common base vector.

Otherwise, S' can be partitioned into sets R_1 and R_2 with

(47.15) $g_1(R_1) + g_2(R_2) < g_1(S')$.

If $g_1(R_1) = f_1(R_1)$ and $g_2(R_2) = f_2(R_2)$, then (47.15) implies

(47.16) $\quad 2f_1(S) > f_1(S_1) + f_2(S_2) + f_1(R_1) + f_2(R_2)$
$\geq f_1(S_1 \cap R_1) + f_1(S_1 \cup R_1) + f_2(S_2 \cap R_2) + f_2(S_2 \cup R_2).$

By symmetry, we can assume that $f_1(S) > f_1(S_1 \cap R_1) + f_2(S_2 \cup R_2)$. However, $S_1 \cap R_1$ and $S_2 \cup R_2$ partition S, contradicting (47.9).

If $g_1(R_1) = f_1(R_1)$ and $g_2(R_2) = f_2(R_2 + s) - \alpha$, then (47.15) implies

(47.17) $\quad f_1(S) - \alpha > f_1(R_1) + f_2(R_2 + s) - \alpha,\ .$

and hence $f_1(S) > f_1(R_1) + f_2(R_2 + s)$, contradicting (47.9).

If $g_1(R_1) = f_1(R_1 + s) - \alpha$ and $g_2(R_2) = f_2(R_2 + s) - \alpha$, then (47.15) implies

(47.18) $\quad f_1(S) - \alpha > f_1(R_1 + s) + f_2(R_2 + s) - 2\alpha,$

implying $f_1(S_1) + f_2(S_2) > f_1(R_1 + s) + f_2(R_2 + s)$, contradicting the minimality of $f_1(S_1) + f_2(S_2)$. This proves (47.14).

By Theorem 47.1, we can find in strongly polynomial time a common base vector x of g_1 and g_2. So $x(S') = g_1(S')$. Extend x to S by defining $x(s) := \alpha$. Then

(47.19) $\quad x$ is a common base vector of f_1 and f_2.

By symmetry, it suffices to show that x is a base vector of f_1. First, x belongs to EP_{f_1}, since for each $U \subseteq S'$ we have

(47.20) $\quad x(U) \leq g_1(U) \leq f_1(U)$ and
$x(U + s) = x(U) + \alpha \leq g_1(U) + \alpha \leq f_1(U + s).$

Next, x is a base vector of f_1, since

(47.21) $\quad x(S) = x(S') + \alpha = g_1(S') + \alpha = f_1(S),$

by (47.12). This proves (47.19).

Moreover,

(47.22) $\quad x(S_i) = f_i(S_i)$

for $i = 1, 2$. Indeed (for $i = 1$),

(47.23) $\quad x(S_1) = x(S) - x(S_2 \setminus \{s\}) \geq f_1(S) - g_2(S_2 \setminus \{s\})$
$\geq f_1(S) - f_2(S_2) + \alpha = f_1(S_1).$

This proves (47.22), which implies that x is a common base vector of f_1 and f_2 maximizing $x(s)$, as for any common base vector z of f_1 and f_2 we have

(47.24) $\quad z(s) = z(S_1) + z(S_2) - z(S) \leq f_1(S_1) + f_2(S_2) - f_1(S)$
$= x(S_1) + x(S_2) - x(S) = x(s).$

So x, S_1, and S_2 have the required properties. ∎

47.3. Weighted polymatroid intersection in polynomial time

It may be shown with the ellipsoid method that the following problem is solvable in polynomial time:

(47.25) given: submodular functions f_1, f_2 on a set S (by value giving oracles) and a function $w : S \to \mathbb{Z}$,
find: a common base vector x of f_1 and f_2 maximizing $w^\mathsf{T} x$, and $w_1, w_2 : S \to \mathbb{Z}$ with $w = w_1 + w_2$ such that, for each $i = 1, 2$, x maximizes $w_i^\mathsf{T} x$ over all base vectors of f_i.

Cunningham and Frank [1985] gave, with the help of Theorem 47.2, a combinatorial polynomial-time algorithm (using submodular function minimization).

In order to describe this, we first give an auxiliary result concerning polymatroids. Let f be a submodular set function on S and let F be a face of EP_f. Define

(47.26) $\quad F^\downarrow := F - \mathbb{R}_+^S$.

Then $F^\downarrow$ is an extended polymatroid again. Moreover, algorithmic properties for $F^\downarrow$ can be deduced from those for EP_f:

Lemma 47.3α. *Let f be a submodular set function on S, let $w : S \to \mathbb{Z}_+$, and let F be the set of vectors x in EP_f maximizing $w^\mathsf{T} x$. Then there is a submodular set function f' on S with $F^\downarrow = EP_{f'}$. Moreover, if f is given by a value giving oracle, $f'(U)$ can be computed in strongly polynomial time, for any $U \subseteq S$.*

Proof. We can assume that $f(\emptyset) = 0$. Let $\emptyset \neq T_1 \subset T_2 \subset \cdots \subset T_{k-1} \subset T_k = S$ be the (unique) sets satisfying

(47.27) $\quad w = \lambda_1 \chi^{T_1} + \cdots + \lambda_k \chi^{T_k}$,

for some $\lambda_1, \ldots, \lambda_{k-1} > 0$ and $\lambda_k \geq 0$. Set $T_0 := \emptyset$, and define f' by:

(47.28) $\quad f'(U) := \sum_{i=1}^{k}(f((U \cap T_i) \cup T_{i-1}) - f(T_{i-1}))$,

for $U \subseteq S$. Then f' is submodular, as it is the sum of k submodular functions. Also,

(47.29) $\quad f' \leq f$,

since for each U we have, by the submodularity of f:

(47.30) $\quad f'(U) = \sum_{i=1}^{k}(f((U \cap T_i) \cup T_{i-1}) - f(T_{i-1}))$
$\leq \sum_{i=1}^{k}(f(U \cap T_i) - f(U \cap T_{i-1})) = f(U)$.

We show:

(47.31) $F^{\downarrow} = EP_{f'}$.

To see $\subseteq$, it suffices to show that $F \subseteq EP_{f'}$. Let $x \in F$. So $x(T_i) = f(T_i)$ for $i = 0, \ldots, k-1$. Hence

(47.32) $$x(U) = \sum_{i=1}^{k} x(U \cap (T_i \setminus T_{i-1})) = \sum_{i=1}^{k} (x((U \cap T_i) \cup T_{i-1}) - x(T_{i-1}))$$
$$\leq \sum_{i=1}^{k} (f((U \cap T_i) \cup T_{i-1}) - f(T_{i-1})) = f'(U)$$

for each $U \subseteq S$. This proves that $x \in EP_{f'}$.

To see $\supseteq$ in (47.31), it suffices to show that any $x \in EP_{f'}$ with $x(S) = f'(S)$ belongs to F. As $f' \leq f$ we know that $x \in EP_f$. So it suffices to show that $x(T_j) = f(T_j)$ for $j = 1, \ldots, k$ (as this implies that x maximizes $w^{\mathsf{T}}x$ over EP_f, by the greedy algorithm). Now, as $f'(S) = f(S)$:

(47.33) $$x(T_j) = x(S) - x(S \setminus T_j) \geq f'(S) - f'(S \setminus T_j)$$
$$= f(S) - \sum_{i=1}^{k} (f((T_i \setminus T_j) \cup T_{i-1}) - f(T_{i-1}))$$
$$= f(S) - \sum_{i=j+1}^{k} (f(T_i) - f(T_{i-1})) = f(T_j).$$

This proves (47.31). ∎

We also will use the following lemma:

Lemma 47.3β. *Let f be a submodular set function on S, let $w : S \to \mathbb{Z}$, and let F be the set of base vectors x of f maximizing $w^{\mathsf{T}}x$. Let $U \subseteq S$ and let x maximize $x(U)$ over F. Then x maximizes $(w + \chi^U)^{\mathsf{T}}x$ over all base vectors of f.*

Proof. Let $w' := w + \chi^U$. As x maximizes $x(U)$ over F, we know that x maximizes $w'^{\mathsf{T}}x$ over F. Also, some $z \in F$ maximizes $w'^{\mathsf{T}}z$ over EP_f, by the greedy method, as there is an ordering of S in which both w and w' are monotonically nondecreasing, and so EP_f has a vertex z maximizing both $w^{\mathsf{T}}z$ and $w'^{\mathsf{T}}z$.

As $w'^{\mathsf{T}}x \geq w'^{\mathsf{T}}z$, x maximizes $w'^{\mathsf{T}}x$ over EP_f. ∎

Now we can derive:

Theorem 47.3. *Problem (47.25) is solvable in polynomial time.*

Proof. We give a polynomial-time algorithm to transform a solution of (47.25) for some w to a solution of (47.25) for $w := w + \chi^s$, for any $s \in S$. This implies a polynomial-time algorithm for (47.25), since we can assume that $w \geq 0$, and since any $w \geq 0$ can be obtained from $w = 0$ by a polynomially bounded number of resettings $w := 2w$ and $w := w + \chi^s$ for $s \in S$. Note that for $w = 0$, (47.25) is trivial, and that a solution x, w_1, w_2 for w yields a solution $x, 2w_1, 2w_2$ for $2w$.

Let $s \in S$. Let x, w_1, w_2 be a solution of (47.25) for some w. For $i = 1, 2$, let F_i be the set of all vectors $x \in EP_{f_i}$ maximizing $w_i^T x$ and let f_i' be a submodular function satisfying $F_i^{\downarrow} = EP_{f_i'}$ (Lemma 47.3α). Applying Theorem 47.2 to f_1', f_2', we find a common base vector x' of f_1' and f_2' maximizing $x'(s)$, and subsets S_1, S_2 of S with $S_1 \cap S_2 = \{s\}$, $S_1 \cup S_2 = S$, and $x'(S_1) = f_1'(S_1)$, $x'(S_2) = f_2'(S_2)$. Then x' maximizes $x'(S_1)$ over $EP_{f_1'}$, and x' maximizes $x'(S_2)$ over $EP_{f_2'}$. Hence, by Lemma 47.3β, x' is a base vector of f_1' maximizing $(w_1 + \chi^{S_1})^T x'$, and also, x' is a base vector of f_2' maximizing $(w_2 + \chi^{S_2})^T x'$. So

(47.34) $\quad x', w_1' := w_1 + \chi^{S_1} - \chi^S, w_2' := w_2 + \chi^{S_2}$,

gives a solution of (47.25) for $w + \chi^s$. ∎

47.4. Weighted polymatroid intersection in strongly polynomial time

A general simultaneous diophantine approximation method of Frank and Tardos [1985,1987] implies that (47.25) is *strongly* polynomial-time solvable. Fujishige, Röck, and Zimmermann [1989] showed that from Theorem 47.3 a combinatorial strongly polynomial-time algorithm can be derived, by extending the method of Tardos [1985a] for the minimum-cost circulation problem.

To prove this, we first show a sensitivity result. Let f_1, f_2 be submodular set functions on S. Call a pair $w_1, w_2 : S \to \mathbb{R}$ *good* if there exists an x that maximizes $w_i^T x$ over EP_{f_i}, for both $i = 1$ and $i = 2$.

Lemma 47.4α. *Let $w : S \to \mathbb{Q}$ and let w_1, w_2 be a good pair with $w = w_1 + w_2$. Then for any $\tilde{w} : S \to \mathbb{Q}$ with $\tilde{w} \geq w$ there exists a good pair $\tilde{w}_1, \tilde{w}_2$ with $\tilde{w} = \tilde{w}_1 + \tilde{w}_2$ and $\|\tilde{w}_i - w_i\|_\infty \leq \|\tilde{w} - w\|_1$ for $i = 1, 2$.*

Proof. We can assume that w and $\tilde{w}$ are integer, and that $\|\tilde{w} - w\|_1 = 1$ (as the general case then follows inductively).

Let F_i be the set of all x maximizing $w_i^T x$ over EP_{f_i}. Let f_i' be a submodular function satisfying $F_i^{\downarrow} = EP_{f_i'}$. Let s be such that $\tilde{w}(s) = w(s) + 1$. By the solvability of problem (47.8), there is a common base vector x of f_1' and f_2' maximizing x_s, and there exist S_1, S_2 with $S_1 \cap S_2 = \{s\}$ and $S_1 \cup S_2 = S$ such that $x(S_1) = f_1'(S_1)$ and $x(S_2) = f_2'(S_2)$. Define

(47.35) $\quad \tilde{w}_1 := w_1 + \chi^{S_1} - \chi^S, \tilde{w}_2 := w_2 + \chi^{S_2}$.

So $\tilde{w} = \tilde{w}_1 + \tilde{w}_2$. By Lemma 47.3$\beta$, x maximizes $\tilde{w}_i^T x$ over EP_{f_i} for $i = 1, 2$. Therefore, the pair $\tilde{w}_1, \tilde{w}_2$ is good. As $\|\tilde{w}_i - w_i\|_\infty \leq 1$ for $i = 1, 2$, this proves the lemma. ∎

Theorem 47.4. *Given submodular functions f_1, f_2 on a set S and $w \in \mathbb{Q}^S$, one can find a common base vector x of f_1 and f_2 maximizing $w^T x$, in strongly polynomial time.*

Proof. Let be given submodular functions f_1, f_2 on a set S and a function $w : S \to \mathbb{Q}$. We may assume that f_1 and f_2 have a common base vector. (This can be checked by Theorem 47.1.)

We keep chains $\mathcal{C}_1, \mathcal{C}_2$ of subsets of S such that for $i = 1, 2$ and each $U \in \mathcal{C}_i$:

(47.36) $x(U) = f_i(U)$ for each common base vector x of f_1 and f_2 maximizing $w^T x$,

and such that $S \in \mathcal{C}_1$ and $S \in \mathcal{C}_2$. Initially we set $\mathcal{C}_i := \{S\}$ for $i = 1, 2$. We describe an iteration that either extends $\mathcal{C}_1$ or $\mathcal{C}_2$, or gives a solution x.

We can assume that, for $i = 1, 2$,

(47.37)   each base vector x of f_i satisfies $x(U) = f_i(U)$ for each $U \in \mathcal{C}_i$.

Indeed, let F_i be the set of vectors x in EP_{f_i} with $x(U) = f_i(U)$ for each $U \in \mathcal{C}_i$. So F_i is equal to the set of $x \in EP_{f_i}$ maximizing $c_i^T x$, where $c_i := \sum_{U \in \mathcal{C}_i} \chi^U$. By Lemma 47.3$\alpha$, we can find f_i' with $F_i^{\downarrow} = EP_{f_i'}$. By (47.36), replacing the f_i by f_i' does not change the set of optimum solutions x of our problem.

Let

(47.38) $L :=$ linear hull of $\{\chi^U \mid U \in \mathcal{C}_1 \cup \mathcal{C}_2\}$.

Determine $y \in L$ minimizing

(47.39) $\|w - y\|_\infty$.

This can be done in strongly polynomial time as follows. For $i = 1, 2$, let $\mathcal{P}_i$ be the partition of S into nonempty classes such that u and v belong to the same class if and only if there is no set in $\mathcal{C}_i$ containing exactly one of u, v. Let D be the directed graph with vertex set $\mathcal{P}_1 \cup \mathcal{P}_2$ such that for each $v \in S$ there is an arc of length $w(v)$ from $U \in \mathcal{P}_1$ to $W \in \mathcal{P}_2$ with $v \in U \cap W$, and an arc of length $-w(v)$ in the reverse direction. Determine the minimum mean-length α of a directed circuit in D (cf. Section 8.5). It is the minimum α for which there exist $p_i : \mathcal{P}_i \to \mathbb{Q}$ such that

(47.40) $-\alpha \leq w(v) + p_1(U) - p_2(W) \leq \alpha$

for each arc as described. Then

(47.41) $y := -\sum_{U \in \mathcal{P}_1} p_1(U) \chi^U + \sum_{W \in \mathcal{P}_2} p_2(W) \chi^W$

minimizes (47.39).

Let α be the value of (47.39). If $\alpha = 0$, then $w \in L$, and so

(47.42) $$w = \sum_{i=1}^{2} \sum_{U \in \mathcal{C}_i} \lambda_i(U) \chi^U$$

for functions $\lambda_i : \mathcal{C}_i \to \mathbb{Q}$. Then for any common base vector x of f_1 and f_2 we have

(47.43) $$w^\mathsf{T} x = \sum_{i=1}^{2} \sum_{U \in \mathcal{C}_i} \lambda_i(U) x(U) = \sum_{i=1}^{2} \sum_{U \in \mathcal{C}_i} \lambda_i(U) f_i(U).$$

So each common base vector is optimum. As we can find any common base vector in strongly polynomial time (by Theorem 47.1), we have solved the problem.

So we can assume that $\alpha > 0$. Define $w' : S \to \mathbb{Z}$ by

(47.44) $$w' := \lfloor \frac{5n^2}{\alpha}(w - y) \rfloor,$$

where $n := |S|$. By definition of α, $\|w'\|_\infty = 5n^2$. Hence by Theorem 47.3, we can find in strongly polynomial time a common base vector x' of f_1 and f_2 and $w'_1, w'_2 : S \to \mathbb{Z}$ with $w' = w'_1 + w'_2$ such that x' is a base vector of f_i maximizing $w'^\mathsf{T}_i x'$, for $i = 1, 2$.

For $i = 1, 2$, we can determine a chain $\mathcal{D}_i$ of subsets of S (with $S \in \mathcal{D}_i$) and a function $\lambda_i : \mathcal{D}_i \to \mathbb{Z}$ such that

(47.45) $$w'_i = \sum_{W \in \mathcal{D}_i} \lambda_i(W) \chi^W$$

and such that $\lambda_i(W) > 0$ if $W \neq S$. We show that

(47.46) there exist $i \in \{1, 2\}$ and $W \in \mathcal{D}_i$ with $\lambda_i(W) > 2n$ and $\chi^W \notin L$.

Suppose not. Let $\mathcal{D}'_i := \{W \in \mathcal{D}_i \mid \chi^W \notin L\}$, and $\mathcal{D}''_i := \mathcal{D}_i \setminus \mathcal{D}'_i$, for $i = 1, 2$. So if $W \in \mathcal{D}'_i$, then $\lambda_i(W) \leq 2n$. This gives the contradiction:

(47.47) $$4n^2 \geq \|\sum_{i=1}^{2} \sum_{W \in \mathcal{D}'_i} \lambda_i(W) \chi^W\|_\infty = \|w' - \sum_{i=1}^{2} \sum_{W \in \mathcal{D}''_i} \lambda_i(W) \chi^W\|_\infty$$
$$> \|\frac{5n^2}{\alpha}(w - y) - \sum_{i=1}^{2} \sum_{W \in \mathcal{D}''_i} \lambda_i(W) \chi^W\|_\infty - 1 \geq 5n^2 - 1.$$

The last inequality holds as y minimizes $\|w - y\|_\infty$ over $y \in L$.

This shows (47.46). We can assume that $W \in \mathcal{D}'_1$ is such that $\lambda_1(W) > 2n$. Then:

(47.48) each optimum common base vector x of f_1 and f_2 satisfies $x(W) = f_1(W)$.

To see this, let

(47.49) $$\tilde{w} := \frac{5n^2}{\alpha}(w-y).$$

Replacing w by $\tilde{w}$ does not change the set of optimum common base vectors, since y belongs to L (implying (by our assumption (47.37)) that $y^\mathsf{T}x$ is the same for all common base vectors x of f_1 and f_2).

By Lemma 47.4α, there exists a good pair $\tilde{w}_1, \tilde{w}_2$ with $\tilde{w} = \tilde{w}_1 + \tilde{w}_2$ and

(47.50) $$\|\tilde{w}_i - w'_i\|_\infty \leq \|\tilde{w} - w'\|_1 < n$$

for $i = 1, 2$. Now for any $v \in W$ and $u \in S \setminus W$ we have $w'_1(v) > w'_1(u) + 2n$, as $\lambda_1(W) > 2n$, and as $\lambda_1(W') \geq 0$ for each $W' \in \mathcal{D}_1 \setminus \{S\}$. Hence, by (47.50), $\tilde{w}_1(v) > \tilde{w}_1(u)$. So (by the greedy method) any base vector x of f_1 maximizing $\tilde{w}_1^\mathsf{T} x$ satisfies $x(W) = f_1(W)$. This shows (47.48).

Let $\mathcal{C}_1 = \{U_1 \subset U_2 \subset \cdots \subset U_t = S\}$. For $j = 1, \ldots, t$, let $W_j := (W \cap U_j) \cup U_{j-1}$, where $U_0 := \emptyset$. Then $x(W_j) = f(W_j)$ for each optimum common base vector x (since W_j arises by taking unions and intersections from W, U_j, and U_{j-1}).

Moreover, $W_j \notin \mathcal{C}_1$ for at least one $j = 1, \ldots, t$, since

(47.51) $$\chi^W = \sum_{j=1}^t (\chi^{W_j} - \chi^{U_{j-1}})$$

while χ^W does not belong to L, implying that not all χ^{W_j} belong to L, and so some W_j does not belong to $\mathcal{C}_1$. So W_j can be added to $\mathcal{C}_1$, and we can iterate. ∎

From an optimum common base vector x, an optimum 'dual solution' w_1, w_2 can be derived, with a method of Cunningham and Frank [1985]. This gives:

Corollary 47.4a. *Problem (47.25) is solvable in strongly polynomial time.*

Proof. By Theorem 47.4, we can find a common base vector x of f_1 and f_2 maximizing $w^\mathsf{T} x$, in strongly polynomial time. Define a directed graph $D = (S, A)$ as follows.

For $i = 1, 2$, let A_i consist of all pairs (u, v) with $u, v \in S$ such that for each $U \subseteq V$:

(47.52)   if $x(U) = f_i(U)$ and $u \in U$ then $v \in U$.

We can find A_i in strongly polynomial time, by finding the minimum of $f_i(U) - x(U)$ over subsets U of S with $u \in U$ and $v \notin U$ (with any strongly polynomial-time submodular function minimization algorithm).

Let D have arc set $A := A_1 \cup A_2^{-1}$ (taking two parallel arcs from u to v in case $(u,v) \in A_1$ and $(v,u) \in A_2$). Define a length function l on A by, for $(u,v) \in A$:

Section 47.4. Strongly polynomial time

(47.53) $\quad l(u,v) := \begin{cases} w(v) - w(u) & \text{for } (u,v) \in A_1, \\ 0 & \text{for } (v,u) \in A_2. \end{cases}$

We claim:

(47.54) D has no negative-length directed circuits.

For suppose that C is a negative-length directed circuit. We take such a C with $|AC|$ smallest. Then two consecutive arcs in C neither both belong to A_1 nor both belong to A_2^{-1}. For suppose that $a = (t,u)$ and $a' = (u,v)$ are in C and that they both belong to A_1. Then $(t,v) \in A_1$ and $l(a)+l(a') = l(t,v)$, contradicting the minimality of $|AC|$. This similarly gives a contradiction if $a, a' \in A_2^{-1}$.

So we can assume that C traverses the vertices $u_0, u_1, \ldots, u_k$ in this order, with $u_0 = u_k$, such that (u_{i-1}, u_i) belongs to A_1 if i is odd, and to A_2^{-1} if i is even. Let $X := \{u_1, u_3, \ldots, u_{k-1}\}$ and $Y := \{u_2, u_4, \ldots, u_k\}$. As C has negative length, we know $l(AC) < 0$, and hence $w(X) < w(Y)$.

By (47.52), for each $i = 1,2$ and for each $U \subseteq V$ with $x(U) = f_i(U)$ we have $|U \cap Y| \geq |U \cap X|$. Hence there exists an $\varepsilon > 0$ such that the vector

(47.55) $\quad x' := x + \varepsilon(\chi^X - \chi^Y)$

belongs to EP_{f_1} and to EP_{f_2}. So, since $x'(S) = x(S)$, x' is again a common base vector of f_1 and f_2. However, $w^\mathsf{T} x' = w^\mathsf{T} x + w(X) - w(Y) > w^\mathsf{T} x$, contradicting the fact that x maximizes $w^\mathsf{T} x$. This proves (47.54).

By Theorem 8.7, we can find a potential $p : S \to \mathbb{Z}$ for D with respect to l, in strongly polynomial time. Then p satisfies

(47.56) $\quad \begin{array}{ll} p(v) - p(u) \leq w(v) - w(u) & \text{for each } (u,v) \in A_1, \\ p(v) - p(u) \geq 0 & \text{for each } (u,v) \in A_2. \end{array}$

Define $w_1 := w - p$ and $w_2 := p$. We show that w_1 and w_2 are as required in (47.25).

(47.56) implies that, for each $i = 1,2$,

(47.57)   if $(u,v) \in A_i$ then $w_i(v) \geq w_i(u)$.

We show that (47.57) implies that, for each $i = 1,2$, x is a base vector of f_i maximizing $w_i^\mathsf{T} x$, as required. We may assume $i = 1$.

Let μ and ν be the minimum and maximum value (respectively) of the entries in w_1. For $j \in \mathbb{Z}$, let $U_j := \{v \in S \mid w_1(v) \geq \mu + j\}$. Then, taking $k := \nu - \mu$,

(47.58) $\quad w_1 = \mu \cdot \chi^S + \sum_{j=1}^{k} \chi^{U_j}.$

Moreover,

(47.59) $\quad x(U_j) = f_1(U_j)$ for each $j = 1, \ldots, k$.

Indeed, for all $s \in U_j$ and $t \in S \setminus U_j$ we have $(s,t) \notin A_1$ (by (47.57), since $w_1(t) < \mu + j \leq w_1(s)$). Hence, by definition of A_1, there is a set $T_{s,t}$ with $s \in T_{s,t}$, $t \notin T_{s,t}$, and $x(T_{s,t}) = f_1(T_{s,t})$. As the collection of sets U with $x(U) = f_1(U)$ is closed under taking unions and intersections, (47.59) follows.

Then for any base vector x' of f_1 we have

$$(47.60) \qquad w_1^\mathsf{T} x' = \mu x'(S) + \sum_{j=1}^{k} x'(U_j) \leq \mu f_1(S) + \sum_{j=1}^{k} f_1(U_j).$$

By (47.59), we here have equality throughout for $x' := x$, which proves that x maximizes $w_1^\mathsf{T} x$ over all base vectors of f_1. ∎

Theorem 47.4 implies for (nonextended) polymatroids (extending a result of Schönsleben [1980] for integer f_1 and f_2 for which there is a fixed K with $P_{f_1} \cap P_{f_2} \subseteq [0,K]^S$):

Corollary 47.4b. *Given submodular set functions f_1, f_2 on S (by value giving oracles) and a weight function $w \in \mathbb{Q}^S$, we can find a maximum-weight vector $x \in P_{f_1} \cap P_{f_2}$ in strongly polynomial time.*

Proof. We can assume that $f_1(\emptyset) = f_2(\emptyset) = 0$ and that f_1 and f_2 are nondecreasing (as we can replace $f_i(U)$ by $\min_{T \supseteq U} f_i(T)$). Extend S with a new element t to a set $S' := S + t$. Define set functions f_1' and f_2' on S' by:

$$(47.61) \qquad f_i'(U) := f_i(U) \text{ and } f_i'(U + t) := 0$$

for $U \subseteq S$ and $i = 1, 2$. Then f_1' and f_2' are submodular (using the nondecreasingness of f_1 and f_2). Moreover, consider any $x' \in \mathbb{R}^{S'}$ with $x'(S') = 0$. Let x be the restriction of x' to S. Then:

$$(47.62) \qquad x' \in EP_{f_i'} \text{ if and only if } x \in P_{f_i}.$$

Indeed, if $x' \in EP_{f_i'}$, then $x'(s) \geq 0$ for each $s \in S$, since $x'(S' - s) \leq f'(S' - s) = 0$ and $x'(S') = 0$, implying that $x(s) = x'(s) \geq 0$. Moreover, for each $U \subseteq S$ one has $x(U) = x'(U) \leq f_i'(U) = f_i(U)$. So $x \in P_{f_i}$.

Conversely, if $x \in P_f$, then for each $U \subseteq S$ one has $x'(U) = x(U) \leq f_i(U) = f_i'(U)$ and $x'(U + t) = x(U) - x(S) = -x(S \setminus U) \leq 0 = f_i'(U + t)$. So $x' \in EP_{f_i'}$. This proves (47.62).

Define $w' \in \mathbb{Q}^{S'}$ by $w'(v) := w(v)$ if $v \in S$, and $w'(t) := 0$. By Theorem 47.4, we can find a common base vector x' of f_1' and f_2' maximizing ${w'}^\mathsf{T} x'$ in strongly polynomial time. Let x be the restriction of x' to S. By (47.62), x maximizes $w^\mathsf{T} x$ over $P_{f_1} \cap P_{f_2}$. ∎

Similarly for maximum-weight common base vectors in (nonextended) polymatroids:

Corollary 47.4c. *Given submodular set functions f_1, f_2 on S (by value giving oracles) and a weight function $w \in \mathbb{Q}^S$, we can find a maximum-weight common base vector x of P_{f_1} and P_{f_2} in strongly polynomial time.*

Proof. Again, we can assume that $f_1(\emptyset) = f_2(\emptyset) = 0$ and that f_1 and f_2 are nondecreasing. Then the present corollary follows directly from Theorem 47.4, since

(47.63) $\quad P_{f_1} \cap P_{f_2} \cap \{x \mid x(S) = f_1(S)\} = EP_{f_1} \cap EP_{f_2} \cap \{x \mid x(S) = f_1(S)\}.$

Indeed, if $x \in EP_{f_i}$ and $x(S) = f_i(S)$, then $x \geq \mathbf{0}$, since for any $s \in S$ one has $x_s = x(S) - x(S - s) \geq f_i(S) - f_i(S - s) \geq 0$. ∎

Back to extended polymatroids, Corollary 47.4b yields that we can optimize over the intersection of two extended polymatroids in strongly polynomial time:

Corollary 47.4d. *Given submodular set functions f_1, f_2 on S (by value giving oracles) and a weight function $w \in \mathbb{Q}_+^S$, we can find a maximum-weight vector $x \in EP_{f_1} \cap EP_{f_2}$ in strongly polynomial time.*

Proof. We may assume that $f_1(\emptyset) = f_2(\emptyset) = 0$. Let

(47.64) $\quad L := \max_{i=1,2}(|f_i(S)| + \sum_{s \in S} |f_i(\{s\})|).$

Then $|f_i(U)| \leq L$ for each $i = 1, 2$ and $U \subseteq S$, since

(47.65) $\quad f_i(U) \leq \sum_{s \in U} f_i(\{s\}) \leq L$

and

(47.66) $\quad f_i(U) \geq f_i(S) - f_i(S \setminus U) \geq f_i(S) - \sum_{s \in S \setminus U} f_i(\{s\}) \geq -L.$

Let $K := |S| \cdot L + 1$. Then for any vertex x of $EP_{f_1} \cap EP_{f_2}$ and any $s \in S$:

(47.67) $\quad x(s) > -K,$

since $x = A^{-1} b$ for some totally unimodular matrix A and some vector b whose entries are values of f_1 and f_2 (as in the proof of Theorem 46.1; observe that the entries of A^{-1} belong to $\{0, \pm 1\}$).

Define $f_i'(U) := f_i(U) + K \cdot |U|$. Then

(47.68) $\quad EP_{f_i'} = K \cdot \mathbf{1} + EP_{f_i}$

for $i = 1, 2$. Hence, by (47.67), all vertices of $EP_{f_1'} \cap EP_{f_2'}$ are nonnegative. So any vector x maximizing $w^T x$ over $P_{f_1'} \cap P_{f_2'}$ also maximizes $w^T x$ over $EP_{f_1'} \cap EP_{f_2'}$. By Corollary 47.4b, x can be found in strongly polynomial time. ∎

47.5. Contrapolymatroids

Similar results hold for intersections of contrapolymatroids, by reduction to polymatroids. Given supermodular set functions g_1 and g_2 on S (by value giving oracles) and a weight function $w \in \mathbb{Q}^S$, we can find in strongly polynomial time:

(47.69)   (i) a minimum-weight vector in $EQ_{g_1} \cap EQ_{g_2}$,
             (ii) a minimum-weight common base vector of EQ_{g_1} and EQ_{g_2},
             (iii) a minimum-weight vector in $Q_{g_1} \cap Q_{g_2}$, and
             (iv) a minimum-weight common base vector of Q_{g_1} and Q_{g_2}.

Here (i) and (ii) follow from Corollary 47.4d and Theorem 47.4 applied to the submodular functions $-g_1$ and $-g_2$. Moreover, (iii) and (iv) follow by application of (i) and (ii) to the supermodular functions $\bar{g}_i$ given by $\bar{g}_i(U) = \max_{T \subseteq U} g_i(T)$ for $U \subseteq S$ and $i = 1, 2$ (assuming without loss of generality $g_1(\emptyset) = g_2(\emptyset) = 0$).

47.6. Intersecting a polymatroid and a contrapolymatroid

Let f be a submodular, and g a supermodular, set function on S. The results on polymatroid intersection also imply that

(47.70)   a maximum-weight vector in the intersection of the extended polymatroid EP_f and the extended contrapolymatroid EQ_g can be found in strongly polynomial time,

assuming that we have value giving oracles for f and g.

To see this, we can assume that $f(\emptyset) = g(\emptyset) = 0$ and $g(S) \leq f(S)$. Let t be a new element. Define submodular set functions f_1 and f_2 on $S + t$ by:

(47.71) $f_1(U) := f(U)$, $f_1(U+t) := f(U) - g(S)$, $f_2(U) := f(S) - g(S \setminus U)$,
             $f_2(U + t) := -g(S \setminus U)$,

for $U \subseteq S$. Reset $f_1(S + t) := 0$. Then for each $x \in \mathbb{R}^S$ and $\lambda \in \mathbb{R}$:

(47.72) (x, λ) is a common base vector of EP_{f_1} and EP_{f_2}
             $\iff \lambda = -x(S)$ and $x \in EP_f \cap EQ_g$.

To see necessity, let (x, λ) be a common base vector of EP_{f_1} and EP_{f_2}. As $f_1(S + t) = 0$, we have $\lambda = -x(S)$. Moreover, for any $U \subseteq S$, we have

(47.73) $x(U) \leq f_1(U) = f(U)$ and
             $x(U) = x(S) - x(S \setminus U) = -\lambda - x(S \setminus U) \geq -f_2((S \setminus U) + t) = g(U)$.

So $x \in EP_f \cap EQ_g$.

To see sufficiency, assume $\lambda = -x(S)$ and $x \in EP_f \cap EQ_g$. Then for each $U \subseteq S$ we have:

(47.74) $x(U) \leq f(U) = f_1(U),$
$x(U+t) = x(U) + \lambda = x(U) - x(S) \leq f(U) - g(S) = f_1(U+t),$
$x(U) = x(S) - x(S \setminus U) \leq f(S) - g(S \setminus U) = f_2(U),$
$x(U+t) = x(U) + \lambda = x(U) - x(S) = -x(S \setminus U) \leq -g(S \setminus U)$
$= f_2(U+t).$

So (x, λ) is a common base vector of EP_{f_1} and EP_{f_2}.

This shows (47.72), which implies that finding a minimum-weight vector in $EP_f \cap EQ_g$ amounts to finding a minimum-weight common base vector of EP_{f_1} and EP_{f_2}.

Similarly, we can find a modular function h satisfying $g \leq h \leq f$ in strongly polynomial time, if $g \leq f$ (Frank's discrete sandwich theorem (Corollary 46.2b)). To see this, let f_1 and f_2 be as above, and find an (x, λ) in $EP_{f_1} \cap EP_{f_2}$ maximizing $x(S) + \lambda$. If $x(S) + \lambda \geq 0$, then $x \in EP_f \cap EQ_g$, that is x gives a modular function h with $g \leq h \leq f$.

47.6a. Further notes

Polymatroid intersection is a special case of submodular flow, as discussed in Chapter 60. We therefore refer for further algorithmic work to the notes in Section 60.3e.

A preflow-push algorithm for finding a maximum common vector in the intersection of two polymatroids was presented by Fujishige and Zhang [1992].

Tardos, Tovey, and Trick [1986] gave an improved version of Cunningham and Frank's polynomial-time algorithm for weighted polymatroid intersection. Fujishige [1978a] gave a (non-polynomial-time) algorithm for weighted polymatroid intersection. Optimizing over the intersection of a base polytope and an affine space was considered by Hartvigsen [1996,1998a,2001a].

Frank [1984c] and Fujishige and Iwata [2000] gave surveys.

Chapter 48

Dilworth truncation

If a submodular function f has $f(\emptyset) < 0$, the associated extended polymatroid is empty, as the conditions $x(U) \leq f(U)$ for all U include $x(\emptyset) < f(\emptyset)$. However, by ignoring the condition for $U = \emptyset$, the obtained polyhedron is yet an extended polymatroid, for a different submodular function, denoted by $\hat{f}$. This function $\hat{f}$ is called the *Dilworth truncation* of f.

48.1. If $f(\emptyset) < 0$

Let f be a submodular set function on S. If $f(\emptyset) < 0$, the associated extended polymatroid EP_f is empty. However, by ignoring the empty set, we yet obtain an extended polymatroid. (The interest in this goes back to Dilworth [1944].)

Consider the system

(48.1) $\qquad x(U) \leq f(U)$ for $U \in \mathcal{P}(S) \setminus \{\emptyset\}$,

and the problem dual to maximizing $w^\mathsf{T} x$ over (48.1), for $w \in \mathbb{R}_+^S$:

(48.2) $\qquad \min\{ \sum_{U \in \mathcal{P}(S) \setminus \{\emptyset\}} y(U) f(U) \mid$
$\qquad\qquad y \in \mathbb{R}_+^{\mathcal{P}(S) \setminus \{\emptyset\}}, \sum_{U \in \mathcal{P}(S) \setminus \{\emptyset\}} y(U) \chi^U = w\}.$

Recall that a collection $\mathcal{F}$ of sets is called *laminar* if

(48.3) $\qquad T \cap U = \emptyset$ or $T \subseteq U$ or $U \subseteq T$ for all $T, U \in \mathcal{F}$.

Then a basic result of Edmonds [1970b] is:

Theorem 48.1. *If f is a submodular set function on S, then (48.2) has an optimum solution y with $\mathcal{F} := \{U \in \mathcal{P}(S) \setminus \{\emptyset\} \mid y(U) > 0\}$ laminar.*

Proof. Let $y : \mathcal{P}(S) \setminus \{\emptyset\} \to \mathbb{R}_+$ achieve this minimum, with

(48.4) $\qquad \sum_{U \in \mathcal{P}(S) \setminus \{\emptyset\}} y(U) |U| |S \setminus U|$

as small as possible. Assume that $\mathcal{F}$ is not laminar, and choose $T, U \in \mathcal{F}$ violating (48.3). Let $\alpha := \min\{y(T), y(U)\}$. Decrease $y(T)$ and $y(U)$ by α, and increase $y(T \cap U)$ and $y(T \cup U)$ by α. Since

(48.5) $\qquad \chi^{T \cap U} + \chi^{T \cup U} = \chi^T + \chi^U,$

y remains a feasible solution of (48.2). As moreover

(48.6) $\qquad f(T \cap U) + f(T \cup U) \leq f(T) + f(U),$

f remains optimum. However, by Theorem 2.1, sum (48.4) decreases, contradicting our assumption. ∎

This implies that system (48.1) is TDI. More generally, it implies the box-total dual integrality of (48.1):

Corollary 48.1a. *For any submodular set function f on S, system (48.1) is box-totally dual integral.*

Proof. Consider some $w : S \to \mathbb{Z}_+$, and problem (48.2) dual to maximizing $w^\mathsf{T} x$ over (48.1). By Theorem 48.1, this minimum is attained by a y with $\mathcal{F} := \{U \in \mathcal{P}(S) \setminus \{\emptyset\} \mid y(U) > 0\}$ laminar. Hence the constraints corresponding to positive entries in y form a totally unimodular matrix (by Theorem 41.11). Therefore, by Theorem 5.35, (48.1) is box-TDI. ∎

Let EP'_f denote the solution set of (48.1). So EP'_f is nonempty for each submodular function f. As for integrality we have:

Corollary 48.1b. *If f is submodular and integer, then EP'_f is integer.*

Proof. Directly from Corollary 48.1a. ∎

In fact, as we shall see in Section 48.2, EP'_f is again an extended polymatroid.

48.2. Dilworth truncation

For each submodular function f, there exists a unique largest submodular function $\hat{f}$ with the property that $\hat{f}(U) \leq f(U)$ for each nonempty $U \subseteq S$, and $\hat{f}(\emptyset) = 0$. This follows from a method of Dilworth [1944].

Let f be a submodular set function on S. The *Dilworth truncation* $\hat{f} : \mathcal{P}(S) \to \mathbb{R}$ of f is given by:

(48.7) $\qquad \hat{f}(U) := \min\{\sum_{P \in \mathcal{P}} f(P) \mid \mathcal{P}$ is a partition of U into nonempty sets$\}$

for $U \subseteq S$. So $\hat{f}(\emptyset) = 0$ (as for $U = \emptyset$, only $\mathcal{P} = \emptyset$ qualifies in (48.7)). Dunstan [1976] showed:

Theorem 48.2. $\hat{f}$ *is submodular.*

Proof. Choose $T, U \subseteq S$, and let $\mathcal{P}$ and $\mathcal{Q}$ be partitions of T and U (respectively) into nonempty sets with

(48.8) $\qquad \hat{f}(T) = \sum_{P \in \mathcal{P}} f(P)$ and $\hat{f}(U) = \sum_{Q \in \mathcal{Q}} f(Q)$.

Consider the family $\mathcal{F}$ made by $\mathcal{P}$ and $\mathcal{Q}$ (taking a set twice if it occurs in both partitions). We can transform $\mathcal{F}$ iteratively into a laminar family, by replacing any $X, Y \in \mathcal{F}$ with $X \cap Y \neq \emptyset$ and $X \not\subseteq Y \not\subseteq X$ by $X \cap Y, X \cup Y$. In each iteration, the sum

(48.9) $\qquad \sum_{Z \in \mathcal{F}} f(Z)$

does not increase (as f is submodular). As at each iteration the sum

(48.10) $\qquad \sum_{Z \in \mathcal{F}} |Z||S \setminus Z|$

decreases (by Theorem 2.1), this process terminates. We end up with a laminar family $\mathcal{F}$.

The inclusionwise maximal sets in $\mathcal{F}$ form a partition $\mathcal{R}$ of $T \cup U$, and the remaining sets form a partition $\mathcal{S}$ of $T \cap U$. Therefore,

(48.11) $\qquad \hat{f}(T \cup U) + \hat{f}(T \cap U) \leq \sum_{X \in \mathcal{R}} f(X) + \sum_{Y \in \mathcal{S}} f(Y)$
$\qquad \leq \sum_{P \in \mathcal{P}} f(P) + \sum_{Q \in \mathcal{Q}} f(Q) = \hat{f}(T) + \hat{f}(U),$

showing that $\hat{f}$ is submodular. ∎

Lovász [1983c] observed that $\hat{f}$ is the unique largest among all submodular set functions g on S with $g(\emptyset) = 0$ and $g(U) \leq f(U)$ for $U \neq \emptyset$. Indeed, each subset U of S can be partitioned into nonempty sets $U_1, \ldots, U_t$ such that

(48.12) $\qquad g(U) \leq \sum_{i=1}^{t} g(U_i) \leq \sum_{i=1}^{t} f(U_i) = \hat{f}(U)$

(the first inequality follows from the submodularity of g, as $g(\emptyset) = 0$).

Trivially, $EP_{\hat{f}} = EP'_f$. In particular, EP'_f is an extended polymatroid. Moreover, by (44.34),

Theorem 48.3.

(48.13) $\qquad \hat{f}(U) = \max\{x(U) \mid x \in EP'_f\}.$

Proof. By (44.34), since $EP'_f = EP_{\hat{f}}$. ∎

$\hat{f}(U)$ can be computed in strongly polynomial time:

Theorem 48.4. *If a submodular set function f on S is given by a value giving oracle, then for each given $U \subseteq S$, $\hat{f}(U)$ can be computed in strongly polynomial time.*

Proof. We can assume that $U = S$. Order $S = \{s_1, \ldots, s_n\}$ arbitrarily. For $i = 1, \ldots, n$, define $U_i := \{s_1, \ldots, s_i\}$. Set $x := \mathbf{0}$. Iteratively, for $i = 1, \ldots, n$, determine

(48.14) $\quad \mu := \min\{f(T) - x(T) \mid s_i \in T \subseteq U_i\}$

(with a submodular function minimization algorithm), and reset $x(s_i) := x(s_i) + \mu$.

We end up with $x \in EP'_f$ and for each $u \in S$ a subset T_u of S with $u \in T_u$ and $x(T_u) = f(T_u)$. As the collection of subsets T of S with $x(T) = f(T)$ is closed under unions and intersections of intersecting sets (cf. Theorem 44.2), we can modify the T_u in such a way that they form a partition $U_1, \ldots, U_k$ of S. Then $\hat{f}(S) = f(U_1) + \cdots + f(U_k)$, as x attains the maximum in (48.13). ∎

As a consequence, given a submodular set function f on S (by a value giving oracle), we can optimize over EP'_f in strongly polynomial time (by Corollary 44.3b, as $EP'_f = EP_{\hat{f}}$ and as we can compute $\hat{f}$).

48.2a. Applications and interpretations

Graphic matroids (Dilworth [1944], also Edmonds [1970b], Dunstan [1976]). Let $G = (V, E)$ be an undirected graph and let for each $F \subseteq E$, $f(F)$ be given by

(48.15) $\quad f(F) := |\bigcup F| - 1$.

It is easily checked that the function f is submodular, and that the function $\hat{f}$ as given by (48.7) satisfies

(48.16) $\quad \hat{f}(F) = |V|$ minus the number of components of the graph (V, F),

i.e., $\hat{f}$ is the rank function of the cycle matroid of G.

Geometric interpretation. The operation of making $\hat{f}$ from f can be interpreted geometrically as follows (Lovász [1977c], Mason [1977,1981]).

Let $\mathcal{F}$ be a collection of flats (subspaces) in a projective space, and define for each subset $\mathcal{F}'$ of $\mathcal{F}$, the rank $r(\mathcal{F}')$ by

(48.17) $\quad r(\mathcal{F}') :=$ the (projective) dimension of $\bigcup \mathcal{F}'$.

One easily checks that r is nondecreasing and submodular and that $r(\emptyset) = 0$. Now let

(48.18) $\quad f(\mathcal{F}') := r(\mathcal{F}') - 1$

for $\mathcal{F}' \subseteq \mathcal{F}$, and consider the function $\hat{f}$. Then $\hat{f}$ can be interpreted geometrically as follows. Let H be some hyperplane 'in general position' in the projective space. Then $\hat{f}(\mathcal{F}')$ is equal to the projective dimension of $H \cap \bigcup \mathcal{F}'$, i.e., $\hat{f}$ is as given by (48.17) if we replace $\mathcal{F}$ by $\{F \cap H \mid F \in \mathcal{F}\}$ (see Lovász [1977c] and Mason [1977, 1981]).

Rigidity. Let $M = (S, \mathcal{I})$ be a loopless matroid, with rank function r. Let d be a natural number. Define the set function f on S by

(48.19) $\quad f(U) := d \cdot r(U) - d + 1$,

for $U \subseteq S$. Again, f is submodular and nondecreasing. Moreover, the function $\hat{f}$ is the rank function of a loopless matroid, as $\hat{f}(\{s\}) = f(\{s\}) = 1$ for all s in S.

Let $M_d = (S, \mathcal{I}_d)$ be this matroid. Since $EP'_f = EP_{\hat{f}}$, a subset I of S is independent in M_d if and only if

(48.20) $\quad |U| \le d \cdot r(U) - d + 1$

for all $U \subseteq I$.

In case M is the cycle matroid of a connected graph $G = (V, E)$, this relates to the following (cf. Crapo [1979] and Crapo and Whiteley [1978]). Let the vertices of G be placed 'in general position' in the d-dimensional Euclidean space. Make the edges 'rigid bars'. Suppose now that the whole graph G is rigid (which only depends on G and not on the embedding, since the vertices are 'in general position'). Then G is called *rigid (in d dimensions)*. It is not difficult to see that the minimal sets F of edges of G for which the subgraph (V, F) is rigid, form the bases of a matroid. For $d = 1$ this matroid is just the cycle matroid of G, as can be checked easily. Laman [1970] (cf. Asimow and Roth [1978,1979]) showed that for $d = 2$, a graph $G = (V, E)$ is a base (i.e., a minimal rigid graph), if and only if

(48.21)   (i) $|E| = 2|V| - 3$,
           (ii) $|E[U]| \le 2|U| - 3$, for each $U \subseteq V$.

Now if M is the cycle matroid of a rigid graph G, with rank function r, then (48.21)(ii) is equivalent to

(48.22) $\quad |F| \le 2r(F) - 1$, for each subset F of E,

that is, by (48.20), to: E is independent in the matroid M_2, as given above. Condition (48.21)(i) implies that M_2 has rank $2r(E) - 1$. Hence, if G is rigid in 2 dimensions, then the bases of M_2 are the minimally rigid subgraphs of G in 2 dimensions.

In general, the matroid of rigid subgraphs of a graph $G = (V, E)$ (in d dimensions) has rank $d|V| - \binom{d+1}{2}$. However, it is not necessarily true that G is minimally rigid in d dimensions if and only if G has $d|V| - \binom{d+1}{2}$ edges and each subgraph (U, F) of G has at most $d|U| - \binom{d+1}{2}$ edges. For instance, if G arises from glueing two copies of the complete graph K_5 together in two vertices, and deleting the edge connecting these two vertices, then G is not rigid in 3 dimensions, but it satisfies the conditions given above for $d = 3$. (These conditions are easily seen to be necessary.)

More on the relation between rigidity and matroid union can be found in Whiteley [1988].

48.3. Intersection

Corollaries 48.1a and 48.1b on submodular functions f not necessarily satisfying $f(\emptyset) \geq 0$, can be extended to pairs of functions. Let f_1 and f_2 be submodular set functions on S, and consider the system

(48.23) $\quad x(U) \leq f_i(U)$ for $U \in \mathcal{P}(S) \setminus \{\emptyset\}$ and $i = 1, 2$.

Then:

Theorem 48.5. *System* (48.23) *is box-totally dual integral.*

Proof. Choose $w \in \mathbb{Z}^S$, and consider the problem dual to maximizing $w^\mathsf{T} x$ over (48.23):

(48.24) $\quad \min\{ \sum_{U \in \mathcal{P}(S) \setminus \{\emptyset\}} (y_1(U)f_1(U) + y_2(U)f_2(U)) \mid$
$\qquad y_1, y_2 \in \mathbb{R}_+^{\mathcal{P}(S) \setminus \{\emptyset\}}, \sum_{U \in \mathcal{P}(S) \setminus \{\emptyset\}} (y_1(U) + y_2(U))\chi^U = w\}.$

Let y_1, y_2 attain the minimum.
For $i \in \{1, 2\}$, define

(48.25) $\quad w_i := \sum_{U \in \mathcal{P}(S) \setminus \{\emptyset\}} y_i(U)\chi^U.$

By Theorem 48.1, for each $i = 1, 2$,

(48.26) $\quad \min\{ \sum_{U \in \mathcal{P}(S) \setminus \{\emptyset\}} y_i(U)f_i(U) \mid$
$\qquad y_i \in \mathbb{R}_+^{\mathcal{P}(S) \setminus \{\emptyset\}}, \sum_{U \in \mathcal{P}(S) \setminus \{\emptyset\}} y_i(U)\chi^U = w_i\}$

has an optimum solution y_i with $\mathcal{F}_i := \{U \mid y_i(U) > 0\}$ laminar.

These (modified) y_1, y_2 again are optimum in (48.24). As the constraints corresponding to positive components of y_1, y_2 give a totally unimodular matrix (by Theorem 41.11), Theorem 5.35 implies that system (48.23) is box-TDI. ∎

Theorem 48.5 implies primal integrality:

Corollary 48.5a. *If f_1 and f_2 are submodular and integer, then $EP'_{f_1} \cap EP'_{f_2}$ is box-integer.*

Proof. Directly from Theorem 48.5. ∎

Given submodular functions f_1 and f_2 (by value giving oracles), we can optimize over $EP'_{f_1} \cap EP'_{f_2}$ in strongly polynomial time (by Corollary 47.4d, as $EP'_{f_1} = EP_{\hat{f}_1}$ and $EP'_{f_2} = EP_{\hat{f}_2}$).

Chapter 49

Submodularity more generally

We now discuss a number of generalizations of submodular functions, namely those defined on a subcollection $\mathcal{C}$ of the collection of all subsets of a set S. The results are similar to those for submodular functions defined on all subsets on S. Often, the corresponding polyhedra form a polymatroid for some derived submodular function defined on all subsets of S.

We consider three kinds of collections, in order of increasing generality: lattice families, intersecting families, and crossing families.

49.1. Submodular functions on a lattice family

We first consider the generalization of submodular functions to those defined on a 'lattice family'.

Let S be a finite set. A family $\mathcal{C}$ of sets is called a *lattice family* if

(49.1) $\quad T \cap U, T \cup U \in \mathcal{C}$ for all $T, U \in \mathcal{C}$.

For a lattice family $\mathcal{C}$, a function $f : \mathcal{C} \to \mathbb{R}$ is called *submodular* if

(49.2) $\quad f(T \cap U) + f(T \cup U) \leq f(T) + f(U)$

for all $T, U \in \mathcal{C}$. Consider the system

(49.3) $\quad x(U) \leq f(U)$ for $U \in \mathcal{C}$,

and the problem dual to maximizing $w^\mathsf{T} x$ over (49.3), for $w \in \mathbb{R}^S$:

(49.4) $\quad \min\{\sum_{U \in \mathcal{C}} y(U) f(U) \mid y \in \mathbb{R}_+^\mathcal{C}, \sum_{U \in \mathcal{C}} y(U) \chi^U = w\}$.

Theorem 49.1. *Let $\mathcal{C}$ be a lattice family, $f : \mathcal{C} \to \mathbb{R}$ a submodular function, and $w \in \mathbb{R}^S$. Then (49.4) has an optimum solution y with $\mathcal{F} := \{U \in \mathcal{C} \mid y(U) > 0\}$ a chain.*

Proof. Let $y : \mathcal{C} \to \mathbb{R}_+$ achieve this minimum, with

(49.5) $\quad \sum_{U \in \mathcal{C}} y(U) |U| |S \setminus U|$

as small as possible. Assume that $\mathcal{F}$ is not a chain, and choose $T, U \in \mathcal{F}$ with $T \not\subseteq U$ and $U \not\subseteq T$. Let $\alpha := \min\{y(T), y(U)\}$. Decrease $y(T)$ and $y(U)$ by α, and increase $y(T \cap U)$ and $y(T \cup U)$ by α. Since

(49.6) $\qquad \chi^{T \cap U} + \chi^{T \cup U} = \chi^T + \chi^U,$

y remains a feasible solution of (49.4). As moreover

(49.7) $\qquad f(T \cap U) + f(T \cup U) \leq f(T) + f(U),$

f remains optimum. However, by Theorem 2.1, sum (49.5) decreases, contradicting our assumption. ∎

This implies the box-total dual integrality of (49.3):

Corollary 49.1a. *If $\mathcal{C}$ is a lattice family and $f : \mathcal{C} \to \mathbb{R}$ is submodular, then system (49.3) is box-TDI.*

Proof. Consider some $w : \mathcal{C} \to \mathbb{Z}$, and problem (49.4) dual to maximizing $w^\mathsf{T} x$ over (49.3). By Theorem 49.1, this minimum is attained by a y with $\mathcal{F} := \{U \in \mathcal{C} \mid y(U) > 0\}$ a chain. So the constraints corresponding to positive components of y form a totally unimodular matrix (by Theorem 41.11). Hence by Theorem 5.35, (49.3) is box-TDI. ∎

For any $\mathcal{C} \subseteq \mathcal{P}(S)$ and $f : \mathcal{C} \to \mathbb{R}$, define:

(49.8) $\qquad P_f := \{x \in \mathbb{R}^S \mid x \geq 0, x(U) \leq f(U) \text{ for each } U \in \mathcal{C}\},$
$\qquad\qquad EP_f := \{x \in \mathbb{R}^S \mid x(U) \leq f(U) \text{ for each } U \in \mathcal{C}\}.$

Then Corollary 49.1a implies:

Corollary 49.1b. *If $\mathcal{C}$ is a lattice family and $f : \mathcal{C} \to \mathbb{R}$ is submodular and integer, then EP_f is box-integer.*

Proof. Directly from Corollary 49.1a. ∎

Another consequence of Theorem 49.1 is that a submodular function f on a lattice family is uniquely determined by EP_f (given the lattice family):

Corollary 49.1c. *If $\mathcal{C}$ is a lattice family and $f : \mathcal{C} \to \mathbb{R}$ is submodular, then*

(49.9) $\qquad f(U) = \max\{x(U) \mid x \in EP_f\}$

for each $U \in \mathcal{C}$.

Proof. Let $w := \chi^U$ and let y attain minimum (49.4), with $\mathcal{F} := \{T \in \mathcal{C} \mid y(T) > 0\}$ a chain. Since

(49.10) $\qquad \chi^U = w = \sum_{T \in \mathcal{F}} y(T) \chi^T,$

we know that $\mathcal{F} = \{U\}$ and $y(U) = 1$. So the maximum in (49.9) is equal to $\sum_{T \in \mathcal{C}} y(T) f(T) = y(U) f(U) = f(U)$. ∎

We note that for any lattice family $\mathcal{C} \subseteq \mathcal{P}(S)$ with $\bigcup \mathcal{C} = S$, and any submodular function $f : \mathcal{C} \to \mathbb{R}$, the polytope P_f is a polymatroid. Indeed, define

(49.11) $\quad f'(U) := \min\{f(T) \mid T \in \mathcal{C}, T \supseteq U\}$.

for $U \subseteq S$. Then f' is submodular, and $P_{f'} = P_f$.

49.2. Intersection

Also the intersection of two of the polyhedra EP_f is tractable. Let S be a finite set. For $i = 1, 2$, let $\mathcal{C}_i$ be a lattice family on S and let $f_i : \mathcal{C}_i \to \mathbb{R}$ be submodular. Consider the system

(49.12) $\quad \begin{array}{l} x(U) \leq f_1(U) \text{ for } U \in \mathcal{C}_1, \\ x(U) \leq f_2(U) \text{ for } U \in \mathcal{C}_2. \end{array}$

Then:

Corollary 49.1d. *System* (49.12) *is box-TDI.*

Proof. Choose $w \in \mathbb{R}^S$, and consider the problem dual to maximizing $w^\mathsf{T} x$ over (49.12):

(49.13) $\quad \min\{ \sum_{U \in \mathcal{C}_1} y_1(U) f_1(U) + \sum_{U \in \mathcal{C}_2} y_2(U) f_2(U) \mid$
$\qquad y_1 \in \mathbb{R}_+^{\mathcal{C}_1}, y_2 \in \mathbb{R}_+^{\mathcal{C}_2}, \sum_{U \in \mathcal{C}_1} y_1(U) \chi^U + \sum_{U \in \mathcal{C}_2} y_2(U) \chi^U = w \}$.

Let y_1, y_2 attain the minimum.
For $i \in \{1, 2\}$, define

(49.14) $\quad w_i := \sum_{U \in \mathcal{C}_i} y_i(U) \chi^U$.

By Theorem 49.1, for each $i = 1, 2$,

(49.15) $\quad \min\{ \sum_{U \in \mathcal{C}_i} y_i(U) f_i(U) \mid y_i \in \mathbb{R}_+^{\mathcal{C}_i}, \sum_{U \in \mathcal{C}_i} y_i(U) \chi^U = w_i \}$

has an optimum solution y_i with $\mathcal{F}_i := \{U \in \mathcal{C}_i \mid y_i(U) > 0\}$ a chain.

These y_1, y_2 again are optimum in (49.13). So, by Theorem 41.11, the constraints corresponding to positive components of y form a totally unimodular matrix. Hence by Theorem 5.35, (49.12) is box-TDI. ∎

This implies primal integrality:

Corollary 49.1e. *If f_1 and f_2 are submodular and integer, then $EP_{f_1} \cap EP_{f_2}$ is box-integer.*

Proof. Directly from Corollary 49.1d. ∎

49.3. Complexity

To find the minimum of a submodular function f defined on a lattice family $\mathcal{C}$ in polynomial time, just an oracle telling if a set U belongs to $\mathcal{C}$, and if so, giving $f(U)$, is not sufficient: if $\mathcal{C} = \{\emptyset, T, S\}$ for some $T \subseteq S$, with $f(\emptyset) = f(S) = 0$ and $f(T) = -1$, we cannot find T by a polynomially bounded number of oracle calls. So we need to have more information on $\mathcal{C}$.

A lattice family $\mathcal{C}$ is fully characterized by the smallest set M and the largest set L in $\mathcal{C}$, together with the pre-order $\preceq$ on S defined by:

(49.16) $\quad u \preceq v \iff$ each $U \in \mathcal{C}$ containing v also contains u.

Then $\preceq$ is a pre-order (that is, it is reflexive and transitive). A subset U of S belongs to $\mathcal{C}$ if and only if $M \subseteq U \subseteq L$ and U is a lower ideal in $\preceq$ (that is, if $v \in U$ and $u \preceq v$, then $u \in U$).

Hence $\mathcal{C}$ has a description of size $O(|S|^2)$, such that for given $U \subseteq S$ one can test in polynomial time if U belongs to $\mathcal{C}$.

For $U \subseteq S$, define

(49.17) $\quad U^{\downarrow} := \{s \in S \mid \exists t \in U : s \preceq t\}$ and
$\quad\quad\quad U^{\uparrow} := \{s \in S \mid \exists t \in U : t \preceq s\}$.

Set

(49.18) $\quad v^{\uparrow} := \{v\}^{\uparrow},\ v^{\downarrow} := \{v\}^{\downarrow},\ \tilde{v} := v^{\uparrow} \cap v^{\downarrow}$.

For any $U \subseteq S$, let $\overline{U}$ be the (unique) smallest set in $\mathcal{C}$ containing $U \cap L$; that is,

(49.19) $\quad \overline{U} = (U \cap L)^{\downarrow} \cup M$.

So having L, M, and $\preceq$, the set $\overline{U}$ can be determined in polynomial time.

Determine a number $\alpha > 0$ such that

(49.20) $\quad \alpha \geq f(S \setminus v^{\uparrow}) - f((S \setminus v^{\uparrow}) \cup \tilde{v})$ and $\alpha \geq f(v^{\downarrow}) - f(v^{\downarrow} \setminus \tilde{v})$

for all $v \in L \setminus M$. Such an α can be found by at most $4|S|$ oracle calls.

Then α satisfies, for any $X, Y \in \mathcal{C}$ with $X \subseteq Y$:

(49.21) $\quad |f(Y) - f(X)| \leq \alpha |Y \setminus X|$.

To show this, we can assume that $Y \setminus X = \tilde{v}$ for some $v \in L \setminus M$. Then $f(Y) - f(X) \leq f(v^{\downarrow}) - f(v^{\downarrow} \setminus \tilde{v}) \leq \alpha$ and $f(Y) - f(X) \geq f((S \setminus v^{\uparrow}) \cup \tilde{v}) - f(S \setminus v^{\uparrow}) \geq -\alpha$, implying (49.21).

Now define a function $\bar{f} : \mathcal{P}(S) \to \mathbb{R}$ by:

(49.22) $\bar{f}(U) := f(\overline{U}) + \alpha |\overline{U} \triangle U|$

for $U \subseteq S$.
  Then:

Theorem 49.2. *For any $\alpha > 0$ satisfying (49.20) for all $v \in L \setminus M$, the function $\bar{f}$ is submodular.*

Proof. First consider $T, U \subseteq L$. Then $T \subseteq \overline{T}$ and $U \subseteq \overline{U}$, and hence:

(49.23) $\quad \bar{f}(T) + \bar{f}(U) = f(\overline{T}) + \alpha |\overline{T} \setminus T| + f(\overline{U}) + \alpha |\overline{U} \setminus U|$
$\qquad \geq f(\overline{T} \cap \overline{U}) + \alpha |(\overline{T} \cap \overline{U}) \setminus (T \cap U)| + f(\overline{T} \cup \overline{U}) + \alpha |(\overline{T} \cup \overline{U}) \setminus (T \cup U)|$
$\qquad \geq f(\overline{T \cap U}) + \alpha |\overline{T \cap U} \setminus (T \cap U)| + f(\overline{T \cup U}) + \alpha |\overline{T \cup U} \setminus (T \cup U)|$
$\qquad = \bar{f}(T \cap U) + \bar{f}(T \cup U).$

(The last inequality uses (49.21), since $\overline{T} \cap \overline{U} \supseteq \overline{T \cap U}$ (while $\overline{T} \cup \overline{U} = \overline{T \cup U}$).)
Hence, for $T, U \subseteq S$ one has:

(49.24) $\quad \bar{f}(T) + \bar{f}(U) = \bar{f}(T \cap L) + \alpha |T \setminus L| + \bar{f}(U \cap L) + \alpha |U \setminus L|$
$\qquad \geq \bar{f}((T \cap L) \cap (U \cap L)) + \bar{f}((T \cap L) \cup (U \cap L)) + \alpha |T \setminus L| + \alpha |U \setminus L|$
$\qquad = \bar{f}((T \cap U) \cap L) + \bar{f}((T \cup U) \cap L) + \alpha |(T \cap U) \setminus L| + \alpha |(T \cup U) \setminus L|$
$\qquad = \bar{f}(T \cap U) + \bar{f}(T \cup U).$

So $\bar{f}$ is submodular. ∎

  The function $\bar{f}$ enables us to reduce optimization problems on submodular functions defined on a lattice family, to those defined on all subsets.

Minimization. By Theorem 45.1, the minimum of $\bar{f}$ can be found in strongly polynomial time. Hence

(49.25) $\quad$ if $\mathcal{C}$ is given by L, M, and $\preceq$, and a submodular function $f : \mathcal{C} \to \mathbb{R}$ is given by a value giving oracle, we can find a $U \in \mathcal{C}$ minimizing $f(U)$ in strongly polynomial time.

Indeed, if $\bar{f}$ attains its minimum at U, then $U \in \mathcal{C}$, since otherwise $\overline{U} \neq U$ and hence $\bar{f}(U) > \bar{f}(\overline{U})$ (as $\alpha > 0$), contradicting the fact that $\bar{f}$ attains its minimum at U. This shows (49.25).

Maximization over EP_f. Given a lattice family $\mathcal{C}$ of subsets of a set S, a submodular function $f : \mathcal{C} \to \mathbb{R}$, and a weight function $w \in \mathbb{Q}^S$, we can maximize $w^\mathsf{T} x$ over EP_f, by adapting the greedy algorithm as follows.
  Note that $\max\{w^\mathsf{T} x \mid x \in EP_f\}$ is finite if and only if $w \geq \mathbf{0}$, $w(s) = 0$ for each $s \in S \setminus L$, and

(49.26) $\quad u \preceq v$ implies $w(u) \geq w(v)$

for all $u, v \in S$. If (49.26) is not the case, the maximum value is infinite, since if $u \preceq v$, then for any $x \in EP_f$, the vector $x + \lambda(\chi^v - \chi^u)$ belongs to EP_f for all $\lambda \geq 0$. Now, if $w(v) > w(u)$, the weight increases to infinity along this line, and therefore the maximum value is ∞.

So we can check in strongly polynomial time if $\max\{w^\mathsf{T} x \mid x \in EP_f\}$ is finite, and therefore we can assume that it is finite. Moreover, we can assume that $L = S$, since $w(s) = 0$ for each $s \in S \setminus L$, and hence we can delete $S \setminus L$. Similarly, we can assume that $\emptyset \in \mathcal{C}$ and $f(\emptyset) = 0$. For if $\emptyset \in \mathcal{C}$ and $f(\emptyset) < 0$, then $EP_f = \emptyset$, and if $f(\emptyset) > 0$, we can reset $f(\emptyset) := 0$, without violating the submodularity and without modifying EP_f. If $\emptyset \notin \mathcal{C}$, then we can add $\emptyset$ to $\mathcal{C}$ and set $f(\emptyset) := 0$, again maintaining submodularity and EP_f. Finally, we can assume that $\preceq$ is a partial order, since if $u \preceq v \preceq u$, then by (49.26), $w(u) = w(v)$, and each set in $\mathcal{C}$ either contains both u and v, or neither of them. So we can merge u and v; and in fact we can merge any set $\tilde{v}$ to one element.

Now let $\leq$ be a linear order such that for any u, v, if $u \preceq v$ or $w(u) > w(v)$, then $u \leq v$. By (49.26), the latter defines a partial order. So $\leq$ is a linear extension of it, and hence can be found in strongly polynomial time.

Let $S = \{s_1, \ldots, s_n\}$ with $s_1 < s_2 < \cdots < s_n$. For $i = 0, \ldots, n$, define $U_i := \{s_1, \ldots, s_i\}$. As $\leq$ is a linear extension of $\preceq$, each U_i is a lower ideal of $\preceq$, and hence each U_i belongs to $\mathcal{C}$. Define $x(s_i) := f(U_i) - f(U_{i-1})$ for $i = 1, \ldots, n$. Then x maximizes $w^\mathsf{T} x$ over EP_f.

To see this, let $\bar{f}$ be defined as above. Then by Theorem 44.3, x belongs to $EP_{\bar{f}}$ (as f and $\bar{f}$ coincide on each U_i), and hence x belongs to EP_f. To see that x is optimum, we have for any $z \in EP_f$:

$$(49.27) \quad w^\mathsf{T} z = \sum_{i=1}^{n-1} z(U_i)(w(s_i) - w(s_{i+1})) + z(S)w(s_n)$$

$$\leq \sum_{i=1}^{n-1} f(U_i)(w(s_i) - w(s_{i+1})) + f(S)w(s_n)$$

$$= \sum_{i=1}^{n} w(s_i)(f(U_i) - f(U_{i-1})) = \sum_{i=1}^{n} w(s_i) x(s_i) = w^\mathsf{T} x.$$

This also gives a dual solution to the corresponding LP-formulation of the problem.

Maximization over intersections. Let $\mathcal{C}_1$ and $\mathcal{C}_2$ be lattice families of subsets of S and let f_1 and f_2 be submodular functions on $\mathcal{C}_1$ and $\mathcal{C}_2$ respectively. Let $\mathcal{C}_i$ be specified by L_i, M_i, and $\preceq_i$.

Find a number $\alpha > 0$ satisfying (49.20) for both $f = f_1$ and $f = f_2$. So by (49.21), $\alpha|S| + \max_{i=1,2} |f_i(L_i)|$ is an upper bound on $|f_i(U)|$ for each $i \in \{1, 2\}$ and each $U \in \mathcal{C}_i$. Define

$$(49.28) \quad K := |S|(\alpha|S| + \max_{i=1,2} |f_i(L_i)|).$$

Now for $i = 1, 2$ and $U \subseteq S$, let $\bar{f}_i(U) := f_i(\overline{U}) + K|\overline{U} \triangle U|$ (where $\overline{U}$ is taken with respect to $\mathcal{C}_i$). So $\bar{f}_1$ and $\bar{f}_2$ are submodular (by Theorem 49.2). Then:

$$(49.29) \quad \max\{w^\mathsf{T} x \mid x \in EP_{f_1} \cap EP_{f_2}\} = \max\{w^\mathsf{T} x \mid x \in EP_{\bar{f}_1} \cap EP_{\bar{f}_2}\},$$

if the first maximum is finite. Clearly $\geq$ holds in (49.29), since $EP_{\tilde{f}_i} \subseteq EP_{f_i}$ for $i = 1, 2$. To see equality, each face of $EP_{f_1} \cap EP_{f_2}$ is determined by equations $x(U) = f_i(U)$ for $i = 1, 2$ and $U \in \mathcal{D}_i$, where $\mathcal{D}_i$ is a chain of sets in $\mathcal{C}_i$. So it is determined by a system of linear equations with totally unimodular constraint set and right-hand sides determined by function values of f_1 and f_2. So each face contains a vector x with $|x_s| \leq K$ for all $s \in S$ (by (49.28), since the inverse of a nonsingular totally unimodular matrix has all its entries in $\{0, \pm 1\}$). But any such x belongs to $EP_{\tilde{f}_1} \cap EP_{\tilde{f}_2}$, since for $i = 1, 2$ and $U \subseteq S$, we have:

(49.30) $\quad x(U) \leq x(\overline{U}) + K|\overline{U} \triangle U| \leq f_i(\overline{U}) + K|\overline{U} \triangle U| = \tilde{f}_i(U)$

(where $\overline{U}$ is taken with respect to $\mathcal{C}_i$). So we have (49.29).

Therefore, by Corollary 47.4d, we can maximize $w^\mathsf{T} x$ over $EP_{f_1} \cap EP_{f_2}$ in strongly polynomial time. Note that for any $w \in \mathbb{Q}^S$, we can decide in strongly polynomial time if the first maximum in (49.29) is finite. For this, we should decide if there exist $w_1, w_2 \in \mathbb{Q}_+^S$ such that $w = w_1 + w_2$ and such that for $i = 1, 2$: $w_i(s) = 0$ for $s \in S \setminus L_i$ and $u \preceq_i v$ implies $w_i(u) \geq w_i(v)$ for all u, v. This can be reduced to checking if a certain digraph with lengths has no negative-length directed circuit.

49.4. Submodular functions on an intersecting family

We next consider functions defined on a broader class of collections, the intersecting families, where the function satisfies a restricted form of submodularity. It yields an extension of the *Dilworth truncation* studied in Chapter 48.

A family $\mathcal{C}$ of sets is called an *intersecting family* if for all $T, U \in \mathcal{C}$ one has:

(49.31) $\quad$ if $T \cap U \neq \emptyset$, then $T \cap U, T \cup U \in \mathcal{C}$.

Let $\mathcal{C}$ be an intersecting family. A function $f : \mathcal{C} \to \mathbb{R}$ is called *submodular on intersecting pairs*, or *intersecting submodular*, if

(49.32) $\quad f(T) + f(U) \geq f(T \cap U) + f(T \cup U)$

for all $T, U \in \mathcal{C}$ with $T \cap U \neq \emptyset$.

Consider the system

(49.33) $\quad x(U) \leq f(U)$ for $U \in \mathcal{C}$,

and the problem dual to maximizing $w^\mathsf{T} x$ over (49.33), for $w \in \mathbb{R}^S$:

(49.34) $\quad \min \{ \sum_{U \in \mathcal{C}} y(U) f(U) \mid y \in \mathbb{R}_+^\mathcal{C}, \sum_{U \in \mathcal{C}} y(U) \chi^U = w \}$.

Recall that a collection $\mathcal{F}$ of sets is called *laminar* if

(49.35) $\quad T\cap U = \emptyset$ or $T\subseteq U$ or $U\subseteq T$, for all $T,U\in\mathcal{F}$.

A basic result (proved with a method due to Edmonds [1970b]) is:

Theorem 49.3. *Let $\mathcal{C}$ be an intersecting family of subsets of a set S, let $f:\mathcal{C}\to\mathbb{R}$ be intersecting submodular and let $w\in\mathbb{R}^S$. Then (49.34) has an optimum solution y with $\mathcal{F}:=\{U\in\mathcal{C}\mid y(U)>0\}$ laminar.*

Proof. Let $y:\mathcal{C}\to\mathbb{R}_+$ achieve this minimum, with

(49.36) $\quad \displaystyle\sum_{U\in\mathcal{C}} y(U)|U||S\setminus U|$

as small as possible. Assume that $\mathcal{F}$ is not laminar, and choose $T,U\in\mathcal{F}$ violating (49.35). Let $\alpha:=\min\{y(T),y(U)\}$. Decrease $y(T)$ and $y(U)$ by α, and increase $y(T\cap U)$ and $y(T\cup U)$ by α. Since

(49.37) $\quad \chi^{T\cap U} + \chi^{T\cup U} = \chi^T + \chi^U,$

y remains a feasible solution of (49.34). As moreover

(49.38) $\quad f(T\cap U) + f(T\cup U) \leq f(T) + f(U),$

f remains optimum. However, by Theorem 2.1, sum (49.36) decreases, contradicting our assumption. ∎

It gives the box-total dual integrality of (49.33):

Corollary 49.3a. *Let $\mathcal{C}$ be an intersecting family of subsets of a set S and let $f:\mathcal{C}\to\mathbb{R}$ be intersecting submodular. Then system (49.33) is box-TDI.*

Proof. Consider problem (49.34) dual to maximizing $w^\top x$ over (49.33). By Theorem 49.3, this minimum is attained by a y with $\mathcal{F}:=\{U\in\mathcal{C}\mid y(U)>0\}$ laminar. As the matrix of constraints corresponding to $\mathcal{F}$ is totally unimodular (Theorem 41.11), Theorem 5.35 gives the corollary. ∎

49.5. Intersection

Again, these results can be extended in a natural way to pairs of functions, by derivation from Theorem 49.3.

Corollary 49.3b. *For $i=1,2$, let $\mathcal{C}_i$ be an intersecting family of subsets of a set S and let $f_i:\mathcal{C}_i\to\mathbb{R}$ be intersecting submodular. Then the system*

(49.39) $\quad x(U) \leq f_1(U)$ *for $U\in\mathcal{C}_1$,*
$\qquad\quad\; x(U) \leq f_2(U)$ *for $U\in\mathcal{C}_2$,*

is box-TDI.

Proof. Choose $w \in \mathbb{R}^S$, and consider the problem dual to maximizing $w^\mathsf{T} x$ over (49.39):

(49.40) $\min\{\sum_{U \in \mathcal{C}_1} y_1(U)f_1(U) + \sum_{U \in \mathcal{C}_2} y_2(U)f_2(U) \mid$
$y_1 \in \mathbb{R}_+^{\mathcal{C}_1}; y_2 \in \mathbb{R}_+^{\mathcal{C}_2}, \sum_{U \in \mathcal{C}_1} y_1(U)\chi^U + \sum_{U \in \mathcal{C}_2} y_2(U)\chi^U = w\}.$

Let y_1, y_2 attain the minimum. For $i \in \{1, 2\}$, define

(49.41) $w_i := \sum_{U \in \mathcal{C}_i} y_i(U)\chi^U.$

By Theorem 49.3,

(49.42) $\min\{\sum_{U \in \mathcal{C}_i} y_i(U)f_i(U) \mid y_i \in \mathbb{R}_+^{\mathcal{C}_i}, \sum_{U \in \mathcal{C}_i} y_i(U)\chi^U = w_i\}$

has an optimum solution y_i with $\mathcal{F}_i := \{U \in \mathcal{C}_i \mid y_i(U) > 0\}$ laminar.

As $\mathcal{F}_1 \cup \mathcal{F}_2$ determine a totally unimodular matrix (by Theorem 41.11), Theorem 5.35 implies that system (49.39) is box-TDI. ∎

This implies the integrality of polyhedra:

Corollary 49.3c. *If f_1 and f_2 are integer, then $EP_{f_1} \cap EP_{f_2}$ is box-integer.*

Proof. Directly from Corollary 49.3b. ∎

49.6. From an intersecting family to a lattice family

Let $\mathcal{C}$ be an intersecting family of subsets of a set S and let $f : \mathcal{C} \to \mathbb{R}$ be submodular on intersecting pairs. Let $\check{\mathcal{C}}$ be the collection of all unions of sets in $\mathcal{C}$. Since $\mathcal{C}$ is closed under unions of intersecting sets, $\check{\mathcal{C}}$ is equal to the collection of *disjoint* unions of nonempty sets in $\mathcal{C}$. It is not difficult to show that $\check{\mathcal{C}}$ is a lattice family and that $\emptyset \in \check{\mathcal{C}}$.

Call a partition *proper* if its classes are nonempty. Define $\check{f} : \check{\mathcal{C}} \to \mathbb{R}$ by:

(49.43) $\check{f}(U) := \min\{\sum_{P \in \mathcal{P}} f(P) \mid \mathcal{P} \subseteq \mathcal{C} \text{ is a proper partition of } U\}.$

for $U \in \check{\mathcal{C}}$. So $\check{f}(\emptyset) = 0$. Then (Dunstan [1976]):

Theorem 49.4. *$\check{f}$ is submodular.*

Proof. Choose $T, U \in \check{\mathcal{C}}$, and let $\mathcal{P}$ and $\mathcal{Q}$ be partitions of T and U (respectively) into nonempty sets in $\mathcal{C}$ with

(49.44) $$\check{f}(T) = \sum_{P \in \mathcal{P}} f(P) \text{ and } \check{f}(U) = \sum_{Q \in \mathcal{Q}} f(Q).$$

Consider the family $\mathcal{F}$ made by the union of $\mathcal{P}$ and $\mathcal{Q}$ (taking a set twice if it occurs in both partitions). We can transform $\mathcal{F}$ iteratively into a laminar family, by replacing any $X, Y \in \mathcal{F}$ with $X \cap Y \neq \emptyset$ and $X \not\subseteq Y \not\subseteq X$ by $X \cap Y, X \cup Y$. In each iteration, the sum

(49.45) $$\sum_{Z \in \mathcal{F}} f(Z)$$

does not increase (as f is submodular on intersecting pairs). As at each iteration the sum

(49.46) $$\sum_{Z \in \mathcal{F}} |Z||S \setminus Z|$$

decreases (by Theorem 2.1), this process terminates. We end up with a laminar family $\mathcal{F}$.

The inclusionwise maximal sets in $\mathcal{F}$ form a partition $\mathcal{R}$ of $T \cup U$, and the remaining sets form a partition $\mathcal{S}$ of $T \cap U$. Therefore,

(49.47) $$\check{f}(T \cup U) + \check{f}(T \cap U) \leq \sum_{X \in \mathcal{R}} f(X) + \sum_{Y \in \mathcal{S}} f(Y)$$
$$\leq \sum_{P \in \mathcal{P}} f(P) + \sum_{Q \in \mathcal{Q}} f(Q) = \check{f}(T) + \check{f}(U),$$

showing that $\check{f}$ is submodular. ∎

Trivially, if $\emptyset \notin \mathcal{C}$ or if $f(\emptyset) \geq 0$, then $EP_{\check{f}} = EP_f$. Hence, by (49.9),

(49.48) $$\check{f}(U) = \max\{x(U) \mid U \in EP_f\}.$$

As we shall see in Section 49.7, this enables us to calculate $\check{f}$ from a value giving oracle, using the greedy algorithm.

49.7. Complexity

The results of the previous section enable us to reduce algorithmic problems on intersecting submodular functions, to those on submodular functions on lattice families.

If $\mathcal{C}$ is an intersecting family on S, then for each $s \in S$, the collection $\mathcal{C}_s := \{U \in \mathcal{C} \mid s \in U\}$ is a lattice family. So (like in Section 49.3) we can assume that $\mathcal{C}$ is given by a representation of $\mathcal{C}_s$ for each $s \in S$, in terms of the pre-order $\preceq_s$ given by: $u \preceq_s v$ if and only if each set in $\mathcal{C}$ containing s and v also contains u, and by $M_s := \bigcap \mathcal{C}_s$ and $L_s := \bigcup \mathcal{C}_s$; next to that we should know if $\emptyset$ belongs to $\mathcal{C}$.

We can derive the information on $\check{\mathcal{C}}$ as follows:

836   Chapter 49. Submodularity more generally

(49.49) $\bigcap \check{\mathcal{C}} = \emptyset$, $\bigcup \check{\mathcal{C}} = \bigcup_{s \in S} L_s$; $u \preceq v$ if and only if $u \in M_v$.

So we can decide in polynomial time if a set U belongs to $\check{\mathcal{C}}$.

For any intersecting submodular function f on $\mathcal{C}$, the restriction f_s of f to $\mathcal{C}_s$ is submodular. So by the results of Section 49.3, we can find a set minimizing f in strongly polynomial time.

For any $U \in \check{\mathcal{C}}$, we can calculate $\check{f}(U)$, as defined in (49.43), in strongly polynomial time, having a value giving oracle for f. To see this, we use (49.48). We can assume that $\emptyset \notin \mathcal{C}$.

Order the elements of U as $t_1, \ldots, t_k$ such that if $L_{t_j} \subset L_{t_i}$, then $j < i$. For $i = 0, \ldots, k$, let $U_i := L_{t_1} \cup \cdots \cup L_{t_i}$. So $U_k = U$.

Initially, set $x(t) := 0$ for each $t \in U$. Next, for $i = 1, \ldots, k$, calculate

(49.50) $\mu := \min\{f(T) - x(T) \mid T \in \mathcal{C}, t_i \in T \subseteq U_i\}$,

and reset $x(t_i) := x(t_i) + \mu$. We prove, by induction on i, that for $i = 0, 1, \ldots, k$ we have, after processing $t_1, \ldots, t_i$:

(49.51)   (i) $x(T) \leq f(T)$ for each $T \in \mathcal{C}$ with $T \subseteq U_i$,
(ii) for each $j = 1, \ldots, i$ there exists a $T \in \mathcal{C}$ with $t_j \in T \subseteq U_i$ and $x(T) = f(T)$.

For $i = 0$ this is trivial. Let $i \geq 1$. Consider any $T \in \mathcal{C}$ with $T \subseteq U_i$. If $t_i \in T$, then $x(T) \leq f(T)$, as at processing t_i we have added μ to $x(t_i)$. If $t_i \notin T$, then $T \subseteq U_{i-1}$. For suppose that there exists a $t_j \in T$ with $t_j \notin U_{i-1}$. So $j > i$ and $t_j \in L_{t_i}$, and therefore $L_{t_j} \subseteq L_{t_i}$, implying $L_{t_j} = L_{t_i}$ (since if $L_{t_j} \subset L_{t_i}$, then $j < i$). But then $t_i \in L_{t_j} \subseteq T$, contradicting the fact that $t_i \notin T$. So $T \subseteq U_{i-1}$. As $t_i \notin T$, $x(T)$ did not change at processing t_i, and hence we know $x(T) \leq f(T)$ by induction. This proves (49.51)(i).

To see (49.51)(ii), choose $j \leq i$. If $j = i$, there exists after processing t_i a T as required, as we have added μ to $x(t_i)$. If $j < i$, by induction there exists a $T \in \mathcal{C}$ with $t_j \in T \subseteq U_{i-1}$ and $x(T) = f(T)$ before processing t_i. If $t_i \notin T$, $x(T) = f(T)$ is maintained at processing t_i. If $t_i \in T$, then $t_i \in U_{i-1}$, and so $t_i \in L_{t_j}$ for some $j < i$. Hence $L_{t_i} \subseteq L_{t_j}$. So, by the choice of the order of U, $L_{t_i} = L_{t_j}$. Hence before processing t_i we have $x(T') \leq f(T')$ for each $T' \subseteq U_i$. So, as $x(T) = f(T)$ and $t_i \in T$, $x(t_i)$ is not modified at processing t_i. Therefore, $x(T) = f(T)$ holds also after processing t_i. This proves (49.51)(ii).

This shows (49.51), which gives, taking $i = k$, that $x(T) \leq f(T)$ for each $T \in \mathcal{C}$ with $T \subseteq U$, and that for each $t \in U$, we have a T containing t with $x(T) = f(T)$. We can replace any two T and T' with $T \cap T' \neq \emptyset$ by $T \cup T'$. We end up with a partition $\mathcal{T}$ of U with $x(U) = \sum_{T \in \mathcal{T}} f(T)$. Hence we know

(49.52) $\check{f}(U) \geq x(U) = \sum_{T \in \mathcal{T}} x(T) = \sum_{T \in \mathcal{T}} f(T) \geq \check{f}(U)$,

and therefore we have equality throughout.

Having this, we can reduce the problem of maximizing $w^\mathsf{T} x$ over EP_f, where f is intersecting submodular, to that of maximizing $w^\mathsf{T} x$ over $EP_{\check{f}}$, which can be done in strongly polynomial time by the results of Section 49.3. Similarly for intersections of two such polyhedra EP_{f_1} and EP_{f_2}.

49.8. Intersecting a polymatroid and a contrapolymatroid

For an intersecting family $\mathcal{C}$, a function $g : \mathcal{C} \to \mathbb{R}$ is called *supermodular on intersecting pairs*, or *intersecting supermodular*, if $-g$ is intersecting submodular.

Let S be a finite set. Let $\mathcal{C}$ and $\mathcal{D}$ be collections of subsets of S and let $f : \mathcal{C} \to \mathbb{R}$ and $g : \mathcal{D} \to \mathbb{R}$. Consider the system

(49.53) $\quad x(U) \leq f(U) \quad \text{for } U \in \mathcal{C},$
$\qquad\quad x(U) \geq g(U) \quad \text{for } U \in \mathcal{D}.$

Theorem 49.5. *If $\mathcal{C}$ and $\mathcal{D}$ are intersecting families, $f : \mathcal{C} \to \mathbb{R}$ is intersecting submodular, and $g : \mathcal{D} \to \mathbb{R}$ is intersecting supermodular, then system (49.53) is box-TDI.*

Proof. Choose $w \in \mathbb{Z}^S$, and consider the dual problem of maximizing $w^\mathsf{T} x$ over (49.53):

(49.54) $\quad \min\{\sum_{U \in \mathcal{C}} y(U)f(U) - \sum_{U \in \mathcal{D}} z(U)g(U) \mid$
$\qquad\qquad y \in \mathbb{R}_+^{\mathcal{C}}, z \in \mathbb{R}_+^{\mathcal{D}}, \sum_{U \in \mathcal{C}} y(U)\chi^U - \sum_{U \in \mathcal{D}} z(U)\chi^U = w\}.$

Let y, z attain this minimum. Define

(49.55) $\quad u := \sum_{U \in \mathcal{C}} y(U)\chi^U \text{ and } v := \sum_{U \in \mathcal{D}} z(U)\chi^U.$

So y attains

(49.56) $\quad \min\{\sum_{U \in \mathcal{C}} y(U)f(U) \mid y \in \mathbb{R}_+^{\mathcal{C}}, \sum_{U \in \mathcal{C}} y(U)\chi^U = u\}$

and z attains

(49.57) $\quad \max\{\sum_{U \in \mathcal{D}} z(U)g(U) \mid z \in \mathbb{R}_+^{\mathcal{D}}, \sum_{U \in \mathcal{D}} z(U)\chi^U = v\}.$

By Theorem 49.3, (49.56) has an optimum solution y with $\mathcal{F} := \{U \in \mathcal{C} \mid y(U) > 0\}$ laminar. Similarly, (49.57) has an optimum solution z with $\mathcal{G} := \{U \in \mathcal{D} \mid z(U) > 0\}$ laminar. Now $\mathcal{F}$ and $\mathcal{G}$ determine a totally unimodular submatrix (by Theorem 41.11), and hence by Theorem 5.35, (49.53) is box-TDI. ∎

49.9. Submodular functions on a crossing family

Finally, we consider submodular functions defined on a crossing family. The results discussed above for submodular functions on intersecting families do not all transfer to crossing families. But certain restricted versions still hold.

A family $\mathcal{C}$ of subsets of a set S is called a *crossing family* if for all $T, U \in \mathcal{C}$ one has:

(49.58)   if $T \cap U \neq \emptyset$ and $T \cup U \neq S$, then $T \cap U, T \cup U \in \mathcal{C}$.

A function $f : \mathcal{C} \to \mathbb{R}$, defined on a crossing family $\mathcal{C}$, is called *submodular on crossing pairs*, or *crossing submodular*, if for all $T, U \in \mathcal{C}$ with $T \cap U \neq \emptyset$ and $T \cup U \neq S$:

(49.59) $f(T) + f(U) \geq f(T \cap U) + f(T \cup U)$.

In general, the system

(49.60) $x(U) \leq f(U)$ for $U \in \mathcal{C}$

is *not* TDI. For instance, if $S = \{1, 2, 3\}$, $\mathcal{C} = \{\{1, 2\}, \{1, 3\}, \{2, 3\}\}$, from S, and $f(U) := 1$ for each $U \in \mathcal{C}$, then (49.60) not even determines an integer polyhedron (as $(\frac{1}{2}, \frac{1}{2}, \frac{1}{2})^\mathsf{T}$ is a vertex of it).

However, for any $k \in \mathbb{R}$, the system

(49.61) $x(U) \leq f(U)$ for $U \in \mathcal{C}$,
$x(S) = k$

is box-TDI. This can be done by reduction to Corollary 49.3a. Similarly for pairs of such functions. This was shown by Frank [1982b,1984a] and Fujishige [1984e].

Let $\mathcal{C}$ be a crossing family of subsets of a set S. Let $\hat{\mathcal{C}}$ be the collection of all nonempty intersections of sets in $\mathcal{C}$ (we allow the intersection of 0 sets, so $S \in \hat{\mathcal{C}}$). Since $\mathcal{C}$ is a crossing family, we know

(49.62) $\hat{\mathcal{C}} = \{U \mid U \neq \emptyset; \exists \mathcal{P} \subseteq \mathcal{C} : \mathcal{P} \text{ is a copartition of } S \setminus U\}$,

where a *copartition* of U is a collection $\mathcal{P}$ of subsets of S such that the collection $\{S \setminus T \mid T \in \mathcal{P}\}$ is a partition of U. We call the copartition *proper* if $T \neq S$ for each $T \in \mathcal{P}$.

Note that, in (49.62), restricting $\mathcal{P}$ to proper copartitions of $S \setminus U$, does not modify $\hat{\mathcal{C}}$. We allow $\mathcal{P} = \emptyset$, so $S \in \hat{\mathcal{C}}$.

Define $\hat{f} : \hat{\mathcal{C}} \to \mathbb{R}$ by:

(49.63) $\hat{f}(U) := \min\{\sum_{P \in \mathcal{P}} f(P) \mid \mathcal{P} \subseteq \mathcal{C} \text{ is a proper copartition of } S \setminus U\}$

for $U \in \hat{\mathcal{C}}$. So $\hat{f}(S) = 0$. Then:

Theorem 49.6. $\hat{\mathcal{C}}$ *is an intersecting family and $\hat{f}$ is submodular on intersecting pairs.*

Proof. Define, for $s \in S$,

(49.64) $\quad \hat{\mathcal{C}}_s := \{U \in \hat{\mathcal{C}} \mid s \in U\}$ and $\mathcal{D} := \{S \setminus U \mid s \in U \in \mathcal{C}\}$.

As $\mathcal{C}$ is crossing, $\mathcal{D}$ is intersecting. Hence $\check{\mathcal{D}}$ is a lattice family. As $\hat{\mathcal{C}}_s = \{S \setminus U \mid U \in \check{\mathcal{D}}\}$, also $\hat{\mathcal{C}}_s$ is a lattice family. As this is true for each $s \in S$, $\hat{\mathcal{C}}$ is intersecting.

To prove that $\hat{f}$ is intersecting submodular, it suffices to show that for each $s \in S$, the restriction $\hat{f}_s$ of $\hat{f}$ to $\hat{\mathcal{C}}_s$ is submodular. Define $g : \mathcal{D} \to \mathbb{R}$ by $g(U) := f(S \setminus U)$ for $U \in \mathcal{D}$. Then g is intersecting submodular (as f is crossing submodular). Hence, by Theorem 49.4, $\check{g}$ is submodular on $\check{\mathcal{D}}$. As $\hat{f}_s(U) = \check{g}(S \setminus U)$ for $U \in \hat{\mathcal{C}}_s$, $\hat{f}_s$ is submodular. ∎

Fujishige [1984e] showed that the following box-TDI result can be derived from Corollary 49.3a:

Theorem 49.7. *Let $\mathcal{C}$ be a crossing family of subsets of S, let $f : \mathcal{C} \to \mathbb{R}$ be crossing submodular, and let $k \in \mathbb{R}$. Then system (49.61) is box-TDI and determines the polyhedron of maximal vectors of $EP_{f'}$ for some submodular function f' defined on a lattice family.*

Proof. We can assume that $k = 0$, since, choosing any $t \in S$ and resetting $f(U) := f(U) - k$ for all $U \in \mathcal{C}$ with $t \in U$, does not change the box-total dual integrality of (49.61). We can also assume that $\emptyset \notin \mathcal{C}$.

The box-total dual integrality of (49.61) follows from that of

(49.65) $\quad x(U) \leq \hat{f}(U) \quad$ for $U \in \hat{\mathcal{C}}$,
$\quad\quad\quad\quad x(S) = 0$,

as (49.65) and (49.61) have the same solution set, and as each constraint in (49.65) is a nonnegative integer combination of constraints in (49.61). The box-total dual integrality of (49.65) follows from Corollary 49.3a (using Theorem 5.25). It also shows that the solution set of (49.61) is the set of maximal vectors of $EP_{f'}$ for some submodular function f' defined on a lattice family. ∎

Frank and Tardos [1984b] observed that this implies a relation with matroids:

Corollary 49.7a. *If $\mathcal{C}$ is a crossing family of subsets of a set S, $f : \mathcal{C} \to \mathbb{Z}$ is crossing submodular, and $k \in \mathbb{Z}$, then the collection*

(49.66) $\quad \{B \subseteq S \mid |B| = k, |B \cap U| \leq f(U) \text{ for each } U \in \mathcal{C}\}$

forms the collection of bases of a matroid (if nonempty).

Proof. Directly from Theorem 49.7, using the observations on (44.43) and (49.11). ∎

Similarly, the box-total dual integrality of pairs of such systems follows:

Theorem 49.8. *For $i = 1, 2$, let C_i be a crossing family of subsets of a set S, let $f_i : C_i \to \mathbb{R}$ be crossing submodular, and let $k \in \mathbb{R}$. Then the system*

(49.67) $\quad x(U) \leq f_1(U) \text{ for } U \in C_1,$
$\qquad\qquad x(U) \leq f_2(U) \text{ for } U \in C_2,$
$\qquad\qquad x(S) = k,$

is box-TDI.

Proof. Similar to the proof of the previous theorem, by reduction to Corollary 49.3b. ∎

This implies the integrality of polyhedra:

Corollary 49.8a. *For $i = 1, 2$, let C_i be a crossing family of subsets of a set S, let $f_i : C_i \to \mathbb{R}$ be crossing submodular, and let $k \in \mathbb{R}$. If f_1, f_2, and k are integer, system (49.67) determines a box-integer polyhedron.*

Proof. Directly from Theorem 49.8. ∎

49.10. Complexity

The reduction given in the proof of Theorem 49.6 also enables us to calculate $\hat{f}(U)$ from a value giving oracle for f, similar to the proof in Section 49.7. We assume that C is given by descriptions of the lattice families $C_{s,t} := \{U \in C \mid s \in U, t \notin U\}$ as in Section 49.3.

Note that

(49.68) $\quad EP_{\hat{f}} \cap \{x \mid x(S) = 0\} = EP_f \cap \{x \mid x(S) = 0\}.$

This follows from the fact that if $x \in EP_f$ and $x(S) = 0$, then for any $U \in \hat{C}$ and any proper copartition $\mathcal{P} = \{U_1, \ldots, U_p\}$ of $S \setminus U$ with $f(U) = \sum_{P \in \mathcal{P}} f(P)$, one has:

(49.69) $\quad x(U) = -x(S \setminus U) = -\sum_{P \in \mathcal{P}} x(S \setminus P) = \sum_{P \in \mathcal{P}} x(P) \leq \sum_{P \in \mathcal{P}} f(P)$
$\qquad\qquad = f(U).$

Having this, the problem of maximizing $w^\mathsf{T} x$ over $EP_f \cap \{x \mid x(S) = 0\}$, where f is crossing submodular, is reduced to the problem of maximizing $w^\mathsf{T} x$ over $EP_{\hat{f}} \cap \{x \mid x(S) = 0\}$. The latter problem can be solved in strongly polynomial time by the results of Section 49.7. Similar results hold for intersections of two such polyhedra:

Theorem 49.9. *For crossing families C_1, C_2 of subsets of a set S, crossing submodular functions $f_1 : C_1 \to \mathbb{Q}$ and $f_2 : C_2 \to \mathbb{Q}$, $k \in \mathbb{Q}$, and $w \in \mathbb{Q}^S$, one* can find an $x \in EP_{f_1} \cap EP_{f_2}$ with $x(S) = k$ and maximizing $w^T x$ in strongly polynomial time.

Proof. From the above. ∎

If $\mathcal{C}$ is a crossing family and $f : \mathcal{C} \to \mathbb{Q}$ is crossing submodular, then we can find its minimum value in polynomial time, as, for each $s, t \in S$, we can minimize f over the lattice family $\{U \in \mathcal{C} \mid s \in U, t \notin U\}$, and take the minimum of all these minima, and of the values in $\emptyset$ and S (if in $\mathcal{C}$).

Hence we can decide in polynomial time if a given vector $x \in \mathbb{Q}^S$ belongs to EP_f, by testing if the minimum value of the crossing submodular function $g : \mathcal{C} \to \mathbb{Q}$ defined by

(49.70) $\quad g(U) := f(U) - x(U)$

for $U \in \mathcal{C}$, is nonnegative.

49.10a. Nonemptiness of the base polyhedron

Let $\mathcal{C}$ be a crossing family of subsets of a set S and let $f : \mathcal{C} \to \mathbb{R}$ be crossing submodular. We give a theorem of Fujishige [1984e] characterizing when EP_f contains a vector x with $x(S) = 0$. If $S \in \mathcal{C}$ and $f(S) = 0$, then the set $EP_f \cap \{x \mid x(S) = 0\}$ is called the *base polyhedron* of f.

To give the characterization, again call a collection $\mathcal{P} \subseteq \mathcal{C}$ a *copartition* of S if the collection $\{S \setminus U \mid U \in \mathcal{P}\}$ is a partition of S.

Theorem 49.10. EP_f *contains a vector* x *satisfying* $x(S) = 0$ *if and only if*

(49.71) $\quad \sum_{U \in \mathcal{P}} f(U) \geq 0$

for each partition or copartition $\mathcal{P} \subseteq \mathcal{C}$ *of* S. *If moreover* f *is integer, there exists an integer such vector* x.

Proof. The condition is necessary, since if $x \in EP_f$ satisfies $x(S) = 0$ and $\mathcal{P} \subseteq \mathcal{C}$ is a partition of S, then

(49.72) $\quad \sum_{U \in \mathcal{P}} f(U) \geq \sum_{U \in \mathcal{P}} x(U) = x(S) = 0$.

Similarly, if $\mathcal{P} \subseteq \mathcal{C}$ is a copartition of S, then

(49.73) $\quad \sum_{U \in \mathcal{P}} f(U) \geq \sum_{U \in \mathcal{P}} x(U) = \sum_{U \in \mathcal{P}} (x(S) - x(S \setminus U)) = |\mathcal{P}|x(S) - x(S) = 0$.

To see sufficiency, let $\mathcal{D} := \hat{\mathcal{C}}$ and $g := \hat{f}$ (cf. (49.62)). By Theorem 49.6, $\mathcal{D}$ is an intersecting family and g is intersecting submodular. Moreover, $S \in \mathcal{D}$. Let $\mathcal{E} := \check{\mathcal{D}}$ and $h := \check{g}$. By Theorem 49.4, $\mathcal{E}$ is a lattice family and h is submodular.

Now if EP_h contains a vector x with $x(S) = 0$, then $x \in EP_g$, and hence $x \in EP_f$ (using $x(S) = 0$). So it suffices to show that EP_h contains a vector x with $x(S) = 0$.

By Corollary 49.1c, in order to show this, it suffices to show that $h(S) \geq 0$. The solution can be taken integer if f (hence h) is integer.

Suppose $h(S) < 0$. Since $h = \check{g}$, $\mathcal{D}$ contains a proper partition $\mathcal{P}$ of S with

(49.74) $$h(S) = \sum_{U \in \mathcal{P}} g(U).$$

Since $g = \hat{f}$, for each $U \in \mathcal{P}$, $\mathcal{C}$ contains a proper copartition $\mathcal{Q}_U$ of $S \setminus U$ such that

(49.75) $$g(U) = \sum_{T \in \mathcal{Q}_U} f(T).$$

Let $\mathcal{F}$ be the family consisting of the union of the $\mathcal{Q}_U$ over $U \in \mathcal{P}$, taking multiplicities into account. Then

(49.76)   (i) all elements of S are contained in the same number of sets in $\mathcal{F}$;
          (ii) $\sum_{T \in \mathcal{F}} f(T) < 0.$

Now apply the following operation as often as possible to $\mathcal{F}$: if $T, W \in \mathcal{F}$ with $T \cap W \neq \emptyset$, $T \cup W \neq S$, and $T \not\subseteq W \not\subseteq T$, replace T and W by $T \cap W$ and $T \cup W$. This maintains (49.76) and decreases $\sum_{T \in \mathcal{F}} |T||S \setminus T|$ (by Theorem 2.1). So the process terminates, and we end up with a *cross-free family*: for all $T, W \in \mathcal{F}$ we have $T \subseteq W$ or $W \subseteq T$ or $T \cap W = \emptyset$ or $T \cup W = S$.

We show that $\mathcal{F}$ contains a partition or copartition $\mathcal{P}$ of S. By (49.71), $\mathcal{F} \setminus \mathcal{P}$ again satisfies (49.76), and hence we can repeat. We end up with $\mathcal{F}$ empty, a contradiction.

To show that $\mathcal{F}$ contains a partition or copartition of S, choose $U \in \mathcal{F}$. If $U = \emptyset$ or $U = S$ we are done (taking $\mathcal{P} := \{U\}$). So we can assume that $\emptyset \neq U \neq S$. Let $\mathcal{X}$ be the collection of inclusionwise maximal subsets of $S \setminus U$ that belong to $\mathcal{F}$. Let $\mathcal{Y}$ be the collection of inclusionwise minimal supersets of $S \setminus U$ that belong to $\mathcal{F}$. Since $\mathcal{F}$ is cross-free and $U \neq \emptyset$, the sets in $\mathcal{X}$ are pairwise disjoint. Similarly, the complements of the sets in $\mathcal{Y}$ are pairwise disjoint.

If $\bigcup \mathcal{X} = S \setminus U$, then $\mathcal{X} \cup \{U\}$ is a partition of S as required. If $\bigcap \mathcal{Y} = S \setminus U$, then $\mathcal{Y} \cup \{U\}$ is a copartition of S as required. So we can assume that there exist $s \in (S \setminus U) \setminus \bigcup \mathcal{X}$ and $t \in U \cap \bigcap \mathcal{Y}$. Since each element of S is contained in the same number of sets in $\mathcal{F}$, and since $s \notin U$, and $t \in U$, there exists a $T \in \mathcal{F}$ with $s \in T$ and $t \notin T$. So $T \not\subseteq U \not\subseteq T$.

Hence $T \cap U = \emptyset$ or $T \cup U = S$. However, if $T \cap U = \emptyset$, then T is contained in some set in $\mathcal{X}$, and hence $s \in T \subseteq \bigcup \mathcal{X}$, a contradiction. If $T \cup U = S$, then T contains some set in $\mathcal{Y}$, and hence $t \notin T \supseteq \bigcap \mathcal{Y}$, again a contradiction. ∎

This theorem will be used in proving Theorem 61.8.

Fujishige and Tomizawa [1983] characterized the vertices of the base polyhedron of a submodular function defined on a lattice family.

49.11. Further results and notes

49.11a. Minimizing a submodular function over a subcollection of a lattice family

In Section 45.7 we saw that the minimum of a submodular function over the odd subsets can be found in strongly polynomial time. A generalization of minimizing a submodular function over the odd subsets (cf. Section 45.7), was given by Grötschel,

Section 49.11a. Minimizing a submodular function over a subcollection

Lovász, and Schrijver [1981,1984a] (the latter paper corrects a serious flaw in the first paper found by A. Frank). This was extended by Goemans and Ramakrishnan [1995] to the following.

Let $\mathcal{C}$ be a lattice family and let $\mathcal{D}$ be a subcollection of $\mathcal{C}$ with the following property:

(49.77) $\qquad$ for all $X, Y \in \mathcal{C} \setminus \mathcal{D}$: $X \cap Y \in \mathcal{D} \iff X \cup Y \in \mathcal{D}$.

Examples are: $\mathcal{D} := \{X \in \mathcal{C} \mid |X| \not\equiv q \pmod{p}\}$ for some natural numbers p, q, and $\mathcal{D} := \mathcal{C} \setminus \mathcal{A}$ for some antichain or some sublattice $\mathcal{A} \subseteq \mathcal{C}$.

To prove that for a submodular function on $\mathcal{C}$, the minimum over $\mathcal{D}$ can be found in strongly polynomial time, Goemans and Ramakrishnan gave the following interesting lemma:

Lemma 49.11α. *Let $\mathcal{C}$ be a lattice family, let f be a submodular function on $\mathcal{C}$, let $\mathcal{D} \subseteq \mathcal{C}$ satisfy (49.77), and let U minimize $f(U)$ over $U \in \mathcal{D}$. If $U \neq \emptyset$, then there exists a $u \in U$ such that $f(W) \geq f(U)$ for each subset W of U with $W \in \mathcal{C}$ and $u \in W$.*

Proof. Suppose not. Then for each $u \in U$ there exists a $W_u \in \mathcal{C}$ satisfying $u \in W_u \subseteq U$ and $f(W_u) < f(U)$. Choose each W_u inclusionwise maximal with this property. Then

(49.78) $\qquad f(\bigcap_{u \in T} W_u) < f(U)$

for each nonempty $T \subseteq U$. To prove this, choose a counterexample T with $|T|$ minimal. Then $|T| > 1$, since $f(W_u) < f(U)$ for each $u \in U$. Choose $t \in T$. Since $\bigcap_{u \in T} W_u \neq \bigcap_{u \in T-t} W_u$ by the minimality of T, we know that $\bigcap_{u \in T-t} W_u \not\subseteq W_t$, and hence W_t is a proper subset of $(\bigcap_{u \in T-t} W_u) \cup W_t$. So by the maximality of W_t, $f((\bigcap_{u \in T-t} W_u) \cup W_t) \geq f(U)$. Hence

(49.79) $\qquad f(U) \leq f(\bigcap_{u \in T} W_u) = f((\bigcap_{u \in T-t} W_u) \cap W_t)$
$\qquad\qquad \leq f(\bigcap_{u \in T-t} W_u) + f(W_t) - f((\bigcap_{u \in T-t} W_u) \cup W_t)$
$\qquad\qquad < f(U) + f(U) - f(U) = f(U)$,

a contradiction.

This shows (49.78), which implies

(49.80) $\qquad \bigcap_{u \in T} W_u \notin \mathcal{D}$

for each nonempty $T \subseteq U$.

This can be extended to:

(49.81) $\qquad X := (\bigcap_{u \in T} W_u) \cap (\bigcup_{u \in V} W_u) \notin \mathcal{D}$

for all disjoint $T, V \subseteq U$ with V nonempty. Suppose to the contrary that $X \in \mathcal{D}$. Choose such X with $|V|$ minimal. By (49.80), $|V| \geq 2$. Choose $v \in V$. The minimality of V gives

(49.82) $\quad (\bigcap_{u \in T} W_u) \cap W_v \not\in \mathcal{D}$ and $(\bigcap_{u \in T} W_u) \cap (\bigcup_{u \in V \setminus \{v\}} W_u) \not\in \mathcal{D}.$

By assumption, the union of these sets belongs to $\mathcal{D}$. Hence, by (49.77), also their intersection belongs to $\mathcal{D}$; that is

(49.83) $\quad (\bigcap_{u \in T \cup \{v\}} W_u) \cap (\bigcup_{u \in V \setminus \{v\}} W_u) \in \mathcal{D}$

This contradicts the minimality of $|V|$.

This proves (49.81), which gives for $T := \emptyset$ and $V := U$ a contradiction, since then $X = U \in \mathcal{D}$. ∎

This lemma is used in proving the following theorem, where we assume that $\mathcal{C}$ is given as in Section 49.3, f is given by a value giving oracle, and $\mathcal{D}$ is given by an oracle telling if any given set in $\mathcal{C}$ belongs to $\mathcal{D}$:

Theorem 49.11. *Given a submodular function f on a lattice family $\mathcal{C}$, and a subcollection $\mathcal{D}$ of $\mathcal{C}$ satisfying (49.77), a set U minimizing $f(U)$ over $U \in \mathcal{D}$ can be found in strongly polynomial time.*

Proof. We describe the algorithm. For all distinct $s, t \in S$ define

(49.84) $\quad \mathcal{C}_{s,t} := \{U \in \mathcal{C} \mid s \in U, t \notin U\}.$

Let $U_{s,t}$ be the inclusionwise minimal set minimizing f over $\mathcal{C}_{s,t}$. ($U_{s,t}$ can be found by minimizing a slight perturbation of f.) Choose in

(49.85) $\quad \{\emptyset, S\} \cup \{U_{s,t} \mid s, t \in S\}$

a $U \in \mathcal{D}$ minimizing f. Then U minimizes f over $\mathcal{D}$.

To see this, we must show that set (49.85) contains a set minimizing f over $\mathcal{D}$. Let W be a set minimizing f over $\mathcal{D}$, with $|W|$ minimal. If $W \in \{\emptyset, S\}$ we are done. So we can assume that $W \notin \{\emptyset, S\}$. By Lemma 49.11$\alpha$ (applied to the function $\tilde{f}(X) := f(S \setminus X)$ for $X \subseteq S$), there exists an element $t \in S \setminus W$ such that each $T \supseteq W$ with $t \notin T$ satisfies $f(T) \geq f(W)$.

The lemma also gives the existence of an $s \in W$ such that each $T \subset W$ with $s \in T$ satisfies $f(T) > f(W)$. Indeed, for small enough $\varepsilon > 0$, W minimizes $f(X) + \varepsilon|X|$ over $X \in \mathcal{D}$. Hence, by Lemma 49.11α, there exists an $s \in W$ such that each $T \subseteq W$ with $s \in T$ satisfies $f(T) + \varepsilon|T| \geq f(W) + \varepsilon|W|$. This implies $f(T) > f(W)$ if $T \neq W$.

We show that $W = U_{s,t}$. Indeed, W minimizes f over $\mathcal{C}_{s,t}$, since

(49.86) $\quad f(U_{s,t}) \geq f(W \cap U_{s,t}) + f(W \cup U_{s,t}) - f(W) \geq f(W) + f(W) - f(W)$
$\quad = f(W).$

Moreover, $W \subseteq U_{s,t}$, as otherwise $W \cap U_{s,t} \subset W$, implying that the second inequality in (49.86) would be strict.

So $f(W) = f(U_{s,t})$, and hence, by the minimality of $U_{s,t}$, we have $W = U_{s,t}$. ∎

It is interesting to note that this algorithm implies that the set $\{\emptyset, S\} \cup \{U_{s,t} \mid s, t \in S\}$ contains a set minimizing f over $\mathcal{D}$, for any nonempty subcollection $\mathcal{D}$ of $\mathcal{C}$ satisfying (49.77).

Goemans and Ramakrishnan showed that if $\mathcal{C}$ and $\mathcal{D}$ are *symmetric* (that is, $U \in \mathcal{C} \iff S \setminus U \in \mathcal{C}$, and similarly for $\mathcal{D}$) and $\emptyset \notin \mathcal{D}$, then (49.77) is equivalent to: if $X, Y \in \mathcal{C} \setminus \mathcal{D}$ are disjoint, then $X \cup Y \in \mathcal{C} \setminus \mathcal{D}$.

Related work was reported by Benczúr and Fülöp [2000].

49.11b. Generalized polymatroids

We now describe a generalization, given by Frank [1984b], that comprises sub- and supermodular functions, and (extended) polymatroids and contrapolymatroids. (Hassin [1978,1982] described the case $\mathcal{C} = \mathcal{D} = \mathcal{P}(S)$.)

Let $\mathcal{C}$ and $\mathcal{D}$ be intersecting families of subsets of a finite set S and let $f : \mathcal{C} \to \mathbb{R}$ and $g : \mathcal{D} \to \mathbb{R}$. We say that the pair (f, g) is *paramodular* if

(49.87)   (i) f is submodular on intersecting pairs,
             (ii) g is supermodular on intersecting pairs,
             (iii) if $T \in \mathcal{C}$ and $U \in \mathcal{D}$ with $T \setminus U \neq \emptyset$ and $U \setminus T \neq \emptyset$, then $T \setminus U \in \mathcal{C}$ and $U \setminus T \in \mathcal{D}$, and

$$f(T \setminus U) - g(U \setminus T) \leq f(T) - g(U).$$

If (f, g) is paramodular, the solution set P of the system (for $x \in \mathbb{R}^S$):

(49.88) $x(U) \leq f(U)$ for $U \in \mathcal{C}$,
             $x(U) \geq g(U)$ for $U \in \mathcal{D}$,

is called a *generalized polymatroid* (*determined by* (f, g)).

Generalized polymatroids generalize polymatroids (where $g(U) = 0$ for each $U \subseteq S$), extended polymatroids (where $\mathcal{D} = \emptyset$), contrapolymatroids (where $\mathcal{C} = \emptyset$ and $g(\{s\}) \geq 0$ for each $s \in S$), and extended contrapolymatroids (where $\mathcal{C} = \emptyset$).

The intersection of a generalized polymatroid with a 'box' $\{x \mid d \leq x \leq c\}$ (for $d, c \in \mathbb{R}^S$) is again a generalized polymatroid: we can add $\{s\}$ to $\mathcal{C}$ and to $\mathcal{D}$ if necessary, and (re)define $f(\{s\}) := w(s)$ and $g(\{s\}) := d(s)$, if necessary. This transformation does not violate the paramodularity of (f, g).

Another transformation is as follows. Let $P \subseteq \mathbb{R}^S$ be a generalized polymatroid and let $\kappa, \lambda \in \mathbb{R}$. Let t be a new element and let $S' := S \cup \{t\}$. Let P' be the polyhedron in $\mathbb{R}^{S'}$ given by

(49.89) $P' := \{(x, \eta) \mid x \in P, \lambda \leq x(S) + \eta \leq \kappa\}$.

Then P' again is a generalized polymatroid, determined by the functions obtained by extending $\mathcal{C}$ and $\mathcal{D}$ with S' and extending f, g with the values $f(S') := \kappa$ and $g(S') := \lambda$.

The class of generalized polymatroids is closed under projections. That is, for any generalized polymatroid $P \subseteq \mathbb{R}^S$ and any $t \in S$, the set

(49.90) $P' := \{x \in \mathbb{R}^{S-t} \mid \exists \eta : (x, \eta) \in P\}$

is again a generalized polymatroid. This will be shown as Corollary 49.13c.

The following theorem will imply that system (49.88) is TDI. Hence, if f and g are integer, then P is integer.

Theorem 49.12. *System* (49.88) *is box-TDI.*

Proof. Let t be a new element. Define

846   Chapter 49. Submodularity more generally

(49.91) $\mathcal{B} := \mathcal{C} \cup \{(S \setminus D) \cup \{t\} \mid D \in \mathcal{D}\}$.

Then $\mathcal{B}$ is a crossing family of subsets of $S \cup \{t\}$. Define $e : \mathcal{B} \to \mathbb{R}$ by: $e(C) := f(C)$ for $C \in \mathcal{C}$ and $e((S \setminus D) \cup \{t\}) := -g(D)$ for $D \in \mathcal{D}$. Then e is crossing submodular. Hence, by Theorem 49.7, system

(49.92) $x(U) \leq e(U)$ for $U \in \mathcal{B}$,
$x(S \cup \{t\}) = 0$,

is box-TDI. Therefore, by Theorem 5.27, system (49.88) is box-TDI.   ∎

This gives for the integrality of generalized polymatroids:

Corollary 49.12a. *If (f, g) is paramodular and f and g are integer, the generalized polymatroid is box-integer.*

Proof. Directly from Theorem 49.12.   ∎

More generally one has the box-total dual integrality of the system

(49.93) $x(U) \leq f_1(U)$ for $U \in \mathcal{C}_1$,
$x(U) \geq g_1(U)$ for $U \in \mathcal{D}_1$,
$x(U) \leq f_2(U)$ for $U \in \mathcal{C}_2$,
$x(U) \geq g_2(U)$ for $U \in \mathcal{D}_2$,

for pairs of paramodular pairs (f_i, g_i):

Corollary 49.12b. *For $i = 1, 2$, let $\mathcal{C}_i$ and $\mathcal{D}_i$ be intersecting families and let $f_i : \mathcal{C}_i \to \mathbb{R}$, $g_i : \mathcal{D}_i \to \mathbb{R}$ form a paramodular pair. Then system (49.93) is box-TDI.*

Proof. Similar to the proof of Theorem 49.12, by reduction to Theorem 49.8.   ∎

This gives for primal integrality:

Corollary 49.12c. *If f_1, g_1, f_2 and g_2 are integer, the intersection of the associated generalized polymatroids is box-integer.*

Proof. Directly from Corollary 49.12b.   ∎

Another consequence is the following box-TDI result of McDiarmid [1978]:

Corollary 49.12d. *Let f_1 and f_2 be submodular set functions on a set S and let $\lambda, \kappa \in \mathbb{R}$. Then the system*

(49.94) $x(U) \leq f_1(U)$ for $U \subseteq S$,
$x(U) \leq f_2(U)$ for $U \subseteq S$,
$\lambda \leq x(S) \leq \kappa$,

is box-TDI.

Proof. Redefine $f_1(S) := \min\{f_1(S), \kappa\}$, and define $g_1 : \{S\} \to \mathbb{R}$ by $g_1(S) := \lambda$, and $g_2 : \emptyset \to \mathbb{R}$. Then (f_1, g_1) and (f_2, g_2) are paramodular pairs, and the box-total

Section 49.11b. Generalized polymatroids

dual integrality of (49.94) is equivalent to the box-total dual integrality of (49.93). ∎

From Corollary 49.12c one can derive that the intersection of two integer generalized polymatroids is integer again. To prove this, we show that for any integer generalized polymatroid P there exists a paramodular pair (f, g) determining P, with f and g integer.

Let P be a generalized polymatroid, determined by the paramodular pair (f, g) of functions $f : \mathcal{C} \to \mathbb{R}$ and $g : \mathcal{D} \to \mathbb{R}$, where $\mathcal{C}$ and $\mathcal{D}$ are intersecting families. For any $U \subseteq S$, define

(49.95) $\quad \tilde{f}(U) := \max\{x(U) \mid x \in P\}$ and $\tilde{g}(U) := \min\{x(U) \mid x \in P\}$.

So $\tilde{f}$ and $\tilde{g}$ are integer if P is integer.

Let

(49.96) $\quad \tilde{\mathcal{C}} := \{U \in \mathcal{C} \mid \tilde{f}(U) < \infty\}$ and $\tilde{\mathcal{D}} := \{U \in \mathcal{D} \mid \tilde{g}(U) > -\infty\}$.

We restrict $\tilde{f}$ and $\tilde{g}$ to $\tilde{\mathcal{C}}$ and $\tilde{\mathcal{D}}$ respectively. We show that $(\tilde{f}, \tilde{g})$ is a paramodular pair determining P.

It is convenient to note that if $w \in \mathbb{R}^S$ with $w = w_1 + w_2$, then

(49.97) $\quad \max\{w^\mathsf{T} x \mid x \in P\} \leq \max\{w_1^\mathsf{T} x \mid x \in P\} + \max\{w_2^\mathsf{T} x \mid x \in P\}$.

Theorem 49.13. *For any generalized polymatroid P, $\tilde{\mathcal{C}}$ and $\tilde{\mathcal{D}}$ are intersecting families, and the pair $(\tilde{f}, \tilde{g})$ is paramodular and determines P.*

Proof. We first show the following. Let $w \in \mathbb{Z}^S$ and let $\lambda > 0$ be such that $w_s \leq \lambda$ for each $s \in S$. Let $U := \{s \in S \mid w(s) = \lambda\}$ and $w' := w - \chi^U$. Then

(49.98) $\quad \max\{w^\mathsf{T} x \mid x \in P\} = \max\{{w'}^\mathsf{T} x \mid x \in P\} + \tilde{f}(U)$.

Here $\leq$ follows from (49.97), by definition of $\tilde{f}$. Equality is proved by induction on $|U|$, the case $U = \emptyset$ being trivial; so let $U \neq \emptyset$.

Let y, z be an optimum solution to the dual of $\max\{w^\mathsf{T} x \mid x \in P\}$:

(49.99) $\quad \min\{\sum_{T \in \mathcal{C}} y_T f(T) - \sum_{T \in \mathcal{D}} z_T g(T) \mid$
$\qquad y \in \mathbb{R}_+^\mathcal{C}, z \in \mathbb{R}_+^\mathcal{D}, \sum_{T \in \mathcal{C}} y_T \chi^T - \sum_{T \in \mathcal{D}} z_T \chi^T = w\}$.

Define $\mathcal{F} := \{T \in \mathcal{C} \mid y_T > 0\}$ and $\mathcal{G} := \{T \in \mathcal{D} \mid z_T > 0\}$. Similarly to Theorem 49.3, we can assume that $\mathcal{F} \cup \mathcal{G}$ is laminar.

Choose $u \in U$, and let W be an inclusionwise minimal set in $\mathcal{F}$ containing u. (Such a set exists, as $w(s) = \lambda > 0$.) Let $\mathcal{H}$ be the collection of inclusionwise maximal sets in $\mathcal{G}$ contained in $W - u$. As $\mathcal{G}$ is laminar, the sets in $\mathcal{H}$ are disjoint. Moreover, each $t \in W \setminus U$ is contained in some set in $\mathcal{H}$: since $w(t) < w(u)$ and since every set in $\mathcal{F}$ containing u also contains t (as $t \in W$), there exists an $X \in \mathcal{G}$ with $t \in X$ and $u \notin X$; as $\mathcal{F} \cup \mathcal{G}$ is laminar, we know that $X \subseteq W - u$.

Now let $Y := W \setminus \bigcup \mathcal{H}$. So Y is a nonempty subset of U. Define $w'' := w - \chi^Y$, let y' be obtained from y by decreasing $y(W)$ by 1, and let z' be obtained from z by decreasing $y(H)$ by 1 for each $H \in \mathcal{H}$. So (since $\chi^Y = \chi^W - \sum_{H \in \mathcal{H}} \chi^H$)

(49.100) $$\sum_{T \in \mathcal{C}} y'(T)\chi^T - \sum_{T \in \mathcal{D}} z'(T)\chi^T = w''$$

and

(49.101) $$\tilde{f}(Y) \leq f(W) - \sum_{H \in \mathcal{H}} g(H).$$

Moreover, setting $U' := U \setminus Y$, we have (by (49.97))

(49.102) $$\tilde{f}(U) \leq \tilde{f}(Y) + \tilde{f}(U'),$$

and by our induction hypothesis, as $|U'| < |U|$,

(49.103) $$\max\{w''^{\mathsf{T}} x \mid x \in P\} = \max\{w'^{\mathsf{T}} x \mid x \in P\} + \tilde{f}(U').$$

Hence

(49.104) $$\begin{aligned}\max\{w^{\mathsf{T}} x \mid x \in P\} &= \sum_{T \in \mathcal{C}} y(T)f(T) - \sum_{T \in \mathcal{D}} z(T)g(T) \\ &= \sum_{T \in \mathcal{C}} y'(T)f(T) - \sum_{T \in \mathcal{D}} z'(T)g(T) + f(W) - \sum_{H \in \mathcal{H}} g(H) \\ &\geq \max\{w''^{\mathsf{T}} x \mid x \in P\} + \tilde{f}(Y) = \max\{w'^{\mathsf{T}} x \mid x \in P\} + \tilde{f}(U') + \tilde{f}(Y) \\ &\geq \max\{w'^{\mathsf{T}} x \mid x \in P\} + \tilde{f}(U),\end{aligned}$$

thus proving (49.98).

We next derive that $\tilde{f}$ is submodular on intersecting pairs. Choose $X, Y \in \tilde{\mathcal{C}}$ with $X \cap Y \neq \emptyset$. Define $w := \chi^X + \chi^Y$. Then by (49.98) and (49.97),

(49.105) $$\tilde{f}(X \cap Y) + \tilde{f}(X \cup Y) = \max\{w^{\mathsf{T}} x \mid x \in P\} \leq \tilde{f}(X) + \tilde{f}(Y).$$

So $\tilde{\mathcal{C}}$ is an intersecting family and $\tilde{f}$ is submodular on intersecting pairs. By symmetry, it follows that $\tilde{\mathcal{D}}$ is an intersecting family and $\tilde{g}$ is supermodular on intersecting pairs.

Finally, to see that (f, g) is paramodular, let $X \in \mathcal{C}$ and $Y \in \mathcal{D}$. Define $w := \chi^X - \chi^Y$. Again, by (49.98) and (49.97),

(49.106) $$\tilde{f}(X \setminus Y) - \tilde{g}(Y \setminus X) = \max\{w^{\mathsf{T}} x \mid x \in P\} \leq \tilde{f}(X) - \tilde{g}(Y).$$

So $\tilde{\mathcal{C}}$ and $\tilde{\mathcal{D}}$ are intersecting families, and the pair $(\tilde{f}, \tilde{g})$ is paramodular. It determines P, since P is determined by upper and lower bounds on $x(U)$ for subsets U of S. ∎

Corollary 49.12a implies:

Corollary 49.13a. *A generalized polymatroid P is integer if and only if there is a paramodular pair (f, g) defining P with f and g integer.*

Proof. Sufficiency follows from Corollary 49.12a. Necessity follows from Theorem 49.13, as P is determined by $(\tilde{f}, \tilde{g})$, where $\tilde{f}$ and $\tilde{g}$ are integer if P is integer. ∎

A second consequence is:

Corollary 49.13b. *The intersection of two integer generalized polymatroids is integer.*

Proof. Directly by combining Corollaries 49.12c and 49.13a. ∎

We should note that the collections $\tilde{\mathcal{C}}$ and $\tilde{\mathcal{D}}$ found in the proof of Theorem 49.13 are lattice families and that $\tilde{f}$ and $\tilde{g}$ are sub- and supermodular respectively. Moreover, $\tilde{f}(T \setminus U) - \tilde{g}(U \setminus T) \leq \tilde{f}(T) - \tilde{g}(U)$ for each pair $T \in \tilde{\mathcal{C}}$, $U \in \tilde{\mathcal{D}}$.

This implies that projections of generalized polymatroids are again generalized polymatroids (Frank [1984b]):

Corollary 49.13c. *Let $P \subseteq \mathbb{R}^S$ be a generalized polymatroid and let $t \in S$. Define $S' := S - t$. Then the projection*

(49.107) $\quad P' := \{x \in \mathbb{R}^{S'} \mid \exists \eta : (x, \eta) \in P\}$

is again a generalized polymatroid.

Proof. We can assume that P is nonempty, that $\mathcal{C}$ and $\mathcal{D}$ are lattice families, and that P is determined by a paramodular pair $(f, g) = (\tilde{f}, \tilde{g})$ as above. Let $\mathcal{C}'$ and $\mathcal{D}'$ be the collections of sets in $\mathcal{C}$ and $\mathcal{D}$ respectively not containing t. Let $f' := f|\mathcal{C}'$ and $g' := g|\mathcal{D}'$.

Trivially, (f', g') is a paramodular pair. We claim that P' is equal to the generalized polymatroid Q determined by (f', g'). Trivially, $P' \subseteq Q$. To see the reverse inclusion, let $x \in Q$. Let η' be the largest real such that $x(T - t) + \eta' \leq f(T)$ for each $T \in \mathcal{C} \setminus \mathcal{C}'$. Let η'' be the smallest real such that $x(U - t) + \eta'' \geq g(U)$ for each $U \in \mathcal{D} \setminus \mathcal{D}'$.

If $x \notin P'$, then $\eta' < \eta''$, and hence there exist $T \in \mathcal{C}$ and $U \in \mathcal{D}$ with $t \in T \cap U$ and $f(T) - x(T - t) < g(U) - x(U - t)$. Hence

(49.108) $\quad x(T \setminus U) - x(U \setminus T) = x(T - t) - x(U - t) > f(T) - g(U)$
$\geq f(T \setminus U) - g(U \setminus T)$.

This contradicts the fact that $x(T \setminus U) \leq f(T \setminus U)$ and $x(U \setminus T) \geq g(U \setminus T)$, as $x \in Q$. ∎

For results on the dimension of generalized polymatroids, see Frank and Tardos [1988], which paper surveys generalized polymatroids and submodular flows. More results on generalized polymatroids are reported by Fujishige [1984b], Nakamura [1988b], Naitoh and Fujishige [1992], and Tamir [1995].

49.11c. Supermodular colourings

A colouring-type of result on supermodular functions was shown by Schrijver [1985]. We give the proof based on generalized polymatroids found by Tardos [1985b].

Theorem 49.14. *Let $\mathcal{C}_1$ and $\mathcal{C}_2$ be intersecting families of subsets of a set S, let $g_1 : \mathcal{C}_1 \to \mathbb{Z}$ and $g_2 : \mathcal{C}_2 \to \mathbb{Z}$ be intersecting supermodular, and let $k \in \mathbb{Z}_+$ with $k \geq 1$. Then S can be partitioned into classes $L_1, \ldots, L_k$ such that*

(49.109) $\quad g_i(U) \leq |\{j \in \{1, \ldots, k\} \mid L_j \cap U \neq \emptyset\}|$

for each $i = 1, 2$ and each $U \in \mathcal{C}_i$ if and only if

(49.110) $\quad g_i(U) \leq \min\{k, |U|\}$

for each $i = 1, 2$ and each $U \in \mathcal{C}_i$.

Proof. Necessity is easy. Sufficiency is shown by induction on k, the case $k = 0$ being trivial. By induction, it suffices to find a subset L of S such that

(49.111) $\qquad |U \setminus L| \geq g_i(U) - 1$ and, if $g_i(U) = k$, then $U \cap L \neq \emptyset$.

Indeed, in that case we can apply induction to the functions $g'_i : \mathcal{C}'_i \to \mathbb{Z}$ on $\mathcal{C}'_i := \{U \setminus L \mid U \in \mathcal{C}\}$, defined by

(49.112) $\qquad g'_i(U \setminus L) := \begin{cases} g_i(U) - 1 & \text{if } U \cap L \neq \emptyset, \\ g_i(U) & \text{if } U \cap L = \emptyset, \end{cases}$

for $U \in \mathcal{C}_i$.

For $i = 1, 2$, consider the system:

(49.113) $\quad$ (i) $\quad 0 \leq x_s \leq 1 \qquad\qquad$ for $s \in S$,
$\qquad\qquad$ (ii) $\quad x(U) \leq |U| - g_i(U) + 1 \quad$ for $U \in \mathcal{C}_i$,
$\qquad\qquad$ (iii) $\quad x(U) \geq 1 \qquad\qquad$ for $U \in \mathcal{C}_i$ with $g_i(U) = k$.

This system determines an integer generalized polymatroid. This can be seen as follows. Let $\mathcal{D}_i$ be the collection of inclusionwise minimal sets in $\{U \in \mathcal{C}_i \mid g_i(U) = k\}$. So $\mathcal{D}_i$ consists of disjoint sets (as $\mathcal{C}_i$ is intersecting and as $g_i(U) \leq k$ for each $U \in \mathcal{C}_i$). Let

(49.114) $\qquad \mathcal{C}'_i := \{U \in \mathcal{C}_i \mid \forall T \in \mathcal{D}_i : U \subseteq T \text{ or } T \cap U = \emptyset\}.$

Then (49.113) has the same solution set as:

(49.115) $\quad$ (i) $\quad 0 \leq x_s \leq 1 \qquad\qquad$ for $s \in S$,
$\qquad\qquad$ (ii) $\quad x(U) \leq |U| - g_i(U) + 1 \quad$ for $U \in \mathcal{C}'_i$,
$\qquad\qquad$ (iii) $\quad x(U) \geq 1 \qquad\qquad$ for $U \in \mathcal{D}_i$.

Indeed, (49.115)(iii) implies (49.113)(iii) (as $x \geq 0$). Moreover, for any $U \in \mathcal{C}_i$ with $T \cap U \neq \emptyset$ for some $T \in \mathcal{D}_i$, one has

(49.116) $\qquad g_i(T \cap U) \geq g_i(T) + g_i(U) - g_i(T \cup U) \geq g_i(U)$

(as $g_i(T \cup U) \leq k = g_i(T)$). So with (49.115)(iii) we have:

(49.117) $\qquad x(U) \leq x(T \cap U) + |U \setminus T| \leq |T \cap U| - g_i(T \cap U) + 1 + |U \setminus T|$
$\qquad\qquad \leq |U| - g_i(U) + 1$

(as $x_s \leq 1$ for all $x \in U \setminus T$). Hence (49.113) and (49.115) have the same solution set.

Now (49.115) is a system defining a generalized polymatroid, as one easily checks (condition (49.87)(iii) follows, since if $T \in \mathcal{C}'_i$ and $U \in \mathcal{D}_i$ with $T \setminus U \neq \emptyset$ and $U \setminus T \neq \emptyset$, then, by definition of $\mathcal{C}'_i$, T and U are disjoint, and then the inequality in (49.87)(iii) is trivial). It is integer, as the right-hand sides in (49.115) are integer.

Also, the intersection of these generalized polymatroids for $i = 1$ and $i = 2$ is nonempty, since the vector $x := k^{-1} \cdot \mathbf{1}$ belongs to it. (49.115)(i) and (iii) hold trivially. To see (ii), we have

(49.118) $\qquad x(U) = \frac{1}{k}|U| = |U| - \frac{k-1}{k}|U| \leq |U| - \frac{k-1}{k}g_i(U) = |U| - g_i(U) + \frac{1}{k}g_i(U)$
$\qquad\qquad \leq |U| - g_i(U) + 1.$

Therefore, the intersection contains an integer vector x, which is, by (49.115)(i), the incidence vector of some subset L of S satisfying (49.111), as required. ∎

This theorem generalizes edge-colouring theorems for bipartite graphs $G = (V, E)$. Let V_1 and V_2 be the colour classes of G. Let $\mathcal{C}_i := \{\delta(v) \mid v \in V_i\}$ for $i = 1, 2$. If we define $g_i(\delta(v)) := |\delta(v)|$ for $v \in V_i$ ($i = 1, 2$), Theorem 49.14 reduces to Kőnig's edge-colouring theorem (Theorem 20.1). If $g_i(\delta(v))$ is set to the minimum degree of G, we obtain Theorem 20.5, and if it is set to the minimum of k and $|\delta(v)|$, we obtain Theorem 20.6.

Theorem 49.14 can also be used in proving the 'disjoint bibranchings theorem' (Theorem 54.11 — see Section 54.7a). Szigeti [1999] gave a generalization of Theorem 49.14.

49.11d. Further notes

Let S and T be disjoint sets. A function $f : \mathcal{P}(S) \times \mathcal{P}(T) \to \mathbb{R}$ is called *bisubmodular* if

(49.119) $f(X_1 \cap X_2, Y_1 \cup Y_2) + f(X_1 \cup X_2, Y_1 \cap Y_2) \leq f(X_1, Y_1) + f(X_2, Y_2)$

for all $X_1, X_2 \subseteq S$ and $Y_1, Y_2 \subseteq T$.

Bisubmodular functions were studied by Kung [1978b] and Schrijver [1978, 1979c]. Most of the results can be obtained from those for submodular functions, by considering the submodular set function f' on $S \cup T$ defined by $f'(X \cup Y) := f((S \setminus X) \cup Y)$ for $X \subseteq S$ and $Y \subseteq T$. Similarly for *bisupermodular* functions, where the inequality sign in (49.119) is reversed.

For an interesting related result of Frank and Jordán [1995b] yielding Győri's theorem, see Section 60.3d.

Fujishige [1984c] gave a framework that includes Theorem 46.2 on the total dual integrality of the intersection of a polymatroid and a contrapolymatroid system, Corollary 46.2b on the existence of a modular function between a sub- and a supermodular function, and Theorem 49.13 on the total dual integrality of the generalized polymatroid constraints (but not the total dual integrality of the intersection of two polymatroids). Fujishige [1984b] described generalized polymatroids as projections of base polyhedra of submodular functions.

Chandrasekaran and Kabadi [1988] introduced the concept of a *generalized submodular function* as a function $f : \mathcal{R} \to \mathbb{R}$, where $\mathcal{R} := \{(T, U) \mid T, U \subseteq S, T \cap U = \emptyset\}$ for some set S, satisfying

(49.120) $f(A, B) + f(C, D)$
$\geq f(A \cap C, B \cap D) + f((A \setminus D) \cup (C \setminus B), (B \setminus C) \cup (D \setminus A))$

for all $(A, B), (C, D) \in \mathcal{R}$. They showed that the system

(49.121) $x(T) - x(U) \leq f(T, U)$ for $(T, U) \in \mathcal{R}$

is box-TDI, and that for any $w \in \mathbb{R}^S$, an x maximizing $w^\mathsf{T} x$ over (49.121) can be found by a variant of the greedy method. Unions of two such systems need not define an integer polyhedron if the functions are integer, as is shown by an example with $|S| = 2$. A similar framework was considered by Nakamura [1990]. More results can be found in Dress and Havel [1986], Bouchet [1987a,1995], Bouchet, Dress, and

Havel [1992], Ando and Fujishige [1996], Fujishige [1997], and Fujishige and Iwata [2001].

It is direct to represent a lattice family on a set S of size n in $O(n^2)$ space (just by giving all pairs (u,v) for which each set in the family containing u also contains v). Gabow [1993b,1995c] gave an $O(n^2)$ representation for intersecting and crossing families. Related results were found by Fleiner and Jordán [1999].

Tardos [1985b] also studied *generalized matroids*, which form the special case of generalized polymatroids with $0,1$ vertices. An instance of it we saw in the proof of Theorem 49.14.

More results on submodularity are given by Fujishige [1980b,1984f,1984g,1988], Nakamura [1988b,1988c,1993], Kabadi and Chandrasekaran [1990], Iwata [1995], Iwata, Murota, and Shigeno [1997], and Murota [1998]. Generalizations were studied by Qi [1988b] and Kashiwabara, Nakamura, and Takabatake [1999].

Part V

Trees, Branchings, and Connectors

Part V: Trees, Branchings, and Connectors

This part focuses on structures that are defined by connecting several pairs of vertices simultaneously, with most basic structure that of a *spanning tree*. A spanning tree can be characterized as a minimal set of edges that connects each pair of vertices by at least one path — that is, a minimal *connector*. Alternatively, it can be characterized as a maximal set of edge that connects each pair of vertices by at most one path — that is, a maximal *forest*.

Finding a shortest spanning tree belongs to classical combinatorial optimization, with lots of applications in planning road, energy, and communication networks, in chip design, and in clustering data in areas like biology, taxonomy, archeology, and, more generally, in any large data base. Spanning trees are well under control polyhedrally and algorithmically, both as to *shortest* and as to *disjoint* spanning trees. They form a prime area of application of matroid theory.

There are several variations and generalizations of the notion of spanning tree that are also well under control, like arborescences, branchings, biconnectors, bibranchings, directed cut covers, and matching forests.

An illustrious variant that is worse under control is the Hamiltonian circuit — in other words, the traveling salesman tour — which (in the directed case) can be considered as a smallest strongly connected subgraph. The traveling salesman problem is NP-complete and no complete polyhedral characterization is known. It implies that more general optimization problems like finding a shortest strong connector or a cheapest connectivity augmentation also are NP-complete. In this part we will however come across some special cases that *are* well-solvable and well-characterized.

In this part we also discuss the powerful framework designed by Edmonds and Giles, based on defining the concept of a *submodular flow* in a directed graph with a submodular function on its vertex set. It unifies several of the results and techniques of the present part and of the previous part on matroids and submodular functions.

Chapters:

50. Shortest spanning trees 855
51. Packing and covering of trees 877
52. Longest branchings and shortest arborescences 893
53. Packing and covering of branchings and arborescences 904
54. Biconnectors and bibranchings 928
55. Minimum directed cut covers and packing directed cuts 946
56. Minimum directed cuts and packing directed cut covers 962
57. Strong connectors 969
58. The traveling salesman problem 981
59. Matching forests 1005
60. Submodular functions on directed graphs 1018
61. Graph orientation 1035
62. Network synthesis 1049
63. Connectivity augmentation 1058

Chapter 50
Shortest spanning trees

In this chapter we consider shortest spanning trees in undirected graphs. We show that the greedy algorithm finds a shortest spanning tree in a graph, and moreover yields min-max relations and polyhedral characterizations. These are special cases of results on matroids discussed in Chapter 40, but deserve special consideration since the graph framework allows a number of additional viewpoints and opportunities.

We recall some terminology and elementary facts. A graph $G = (V, E)$ is called a *tree* if G is connected and contains no circuit. For any graph $G = (V, E)$, a subset F of E is called:

- a *spanning tree* if (V, F) is a tree,
- a *forest* if F contains no circuit,
- a *maximal forest* if F is an inclusionwise maximal forest,
- a *connector* if (V, F) is connected.

A graph G has a spanning tree if and only if G is connected. For any connected graph $G = (V, E)$, each of the following characterizes a subset F of E as a spanning tree:

- F is a maximal forest;
- F is an inclusionwise minimal connector;
- F is a forest with $|F| = |V| - 1$;
- F is a connector with $|F| = |V| - 1$.

In any graph $G = (V, E)$, a maximal forest has $|V| - k$ edges, where k is the number of components of G; it forms a spanning tree in each of the components of G. So each inclusionwise maximal forest is a maximum-size forest; that is, each forest is contained in a maximum-size forest. Similarly, each connector contains a minimum-size connector.

50.1. Shortest spanning trees

Let $G = (V, E)$ be a connected graph and let $l : E \to \mathbb{R}$ be a function, called the *length* function. For any subset F of E, the *length* $l(F)$ of F is, by definition:

(50.1) $\qquad l(F) := \sum_{e \in F} l(e).$

856   Chapter 50. Shortest spanning trees

In this section we consider the problem of finding a shortest spanning tree in G — that is, one of minimum length.

While this is a special case of finding a minimum-weight base in a matroid, and hence can be solved with the greedy algorithm (Section 40.1), spanning trees allow some variation on the method, essentially because we can exploit the presence of the vertex set (graphic matroids are defined on the edge set only).

Also these variants of the greedy method will be called greedy. Such methods go back to Borůvka [1926a]. The correctness of each of the variants follows from the following basic phenomenon.

Call a forest F *good* if there exists a shortest spanning tree T of G that contains F. (So we are out for a good spanning tree.) Then:

Theorem 50.1. *Let F be a good forest and let e be an edge not in F. Then $F \cup \{e\}$ is a good forest if and only if*

(50.2)   *there exists a cut C disjoint from F such that e is shortest among the edges in C.*

Proof. To see necessity, let T be a shortest spanning tree containing $F \cup \{e\}$. Let C be the unique cut disjoint from $T \setminus \{e\}$. Then e is shortest in C, since if $f \in C$, then $T' := (T \setminus \{e\}) \cup \{f\}$ is again a spanning tree. As $l(T') \geq l(T)$ we have $l(f) \geq l(e)$.

To see sufficiency, let T be a shortest spanning tree containing F. Let P be the path in T between the ends of e. Then P contains at least one edge f that belongs to C. Then $T' := (T \setminus \{f\}) \cup \{e\}$ is a spanning tree again. By assumption, $l(e) \leq l(f)$ and hence $l(T') \leq l(T)$. Hence T' is a shortest spanning tree again. As $F \cup \{e\}$ is contained in T', it is a good forest. ∎

(The idea of this proof is in Jarník [1930].)

This theorem offers us a framework for an algorithm: starting with $F := \emptyset$, iteratively extend F by an edge e satisfying (50.2). We end up with a shortest spanning tree.

Rule (50.2) was formulated by Tarjan [1983], and is the most liberal rule in obtaining greedily a shortest spanning tree. The variants of the greedy method are obtained by specifying how to choose edge e.

The first variant, the *tree-growing method*, was given by Jarník [1930] (and by Kruskal [1956], Prim [1957], Dijkstra [1959]). It is also called the *Jarník-Prim method* or *Prim's method* (Prim was the first giving an $O(n^2)$ implementation):

(50.3)   Fix a vertex r. Set $F := \emptyset$. As long as F is not a spanning tree, let K be the component of F containing r, let e be a shortest edge leaving K, and reset $F := F \cup \{e\}$.

Corollary 50.1a. *Prim's method yields a shortest spanning tree.*

Proof. Directly from Theorem 50.1, by taking $C := \delta(K)$. ∎

A second variant, the *forest-merging method* or *Kruskal's method*, is due to Kruskal [1956] (and to Loberman and Weinberger [1957] and Prim [1957]):

(50.4)   Set $F := \emptyset$. As long as F is not a spanning tree, choose a shortest edge e for which $F \cup \{e\}$ is a forest, and reset $F := F \cup \{e\}$.

(So this version is the true specialization of the greedy algorithm for matroids to graphs.)

Corollary 50.1b. *Kruskal's method yields a shortest spanning tree.*

Proof. Again directly from Theorem 50.1, as e is shortest in the cut $\delta(K)$ for each of the two components K of F incident with e. ∎

Prim [1957] and Loberman and Weinberger [1957] observed that the optimality of the greedy method implies that each length function which gives the same order of the edges (like the logarithm or square of the lengths), has the same collection of shortest spanning trees. Similarly, the shortest spanning tree minimizes the *product* of the lengths (if nonnegative).

In a similar way one finds a *longest* spanning tree. The maximum length of a forest and the minimum length of a connector can also be found with the greedy method.

Note that the greedy method is flexible: We can change our rule of choosing the new edge e at any time throughout the algorithm, as long as at any choice of e, (50.2) is satisfied.

As Prim [1957] and Dijkstra [1959] remark, the value of any variant of the greedy method depends on its implementation. One should have efficient ways to store and update information on the components of F, and on finding an edge satisfying (50.2). We now consider such implementations for Prim's and for Kruskal's method.

50.2. Implementing Prim's method

Prim [1957] and Dijkstra [1959] described implementations of Prim's method that run in time $O(n^2)$. (Here we assume without loss of generality that the graph is simple.)

To this end, we indicate at any vertex v, whether or not v belongs to the component K containing r of the current forest F, and in case $v \notin K$, we store at v a shortest edge e_v connecting v with K (void if there is no such edge). Then at each iteration, we scan all vertices, and select one, v say, for which $v \notin K$ and e_v is shortest. We add e_v to F, and v to K, and for each edge vu incident with v, we replace e_u by vu if $u \notin K$ and vu is shorter than e_u (or if e_u is void).

As each iteration takes $O(n)$ time and as there are $n-1$ iterations we have the result stated by Dijkstra [1959]:

Theorem 50.2. *A shortest spanning tree can be found in time $O(n^2)$.*

Proof. See above. ∎

In fact, by applying 2-heaps (Section 7.3) one can obtain a running time bound of $O(m \log n)$ (E.L. Johnson, cf. Kershenbaum and Van Slyke [1972]), and with Fibonacci heaps (Section 7.4) one obtains (Fredman and Tarjan [1984,1987]):

Theorem 50.3. *A shortest spanning tree can be found in time $O(m + n \log n)$.*

Proof. Directly by applying Fibonacci heaps as described in Section 7.4. ∎

50.3. Implementing Kruskal's method

Bottleneck in implementing Kruskal's method is the necessity to scan the edges sorted by length. As the best bound for sorting is $O(m \log n)$, we cannot hope for implementations of Kruskal's method faster than that.

However, the bound $O(m \log n)$ is easy to achieve. In fact, as was noticed by Kershenbaum and Van Slyke [1972] (using ideas of Van Slyke and Frank [1972]), it is easy to implement Kruskal's method such that the time *after sorting* is $O(m + n \log n)$. This can be done with elementary data-structures like lists; no heaps are needed.

Indeed, it is not hard to design a simple data structure that tests in constant time if the ends of any edge belong to different components of the current forest F, and that merges components in time linear in the size of the smaller component[1].

Then the iterations take $O(m + n \log n)$ time, since checking if the ends of an edge belong to different components takes $O(m)$ time overall, while merging takes $O(n \log n)$ time overall: any vertex v belongs at most $\log_2 n$ times to the smaller component when merging, as, at any such event, the component containing v at least doubles in size.

---

[1] Consider any forest F. Represent each component K by a (singly) linked list. For any vertex v, let $r(v)$ be the first vertex of the list L_v containing v.

Initially, for each vertex v, $r(v) = v$, as $L_v = \{v\}$. At any iteration, the edge $e = uv$ considered connects different components of F if and only if $r(u) \neq r(v)$. Checking this takes constant time.

If $r(u) \neq r(v)$, we can determine which of the lists L_u, L_v is smallest in time $O(\min\{|L_u|, |L_v|\})$ (by scanning them in parallel, starting at $r(u)$ and $r(v)$). Assume without loss of generality that $|L_u| \leq |L_v|$. Then we reset $r(u') := r(v)$ for all u' in L_u, and we insert L_u into L_v directly after v. This can be done in time $O(|L_u|)$.

Tarjan [1983] showed that if the edges are presorted, a minimum spanning tree can be found in time $O(m\alpha(m,n))$ (where $\alpha(m,n)$ is the 'inverse Ackermann function — see Section 50.6a).

50.3a. Parallel forest-merging

A variant that suggests parallel implementation was given by Borůvka [1926a,1926b] — the *parallel forest-merging method* or *Borůvka's method*. (This method was also given by Choquet [1938] (without proof) and Florek, Łukaszewicz, Perkal, Steinhaus, and Zubrzycki [1951a].) It assumes that all edge lengths are different:

(50.5)   Set $F := \emptyset$. As long as F is not a spanning tree do the following: choose for each component K of F the shortest edge leaving K, and add all chosen edges to F.

Theorem 50.4. *Assuming that all edge lengths are different, the parallel forest-merging variant yields a shortest spanning tree.*

Proof. We show that F remains a good forest throughout the iterations. Consider some iteration, and let F be the good forest at the start of the iteration. Let $e_1, \ldots, e_k$ be the edges added in the iteration, indexed such that $l(e_1) < l(e_2) < \cdots < l(e_k)$. By the selection rule (50.5), for each $i = 1, \ldots, k$, e_i is the shortest edge leaving some component K of F. Then K is also a component of $F \cup \{e_1, \ldots, e_{i-1}\}$, as none of $e_1, \ldots, e_{i-1}$ leave K (since e_i is shortest leaving K). Hence for each $i = 1, \ldots, k$, $F \cup \{e_1, \ldots, e_i\}$ is a good forest (by induction on i). Concluding, the iteration yields a good forest. ∎

50.3b. A dual greedy algorithm

We can consider a dual approach by iteratively decreasing a connector, instead of iteratively growing a forest. The analogy can be exhibited as follows.

Let $G = (V, E)$ be a connected graph and let $l : E \to \mathbb{R}$ be a length function. Call a connector $K \subseteq E$ *good* if K contains a shortest spanning tree. Then we have:

Theorem 50.5. *Let K be a good connector and let $e \in K$. Then $K \setminus \{e\}$ is a good connector if and only if*

(50.6) K *contains a circuit C such that e is a longest edge in C.*

Proof. To see necessity, let T be a shortest spanning tree contained in $K \setminus \{e\}$. Let C be the unique circuit contained in $T \cup \{e\}$. Then e is longest in C, since if $f \in C$, then $T' := (T \setminus \{f\}) \cup \{e\}$ is again a spanning tree. As $l(T') \geq l(T)$ we have $l(e) \geq l(f)$.

To see sufficiency, let T be a shortest spanning tree contained in K. If $e \notin T$, then also $K \setminus \{e\}$ contains T, and hence $K \setminus \{e\}$ is a good connector. So we can assume that $e \in T$. Let D be the cut determined by $T - e$. Then the circuit C contains at least one edge $f \neq e$ that belongs to D. So $T' := (T \setminus \{e\}) \cup \{f\}$ is a spanning tree again. By assumption, $l(e) \geq l(f)$ and hence $l(T') \leq l(T)$. Hence T'

is a shortest spanning tree again. It is contained in $K \setminus \{e\}$, which therefore is a good connector. ∎

So we can formulate the *dual greedy algorithm*: starting with $K := E$, iteratively remove from K an edge e satisfying (50.6). We end up with a shortest spanning tree.

A special case is the following algorithm, proposed by Kruskal [1956]: iteratively delete a longest edge e that is not a bridge. We end up with a shortest spanning tree.

50.4. The longest forest and the forest polytope

The greedy algorithm can be easily adapted so as to give:

Theorem 50.6. *A longest forest can be found in strongly polynomial time.*

Proof. It suffices to find a longest spanning tree in any component. This can be done with the greedy method. ∎

As Edmonds [1971] noticed, it is easy to derive with the greedy method a min-max relation for the maximum length of a forest in a graph $G = (V, E)$. This is similar to the results of Section 40.2.

Theorem 50.7. *Let $G = (V, E)$ be a graph and let $l \in \mathbb{Z}_+^E$. Then the maximum length of a forest is equal to the minimum value of*

(50.7) $$\sum_{U \in \mathcal{P}(V) \setminus \{\emptyset\}} y_U (|U| - 1),$$

where $y \in \mathbb{Z}_+^{\mathcal{P}(V) \setminus \{\emptyset\}}$ satisfies

(50.8) $$\sum_{U \in \mathcal{P}(V) \setminus \{\emptyset\}} y_U \chi^{E[U]} \geq l.$$

Proof. The maximum cannot be larger than the minimum, since for any forest F and any $y \in \mathbb{Z}_+^{\mathcal{P}(V) \setminus \{\emptyset\}}$ satisfying (50.8) one has:

(50.9) $$l(F) \leq \sum_{U \in \mathcal{P}(V) \setminus \{\emptyset\}} y_U |E[U] \cap F| \leq \sum_{U \in \mathcal{P}(V) \setminus \{\emptyset\}} y_U (|U| - 1).$$

To see equality, let $k := \max\{l(e) \mid e \in E\}$, and let E_i be the set of edges e with $l(e) \geq i$, for $i = 0, 1, \ldots, k$. For each $U \in \mathcal{P}(V) \setminus \{\emptyset\}$, let y_U be the number of $i \in \{1, \ldots, k\}$ such that U is a component of the graph (V, E_i). Then it is easy to see that y satisfies (50.8).

We can find a sequence of forests $F_k \subseteq \cdots \subseteq F_1 \subseteq F_0$, where for $i = 0, 1, \ldots, k$, F_i is a maximal forest in (V, E_i) containing F_{i+1}, setting $F_{k+1} := \emptyset$.

Then for $F := F_0$ we have:

Section 50.4. The longest forest and the forest polytope

$$(50.10) \quad l(F) = \sum_{i=0}^{k} i |F_i \setminus F_{i+1}| = \sum_{i=1}^{k} |F_i| = \sum_{i=1}^{k} (|V| - \kappa(V, E_i))$$
$$= \sum_{U \in \mathcal{P}(V) \setminus \{\emptyset\}} y_U (|U| - 1),$$

where $\kappa(V, E_i)$ denotes the number of components of the graph (V, E_i). ∎

(The series of forests $F_k \subseteq F_{k-1} \subseteq \cdots \subseteq F_1 \subseteq F_0$, corresponds to the greedy method.)

Note that this theorem gives, if G is connected, a min-max relation for the maximum length of a spanning tree.

For any graph $G = (V, E)$, let the *forest polytope* of G, denoted by $P_{\text{forest}}(G)$, be the convex hull of the incidence vectors (in $\mathbb{R}^E$) of the forests of G. The following characterization of the forest polytope is (in matroid terms) due to Edmonds [1971] (announced in Edmonds [1967a]):

Corollary 50.7a. *The forest polytope of a graph G is determined by*

$$(50.11) \quad \begin{array}{lll} \text{(i)} & x_e \geq 0 & \text{for } e \in E, \\ \text{(ii)} & x(E[U]) \leq |U| - 1 & \text{for nonempty } U \subseteq V. \end{array}$$

Proof. Trivially, the incidence vector of any forest satisfies (50.11), and hence the forest polytope is contained in the polytope determined by (50.11). Suppose now that the latter polytope is larger. Then (since both polytopes are rational and down-monotone in $\mathbb{R}_+^E$) there exists a vector $l \in \mathbb{Q}_+^E$ such that the maximum value of $l^\top x$ over (50.11) is larger than the maximum of $l(F)$ over forests F. We can assume that l is integer. However, by Theorem 50.7, the maximum of $l(F)$ is at least the minimum value of the problem dual to maximizing $l^\top x$ over (50.11), a contradiction. ∎

Theorem 50.7 can be stated equivalently in TDI terms as follows:

Corollary 50.7b. *System (50.11) is totally dual integral.*

Proof. This follows from Theorem 50.7, by the definition of total dual integrality. ∎

Having a description of the forest polytope, we can derive a description of the *spanning tree polytope* $P_{\text{spanning tree}}(G)$ of a graph $G = (V, E)$, which is the convex hull of the incidence vectors of the spanning trees in G.

Corollary 50.7c. *The spanning tree polytope of a graph $G = (V, E)$ is determined by*

(50.12) (i) $x_e \geq 0$ for $e \in E$,
          (ii) $x(E[U]) \leq |U| - 1$ for nonempty $U \subseteq V$,
          (iii) $x(E) = |V| - 1$.

Proof. Directly from Corollary 50.7a, since the spanning trees are exactly the forests of size $|V| - 1$, and since there exist no forests larger than that. ∎

One also directly has a TDI result:

Corollary 50.7d. *System (50.12) is totally dual integral.*

Proof. Directly from Corollary 50.7b, since (50.12) arises from (50.11) by setting an inequality to equality (cf. Theorem 5.25). ∎

Theorem 40.5 implies that (if G is loopless) an inequality (50.12)(ii) is facet-inducing if and only if $|U| \geq 2$ and U induces a 2-connected subgraph of G (cf. Grötschel [1977a]).

In Section 51.4 we consider the problem of testing membership of the forest polytope.

50.5. The shortest connector and the connector polytope

The greedy method also provides a min-max relation for the minimum length of a connector in a graph $G = (V, E)$. Let Π denote the collection of partitions of V into nonempty subsets. For any partition $\mathcal{P}$ of V, let $\delta(\mathcal{P})$ denote the set of edges connecting two different classes of $\mathcal{P}$. So any connector contains at least $|\mathcal{P}| - 1$ edges in $\delta(\mathcal{P})$.

Theorem 50.8. *Let $G = (V, E)$ be a connected graph and let $l \in \mathbb{Z}_+^E$. Then the minimum length of a spanning tree is equal to the maximum value of*

(50.13) $$\sum_{\mathcal{P} \in \Pi} y_\mathcal{P}(|\mathcal{P}| - 1),$$

where $y \in \mathbb{Z}_+^\Pi$ such that

(50.14) $$\sum_{\mathcal{P} \in \Pi} y_\mathcal{P} \chi^{\delta(\mathcal{P})} \leq l.$$

Proof. The minimum cannot be smaller than the maximum, since for any spanning tree T and any $y \in \mathbb{Z}_+^\Pi$ satisfying (50.14) one has:

(50.15) $$l(T) \geq \sum_{\mathcal{P} \in \Pi} y_\mathcal{P} \chi^{\delta(\mathcal{P})}(T) = \sum_{\mathcal{P} \in \Pi} y_\mathcal{P} |\delta(\mathcal{P}) \cap T| \geq \sum_{\mathcal{P} \in \Pi} y_\mathcal{P}(|\mathcal{P}| - 1).$$

To see equality, define $k := \max\{l(e) \mid e \in E\}$ and for $i = 0, 1, \ldots, k$, let E_i be the set of edges e with $l(e) \leq i$. For each $\mathcal{P} \in \Pi$, let $y_\mathcal{P}$ be the number of

$i \in \{1, \ldots, k\}$ such that $\mathcal{P}$ is the collection of components of (V, E_i). Then it is easy to see that y satisfies (50.14).

We can find a sequence of forests $F_0 \subseteq F_1 \subseteq \cdots F_{k-1} \subseteq F_k$, where F_0 is a maximal forest in (V, E_0), and where for $i = 0, \ldots, k$, F_i is a maximal forest in (V, E_i) containing F_{i-1}, setting $F_{-1} := \emptyset$.

Then for $T := F_k$ we have:

$$(50.16) \quad l(T) = \sum_{i=0}^{k} i|F_i \setminus F_{i-1}| = k|T| - \sum_{i=0}^{k-1} |F_i| = \sum_{i=0}^{k-1} (|V| - 1 - |F_i|)$$

$$= \sum_{i=1}^{k-1} (\kappa(V, E_i) - 1) = \sum_{\mathcal{P} \in \Pi} y_{\mathcal{P}}(|\mathcal{P}| - 1),$$

where $\kappa(V, E_i)$ denotes the number of components of the graph (V, E_i). ∎

For any graph $G = (V, E)$, let the *connector polytope* of G, denoted by $P_{\text{connector}}(G)$, be the convex hull of the incidence vectors (in $\mathbb{R}^E$) of the connectors of G. The following characterization can be derived from Edmonds [1970b], and was stated explicitly by Fulkerson [1970b]:

Corollary 50.8a. *The connector polytope of a graph G is determined by*

$$(50.17) \quad \begin{array}{lll} \text{(i)} & 0 \le x_e \le 1 & \text{for } e \in E, \\ \text{(ii)} & x(\delta(\mathcal{P})) \ge |\mathcal{P}| - 1 & \text{for } \mathcal{P} \in \Pi. \end{array}$$

Proof. Trivially, the incidence vector of any connector satisfies (50.17), and hence the connector polytope is contained in the polytope determined by (50.17). Suppose now that the latter polytope is larger. Then (since both polytopes are rational and up-monotone in $[0, 1]^E$) there exists a vector $l \in \mathbb{Q}_+^E$ such that the minimum value of $l^\mathsf{T} x$ over (50.17) is smaller than the minimum of $l(C)$ over connectors C. We can assume that l is integer. However, by Theorem 50.8, the minimum of $l(C)$ is at most the maximum value of the problem dual to minimizing $l^\mathsf{T} x$ over (50.17), a contradiction. ∎

Theorem 50.8 can be stated equivalently in TDI terms as follows:

Corollary 50.8b. *System (50.17) is totally dual integral.*

Proof. This follows from Theorem 50.8, by the definition of total dual integrality. ∎

Chopra [1989] described the facets of the connector polytope. In Section 51.4 we consider the problem of testing membership of the connector polytope.

50.6. Further results and notes

50.6a. Complexity survey for shortest spanning tree

	$O(nm)$	Jarník [1930]
	$O(n^2)$	Prim [1957], Dijkstra [1959]
	$O(m \log n)$	Kershenbaum and Van Slyke [1972], E.L. Johnson (cf. Kershenbaum and Van Slyke [1972])
	$O(m \log_{m/n} n)$	D.B. Johnson [1975b]
	$O(m\sqrt{\log n})$	R.E. Tarjan (cf. Yao [1975])
	$O(m \log \log n)$	Yao [1975]
	$O(m \log \log_{m/n} n)$	Cheriton and Tarjan [1976], Tarjan [1983]
*	$O((m + n \log L) \log \log L)$	D.B. Johnson [1977b]
	$O(m + n \log n)$	Fredman and Tarjan [1984,1987]
	$O(m\beta(m, n))$	Fredman and Tarjan [1984,1987]
	$O(m \log \beta(m, n))$	Gabow, Galil, Spencer, and Tarjan [1986] (cf. Gabow, Galil, and Spencer [1984])
	$O(m(\log_n L + \alpha(m, n)))$	Gabow [1983b,1985b]
	$O(m\alpha(m, n) \log \alpha(m, n))$	Chazelle [1997]
*	$O(m\alpha(m, n))$	Chazelle [2000]

As before, * indicates an asymptotically best bound in the table. Moreover, $\beta(m, n) := \min\{i \mid \log_2^{(i)} n \leq m/n\}$ and $L := \max\{l(e) \mid e \in E\}$ (assuming l nonnegative integer). The function $\alpha(m, n)$ is the *inverse Ackermann function*, defined as follows. For $i, j \geq 1$, the *Ackermann function* $A(i, j)$ is defined recursively by:

(50.18) $\quad A(1, j) = 2^j \quad$ for $j \geq 1$,
$\qquad A(i, 1) = A(i-1, 2) \quad$ for $i \geq 2$,
$\qquad A(i, j) = A(i-1, A(i, j-1)) \quad$ for $i, j \geq 2$.

Next, for $m \geq n \geq 1$,

(50.19) $\quad \alpha(m, n) := \min\{i \geq 1 \mid A(i, \lfloor m/n \rfloor) > \log_2 n\}$.

The function $\alpha(m, n)$ is extremely slowly growing.

Fredman and Willard [1990,1994] gave a 'strongly trans-dichotomous' linear-time minimum spanning tree algorithm (where capabilities of random access machines, like addressing, can be used). Based on sampling, Karger [1993,1998] found a simple linear-time approximative spanning tree algorithm, and an $O(m+n \log n)$-time minimum spanning tree algorithm not using Fibonacci heaps.

Katoh, Ibaraki, and Mine [1981] gave an algorithm to find the Kth shortest spanning tree in time $O(Km + \min\{n^2, m \log \log n\})$ (improving slightly Gabow

[1977]). They also gave an algorithm to find the second shortest spanning tree in time $O(\min\{n^2, m\alpha(m,n)\})$.

Pettie and Ramachandran [2000,2002a] showed that a shortest spanning tree can be found in time $O(\mathcal{T}^*(m,n))$, where $\mathcal{T}^*(m,n)$ is the minimum number of edge length comparisons needed to determine the solution.

Frederickson [1983a,1985] gave an $O(\sqrt{m})$-time algorithm to update a shortest spanning tree (and the data-structure) if one edge changes length. Spira and Pan [1973,1975] and Chin and Houck [1978] gave fast algorithms to update a shortest spanning tree if vertices are added or removed. More on sensitivity and most vital edges can be found in Tarjan [1982], Hsu, Jan, Lee, Hung, and Chern [1991], Dixon, Rauch, and Tarjan [1992], Iwano and Katoh [1993], Lin and Chern [1993], and Frederickson and Solis-Oba [1996,1999].

Tarjan [1979] showed that the minimality of a given spanning tree can be checked in time $O(m\alpha(m,n))$ (cf. Dixon, Rauch, and Tarjan [1992]). Komlós [1984, 1985] showed that the minimality of a given spanning tree can be checked by $O(m)$ comparisons of edge lengths. King [1997] gave a linear-time implementation in the unit-cost RAM model. A randomized linear-time algorithm was given by Klein and Tarjan [1994], and Karger, Klein, and Tarjan [1995].

Gabow and Tarjan [1984] (cf. Gabow and Tarjan [1979]) showed that the problem of finding a shortest spanning tree with a prescribed number of edges incident with a (one) given vertex r, is linear-time equivalent to the (unconstrained) shortest spanning tree problem. They also showed that if the edges of a graph are coloured red and blue, a shortest spanning tree having exactly k red edges (for given k) can be found in time $O(m \log \log_{2+\frac{m}{n}} n + n \log n)$.

Brezovec, Cornuéjols, and Glover [1988] gave an efficient algorithm to find a shortest spanning tree in a coloured graph with, for each colour, an upper and a lower bound on the number of edges in the tree of that colour.

Camerini [1978] showed that a spanning tree minimizing $\max_{e \in T} l(e)$ can be found in $O(m)$ time.

Geometric spanning trees (on vertices in Euclidean space, with Euclidean distance as length function) were considered by Bentley, Weide, and Yao [1980], Yao [1982], Supowit [1983], Clarkson [1984,1989], and Agarwal, Edelsbrunner, Schwarzkopf, and Welzl [1991].

50.6b. Characterization of shortest spanning trees

The following theorem is implicit in Kalaba [1960]:

Theorem 50.9. *Let $G = (V, E)$ be a graph, let $l \in \mathbb{R}^E$ be a length function, and let T be a spanning tree in G. Then T is a shortest spanning tree if and only if $l(f) \geq l(e)$ for all $e \in T$ and $f \in E \setminus T$ with $T - e + f$ a spanning tree.*

Proof. Necessity being trivial, we show sufficiency. Let the condition be satisfied, and suppose that T is not a shortest spanning tree. Choose a shorter spanning tree T' with $|T' \setminus T|$ minimal. Let $f \in T' \setminus T$. Let e be an edge on the circuit in $T \cup \{f\}$ with $e \neq f$, such that e connects the two components of $T' \setminus \{f\}$. Then $(T\setminus\{e\})\cup\{f\}$ is a spanning tree, and hence $l(f) \geq l(e)$. Define $T'' := (T'\setminus\{f\})\cup\{e\}$. Then $l(T'') \leq l(T') < l(T)$ and $|T'' \setminus T| < |T' \setminus T|$, contradicting our minimality assumption. ∎

This theorem gives a good characterization of the minimum length of a spanning tree. (As Kalaba [1960] pointed out, it also gives an algorithm to find a shortest spanning tree (by iteratively exchanging one edge for another if it makes the tree shorter), but it is not polynomial-time.)

Recall that a forest is called *good* if it is contained in a shortest spanning tree.

Corollary 50.9a. *Let $G = (V, E)$ be a connected graph, let $l \in \mathbb{R}^E$ be a length function, and let F be a forest. Then F is good if and only if for each $e \in F$ there exists a cut C with $C \cap F = \{e\}$ and with e shortest in C.*

Proof. To see necessity, let F be good and let $e \in F$. So there exists a shortest spanning tree T containing F. By Theorem 50.9, e is a shortest edge connecting the two components of $T - e$. This gives the required cut C.

Sufficiency is shown by induction on $|F|$, the case $F = \emptyset$ being trivial. Choose $e \in F$. By induction, $F \setminus \{e\}$ is good (as the condition is maintained for $F \setminus \{e\}$). The condition implies that (50.2) is satisfied, and hence F is good by Theorem 50.1. ∎

50.6c. The maximum reliability problem

Often, in designing a network, one is not primarily interested in minimizing the total length, but rather in maximizing 'reliability' (for instance when designing energy or communication networks).

Let $G = (V, E)$ be a connected graph and let $r : E \to \mathbb{R}_+$ be a function. Let us call $r(e)$ the *reliability* of edge e. For any path P in G, the *reliability* of P is, by definition, the minimum reliability of the edges occurring in P. The *reliability* $r_G(s,t)$ of two vertices s and t is equal to the maximum reliability of P where P ranges over all $s - t$ paths. That is,

(50.20) $\qquad r_G(s,t) := \max_P \min_{e \in EP} r(e),$

where the maximum ranges over all $s - t$ paths P. (The value of $r_G(s,t)$ can be found with the method described in Section 8.6e.)

The problem now is to find a minimal subgraph H of G having the same reliability as G; that is, with $r_H = r_G$. Hu [1961] observed that there is a spanning tree carrying the reliability of G. More precisely, Hu showed that any spanning tree T of maximum total reliability is such a tree (also shown by Kalaba [1964]):

Corollary 50.9b. *Let $G = (V, E)$ be a graph, let $r \in \mathbb{R}^E$, and let T be any spanning tree. Then $r_T(s,t) = r_G(s,t)$ for all s,t if and only if T is a spanning tree in G maximizing $r(T)$.*

Proof. To see sufficiency, let T maximize $r(T)$. Choose $s, t \in V$, and let P be a path in G attaining maximum (50.20). Let e be an edge on the $s - t$ path in T with minimum $r(e)$. Then P contains an edge f connecting the two components of $T - e$. As T maximizes $r(T)$ we have $r(f) \le r(e)$. Hence

(50.21) $\qquad r_T(s,t) = r(e) \ge r(f) \ge r_G(s,t).$

Since trivially $r_T(s,t) \leq r_G(s,t)$, this shows sufficiency.

To see necessity, we apply Theorem 50.9. Choose $e \in T$, and suppose that there is an edge f connecting the components of $T - e$, with $r(f) > r(e)$. Then for the ends s, t of f we have

(50.22) $\qquad r_G(s,t) \geq r(f) > r(e) \geq r_T(s,t),$

a contradiction. ∎

Corollary 50.9b implies:

Corollary 50.9c. *Let $G = (V, E)$ be a complete graph and let $l : E \to \mathbb{R}_+$ be a length function satisfying*

(50.23) $\qquad l(uw) \geq \min\{l(uv), l(vw)\}$

for all distinct $u, v, w \in V$. Let T be a longest spanning tree in G. Then for all $u, w \in V$, $l(uw)$ is equal to the minimum length of the edges in the $u - w$ path in T.

Proof. Note that (50.23) implies that $l(uw)$ is equal to the reliability $r_G(u, w)$ of u and w, taking $r := l$. So the corollary follows from Corollary 50.9b. ∎

This implies the following. Let $G = (V, E)$ be a graph and let $c : E \to \mathbb{R}_+$ be a capacity function. Let K be the complete graph on V. For each edge st of K, let the length $l(st)$ be the minimum capacity of any $s - t$ cut in G. (An $s - t$ cut is any subset $\delta(W)$ with $s \in W, t \notin W$.)

Let T be a longest spanning tree in K. Then for all $s, t \in V$, $l(st)$ is equal to the minimum length of the edges of T in the $s - t$ path in T.

(This tree need not be a Gomory-Hu tree, as is shown by the complete graph on vertices $1, 2, 3$ and $c(12) = 1$ and $c(13) = c(23) = 2$. Then edges 12 and 13 form a tree as above, but it is not a Gomory-Hu tree.)

50.6d. Exchange properties of forests

The following fundamental property of forests in fact is the basis of most theorems in this chapter. It is the 'exchange property' that makes the collection of forests into a matroid.

Theorem 50.10. *Let $G = (V, E)$ be a graph and let F and F' be forests with $|F| < |F'|$. Then $F \cup \{e\}$ is a forest for some $e \in F' \setminus F$.*

Proof. We can assume that $E = F \cup F'$. If no such edge e exists, then F is a maximal forest in G. This however implies that $|F| \geq |F'|$, a contradiction. ∎

Call a forest F *extreme* if $l(F') \geq l(F)$ for each forest F' satisfying $|F'| = |F|$. The forests made iteratively in Kruskal's method all are extreme, since:

Corollary 50.10a. *Let F be an extreme forest and let e be a shortest edge with $e \notin F$ and $F \cup \{e\}$ a forest. Then $F \cup \{e\}$ is extreme again.*

Proof. Let F' be an extreme forest with $|F'| = |F| + 1$. By Theorem 50.10, there exists an $e' \in F' \setminus F$ such that $F \cup \{e'\}$ is a forest. As F is extreme we have $l(F' \setminus \{e'\}) \geq l(F)$. Hence $l(F \cup \{e'\}) \leq l(F')$. Also, by the choice of e, $l(e) \leq l(e')$. So $l(F \cup \{e\}) \leq l(F')$. Concluding, $F \cup \{e\}$ is extreme (as F' is extreme). ∎

The following corollary is due to Florek, Łukaszewicz, Perkal, Steinhaus, and Zubrzycki [1951a]. Recall that a forest is called *good* if it is contained in a shortest spanning tree.

Corollary 50.10b. *Each extreme forest is good.*

Proof. Directly from Corollary 50.10a, since it implies that each extreme forest is contained in an extreme maximal forest, and hence in a shortest maximal forest; so it is good. ∎

We also can derive a 'slice-integrality' result:

Corollary 50.10c. *Let $G = (V, E)$ be a graph and let $k, l \in \mathbb{Z}_+$. Then the convex hull of the incidence vectors of forests F with $k \leq |F| \leq l$ is equal to the intersection of the forest polytope of G with $\{x \in \mathbb{R}^E \mid k \leq x(E) \leq l\}$.*

Proof. Let x be in the forest polytope with $k \leq x(E) \leq l$. Let $x = \sum_F \lambda_F \chi^F$, where F ranges over all forests and where the λ_F are nonnegative reals with $\sum_F \lambda_F = 1$. Choose the λ_F with

(50.24) $$\sum_F \lambda_F |F|^2$$

minimal. Then

(50.25) $\quad |F'| \leq |F| + 1$ for all F, F' with $\lambda_F > 0$ and $\lambda_{F'} > 0$.

Otherwise we can choose $e \in F' \setminus F$ such that $F \cup \{e\}$ is a forest (by Theorem 50.10). Let $\alpha := \min\{\lambda_F, \lambda_{F'}\}$. Then decreasing λ_F and $\lambda_{F'}$ by α and increasing $\lambda_{F \cup \{e\}}$ and $\lambda_{F' \setminus \{e\}}$ by α, decreases sum (50.24). This contradicts our assumption, and proves (50.25).

It implies that $k \leq |F| \leq l$ for each F with $\lambda_F > 0$, and we have the corollary. ∎

50.6e. Uniqueness of shortest spanning tree

Kotzig [1961b] characterized when there is a unique shortest spanning tree:

Theorem 50.11. *Let $G = (V, E)$ be a graph, let $l \in \mathbb{R}^E$ be a length function, and let T be a spanning tree in G. Then T is a unique shortest spanning tree if and only if $l(f) > l(e)$ for all $e \in T$ and $f \in E \setminus T$ such that $T - e + f$ is a spanning tree.*

Proof. As the proof of Theorem 50.9. ∎

This implies a sufficient condition given by Borůvka [1926a]:

Corollary 50.11a. *Let $G = (V, E)$ be a graph and let $l \in \mathbb{R}^E$ be a length function with $l(e) \neq l(f)$ if $e \neq f$. Then there is a unique shortest spanning tree.*

Proof. Directly from Theorem 50.11. ∎

Let $G = (V, E)$ be a connected graph and let $l \in \mathbb{R}^E$ be a length function, with $l(e) \neq l(f)$ if $e \neq f$. Define

(50.26) $\qquad T := \{e \in E \mid \exists \text{ cut } C \text{ such that } e \text{ is the shortest edge of } C\}$.

Then

(50.27) $\qquad E \setminus T = \{e \in E \mid \exists \text{ circuit } D \text{ such that } e \text{ is the longest edge in } D\}$.

This is easy, since if some edge e is contained in some cut C and some circuit D, then there exists an edge $f \neq e$ in $C \cap D$. If $l(f) < l(e)$, then e is not shortest in C, and if $l(f) > l(e)$, then e is not longest in D. Moreover, for any $e \in E$, if no circuit D as in (50.27) exists, then each circuit D containing e contains an edge f with $l(f) > l(e)$. Hence the set of edges f with $l(f) \geq l(e)$ contains a cut C containing e. This C is as in (50.26).

Now (Dijkstra [1960], Rosenstiehl [1967]):

(50.28) $\qquad T$ is the unique shortest spanning tree in G.

Indeed, T is a forest, since each circuit D intersects $E \setminus T$ (namely, in the longest edge of D). Moreover, T is a connector, since each cut C intersects T (namely, in the shortest edge of C). T is the unique shortest spanning tree. This follows from Theorem 50.11, since for each $e \in T$ and each $f \notin T$, if $(T \setminus \{e\}) \cup \{f\}$ is a spanning tree, then $l(e) < l(f)$ as e is the shortest edge in the cut determined by $T - e$.

50.6f. Forest covers

Let $G = (V, E)$ be an undirected graph. A subset F of E is called a *forest cover* if F is both a forest and an edge cover. Forest covers turn out to be interesting algorithmically and polyhedrally.

As Gamble and Pulleyblank [1989] point out, White [1971] showed:

Theorem 50.12. *Given a graph $G = (V, E)$ and a weight function $w \in \mathbb{Q}^E$, a minimum-weight forest cover can be found in strongly polynomial time.*

Proof. Let E_- be the set of edges of negative weight and let V_- be the set of vertices covered by E_-. Let $V_+ := V \setminus V_-$. First find a subset F' of $E[V_+] \cup \delta(V_+)$ covering V_+, of minimum weight. This can be done in strongly polynomial time, by a variation of the strongly polynomial-time algorithm for the minimum weight edge cover problem. (In fact, it is a special case of Theorem 34.4.)

Next find a forest F'' in $E[V_-]$ of minimum weight. Again, this can be done in strongly polynomial time, by Theorem 50.6.

We can assume that any proper subset of F' does not cover V_+. It implies that F' is a forest and that for any vertex $v \in V_+$ incident with some edge e in F' with $e \in \delta(V_+)$, e is the only edge in F' incident with v.

This implies that $F' \cup F''$ is a forest. Moreover, it is an edge cover, since F' covers V_+ and F'' covers V_-, since any vertex in V_- is incident with an edge of negative weight.

So $F' \cup F''$ is a forest cover. To see that it has minimum weight, let $B \subseteq E$ be any forest cover. Let $B'' := B \cap E[V_-]$ and $B' := B \setminus B''$. Then $w(B') \geq w(F')$, since B' covers V_+. Also, $w(B'') \geq w(F'')$, since B'' is a forest. So $w(B) \geq w(F)$. ∎

Gamble and Pulleyblank [1989] showed that White's method implies a characterization of the *forest cover polytope* $P_{\text{forest cover}}(G)$ of a graph G, which is the convex hull of the incidence vectors of forest covers in G. It turns out to be equal to the intersection of the forest polytope (characterized in Corollary 50.7a) and the edge cover polytope (characterized in Corollary 27.3a):

Theorem 50.13. *For any undirected graph $G = (V, E)$:*

(50.29) $\qquad P_{\text{forest cover}}(G) = P_{\text{forest}}(G) \cap P_{\text{edge cover}}(G).$

Proof. The inclusion $\subseteq$ is trivial, as any forest cover is both a forest and an edge cover. Suppose that the reverse inclusion does not hold, and let x be a vertex of $P_{\text{forest}}(G) \cap P_{\text{edge cover}}(G)$ which is not in $P_{\text{forest cover}}(G)$. Let $w \in \mathbb{Q}^E$ be a weight function such that x uniquely minimizes $w^\mathsf{T} x$ over $P_{\text{forest}}(G) \cap P_{\text{edge cover}}(G)$. We can assume that $w(e) \neq 0$ for each edge e (as we can perturb w slightly).

Again let E_- be the set of edges of negative weight, V_- be the set of vertices covered by E_-, and $V_+ := V \setminus V_-$. Since x is in the edge cover polytope, there exists a subset F' of $E[V_+] \cup \delta(V_+)$ covering V_+ with

(50.30) $\qquad w(F') \leq \sum_{e \in E[V_+] \cup \delta(V_+)} w(e) x_e.$

Similarly, since x is in the forest polytope, there is a forest F'' in $E[V_-]$ with

(50.31) $\qquad w(F'') \leq \sum_{e \in E[V_-]} w(e) x_e.$

Now, as in the proof of Theorem 50.12, $F := F' \cup F''$ is a forest cover. Since $w(F) \leq w^\mathsf{T} x$, this contradicts our assumptions on x and w. ∎

White [1971] also considered the problem of finding a minimum weight forest cover of given size k. Gamble and Pulleyblank [1989] showed that the convex hull of the incidence vectors of forest covers of size k is equal to the intersection of the forest cover polytope with the hyperplane $\{x \in \mathbb{R}^E \mid x(E) = k\}$.

Cerdeira [1994] related forest covers to matroid intersection.

50.6g. Further notes

Let $G = (V, E)$ be a graph. Call a subset U of V *circuit-free* if U spans no circuit; that is, it induces a forest as subgraph of G. Ding and Zang [1999] characterized the graphs G for which the convex hull of the incidence vectors of circuit-free sets is determined by

(50.32) $0 \leq x_v \leq 1$ for each vertex v,
           $x(VC) \leq |VC| - 1$ for each circuit C.

Their characterization implies that (50.32) is totally dual integral as soon as it determines an integer polytope.

Goemans [1992] studied the convex hull of the incidence vectors of (not necessarily spanning) subtrees of a graph.

Brennan [1982] reported on good experimental results with an implementation of Kruskal's method by only partially sorting the edges until the successive shortest edges to be added to the current forest can be identified.

Győri [1978] and Lovász [1977a] showed that if $G = (V, E)$ is k-connected and $v_1, \ldots, v_k$ are distinct vertices, and $n_1, \ldots, n_k$ are positive integers with $n_1 + \cdots + n_k = |V|$, then G contains a forest F such that the component containing v_i has size n_i ($i = 1, \ldots, k$). For $k = 2$, Győri's proof gives an $O(nm)$-time algorithm. A linear-time algorithm for $k = 2$ was given by Suzuki, Takahashi, and Nishizeki [1990]. More can be found in Győri [1981].

Khuller, Raghavachari, and Young [1993,1995b] considered spanning trees that belance between shortest spanning trees and shortest paths trees.

Books covering shortest spanning trees include Even [1973,1979], Christofides [1975], Lawler [1976b], Minieka [1978], Hu [1982], Papadimitriou and Steiglitz [1982], Smith [1982], Aho, Hopcroft, and Ullman [1983], Sysło, Deo, and Kowalik [1983], Tarjan [1983], Gondran and Minoux [1984], Nemhauser and Wolsey [1988], Chen [1990], Cormen, Leiserson, and Rivest [1990], Lengauer [1990]. Ahuja, Magnanti, and Orlin [1993], Cook, Cunningham, Pulleyblank, and Schrijver [1998], Jungnickel [1999], and Korte and Vygen [2000]. Pierce [1975] and Golden and Magnanti [1977] gave bibliographies on algorithms for shortest spanning tree.

50.6h. Historical notes on shortest spanning trees

We refer to Graham and Hell [1985] for an extensive historical survey of shortest tree algorithms, with several quotes (with translations) from old papers. Our notes below have profited from their investigations.

We recall some terminology for a shortest spanning tree algorithm. We call it *tree-growing* if we keep a tree on a subset of the vertices, and iteratively extend it by adding an edge joining the tree with a vertex outside of the tree. It is *forest-merging* if we keep a forest, and iteratively merge two components by joining them by an edge. It is called *parallel forest-merging* if forest-merging is performed in parallel, by connecting each component to its nearest neighbouring component (assuming all lengths are distinct).

Borůvka: parallel forest-merging

Borůvka [1926a] described the problem of finding a shortest spanning tree as follows (the paper is in Czech; we quote from its German summary; for quotes from Czech with translation, see Graham and Hell [1985]):

> In dieser Arbeit löse ich folgendes Problem:
> Es möge eine Matrix der bis auf die Bedingungen $r_{\alpha\alpha} = 0$, $r_{\alpha\beta} = r_{\beta\alpha}$ positiven und von einander verschiedenen Zahlen $r_{\alpha\beta}$ ($\alpha, \beta = 1, 2, \ldots n; n \geq 2$) gegeben sein.

Aus dieser ist eine Gruppe von einander und von Null verschiedener Zahlen auszuwählen, so dass
1° in ihr zu zwei willkürlich gewählten natürlichen Zahlen p_1, p_2 ($\leq n$) eine Teilgruppe von der Gestalt

$$r_{p_1 c_2}, r_{c_2 c_3}, r_{c_3 c_4}, \ldots r_{c_{q-2} c_{q-1}}, r_{c_{q-1} p_2}$$

existiere,
2° die Summe ihrer Glieder kleiner sei als die Summe der Glieder irgendeiner anderen, der Bedingung 1° genügenden Gruppe von einander und von Null verschiedenen Zahlen.[2]

So Borůvka stated that the spanning tree found is the unique shortest. He assumed that all edge lengths are different.

Borůvka next described parallel forest-merging, in a somewhat complicated way. (He did not have the language of graph theory at hand.) The idea is to update a number of vertex-disjoint paths $P_1, \ldots, P_k$ (initially $k = 0$). Along any P_i, the edge lengths are decreasing. Let v be the last vertex of P_k and let e be the edge of shortest length incident with v. If the other end vertex of e is not yet covered by any P_i, we extend P_k with e, and iterate. Otherwise, if not all vertices are covered yet by the P_i, we choose such a vertex v, and start a new path P_{k+1} at v. If all vertices are covered by the P_i, we shrink each of the P_i to one vertex, and iterate. At the end, the edges chosen throughout the iterations form a shortest spanning tree. It is easy to see that this in fact is 'parallel forest-merging'.

The interest of Borůvka in this problem came from a question of the Electric Power Company of Western Moravia in Brno, at the beginning of the 1920s, asking for the most economical construction of an electric power network (see Borůvka [1977]).

In a follow-up paper, Borůvka [1926b] gave a simple explanation of the method by means of an example. We refer to Nešetřil, Milková, and Nešetřilova [2001] for translations of and comments on the two papers of Borůvka.

Jarník: tree-growing

In a reaction to Borůvka's work, Jarník wrote on 12 February 1929 a letter to Borůvka in which he described a 'new solution of a minimal problem discussed by Mr Borůvka'. This 'new solution' is the tree-growing method. An extract of the letter was published as Jarník [1930]. We quote from the German summary:

a_1 ist eine beliebige unter den Zahlen $1, 2, \ldots, n$.
a_2 ist durch

---

[2] In this work, I solve the following problem:
A matrix may be given of positive distinct numbers $r_{\alpha\beta}$ ($\alpha, \beta = 1, 2 \ldots n$; $n \geq 2$), up to the conditions $r_{\alpha\alpha} = 0$, $r_{\alpha\beta} = r_{\beta\alpha}$.
From this, a group of numbers, different from each other and from zero, should be selected such that
1° for arbitrarily chosen natural numbers p_1, p_2 ($\leq n$) a subgroup of it exists of the form

$$r_{p_1 c_2}, r_{c_2 c_3}, r_{c_3 c_4}, \ldots r_{c_{q-2} c_{q-1}}, r_{c_{q-1} p_2},$$

2° the sum of its members be smaller than the sum of the members of any other group of numbers different from each other and from zero, satisfying condition 1°.

Section 50.6h. Historical notes on shortest spanning trees

$$r_{a_1,a_2} = \left(\begin{array}{c} \min \\ l = 1, 2, \ldots, n \\ l \neq a_1 \end{array} \right) r_{a_1,l}$$

definiert.
Wenn $2 \leq k < n$ und wenn $[a_1, a_2], \ldots, [a_{2k-3}, a_{2k-2}]$ bereits bestimmt sind, so wird $[a_{2k-1}, a_{2k}]$ durch

$$r_{a_{2k-1},a_{2k}} = \min r_{i,j},$$

definiert, wo i alle Zahlen $a_1, a_2, \ldots, a_{2k-2}$, j aber alle übrigen von den Zahlen $1, 2, \ldots, n$ durchläuft.[3]

Again, Jarník assumed that all lengths are distinct and showed that then the shortest spanning tree is unique. For a detailed discussion and translation of the article of Jarník [1930] (and of Jarník and Kössler [1934] on the Steiner tree problem), see Korte and Nešetřil [2001].

Other discoveries of parallel forest-merging

Parallel forest-merging was described also by Choquet [1938] (without proof), who gave as motivation the construction of road systems:

> Étant donné n villes du plan, il s'agit de trouver un réseau de routes permettant d'aller d'une quelconque de ces villes à une autre et tel que:
> 1° la longueur globale du réseau soit minimum;
> 2° exception faite des villes, on ne peut partir d'aucun point dans plus de deux directions, afin d'assurer la sûreté de la circulation; ceci entraîne, par exemple, que lorsque deux routes semblent se croiser en un point qui n'est pas une ville, elles passent en fait l'une au-dessus de l'autre et ne communiquent pas entre elles en ce point, qu'on appellera faux-croisement.[4]

He was one of the first concerned on the complexity of the method:

> Le réseau cherché sera tracé après $2n$ opérations élémentaires au plus, en appelant opération élémentaire la recherche du continu le plus voisin d'un continu donné.[5]

---

[3] a_1 is an arbitrary one among the numbers $1, 2, \ldots, n$.
a_2 is defined by

$$r_{a_1,a_2} = \left(\begin{array}{c} \min \\ l = 1, 2, \ldots, n \\ l \neq a_1 \end{array} \right) r_{a_1,l}.$$

If $2 \leq k < n$ and if $[a_1, a_2], \ldots, [a_{2k-3}, a_{2k-2}]$ are determined already, then $[a_{2k-1}, a_{2k}]$ is defined by

$$r_{a_{2k-1},a_{2k}} = \min r_{i,j},$$

where i runs through all numbers $a_1, a_2, \ldots, a_{2k-2}$, j however through all remaining of the numbers $1, 2, \ldots, n$.

[4] Being given n cities of the plane, the point is to find a network of routes allowing to go from an arbitrary of these cities to another and such that:
1° the global length of the network be minimum;
2° except for the cities, one cannot depart from any point in more than two directions, in order to assure the certainty of the circulation; this entails, for instance, that when two routes seem to cross each other in a point which is not a city, they pass in fact one above the other and do not communicate among them in this point, which we shall call a false crossing.

[5] The network looked for will be traced after at most $2n$ elementary operations, calling the search for the continuum closest to a given continuum an elementary operation.

Also Florek, Łukaszewicz, Perkal, Steinhaus, and Zubrzycki [1951a,1951b] described parallel forest-merging. They were motivated by clustering in anthropology, taxonomy, etc. In the latter paper, they apply the method to:

> 1° the capitals of Poland's provinces, 2° two collections of excavated skulls, 3° 42 archeological finds, 4° the liverworts of Silesian Beskid mountains with forests as their background, and to the forests of Silesian Beskid mountains with the liverworts appearing in them as their background.

Kruskal

Kruskal [1956] was motivated by Borůvka's first paper and by the application to the traveling salesman problem, described as follows (where [1] refers to Borůvka [1926a]):

> Several years ago a typewritten translation (of obscure origin) of [1] raised some interest. This paper is devoted to the following theorem: If a (finite) connected graph has a positive real number attached to each edge (the *length* of the edge), and if these lengths are all distinct, then among the spanning trees (German: Gerüst) of the graph there is only one, the sum of whose edges is a minimum; that is, the shortest spanning tree of the graph is unique. (Actually in [1] this theorem is stated and proved in terms of the "matrix of lengths" of the graph, that is, the matrix $\|a_{ij}\|$ where a_{ij} is the length of the edge connecting vertices i and j. Of course, it is assumed that $a_{ij} = a_{ji}$ and that $a_{ii} = 0$ for all i and j.) The proof in [1] is based on a not unreasonable method of constructing a spanning subtree of minimum length. It is in this construction that the interest largely lies, for it is a solution to a problem (Problem 1 below) which on the surface is closely related to one version (Problem 2 below) of the well-known traveling salesman problem.
> PROBLEM 1. Give a practical method for constructing a spanning subtree of minimum length.
> PROBLEM 2. Give a practical method for constructing an unbranched spanning subtree of minimum length.
> The construction in [1] is unnecessarily elaborate. In the present paper I give several simpler constructions which solve Problem 1, and I show how one of these constructions may be used to prove the theorem of [1]. Probably it is true that any construction which solves Problem 1 may be used to prove this theorem.

Kruskal described three algorithms: Construction A: iteratively choose the shortest edge that can be added (forest-merging); Construction B: fix a nonempty set U of vertices, and choose iteratively the shortest edge leaving some component intersecting U (a generalization of tree-growing); Construction A': iteratively remove the longest edge that can be removed without making the graph disconnected. He proved that Construction A implies the uniqueness of shortest spanning tree if all lengths are distinct.

In his reminiscences, Kruskal [1997] wrote about Borůvka's method:

> In one way, the method of construction was very elegant. In another way, however, it was unnecessarily complicated. A goal which has always been important to me is to find simpler ways to describe complicated ideas, and that is all I tried to do here. I simplified the construction down to its essence, but it seems to me that the idea of Professor Borůvka's method is still present in my version.

Prim

Prim [1957] gave the following motivation:

> A problem of inherent interest in the planning of large-scale communication, distribution and transportation networks also arises in connection with the current rate structure for Bell System leased-line services.

He described the following algorithm: choose a component of the current forest, and connect it to the nearest component. He observed that Kruskal's constructions A and B are special cases of this.

Prim noticed that in fact only the order of the lengths determines if a spanning tree is shortest:

> The *shortest spanning subtree* of a connected labelled graph also minimizes all increasing symmetric functions, and maximizes all decreasing symmetric functions, of the edge "lengths."

Prim preferred starting at a vertex and growing a tree for computational reasons:

> This computational procedure is easily programmed for an automatic computer so as to handle quite large-scale problems. One of its advantages is its avoidance of checks for closed cycles and connectedness. Another is that it never requires access to more than two rows of distance data at a time — no matter how large the problem.

The implementation described by Prim has $O(n^2)$ running time.

Loberman and Weinberger

Loberman and Weinberger [1957] gave minimizing wire connections as motivation:

> In the construction of a digital computer in which high-frequency circuitry is used, it is desirable and often necessary when making connections between terminals to minimize the total wire length in order to reduce the capacitance and delay-line effects of long wire leads.

They described two methods: tree-growing and forest-merging. Only after they had designed their algorithms, they discovered that their algorithms were given earlier by Kruskal [1956].

> However, it is felt that the more detailed implementation and general proofs of the procedures justify this paper.

They next described how to implement Kruskal's method, in particular, how to merge forests. They also observed that the minimality of a spanning tree depends only on the order of the lengths, and not on their specific values:

> After the initial sorting into a list where the branches are of monotonically increasing length, the actual value of the length of any branch no longer appears explicitly in the subsequent manipulations. As a result, some other parameter such as the square of the length could have been used. More generally, the same minimum tree will persist for all variations in branch lengths that do not disturb the original relative order.

Dijkstra

Dijkstra [1959] gave again the tree-growing method, which he preferred (for computational reasons) above the forest-merging method of Kruskal and Loberman and Weinberger (overlooking the fact that these authors also gave the tree-growing method):

> The solution given here is to be preferred to the solution given by J.B. KRUSKAL [1] and those given by H. LOBERMAN and A. WEINBERGER [2]. In their solutions all the — possibly $\frac{1}{2}n(n-1)$ — branches are first of all sorted according to length. Even if the length of the branches is a computable function of the node coordinates, their methods demand that data for all branches are stored simultaneously. Our method requires the simultaneous storing of the data for at most n branches,
> ...

Dijkstra described an $O(n^2)$ implementation.

Dijkstra [1960] gave the following alternative shortest spanning tree method: order edges arbitrarily, find the first edge that forms a circuit with previous edges; delete the longest edge from this circuit, and continue. (This method was also found by Rosenstiehl [1967].) This generalizes both forest-merging and tree-growing, by choosing the order appropriately.

Further work

Kalaba [1960] proposed the method of first choosing a spanning tree arbitrarily, and next adding, iteratively, an edge and removing the longest edge in the circuit arising.

Kotzig [1961b] gave again Kruskal's Algorithm A' (a referee pointed Kruskal's work out to Kotzig). Kotzig moreover showed that there is a unique minimum spanning tree T if and only if for each edge e not in T, e is the unique longest edge in the circuit in $T \cup \{e\}$.

As mentioned, Graham and Hell [1985] give an extensive survey on the history of the minimum spanning tree (and minimum Steiner tree) problem. See also Nešetřil [1997] for additional notes on the history of the minimum spanning tree problem.

Chapter 51

Packing and covering of trees

The basic facts on packing and covering of trees follow directly from those on matroid union. In this chapter we check what these results amount to in terms of graphs, and we give some more direct algorithms.

51.1. Unions of forests

For any graph $G = (V, E)$ and any partition $\mathcal{P}$ of V, let $\delta(\mathcal{P})$ denote the set of edges connecting distinct classes of $\mathcal{P}$. From the following consequence of the matroid union theorem we will derive other results on tree packing and covering:

Theorem 51.1. *Let $G = (V, E)$ be an undirected graph and let $k \in \mathbb{Z}_+$. Then the maximum size of the union of k forests is equal to the minimum value of*

(51.1) $|\delta(\mathcal{P})| + k(|V| - |\mathcal{P}|)$

taken over all partitions $\mathcal{P}$ of V into nonempty classes.

Proof. This follows directly from the matroid union theorem (Corollary 42.1a) applied to the cycle matroid M of G. Indeed, by Corollary 42.1b, the maximum size of the union of k forests is equal to the minimum value of

(51.2) $|E \setminus F| + k r_M(F),$

where $r_M(F)$ is the maximum size of a forest contained in F. We can assume that each component of (V, F) is an induced subgraph of G. So taking $\mathcal{P}$ equal to the set of components of (V, F), we see that $r_M(F) = |V| - |\mathcal{P}|$, and hence that the minimum of (51.2) is equal to the minimum of (51.1). ∎

51.2. Disjoint spanning trees

Theorem 51.1 has a number of consequences. First we have the following tree packing result of Tutte [1961a] and Nash-Williams [1961b]:

Corollary 51.1a (Tutte-Nash-Williams disjoint trees theorem). *A graph $G = (V, E)$ contains k edge-disjoint spanning trees if and only if*

(51.3) $\quad |\delta(\mathcal{P})| \geq k(|\mathcal{P}| - 1)$

for each partition $\mathcal{P}$ of V into nonempty classes.

Proof. To see necessity of (51.3), each spanning tree contains at least $|\mathcal{P}|-1$ edges in $\delta(\mathcal{P})$. To show sufficiency, it is equivalent to show that there exist $k(|V|-1)$ edges that can be covered by k forests. By Theorem 51.1, this is indeed possible, since

(51.4) $\quad |\delta(\mathcal{P})| + k(|V| - |\mathcal{P}|) \geq k(|\mathcal{P}| - 1) + k(|V| - |\mathcal{P}|) = k(|V| - 1)$

for each partition $\mathcal{P}$ of V into nonempty sets. ∎

Gusfield [1983] observed that the Tutte-Nash-Williams disjoint trees theorem (Corollary 51.1a) implies that each $2k$-edge-connected undirected graph has k edge-disjoint spanning trees (since $|\delta(\mathcal{P})| \geq k|\mathcal{P}| \geq k(|\mathcal{P}|-1)$).

Similarly to the line pursued in Section 42.2, Corollary 51.1a can be formulated equivalently in polyhedral terms:

Corollary 51.1b. *The connector polytope of a graph has the integer decomposition property.*

Proof. Similar to the proof of Corollary 42.1e. ∎

For any connected graph $G = (V, E)$, define the *strength* of G by:

(51.5) $\quad$ strength$(G) := \max\{\lambda \mid \mathbf{1} \in \lambda \cdot P_{\text{connector}}(G)\}$
$= \max\{\sum_T \lambda_T \mid \lambda_T \geq 0, \sum_T \lambda_T \chi^T \leq \mathbf{1}\}$,

where T ranges over the spanning trees of G, and where $\mathbf{1}$ denotes the all-1 vector in $\mathbb{R}^E$.

The Tutte-Nash-Williams disjoint trees theorem is equivalent to: the maximum number of disjoint spanning trees in a graph $G = (V, E)$ is equal to $\lfloor \text{strength}(G) \rfloor$. Similarly, the capacitated version of the Tutte-Nash-Williams theorem is equivalent to the integer rounding property of the system (cf. Section 42.2):

(51.6) $\quad x_e \geq 0 \quad$ for $e \in E$,
$\quad\quad\quad\;\; x(T) \geq 1 \quad$ for each spanning tree T.

51.3. Covering by forests

Dual to Corollary 51.1a is the following forest covering theorem of Nash-Williams [1964], where $E[U]$ denotes the set of edges contained in U. (The theorem is also a consequence of a theorem of Horn [1955] on covering vector sets by linearly independent sets, since each graphic matroid is linear.)

Corollary 51.1c (Nash-Williams' covering forests theorem). *The edge set of a graph $G = (V, E)$ can be covered by k forests if and only if*

(51.7) $\qquad |E[U]| \leq k(|U| - 1)$

for each nonempty subset U of V.

Proof. Since any forest has at most $|U| - 1$ edges contained in U, we have necessity of (51.7). To see sufficiency, notice that (51.7) implies

(51.8) $\qquad |E| - |\delta(\mathcal{P})| = \sum_{U \in \mathcal{P}} |E[U]| \leq \sum_{U \in \mathcal{P}} k(|U| - 1) = k(|V| - |\mathcal{P}|)$

for any partition $\mathcal{P}$ of V into nonempty sets. So $|\delta(\mathcal{P})| + k(|V| - |\mathcal{P}|) \geq |E|$, and hence Theorem 51.1 implies that there exist k forests covering E. ∎

(Nash-Williams [1964] derived Corollary 51.1c from Corollary 51.1a.)

Again, this corollary can be formulated in terms of the integer decomposition property:

Corollary 51.1d. *For any graph G, the forest polytope has the integer decomposition property.*

Proof. Similar to the proof of Corollary 42.1e. ∎

These results are equivalent to: the minimum number of forests needed to cover the edges of a graph $G = (V, E)$ is equal to

(51.9) $\qquad \lceil \min\{\lambda \mid \mathbf{1} \in \lambda \cdot P_{\text{forest}}(G)\} \rceil$,

where $\mathbf{1}$ denotes the all-one vector in $\mathbb{R}^E$. A similar relation holds for the capacitated case, which is equivalent to the integer rounding property of the system:

(51.10) $\qquad x_e \geq 0 \qquad$ for $e \in E$,
$\qquad\qquad\quad x(F) \leq 1 \quad$ for each forest F.

The minimum number of forests needed to cover the edges of a graph G is called the *arboricity* of G.

51.4. Complexity

The complexity results on matroid union in Sections 40.3, 42.3 and 42.4 imply that these packing and covering problems for forests and trees are solvable in polynomial time:

Theorem 51.2. *For any graph $G = (V, E)$, a maximum number of edge-disjoint spanning trees and a minimum number of forests covering E can be found in polynomial time.*

Proof. See Section 42.3.   ∎

Also weighted versions of it can be solved in strongly polynomial time, for instance, finding a maximum-weight union of k forests in a graph. We give in this section some direct proofs.

To study the complexity of the capacitated and fractional cases, we first observe the following auxiliary result, that is (when applied to undirected graphs) at the base of several algorithms on the forest and connector polytopes, and was observed by Rhys [1970], Picard and Ratliff [1975,1978], Picard [1976], Trubin [1978], Picard and Queyranne [1982a], Padberg and Wolsey [1983], and Cunningham [1985a]. (It also follows from the strong polynomial-time solvability of submodular function minimization, but there is an easier direct method.)

Theorem 51.3. *Given a digraph $D = (V, A)$, $x \in \mathbb{Q}_+^A$, $y \in \mathbb{Q}^V$, and disjoint subsets S and T, we can find a set U with $T \subseteq U \subseteq V \setminus S$ minimizing*

(51.11) $x(\delta^{\text{in}}(U)) + y(U)$

in strongly polynomial time.

Proof. Extend D by two new vertices s and t, and arcs (s, v) for $v \in V$ with $y_v > 0$ and (v, t) for $v \in V$ with $y_v < 0$. This gives the digraph $D' = (V \cup \{s, t\}, A')$. Define a capacity function c on A' by:

(51.12) $\begin{aligned} c(u, v) &:= x(u, v) &&\text{for } (u, v) \in A, \\ c(s, v) &:= y_v &&\text{if } (s, v) \in A', \\ c(v, t) &:= -y_v &&\text{for } (v, t) \in A'. \end{aligned}$

Let $\kappa := -c(\delta_{A'}^{\text{in}}(t))$ (the sum of the negative y_v's). Then

(51.13) $c(\delta_{A'}^{\text{in}}(U \cup \{t\})) = x(\delta_A^{\text{in}}(U)) + \sum_{\substack{v \in U \\ y_v > 0}} y_v - \sum_{\substack{v \in V \setminus U \\ y_v > 0}} y_v$

$= x(\delta_A^{\text{in}}(U)) + \sum_{v \in U} y_v - \sum_{\substack{v \in V \\ y_v < 0}} y_v = x(\delta_A^{\text{in}}(U)) + y(U) - \kappa$

for any $U \subseteq V$. Thus minimizing $x(\delta_A^{\text{in}}(U)) + y(U)$ is reduced to finding a minimum-capacity $(S \cup \{s\}) - (T \cup \{t\})$ cut in D'.   ∎

Testing membership and finding most violated inequalities

A first consequence of Theorem 51.3 is that we can test membership, and find a most violated inequality, for the forest polytope (Picard and Queyranne [1982b] (suggested by W.H. Cunningham) and Padberg and Wolsey [1983]).

Corollary 51.3a. *Given a graph $G = (V, E)$ and $x \in \mathbb{Q}_+^E$, we can decide if x belongs to $P_{\text{forest}}(G)$, and if not, find a most violated inequality among (50.11), in strongly polynomial time.*

Proof. Define $y_v := 2 - x(\delta(v))$ for $v \in V$. Then

(51.14) $\quad 2(x(E[U]) - |U|) = \sum_{v \in U} x(\delta(v)) - x(\delta(U)) - 2|U|$

$\qquad\qquad\qquad\qquad = -x(\delta(U)) - y(U).$

So any nonempty $U \subseteq V$ minimizing $x(\delta(U)) + y(U)$, maximizes $x(E[U]) - |U|$. By Theorem 51.3, we can find such a U in strongly polynomial time. If $x(E[U]) \leq |U| - 1$, x belongs to $P_{\text{forest}}(G)$, and otherwise U gives a most violated inequality. ∎

A similar result holds for the up hull of the connector polytope, which we show with a method of Jünger and Pulleyblank [1995]:

Corollary 51.3b. *Given a graph $G = (V, E)$ and $x \in \mathbb{Q}_+^E$, we can find a partition $\mathcal{P}$ of V into nonempty sets minimizing*

(51.15) $\quad x(\delta(\mathcal{P})) - |\mathcal{P}|$

in strongly polynomial time.

Proof. We first construct a vector $y \in \mathbb{Q}^V$, by updating a vector y. Throughout, y satisfies

(51.16) $\quad y(U) \leq x(\delta(U)) - 2$ for each nonempty $U \subseteq V$.

Start with $y_v := -2$ for all $v \in V$. Successively, for each $v \in V$, reset y_v to $y_v + \alpha$, where α is the minimum value of

(51.17) $\quad x(\delta(U)) - 2 - y(U)$

taken over all $U \subseteq V$ containing v. Such a U can be found in strongly polynomial time by Theorem 51.3.

We end up with a y satisfying (51.16). Moreover, each $v \in V$ is contained in some set U with $y(U) = x(\delta(U)) - 2$.

Let $\mathcal{P}$ be the inclusionwise maximal sets U satisfying $y(U) = x(\delta(U)) - 2$. Then $\mathcal{P}$ is a partition of V, since if $T, U \in \mathcal{P}$ and $T \cap U \neq \emptyset$, then (by the submodularity of $x(\delta(Y))$) $y(T \cup U) = x(\delta(T \cup U)) - 2$, and hence $T = U = T \cup U$.

This $\mathcal{P}$ is as required, since for each partition $\mathcal{Q}$ of V into nonempty sets we have

(51.18) $\quad 2x(\delta(\mathcal{Q})) - 2|\mathcal{Q}| = \sum_{U \in \mathcal{Q}} (x(\delta(U)) - 2) \geq \sum_{U \in \mathcal{Q}} y(U) = y(V),$

with equality if $\mathcal{Q} = \mathcal{P}$. ∎

(This method is analogous to calculating the Dilworth truncation as discussed in Theorem 48.4.)

Corollary 51.3b implies for finding the most violated inequality:

Corollary 51.3c. *Given a graph $G = (V, E)$ and $x \in \mathbb{Q}_+^E$, we can decide if x belongs to $P_{\text{connector}}^\uparrow(G)$, and if not, find a most violated inequality among (50.17)(ii), in strongly polynomial time.*

Proof. By Corollary 51.3b, we can find a partition $\mathcal{P}$ of V into nonempty sets, minimizing $x(\delta(\mathcal{P})) - |\mathcal{P}|$. If this value is at least -1, then x belongs to the up hull of the connector polytope, while otherwise $\mathcal{P}$ gives a most violated inequality among (50.17)(ii). ∎

Barahona [1992] showed that membership in the connector polytope can be tested by solving $O(n)$ maximum flow computations (improving Cunningham [1985c]).

Fractional decomposition into trees

By definition, any vector in $P_{\text{forest}}(G)$ or $P_{\text{connector}}(G)$ can be decomposed as a convex combination of incidence vectors of forests or of connectors. These decompositions can be found in strongly polynomial time, a result due to Cunningham [1984] and Padberg and Wolsey [1984] (for the forest polytope).

In order to decompose a vector in the forest polytope as a convex combination of forests, by the following theorem it suffices to have a method to decompose a vector in the spanning tree polytope as a convex combination of spanning trees:

Theorem 51.4. *Given a connected graph $G = (V, E)$ and $x \in P_{\text{forest}}(G)$, we can find a $z \in P_{\text{spanning tree}}(G)$ with $x \leq z$ in strongly polynomial time.*

Proof. We reset x successively for each edge $e = uv$ of G as follows. Reset x_e to $x_e + \alpha$, where α is the largest value such that x remains to belong to $P_{\text{forest}}(G)$. That is, α equals the minimum value of

(51.19) $\quad |U| - 1 - x(E[U]) = |U| - 1 - \frac{1}{2}\sum_{v \in U} x(\delta(v)) + \frac{1}{2}x(\delta(U)),$

taken over subsets U of V with $u, v \in U$. Such a U can be found in strongly polynomial time by Theorem 51.3.

As $P_{\text{forest}}(G) = P_{\text{spanning tree}}^\downarrow(G) \cap \mathbb{R}_+^E$, the final x is a z as required. ∎

Hence, to decompose a vector in the forest polytope, we can do with decomposing vectors in the spanning tree polytope:

Section 51.4. Complexity

Theorem 51.5. *Given a graph $G = (V, E)$ and $y \in P_{\text{spanning tree}}(G)$, we can find spanning trees $T_1, \ldots, T_k$ and $\lambda_1, \ldots, \lambda_k \geq 0$ satisfying*

(51.20) $\qquad y = \lambda_1 \chi^{T_1} + \cdots + \lambda_k \chi^{T_k}$

and $\lambda_1 + \cdots + \lambda_k = 1$, in strongly polynomial time.

Proof. Iteratively resetting y, we keep an integer weight function w such that y maximizes $w^\mathsf{T} y$ over the spanning tree polytope. Initially, $w := \mathbf{0}$. We describe the iteration.

Let T be a spanning tree in G with $T \subseteq \text{supp}(y)$, maximizing $w(T)$. Let $a := y - \chi^T$. If $a = \mathbf{0}$ we stop; then $y = \chi^T$. If $a \neq \mathbf{0}$, let λ be the largest rational such that

(51.21) $\qquad \chi^T + \lambda \cdot a$

belongs to $P^\uparrow_{\text{spanning tree}}(G)$.

We describe an inner iteration to find λ. We iteratively consider vectors y along the halfline $L := \{\chi^T + \lambda \cdot a \mid \lambda \geq 0\}$. Note that the function $w^\mathsf{T} x$ is constant on L. First we let λ be the largest rational such that $\chi^T + \lambda \cdot a$ is nonnegative, and set $z := \chi^T + \lambda \cdot a$.

We iteratively reset z. We check if z belongs to the spanning tree polytope, and if not, we find a constraint among (50.12) most violated by z. That is, we find a nonempty subset U of V minimizing $|U| - 1 - z(E[U])$. Let z' be the vector on L attaining $x(E[U]) \leq |U| - 1$ with equality.

Consider any inequality $x(E[U']) \leq |U'| - 1$ violated by z'. Then

(51.22) $\qquad |U'| - 1 - |T \cap E[U']| < |U| - 1 - |T \cap E[U]|.$

This can be seen by considering the function $d(x) := (|U| - 1 - x(E[U])) - (|U'| - 1 - x(E[U']))$. We have $d(z) \leq 0$ and $d(z') > 0$, and hence, as d is linear, $d(\chi^T) > 0$; that is, we have (51.22). So resetting $z := z'$, there are at most $|V|$ inner iterations.

Let y' be the final z found. Since $\lambda \geq 1$ (as $y \in P_{\text{spanning tree}}(G)$) and $y = \lambda^{-1} \cdot y' + (1 - \lambda^{-1}) \cdot \chi^T$, any convex decomposition of y' into incidence vectors of spanning trees, yields such a decomposition of y. We show that this recursion terminates.

If we apply no iteration, then $\text{supp}(y') \subset \text{supp}(y)$. So replacing y, w by y', w gives a reduction.

If we do at least one iteration, we find a U such that y' satisfies $y'(E[U]) = |U| - 1$ while $|T \cap E[U]| < |U| - 1$. In this case we replace y, w by $y', w' := 2w + \chi^{E[U]}$.

Then y' maximizes $w'^\mathsf{T} x$ over the spanning tree polytope. Indeed, for any x in the spanning tree polytope, we have

(51.23) $\qquad w'^\mathsf{T} x = 2w^\mathsf{T} x + x(E[U]) \leq 2w^\mathsf{T} y + |U| - 1 = 2w^\mathsf{T} y' + y'(E[U])$
$\qquad\qquad = w'^\mathsf{T} y'.$

Moreover, each tree T' maximizing $w'(T')$ also maximizes $w(T')$ (by the greedy method: for any ordering of V for which w' is nondecreasing, also w is nondecreasing). However, T does not maximize $w'(T)$, since $w'(T) = 2w(T) + |T \cap E[U]| < 2w(T) + |U| - 1 = 2w^\mathsf{T}y + |U| - 1 = w'^\mathsf{T}y'$. So the dimension of the face of vectors x maximizing $w'^\mathsf{T}x$ is less than the dimension of the face of vectors x maximizing $w^\mathsf{T}x$.

So the number of iterations is at most $|E|$. This shows that the method is strongly polynomial-time. ∎

Now we can derive from the previous two theorems, an algorithmic result for fractional forest decomposition:

Corollary 51.5a. *Given a graph $G = (V, E)$ and $y \in P_{\text{forest}}(G)$, we can find forests $F_1, \ldots, F_k$ and $\lambda_1, \ldots, \lambda_k \geq 0$ satisfying*

(51.24) $\qquad y = \lambda_1 \chi^{F_1} + \cdots + \lambda_k \chi^{F_k}$

and $\lambda_1 + \cdots + \lambda_k = 1$, in strongly polynomial time.

Proof. We can assume that G is connected, as we can consider each component of G separately. By Theorem 51.4, we can find a $z \in P_{\text{spanning tree}}(G)$ with $y \leq z$ in strongly polynomial time. By Theorem 51.5, we can decompose z as a convex combination of incidence vectors of spanning trees in strongly polynomial time. By restricting the spanning trees to subforests if necessary, we obtain a convex decomposition of y into incidence vectors of forests. ∎

We can proceed similarly for decomposing a vector in the connector polytope. To this end, we show the analogue for connectors of Theorem 51.4:

Theorem 51.6. *Given a graph $G = (V, E)$ and $x \in P^{\uparrow}_{\text{connector}}(G)$, we can find a $z \in P_{\text{spanning tree}}(G)$ with $x \geq z$, in strongly polynomial time.*

Proof. The method described in the proof of Corollary 51.3b gives a vector $y \in \mathbb{Q}^V$ satisfying

(51.25) $\qquad y(U) \leq x(\delta(U)) - 2$ for each nonempty $U \subseteq V$,

and a partition $\mathcal{P}$ of V into nonempty sets with $y(U) = x(\delta(U)) - 2$ for each $U \in \mathcal{P}$. Hence

(51.26) $\qquad y(V) = \sum_{U \in \mathcal{P}} (x(\delta(U)) - 2) = 2x(\delta(\mathcal{P})) - 2|\mathcal{P}| \geq -2$.

By decreasing components of y appropriately, we can achieve that $y(V) = -2$, while maintaining (51.25).

We are going to modify y and x, maintaining (51.25) and $y(V) = -2$. For each $u, v \in V$ with $e = uv \in E$, we do the following. Let α be the minimum value of $x(\delta(U)) - 2 - y(U)$ taken over subsets U of V with $u \in U$, $v \notin U$. So $\alpha \geq 0$. Let $\beta := \min\{x_e, \tfrac{1}{2}\alpha\}$ and reset

(51.27) $x_e := x_e - \beta, y_u := y_u + \beta, y_v := y_v - \beta.$

Then (51.25) is maintained, and the collection $\mathcal{C}$ of subsets U having equality in (51.25) is not reduced. Moreover, in the new situation, $x_e = 0$ or there is a $U \in \mathcal{C}$ with $u \in U$ and $v \notin U$. Also, the new x belongs to $P^{\uparrow}_{\text{connector}}(G)$, as for any partition $\mathcal{Q}$ of V into nonempty sets we have

(51.28) $\sum_{U \in \mathcal{Q}} x(\delta(U)) \geq y(V) + 2|\mathcal{Q}| = 2|\mathcal{Q}| - 2.$

Doing this for each edge e (in both directions), we end up with x, y satisfying (51.25) such that

(51.29)   for all adjacent u, v, if $x_{uv} > 0$, then there is a $U \in \mathcal{C}$ with $u \in U$ and $v \notin U$.

This implies that

(51.30) $y_u = x(\delta(u)) - 2$

for each $u \in V$. Indeed, $\mathcal{C}$ is closed under unions and intersections of intersecting sets. Let U be the smallest set in $\mathcal{C}$ containing u. (This exists, since $V \in \mathcal{C}$.) To show (51.30), we must show $U = \{u\}$. Suppose therefore that $U \neq \{u\}$. By (51.29), there is no edge e connecting u and $U \setminus \{u\}$ with $x_e > 0$. Hence

(51.31) $y(U) = y_u + y(U \setminus \{u\}) \leq x(\delta(u)) - 2 + x(\delta(U \setminus \{u\})) - 2$
           $= x(\delta(U)) - 4 < x(\delta(U)) - 2,$

contradicting the fact that $U \in \mathcal{C}$. This proves (51.30).

Hence

(51.32) $2x(E) = \sum_{u \in V} x(\delta(u)) = y(V) + 2|V| = 2(|V| - 1),$

and so $x(E) = |V| - 1$. This implies that x belongs to the spanning tree polytope. ∎

This implies for fractional connector decomposition:

Corollary 51.6a. *Given a graph* $G = (V, E)$ *and* $x \in P_{\text{connector}}(G)$, *we can find connectors* $C_1, \ldots, C_k$ *and* $\lambda_1, \ldots, \lambda_k \geq 0$ *satisfying*

(51.33) $x = \lambda_1 \chi^{C_1} + \cdots + \lambda_k \chi^{C_k}$

and $\lambda_1 + \cdots + \lambda_k = 1$, *in strongly polynomial time.*

Proof. By Theorem 51.6, we can find a $z \in P_{\text{spanning tree}}(G)$ with $x \geq z$ in strongly polynomial time. By Theorem 51.5, we can decompose z as a convex combination of incidence vectors of spanning trees in strongly polynomial time. This gives a decomposition as required. ∎

Fractionally packing and covering trees and forests

We now consider the problem of finding a maximum fractional packing of spanning trees subject to a given capacity function, and its dual, finding a minimum fractional covering by forests of a given demand function.

Since we have proved above that convex decompositions can be found in strongly polynomial time, we only need to give a method to find the optimum *values* of the fractional packing and covering.

The method is a variant of a 'fractional programming method' initiated by Isbell and Marlow [1956], and developed by Dinkelbach [1967], Schaible [1976], Picard and Queyranne [1982a], Padberg and Wolsey [1984], and Cunningham [1985c].

It implies the following result of Picard and Queyranne [1982a] and Padberg and Wolsey [1984]:

Theorem 51.7. *Given a graph $G = (V, E)$ and $y \in \mathbb{Q}_+^E$, we can find the minimum λ such that $y \in \lambda \cdot P_{\text{forest}}(G)$, in strongly polynomial time.*

Proof. We can assume that y does not belong to the forest polytope. (Otherwise multiply y by a sufficiently large scalar.) Let L be the line through $\mathbf{0}$ and y. We iteratively reset y as follows. Find a nonempty subset U of V minimizing $|U| - 1 - y(E[U])$. Let y' be the vector on L with $|U| - 1 - y'(E[U]) = 0$.

Now if y' violates $x(E[U']) \leq |U'| - 1$ for some U', then $|U'| < |U|$, since the function $d(x) := (|U| - 1 - x(E[U])) - (|U'| - 1 - x(E[U']))$ is nonpositive at y and positive at y', implying that it is positive at $\mathbf{0}$ (as d is linear in x).

We reset $y := y'$ and iterate, until y belongs to $P_{\text{forest}}(G)$. So after at most $|V|$ iterations the process terminates, with a y on the boundary of $P_{\text{forest}}(G)$. Comparing the final y with the original y gives the required λ. ∎

Hence we have for fractional forest covering:

Corollary 51.7a. *Given a graph $G = (V, E)$ and $y \in \mathbb{Q}_+^E$, we can find forests $F_1, \ldots, F_k$ and rationals $\lambda_1, \ldots, \lambda_k \geq 0$ such that*

$$(51.34) \qquad y = \lambda_1 \chi^{F_1} + \cdots + \lambda_k \chi^{F_k}$$

with $\lambda_1 + \ldots + \lambda_k$ minimal, in strongly polynomial time.

Proof. By Theorem 51.7, we can find the minimum value of λ such that y belongs to $\lambda \cdot P_{\text{forest}}(G)$. If $\lambda = 0$, then $y = \mathbf{0}$, and (51.34) is trivial. If $\lambda > 0$, then by Corollary 51.5a we can decompose $\lambda^{-1} \cdot y$ as a convex combination of incidence vectors of forests. This gives the required decomposition of y. ∎

Similar results holds for fractional tree packing (Cunningham [1984, 1985c]). First one has:

Theorem 51.8. *Given a connected graph $G = (V, E)$ and $y \in \mathbb{Q}_+^E$, we can find the maximum λ such that $y \in \lambda \cdot P_{\text{connector}}(G)$, in strongly polynomial time.*

Proof. If supp(y) is not a connector, then $\lambda = 0$. So we may assume that supp(y) is a connector. We can also assume that $y \notin P_{\text{connector}}(G)$. Let L be the line through $\mathbf{0}$ and y. We iteratively reset y as follows. Find a partition $\mathcal{P}$ of V into nonempty sets minimizing $y(\delta(\mathcal{P})) - (|\mathcal{P}| - 1)$ (by Corollary 51.3b). Let y' be the vector on L with $y'(\delta(\mathcal{P})) = |\mathcal{P}| - 1$.

Now if y' violates $x(\delta(\mathcal{P}')) \geq |\mathcal{P}'| - 1$ for some partition $\mathcal{P}'$ of V into nonempty sets, then $|\mathcal{P}'| < |\mathcal{P}|$, since the function $d(x) := (x(\delta(\mathcal{P})) - |\mathcal{P}| + 1) - (x(\delta(\mathcal{P}')) - |\mathcal{P}'| + 1)$ is nonpositive at y and positive at y', implying that it is negative at $\mathbf{0}$ (as d is linear in x).

We reset $y := y'$ and iterate, until y belongs to $P_{\text{connector}}(G)$. So after at most $|V|$ iterations the process terminates, in which case we have a y on the boundary of $P_{\text{connector}}(G)$. Comparing the final y with the original y gives the required λ. ∎

This implies for fractional tree packing:

Corollary 51.8a. *Given a connected graph $G = (V, E)$ and $x \in \mathbb{Q}_+^E$, we can find spanning trees $T_1, \ldots, T_k$ and rationals $\lambda_1, \ldots, \lambda_k \geq 0$ such that*

(51.35) $\quad x \geq \lambda_1 \chi^{T_1} + \cdots + \lambda_k \chi^{T_k}$

with $\lambda_1 + \ldots + \lambda_k$ maximal, in strongly polynomial time.

Proof. By Theorem 51.8, we can find the maximum value of λ such that x belongs to $\lambda \cdot P_{\text{connector}}(G)$. If $\lambda = 0$, we take $k = 0$. If $\lambda > 0$, by Corollary 51.6a we can decompose $\lambda^{-1} \cdot x$ as a convex combination of incidence vectors of connectors. This gives the required decomposition of x. ∎

Integer packing and covering of trees

It is not difficult to derive integer versions of the above algorithms, but they are not strongly polynomial-time, as we round numbers in it. In fact, an integer packing or covering cannot be found in strongly polynomial time, as it would imply a strongly polynomial-time algorithm for testing if an integer k is even (which algorithm does not exist[6]): k is even if and only if K_3 has $\frac{3}{2}k$ spanning trees containing each edge at most k times.

---

[6] For any strongly polynomial-time algorithm with one integer k as input, there is a number L and a rational function $q : \mathbb{Z} \to \mathbb{Q}$ such that if $k > L$, then the output equals $q(k)$. (This can be proved by induction on the number of steps of the algorithm.) However, there do not exist a rational function q and a number L such that for $k > L$, $q(k) = 0$ if k is even, and $q(k) = 1$ if k is odd.

888   Chapter 51. Packing and covering of trees

Weakly polynomial-time algorithms follow directly from the fractional case with the help of the theorems of Nash-Williams and Tutte on disjoint trees and covering forests.

Theorem 51.9. *Given a graph $G = (V, E)$ and $y \in \mathbb{Z}_+^E$, we can find forests $F_1, \ldots, F_t$ and integers $\lambda_1, \ldots, \lambda_t \geq 0$ such that*

(51.36) $\quad y = \lambda_1 \chi^{F_1} + \cdots + \lambda_t \chi^{F_t}$

with $\lambda_1 + \ldots + \lambda_t$ minimal, in polynomial time.

Proof. First find $F_1, \ldots, F_k$ and $\lambda_1, \ldots, \lambda_k$ as in Corollary 51.7a. We can assume that $k \leq |E|$ (by Carathéodory's theorem). Let

(51.37) $\quad y' := \sum_{i=1}^{k}(\lambda_i - \lfloor\lambda_i\rfloor)\chi^{F_i} = y - \sum_{i=1}^{k}\lfloor\lambda_i\rfloor\chi^{F_i}.$

So y' is integer.

Replace each edge e by y'_e parallel edges, making G'. By Theorem 51.2, we can find a minimum number of forests partitioning the edges of G', in polynomial time (as $y'_e \leq |E|$ for each $e \in E$). This gives forests $F_{k+1}, \ldots, F_t$ in G.

Setting $\lambda_i := 1$ for $i = k+1, \ldots, t$, we show that this gives a solution of our problem. Trivially, (51.36) is satisfied (with λ_i replaced by $\lfloor\lambda_i\rfloor$). By Nash-Williams' covering forests theorem (Theorem 51.1c), using (51.37),

(51.38) $\quad t - k \leq \lceil \sum_{i=1}^{k}(\lambda_i - \lfloor\lambda_i\rfloor) \rceil.$

Therefore,

(51.39) $\quad \sum_{i=1}^{t}\lfloor\lambda_i\rfloor = (t-k) + \sum_{i=1}^{k}\lfloor\lambda_i\rfloor \leq \lceil \sum_{i=1}^{k}\lambda_i \rceil,$

proving that the decomposition is optimum. ∎

One similarly shows for tree packing:

Theorem 51.10. *Given a connected graph $G = (V, E)$ and $y \in \mathbb{Z}_+^E$, we can find spanning trees $T_1, \ldots, T_t$ and integers $\lambda_1, \ldots, \lambda_t \geq 0$ such that*

(51.40) $\quad y \geq \lambda_1 \chi^{T_1} + \cdots + \lambda_t \chi^{T_t}$

with $\lambda_1 + \ldots + \lambda_t$ maximal, in polynomial time.

Proof. First find $T_1, \ldots, T_k$ and $\lambda_1, \ldots, \lambda_k$ as in Corollary 51.8a. We can assume that $k \leq |E|$ (by Carathéodory's theorem). Let

(51.41) $\quad y' := \lceil \sum_{i=1}^{k}(\lambda_i - \lfloor\lambda_i\rfloor)\chi^{T_i} \rceil.$

Replace each edge e by y'_e parallel edges, making G'. By Theorem 51.2, we can find a maximum number of edge-disjoint spanning trees in G', in polynomial time (as $y'_e \leq |E|$ for each $e \in E$). This gives spanning trees $T_{k+1}, \ldots, T_t$ in G.

Setting $\lambda_i := 1$ for $i = k+1, \ldots, t$, we show that this gives a solution of our problem. Trivially, (51.40) is satisfied (with λ_i replaced by $\lfloor \lambda_i \rfloor$). By the Tutte-Nash-Williams disjoint trees theorem using (51.41),

$$(51.42) \qquad t - k \geq \lfloor \sum_{i=1}^{k} (\lambda_i - \lfloor \lambda_i \rfloor) \rfloor.$$

Therefore,

$$(51.43) \qquad \sum_{i=1}^{t} \lfloor \lambda_i \rfloor = (t-k) + \sum_{i=1}^{k} \lfloor \lambda_i \rfloor \geq \lfloor \sum_{i=1}^{k} \lambda_i \rfloor,$$

proving that the decomposition is optimum. ∎

51.5. Further results and notes

51.5a. Complexity survey for tree packing and covering

Complexity survey for finding a maximum number of (or k) disjoint spanning trees (∗ indicates an asymptotically best bound in the table):

	$O(m^2 \log n)$	Imai [1983a]
	$O(m^2)$	Roskind and Tarjan [1985] (announced by Tarjan [1976]) *for simple graphs*
∗	$O(m\sqrt{\frac{m}{n}}(m + n \log n) \log \frac{m}{n})$	Gabow and Westermann [1988,1992]
∗	$O(nm \log \frac{m}{n})$	Gabow and Westermann [1988,1992]
∗	$O(kn\sqrt{m} + kn \log n)$	Gabow [1991a] (announced)

Complexity survey for finding a minimum number of forests covering all edges of the graph:

$O(n^4)$	Picard and Queyranne [1982a] (finding the number) *for simple graphs*
$O(n^2 m \log^2 n)$	Picard and Queyranne [1982a] (finding the number) *for simple graphs*

continued

$O(m^2)$	Imai [1983a], Roskind and Tarjan [1985] (announced by Tarjan [1976]) *for simple graphs*
* $O(nm \log n)$	Gabow and Westermann [1988,1992]
$O(m(m(m + n \log n) \log m)^{1/3})$	Gabow and Westermann [1988,1992]
* $O(m^{3/2} \log(n^2/m))$	Gabow [1995b,1998]

Liu and Wang [1988] gave an $O(k^2 n^2 m(m + kn^2))$-time algorithm to find a minimum-weight union F of k edge-disjoint spanning trees in a graph $G = (V, E)$, where E is partitioned into classes $E_1, \ldots, E_t$, such that $a_i \leq |F \cap E_i| \leq b_i$ for each i, given a partition $E_1, \ldots, E_t$ of E and numbers a_i and b_i for all i.

Complexity survey for finding a maximum-size union of k forests:

$O(k^2 n^2)$	Imai [1983a], Roskind and Tarjan [1985] (announced by Tarjan [1976]) *for simple graphs*
$O(k^{3/2} \sqrt{nm(m + n \log n)})$	Gabow and Stallmann [1985]
* $O(k^{3/2} n \sqrt{m + n \log n})$	Gabow and Westermann [1988,1992]
* $O(k^{1/2} m \sqrt{m + n \log n})$	Gabow and Westermann [1988,1992]
* $O(kn^2 \log k)$	Gabow and Westermann [1988,1992]
* $O(\frac{m^2}{k} \log k)$	Gabow and Westermann [1988,1992]

Algorithms for finding a maximum-size union of two forests were given by Kishi and Kajitani [1967,1968,1969] and Kameda and Toida [1973].

Complexity survey for finding a maximum-weight union of k forests:

$O(k^2 n^2 + m \log m)$	Roskind and Tarjan [1985] *for simple graphs*
* $O(kn^2 \log k + m \log m)$	Gabow and Westermann [1988,1992]
* $O(\frac{m^2}{k} \log k + m \log m)$	Gabow and Westermann [1988,1992]

Roskind and Tarjan [1985] (cf. Clausen and Hansen [1980]) gave an $O(k^2 n^2 + m \log m)$-time algorithm for finding a maximum-weight union of k disjoint spanning trees, in a simple graph.

As for **the capacitated case**, the methods given in Section 51.4 indicate that packing and covering problems on forests and trees can be solved by a series of minimum-capacity cut problem (as they reduce to Theorem 51.3). A parametric minimum cut method designed by Gallo, Grigoriadis, and Tarjan [1989] allows to combine several consecutive minimum cut computations, improving the efficiency of the corresponding tree packing and covering problem, as was done by Gusfield [1991].

The published algorithms for integer packing and coverings of trees all are based on rounding the fractional version, not increasing the complexity of the problem,

Section 51.5a. Complexity survey for tree packing and covering

except that rounding is included as an operation. This blocks these algorithms from being strongly polynomial-time: as we saw in Section 51.4, it can be proved that there exists no strongly polynomial-time algorithm for finding an optimum integer packing of spanning trees under a given capacity (similarly, for integer covering by forests).

The following table gives a complexity survey for finding a maximum fractional packing of spanning trees subject to a given integer capacity function c, or a minimum fractional covering by forests subject to a given demand function c. Here it seems that the optimum value can be found faster than an explicit fractional packing or covering. The problems of finding an optimum fractional packing of trees is close to that of finding an optimum fractional covering of forests (or trees), so we present their complexity in one survey.

For any graph $G = (V, E)$ and $c : E \to \mathbb{R}_+$, the *strength* is the maximum value of λ such that c belongs to $\lambda \cdot P_{\text{connector}}(G)$. It is equal to the maximum size of a fractional packing of spanning trees subject to c. The *fractional arboricity* is the minimum value of λ such that c belongs to $\lambda \cdot P_{\text{forest}}(G)$. This is equal to the minimum size of a fractional c-covering by forest.

	$O(nm^8)$	Cunningham [1984]: finding an optimum fractional packing of trees
	$O(nm \cdot \text{MF}(n, n^2))$	Cunningham [1985c]: computing strength
	$O(n^4 m^2 \log^2 C)$	Gabow [1991a] (announced): computing strength
*	$O(n^3 m)$	Gusfield [1991]: computing strength
	$O(nm^2 \log(n^2/m))$	Gusfield [1991]: computing strength
	$O(n^3 \cdot \text{MF}(n, m))$	Trubin [1991]: finding an optimum fractional packing of trees
	$O(n^2 \cdot \text{MF}(n, n^2))$	Barahona [1992]: computing strength
	$O(n^2 \cdot \text{MF}(n, n^2))$	Barahona [1995]: finding optimum fractional packing of trees
*	$O(n \cdot \text{MF}(n, m))$	Cheng and Cunningham [1994]: computing strength
*	$O(n \cdot \text{MF}(n, m))$	Gabow [1995b,1998]: computing strength and fractional arboricity
*	$O(n^3 m \log(n^2/m))$	Gabow and Manu [1995,1998]: finding an optimum fractional packing of trees and an optimum fractional covering by forests
*	$O(n^2 m \log C \log(n^2/m))$	Gabow and Manu [1995,1998]: finding an optimum fractional packing of trees and an optimum fractional covering by forests

Here $\text{MF}(n, m)$ is the complexity of finding a maximum-value $s - t$ flow subject to c in a digraph with n vertices and m arcs, and $C := \|c\|_{\max}$ (assuming c integer).

51.5b. Further notes

A special case of a question asked by A. Frank (cf. Schrijver [1979b], Frank [1995]) amounts to the following:

(51.44) (?) Let $G = (V, E)$ be an undirected graph and let $s \in V$. Suppose that for each vertex $t \neq s$, there exist k internally vertex-disjoint $s-t$ paths. Then G has k spanning trees such that for each vertex $t \neq s$, the $s-t$ paths in these trees are internally vertex-disjoint. (?)

(The spanning trees need not be edge-disjoint — otherwise $G = K_3$ would form a counterexample.) For $k = 2$, (51.44) was proved by Itai and Rodeh [1984,1988], and for $k = 3$ by Cheriyan and Maheshwari [1988] and Zehavi and Itai [1989].

Peng, Chen, and Koh [1991] showed that for any undirected graph $G = (V, E)$ and any $p, k \in \mathbb{Z}_+$, there exist k disjoint forests each with p components if and only if

(51.45) $\quad |\delta(\mathcal{P})| \geq k(|\mathcal{P}| - p)$

for each partition $\mathcal{P}$ of V into nonempty sets. This in fact is the matroid base packing theorem (Corollary 42.1d) applied to the $(|V| - p)$-truncation of the cycle matroid of G.

Theorem 42.10 of Seymour [1998] implies that if the edges of a graph $G = (V, E)$ can be partitioned into k forests and if for each $e \in E$ a subset L_e of $\{1, 2, \ldots\}$ with $|L_e| = k$ is given, then we can partition E into forests $F_1, F_2, \ldots$ such that $j \in L_e$ for each $j \geq 1$ and each $e \in F_j$.

Henneberg [1911] and Laman [1970] characterized those graphs which have, after adding any edge, two edge-disjoint spanning trees. This was extended to k edge-disjoint spanning trees by Frank and Szegő [2001].

Farber, Richter, and Shank [1985] showed the following. Let $G = (V, E)$ be an undirected graph. Let $\mathcal{V}$ be the collection of pairs (T_1, T_2) of edge-disjoint spanning trees T_1 and T_2 in G. Call two pairs (T_1, T_2) and (S_1, S_2) in $\mathcal{V}$ adjacent if $|(T_1 \cup T_2) - (S_1 \cup S_2)| = 2$. Then this determines a connected graph on $\mathcal{V}$.

Cunningham [1985c] gave a strongly polynomial-time algorithm ($O(nm \min\{n^2, m \log n\})$) to find a minimum-cost set of capacities to be added to a capacitated graph so as to create the existence of k edge-disjoint spanning trees; that is, given $G = (V, E)$ and $c, k \in \mathbb{Z}_+^E$, solving

(51.46) $\quad \displaystyle\sum_{e \in E} k(e) x_e$

where $x \in \mathbb{Z}_+^E$ satisfies

(51.47) $\quad (c + x)(\delta(\mathcal{P})) \geq k(|\mathcal{P}| - 1)$

for each partition $\mathcal{P}$ of V into nonempty sets. (It amounts to finding a minimum-cost integer vector in a contrapolymatroid.) Related work can be found in Baïou, Barahona, and Mahjoub [2000].

Chapter 52

Longest branchings and shortest arborescences

We next consider trees in directed graphs. We recall some terminology and facts. Let $D = (V, A)$ be a digraph. A *branching* is a subset B of A such that B contains no undirected circuit and such that for each vertex v there is at most one arc in B entering v. A *root* of B is a vertex not entered by any arc in B. For any branching B, each weak component of (V, B) contains a unique root.

A branching B is called an *arborescence* if the digraph (V, B) is weakly connected; equivalently, if (V, B) is a rooted tree. So each arborescence B has a unique root r. We say that B is *rooted at r*, and we call B an *r-arborescence*. An r-arborescence can be characterized as a directed spanning tree B such that each vertex is reachable in B from r. A digraph $D = (V, A)$ contains an r-arborescence if and only if each vertex of D is reachable from r.

52.1. Finding a shortest r-arborescence

Let be given a digraph $D = (V, A)$, a vertex r, and a length function $l : A \to \mathbb{Q}_+$. We consider the problem of finding a shortest (= minimum-length) r-arborescence.

We cannot apply here the greedy method of starting at the root r and iteratively extending an r-arborescence on a subset U of V, by the shortest arc leaving U. This is shown by the example of Figure 52.1.

The following algorithm was given by Chu and Liu [1965], Edmonds [1967a], and Bock [1971]:

Algorithm to find a shortest r-arborescence. Let $A_0 := \{a \in A \mid l(a) = 0\}$. If A_0 contains an r-arborescence B, then B is a shortest r-arborescence. If A_0 contains no r-arborescence, there is a strong component K of (V, A_0) with $r \notin K$ and with $l(a) > 0$ for each $a \in \delta^{\text{in}}(K)$. Let $\alpha := \min\{l(a) \mid a \in \delta^{\text{in}}(K)\}$. Set $l'(a) := l(a) - \alpha$ if $a \in \delta^{\text{in}}(K)$ and $l'(a) := l(a)$ otherwise.

Find (recursively) a shortest r-arborescence B with respect to l'. As K is a strong component of (V, A_0), we can choose B such that $|B \cap \delta^{\text{in}}(K)| = 1$

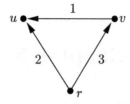

Figure 52.1
In a greedy method one would first choose the shortest arc leaving r, which is (r, u). This arc however is not contained in the shortest r-arborescence.

(since if $|B \cap \delta^{\text{in}}(K)| \geq 2$, then there exists an $a \in B \cap \delta^{\text{in}}(K)$ such that the set $(B \setminus \{a\}) \cup A_0$ contains an r-arborescence, say B', with $l'(B') \leq l'(B) - l'(a) \leq l'(B)$).

Then B is also a shortest r-arborescence with respect to l, since for any r-arborescence B':

(52.1) $\quad l(B') = l'(B') + \alpha|B' \cap \delta^{\text{in}}(K)| \geq l'(B') + \alpha \geq l'(B) + \alpha = l(B).$

Since the number of iterations is at most m (as in each step A_0 increases), we have:

Theorem 52.1. *A shortest r-arborescence can be found in strongly polynomial time.*

Proof. See above. ∎

In fact, direct analysis gives the following result of Chu and Liu [1965], Edmonds [1967a], and Bock [1971]:

Theorem 52.2. *A shortest r-arborescence can be found in time $O(nm)$.*

Proof. First note that there are at most $2n$ iterations. This can be seen as follows. Let k be the number of strong components of (V, A_0), and let k_0 be the number of strong components K of (V, A_0) with $d^{\text{in}}_{A_0}(K) = 0$. Then at any iteration, the number $k + k_0$ decreases. Indeed, if the strong component K selected remains a strong component, then $d^{\text{in}}_{A_0}(K) \neq 0$ in the next iteration; so k_0 decreases. Otherwise, k decreases. Hence there are at most $2n$ iterations.

Next, each iteration can be performed in time $O(m)$. Indeed, in time $O(m)$ we can identify the set U of vertices not reachable in (V, A_0) from r. Next, by Theorem 6.6 one can identify the strong components of the subgraph of (V, A_0) induced by U, in time $O(m)$. Moreover, by Theorem 6.5 we can order the vertices in U pre-topologically. Then the first vertex in this order belongs to a strong component K such that each arc a entering K has $l(a) > 0$. ∎

Tarjan [1977] showed that this algorithm has an $O(\min\{n^2, m \log n\})$-time implementation.

52.1a. r-arborescences as common bases of two matroids

Let $D = (V, A)$ be a digraph and let $r \in V$. The r-arborescences can be considered as the common bases in two matroids on A: M_1 is the cycle matroid of the underlying undirected graph, and M_2 is the partition matroid on A induced by the sets $\delta^{\text{in}}(v)$ for $v \in V \setminus \{r\}$. We assume without loss of generality that no arc of D enters r.

Then the r-arborescences are exactly the common bases of M_1 and M_2. This gives us a reduction of polyhedral and algorithmic results to matroid intersection. In particular, Theorem 52.1 follows from the strong polynomial-time solvability of weighted matroid intersection.

52.2. Related problems

The complexity results of Section 52.1 immediately imply similar results for finding optimum branchings and arborescences without specifying a root. First we note:

Corollary 52.2a. *Given a digraph $D = (V, A)$, $r \in V$, and a length function $l : A \to \mathbb{Q}$, a longest r-arborescence can be found in $O(nm)$ time.*

Proof. Define $L := \max\{l(a) \mid a \in A\}$ and $l'(a) := L - l(a)$ for each $a \in A$. Then an r-arborescence B minimizing $l'(B)$ is an r-arborescence maximizing $l(B)$. ∎

Then we have for longest branching:

Corollary 52.2b. *Given a digraph $D = (V, A)$ and a length function $l \in \mathbb{Q}^A$, a longest branching can be found in time $O(nm)$.*

Proof. We can assume that l is nonnegative, by deleting all arcs of negative length. Extend D by a new vertex r and new arcs (r, v) for all $v \in V$, each of length 0. Let B be a longest r-arborescence in D' (this can be found in $O(nm)$-time by Corollary 52.2a). Then trivially $B \cap A$ is a longest branching in D. ∎

Similarly, for finding a shortest arborescence, without prescribing a root:

Corollary 52.2c. *Given a digraph $D = (V, A)$ and a length function $l \in \mathbb{Q}_+^A$, a shortest arborescence can be found in time $O(nm)$.*

Proof. Extend D by a new vertex r and arcs (r, v) for each $v \in V$, giving digraph D'. Let $l(r, v) := Ln$, where $L := \max\{l(a) \mid a \in A\}$. If D has an arborescence, then a shortest r-arborescence in D' has only one arc leaving r, and deleting this arc gives a shortest arborescence in D. ∎

52.3. A min-max relation for shortest r-arborescences

We now characterize the minimum length of an r-arborescence. Let $D = (V, A)$ be a digraph and let $r \in V$. Call a set C of arcs an r-cut if there exists a nonempty subset U of $V \setminus \{r\}$ with

(52.2) $\quad C = \delta^{\text{in}}(U)$.

It is not difficult to show that

(52.3) the collection of inclusionwise minimal arc sets intersecting each r-arborescence is equal to the collection of inclusionwise minimal r-cuts,

and

(52.4) the collection of inclusionwise minimal arc sets intersecting each r-cut is equal to the collection of r-arborescences.

The following theorem follows directly from the method of Edmonds [1967a], and was stated explicitly by Bock [1971] (and also by Fulkerson [1974]):

Theorem 52.3 (optimum arborescence theorem). *Let $D = (V, A)$ be a digraph, let $r \in V$, and let $l : A \to \mathbb{Z}_+$. Then the minimum length of an r-arborescence is equal to the maximum size of a family of r-cuts such that each arc a is in at most $l(a)$ of them.*

Proof. Clearly, the maximum is not more than the minimum, as each r-cut intersects each r-arborescence.

We prove the reverse inequality by induction on $\sum_{a \in A} l(a)$. Let $A_0 := \{a \in A \mid l(a) = 0\}$. If A_0 contains an r-arborescence, the minimum is 0, while the maximum is at least 0.

If A_0 contains no r-arborescence, there exists a strong component K of the digraph (V, A_0) with $r \notin K$ and with $l(a) > 0$ for each $a \in \delta^{\text{in}}(K)$. Define $l' := l - \chi^{\delta^{\text{in}}(K)}$. By induction there exist an r-arborescence B and r-cuts $C_1, \ldots, C_t$ such that each arc a is in at most $l'(a)$ of the C_i and such that $l'(B) = t$. We may assume that $|B \cap \delta^{\text{in}}(K)| = 1$, since if $|B \cap \delta^{\text{in}}(K)| \geq 2$, then for each $a \in B \cap \delta^{\text{in}}(K)$, $(B \setminus \{a\}) \cup A_0$ contains an r-arborescence, say B', with $l'(B') \leq l'(B) - l'(a) \leq l'(B)$.

It follows that $l(B) = t + 1$. Moreover, taking $C_{t+1} := \delta^{\text{in}}(K)$, each arc a is in at most $l(a)$ of the $C_1, \ldots, C_{t+1}$. ∎

Note that if B is a shortest r-arborescence, then $|B\cap C|=1$ for any r-cut C in the maximum-size family. Moreover, for any $a\in B$, $l(a)$ is equal to the number of r-cuts C chosen with $a\in C$.

52.4. The r-arborescence polytope

Given a digraph $D=(V,A)$ and a vertex $r\in V$, the *r-arborescence polytope* is defined as the convex hull of the incidence vectors (in $\mathbb{R}^A$) of the r-arborescences; that is,

(52.5) $\quad P_{r\text{-arborescence}}(D):=\text{conv.hull}\{\chi^B\mid B\ r\text{-arborescence}\}.$

Theorem 52.3 implies that the r-arborescence polytope of D is determined by:

(52.6)
- (i) $x_a\geq 0$ for $a\in A$,
- (ii) $x(C)\geq 1$ for each r-cut C,
- (iii) $x(\delta^{\text{in}}(v))=1$ for $v\in V\setminus\{r\}$.

To prove this, we first characterize the up hull of the r-arborescence polytope, where as usual the up hull of the r-arborescence polytope is defined as

(52.7) $\quad P^{\uparrow}_{r\text{-arborescence}}(D):=P_{r\text{-arborescence}}(D)+\mathbb{R}^A_+.$

Corollary 52.3a. $P^{\uparrow}_{r\text{-arborescence}}(D)$ *is determined by*

(52.8)
- (i) $x_a\geq 0$ for $a\in A$,
- (ii) $x(C)\geq 1$ for each r-cut C.

Proof. The incidence vector of any r-arborescence trivially satisfies (52.8); hence $P^{\uparrow}_{r\text{-arborescence}}(D)$ is contained in the polyhedron Q determined by (52.8).

Suppose that the reverse inclusion does not hold. Then there exists a rational length function $l\in\mathbb{Q}^A_+$ such that the minimum value of $l^{\top}x$ over Q is less than the minimum length of an r-arborescence. We can assume that l is integer. However, the minimum value of $l^{\top}x$ over Q cannot be less than the maximum described in Theorem 52.3. So we have a contradiction. ∎

Since the r-arborescence polytope is a face of its up hull, this implies:

Corollary 52.3b. *The r-arborescence polytope is determined by* (52.6).

Proof. Directly from Corollary 52.3a. ∎

Corollary 52.3a also implies for the restriction to the unit cube:

Corollary 52.3c. *The convex hull of incidence vectors of arc sets containing an r-arborescence is determined by*

(52.9) (i) $0 \leq x_a \leq 1$ for $a \in A$,
        (ii) $x(C) \geq 1$ for each r-cut C.

Proof. Directly from Corollary 52.3a with Theorem 5.19. ∎

Theorem 52.3 can be reformulated in TDI terms as:

Corollary 52.3d. *System (52.8) is TDI.*

Proof. Choose a length function $l \in \mathbb{Z}_+^A$, and consider the dual problem of minimizing $l^\mathsf{T} x$ over (52.8). For each r-cut C, let y_C be the number of times C is chosen in the maximum family in Theorem 52.3. Moreover, let B be a shortest r-arborescence. Then by Theorem 52.3, $x := \chi^B$ and the y_C form a dual pair of optimum solutions. As the y_C are integer, it follows that (52.8) is TDI. ∎

This in turn implies for the r-arborescence polytope:

Corollary 52.3e. *System (52.6) is TDI.*

Proof. Directly from Corollary 52.3d, with Theorem 5.25, since (52.6) arises from (52.8) by setting some of the inequalities to equality. ∎

For the intersection with the unit cube it gives:

Corollary 52.3f. *System (52.9) is TDI.*

Proof. Directly from Corollary 52.3d, with Theorem 5.23. ∎

In fact, (poly)matroid intersection theory gives the box-total dual integrality of (52.8):

Theorem 52.4. *System (52.8) is box-TDI.*

Proof. Let M_1 be the cycle matroid of the undirected graph underlying $D = (V, A)$, and let M_2 be the partition matroid induced by the sets $\delta^{\text{in}}(v)$ for $v \in V \setminus \{r\}$. By Corollary 46.1d, the system

(52.10) $x(B) \geq |V| - 1 - r_{M_i}(A \setminus B)$ for $i = 1, 2$ and $B \subseteq A$,

is box-TDI. Now any inequality in (52.10) is a nonnegative integer combination of inequalities (52.8).

Indeed, if $i = 1$, then $r_{M_1}(A \setminus B)$ is equal to $|V|$ minus the number of weak components of the digraph $(V, A \setminus B)$. So the inequality in (52.10) states that $x(B)$ is at least the number of weak components of $(V, A \setminus B)$ not containing r.

Hence it is a sum of the inequalities $x(\delta^{\text{in}}(K)) \geq 1$ for each weak component K of $(V, A \setminus B)$ not containing r, and of $x_a \geq 0$ for all $a \in B$ not entering any of these components.

If $i = 2$, then $r_{M_2}(A \setminus B)$ is equal to the number of $v \neq r$ entered by at least one arc in $A \setminus B$. So the inequality in (52.10) states that $x(B)$ is at least the number of $v \neq r$ with $\delta^{\text{in}}(v) \subseteq B$. It therefore is a sum of the inequalities $x(\delta^{\text{in}}(v)) \geq 1$ for these v, and $x_a \geq 0$ for all $a \in B$ not entering any of these vertices.

So Corollary 46.1d implies that (52.8) is box-TDI. ∎

52.4a. Uncrossing cuts

Edmonds and Giles [1977] and Frank [1979b] gave the following procedure of proving that system (52.8) is box-TDI (cf. Corollary 52.3b). The proof is longer than that given above, but it is a special case of a far more general approach (to be discussed in Chapter 60), and is therefore worth noting at this point.

System (52.8) is equivalent to:

(52.11)   (i) $x_a \geq 0$ for $a \in A$,
                (ii) $x(\delta^{\text{in}}(U)) \geq 1$ for $\emptyset \neq U \subseteq V \setminus \{r\}$

Consider any length function $l \in \mathbb{R}_+^A$. Let y_U form an optimum solution to the problem dual to minimizing $l^\top x$ over (52.11):

(52.12)   maximize $\sum_U y_U$
               subject to $y_U \geq 0$ for all U,
                             $\sum_U y_U \chi^{\delta^{\text{in}}(U)} \leq l$,

where U ranges over the nonempty subsets of $V \setminus \{r\}$.

Choose the y_U in such a way that

(52.13) $\sum_U y_U |U||V \setminus U|$

is as small as possible. Then the collection

(52.14) $\mathcal{F} := \{U \mid y_U > 0\}$

is laminar; that is,

(52.15) $U \cap W = \emptyset$ or $U \subseteq W$ or $W \subseteq U$ for all $U, W \in \mathcal{F}$.

For suppose not. Let $\alpha := \min\{y_U, y_W\}$. Decrease y_U and y_W by α, and increase $y_{U \cap W}$ and $y_{U \cup W}$ by α. Then y remains a feasible dual solution, since

(52.16) $\chi^{\delta^{\text{in}}(U \cap W)} + \chi^{\delta^{\text{in}}(U \cup W)} \leq \chi^{\delta^{\text{in}}(U)} + \chi^{\delta^{\text{in}}(W)}$.

Moreover, y remains trivially optimum. However, sum (52.13) decreases (by Theorem 2.1), contradicting our assumption. So $\mathcal{F}$ is laminar.

Now the $\mathcal{F} \times A$ matrix M with

(52.17) $M_{U,a} := \begin{cases} 1 & \text{if } a \in \delta^{\text{in}}(U), \\ 0 & \text{otherwise,} \end{cases}$

is totally unimodular. In fact, it is a network matrix. For make a directed tree T as follows. The vertex set of T is the set $\mathcal{F}' := \mathcal{F} \cup \{V\}$, while for each $U \in \mathcal{F}$ there is an arc a_U from W to U where W is the smallest set in $\mathcal{F}'$ with $W \supset U$. This is in fact an arborescence with root V.

We also define a digraph $\tilde{D} = (\mathcal{F}', \tilde{A})$. For each arc $a = (u,v)$ of D, let $\tilde{a}$ be an arc from the smallest set in $\mathcal{F}'$ containing both u and v, to the smallest set in $\mathcal{F}'$ containing v. Let $\tilde{A} := \{\tilde{a} \mid a \in A\}$.

Identifying any set U in $\mathcal{F}$ with the arc a_U of T, the network matrix generated by directed tree T and digraph $\tilde{D}$ is an $\mathcal{F} \times \tilde{A}$ matrix which is the same as M. So M is totally unimodular. Therefore, by Theorem 5.35, (52.11) is box-TDI.

52.5. A min-max relation for longest branchings

We now consider longest branchings. Characterizing the maximum size of a branching is easy:

Theorem 52.5. *Let $D = (V, A)$ be a digraph. Then the maximum size of a branching is equal to $|V|$ minus the number of strong components K of D with $d_A^{in}(K) = 0$.*

Proof. The theorem follows directly from: (i) each branching has at least one root in any strong component K of D with $d_A^{in}(K) = 0$, and (ii) if a set R intersects each such K, then there is a branching with root set R (since each vertex of D is reachable from R). ∎

From Theorem 52.3 one can derive a min-max relation for the maximum length of a branching in a digraph. The reduction is similar to the reduction of the algorithmic problem of finding a longest branching to that of finding a shortest r-arborescence.

However, a direct proof can be derived from matroid intersection. Consider the system:

(52.18) (i) $x_a \geq 0$ for $a \in A$,
         (ii) $x(\delta^{in}(v)) \leq 1$ for $v \in V$,
         (iii) $x(A[U]) \leq |U| - 1$ for $U \subseteq V$, $U \neq \emptyset$.

Theorem 52.6. *System (52.18) is TDI.*

Proof. Directly from Theorem 41.12, applied to the cycle matroid M_1 of the undirected graph underlying $D = (V, A)$, and the partition matroid M_2 induced by the sets $\delta^{in}(v)$ for $v \in V$. Then each inequality $x(B) \leq r_{M_1}(B)$ is the sum of the inequalities $x(A[U]) \leq |U| - 1$ for the weak components U of (V, B), and $-x_a \leq 0$ for those arcs $a \in A \setminus B$ contained in any weak component of (V, B). Each inequality $x(B) \leq r_{M_2}(B)$ is the sum of the inequalities $x(\delta^{in}(v)) \leq 1$ for those v entered by at least one arc in B, and

$-x_a \leq 0$ for those arcs $a \in A \setminus B$ that enter a vertex v entered by at least one arc in B. ∎

52.6. The branching polytope

The previous corollary immediately implies a description of the *branching polytope* $P_{\text{branching}}(D)$ of D, which is the convex hull of the incidence vectors of branchings in D (stated by Edmonds [1967a]):

Corollary 52.6a. *The branching polytope of $D = (V, A)$ is determined by (52.18).*

Proof. Directly from Theorem 52.6, since the integer solutions of (52.18) are the incidence vectors of the branchings. ∎

Also the following theorem of Edmonds [1967a] follows from matroid intersection theory:

Corollary 52.6b. *Let $D = (V, A)$ be a digraph and let $k \in \mathbb{Z}_+$. Then the convex hull of the incidence vectors of branchings of size k is equal to the intersection of the branching polytope of D with the hyperplane $\{x \mid x(A) = k\}$.*

Proof. This is the common base polytope of the k-truncations of the matroids M_1 and M_2 defined in the proof of Theorem 52.6. ∎

In Corollary 53.3a we shall see that the convex hull of the incidence vectors of branchings of size k has the integer decomposition property (McDiarmid [1983]).

Giles and Hausmann [1979] characterized which pairs of branchings give adjacent vertices of the branching polytope, and Giles [1975,1978b] and Grötschel [1977a] characterized the facets of the branching polytope.

52.7. The arborescence polytope

The results on branchings in the previous section can be specialized to arborescences (without prescribed root). Given a digraph $D = (V, A)$, the *arborescence polytope* of D, denoted by $P_{\text{arborescence}}(D)$, is the convex hull of the incidence vectors of arborescences.

Corollary 52.6c. *The arborescence polytope is determined by*

(52.19) (i) $x_a \geq 0$ for $a \in A$,
         (ii) $x(\delta^{\text{in}}(v)) \leq 1$ for $v \in V$,
         (iii) $x(A[U]) \leq |U| - 1$ for $U \subseteq V$, $U \neq \emptyset$,
         (iv) $x(A) = |V| - 1$.

Proof. Directly from Corollary 52.6a, since $P_{\text{arborescence}}(D)$ is the face of $P_{\text{branching}}(D)$ determined by the hyperplane $x(A) = |V| - 1$. ∎

One similarly obtains from Theorem 52.6 the following, which yields a min-max relation for the minimum length of an arborescence:

Corollary 52.6d. *System* (52.19) *is TDI.*

Proof. From Theorem 52.6, with Theorem 5.25. ∎

52.8. Further results and notes

52.8a. Complexity survey for shortest r-arborescence

	$O(nm)$	Chu and Liu [1965], Edmonds [1967a], Bock [1971]
	$O(n^2)$	Tarjan [1977] (cf. Camerini, Fratta, and Maffioli [1979])
	$O(m \log n)$	Tarjan [1977] (cf. Camerini, Fratta, and Maffioli [1979])
	$O(n \log n + m \log \log \log_{m/n} n)$	Gabow, Galil, and Spencer [1984]
*	$O(m + n \log n)$	Gabow, Galil, Spencer, and Tarjan [1986]

As before, ∗ indicates an asymptotically best bound in the table.

X. Guozhi (see Guan [1979]), Gabow and Tarjan [1979,1984], and Gabow, Galil, Spencer, and Tarjan [1986] studied the problem of finding a shortest r-arborescence with exactly k arcs leaving r, yielding an $O(m+n \log n)$-time algorithm. Hou [1996] gave an $O(k^3 m^3)$-time algorithm to find the k shortest r-arborescences in a digraph.

Gabow and Tarjan [1988a] gave $O(m+n \log n)$- and $O(m \log^* n)$-time algorithms for the *bottleneck* r-arborescence problem (that is, minimizing the maximum arc cost), improving the $O(m \log n)$-time algorithm of Camerini [1978]. (Here $\log^* n$ is the minimum i with $\log_2^{(i)} n \leq 1$.)

52.8b. Concise LP-formulation for shortest r-arborescence

Wong [1984] and Maculan [1986] observed that the problem of finding a shortest r-arborescence can be formulated as a concise linear programming problem. In fact, the dominant $P^\uparrow_{r\text{-arborescence}}(D)$ of the r-arborescence polytope is the projection of a polyhedron in nm dimensions determined by at most $n(2m+n)$ constraints.

Theorem 52.7. *Let $D = (V, A)$ be a digraph and let $r \in V$. Then $P^\uparrow_{r\text{-arborescence}}(D)$ is equal to the set Q of all vectors $x \in \mathbb{R}^A_+$ such that for each $u \in V \setminus \{r\}$ there exists an $r - u$ flow f_u of value 1 satisfying $f_u \leq x$.*

Proof. Since the incidence vector $x = \chi^B$ of any r-arborescence satisfies the constraints, we know that $P^\uparrow_{r\text{-arborescence}}(D)$ is contained in Q.

To see the reverse inclusion, let $x \in Q$. Then for each nonempty subset U of $V \setminus \{r\}$ one has

(52.20) $\quad x(\delta^{\text{in}}(U)) \geq f_u(\delta^{\text{in}}(U)) \geq 1$,

where u is any vertex in U and where f_u is an $r - u$ flow of value 1 with $f_u \leq x$. So by Corollary 52.3a, x belongs to $P^\uparrow_{r\text{-arborescence}}(D)$. ∎

This implies that a shortest r-arborescence can be found by solving a linear programming problem of polynomial size:

Corollary 52.7a. *Let $D = (V, A)$ be a digraph and let $r \in V$ and $l \in \mathbb{R}^A_+$. Then the length of a shortest r-arborescence is equal to the minimum value of*

(52.21) $\quad \displaystyle\sum_{a \in A} l(a) x_a$,

where $x \in \mathbb{R}^A$ is such that for each $u \in V \setminus \{r\}$ there exists an $r - u$ flow f_u of value 1 with $f_u \leq x$.

Proof. Directly from Theorem 52.7. ∎

52.8c. Further notes

Frank [1979b] showed the following. Let $D = (V, A)$ be a digraph and let $r \in V$. Then a subset A' of A is contained in an r-arborescence if and only if $|\mathcal{U}| \leq |V| - 1 - |A'|$ for each laminar collection $\mathcal{U}$ of nonempty subsets of $V \setminus \{r\}$ such that each arc of D enters at most one set in $\mathcal{U}$ and no arc in A' enters any set in $\mathcal{U}$.

Goemans [1992,1994] studied the convex hull of (not necessarily spanning) partial r-arborescences.

Karp [1972a] gave a shortening of the proof of Edmonds [1967a] of the correctness of the shortest r-arborescence algorithm.

Books covering shortest arborescences include Minieka [1978], Papadimitriou and Steiglitz [1982], and Gondran and Minoux [1984].

Chapter 53

Packing and covering of branchings and arborescences

Packing arborescences is a special case of packing common bases in two matroids. However, no general matroid theorem is known that covers this case. In Section 42.6c the maximum number of common bases in two strongly base orderable matroids was characterized, but this does not apply to packing arborescences, as graphic matroids are generally not strongly base orderable. Yet, min-max relations and polyhedral characterizations can be proved for packing arborescences, and similarly for covering by branchings.

53.1. Disjoint branchings

Edmonds [1973] gave the following characterization of the existence of disjoint branchings in a given directed graph $D = (V, A)$. We give the proof of Lovász [1976c]. The *root set* of a branching B is the set of roots of B, that is, the set of sources of the digraph (V, B).

Theorem 53.1 (Edmonds' disjoint branchings theorem). *Let $D = (V, A)$ be a digraph and let $R_1, \ldots, R_k$ be subsets of V. Then there exist disjoint branchings $B_1, \ldots, B_k$ such that B_i has root set R_i (for $i = 1, \ldots, k$) if and only if*

(53.1) $\quad d^{\text{in}}(U) \geq |\{i \mid R_i \cap U = \emptyset\}|,$

for each nonempty subset U of V.

Proof. Necessity being trivial, we show sufficiency, by induction on $|V \setminus R_1| + \cdots + |V \setminus R_k|$. If $R_1 = \cdots = R_k = V$, the theorem is trivial, so we can assume that $R_1 \neq V$. For each $U \subseteq V$, define

(53.2) $\quad g(U) := |\{i \mid R_i \cap U = \emptyset\}|.$

Let W be an inclusionwise minimal set with the properties that $W \cap R_1 \neq \emptyset$, $W \setminus R_1 \neq \emptyset$, and $d^{\text{in}}(W) = g(W)$. Such a set exists, since $W = V$ would qualify.

Then

(53.3) $\quad d^{in}(W \setminus R_1) \geq g(W \setminus R_1) > g(W) = d^{in}(W),$

and hence there exists an arc $a = (u,v)$ in A with $u \in W \cap R_1$ and $v \in W \setminus R_1$. It suffices to show that (53.1) is maintained after resetting $A := A \setminus \{a\}$ and $R_1 := R_1 \cup \{v\}$, since after resetting we can apply induction, and assign a to B_1.

To see that (53.1) is maintained, suppose that to the contrary there is a $U \subseteq V$ violating the condition after resetting. Then in resetting, $d^{in}(U)$ decreases by 1 while $g(U)$ is unchanged. So a enters U, and, before resetting we had $d^{in}(U) = g(U)$ and $U \cap R_1 \neq \emptyset$. This implies (before resetting):

(53.4) $\quad d^{in}(U \cap W) \leq d^{in}(U) + d^{in}(W) - d^{in}(U \cup W)$
$\qquad \leq g(U) + g(W) - g(U \cup W) \leq g(U \cap W).$

So we have equality throughout. Hence $d^{in}(U \cap W) = g(U \cap W)$ and $R_1 \cap (U \cap W) \neq \emptyset$ (as $R_1 \cap W \neq \emptyset$ and $R_1 \cap U \neq \emptyset$, and $g(U \cap W) = g(U) + g(W) - g(U \cup W)$). Also $(U \cap W) \setminus R_1 \neq \emptyset$ (since $v \in U \cap W$) and $U \cap W \subset W$ (as $u \notin U \cap W$). This contradicts the minimality of W. ∎

(Also the method of Tarjan [1974a] is based on the existence of an arc a as in this proof. Fulkerson and Harding [1976] gave another proof of the existence of such ar arc (more complicated than that of Lovász given above).)

53.2. Disjoint r-arborescences

The previous theorem implies a characterization of the existence of disjoint arborescences with prescribed roots:

Corollary 53.1a. *Let $D = (V, A)$ be a digraph and let $r_1, \ldots, r_k \in V$. Then there exist k disjoint arborescences $B_1, \ldots, B_k$, where B_i has root r_i (for $i = 1, \ldots, k$) if and only if each nonempty subset U of V is entered by at least as many arcs as there exist i with $r_i \notin U$.*

Proof. Directly from Edmonds' disjoint branchings theorem (Theorem 53.1) by taking $R_i := \{r_i\}$ for all i. ∎

If all roots are equal, we obtain the following min-max relation, announced by Edmonds [1970b]. Recall that an r-cut is a cut $\delta^{in}(U)$ where U is a nonempty subset of $V \setminus \{r\}$.

Corollary 53.1b (Edmonds' disjoint arborescences theorem). *Let $D = (V, A)$ be a digraph and let $r \in V$. Then the maximum number of disjoint r-arborescences is equal to the minimum size of an r-cut.*

Proof. Directly from Corollary 53.1a by taking k equal to the minimum size of an r-cut and $r_i := r$ for $i = 1, \ldots, k$. ∎

Note that Edmonds' disjoint arborescences theorem implies Menger's theorem: for any digraph $D = (V, A)$ and $r, s \in V$, if k is the minimum size of an $r - s$ cut, we can extend D by k parallel arcs from s to v, for each vertex $v \neq s$; in the extended graph, the minimum size of an r-cut is k, and hence it contains k arc-disjoint r-arborescences. This gives k arc-disjoint $r - s$ paths in the original graph D.

One can reformulate Edmonds' disjoint arborescences theorem in a number of ways (Edmonds [1975]):

Corollary 53.1c. *Let $D = (V, A)$ be a digraph and let $r \in V$. Then for each $k \in \mathbb{Z}_+$ the following are equivalent:*

(53.5)   (i) *there exist k disjoint r-arborescences;*
           (ii) *for each nonempty $U \subseteq V \setminus \{r\}$, $d^{\text{in}}(U) \geq k$;*
           (iii) *for each $s \neq r$ there exist k arc-disjoint $r - s$ paths in D;*
           (iv) *there exist k edge-disjoint spanning trees in the underlying undirected graph such that for each $s \neq r$ there are exactly k arcs entering s covered by these trees.*

Proof. The equivalence of (i) and (ii) follows from Edmonds' disjoint arborescences theorem (Theorem 53.1b), and the equivalence of (ii) and (iii) is a direct consequence of Menger's theorem.

The implication (i) $\Rightarrow$ (iv) is trivial. To prove (iv) $\Rightarrow$ (ii), suppose that (iv) holds, and let U be a nonempty subset of $V \setminus \{r\}$. Each spanning tree has at most $|U| - 1$ arcs contained in U. So the spanning trees of (iv) together have at most $k(|U| - 1)$ arcs contained in U. Moreover, they have exactly $k|U|$ arcs with head in U. Hence, at least k arcs enter U. ∎

An interesting consequence of Edmonds' disjoint arborescences theorem was observed by Shiloach [1979a] and concerns the arc-connectivity of a directed graph:

Corollary 53.1d. *A digraph $D = (V, A)$ is k-arc-connected if and only if for all $s_1, t_1, \ldots, s_k, t_k \in V$ there exist arc-disjoint paths $P_1, \ldots, P_k$, where P_i runs from s_i to t_i $(i = 1, \ldots, k)$.*

Proof. Sufficiency follows by taking $s_1 = \cdots = s_k$ and $t_1 = \cdots = t_k$. To see necessity, extend D by a vertex r and arcs (r, s_i) for $i = 1, \ldots, k$. By Edmonds' disjoint arborescences theorem (Corollary 53.1b), the extended digraph has k disjoint r-arborescences, since each nonempty subset U of V is entered by at least k arcs of D'. Choosing the $s_i - t_i$ path in the r-arborescence containing (r, s_i), for $i = 1, \ldots, k$, we obtain paths as required. ∎

53.3. The capacitated case

The capacitated version of the min-max relation for disjoint r-arborescences reads:

Corollary 53.1e. *Let $D = (V, A)$ be a digraph, let $r \in V$, and let $c \in \mathbb{Z}_+^A$ be a capacity function. Then the minimum capacity of an r-cut is equal to the maximum value of $\sum_B \lambda_B$, where λ_B is a nonnegative integer for each r-arborescence B such that*

(53.6) $\quad \sum_B \lambda_B \chi^B \leq c.$

Proof. Directly from Corollary 53.1b by replacing each arc a by $c(a)$ parallel arcs. ∎

One can equivalently formulate this in term of total dual integrality. To see this, consider the r-cut polytope $P_{r\text{-cut}}(D)$ of D, defined as the convex hull of the incidence vectors of the r-cuts in D. In particular, consider the up hull

(53.7) $\quad P_{r\text{-cut}}^{\uparrow}(D) := P_{r\text{-cut}}(D) + \mathbb{R}_+^A$

of the r-cut polytope.

In Corollary 52.3a we saw that the up hull $P_{r\text{-arborescence}}^{\uparrow}(D)$ of the r-arborescence polytope of D is determined by:

(53.8)   (i) $x_a \geq 0$ for each arc a,
             (ii) $x(C) \geq 1$ for each r-cut C.

By the theory of blocking polyhedra, this implies that $P_{r\text{-cut}}^{\uparrow}(D)$ is determined by:

(53.9)   (i) $x_a \geq 0$ for each arc a,
             (ii) $x(B) \geq 1$ for each r-arborescence B.

In fact:

Corollary 53.1f. *System (53.9) determines $P_{r\text{-cut}}^{\uparrow}(D)$ and is TDI.*

Proof. The first part follows from the theory of blocking polyhedra applied to Corollary 52.3a, and the second part is equivalent to Corollary 53.1e. ∎

Another equivalent form is:

(53.10)   For any digraph $D = (V, A)$ and $r \in V$, the r-arborescence polytope has the integer decomposition property.

By Theorem 5.30, the number of r-arborescences B with $\lambda_B \geq 1$ in Corollary 53.1e can be taken to be at most $2|A| - 1$. (This improves a result of Pevzner [1979a] giving an $O(nm)$ upper bound.) Gabow and Manu [1995, 1998] showed an upper bound of $|V| + |A| - 2$.

53.4. Disjoint arborescences

Frank [1979a,1981c] derived from Corollary 53.1a the following min-max relation for disjoint arborescences without a prescribed root. (A *subpartition* of V is a partition of a subset of V.)

Corollary 53.1g. *Let $D = (V, A)$ be a digraph and let $k \in \mathbb{Z}_+$. Then A contains k disjoint arborescences if and only if*

$$(53.11) \qquad \sum_{U \in \mathcal{P}} d^{\mathrm{in}}(U) \geq k(|\mathcal{P}| - 1)$$

for each subpartition $\mathcal{P}$ of V with nonempty classes.

Proof. Necessity being easy, we show sufficiency. Choose $x \in \mathbb{Z}_+^V$ such that

$$(53.12) \qquad x(U) \geq k - d^{\mathrm{in}}(U)$$

for each nonempty subset U of V, with $x(V)$ as small as possible. We show that $x(V) = k$. Since $x(V) \geq k$ by (53.12), it suffices to show $x(V) \leq k$.

Let $\mathcal{P}$ be the collection of inclusionwise maximal nonempty sets having equality in (53.12). Then $\mathcal{P}$ is a subpartition, for suppose that $U, W \in \mathcal{P}$ with $U \cap W \neq \emptyset$. Then

$$(53.13) \qquad \begin{aligned} x(U \cup W) &= x(U) + x(W) - x(U \cap W) \\ &\leq (k - d^{\mathrm{in}}(U)) + (k - d^{\mathrm{in}}(W)) - (k - d^{\mathrm{in}}(U \cap W)) \\ &\leq (k - d^{\mathrm{in}}(U \cup W)), \end{aligned}$$

and hence $U \cup W \in \mathcal{P}$. So $U = W$.

Now for each $v \in V$ with $x_v > 0$ there exists a set U in $\mathcal{P}$ containing v, since otherwise we could decrease x_v. Hence

$$(53.14) \qquad x(V) = \sum_{U \in \mathcal{P}} x(U) = \sum_{U \in \mathcal{P}} (k - d^{\mathrm{in}}(U)) \leq k,$$

by (53.11).

So $x(V) = k$. Now let $r_1, \ldots, r_k$ be vertices such that any vertex v occurs x_v times among the r_i. Then by Corollary 53.1a there exist disjoint arborescences $B_1, \ldots, B_k$, where B_i has root r_i. This shows the corollary. ∎

53.5. Covering by branchings

Let $A[U]$ denote the set of arcs in A with both ends in U. Frank [1979a] observed that the following min-max relation for covering by branchings can be derived from Edmonds' disjoint arborescences theorem:

Corollary 53.1h. *Let $D = (V, A)$ be a digraph and let $k \in \mathbb{Z}_+$. Then A can be covered by k branchings if and only if*

(53.15)   (i) $\deg^{\text{in}}(v) \leq k$ for each $v \in V$,
(ii) $|A[U]| \leq k(|U| - 1)$ for each nonempty subset U of V.

Proof. Necessity being trivial, we show sufficiency. Extend D by a new vertex r, and for each $v \in V$, $k - \deg^{\text{in}}(v)$ parallel arcs from r to v. Let D' be the digraph thus arising. So each vertex in V is entered by exactly k arcs of D', and D' has $k|V|$ arcs.

Now each nonempty subset U of V is entered by at least k arcs of D', since exactly $k|U|$ arcs have their head in U and at most $k(|U|-1)$ arcs have both ends in U. So by Edmonds' disjoint arborescences theorem (Theorem 53.1b), D' has k disjoint r-arborescences. Since D' has exactly $k|V|$ arcs, these arborescences partition the arc set of D'. Hence restricting them to the arcs of the original graph D, we obtain k branchings partitioning A. ∎

(This was also shown by Markosyan and Gasparyan [1986].)

Corollary 53.1h is equivalent to:

Corollary 53.1i. *Let $D = (V, A)$ be a digraph and let $k \in \mathbb{Z}_+$. Then A can be covered by k branchings if and only if $\deg^{\text{in}}(v) \leq k$ for each $v \in V$ and A can be covered by k forests of the underlying undirected graph.*

Proof. Directly from Corollary 53.1h with Corollary 51.1c. ∎

Corollary 53.1h implies a polyhedral result of Baum and Trotter [1981] (attributing the proof to R. Giles):

Corollary 53.1j. *The branching polytope of a digraph $D = (V, A)$ has the integer decomposition property.*

Proof. Let $k \in \mathbb{Z}_+$ and let x be an integer vector in $k \cdot P_{\text{branching}}(D)$. Let $D' = (V, A')$ be the digraph obtained from D by replacing any arc $a = (u, v)$ by x_a parallel arcs from u to v. Then by Corollary 53.1h, A' can be partitioned into k branchings. This gives a decomposition of x as a sum of the incidence vectors of k branchings in D. ∎

53.6. An exchange property of branchings

We derive an exchange property of branchings from Edmonds' disjoint branchings theorem (Theorem 53.1). It implies that the branchings in an optimum covering can be taken of almost equal size. It will also be used in Section 59.5 on the total dual integrality of the matching forest constraints.

We first show a lemma. For any branching B, let $R(B)$ denote the set of roots of B.

Lemma 53.2α. *Let B_1 and B_2 be branchings partitioning the arc set A of a digraph $D = (V, A)$. Let R_1 and R_2 be sets with $R_1 \cup R_2 = R(B_1) \cup R(B_2)$ and $R_1 \cap R_2 = R(B_1) \cap R(B_2)$. Then A can be split into branchings B_1' and B_2' with $R(B_i') = R_i$ for $i = 1, 2$ if and only if each strong component K of D with $d^{\text{in}}(K) = 0$ intersects both R_1 and R_2.*

Proof. Necessity is easy, since the root set of any branching intersects any strong component K with $d^{\text{in}}(K) = 0$.

To see sufficiency, by Edmonds' disjoint branchings theorem (Theorem 53.1), branchings B_1' and B_2' as required exist if and only if

(53.16) $\quad d^{\text{in}}(U) \geq |\{i \in \{1,2\} \mid U \cap R_i = \emptyset\}|$

for each nonempty $U \subseteq V$. (Actually, Edmonds' theorem gives the existence of disjoint branchings B_1' and B_2' satisfying $R(B_i') = R_i$ for $i = 1, 2$. That $B_1' \cup B_2' = A$ follows from the fact that $|B_1'| + |B_2'| = |B_1| + |B_2|$, as $|R(B_1')| + |R(B_2')| = |R(B_1)| + |R(B_2)|$.)

Suppose that inequality (53.16) does not hold. Then the right-hand side is positive. If it is 2, then U is disjoint from both R_1 and R_2, and hence from both $R(B_1)$ and $R(B_2)$ (since $R_1 \cup R_2 = R(B_1) \cup R(B_2)$), implying that both B_1 and B_2 enter U, and so $d^{\text{in}}(U) \geq 2$.

So the right-hand side is 1, and hence the left-hand side is 0. We can assume that U is an inclusionwise minimal set with this property. It implies that U is a strong component of D. Then by the condition, U intersects both R_1 and R_2, contradicting the fact that the right-hand side in (53.16) is 1. ∎

First, this implies the following exchange property of branchings:

Theorem 53.2. *Let B_1 and B_2 be branchings in a digraph $D = (V, A)$. Let s be a root of B_2 and let r be the root of the arborescence in B_1 containing s. Then D contains branchings B_1' and B_2' satisfying*

(53.17) $\quad B_1' \cup B_2' = B_1 \cup B_1,\ B_1' \cap B_2' = B_1 \cap B_2,$
$\quad \text{and } R(B_1') = R(B_1) \cup \{s\} \text{ or } R(B_1') = (R(B_1) \setminus \{r\}) \cup \{s\}.$

Proof. We may assume that B_1, B_2 partition A, since we can delete all arcs not occurring in $B_1 \cup B_2$, and add parallel arcs for those in $B_1 \cap B_2$. We may also assume that $s \neq r$ (since the theorem is trivial if $s = r$).

Let K be the strong component of D containing s. If no arc of D enters K, then $r \in K$ (as B_1 contains a directed path from r to s), and hence r is not a root of B_2 (as otherwise no arc enters r while K is strongly connected); define $R_1 := (R(B_1) \setminus \{r\}) \cup \{s\}$ and $R_2 := (R(B_2) \setminus \{s\}) \cup \{r\}$.

Alternatively, if some arc of D enters K, define $R_1 := R(B_1) \cup \{s\}$ and $R_2 := R(B_2) \setminus \{s\}$. Then Lemma 53.2α implies that A can be split into branchings B_1' and B_2' with $R(B_i') = R_i$ for $i = 1, 2$. ∎

The lemma also implies that a packing of branchings can be balanced in the following sense:

Theorem 53.3. *Let $D = (V, A)$ be a digraph. If A can be covered by k branchings, then A can be covered by k branchings each of size $\lfloor |A|/k \rfloor$ or $\lceil |A|/k \rceil$.*

Proof. Consider any two branchings B_1, B_2 in the covering which differ in size by at least 2. Consider the digraph $D' = (V, B_1 \cup B_2)$. We can find subsets R_1 and R_2 of V with $R_1 \cup R_2 = R(B_1) \cup R(B_2)$ and $R_1 \cap R_2 = R(B_1) \cap R(B_2)$, such that each strong component K of D' with $d_{D'}^{\text{in}}(K) = 0$ intersects both R_1 and R_2, and such that R_1 and R_2 differ by at most 1 in size. (We can first include, for any such component K, one element in $K \cap R(B_1)$ in R_1, and one element in $K \cap R(B_2)$ in R_2; next we distribute the remaining elements in $R(B_1)$ and $R(B_2)$ almost equally over R_1 and R_2).

Then, by Lemma 53.2α, $B_1 \cup B_2$ can be partitioned into branchings B_1' and B_2' with $R(B_i') = R_i$ for $i = 1, 2$. Then B_1' and B_2' differ by at most 1 in size. Replacing B_1 and B_2 in the covering by B_1' and B_2', and iterating this, we end up with a covering by k branchings, any two of which differ in size by at most 1. This is a covering as required. ∎

This theorem implies the integer decomposition property of the convex hull of branchings of size k (McDiarmid [1983]):

Corollary 53.3a. *Let $D = (V, A)$ be a digraph and let $k \in \mathbb{Z}_+$. Then the convex hull of the incidence vectors of the branchings of size k has the integer decomposition property.*

Proof. Choose $p \in \mathbb{Z}_+$, and let x be an integer vector in $p \cdot \text{conv.hull}\{\chi^B \mid B \text{ branching}, |B| = k\}$. By Corollary 53.1j, x is a sum of the incidence vectors of p branchings. Let $D' = (V, A')$ be the digraph arising from D by replacing any arc a by x_a parallel arcs. Then A' can be partitioned into p branchings. Now $|A'|/p = x(A)/p = k$. So, by Theorem 53.3, we can take these branchings all of size k. Hence x is the sum of the incidence vectors of p branchings each of size k. ∎

53.7. Covering by r-arborescences

Vidyasankar [1978a] proved the following covering analogue of Edmonds' disjoint branchings theorem. (A weaker version was shown by Frank [1979a] (cf. Frank [1979b]).) For any digraph $D = (V, A)$ and $U \subseteq V$, let $H(U)$ denote the set of outneighbours of $V \setminus U$; that is, the set of the heads of the arcs entering U. So $H(U) \subseteq U$.

Theorem 53.4. *Let $D = (V, A)$ be a digraph, let $r \in V$, and let $k \in \mathbb{Z}_+$. Then A can be covered by k r-arborescences if and only if*

(53.18) $\qquad \deg^{\text{in}}(v) \leq k$ *for each* $v \in V$, *and* $\deg^{\text{in}}(r) = 0$,

and

(53.19) $\qquad \displaystyle\sum_{v \in H(U)} (k - \deg^{\text{in}}(v)) \geq k - d^{\text{in}}(U)$

for each nonempty subset U of $V \setminus \{r\}$.

Proof. Necessity of (53.18) is trivial. To see necessity of (53.19), let U be a nonempty subset of $V \setminus \{r\}$. Then each r-arborescence B intersects the set

(53.20) $\qquad \displaystyle\bigcup_{v \in H(U)} \delta^{\text{in}}(v) \setminus \delta^{\text{in}}(U)$

in at most $|H(U)| - 1$ arcs, since at least one arc of B should enter U. Hence if A can be covered by k r-arborescences, the size of set (53.20) is at most $k(|H(U)| - 1)$, implying (53.19).

To see sufficiency, we can assume that for any arc a of D, if we would add a parallel arc to a, then (53.18) or (53.19) is violated (since deleting parallel arcs does not increase the minimum number of r-arborescences needed to cover the arcs).

If $\deg^{\text{in}}(v) = k$ for each vertex $v \neq r$, then A can be decomposed into k r-arborescences by Edmonds' disjoint arborescences theorem (Corollary 53.1b), since then (53.19) implies that $d^{\text{in}}(U) \geq k$ for each nonempty subset U of $V \setminus \{r\}$.

So we can assume that there exists a vertex $u \neq r$ with $\deg^{\text{in}}(u) < k$. Consider the collection $\mathcal{C}$ of nonempty subsets U of $V \setminus \{r\}$ having equality in (53.19) and with $u \in H(U)$. Then $\mathcal{C}$ is closed under taking union and intersection. Indeed, let U and W be in $\mathcal{C}$. Then

(53.21) $\qquad \displaystyle\sum_{v \in H(U \cap W)} (k - \deg^{\text{in}}(v)) + \sum_{v \in H(U \cup W)} (k - \deg^{\text{in}}(v))$
$\qquad \leq \displaystyle\sum_{v \in H(U)} (k - \deg^{\text{in}}(v)) + \sum_{v \in H(W)} (k - \deg^{\text{in}}(v))$
$\qquad = (k - d^{\text{in}}(U)) + (k - d^{\text{in}}(W))$
$\qquad \leq (k - d^{\text{in}}(U \cap W)) + (k - d^{\text{in}}(U \cup W))$.

The first inequality follows from

(53.22) $\qquad H(U \cap W) \cap H(U \cup W) \subseteq H(U) \cap H(W)$ and
$\qquad H(U \cap W) \cup H(U \cup W) \subseteq H(U) \cup H(W)$,

as one easily checks.

By (53.19), (53.21) implies that we have equality throughout, As we have equality in the first inequality in (53.21), and as $k - \deg^{\text{in}}(u) > 0$, we know that $u \in H(U \cap W) \cap H(U \cup W)$. So $U \cap W$ and $U \cup W$ belong to $\mathcal{C}$.

Now for each arc a entering u, if we would add an arc parallel to a, (53.19) is violated for some U. This implies that for each arc a entering u there exists a $U \in \mathcal{C}$ such that the tail of a is in U. We can take for U the largest set in $\mathcal{C}$. Hence for each arc a entering u, the tail of a is in U. This contradicts the fact that $u \in H(U)$. ∎

Frank [1979b] showed the following consequence of this result:

Corollary 53.4a. *Let $D = (V, A)$ be a digraph, let $r \in V$, and let $k \in \mathbb{Z}_+$. Then A can be covered by k r-arborescences if and only if*

(53.23) $\quad k \cdot s(A') \geq |A'|$

for each $A' \subseteq A$. Here $s(A')$ denotes the maximum of $|B \cap A'|$ over r-arborescences B.

Proof. As necessity is trivial, we show sufficiency, by showing that (53.23) implies (53.18) and (53.19). To see (53.18), apply (53.23) to $A' := \delta^{\text{in}}(v)$. To see (53.19), apply (53.23) to A' equal to the set (53.20). ∎

Note that for *acyclic* digraphs, the minimum number of r-arborescences needed to cover all arcs is easily characterized (Vidyasankar [1978a]):

Theorem 53.5. *Let $D = (V, A)$ be an acyclic digraph and let $r \in V$. Then A can be covered by k r-arborescences if and only if r is the only source of D and each indegree is at most k.*

Proof. Necessity being easy, we show sufficiency. Trivially, we can cover A by sets $B_1, \ldots, B_k$ such that each B_i enters each $v \neq r$ precisely once. As D is acyclic, each B_i is an r-arborescence. ∎

53.8. Minimum-length unions of k r-arborescences

Let $D = (V, A)$ be a digraph, let $r \in V$, and let $k \in \mathbb{Z}_+$. Consider the following system in the variable $x \in \mathbb{R}^A$:

(53.24)   (i) $x_a \geq 0$ for each $a \in A$,
            (ii) $x(\delta^{\text{in}}(U)) \geq k$ for each nonempty $U \subseteq V \setminus \{r\}$.

The following basic result of Frank [1979b] follows from Theorem 52.4.

Theorem 53.6. *System (53.24) is box-TDI.*

Proof. Directly from Theorem 52.4, since if a system $Ax \leq b$ is box-TDI, then for any $k \geq 0$, the system $Ax \leq k \cdot b$ is box-TDI. ∎

This theorem has several consequences. First consider the system

(53.25) (i) $0 \leq x_a \leq 1$ for each $a \in A$,
(ii) $x(\delta^{\text{in}}(U)) \geq k$ for each nonempty $U \subseteq V \setminus \{r\}$.

The following (cf. Frank [1979b]) implies a min-max relation for the minimum length of the union of k disjoint r-arborescences in D:

Corollary 53.6a. *System (53.25) is TDI, and determines the convex hull of subsets of A containing k disjoint r-arborescences.*

Proof. Directly from Theorem 53.6 and Edmonds' disjoint arborescences theorem (Corollary 53.1b). ∎

Another consequence of Theorem 53.6 is as follows. Let $D = (V, A)$ and $D' = (V, A')$ be digraphs, let $r \in V$, and let $k \in \mathbb{Z}_+$. Consider the system in the variable $x \in \mathbb{R}^A$:

(53.26) (i) $x_a \geq 0$ for each $a \in A$,
(ii) $x(\delta_A^{\text{in}}(U)) \geq k - d_{A'}^{\text{in}}(U)$ for each nonempty $U \subseteq V \setminus \{r\}$.

Then:

Corollary 53.6b. *System (53.26) is box-TDI.*

Proof. Choose $d, c \in \mathbb{Z}_+^A$. We must show that the system

(53.27) (i) $d(a) \leq x_a \leq c(a)$ for each $a \in A$,
(ii) $x(\delta_A^{\text{in}}(U)) \geq k - d_{A'}^{\text{in}}(U)$ for each nonempty $U \subseteq V \setminus \{r\}$

is TDI. Let $D'' = (V, A'')$ be the digraph with $A'' := A \cup A'$ (taking arcs multiple if they occur both in A and A'). By Theorem 53.6, the following system in the variable $x \in \mathbb{R}^{A''}$ is TDI:

(53.28) (i) $d(a) \leq x_a \leq c(a)$ for each $a \in A$,
(ii) $1 \leq x_a \leq 1$ for each $a \in A'$,
(iii) $x(\delta_{A''}^{\text{in}}(U)) \geq k$ for each nonempty $U \subseteq V \setminus \{r\}$.

This implies the total dual integrality of (53.27) by Corollary 5.27a. ∎

Frank [1979a] derived the following 'rank' formula for coverings by k r-arborescences:

Corollary 53.6c. *Let $D = (V, A)$ be a digraph, let $r \in V$, and let $A' \subseteq A$. Then the maximum number of arcs in A' that can be covered by k r-arborescences is equal to the minimum value of*

(53.29) $\quad k(|V| - 1) + \sum_{i=1}^{t} (d_{A'}^{\text{in}}(V_i) - k),$

where $V_1, \ldots, V_t$ is a laminar collection of nonempty subsets of $V \setminus \{r\}$ such that each arc in A enters at most one of these sets.

Section 53.8. Minimum-length unions of k r-arborescences 915

Proof. Let μ be the maximum number of arcs in A' that can be covered by k r-arborescences. Consider the system (in $x \in \mathbb{R}^A$)

(53.30) (i) $x_a \geq 0$ for $a \in A$,
(ii) $x(\delta_A^{\text{in}}(U)) \geq k - d_{A'}^{\text{in}}(U)$ for each nonempty $U \subseteq V \setminus \{r\}$.

By Corollary 53.6b, this system is TDI. Let x be an integer vector attaining the minimum of $x(A)$ over (53.30). Then

(53.31) $\mu = k(|V| - 1) - x(A)$.

Indeed, by (53.30) and by Edmonds' disjoint arborescences theorem, there exist k r-arborescences $B_1, \ldots, B_k$ with

(53.32) $x + \chi^{A'} \geq \chi^{B_1} + \cdots + \chi^{B_k}$.

Let A'' be the set of arcs in A' covered by no B_i. By the minimality of $x(A)$, we have that $x(A) + |A'| = k(|V| - 1) + |A''|$. As $\mu \geq |A'| - |A''|$ we have $\geq$ in (53.31). Since we can reverse this construction (starting from a set of k r-arborescences covering μ arcs in A', and making x), we have the equality in (53.31).

By the total dual integrality of (53.30), $x(A)$ is equal to the maximum value of

(53.33) $\sum_{i=1}^{t}(k - d_{A'}^{\text{in}}(V_i))$,

taken over nonempty subsets $V_1, \ldots, V_t$ of $V \setminus \{r\}$ such that each arc in A enters at most one of these sets. If, say, $V_1 \cap V_2 \neq \emptyset$ and $V_1 \not\subseteq V_2 \not\subseteq V_1$, we can replace V_1 and V_2 by $V_1 \cap V_2$ and $V_1 \cup V_2$ without violating these conditions. Such replacements terminate by Theorem 2.1. We end up with $V_1, \ldots, V_t$ laminar as required. Therefore, with (53.31) we have the corollary. ∎

Taking $A' = A$, we get (Frank [1979b]):

Corollary 53.6d. *Let $D = (V, A)$ be a digraph and let $r \in V$. Then the maximum number of arcs that can be covered by k r-arborescences is equal to the minimum value of*

(53.34) $k(|V| - 1) + \sum_{i=1}^{t}(d^{\text{in}}(V_i) - k)$,

where $V_1, \ldots, V_t$ form a laminar collection of nonempty subsets of $V \setminus \{r\}$ such that each arc enters at most one of these sets.

Proof. This is the case $A' = A$ in Corollary 53.6c. ∎

This directly implies a min-max characterization for the minimum number of r-arborescences needed to cover all arcs. However, Theorem 53.4 gives a stronger relation.

As for unions of k branchings, Frank [1979a] derived from Corollary 53.6c:

Corollary 53.6e. *Let $D = (V, A)$ be a digraph and let $k \in \mathbb{Z}_+$. The maximum number of arcs of D that can be covered by k branchings is equal to the minimum value of*

(53.35) $$k(|V| - |\mathcal{P}|) + \sum_{U \in \mathcal{P}} d^{\text{in}}(U)$$

taken over all subpartitions $\mathcal{P}$ of V with nonempty classes.

Proof. Let D' be the digraph obtained from D by adding a new vertex r and arcs (r, v) for each $v \in V$. Then the maximum number of arcs of D that can be covered by k branchings in D is equal to the maximum number of arcs in A that can be covered by k r-arborescences in D'. So Corollary 53.6c gives a min-max relation for this.

The subsets V_i form a subpartition of V since if V_i and V_j would intersect in a vertex v say, then the arc (r, v) of D' enters two sets among the V_i, contradicting the condition. ∎

As for unions of k arborescences without prescribed root, Frank [1979a] derived:

Corollary 53.6f. *Let $D = (V, A)$ be a digraph and let $k \in \mathbb{Z}_+$. Then A can be covered by k arborescences if and only if*

(53.36) $$k(|V| - 1 + \lambda) \geq |A| + \sum_{i=1}^{t}(k - d^{\text{in}}(V_i))$$

for each laminar family $(V_1, \ldots, V_t)$ of nonempty sets such that no arc enters more than one of the V_i. Here λ denotes the maximum number of V_i's having nonempty intersection.

Proof. Necessity can be seen as follows. Let A be covered by arborescences $B_1, \ldots, B_k$. For each $v \in V$, let $r(v)$ be the number of B_i having v as root. So $r(V) = k$. For each $a \in A$, let $s(a)$ be the number of B_i containing a. So $s(a) \geq 1$ for each $a \in A$. Moreover, $s(\delta^{\text{in}}(V_i)) + r(V_i) \geq k$ for each i. Hence

(53.37) $$|A| + \sum_{i=1}^{t}(k - d^{\text{in}}(V_i)) \leq |A| + \sum_{i=1}^{t}\left(r(V_i) + s(\delta^{\text{in}}(V_i)) - d^{\text{in}}(V_i)\right)$$

$$\leq |A| + \sum_{a \in A}(s(a) - 1) + \sum_{i=1}^{t} r(V_i) = \sum_{a \in A} s(a) + \sum_{i=1}^{t} r(V_i)$$

$$= k(|V| - 1) + \sum_{i=1}^{t} r(V_i) \leq k(|V| - 1) + k\lambda.$$

The second inequality holds as each arc enters at most one of the V_i. For the last inequality, we use that $r(V) = k$ and that $V_1, \ldots, V_t$ can be partitioned into λ collections, each consisting of disjoint sets. (53.37) shows necessity.

To see sufficiency, extend D by a vertex r and by the arc set $A' := \{(r, v) \mid v \in V\}$, yielding the digraph $D' = (V \cup \{r\}, A \cup A')$. Consider the following constraints for $x \in \mathbb{R}^{A \cup A'}$:

(53.38) (i) $x_a \geq 0$ for each $a \in A \cup A'$,
(ii) $x(\delta^{\text{in}}_{D'}(U)) \geq k - d^{\text{in}}_D(U)$ for each nonempty $U \subseteq V$,
(iii) $x(\delta^{\text{in}}_{D'}(V)) = k$.

Let x attain the minimum of $x(A)$ over (53.38). Since system (53.38) is TDI by Corollary 53.6b (with Theorem 5.25), we can assume that x is integer. We show

(53.39) $\quad x(A) = k(|V| - 1) - |A|$.

First, $x(A) \geq k(|V| - 1) - |A|$, since

(53.40) $\quad x(A) + |A| + k = x(A) + |A| + x(\delta^{\text{in}}_{A'}(V))$
$= \sum_{v \in V} \left(x(\delta^{\text{in}}_{A'}(v)) + x(\delta^{\text{in}}_{A}(v)) + d^{\text{in}}_A(v) \right) \geq k|V|.$

To see the reverse inequality, $x(A)$ is equal to the optimum value μ of the problem dual to the above minimization problem: maximize

(53.41) $\quad \sum_{U \in \mathcal{P}(V) \setminus \{\emptyset\}} z_U(k - d^{\text{in}}_A(U))$

where $z \in \mathbb{R}_+^{\mathcal{P}(V) \setminus \{\emptyset\}}$ such that

(53.42) $\quad \sum_U z_U \chi^{\delta^{\text{in}}_{D'}(U)} \leq \chi^A.$

So we should prove that $\mu \leq k(|V| - 1) - |A|$.

Now let $\mathcal{U}$ be the collection of nonempty proper subsets U of V with $z_U = 1$. We may assume that $\mathcal{U}$ is laminar. Let λ be the maximum number of $U \in \mathcal{U}$ containing any vertex. Then (53.42) implies that $\lambda \leq -z_V$ (since $\chi^A(a) = 0$ for each $a = (r, v)$). Hence

(53.43) $\quad \mu = k \cdot z_V + \sum_{U \in \mathcal{U}} (k - d^{\text{in}}_A(U)) \leq -k\lambda + \sum_{U \in \mathcal{U}} (k - d^{\text{in}}_A(U))$
$\leq k(|V| - 1) - |A|$

by (53.36), and we have the required inequality. This proves (53.39).

Then the vector $y := x + \chi^A$ satisfies:

(53.44) $\quad y(\delta^{\text{in}}_{D'}(U)) \geq k$ for each nonempty $U \subseteq V$,
$y(\delta^{\text{in}}_{D'}(V)) = k$,
$y(A \cup A') = x(A) + x(A') + |A| = k(|V| - 1) - |A| + k + |A| = k|V|.$

So by Edmonds' disjoint arborescences theorem (Corollary 53.1b), y is the sum of the incidence vectors of k r-arborescences, each with exactly one arc leaving r. Hence (by the definition of y) A can be covered by k arborescences. ∎

53.9. The complexity of finding disjoint arborescences

By Edmonds' disjoint arborescences theorem, the maximum number of disjoint r-arborescences can be calculated in polynomial time, just by determining the minimum size of an r-cut. This can be done by determining, for each $v \in V \setminus \{r\}$, the minimum size of an $r - v$ cut, and taking the minimum of these values.

Lovász [1976c] and Tarjan [1974a] showed that actually also a maximum collection of disjoint r-arborescences can be found in polynomial time.

The proof (due to Lovász [1976c]) of Theorem 53.1 described above gives such a polynomial-time algorithm. In fact, Lovász observed that it implies the following result (obtained also by Tarjan [1974a]). Call a subset B of the arc set A of a digraph $D = (V, A)$ a *partial r-arborescence* if B is an r-arborescence for the subgraph of D induced by the set $V(B)$ of vertices covered by B. We take $V(B) := \{r\}$ if B is empty.

Theorem 53.7. *Given a digraph $D = (V, A)$ and a vertex $r \in V$, a maximum number k of disjoint r-arborescences can be found in time $O(k^2 m^2)$.*

Proof. First, the number k can be determined in time $O(knm)$. Since k is equal to the minimum size of a cut $d^{\text{in}}(U)$ over nonempty subsets U of $V \setminus \{r\}$, we can determine for each $v \in V \setminus \{r\}$ a maximum set of arc-disjoint $r - v$ paths, by the augmenting path method described in Section 9.2. Actually, for $i = 1, \ldots, k$, we determine the ith augmenting paths for all $v \in V \setminus \{r\}$, before searching for the $(i+1)$th augmenting paths. In this way we can stop if for some $v \in V \setminus \{r\}$ no augmenting path exists. So in total we do at most $(n-1)(k+1)$ augmenting path searches. Thus it takes $O(knm)$ time to determine k.

Next, we can find an r-arborescence B such that

(53.45) $\quad d^{\text{in}}_{A \setminus B}(U) \geq k - 1$ for each nonempty $U \subseteq V \setminus \{r\}$,

in time $O(km^2)$. This recursively implies the theorem.

To find B, as in the proof of Theorem 53.1, we can grow a partial r-arborescence B satisfying (53.45), starting with $B = \emptyset$. By the proof of Theorem 53.1, if $V(B) \neq V$, there exists an arc a leaving $V(B)$ such that resetting $B := B \cup \{a\}$ maintains (53.45). For any given arc a leaving $V(B)$ it amounts to testing if there exists a set $U \subseteq V \setminus \{r\}$ such that $a \in \delta^{\text{in}}(U)$ and $d^{\text{in}}_{A \setminus B}(U) = k - 1$. This can be done in $O(km)$ time with a minimum cut algorithm.

Now it is important to observe that for each arc a we need to do this test at most once: if the test result is negative, then in growing B we never have to consider arc a anymore; if the result is positive, a is added to B, and again we will not consider a again.

So to obtain an r-arborescence, we determine at most m minimum cuts, and so finding the r-arborescence B takes $O(km^2)$ time. ∎

Tong and Lawler [1983] observed that the following quite easily follows from Edmonds' disjoint arborescences theorem:

Theorem 53.8. *Given a digraph $D = (V, A)$ and a vertex $r \in V$, we can find in time $O(knm)$ a set of arcs that is the union of a maximum number k of disjoint r-arborescences.*

Proof. As in the proof of Theorem 53.7 we can determine the number k in time $O(knm)$. Now consider any vertex $v \in V$. Find k arc-disjoint $r - v$ paths in D, and delete from D each arc entering v that is on none of these paths. After that we still have $d^{\text{in}}(U) \geq k$ for any nonempty $U \subseteq V \setminus \{r\}$, since if $v \notin U$, then no arc entering U has been deleted, and if $v \in U$, then k arcs entering U are maintained, as after deletion there are still k arc-disjoint $r - v$ paths in D.

Doing this successively for all vertices $v \in V$, we are left with a digraph D with $\deg^{\text{in}}(v) = k$ if $v \neq r$ and $\deg^{\text{in}}(r) = 0$, and with $d^{\text{in}}(U) \geq k$ for each nonempty $U \subseteq V \setminus \{r\}$. So the remaining arc set is the union of k disjoint r-arborescences. As k arc-disjoint $r - v$-paths can be found in time $O(km)$, we have the required result. ∎

This implies with Theorem 53.7 a sharpening of Theorem 53.7:

Corollary 53.8a. *Given a digraph $D = (V, A)$ and a vertex $r \in V$, a maximum number k of disjoint r-arborescences can be found in time $O(knm + k^4n^2)$.*

Proof. By Theorem 53.8, we can find in time $O(knm)$ a set A' that is the union of k disjoint r-arborescences. So $m' := |A'| = k(n - 1)$. Then by Theorem 53.7 we can find k disjoint r-arborescences in A', in time $O(k^2 m'^2)$. Since $O(k^2 m'^2) = O(k^4 n^2)$, the corollary follows. ∎

Tong and Lawler [1983] in fact showed that the method of Lovász [1976c] has an $O(k^2 nm)$-time implementation, yielding with Theorem 53.8 an $O(knm + k^3n^2)$-time algorithm for finding k disjoint r-arborescences.

Also the capacitated case can be solved in strongly polynomial time (Gabow [1991a,1995a]), as can be shown with the help of Edmonds' disjoint branchings theorem. (Pevzner [1979a] proved that it can be solved in *semi-strongly polynomial time*, that is, by taking rounding as one arithmetic step.)

Theorem 53.9. *Given a digraph $D = (V, A)$, $r \in V$, and a capacity function $c : A \to \mathbb{Z}_+$, we can find r-arborescences $B_1, \ldots, B_k$ and integers $\lambda_1, \cdots, \lambda_k \geq 0$ with $\sum_{i=1}^{k} \lambda_i \chi^{B_i} \leq c$ and with $\sum_{i=1}^{k} \lambda_i$ maximized, in strongly polynomial time.*

Proof. We can find the maximum value in strongly polynomial time, as it is equal to the minimum capacity of an r-cut. To find the λ_i explicitly, we show more generally that the following problem is solvable in strongly polynomial time (where $R(B)$ denotes the set of roots of B):

(53.46) given: a digraph $D = (V, A)$, a capacity function $c : A \to \mathbb{Z}_+$, a collection $\mathcal{R}$ of nonempty subsets of V, and a demand function $d : \mathcal{R} \to \mathbb{Z}_+$,
find: a collection $\mathcal{B}$ of branchings and a function $\lambda : \mathcal{B} \to \mathbb{Z}_+$, with $\sum_{B \in \mathcal{B}} \lambda_B \chi^B \leq c$ and $\sum(\lambda_B \mid B \in \mathcal{B}, R(B) = R) = d(R)$ for each $R \in \mathcal{R}$.

For any $U \subseteq V$, define

(53.47) $g(U) := \sum(d(R) \mid R \in \mathcal{R}, R \cap U = \emptyset)$.

By replacing each arc a by $c(a)$ parallel arcs, it follows from Edmonds' disjoint branchings theorem (Theorem 53.1) that a necessary and sufficient condition for the existence of a solution of (53.46) is that

(53.48) $c(\delta^{\text{in}}(U)) \geq g(U)$

for each nonempty $U \subseteq V$.

We can assume that $c(a) > 0$ for each $a \in A$ and $d(R) > 0$ for each $R \in \mathcal{R}$, and that we have an $R_1 \in \mathcal{R}$ with $R_1 \neq V$.

We may also assume that (53.46) has a solution. This implies that there exists an arc $a = (u, v) \in A$ leaving R_1 and a $\mu \geq 1$ such that resetting $d(R_1) := d(R_1) - \mu$, $d(R_1 \cup \{v\}) := d(R_1 \cup \{v\}) + \mu$, $c(a) := c(a) - \mu$, maintains feasibility of (53.46). (If $R_1 \cup \{v\}$ did not belong to $\mathcal{R}$, we add it to $\mathcal{R}$.) We apply this for the maximum possible μ. This value of μ can be calculated in strongly polynomial time, as it satisfies

(53.49) $\mu = \min\{c(a), \min\{c(\delta^{\text{in}}(W)) - g(W) \mid a \in \delta^{\text{in}}(W), W \cap R_1 \neq \emptyset\}\}$

(for the original c and g).

To minimize $c(\delta^{\text{in}}(W)) - g(W)$ over W with $a \in \delta^{\text{in}}(W)$ and $W \cap R_1 \neq \emptyset$, add, for each $R \in \mathcal{R}$, a new vertex v_R and, for each $v \in R$, a new arc (v_R, v) of capacity $d(R)$. Moreover, add a new vertex r, and for each $R \in \mathcal{R}$, a new arc (r, v_R) of capacity $d(R)$. Let $D' = (V', A')$ be the extended digraph. With a minimum cut algorithm we can find a subset W' of $V' \setminus \{r\}$ with $a \in \delta^{\text{in}}(W)$

and $W \cap R_1 \neq \emptyset$, minimizing the capacity of $\delta_{A'}^{\text{in}}(W)$. Then $W := W' \cap V$ is a set as required.

We next apply the algorithm recursively. This describes the algorithm.

Running time. In each iteration, the number of arcs a with $c(a) > 0$ decreases or the collection $\mathcal{C} := \{U \subseteq V \mid U \neq \emptyset, c(\delta^{\text{in}}(U)) = g(U)\}$ increases. As $\mathcal{C}$ is an intersecting family, the number of times $\mathcal{C}$ increases is at most $|V|^3$ (since for each $v \in V$, the collection $\mathcal{C}_v := \{U \in \mathcal{C} \mid v \in U\}$ is a lattice family, and since each lattice family $\mathcal{L}$ is determined by the preorder $\preceq$ given by: $s \preceq t \iff$ each set in $\mathcal{L}$ containing t contains s; if $\mathcal{L}$ increases, then $\preceq$ decreases, which can happen at most $|V|^2$ times.)

So the number of iterations is at most $|A| + |V|^3$. ∎

With the reductions given earlier, this implies that the capacitated versions of packing arborescences and covering by branchings also can be solved in strongly polynomial time.

Edmonds [1975] observed that matroid intersection and union theory implies:

Theorem 53.10. *Given a digraph $D = (V, A)$, $r \in V$, $k \in \mathbb{Z}_+$, and a length function $l \in \mathbb{Q}^A$, we can find k disjoint r-arborescences $B_1, \ldots, B_k$ minimizing $l(B_1) + \cdots + l(B_k)$ in strongly polynomial time.*

Proof. This follows, with Corollary 53.1c and Theorem 53.7, from Theorem 41.8 applied to the intersection of two matroids: one being the union of k times the cycle matroid of the undirected graph underlying D; the other being the matroid in which a subset B of A is independent if and only if any $v \in V \setminus \{r\}$ is entered by at most k arcs in B. ∎

This implies:

Corollary 53.10a. *Given a digraph $D = (V, A)$, $r \in V$, $k \in \mathbb{Z}_+$, and a length function $l \in \mathbb{Q}^A$, we can find a minimum-length subset B of A with $\delta_B^{\text{in}}(U) \geq k$ for each nonempty $U \subseteq V \setminus \{r\}$ in strongly polynomial time.*

Proof. Directly from Theorem 53.10, with Edmonds' disjoint arborescences theorem (Corollary 53.1b). ∎

53.10. Further results and notes

53.10a. Complexity survey for disjoint arborescences

Finding k disjoint r-arborescences in an uncapacitated digraph (∗ indicates an asymptotically best bound in the table):

$O(k^2 m^2)$	Lovász [1976c], Tarjan [1974a]
$O(knm + k^3 n^2)$	Tong and Lawler [1983]
* $O(k^2 n^2 + m)$	Gabow [1991a,1995a]

(As noticed by Tong and Lawler [1983], the paper of Shiloach [1979a] claiming an $O(k^2 nm)$ bound, contains an essential error (the set A constructed on page 25 of Shiloach [1979a] need not have the desired properties: it is maximal under the condition that $y \notin A$, while it should be maximal under the condition that $A \cup V(T) \neq V$).)

The $O(k^2 m^2)$ bound for finding k pairwise disjoint r-arborescences implies the $O(n^2 \Delta^4 \log \Delta)$ bound of Markosyan and Gasparyan [1986] for finding a minimum number of branchings covering all arcs (where Δ is the maximum indegree of the vertices), by the construction given in the proof of Corollary 53.1h (as we can take $m \leq n\Delta$ and $k \leq 2\Delta$).

Tarjan [1974c] gave an $O(m + n \log n)$-time algorithm to find two disjoint r-arborescences (actually, to find two r-arborescences with smallest intersection). This was improved to $O(m\alpha(m,n))$ by Tarjan [1976] (where $\alpha(m,n)$ is the inverse Ackermann function), and to $O(m)$ by Gabow and Tarjan [1985].

Clearly, each of the bounds in the table above implies a complexity bound for the capacitated case, by replacing arcs by multiple arcs. However, this can increase the number m of arcs dramatically, and does not lead to a polynomial-time algorithm. Better bounds are given in the following table:

$O(n^3 \cdot \mathrm{MF}(n, m))$	Pevzner [1979a] *taking rounding as one arithmetic step*
* $O(k^2 n^2 + m)$	Gabow [1991a,1995a]
* $O(n^3 m \log \frac{n^2}{m})$	Gabow and Manu [1995,1998]
* $O(n^2 m \log C \log \frac{n^2}{m})$	Gabow and Manu [1995,1998]

In these bounds, m is the number of arcs in the original graph, $\mathrm{MF}(n,m)$ denotes the time needed to solve a maximum flow problem in a digraph with n vertices and m arcs, and C is the maximum capacity (for integer capacity function).

The bounds of Gabow [1991a,1995a] and Gabow and Manu [1995,1998] in these tables also apply to the problem considered in Edmonds' disjoint branching theorem (Theorem 53.1): finding k disjoint branchings $B_1, \ldots, B_k$ where B_i has a given root set R_i ($i = 1, \ldots, k$), finding minimum coverings by branchings, and related problems. Gabow and Manu [1995,1998] also gave an $O(n^3 m \log \frac{n^2}{m})$ fractional packing algorithm of r-arborescences.

Gabow [1991a,1995a] announced $O(kn(m+n\log n)\log n)$- and $O(k\sqrt{n\log n}(m + kn\log n)\log(nK))$-time algorithms to find a minimum-cost union of k disjoint r-arborescences (where K is the maximum cost, with integer cost function).

53.10b. Arborescences with roots in given subsets

Let $D = (V, A)$ be a digraph. Call a vector $x \in \mathbb{Z}_+^V$ a *root vector* if there exist disjoint arborescences such that for each $v \in V$, exactly x_v of these arborescences have root v. By Corollary 53.1a, root vectors are the integer solutions of the following system:

(53.50)   (i) $x_v \geq 0$ for $v \in V$,
             (ii) $x(U) \leq d^{\text{out}}(U)$ for each $U \subset V$.

This system generally does not define an integer polytope P, as is shown by the digraph with vertices u, v, w and arcs $(u, v), (v, w),$ and (w, u), where $\frac{1}{2} \cdot \mathbf{1}$ is in P, but each integer vector x in P satisfies $\mathbf{1}^\mathsf{T} x \leq 1$.

Moreover, sets R of vertices for which there exist $|R|$ disjoint arborescences, rooted at distinct vertices in R, do not form the independent sets of a matroid, as is shown by the graph of Figure 53.1.

Figure 53.1

However, for any $k \in \mathbb{Z}_+$, the system

(53.51) $x(U) \geq k - d^{\text{in}}(U)$ for each nonempty $U \subseteq V$,

is box-TDI, since the right-hand side in (ii) is intersecting supermodular (cf. Sections 44.5 and 48.1).

Cai [1983] proved the following result, with a method (described below) of Frank [1981c] for proving a special case (Corollary 53.11a):

Theorem 53.11. *Let $D = (V, A)$ be a digraph such that D has k arc-disjoint arborescences. Let $l, u \in \mathbb{Z}_+^V$ with $l \leq u$. Then D has k arc-disjoint arborescences such that, for each $v \in V$, at least $l(v)$ and most $u(v)$ of these arborescences are rooted at v if and only if*

(53.52) $u(U) + d^{\text{in}}(U) \geq k$ and $l(U) + \sum_{W \in \mathcal{P}} (k - d^{\text{in}}(W)) \leq k$

for each nonempty subset U of V and each partition $\mathcal{P}$ of $V \setminus U$ into nonempty sets.

Proof. Necessity being easy, we show sufficiency. Choose $x \in \mathbb{Z}_+^V$ such that $l \leq x \leq u$ and such that (53.51) holds, with $x(V)$ as small as possible. (Such an x exists since $u(U) \geq k - d^{\text{in}}(U)$ for each nonempty subset U of V.)

We show that $x(V) = k$. Since $x(V) \geq k$ by (53.51), it suffices to show $x(V) \leq k$. Let $\mathcal{P}$ be the collection of inclusionwise maximal sets having equality in (53.51). Then $\mathcal{P}$ is a subpartition, for suppose that $U, W \in \mathcal{P}$ with $U \cap W \neq \emptyset$. Then

(53.53) $x(U \cup W) = x(U) + x(W) - x(U \cap W)$
                  $\leq (k - d^{\text{in}}(U)) + (k - d^{\text{in}}(W)) - (k - d^{\text{in}}(U \cap W))$
                  $\leq (k - d^{\text{in}}(U \cup W))$,

and hence $U \cup W \in \mathcal{P}$. So $U = W$.

Now for each $v \in V$ with $x_v > l(v)$ there exists a set W in $\mathcal{P}$ containing v, since otherwise we could decrease x_v. Hence

$$(53.54) \quad x(V) - l(V) = \sum_{W \in \mathcal{P}} (x(W) - l(W)) = \sum_{W \in \mathcal{P}} (k - d^{\text{in}}(W) - l(W))$$
$$\leq k - l(V),$$

by (53.52).

So $x(V) = k$. Now let $r_1, \ldots, r_k$ be vertices such that each vertex v occurs x_v times among the r_i. Then by Corollary 53.1a there exist disjoint arborescences $B_1, \ldots, B_k$, where B_i has root r_i. This shows the theorem. ∎

(In this proof we did not use the box-total dual integrality of (53.51), but we applied a similar argument.)

This has as special case the following result of Frank [1981c]:

Corollary 53.11a. *Let $D = (V, A)$ be a digraph such that D has k arc-disjoint arborescences. Let $u \in \mathbb{Z}_+^V$. Then D has k arc-disjoint arborescences such that, for each $v \in V$, at most $u(v)$ of these arborescences have their root in v if and only if*

$$(53.55) \quad u(U) + d^{\text{in}}(U) \geq k$$

for each nonempty subset U of V.

Proof. Directly from Theorem 53.11. ∎

A related theorem is:

Theorem 53.12. *Let $D = (V, A)$ be a digraph and let $R_1, \ldots, R_k$ be subsets of V. Then there exist disjoint arborescences $B_1, \ldots, B_k$, where B_i has its root in R_i (for $i = 1, \ldots, k$) if and only if*

$$(53.56) \quad \sum_{U \in \mathcal{P}} (k - d^{\text{in}}(U)) \leq |\{i \mid R_i \cap \bigcup \mathcal{P} \neq \emptyset\}|$$

for each subpartition $\mathcal{P}$ of V with nonempty classes.

Proof. Necessity is easy, since if the B_i exist, with roots $r_i \in R_i$, then for each $U \in \mathcal{P}$ one has that $r_i \in U$ or B_i contains at least one arc entering U. That is,

$$(53.57) \quad |\{r_i\} \cap U| + d_{B_i}^{\text{in}}(U) \geq 1.$$

Summing this inequality over $U \in \mathcal{P}$ and over $i = 1, \ldots, k$ we obtain (53.56), with R_i replaced by $\{r_i\}$. This implies (53.56) for the original R_i.

To see sufficiency, first observe that the condition implies that the R_i are nonempty (by taking $\mathcal{P} := \{V\}$). If the R_i are singletons, the theorem is equivalent to Corollary 53.1a. So we can assume that $|R_1| \geq 2$. Choose distinct vertices $u, w \in R_1$.

If the condition is maintained after replacing R_1 by $R_1 \setminus \{u\}$, the theorem follows by induction. So we can assume that this violates the condition. That is, there exists a subpartition $\mathcal{P}$ of V into nonempty classes such that (setting $X := \bigcup \mathcal{P}$):

(53.58) $$\sum_{U \in \mathcal{P}} (k - d^{\text{in}}(U)) = |\{i \mid R_i \cap X \neq \emptyset\}|$$

and such that $X \cap R_1 = \{u\}$ (for the original R_1). Similarly we can assume that there exists a subpartition $\mathcal{Q}$ of V into nonempty classes such that (setting $Y := \bigcup \mathcal{Q}$):

(53.59) $$\sum_{U \in \mathcal{Q}} (k - d^{\text{in}}(U)) = |\{i \mid R_i \cap Y \neq \emptyset\}|$$

and such that $Y \cap R_1 = \{w\}$.

Let $\mathcal{F}$ be the union of $\mathcal{P}$ and $\mathcal{Q}$ (any set occurring both in $\mathcal{P}$ and in $\mathcal{Q}$ occurs twice in $\mathcal{F}$). Now iteratively replace any $T, U \in \mathcal{F}$ with $T \cap U \neq \emptyset$ and $T \not\subseteq U \not\subseteq T$ by $T \cap U$ and $T \cup U$. Then the final family $\mathcal{F}$ is laminar. Let $\mathcal{R}$ be the collection of inclusionwise minimal sets in $\mathcal{F}$ and let $\mathcal{S}$ be the collection of inclusionwise maximal sets in $\mathcal{F}$. Then $\mathcal{R}$ and $\mathcal{S}$ are subpartitions of V into nonempty classes, and $\bigcup \mathcal{R} = X \cap Y$ and $\bigcup \mathcal{S} = X \cup Y$. Moreover

(53.60) $$\sum_{U \in \mathcal{R}} (k - d^{\text{in}}(U)) + \sum_{U \in \mathcal{S}} (k - d^{\text{in}}(U)) = \sum_{U \in \mathcal{F}} (k - d^{\text{in}}(U))$$
$$\geq \sum_{U \in \mathcal{P}} (k - d^{\text{in}}(U)) + \sum_{U \in \mathcal{Q}} (k - d^{\text{in}}(U))$$
$$= |\{i \mid R_i \cap X \neq \emptyset\}| + |\{i \mid R_i \cap Y \neq \emptyset\}|$$
$$> |\{i \mid R_i \cap (X \cap Y) \neq \emptyset\}| + |\{i \mid R_i \cap (X \cup Y) \neq \emptyset\}|.$$

The first inequality follows from the submodularity of $d^{\text{in}}(U)$. The last inequality holds as (i) if R_i intersects $X \cup Y$, then it intersects X or Y, (ii) if R_i intersects $X \cap Y$, then it intersects X and Y, and (iii) R_1 intersects X and Y but not $X \cap Y$, since $R_1 \cap X = \{u\}$ and $R_1 \cap Y = \{w\}$.

However, (53.60) contradicts (53.56). ∎

53.10c. Disclaimers

The equivalence of (i) and (iii) in Corollary 53.1c suggests the following question, raised by A. Frank (cf. Schrijver [1979b], Frank [1995]; it generalizes a similar question for the undirected case, described in Section 51.5b):

(53.61) (?) Let $D = (V, A)$ be a k-arc-connected digraph and let $r \in V$. Suppose that for each $s \in V$ there exist k internally vertex-disjoint $r - s$ paths in D. Then there exist k r-arborescences such that, for any vertex s, the k $r - s$ paths determined by the respective r-arborescences are internally vertex-disjoint. (?)

For $k = 2$ this was proved by Whitty [1987]. However, for $k = 3$, a counterexample was found by Huck [1995].

Two potential generalizations of Edmonds' disjoint arborescences theorem have been raised, neither of which holds however. For vertices s, t, let $\lambda(s, t)$ denote the maximum number of arc-disjoint $s - t$ paths. It is not true that for any digraph $D = (V, A)$, $r \in V$, and $T \subseteq V \setminus \{r\}$, there exist k disjoint subsets $A_1, \ldots, A_k$ of A such that each A_i contains an $r - t$ path for each $t \in T$ if and only if $\lambda(r, t) \geq k$ for each $t \in T$ (see Figure 53.2, for $k = 2$).

N. Robertson raised the question if it is true that in any digraph $D = (V, A)$ and any $r \in V$, there exist partial r-arborescences $B_1, B_2, \ldots$ such that each vertex $v \in V \setminus \{r\}$ is in exactly $\lambda(r, v)$ of them. Lovász [1973b] showed that Figure 53.3 is a counterexample. (Related work is reported by Bang-Jensen, Frank, and Jackson [1995] and Gabow [1996].)

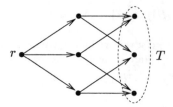

Figure 53.2

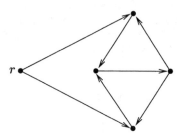

Figure 53.3

53.10d. Further notes

Frank [1981c] gave the following results for mixed graphs. Let $G = (V, E, A)$ be a mixed graph (that is, (V, E) is an undirected graph and (V, A) is a directed graph). A *mixed branching* is a subset B of $E \cup A$ such that the undirected edges in B can be oriented such that B becomes a branching. Then E can be covered by k mixed branchings if and only if

(53.62)   (i) $d_A^{\text{in}}(U) + |E[U]| \leq k|U|$ for each $U \subseteq V$,
           (ii) $|A[U]| + |E[U]| \leq k(|U| - 1)$ for each $\emptyset \neq U \subseteq V$.

Similarly, a *mixed r-arborescence* is a subset B of $E \cup A$ such that the undirected edges in B can be oriented such that B becomes an r-arborescence. Then for any $r \in V$, G has k disjoint mixed r-arborescences if and only if for each subpartition $\mathcal{P}$ of $V \setminus \{r\}$ with nonempty classes, the number of edges (directed or not) entering any class of $\mathcal{P}$, is at least $k|\mathcal{P}|$.

Cai [1989] characterized when, for given digraphs $D_1 = (V, A_1)$ and $D_2 = (V, A_2)$, $a, b \in \mathbb{Z}_+^V$, and $k \in \mathbb{Z}_+$, there exists an $r \in \mathbb{Z}_+^V$ with $a \leq r \leq b$ and there exist, for $i = 1, 2$, k disjoint arborescences in D_i such that for each $v \in V$, $r(v)$ of these arborescences have root v. This can be proved using polymatroid intersection theory, in particular the box-total dual integrality of

(53.63) $x(U) \geq k - d_{A_i}^{\text{in}}(U)$ for $i = 1, 2$ and nonempty $U \subseteq V$

(Theorem 48.5). (For a generalization, see Cai [1990a,1993].)

Cai [1990b] showed, for given digraph $D = (V, A)$, $r \in V$, $d, c \in \mathbb{Z}_+^A$, and $k \in \mathbb{Z}_+$: there exist k r-arborescences such that each arc is covered at least $d(a)$ times and at most $c(a)$ times, if and only if $d(\delta^{\text{in}}(v)) \leq k$, for each $v \in V \setminus \{r\}$, and

(53.64) $$\sum_{v \in U} \min\{k - d(\delta^{\text{in}}(v) \setminus \delta^{\text{in}}(U)), c(\delta^{\text{in}}(v) \cap \delta^{\text{in}}(U))\} \geq k$$

for each nonempty subset U of $V \setminus \{r\}$.

Chapter 54

Biconnectors and bibranchings

The concept of biconnector is a generalization of that of a connector. Let $G = (V, E)$ be an undirected graph and let V be partitioned into classes R and S. An $R-S$ *biconnector* is a subset F of E such that each component of (V, F) intersects both R and S. So contracting R or S gives a connector. If R is a singleton, $R-S$ biconnectors are precisely the connectors.

For biconnectors, min-max relations, polyhedral characterizations, and complexity results similar to those for connectors hold.

In this chapter we also consider the forest analogue of biconnector, the biforest. An $R-S$ *biforest* is a forest F such that each component of (V, F) has at most one edge in the cut $\delta(R)$. So contracting R or S gives a forest. Also biforests show good polyhedral and algorithmical behaviour.

Similar results hold for the directed analogues of biconnectors and biforests, the bibranchings and the bifurcations. An $R-S$ *bibranching* is a set B of arcs such that for each $s \in S$, B contains an $R-s$ path and for each $r \in R$, B contains an $r-S$ path. Bibranchings form a generalization of arborescences, and give rise to similar min-max relations and polyhedral characterizations. An $R-S$ *bifurcation* is a set B of arcs containing no undirected circuit, such that each vertex in R is left by at most one arc in B, each vertex in S is entered by at most one arc in B, and B contains no arcs from S to R.

Theorem 54.11 on disjoint bibranchings will be the only result of this chapter that will be used later in this book, namely in Chapter 56 to obtain a dual form of the Lucchesi-Younger theorem, on packing directed cut covers in a source-sink connected digraph. The proof of Theorem 54.11 uses no other results from this chapter.

54.1. Shortest $R-S$ biconnectors

Let $G = (V, E)$ be a graph and let V be partitioned into two sets R and S. A subset F of E is called an $R-S$ *biconnector* if each component of the graph (V, F) intersects both R and S. So F is an $R-S$ biconnector if and only if each component of (V, F) has at least one edge in $\delta(R)$.

A min-max relation for the minimum size of an $R-S$ biconnector can be derived easily from the Kőnig-Rado edge cover theorem:

Section 54.1. Shortest $R-S$ biconnectors 929

Theorem 54.1. *Let $G = (V,E)$ be a graph and let V be partitioned into sets R and S such that each component of G intersects both R and S. Then the minimum size of an $R-S$ biconnector is equal to the maximum size of a subset of V spanning no edge connecting R and S.*

Proof. To see that the minimum is not less than the maximum, let F be a minimum $R-S$ biconnector and let U attain the maximum. Then F is a forest. For each $r \in U \cap R$, let $\phi(r)$ be the first edge in any $r-S$ path in F; and for each $s \in U \cap S$ let $\phi(s)$ be the first edge in any $s-R$ path in F. Then ϕ is injective from U to F (as U spans no edge in $\delta(R)$). Hence $|U| \leq |F|$.

To see equality, let $H := (N(R) \cup N(S), \delta(R))$, where $N(R)$ and $N(S)$ are the sets of neighbours of R and of S respectively. (So $N(R) \subseteq S$ and $N(S) \subseteq R$, and H is bipartite.) Let U' be a maximum-size stable set in H. Let F' be a minimum-size edge cover in H. By the König-Rado edge cover theorem (Theorem 19.4) we know $|F'| = |U'|$. Let $U := U' \cup (V \setminus (N(R) \cup N(S)))$. Then U spans no edge connecting R and S. By adding $|V \setminus (N(R) \cup N(S))|$ edges to F' we obtain an $R-S$ biconnector F with

(54.1) $\qquad |F| = |U'| + |V \setminus (N(R) \cup N(S))| = |U|.$

This shows the required equality. ∎

To obtain a min-max relation for the minimum length of an $R-S$ biconnector (given a length function on the edges), consider the system

(54.2) $\qquad$ (i) $\;\; 0 \leq x_e \leq 1 \qquad$ for each $e \in E$,
$\qquad\qquad$ (ii) $\;\; x(\delta(\mathcal{P})) \geq |\mathcal{P}| \qquad$ for each subpartition $\mathcal{P}$ of R or S with nonempty classes.

Here a *subpartition* of a set X is a partition of a subset of X (that is, a collection of disjoint subsets of X). $\delta(\mathcal{P})$ denotes the set of edges incident with but not spanned by any set in $\mathcal{P}$. Then system (54.2) determines the $R-S$ *biconnector polytope* — the convex hull of the incidence vectors of $R-S$ biconnectors:

Theorem 54.2. *System (54.2) is box-totally dual integral and determines the $R-S$ biconnector polytope.*

Proof. This follows from matroid intersection theory, applied to the matroids M_1 and M_2 on E, where M_1 is obtained from the cycle matroid $M(G)$ of G by contracting R to one vertex, making all edges spanned by R to a loop, and where M_2 is obtained similarly from $M(G)$ by contracting S to one vertex, making all edges spanned by S to a loop.

So the spanning sets of M_1 are the subsets F of E such that each component of (V, F) intersects R. Similarly, the spanning sets of M_2 are the subsets F of E such that each component of (V, F) intersects S. Hence the common spanning sets are precisely the $R-S$ biconnectors. Therefore, by Corollaries

41.12f and 50.8a, system (54.2) determines the convex hull of the incidence vectors of $R-S$ biconnectors. To see that the system is box-TDI, we use Corollary 41.12g and the fact that for each $F \subseteq E$, the inequality

(54.3) $\qquad x(F) \geq r_{M_i}(E) - r_{M_i}(E \setminus F)$

is a nonnegative integer combination of the inequalities (54.2). Indeed (for $i=1$), if $\mathcal{P}$ denotes the collection of the components of $(V, E \setminus F)$ contained in S, then

(54.4) $\qquad x(F) \geq x(\delta(\mathcal{P})) \geq |\mathcal{P}| \geq r_{M_1}(E) - r_{M_1}(E \setminus F),$

as $r_{M_1}(E) \leq |S|$ and $r_{M_1}(E \setminus F) = |S| - |\mathcal{P}|$. ∎

This implies a min-max relation for the minimum length of an $R-S$ biconnector. The reduction to matroid intersection also immediately implies that one can find a shortest $R-S$ biconnector in strongly polynomial time.

54.2. Longest $R-S$ biforests

Again, let $G = (V, E)$ be a graph and let V be partitioned into two sets R and S. Call a subset F of E an $R-S$ *biforest* if F is a forest and each component of F contains at most one edge in $\delta(R)$.

A min-max relation for the maximum size of an $R-S$ biforest can be derived easily from Kőnig's matching theorem:

Theorem 54.3. *Let $G = (V, E)$ be a graph and let V be partitioned into sets R and S. Then the maximum size of an $R-S$ biforest is equal to the minimum value of $|V| - |\mathcal{U}|$, where $\mathcal{U}$ is a collection of components of $G - R$ and $G - S$ such that no edge connects any two sets in $\mathcal{U}$.*

Proof. We may assume that G has no loops. To see that the maximum is not more than the minimum, consider any $R-S$ biforest F and any collection $\mathcal{U}$ as in the theorem. Then F contains no path connecting two distinct sets in $\mathcal{U}$. Hence $|F| \leq |V| - |\mathcal{U}|$.

The reverse inequality is proved by induction on the number of edges not in $\delta(R)$.

If $E = \delta(R)$, then G is bipartite, and $R-S$ biforests coincide with matchings. Then the theorem is equivalent to Kőnig's matching theorem (Theorem 16.2).

If $E \neq \delta(R)$, choose an edge $e = uv$ in $E \setminus \delta(R)$. If we contract e, the minimum value in the theorem reduces by precisely 1. Moreover, the maximum reduces by at least 1, since any $R-S$ forest in the contracted graph gives with e an $R-S$ forest in the original graph. So we are done by induction. ∎

To obtain a min-max relation for the maximum length of an $R-S$ biforest (given a length function on the edges), consider the following system:

(54.5)   (i) $0 \leq x_e \leq 1$ for each $e \in E$,
  (ii) $x(E[U]) \leq |U| - 1$ for each nonempty subset U of R or S,
  (iii) $x(E[U] \cup (\delta(U) \cap \delta(R))) \leq |U|$
    for each subset U of R or S.

This determines the $R-S$ *biforest polytope* — the convex hull of the incidence vectors of $R - S$ biforests:

Theorem 54.4. *System* (54.5) *is box-totally dual integral and determines the $R - S$ biforest polytope.*

Proof. This can be reduced to matroid intersection theory, similar to the proof of Theorem 54.3. ∎

Again, this theorem implies a min-max relation for the maximum length of an $R - S$ biforest, and the reduction to matroid intersection also implies that a longest $R - S$ biforest can be found in strongly polynomial time.

54.3. Disjoint $R - S$ biconnectors

We give a min-max relation for the maximum number of disjoint $R - S$ biconnectors (Keijsper and Schrijver [1998]). It generalizes the Tutte-Nash-Williams disjoint trees theorem (Corollary 51.1a) — which theorem however is used in the proof — and the disjoint edge covers theorem for bipartite graphs (Theorem 20.5).

We follow the (algorithmic) proof method of Keijsper [1998a], based on the following lemma:

Lemma 54.5α. *Let $T_1 = (V, E_1)$ and $T_2 = (V, E_2)$ be edge-disjoint spanning trees and let $r \in V$. For each $e = rv \in \delta_{T_1}(r)$, let $\phi(e)$ be the first edge of the $v - r$ path in T_2 that leaves the component of $T_1 - e$ containing v. Let $B \subseteq \delta_{T_1}(r)$ be such that $\phi(B)$ contains at most one edge not in $\delta_{T_2}(r)$. Then $(E_1 \setminus B) \cup \phi(B)$ and $(E_2 \setminus \phi(B)) \cup B$ are spanning trees again.*

Proof. By induction on $|B|$, the case $|B| \leq 1$ being easy. Let $|B| \geq 2$. Then there exists an edge $f = rw \in B$ with $\phi(f) \in \delta_{T_2}(r)$ (by the condition given in the theorem). Define

(54.6) $T_1' := (T_1 \setminus \{f\}) \cup \{\phi(f)\}$ and $T_2' := (T_2 \setminus \{\phi(f)\}) \cup \{f\}$.

Let $B' := B \setminus \{f\}$. Then for each $e = rv \in B'$,

(54.7) $\phi(e)$ is equal to the first edge of the $v - r$ path in T_2' that leaves the component K of $T_1' - e$ containing v.

To see this, let L be the component of $T_1 - f$ containing w. Since $\phi(f)$ connects L and r, K is equal to the component of $T_1 - e$ containing v. Moreover, the

$v-r$ path P in T_2' does not differ from the $v-r$ path in T_2 before entering L, and hence the first edge of P leaving K equals $\phi(e)$. This shows (54.7).

Now $(T_1 \setminus B) \cup \phi(B) = (T_1' \setminus B') \cup \phi(B')$ and $(T_2 \setminus \phi(B)) \cup B = (T_2' \setminus \phi(B')) \cup B'$, and by induction, they are spanning trees. ∎

Notice that the function $\phi : E_1 \cap \delta(r) \to E_2$ defined in the lemma is injective.

In the following lemma, we consider forests as edge sets. We recall that G/R denotes the graph obtained from G by contracting all vertices in R to one new vertex, denoted by R. The edges in the contracted graph are named after the edges in the original graph.

Lemma 54.5β. *Let $G = (V, E)$ be a graph and let V be partitioned into sets R and S. Let X_1 and X_2 be disjoint forests in G/R and let Y_1 and Y_2 be disjoint forests in G/S. Then there exist disjoint forests X_1' and X_2' in G/R and disjoint forests Y_1' and Y_2' in G/S with $X_1' \cup X_2' = X_1 \cup X_2$, $Y_1' \cup Y_2' = Y_1 \cup Y_2$, $X_1' \cap Y_2' = \emptyset$, and $X_2' \cap Y_1' = \emptyset$.*

Proof. By adding new edges spanned by S, we can assume that the X_i are spanning trees in G/R. Similarly, we can assume that the Y_i are spanning trees in G/S. (At the conclusion, we delete the new edges from the X_i' and Y_i'.)

If $X_1 \cap Y_2 = \emptyset$ and $X_2 \cap Y_1 = \emptyset$, we are done. So, by symmetry, we can assume that $X_1 \cap Y_2 \neq \emptyset$.

For each $e = rs \in X_1 \cap \delta(R)$, with $r \in R, s \in S$, let $\phi(e)$ be the first edge on the $s - R$ path in X_2 that leaves the component of $X_1 - e$ containing s. For each $e = rs \in Y_2 \cap \delta(R)$, with $r \in R, s \in S$, let $\psi(e)$ be the first edge on the $r - S$ path in Y_1 that leaves the component of $Y_2 - e$ containing r.

This gives injective functions

(54.8) $\phi : X_1 \cap \delta(R) \to X_2$ and $\psi : Y_2 \cap \delta(R) \to Y_1$.

Observe that $X_1 \cap (Y_1 \cup Y_2) \subseteq \delta(R)$ and $Y_2 \cap (X_1 \cup X_2) \subseteq \delta(R)$. Consider the directed graph with vertex set E and arc set

(54.9) $A := \{(e, \phi(e)) \mid e \in X_1 \cap \delta(R)\} \cup \{(e, \psi(e)) \mid e \in Y_2 \cap \delta(R)\}$.

Choose $e_0 \in X_1 \cap Y_2$ and set $e_1 := \phi(e_0)$. Then D contains a unique directed path $e_0, e_1, \ldots, e_h$ such that $e_0, \ldots, e_{h-1} \in X_1 \cup Y_2$ and $e_h \notin X_1 \cup Y_2$. (This because each vertex in $X_1 \cap Y_2$ has outdegree 2 and indegree 0 in D, and each vertex in $(X_1 \cup Y_2) \setminus (X_1 \cap Y_2)$ has outdegree 1 and indegree at most 1.)

It follows that for each $j < h$ one has $e_{j+1} = \phi(e_j)$ if j is even and $e_{j+1} = \psi(e_j)$ if j is odd. Define

(54.10) $B := \{e_j \mid 0 \leq j < h, j \text{ even}\}$ and $C := \{e_j \mid 1 \leq j < h, j \text{ odd}\}$.

Then by Lemma 54.5α,

(54.11) $\quad X_1' := (X_1 \setminus B) \cup \phi(B), X_2' := (X_2 \setminus \phi(B)) \cup B,$
$\quad\quad\quad Y_1' := (Y_1 \setminus \psi(C)) \cup C, Y_2' := (Y_2 \setminus C) \cup \psi(C),$

are again spanning tree of G/R and G/S respectively. Note that $X_1' \cap X_2' = \emptyset$, $Y_1' \cap Y_2' = \emptyset$, $X_1' \cup X_2' = X_1 \cup X_2$ and $Y_1' \cup Y_2' = Y_1 \cup Y_2$.

Now $\phi(B) \cap \psi(C) = \emptyset$, $\phi(B) \cap (Y_2 \setminus C) = \emptyset$ (since $\phi(B) \cap Y_2 \subseteq C$, as $e_h \notin Y_2$), and $\psi(C) \cap (X_1 \setminus B) = \emptyset$ (since $\psi(C) \cap X_1 \subseteq B$, as $e_h \notin X_1$). So $X_1' \cap Y_2' \subseteq (X_1 \cap Y_2) \setminus \{e_0\}$ (since $e_0 \notin X_1'$).

Moreover, $B \cap C = \emptyset$, $B \cap (Y_1 \setminus \psi(C)) = \emptyset$ (since $B \cap Y_1 \subseteq \psi(C)$, as $e_0 \notin Y_1$), and $C \cap (X_2 \setminus \phi(B)) = \emptyset$ (since $C \cap X_2 \subseteq \phi(B)$, as $e_0 \notin X_2$). So $X_2' \cap Y_1' \subseteq X_2 \cap Y_1$.

Concluding, $|X_1' \cap Y_2'| + |X_2' \cap Y_1'| < |X_1 \cap Y_2| + |X_2 \cap Y_1|$. Therefore, iterating this, we obtain trees as required. ∎

Now a min-max relation for disjoint $R - S$ biconnectors can be deduced:

Theorem 54.5. *Let $G = (V, E)$ be a graph, let V be partitioned into sets R and S, and let $k \in \mathbb{Z}_+$. Then there exist k disjoint $R - S$ biconnectors if and only if $|\delta(\mathcal{P})| \geq k|\mathcal{P}|$ for each subpartition $\mathcal{P}$ of R or S with nonempty classes.*

Proof. Necessity being easy, we show sufficiency. By Corollary 51.1a, the graph G/R (obtained from G by contracting R) has k disjoint spanning trees $X_1, \ldots, X_k$. Similarly, the graph G/S has k disjoint spanning trees $Y_1, \ldots, Y_k$. Then $X_i \cap Y_j$ is a subset of $\delta(R)$, for all i, j. Choose the X_i and Y_i in such a way that

(54.12) $\quad \sum_{i=1}^{k} |X_i \cap Y_i|$

is as large as possible.

Then $X_i \cap Y_j = \emptyset$ for all distinct i, j, for if, say, $X_1 \cap Y_2 \neq \emptyset$, we can replace X_1, X_2, Y_1, Y_2 by X_1', X_2', Y_1', Y_2' as in Lemma 54.5β. Then we have

(54.13) $\quad |X_1' \cap Y_1'| + |X_2' \cap Y_2'| = |(X_1' \cup X_2') \cap (Y_1' \cup Y_2')|$
$\quad\quad\quad = |(X_1 \cup X_2) \cap (Y_1 \cup Y_2)| > |X_1 \cap Y_1| + |X_2 \cap Y_2|.$

This contradicts the maximality of sum (54.12).

Hence $X_1 \cup Y_1, \ldots, X_k \cup Y_k$ form k disjoint $R - S$ biconnectors as required. ∎

This proof gives a polynomial-time algorithm to find a maximum number of disjoint $R - S$ biconnectors. Keijsper [1998a] gave an $O(\mathrm{DT}(n, m) + nm)$-time algorithm for this problem, where $\mathrm{DT}(n, m)$ denotes the time needed to find a maximum number of disjoint spanning trees in an undirected graph with n vertices and m edges.

By replacing edges by parallel edges, one obtains a capacitated version of Theorem 54.5. The corresponding optimization problem can be solved in polynomial time, by a straightforward adaptation of the methods described in the proofs of Theorems 51.8 and 51.10 and Corollary 51.8a. However, the capacitated problem cannot be solved in strongly polynomial time if we do not allow rounding (cf. the argument given in Section 51.4).

A generalization of Theorem 54.5 is given by Keijsper [1998a].

54.4. Covering by $R - S$ biforests

With the foregoing two lemmas, one can also derive a min-max relation for the minimum number of $R - S$ biforests that cover all edges (Keijsper [1998b]). It generalizes the Nash-Williams' covering forests theorem (Corollary 51.1c) — which theorem however is used in the proof — and Kőnig's edge-colouring theorem for bipartite graphs (Theorem 20.1).

Theorem 54.6. *Let $G = (V, E)$ be a graph, let V be partitioned into sets R and S, and let $k \in \mathbb{Z}_+$. Then E can be covered by k $R - S$ biforests if and only if*

(54.14) $\qquad |E[U]| \leq k(|U| - 1)$ *and* $|E[U]| + |\delta(U) \cap \delta(R)| \leq k|U|$

for each nonempty subset U of R or S.

Proof. Necessity being easy, we show sufficiency. We can assume that G is connected, as otherwise we can consider any component of G separately.

By Corollary 51.1c, the edges of the graph G/R can be partitioned into k forests $X_1, \ldots, X_k$. Similarly, the edges of the graph G/S can be partitioned into k forests $Y_1, \ldots, Y_k$. So $X_i \cap Y_j \subseteq \delta(R)$, for all i, j. Choose the X_i and Y_i in such a way that sum (54.12) is as large as possible. Then, as in the proof of Theorem 54.5, $X_i \cap Y_j = \emptyset$ for distinct i, j. Hence each $e \in \delta(R)$ belongs to $X_i \cap Y_i$ for some $i = 1, \ldots, k$. Concluding, $X_1 \cup Y_1, \ldots, X_k \cup Y_k$ form $R - S$ biforests as required. ∎

This proof gives a polynomial-time algorithm for finding a minimum covering by $R - S$ biforests. The methods of Section 51.4 can be extended to imply the polynomial-time solvability of the corresponding capacitated version, while strong polynomial-time solvability is again impossible.

54.5. Minimum-size bibranchings

We now turn to the directed analogues of biconnectors and biforests. Let $D = (V, A)$ be a digraph and let V be partitioned into two sets R and S. Call a subset B of A an $R - S$ *bibranching* if in the graph (V, B), each vertex in S is reachable from R, and each vertex in R reaches S.

Similarly to minimum $R - S$ biconnectors, a min-max relation for the minimum size of an $R - S$ bibranching follows easily from the Kőnig-Rado edge cover theorem.

Theorem 54.7. *Let $D = (V, A)$ be a graph and let V be partitioned into sets R and S such that each vertex in R can reach S and such that each vertex in S is reachable from R. Then the minimum size of an $R - S$ bibranching is equal to the maximum size of a subset of V spanning no arc in $\delta^{\text{out}}(R)$.*

Proof. To see that the minimum is not less than the maximum, let B be a minimum-size $R - S$ bibranching and let U attain the maximum. For each $r \in U \cap R$, let $\phi(r)$ be any arc in B leaving r, and for each $s \in U \cap S$ let $\phi(s)$ be any arc in B entering s. Then ϕ is injective from U to B, and hence $|U| \leq |B|$.

To see equality, let U' be a maximum stable set in the bipartite graph H with colour classes $N^{\text{in}}(S) \subseteq R$ and $N^{\text{out}}(R) \subseteq S$, with $r \in N^{\text{in}}(S)$ and $s \in N^{\text{out}}(R)$ adjacent if and only if D has an arc from r to s. (Here $N^{\text{out}}(X)$ and $N^{\text{in}}(X)$ are the sets of outneighbours and of inneighbours of X, respectively.)

Let B' be a minimum-size edge cover in H. By the Kőnig-Rado edge cover theorem (Theorem 19.4) we know $|B'| = |U'|$. Now by adding $|V \setminus (N^{\text{out}}(R) \cup N^{\text{in}}(S))|$ arcs to B' we obtain an $R - S$ bibranching B with

(54.15) $\quad |B| = |U'| + |V \setminus (N^{\text{out}}(R) \cup N^{\text{in}}(S))| = |U|,$

where $U := U' \cup (V \setminus (N^{\text{out}}(R) \cup N^{\text{in}}(S)))$. This shows the required equality. ∎

If each arc of D belongs to $\delta^{\text{out}}(R)$, then Theorem 54.7 reduces to the Kőnig-Rado edge cover theorem (Theorem 19.4).

The proof gives a polynomial-time algorithm to find a minimum-size $R-S$ bibranching (as we can find a minimum-size edge cover in a bipartite graph in polynomial time (Corollary 19.3a)).

54.6. Shortest bibranchings

To obtain a min-max relation for the minimum length of an $R-S$ bibranching (given a length function on the arcs), define a set of arcs C to be an $R - S$ *bicut* if $C = \delta^{\text{in}}(U)$ for some nonempty proper subset U of V satisfying $U \subseteq S$ or $S \subseteq U$.

Consider the system:

(54.16)   (i) $x_a \geq 0$ for each $a \in A$,
               (ii) $x(C) \geq 1$ for each $R - S$ bicut C.

Then the following implies a min-max relation for the minimum length of an $R - S$ bibranching.

Theorem 54.8. *System* (54.16) *is box-TDI.*

Proof. Let $w : A \to \mathbb{R}_+$. Let $\mathcal{U}$ be the collection of nonempty proper subsets U of V satisfying $U \subseteq S$ or $S \subseteq U$. Consider the maximum value of

(54.17) $$\sum_{U \in \mathcal{U}} y_U$$

where $y : \mathcal{U} \to \mathbb{R}_+$ satisfies

(54.18) $$\sum_{U \in \mathcal{U}} y_U \chi^{\delta^{\text{in}}(U)} \leq w.$$

Choose $y : \mathcal{U} \to \mathbb{R}_+$ attaining the maximum, such that

(54.19) $$\sum_{U \in \mathcal{U}} y_U |U||V \setminus U|$$

is minimized. We show that the collection $\mathcal{F} := \{U \in \mathcal{U} \mid y_U > 0\}$ is cross-free; that is, for all $T, U \in \mathcal{F}$ one has

(54.20) $\quad T \subseteq U$ or $U \subseteq T$ or $T \cap U = \emptyset$ or $T \cup U = V$.

Suppose that this is not true. Let $\alpha := \min\{y_T, y_U\}$. Decrease y_T and y_U by α, and increase $y_{T \cap U}$ and $y_{T \cup U}$ by α. Now (54.18) is maintained, and (54.17) did not change. However, (54.19) decreases (by Theorem 2.1), contradicting our minimality assumption.

So $\mathcal{F}$ is cross-free. Now the $\mathcal{F} \times A$ matrix M with

(54.21) $$M_{U,a} := \begin{cases} 1 & \text{if } a \in \delta^{\text{in}}(U), \\ 0 & \text{otherwise,} \end{cases}$$

is totally unimodular. To see this, let $T = (W, B)$ and $\pi : V \to W$ form a tree-representation of $\mathcal{F}$ (see Section 13.4). That is, T is a directed tree and $\mathcal{F} = \{V_b \mid b \in B\}$, where

(54.22) $\quad V_b := \{v \in V \mid \pi(v)$ belongs to the same component of $T - b$ as the head of $b\}$.

Then for any arc $a = (u, v)$ of D, the set of forward arcs in the undirected $\pi(u) - \pi(v)$ path in T is contiguous, that is, forms a directed path, say from u' to v'. This follows from the fact that there exist no arcs b, c, d in this order on the path with b and d forward and c backward.

Define $a' := (u', v')$, and let $D' = (W, A')$ be the digraph with $A' := \{a' \mid a \in A\}$. Then M is equal to the network matrix generated by T and D' (identifying $b \in B$ with the set V_b in $\mathcal{F}$ determined by b). Hence by Theorem 13.20, M is totally unimodular.

This implies with Theorem 5.35 that (54.16) is box-TDI. ∎

This implies that the $R-S$ bibranching polytope — the convex hull of the incidence vectors of $R-S$ bibranchings — can be described as follows:

Corollary 54.8a. *The $R-S$ bibranching polytope is determined by*

(54.23)   (i) $0 \leq x_a \leq 1$ *for each $a \in A$,*
           (ii) $x(C) \geq 1$ *for each $R-S$ bicut C.*

Proof. By Theorem 54.8, (54.23) determines an integer polytope. Necessarily, each vertex of it is the incidence vector of an $R-S$ branching. ∎

The box-total dual integrality of (54.16) has as special case the total dual integrality of (54.23), which is equivalent to:

Corollary 54.8b (optimum bibranching theorem). *Let $D = (V, A)$ be a digraph, let V be partitioned into sets R and S, and let $l : A \to \mathbb{Z}_+$ be a length function. Then the minimum length of an $R-S$ bibranching is equal to the maximum size of a family of $R-S$ dicuts, such that each arc a is in at most $l(a)$ of them.*

Proof. This is a reformulation of the total dual integrality of (54.23), which follows from Theorem 54.8. ∎

We also note that Theorem 54.8 implies that for each $k \in \mathbb{Z}_+$ the system

(54.24)   (i) $x_a \geq 0$ for each $a \in A$,
           (ii) $x(C) \geq k$ for each $R-S$ bicut C,

is box-TDI (since if $Ax \leq b$ is box-TDI, then for each $k > 0$, $Ax \leq k \cdot b$ is box-TDI).

Keijsper and Pendavingh [1998] gave an $O(n'(m + n \log n))$ algorithm to find a shortest bibranching, where $n' := \min\{|R|, |S|\}$. The strong polynomial-time solvability follows also from the strong polynomial-time solvability of finding a minimum-length strong connector for a source-sink connected digraph, which by the method of Theorem 57.3 can be reduced to finding a minimum-length directed cut cover, which is a special case of weighted matroid intersection (Section 55.5).

54.6a. Longest bifurcations

Let $D = (V, A)$ be a digraph and let V be partitioned into two sets R and S. Call a subset B of A an $R-S$ *bifurcation* if B contains no undirected circuits, each vertex in R is left by at most one arc in B, each vertex in S is entered by at most one arc in B, and B contains no arcs from S to R. So B is an $R-S$ bifurcation if and only if contracting R gives a branching and contracting S gives a cobranching. (A *cobranching* is a set B of arcs whose reversal B^{-1} is a branching.)

Similarly to maximum $R-S$ biforests, a min-max relation for the maximum size of an $R-S$ bifurcation follows from Kőnig's matching theorem:

938   Chapter 54. Biconnectors and bibranchings

Theorem 54.9. *Let $D = (V, A)$ be a graph and let V be partitioned into sets R and S, with $\delta^{\text{in}}(R) = \emptyset$. Then the maximum size of an $R - S$ bifurcation is equal to the minimum size of $|V| - |\mathcal{L}|$, where $\mathcal{L}$ is a collection of strong components K of D with either $K \subseteq R$ and $\delta^{\text{out}}(K) \subseteq \delta^{\text{out}}(R)$, or $K \subseteq S$ and $\delta^{\text{in}}(K) \subseteq \delta^{\text{in}}(S)$, such that no arc connects two components in $\mathcal{L}$.*

Proof. To see that the minimum is not less than the maximum, let B be a maximum-size $R - S$ bifurcation and let $\mathcal{L}$ attain the minimum. Let U be the set of vertices v with $v \in R$ and $\delta_B^{\text{out}}(v) = \emptyset$, or $v \in S$ and $\delta_B^{\text{in}}(v) = \emptyset$. Then

(54.25) $\quad |B| = |V| - |U| - |B \cap \delta^{\text{out}}(R)| \leq |V| - |\mathcal{L}|,$

since each $K \in \mathcal{L}$ contains a vertex in U or is entered or left by an arc in $B \cap \delta^{\text{out}}(R)$.

To see equality, consider the following bipartite graph H. H has vertex set the set $\mathcal{K}$ of strong components K of D with either $K \subseteq R$ and $\delta^{\text{out}}(K) \subseteq \delta^{\text{out}}(R)$, or $K \subseteq S$ and $\delta^{\text{in}}(K) \subseteq \delta^{\text{in}}(S)$. Two sets $K, L \in \mathcal{K}$ are adjacent if and only if there is an arc connecting K and L. (This implies that one of K, L is contained in R, the other in S.) Let $\mathcal{L}$ be a maximum-size stable set in H and let B' be a maximum-size matching in H. By Kőnig's matching theorem (Theorem 16.2), $|B'| + |\mathcal{L}| = |\mathcal{K}|$. Now by adding $|V| - |\mathcal{K}|$ arcs to the arc set in D corresponding to B', we can obtain an $R - S$ bifurcation of size $|B'| + |V| - |\mathcal{K}| = |V| - |\mathcal{L}|$. ∎

If each arc of D belongs to $\delta^{\text{out}}(R)$, then Theorem 54.9 reduces to Kőnig's matching theorem (Theorem 16.2).

We next give a min-max relation for the maximum length of an $R-S$ bifurcation, by reduction to Theorem 54.8 on minimum-length bibranching:

Theorem 54.10. *Let $D = (V, A)$ be a digraph and let V be partitioned into R and S such that there are no arcs from S to R. Let $l \in \mathbb{Z}_+^A$ be a length function. Then the maximum length of an $R - S$ bifurcation is equal to the minimum value of*

(54.26) $\quad \displaystyle\sum_{v \in V} y_v + \sum_{U \in \mathcal{U}} z_U(|U| - 1)$

where $y \in \mathbb{Z}_+^V$ and $z \in \mathbb{Z}_+^\mathcal{U}$, with $\mathcal{U} := \{U \mid U \neq \emptyset, U \subseteq R \text{ or } U \subseteq S\}$, such that

(54.27) $\quad \displaystyle\sum_{v \in R} y_v \chi^{\delta^{\text{out}}(v)} + \sum_{v \in S} y_v \chi^{\delta^{\text{in}}(v)} + \sum_{U \in \mathcal{U}} z_U \chi^{A[U]} \geq l.$

Proof. To see that the maximum is not more than the minimum, let B be any $R - S$ bifurcation and let y_v, z_U satisfy (54.27). Then

(54.28) $\quad l(B) = \displaystyle\sum_{a \in B} l(a) \leq \sum_{a \in B}\Big(\sum_{\substack{v \in R \\ a \in \delta^{\text{out}}(v)}} y_v + \sum_{\substack{v \in S \\ a \in \delta^{\text{in}}(v)}} y_v + \sum_{\substack{U \in \mathcal{U} \\ a \in A[U]}} z_U\Big)$

$\quad = \displaystyle\sum_{v \in R} y_v |B \cap \delta^{\text{out}}(v)| + \sum_{v \in S} y_v |B \cap \delta^{\text{in}}(v)| + \sum_{U \in \mathcal{U}} z_U |B \cap A[U]|$

$\quad \leq \displaystyle\sum_{v \in V} y_v + \sum_{U \in \mathcal{U}} z_U(|U| - 1).$

To see equality, extend D by two new vertices, r and s, and by arcs (r, v) for each $v \in S \cup \{s\}$ and (v, s) for each $v \in R$. This makes the digraph $D' = (V', A')$. Define $R' := R \cup \{r\}$ and $S' := S \cup \{s\}$. Let $L := \max\{l(a) \mid a \in A\} + 1$. Define $l' \in \mathbb{Z}_+^{A'}$ by:

Section 54.6a. Longest bifurcations   939

$$
(54.29) \qquad l'(a) := \begin{cases} L - l(a) & \text{for each } a \in A[R] \cup A[S], \\ 2L - l(a) & \text{for each } a \in \delta^{\text{out}}(R), \\ L & \text{for each } a = (r,v) \text{ with } v \in S \text{ and } a = (v,s) \\ & \text{with } v \in R, \\ 0 & \text{for } a = (r,s). \end{cases}
$$

Let $\mathcal{U}'$ be the collection of nonempty subsets U of R' or S'. By Theorem 54.8, applied to D', there exists an $R' - S'$ bibranching B' in D' and a $z : \mathcal{U}' \to \mathbb{Z}_+$ such that

$$
(54.30) \qquad l'(B') = \sum_{U \in \mathcal{U}'} z_U,
$$

and

$$
(54.31) \qquad \sum_{\substack{U \in \mathcal{U}' \\ U \subseteq R'}} z_U \chi^{\delta^{\text{out}}(U)} + \sum_{\substack{U \in \mathcal{U}' \\ U \subseteq S'}} z_U \chi^{\delta^{\text{in}}(U)} \leq l'.
$$

Since $l'(r,s) = 0$ we know that $z_U = 0$ if r or s belongs to U. That is, $z_U = 0$ if $U \in \mathcal{U}' \setminus \mathcal{U}$.

For each $v \in V$, define

$$
(54.32) \qquad y_v := L - \sum_{\substack{U \in \mathcal{U} \\ v \in U}} z_U.
$$

Then $y_v \geq 0$ for each $v \in V$, as

$$
(54.33) \qquad y_v = L - \sum_{\substack{U \in \mathcal{U} \\ v \in U}} z_U \geq L - l'(r,v) = 0
$$

if $v \in S$, and similarly $y_v \geq 0$ if $v \in R$.

Also, y and z satisfy (54.27), since for any arc $a = (u,v)$ one has, if $u,v \in R$:

$$
(54.34) \qquad y_u + \sum_{\substack{U \in \mathcal{U} \\ a \in A[U]}} z_U = L - \sum_{\substack{U \in \mathcal{U} \\ u \in U}} z_U + \sum_{\substack{U \in \mathcal{U} \\ a \in A[U]}} z_U = L - \sum_{\substack{U \in \mathcal{U} \\ a \in \delta^{\text{out}}(U)}} z_U
$$
$$
\geq L - l'(a) = l(a).
$$

Similarly, if $u, v \in S$, then

$$
(54.35) \qquad y_v + \sum_{\substack{U \in \mathcal{U} \\ a \in A[U]}} z_U \geq l(a).
$$

Finally, if $u \in R$ and $v \in S$, then:

$$
(54.36) \qquad y_u + y_v = 2L - \sum_{\substack{U \in \mathcal{U} \\ u \in U}} z_U - \sum_{\substack{U \in \mathcal{U} \\ v \in U}} z_U
$$
$$
= 2L - \sum_{\substack{U \in \mathcal{U} \\ a \in \delta^{\text{out}}(U)}} z_U - \sum_{\substack{U \in \mathcal{U} \\ a \in \delta^{\text{in}}(U)}} z_U \geq 2L - l'(a) = l(a).
$$

So y and z satisfy (54.27).

Note that each $u \in R$ is left by a unique arc in B', since if there is more than one, all arcs leaving u should have their heads in S' (since if $(u,v), (u',v') \in B'$, $v \neq v'$, $v \notin S'$, then $B' \setminus \{(u,v)\}$ is again an $R' - S'$ bibranching). Then replacing one outgoing arc $(u,v) \in B'$ by the arc (r,v) keeps B' an $R - S$ bibranching, however of smaller length. This contradicts our assumption. So each vertex in R is left by exactly one arc in B', and similarly, each vertex in S is entered by exactly one arc in B'. This implies that $B := B' \cap A$ is an $R - S$ bifurcation.

We finally show that equality holds throughout in (54.28). Indeed, if $a \in B$, then $a \in B'$, and hence we have equality in (54.34), implying that the first inequality in (54.28) is satisfied with equality. Moreover, if $y_v > 0$ and $v \in S$, then we have strict inequality in (54.33), and hence $(r,v) \notin B'$. Therefore $|B \cap \delta^{\text{in}}(v)| = 1$. Similarly, $y_v > 0$ and $v \in R$ implies $|B \cap \delta^{\text{out}}(v)| = 1$. Finally, if $z_U > 0$ and (say) $U \subseteq R$, then $|B' \cap \delta^{\text{out}}(U)| = 1$, and hence $|B' \cap A[U]| = |U| - 1$ (since each $v \in R$ is left by precisely one arc in B'), implying $|B \cap A[U]| = |U| - 1$. This shows that also the second inequality in (54.28) is satisfied with equality. ∎

Theorem 54.10 is equivalent to the total dual integrality of the following system:

(54.37)   (i) $x_a \geq 0$ for each $a \in A$,
                (ii) $x(\delta^{\text{out}}(v)) \leq 1$ for each $v \in R$,
                (iii) $x(\delta^{\text{in}}(v)) \leq 1$ for each $v \in S$,
                (iv) $x(A[U]) \leq |U| - 1$ for each nonempty U with $U \subseteq R$ or $U \subseteq S$.

It yields a description of the $R - S$ *bifurcation polytope* — the convex hull of the incidence vectors of the $R - S$ bifurcations in D.

Corollary 54.10a. *System (54.37) is TDI and determines the $R - S$ bifurcation polytope.*

Proof. This is equivalent to Theorem 54.10. ∎

As for the *complexity*, the reduction given in Theorem 54.10 also implies that a maximum-length $R - S$ bifurcation can be found in strongly polynomial time (since a minimum-length $R - S$ bibranching can be found in strongly polynomial time).

54.7. Disjoint bibranchings

Consider the system

(54.38)   (i) $0 \leq x_a \leq 1$ for each $a \in A$,
                (ii) $x(B) \geq 1$ for each $R - S$ bibranching B.

By the theory of blocking polyhedra, Corollary 54.8a implies:

Corollary 54.10b. *System (54.38) determines the convex hull of the incidence vectors of arc sets containing an $R - S$ bicut.*

Proof. Directly from Corollary 54.8a with the theory of blocking polyhedra. ∎

System (54.38) in fact is TDI, which is equivalent to the following statement:

Theorem 54.11 (disjoint bibranchings theorem). *Let $D = (V, A)$ be a digraph and let V be partitioned into sets R and S. Then the maximum number of disjoint $R-S$ bibranchings is equal to the minimum size of an $R-S$ bicut.*

Proof. Let k be the minimum size of an $R - S$ bicut. Clearly, there are at most k disjoint $R - S$ bibranchings. We show equality. For any digraph $D = (V, A)$ and $r \in V$, call a subset B of A an *r-coarborescence* if the set B^{-1} of reverse arcs of B is an r-arborescence.

By Edmonds' disjoint arborescences theorem (Corollary 53.1b), the graph D/R (obtained from D by contracting R to one vertex) has k disjoint R-arborescences $B_1, \ldots, B_k$. Similarly, the graph D/S has k disjoint S-coarborescences $B'_1, \ldots, B'_k$. Choose the B_i and B'_i such that the sum

$$(54.39) \qquad \sum_{i=1}^{k} \sum_{\substack{j=1 \\ j \neq i}}^{k} |B_i \cap B'_j|$$

is as small as possible. If the sum is 0, then

$$(54.40) \qquad B_1 \cup B'_1, \ldots, B_k \cup B'_k$$

are k disjoint $R - S$ bibranchings in D as required. So we can assume that the sum is positive. Without loss of generality, $B_1 \cap B'_2 \neq \emptyset$.

Define

$$(54.41) \qquad X := (B_1 \cup B_2) \cap A[S], \ X' := (B'_1 \cup B'_2) \cap A[R],$$
$$Y := (B_1 \cup B_2) \cap \delta^{\text{out}}(R), \ Y' := (B'_1 \cup B'_2) \cap \delta^{\text{out}}(R).$$

Let $\mathcal{K}$ be the collection of strong components K of the digraph (S, X) with $\delta_X^{\text{in}}(K) = \emptyset$. Similarly, let $\mathcal{K}'$ be the collection of strong components K of the digraph (R, X') with $\delta_{X'}^{\text{out}}(K) = \emptyset$.

Now $d_Y^{\text{in}}(K) = d_{B_1 \cup B_2}^{\text{in}}(K) \geq 2$ for each $K \in \mathcal{K}$, and similarly $d_{Y'}^{\text{out}}(K) \geq 2$ for each $K \in \mathcal{K}'$. Then we can split Y into Y_1 and Y_2 and Y' into Y'_1 and Y'_2 such that

$$(54.42) \qquad \begin{array}{l} d_{Y_i}^{\text{in}}(K) \geq 1 \text{ for each } K \in \mathcal{K} \text{ and } i = 1, 2, \\ d_{Y'_i}^{\text{out}}(K) \geq 1 \text{ for each } K \in \mathcal{K}' \text{ and } i = 1, 2, \\ \text{and } Y_1 \cap Y'_2 = \emptyset \text{ and } Y_2 \cap Y'_1 = \emptyset. \end{array}$$

This can be seen as follows. Select for each $U \in \mathcal{K}$ a pair e_U from $\delta_Y^{\text{in}}(U)$. Similarly, select for each $U \in \mathcal{K}'$ a pair e_U from $\delta_{Y'}^{\text{out}}(U)$. So the e_U for $U \in \mathcal{K}$ are disjoint, and the e_U for $U \in \mathcal{K}'$ are disjoint. Hence the e_U for $U \in \mathcal{K} \cup \mathcal{K}'$ form a bipartite graph on $Y \cup Y'$ (in fact, a set of vertex-disjoint paths and even circuits). The two colour classes of this bipartite graph give the partitions of Y and Y' as required.

Then by Lemma 53.2α, X can be split into two branchings X_1 and X_2 such that the set of roots of X_i is equal to the set of heads of Y_i ($i = 1, 2$). Similarly, X' can be split into two cobranchings X_1' and X_2' such that the set of coroots of X_i' is equal to the set of tails of Y_i' ($i = 1, 2$). (A *cobranching* is a set B of arcs whose reversal B^{-1} is a branching. A *coroot* of B is a root of B^{-1}.)

Define

(54.43) $\quad \widetilde{B}_i := X_i \cup Y_i$ and $\widetilde{B}_i' := X_i' \cup Y_i'$

for $i = 1, 2$. Since $\widetilde{B}_1 \cap \widetilde{B}_2' = \emptyset$ and $\widetilde{B}_2 \cap \widetilde{B}_1' = \emptyset$, replacing B_1, B_2, B_1', B_2' by $\widetilde{B}_1, \widetilde{B}_2, \widetilde{B}_1', \widetilde{B}_2'$ decreases sum (54.39), contradicting the minimality assumption. ∎

The capacitated case can be derived as a consequence:

Corollary 54.11a. *Let $D = (V, A)$ be a digraph, let V be partitioned into sets R and S, and let $c \in \mathbb{Z}_+^A$ be a capacity function. Then the maximum number of $R - S$ bibranchings such that no arc a is in more than $c(a)$ of these bibranchings is equal to the minimum capacity of an $R - S$ bicut.*

Proof. This follows from Theorem 54.11 by replacing any arc a by $c(a)$ parallel arcs. ∎

Equivalently, in TDI terms:

Corollary 54.11b. *System (54.38) is totally dual integral.*

Proof. This is a reformulation of Corollary 54.11a. ∎

Another consequence is:

(54.44) For any digraph $D = (V, A)$ and any partition of V into R and S, the $R - S$ bibranching polytope has the integer decomposition property.

As for the *complexity*, the proof of Theorem 54.11 gives a polynomial-time algorithm for finding a maximum number of disjoint $R - S$ bibranchings. For the capacitated case there is a semi-strongly polynomial-time algorithm (that is, where rounding takes one arithmetic step): first find a fractional dual solution, then round (Grötschel, Lovász, and Schrijver [1988]). A combinatorial semi-strongly polynomial-time algorithm follows from the results in Section 57.5.

54.7a. Proof using supermodular colourings

We show how to derive Theorem 54.11 on disjoint bibranchings from Edmonds' disjoint branchings theorem (Theorem 53.1) and Theorem 49.14 on supermodular colourings.

Let $D = (V, A)$ be a digraph and let V be partitioned into R and S. Let $k \in \mathbb{Z}_+$. Define $H := \delta^{\text{out}}(R)$, and define the following collections of subsets of H:

(54.45) $\quad \mathcal{C}_1 := \{\delta_H^{\text{in}}(U) \mid \emptyset \neq U \subseteq S\}$ and $\mathcal{C}_2 := \{\delta_H^{\text{out}}(U) \mid \emptyset \neq U \subseteq R\}$.

Then $\mathcal{C}_1$ and $\mathcal{C}_2$ are intersecting families on H. Define $g_j : \mathcal{C}_j \to \mathbb{Z}$ for $j = 1, 2$ by:

(54.46) $\quad g_1(B) := \max\{k - d_{A[S]}^{\text{in}}(U) \mid \emptyset \neq U \subseteq S, B = \delta_H^{\text{in}}(U)\}$ for $B \in \mathcal{C}_1$,
$\quad g_2(B) := \max\{k - d_{A[R]}^{\text{out}}(U) \mid \emptyset \neq U \subseteq R, B = \delta_H^{\text{out}}(U)\}$ for $B \in \mathcal{C}_2$.

Then g_1 and g_2 are intersecting supermodular. Moreover, if U attains the maximum in (54.46), then

(54.47) $\quad g_1(B) = k - d_{A[S]}^{\text{in}}(U) \leq d_A^{\text{in}}(U) - d_{A[S]}^{\text{in}}(U) = d_H^{\text{in}}(U) = |B|$ if $U \subseteq S$
and
$\quad g_2(B) = k - d_{A[R]}^{\text{out}}(U) \leq d_A^{\text{out}}(U) - d_{A[R]}^{\text{out}}(U) = d_H^{\text{out}}(U) = |B|$ if $U \subseteq R$.

Since $g_j(B) \leq k$ for $j = 1, 2$ and $B \in \mathcal{C}_j$, by Theorem 49.14 we can partition H into classes $H_1, \ldots, H_k$ such that:

(54.48) $\quad$ (i) if $\emptyset \neq U \subseteq S$, then U is entered by at least $k - d_{A[S]}^{\text{in}}(U)$ of the classes H_i, and
$\quad$ (ii) if $\emptyset \neq U \subseteq R$, then U is left by at least $k - d_{A[R]}^{\text{out}}(U)$ of the classes H_i.

By Edmonds' disjoint branchings theorem, (i) implies that $A[S]$ contains disjoint branchings $B_1, \ldots, B_k$ such that, for each $i = 1, \ldots, k$, the root set of B_i is equal to the set of heads of the arcs in H_i; that is, each vertex in S is entered by at least one arc in $B_i \cup H_i$. Similarly, $A[R]$ contains disjoint cobranchings (= branchings if all orientations are reversed) $B'_1, \ldots, B'_k$ such that, for each $i = 1, \ldots, k$, each vertex in R is left by at least one arc in $B'_i \cup H_i$. Then the $B_i \cup H_i \cup B'_i$ form disjoint $R - S$ bibranchings.

54.7b. Covering by bifurcations

Theorem 54.11 also implies the following characterization of the minimum number of $R - S$ bifurcations needed to cover all arcs (Keijsper [1998b]):

Corollary 54.11c. *Let $D = (V, A)$ be a digraph and let V be partitioned into sets R and S, with no arc from S to R. Then A can be covered by k $R - S$ bifurcations if and only if*

(54.49) $\quad$ (i) $\deg^{\text{out}}(v) \leq k$ for each $v \in R$;
$\quad$ (ii) $\deg^{\text{in}}(v) \leq k$ for each $v \in S$;
$\quad$ (iii) $|A[U]| \leq k(|U| - 1)$ for each nonempty subset U of R or S.

Proof. Necessity being easy, we show sufficiency. Extend D by two new vertices r and s, for each $v \in S$ by $k - \deg^{\text{in}}(v)$ parallel arcs from r to v, for each $v \in R$ by $k - \deg^{\text{out}}(v)$ parallel arcs from v to s, and by k parallel arcs from r to s. Let D' be the graph arising in this way. So in D', each $v \in R$ has outdegree k, and each $v \in S$ has indegree k. Define $R' := R \cup \{r\}$ and $S' := S \cup \{s\}$.

Then by Theorem 54.11, D' has k disjoint $R' - S'$ bibranchings. Indeed, any nonempty subset U of R' is left by $k|U| - |A[U]| \geq k$ arcs of D' if $r \notin U$ (since each vertex in R has outdegree k in D'), and by at least k arcs of D' if $r \in U$. Similarly, any nonempty subset of S' is entered by at least k arcs of D'.

Now each of these bibranchings leaves any $v \in R$ exactly once (as v has outdegree k in D'), and (similarly) enters any $v \in S$ exactly once. Moreover, these bibranchings cover A. Hence restricted to A we obtain a covering of A by k $R - S$ bifurcations. ∎

An equivalent way of saying this is (using Corollary 54.10a):

(54.50)   For any digraph $D = (V, A)$ and any partition of V into R and S, the $R - S$ bifurcation polytope has the integer decomposition property.

As for the *complexity*, the reduction given in the proof of Corollary 54.11c implies a polynomial-time algorithm to find a minimum number of $R - S$ bifurcations covering the arc set (by reduction to finding a maximum number of disjoint bibranchings). The capacitated version can be solved in semi-strongly polynomial time, with the help of the ellipsoid method, by first finding a fractional packing, and next round (like in Section 51.4).

54.7c. Disjoint $R - S$ biconnectors and $R - S$ bibranchings

As in Keijsper and Schrijver [1998], one can derive Theorem 54.5 on disjoint $R-S$ biconnectors (in an undirected graph) from Theorem 54.11 on disjoint $R-S$ bibranchings (in a directed graph), with the help of the Tutte-Nash-Williams disjoint trees theorem (Corollary 51.1a).

Indeed, the condition in Theorem 54.5 gives, with the Tutte-Nash-Williams disjoint trees theorem, that the graph G/R obtained from G by contracting R to one vertex, has k edge-disjoint spanning trees.

By orienting the edges in these trees appropriately, we see that G/R has an orientation such that any nonempty $U \subseteq S$ is entered by at least k arcs, and such that each edge incident with R is oriented away from R. Similarly, G/S has an orientation such that any nonempty $U \subseteq R$ is left by at least k arcs, and such that each edge incident with S is oriented towards S.

Combining the two orientations, we obtain an orientation $D = (V, A)$ of G such that each $R - S$ bicut has size at least k. Hence, by Theorem 54.11, D has k disjoint $R - S$ bibranchings, and hence, G has k disjoint $R - S$ biconnectors.

54.7d. Covering by $R - S$ biforests and by $R - S$ bifurcations

Similarly, one can derive Theorem 54.6 on covering $R - S$ biforests from Corollary 54.11c on covering $R - S$ bifurcations, with the help of Nash-Williams' covering forests theorem (Corollary 51.1c). Indeed, the condition in Theorem 54.6 gives,

Section 54.7d. Covering by $R-S$ biforests and by $R-S$ bifurcations   945

with Nash-Williams' covering forests theorem, that the edges of the graph G/R obtained from G by contracting R to one vertex, can be covered by k forests. Hence G/R has an orientation such that any vertex in S is entered by at most k arcs, and such that R is only left by arcs. Similarly, G/S has an orientation such that any vertex in R is left by at most k arcs, and such that S is only entered by arcs.

Combining the two orientations, we obtain an orientation $D = (V, A)$ of G satisfying the condition in Corollary 54.11c. Hence the arcs of D can be covered by k $R-S$ bifurcations, and hence the edges of G can be covered by k $R-S$ biforests.

Chapter 55

Minimum directed cut covers and packing directed cuts

A *directed cut* in a directed graph $D = (V, A)$ is a set of arcs $\delta^{in}(U)$ for some nonempty proper subset U of V with $\delta^{out}(U) = \emptyset$. A *directed cut cover* is a set of arcs intersecting each directed cut — equivalent, it is a set of arcs such that their contraction makes the graph strongly connected. For planar digraphs, a directed cut cover corresponds to a *feedback arc set* in the dual digraph — a set of arcs whose removal makes the digraph acyclic. Lucchesi and Younger showed that the minimum size of a directed cut cover is equal to the maximum number of disjoint directed cuts. This min-max relation is the basis for several other results on *shortest* directed cut covers, which we survey in this chapter. In the next chapter we consider the, less tractable, *disjoint* directed cut covers.

55.1. Minimum directed cut covers and packing directed cuts

Let $D = (V, A)$ be a digraph. A subset C of A is called a *directed cut* if there exists a nonempty proper subset U of V with $\delta^{in}(U) = C$ and $\delta^{out}(U) = \emptyset$. A *directed cut cover* is a set of arcs intersecting each directed cut.

It is easy to show that for any subset B of A the following are equivalent:

(55.1)   (i) B is a directed cut cover;
(ii) adding to D all arcs (u, v) with $(v, u) \in B$ makes the digraph strongly connected;
(iii) contracting all arcs in B makes the digraph strongly connected.

So a minimum directed cut cover gives a minimum number of arcs in D such that making them two-way we obtain a strongly connected digraph.

Moreover, A. Frank (cf. Lovász [1979a] p. 271) showed:

Theorem 55.1. *Let $D = (V, A)$ be a weakly connected digraph without cut arcs and let $B \subseteq A$. Then B is an inclusionwise minimal directed cut cover if and only if B is an inclusionwise minimal set such that if we invert the orientations of all arcs in B, the digraph becomes strongly connected.*

Proof. Define $\widetilde{A} := (A \setminus B) \cup B^{-1}$, where $B^{-1} := \{a^{-1} \mid a \in B\}$, and where a^{-1} is the arc arising from a by inverting its orientation.

Trivially, if $(V, \widetilde{A})$ is strongly connected, then B is a directed cut cover. Hence it suffices to show that if B is an inclusionwise minimal directed cut cover, then $\widetilde{D} = (V, \widetilde{A})$ is strongly connected.

Suppose that $\widetilde{D}$ is not strongly connected. Let K be a strong component of $\widetilde{D}$ with $\delta^{\text{in}}_{\widetilde{A}}(K) = \emptyset$. Let $\delta^{\text{in}}_A(K) = \{a_1, \ldots, a_t\}$. So $a_1, \ldots, a_t$ belong to B. Hence, as B is an inclusionwise minimal directed cut cover, for each $i = 1, \ldots, t$ there exists a subset U_i of V with $\delta^{\text{in}}_A(U_i) = \emptyset$ and $\delta^{\text{out}}_B(U_i) = \{a_i\}$.

Then for each i, $U_i \cap K = \emptyset$. For suppose that $U_i \cap K \neq \emptyset$. As the head of a_i does not belong to U_i, U_i splits K. Hence some arc $a \in \widetilde{A}$ enters U_i, with a spanned by K. As $\delta^{\text{in}}_A(U_i) = \emptyset$, we know $a \in B^{-1}$, and therefore $a^{-1} \in \delta^{\text{out}}_B(U_i)$ while $a \neq a_i$, a contradiction.

Also, $U_i \cap U_j = \emptyset$ for $i \neq j$, as $\delta^{\text{in}}_A(U_i \cap U_j) = \emptyset$ and

(55.2) $\qquad d^{\text{out}}_B(U_i \cap U_j) \leq d^{\text{out}}_B(U_i) + d^{\text{out}}_B(U_j) - d^{\text{out}}_B(U_i \cup U_j) = 0$,

since both a_i and a_j leave $U_i \cup U_j$.

So $U_1, \ldots, U_t$ are disjoint subsets of $V \setminus K$. As D has no cut arcs, $d^{\text{out}}_A(U_i) \geq 2$ for each i. Hence, as no arc in A enters any U_i, and only one arc (namely a_i) leaves U_i to enter K, the set $W := V \setminus (K \cup U_1 \cup \cdots \cup U_t)$ is nonempty. Also, $\delta^{\text{out}}_A(W) = \emptyset$, and so $\delta^{\text{out}}_B(W) \neq \emptyset$, that is $\delta^{\text{out}}_B(K \cup U_1 \cup \cdots \cup U_t) \neq \emptyset$. However, $\delta^{\text{out}}_B(K) = \emptyset$ (since $\delta^{\text{in}}_{\widetilde{A}}(K) = \emptyset$) and $\delta^{\text{out}}_B(U_i) = \{a_i\}$, implying $\delta^{\text{out}}_B(K \cup U_i) = \emptyset$ for each i, a contradiction. ∎

55.2. The Lucchesi-Younger theorem

Lucchesi and Younger [1978] proved the following min-max relation for the minimum size of a directed cut cover, which was conjectured by N. Robertson and by Younger [1965,1969] (for planar graphs by Younger [1963a], inspired by a question suggested by J.P. Runyan to Seshu and Reed [1961]).

The proof below is a variant of the proof of Lovász [1976c] (cf. Lovász [1979b]).

Theorem 55.2 (Lucchesi-Younger theorem). *Let $D = (V, A)$ be a weakly connected digraph. Then the minimum size of a directed cut cover is equal to the maximum number of disjoint directed cuts.*

Proof. For any digraph D, let $\nu(D)$ be the maximum number of disjoint directed cuts in D and let $\tau(D)$ be the minimum size of a directed cut cover. Choose a counterexample $D = (V, A)$ with a minimum number of arcs.

For any $B \subseteq A$, let D_B be the graph obtained from D by replacing each arc (u, v) in B by a directed $u - v$ path of length 2 (the intermediate vertex being new). Choose an inclusionwise maximal subset B of A with $\nu(D_B) = \nu(D)$. Then $B \neq A$, as $\nu(D_A) \geq 2\nu(D) > \nu(D)$.

Choose $b \in A \setminus B$. So $\nu(D_{B\cup\{b\}}) > \nu(D)$. Moreover, as D is a smallest counterexample, the graph D' obtained from D by contracting b satisfies $\nu(D') = \tau(D') \geq \tau(D) - 1 \geq \nu(D)$. Combining a maximum-size packing of directed cuts in D' and one in $D_{B\cup\{b\}}$, we obtain a family $\mathcal{F}$ of nonempty proper subsets of the vertex set V' of D_B with the property that

(55.3) $|\mathcal{F}| = 2\nu(D) + 1$, and the $\delta^{\text{in}}(U)$ for $U \in \mathcal{F}$ are directed cuts in D_B covering any arc of D_B at most twice.

Now we choose $\mathcal{F}$ satisfying (55.3) such that

(55.4) $$\sum_{U \in \mathcal{F}} |U||V' \setminus U|$$

is minimized. Then $\mathcal{F}$ is a cross-free family. Indeed, if $X, Y \in \mathcal{F}$ with $X \not\subseteq Y \not\subseteq X$, $X \cap Y \neq \emptyset$ and $X \cup Y \neq V'$, we can replace X and Y by $X \cap Y$ and $X \cup Y$, while not violating (55.3) but decreasing sum (55.4) (by Theorem 2.1), contradicting its minimality.

So $\mathcal{F}$ is cross-free. For each $X \in \mathcal{F}$, define

(55.5) $\beta(X) := \{U \in \mathcal{F} \mid U \subseteq X \text{ or } U \cap X = \emptyset\}$.

Let $\mathcal{F}_2$ be the collection of sets occurring twice in $\mathcal{F}$ and let $\mathcal{F}_1$ be the collection of sets occurring precisely once in $\mathcal{F}$. Then

(55.6) if X and Y are distinct sets in $\mathcal{F}_1$ with $|\beta(X)| \equiv |\beta(Y)|$ (mod 2), then no arc of D enters both X and Y.

Suppose that to the contrary arc a enters both X and Y. As $\mathcal{F}$ is cross-free, we can assume that $X \subset Y$.

If $|\beta(Y)| \leq |\beta(X)|$, then (as $Y \in \beta(Y) \setminus \beta(X)$) there exists a Z in $\beta(X) \setminus \beta(Y)$. So $Z \not\subseteq Y$ and $Z \cap Y \neq \emptyset$. Hence $Z \not\subseteq X$, and so $Z \cap X = \emptyset$. So $Y \not\subseteq Z$, and hence (as $\mathcal{F}$ is cross-free) $Z \cup Y = V'$. So a leaves Z, a contradiction (since no arc leaves any set in $\mathcal{F}$).

If $|\beta(Y)| \geq |\beta(X)|+2$, then there exists a $Z \neq Y$ with $Z \in \beta(Y) \setminus \beta(X)$. So $Z \not\subseteq X$ and $Z \cap X \neq \emptyset$. Hence $Z \cap Y \neq \emptyset$, and so $Z \subseteq Y$. So $Z \cup X \neq V'$, and hence (as $\mathcal{F}$ is cross-free) $X \subset Z$. So a enters X, Y, and Z, a contradiction. This proves (55.6).

It follows that for some $j \in \{0, 1\}$, the collection

(55.7) $\mathcal{F}_2 \cup \{X \in \mathcal{F}_1 \mid |\beta(X)| \equiv j \pmod{2}\}$

has size at least $\nu(D) + 1$. By (55.6), it gives $\nu(D) + 1$ disjoint directed cuts in D_B, contradicting our assumption. ∎

Equivalent to the Lucchesi-Younger theorem is the following weighted version of it:

Corollary 55.2a. *Let $D = (V, A)$ be a digraph and let $l : A \to \mathbb{Z}_+$ be a length function. Then the minimum length of a directed cut cover is equal to* the maximum number of directed cuts such that each arc a is in at most $l(a)$ of them.

Proof. Replace any arc a by a path of length $l(a)$ (contracting a if $l(a) = 0$). Then the Lucchesi-Younger theorem applied to the new graph gives the present corollary. ∎

This can be formulated in terms of the total dual integrality of the following system:

(55.8)   (i) $0 \leq x_a \leq 1$ for each $a \in A$,
            (ii) $x(C) \geq 1$ for each directed cut C.

Define the *directed cut cover polytope* of D as the convex hull of the incidence vectors of directed cut covers. Then:

Corollary 55.2b. *System (55.8) is TDI and determines the directed cut cover polytope of D.*

Proof. The total dual integrality is a reformulation of Corollary 55.2a. The total dual integrality of (55.8) implies that it determines an integer polytope. Hence the second part of the corollary follows. ∎

55.3. Directed cut k-covers

In fact, system (55.8) is box-TDI, and more generally, the following system is box-TDI, as was shown by Edmonds and Giles [1977]:

(55.9) $x(C) \geq 1$ for each directed cut C.

Edmonds and Giles' proof gives the following alternative way of proving the Lucchesi-Younger theorem.

Theorem 55.3. *System (55.9) is box-TDI.*

Proof. Let $\mathcal{U}$ be the collection of nonempty proper subsets U of V with $\delta^{\text{out}}(U) = \emptyset$. So $\{\delta^{\text{in}}(U) \mid U \in \mathcal{U}\}$ is the collection of all directed cuts.

Choose $w \in \mathbb{R}^A$, and let y achieve the maximum in the dual of minimizing $w^T x$ over (55.9), that is, in:

(55.10) $\max\{\sum_{U \in \mathcal{U}} y_U \mid y \in \mathbb{R}_+^\mathcal{U}, \sum_{U \in \mathcal{U}} y_U \chi^{\delta^{\text{in}}(U)} = w\}$,

such that

(55.11) $\sum_{U \in \mathcal{U}} y_U |U||V \setminus U|$

is as small as possible. Let $\mathcal{F} := \{U \in \mathcal{U} \mid y_U > 0\}$. Then $\mathcal{F}$ is cross-free. Suppose to the contrary that $T, U \in \mathcal{F}$ with $T \not\subseteq U \not\subseteq T$, $T \cap U \neq \emptyset$, $T \cup U \neq V$. Let $\alpha := \min\{y_T, y_U\} > 0$. Then decreasing y_T and y_U by α, and increasing $y_{T \cap U}$ and $y_{T \cup U}$ by α, maintains feasibility of y, while its value is not changed; so it remains an optimum solution. However, sum (55.11) decreases (by Theorem 2.1). This contradicts the minimality of (55.11).

So $\mathcal{F}$ is cross-free, and hence the constraints corresponding to $\mathcal{F}$ form a totally unimodular matrix (Corollary 13.21a). Hence, by Theorem 5.35, system (55.9) is box-TDI. ∎

This implies the box-total dual integrality of (for $k \geq 0$):

(55.12) $x(C) \geq k$ for each directed cut C.

Corollary 55.3a. *For each $k \in \mathbb{R}_+$, system (55.12) is box-TDI.*

Proof. Directly from Theorem 55.3, since if a system $Ax \leq b$ is box-TDI, then also $Ax \leq k \cdot b$ is box-TDI. ∎

This has the following consequences. Call a subset C of the arc set A of a digraph $D = (V, A)$ a *directed cut k-cover* if C intersects each directed cut in at least k arcs. Consider the system:

(55.13) $0 \leq x_a \leq 1$ for $a \in A$,
         $x(C) \geq k$ for each directed cut C.

Then:

Corollary 55.3b. *System (55.13) is TDI and determines the convex hull of the incidence vectors of the directed cut k-covers.*

Proof. Directly from Corollary 55.3a. ∎

From this, a min-max relation for the minimum size of a directed cut k-cover can be derived:

Corollary 55.3c. *Let $D = (V, A)$ be a digraph and let $k \in \mathbb{Z}_+$, such that each directed cut has size at least k. Then the minimum size of a directed cut k-cover is equal to the maximum value of*

(55.14) $|\bigcup \mathcal{C}| + k|\mathcal{C}| - \sum_{C \in \mathcal{C}} |C|$

taken over all collections $\mathcal{C}$ of directed cuts.

Proof. By Corollary 55.3b, the minimum size of a directed cut k-cover is equal to the minimum value of $\mathbf{1}^T x$ over (55.13). Hence, as (55.12) is TDI, the minimum size of a directed cut k-cover is equal to the maximum value of

(55.15) $$k \sum_C y_C - z(A),$$

where $y_C \in \mathbb{Z}_+$ for each directed cut C and where $z \in \mathbb{Z}_+^A$ such that

(55.16) $$\sum_C y_C \chi^C - z \leq 1.$$

Now we can assume that $y_C \in \{0, 1\}$ for each C, since if $y_C \geq 2$, then $z_a \geq 1$ for each $a \in C$ (by (55.16)). Hence decreasing y_C by 1 and decreasing z_a by 1 for each $a \in C$, maintains (55.16) while (55.15) is not decreased (as $|C| \geq k$ by assumption).

Let $\mathcal{C} := \{C \mid y_C = 1\}$. As $z(A)$ is minimized, we have

(55.17) $$z = \sum_{C \in \mathcal{C}} \chi^C - \chi^{\bigcup \mathcal{C}},$$

and hence that $z(A)$ is equal to $\sum_{C \in \mathcal{C}} |C| - |\bigcup \mathcal{C}|$. This proves the corollary. ∎

55.4. Feedback arc sets

The Lucchesi-Younger theorem implies a min-max relation for the minimum size of a feedback arc set in a planar digraph. A *feedback arc set* in a digraph $D = (V, A)$ is a set of arcs intersecting every directed circuit.

In fact, if D has no loops, then a set A' is an inclusionwise minimal feedback arc set if and only if A' is an inclusionwise minimal set of arcs such that inverting all arcs in A' makes the digraph acyclic (Grinberg and Dambit [1966], Gallai [1968a]).

E.L. Lawler and R.M. Karp (see Karp [1972b]) showed that finding a minimum-size feedback arc set in a digraph, is NP-complete. For planar digraphs one has however:

Theorem 55.4. *Let $D = (V, A)$ be a planar digraph. Then the minimum size of a feedback arc set is equal to the maximum number of arc-disjoint directed circuits.*

Proof. Consider the dual digraph D^* of D. Then a set of arcs of D forms a directed circuit if and only if the set of dual arcs forms a directed cut in D^*. Hence the corollary follows immediately from the Lucchesi-Younger theorem (Theorem 55.2). ∎

Notes. Figure 55.1 (from Younger [1965]) shows that we cannot drop the planarity condition. This is a counterexample with a smallest number of vertices, since Barahona, Fonlupt, and Mahjoub [1994] showed that in a digraph with no $K_{3,3}$ minor, the minimum size of a feedback arc set is equal to the maximum number of disjoint

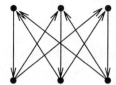

Figure 55.1
The minimum size of a feedback arc set equals 2, while there are no two disjoint directed cuts.

directed circuits. The proof is based on a theorem of Wagner [1937b] on decomposing graphs without $K_{3,3}$ minor into planar graphs and copies of K_5. (Nutov and Penn [1995] gave a similar proof. Related work is done is reported in Nutov and Penn [2000].)

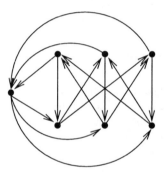

Figure 55.2
An Eulerian digraph where the minimum size of a feedback arc set equals 5, while there are no 5 disjoint directed cuts.

Moreover, Borobia, Nutov, and Penn [1996] showed that in an Eulerian digraph with at most 6 vertices, the minimum size of a feedback arc set is equal to the maximum value of a *fractional* packing of directed circuits. This is not the case if there are more than 6 vertices, as is shown by the graph in Figure 55.2.

Guenin and Thomas [2001] characterized the digraphs D that have the property that for every subhypergraph D' of D, the maximum number of disjoint circuits in D' is equal to the minimum size of a feedback arc set in D'.

More on the polytope determined by the feedback arc sets, equivalently on the *acyclic subgraph polytope* (the convex hull of the incidence vectors of arc sets containing no directed circuit) is presented in Young [1978], Grötschel, Jünger, and Reinelt [1984,1985a,1985b], Reinelt [1993], Leung and Lee [1994], Goemans and Hall [1996], and Bolotashvili, Kovalev, and Girlich [1999]. (Bowman [1972] wrongly claimed to give a system determining the acyclic subgraph polytope.)

The problem of finding a minimum-weight feedback arc set is equivalent to the *linear ordering problem*: given a matrix M, find a permutation matrix P such that the sum of the elements below the main diagonal of $P^\mathsf{T} MP$ is minimized. More on this can be found in Younger [1963b], Jünger [1985], Reinelt [1985], Berger and Shor [1990,1997], Arora, Frieze, and Kaplan [1996,2002], Fernandez de la Vega [1996], Frieze and Kannan [1996,1999], and Newman and Vempala [2001].

For feedback arc sets in linklessly embeddable graphs, see Section 55.6b. For feedback vertex sets, see Section 55.6c.

55.5. Complexity

It was shown by Lucchesi [1976], Karzanov [1979c,1981], and Frank [1981b] that a minimum-size directed cut cover and a maximum packing of directed cuts can be found in polynomial time. Lucchesi [1976] also gave a weakly polynomial-time algorithm for the weighted versions of these problems, and Frank [1981b] gave a strongly polynomial-time algorithm for these problems.

Frank and Tardos [1984b] showed that finding a minimum-length directed cut k-cover in fact can be reduced to a weighted matroid intersection problem. Thus all ingredients for a strongly polynomial-time algorithm are ready at hand.

We describe the reduction. Let $D = (V, A)$ be a digraph, let $l : A \to \mathbb{Q}_+$ be a length function, and let $k \in \mathbb{Z}_+$. We want to find a directed cut k-cover of minimum length.

Let $D^{-1} = (V, A^{-1})$ be the reverse digraph of D, where $A^{-1} := \{a^{-1} \mid a \in A\}$ and $a^{-1} = (v, u)$ if $a = (u, v)$. We will define matroids M_1 and M_2 on $A \cup A^{-1}$.

M_1 is easy: it is the partition matroid induced by the sets $\{a, a^{-1}\}$ for $a \in A$. To define M_2, let $\mathcal{U}$ be the collection of nonempty proper subsets U of V with $\delta_A^{\text{in}}(U) = \emptyset$. Define

(55.18) $\quad P := \{x \in \mathbb{Z}_+^V \mid x(V) = |A| \text{ and } x(U) \geq |A[U]| + k \text{ for each } U \in \mathcal{U}\}$.

Then:

(55.19) $\quad$ for $x, y \in P$ and $u \in V$ with $x_u < y_u$, there exists a $v \in V$ with $x_v > y_v$ and $x + \chi^u - \chi^v \in P$.

Indeed, let $\mathcal{K}$ be the collection of inclusionwise maximal subsets U of $V \setminus \{u\}$ with $U \in \mathcal{U}$ and $x(U) = |A[U]| + k$. As sets with this property are closed under unions of intersecting sets, $\mathcal{K}$ consists of disjoint sets, and no two of them are connected by an arc. Hence for $W := V \setminus \bigcup \mathcal{K}$, we have

(55.20) $\quad y(W) = y(V) - \sum_{U \in \mathcal{K}} y(U) \leq |A| - \sum_{U \in \mathcal{K}} (|A[U]| + k) = x(W)$.

As $x_u < y_u$ and $u \in W$, we know that $x_v > y_v$ for some $v \in W$. Also, $x + \chi^u - \chi^v \in P$, since there is no subset U of $V \setminus \{u\}$ with $\delta^{\text{in}}(U) = \emptyset$, $x(U) = |A[U]| + k$, and $v \in U$.

This shows (55.19), which implies that

(55.21) $\quad \mathcal{B} := \{B \subseteq A \cup A^{-1} \mid \deg_B^{\text{in}} \in P\}$

forms the collection of bases of a matroid M_2 on $A \cup A^{-1}$, provided that $\mathcal{B}$ is nonempty; equivalently, provided that each directed cut in D has size at least k. (That M_2 is a matroid can also be derived from Corollary 49.7a.)

To test independence in M_2, it suffices to have one base of M_2 (which we have: A^{-1}), and to have a test of being a base. Equivalently, we should be able to test membership of P. Let $x \in \mathbb{Z}_+^V$ with $x(V) = |A|$. By Theorem 51.3, we can find a nonempty proper subset U of V minimizing

(55.22) $\quad x(U) - |A[U]| + (k + |A|)d^{\text{in}}(U)$
$= x(U) - \sum_{v \in U} \deg^{\text{out}}(v) + d^{\text{out}}(U) + (k + |A|)d^{\text{in}}(U),$

in strongly polynomial time. If this minimum is at least k, then x belongs to P. If this minimum is less than k, then $d^{\text{in}}(U) = 0$, and hence $x(U) < |A[U]| + k$, implying that x does not belong to P.

Now a subset C of A is a directed cut k-cover if and only if $B := (A \setminus C) \cup C^{-1}$ is a common base of M_1 and M_2. Hence:

Theorem 55.5. *Given a digraph $D = (V, A)$, a length function $l : A \to \mathbb{Q}_+$, and $k \in \mathbb{Z}_+$, a minimum-length directed cut k-cover can be found in strongly polynomial time.*

Proof. Directly from the above, with Theorem 41.8. We apply the weighted matroid intersection algorithm to find a maximum-length common base B in the matroids M_1 and M_2 on $A \cup A^{-1}$, defining $l(a^{-1}) := 0$ for $a \in A$. Then $A \setminus B$ is a minimum-length directed cut cover. ∎

55.5a. Finding a dual solution

Also a maximum packing of directed cuts can be found in polynomial time. Let B be the maximum-length base found and let C be the directed cut k-cover with $B = (A \setminus C) \cup C^{-1}$.

The weighted matroid intersection algorithm also yields a dual solution. Indeed, if l is integer-valued, it gives length functions $l_1, l_2 : A \cup A^{-1} \to \mathbb{Z}$ such that $l = l_1 + l_2$ and such that B is an l_i-maximal base of M_i, for $i = 1, 2$ (cf. Section 41.3a).

Define

(55.23) $\quad \mathcal{F} := \{U \subseteq V \mid d_A^{\text{in}}(U) = 0, d_C^{\text{out}}(U) = k\},$

and define a pre-order $\preceq$ on V by:

(55.24) $\quad u \preceq v \iff$ each $U \in \mathcal{F}$ containing u also contains v, for $u, v \in V$. It can be checked in polynomial time whether $u \preceq v$ holds, since it is equivalent to: $\deg_B^{\text{in}} - \chi^u + \chi^v \in P$. Indeed, $\deg_B^{\text{in}} - \chi^u + \chi^v$ belongs to P if and only if $v \in U$ for each $U \in \mathcal{U}$ satisfying $u \in U$ and $\sum_{s \in U} \deg_B^{\text{in}}(s) = |A[U]| + k$. Now $\sum_{s \in U} \deg_B^{\text{in}}(s) = |A[U]| + d_C^{\text{out}}(U)$. So it is equivalent to: $u \preceq v$.

Next define for each $u \in V$:

(55.25) $\quad p(u) := \max\{l_2(a) \mid \exists v \succeq u : a^{-1} \in \delta_B^{\text{out}}(v)\}$.

Let $h_0 < \cdots < h_t$ be the elements of the set $\{p(u) \mid u \in V\}$. For $j = 1, \ldots, t$, define $V_j := \{u \mid p(u) \geq h_j\}$. Let $\mathcal{K}$ be the collection of all weak components of $D - V_j$, over all $j = 1, \ldots, t$, and for each $K \in \mathcal{K}$, let

(55.26) $\quad y_K := \sum (h_j - h_{j-1} \mid j = 1, \ldots, t;\ K \text{ is a weak component of } D - V_j)$.

So

(55.27) $\quad P = h_0 \chi^V + \sum_{K \in \mathcal{K}} y_K \chi^K$.

Then:

Theorem 55.6. *Each $K \in \mathcal{K}$ belongs to $\mathcal{F}$. Moreover, for each $a = (u, v) \in A$:*

(55.28) $\quad$ (i) $\quad \sum(y_K \mid K \in \mathcal{K}, a \in \delta^{\text{out}}(K)) \leq l(a) \quad$ *if $a \in A \setminus C$,*
$\qquad\quad$ (ii) $\quad \sum(y_K \mid K \in \mathcal{K}, a \in \delta^{\text{out}}(K)) \geq l(a) \quad$ *if $a \in C$.*

Proof. Consider any $j = 1, \ldots, t$. By definition of $p(u)$, we know that V_j is a lower ideal with respect to $\preceq$. That is, if $v \in V_j$ and $u \preceq v$, then $u \in V_j$. (Indeed, if $v \in V_j$, then $p(v) \geq h_j$, hence $l_2(a) \geq h_j$ for some a with $a^{-1} \in \delta_B^{\text{out}}(w)$ for some $w \succeq v$. Since $w \succeq u$ we have $p(u) \geq l_2(a) \geq h_j$.)

Hence, for each $v \in V_j$ and $u \notin V_j$ we have $u \not\preceq v$. Therefore, there is a $U \in \mathcal{F}$ with $u \in U$ and $v \notin U$. This implies, as $\mathcal{F}$ is a crossing family, that there is a partition of $V \setminus V_j$ into sets in $\mathcal{F}$. As $d_A^{\text{in}}(U) = 0$ and $d_C^{\text{out}}(U) = k$, it follows that this partition is equal to the collection of weak components of the digraph $D - V_j$. So each weak component K of $D - V_j$ satisfies $d_A^{\text{in}}(K) = 0$ and $d_C^{\text{out}}(K) = k$; that is, K belongs to $\mathcal{F}$.

Consider any arc $a = (v, u) \in B$. As B is an l_1-maximal base of M_1, we have $l_1(a^{-1}) \leq l_1(a)$. Let $p(u) = l_2(b)$ for some $b^{-1} \in \delta_B^{\text{out}}(w)$ and some $w \succeq u$. Since $u \preceq w$, we know that $\deg_B^{\text{in}} - \chi^u + \chi^w \in P$. So $(B \cup \{b\}) \setminus \{a\}$ is again a base of M_2. Hence we have (as B is an l_2-maximal base of M_2) $l_2(b) \leq l_2(a)$. So $l_2(a) \geq l_2(b) \geq p(u)$. Also $p(v) \geq l_2(a^{-1})$, by definition of $p(v)$. Hence

(55.29) $\quad l(a) - l(a^{-1}) = l_1(a) + l_2(a) - l_1(a^{-1}) - l_2(a^{-1}) \geq l_2(a) - l_2(a^{-1})$
$\qquad\qquad \geq p(u) - p(v)$.

If $a \in A \setminus C$, we have $l(a^{-1}) = 0$, and obtain (55.28)(i), since a enters no $K \in \mathcal{K}$ and so $p(u) \geq p(v)$. Hence

(55.30) $\quad l(a) \geq p(u) - p(v) = \sum(y_K \mid K \in \mathcal{K}, a \in \delta^{\text{out}}(K))$.

If $a \in C^{-1}$, we have $l(a) = 0$ and obtain (55.28)(ii), since a^{-1} enters no $K \in \mathcal{K}$, and so $p(u) \leq p(v)$. Hence

(55.31) $\quad l(a^{-1}) \leq p(v) - p(u) = \sum(y_K \mid K \in \mathcal{K}, a^{-1} \in \delta^{\text{out}}(K))$. ∎

For each $a \in C$, let $s(a)$ be the difference of the two terms in (55.28)(ii), and for $a \in A \setminus C$ let $s(a) := 0$. Then

(55.32) $$\sum_{K \in \mathcal{K}} y_K \chi^{\delta_A^{\text{out}}(K)} - s \leq l$$

and

(55.33) $$k \sum_{K \in \mathcal{K}} y_K - s(A) = \sum_{K \in \mathcal{K}} y_K |\delta_A^{\text{out}}(K) \cap C| - s(A)$$
$$= \sum_{a \in C} \sum (y_K \mid K \in \mathcal{K}, a \in \delta^{\text{out}}(K)) - s(A) = l(C).$$

Thus we have an integer optimum dual solution to maximizing $l^T x$ over the system $0 \leq x \leq 1$, $x(Y) \geq k$ (Y directed cut). If $k = 1$, we can do with $s = 0$, and obtain an integer optimum packing of directed cuts subject to l.

So we have:

Theorem 55.7. *Given a digraph $D = (V, A)$ and a length function $l : A \to \mathbb{Z}_+$, an optimum packing of directed cuts subject to l can be found in strongly polynomial time.*

Proof. See above. ∎

55.6. Further results and notes

55.6a. Complexity survey for minimum-size directed cut cover

	$O(n^5 \log n)$	Lucchesi [1976]
	$O(n^3 m)$	Frank [1981b]
	$O(n^2 M(n))$	Frank [1981b]
*	$O(n^2 m)$	Gabow [1993b,1995c]
*	$O(nM(n))$	Gabow [1993b,1995c]

As before, * indicates an asymptotically best bound in the table. $M(n)$ denotes the time to multiply $n \times n$ matrices. Also Karzanov [1979c,1981] gave a polynomial-time algorithm to find a minimum-size directed cut cover. Lucchesi [1976] gave also a polynomial-time algorithm to find a minimum-weight directed cut cover, and Frank [1981b] and Gabow [1993a,1993b,1995c] gave strongly polynomial-time algorithms for this.

55.6b. Feedback arc sets in linklessly embeddable digraphs

An undirected graph is called *linklessly embeddable* if it can be embedded in $\mathbb{R}^3$ such that any two vertex-disjoint circuits C_1 and C_2 are unlinked (that is, there is a topological sphere S such that C_1 is in the interior of S and C_2 is in the

Section 55.6b. Feedback arc sets in linklessly embeddable digraphs 957

exterior of S). A digraph is called linklessly embeddable if its underlying undirected graph is linklessly embeddable. (Linklessly embeddable graphs are characterized by Robertson, Seymour, and Thomas [1995] in terms of 'forbidden minors'.)

Seymour [1996] showed that in an Eulerian linklessly embeddable directed graph, the minimum-size of a feedback arc set is equal to the maximum number of arc-disjoint directed circuits. We sketch the proof.

The basic combinatorial-topological part of the proof consists of showing:

Theorem 55.8. *Let D be an Eulerian linklessly embeddable digraph. Suppose that there exist $2k+1$ directed circuits such that any arc is in at most two of them. Then there exist $k+1$ arc-disjoint directed circuits.*

Sketch of proof. By a theorem of Robertson, Seymour, and Thomas [1995], D can be embedded in $\mathbb{R}^3$ such that for each undirected circuit C in D there exists an open disk B in $\mathbb{R}^3$ with boundary C and disjoint from D. (We identify D with its embedding.)

Let $C_1, \ldots, C_t$ be a maximum number of directed circuits in D such that any arc of D is in at most two of them. So $t \geq 2k + 1$. Moreover, each arc of D is contained in *exactly* two of the C_i. Otherwise, the arcs not covered twice contain a directed circuit (as D is Eulerian). This contradicts the maximality of t.

For each $i = 1, \ldots, t$, let B_i by an open disk with boundary C_i and disjoint from D. We can assume that the B_i are pairwise disjoint, as can be seen as follows. First, we can assume that the B_i are tame and in general position. In particular, no point is in four of the B_i. Any point p in three of the B_i is the intersection point of three of the B_i, pairwise crossing at p. Any point p in two of the B_i is the intersection point of two of the B_i, crossing at p. Moreover, any two distinct B_i and B_j intersect each other in a finite number of closed and open curves, each representing crossings of B_i and B_j. Let $c(B_i, B_j)$ denote the number of such components.

We choose the C_i and B_i such that the sum of the $c(B_i, B_j)$ for $i \neq j$ is minimized.

Now it is elementary combinatorial topology to prove that there exist for any distinct i, j, with $B_i \cap B_j \neq \emptyset$, directed circuits C'_i and C'_j in D with

(55.34) $\chi^{AC'_i} + \chi^{AC'_j} = \chi^{AC_i} + \chi^{AC_j}$

and open disks B'_i and B'_j with boundaries C'_i and C'_j respectively, such that $B'_i \cap B'_j = \emptyset$ and $c(B'_i, B_h) + c(B'_j, B_h) \leq c(B_i, B_h) + c(B_j, B_h)$ for all $h \neq i, j$.

Hence, by the minimality of the sum of the $c(B_i, B_j)$, it follows that the B_i are disjoint. So D, together with the B_i forms the union of a number of compact surfaces, certain points of which are identified. As these surfaces are orientable (since they are embedded in $\mathbb{R}^3$), the B_i fall apart into two classes: those with boundary oriented clockwise, and those with boundary oriented counter-clockwise. Each of these classes have arc-disjoint boundaries, and at least one of these classes has size at least $k+1$. This proves the theorem. ∎

Seymour [1996] next continues by deriving (for linklessly embeddable graphs) the total dual integrality of the following system in $x \in \mathbb{R}^A$:

(55.35) $x(C) \geq 1$ for each directed circuit C in D.

Note that nonnegativity of x is not required here.

Corollary 55.8a. *For any linklessly embeddable digraph $D = (V, A)$, system (55.35) is totally dual integral.*

Proof. Let $w \in \mathbb{Z}^A$ be such that the minimum of $w^\mathsf{T} x$ over (55.35) is finite. Let $\mathcal{C}$ be the collection of directed circuits in D. By Theorem 5.29, it suffices to show that the maximum value μ of $\sum_{C \in \mathcal{C}} y(C)$ taken over $y : \mathcal{C} \to \frac{1}{2}\mathbb{Z}_+$ satisfying

(55.36) $$\sum_{C \in \mathcal{C}} y_C \chi^{AC} \leq w$$

is attained by an integer-valued y.

Now (as the minimum is finite) w belongs to the cone generated by the incidence vectors of directed circuits, and hence w is a nonnegative circulation. Replace any arc $a = (u, v)$ by $w(a)$ parallel arcs from u to v, giving the Eulerian digraph $D' = (V, A')$. Then μ is equal to half of the maximum number μ' of directed circuits in D' such that any arc of D' is in at most two of these circuits. By Theorem 55.8, D' contains at least $\lceil \frac{1}{2}\mu' \rceil$ arc-disjoint directed circuits. Since $\lceil \frac{1}{2}\mu' \rceil \geq \mu$, this gives in D an integer vector $y : \mathcal{C} \to \mathbb{Z}_+$ as required. ∎

This finally gives:

Theorem 55.9. *The minimum size of a feedback arc set in an Eulerian linklessly embeddable digraph $D = (V, A)$ is equal to the maximum number of arc-disjoint directed circuits.*

Proof. Consider the LP-duality relation for maximizing $x(U)$ over (55.35):

(55.37) $\min\{x(A) \mid x(C) \geq 1 \text{ for each directed circuit } C\}$
$= \max\{\sum_C y_C \mid y_C \geq 0, \sum_C y_C \chi^{AC} = \mathbf{1}\},$

where C ranges over all directed circuits. By Corollary 55.8a and the theory of total dual integrality (Theorem 5.22), both optima have an integer optimum solution. So the maximum is equal to the maximum number of arc-disjoint directed circuits. Let x attain the minimum. By Theorem 8.2, there exists a ('potential') $p : V \to \mathbb{Z}$ with $x_a \geq p(v) - p(u)$ for each arc $a = (u, v)$ of D. Define $x'(a) := x_a - p(v) + p(u)$ for each arc $a = (u, v)$. Then $x' \in \mathbb{Z}_+^A$, $x'(C) = x(C) \geq 1$ for each directed circuit C, and $x'(A) = x(A)$ (since D is Eulerian). Hence the set of arcs a with $x'(a) \geq 1$ forms a feedback arc set of size at most $x(A)$, proving the theorem. ∎

System (55.35) can be tested in polynomial time, for any digraph (with the Bellman-Ford method). It implies that in an Eulerian linklessly embeddable digraph, a minimum-size feedback arc set can be found in polynomial time (with the ellipsoid method).

55.6c. Feedback vertex sets

A *feedback vertex set* in a digraph $D = (V, A)$ is a subset U of V with $D - U$ acyclic — that is, U intersects every directed circuit. Reed, Robertson, Seymour, and Thomas [1996] proved:

(55.38)   for each integer $k \geq 0$ there exists an integer $n_k \geq 0$ such that each digraph $D = (V, A)$ having no k vertex-disjoint directed circuits, has a feedback vertex set of size at most n_k.

(For $k = 2$ this answers a question of Gallai [1968b] and Younger [1973].)

Reed, Robertson, Seymour, and Thomas also showed that for each fixed integer k, there is a polynomial-time algorithm to find k vertex-disjoint directed circuits in a digraph if they exist.

Earlier, progress on (55.38) was made by McCuaig [1993], who proved it for $k = 2$, where $n_2 = 3$, by Seymour [1995b], who proved a fractional version of it (if there is no fractional packing of k directed circuits, then there is a feedback vertex set of size at most n_k), and by Reed and Shepherd [1996] for planar graphs.

According to Reed, Robertson, Seymour, and Thomas [1996], N. Alon proved that n_k is at least $Ck \log k$ for some constant C.

Cai, Deng, and Zang [1999,2002] characterized for which orientations $D = (V, A)$ of a complete bipartite graph the system

(55.39) $x_v \geq 0$ for $v \in V$,
          $x(VC) \geq 1$ for each directed circuit C,

is totally dual integral. (Related results can be found in Cai, Deng, and Zang [1998].)

Guenin [2001b] gave a characterization of digraphs $D = (V, A)$ for which the linear system in $\mathbb{R}^{V \cup A}$ for feedback arc and vertex sets:

(55.40) $x_v \geq 0$ for $v \in V$,
          $x_a \geq 0$ for $a \in A$,
          $x(VC) + x(AC) \geq 1$ for each directed circuit C,

is totally dual integral.

The undirected analogue of (55.38) was proved for $k = 2$ by Bollobás [1963], and for general k by Erdős and Pósa [1965]. Ding and Zang [1999] characterized the undirected graphs $G = (V, E)$ for which the system

(55.41) $x_v \geq 0$ for $v \in V$,
          $x(VC) \geq 1$ for each circuit C,

is totally dual integral. Their characterization implies that (55.41) is totally dual integral if and only if it defines an integer polyhedron.

A polyhedral approach to the feedback vertex set problem was investigated by Funke and Reinelt [1996]. Approximation algorithms for feedback problems were given by Monien and Schulz [1982], Eades, Lin, and Smyth [1993], Bar-Yehuda, Geiger, Naor, and Roth [1994,1998], Becker and Geiger [1994,1996], Bafna, Berman, and Fujito [1995,1999], Even, Naor, Schieber, and Sudan [1995,1998], Even, Naor, and Zosin [1996,2000], Goemans and Williamson [1996,1998], Chudak, Goemans, Hochbaum, and Williamson [1998], Bar-Yehuda [2000], and Cai, Deng, and Zang [2001]. More on the feedback vertex set problem was presented by Smith and Walford [1975], Kevorkian [1980], Rosen [1982], Speckenmeyer [1988], Stamm [1991], Hackbusch [1997], and Pardalos, Qian, and Resende [1999].

55.6d. The bipartite case

McWhirter and Younger [1971] (cf. Younger [1970], Vidyasankar [1978b]) proved the Lucchesi-Younger theorem in case the arcs of D form a directed cut; that is, in case the underlying undirected graph is bipartite, while all arcs are oriented from one colour class to the other. It amounts to the following:

Theorem 55.10. *Let $G = (V, E)$ be a connected bipartite graph and let $\mathcal{F}$ be the collection of subsets $E[U]$ of E for which U is a vertex cover with $E[U]$ nonempty. Then the minimum size of a set of edges intersecting each set in $\mathcal{F}$ is equal to the maximum number of disjoint sets in $\mathcal{F}$.*

Proof. Let U and W be the colour classes of G and let digraph D be obtained from G by orienting each edge from U to W. Then a set of edges belongs to $\mathcal{F}$ if and only if it is a directed cut of D. Hence the theorem follows from the Lucchesi-Younger theorem. ∎

D.H. Younger (cf. Frank [1993b]) showed that the maximum number of disjoint nonempty cuts in a bipartite graph G is equal to the maximum number of disjoint directed cuts in the directed graph obtained from G by orienting all edges from one colour class to the other (cf. Corollary 29.13b). (Vidyasankar [1978b] showed that a set of edges J intersecting each set in $\mathcal{F}$ attains the minimum in Theorem 55.10 if and only if J intersects any circuit C of G in at most $\frac{1}{2}|EC|$ edges; that is, if and only if J is a join — cf. Section 29.11d.)

As noted by Younger [1979], the Lucchesi-Younger theorem, in the form of Corollary 55.2b, implies the Kőnig-Rado edge cover theorem (Theorem 19.4): the minimum size of an edge cover in a bipartite graph $G = (V, E)$ is equal to the maximum size of a stable set in G. To obtain this as a consequence, let U and W be the colour classes of G and let $D = (V, A)$ be the directed graph with vertex set V and arcs all pairs (u, v) with $u \in U$ and $v \in W$. Define a weight function $w : A \to \mathbb{Z}$ by $w(u, v) := 1$ if $uv \in E$, and ∞ otherwise. Then the minimum weight of a directed cut cover in D is equal to the minimum size of an edge cover in G. With this correspondence, Corollary 55.2b gives the Kőnig-Rado edge cover theorem.

55.6e. Further notes

Frank, Tardos, and Sebő [1984] showed that the Lucchesi-Younger theorem implies that in a digraph $D = (V, A)$, the minimum size of a directed cut cover is equal to the maximum value of

(55.42) $\quad \sum_{i=1}^{k}$ number of weak components of $D - V_i$,

where $\emptyset \neq V_1 \subset V_2 \subset \cdots \subset V_k \subset V$ are such that no arc leaves any V_i and enters at most one of the V_i.

Frank and Tardos [1989] showed that a weakly connected digraph $D = (V, A)$ has a branching that intersects all directed cuts if and only if for each nonempty $U \subseteq V$, the number of weak components K of $D - U$ with $d^{\text{in}}(K) = 0$, is at most $|U|$.

Younger [1965] proved the Lucchesi-Younger theorem for digraphs having an arborescence, and, more generally, Younger [1979] proved it for source-sink connected digraphs (that is, each strong component not left by any arc is reachable by a directed path from each strong component not entered by any arc).

Tuza [1994] showed that for any planar directed graph $D = (V, A)$ and any collection $\mathcal{T}$ of directed triangles in D, the minimum number of arcs intersecting each triangle in $\mathcal{T}$ is equal to the maximum number of arc-disjoint triangles in $\mathcal{T}$.

Chapter 56

Minimum directed cuts and packing directed cut covers

A minimum-capacity directed cut can be found in strongly polynomial time, by applying the minimum-capacity $s-t$ cut algorithm, for all s,t, in some modified digraph.

As for packing directed cut covers it is unknown if it is polynomial-time tractable. Also it is unknown if the maximum number of disjoint directed cut covers is equal to the minimum size of a directed cut — this is Woodall's conjecture. But the capacitated version of it does not hold.

In this chapter we consider a few cases where Woodall's conjecture has been proved, in particular for the *source-sink connected* digraphs.

56.1. Minimum directed cuts and packing directed cut covers

The Lucchesi-Younger theorem states that in a digraph $D = (V,A)$, the minimum size of a directed cut cover is equal to the maximum number of disjoint directed cuts. Woodall [1978a,1978b] ventured the conjecture that this min-max relation would be maintained after interchanging the terms directed cut and directed cut cover:

Conjecture (Woodall's conjecture). In a digraph, the minimum size of a directed cut is equal to the maximum number of disjoint directed cut covers.

This conjecture is open.

A capacitated version of Woodall's conjecture (conjectured by Edmonds and Giles [1977] and D.H. Younger) is however not true. Note that the Lucchesi-Younger theorem is equivalent to its weighted version, by replacing arcs by directed paths of length $l(a)$ if $l(a) \geq 1$, and contracting an arc a if $l(a) = 0$. We could attempt this approach to obtain an equivalent capacitated version from Woodall's conjecture, by replacing any arc a by $c(a)$ parallel arcs, but there is a problem here: if $c(a) = 0$, we delete a and can create new directed cuts.

A capacitated version with capacities 0 and 1 amounts to the statement that each directed cut k-cover can be partitioned into k directed cut covers.

Section 56.1. Minimum directed cuts and packing directed cut covers

Figure 56.1 gives a counterexample to this for the case $k = 2$ (Schrijver [1980a]). Note that the counterexample is planar, and that therefore the 'planar dual' assertion (on packing feedback arc sets) also does not hold.

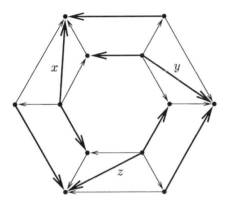

Figure 56.1
A directed cut 2-cover that cannot be split into two directed cut covers. Let C be the set of heavy arcs. Then C is a directed cut 2-cover, since for each arc $c \in C$, the set $C \setminus \{c\}$ is a directed cut cover, which is easy to check since up to symmetry there are only two types of arcs in C.

However, C cannot be split into directed cut covers C_1 and C_2. To see this, observe that each of these C_i must contain exactly one of the two arcs in C meeting any source or sink. Moreover, each C_i must contain at least one of the arcs labeled x, y, z, since the set of arcs from the inner hexagon to the outer hexagon forms a directed cut. Hence we may assume without loss of generality that C_1 contains the arcs x and y, but not z. But then C_1 does not intersect the directed cut of those arcs going from the right half of the figure to the left half.

To interpret this polyhedrally, consider the system:

(56.1) (i) $0 \leq x_a \leq 1$ for each $a \in A$,
        (ii) $x(B) \geq 1$ for each directed cut cover B.

With Corollary 55.2b, the theory of blocking polyhedra gives that system (56.1) determines the convex hull of the incidence vectors of arc sets containing a directed cut. However, by the example in Figure 56.1, system (56.1) generally is not TDI, as total dual integrality amounts to the capacitated version of Woodall's conjecture.

In a number of special cases, Woodall's conjecture, and its capacitated extension, have been proved. In the remainder of this chapter we will consider such cases.

Two more counterexamples to the conjecture of Edmonds and Giles were given by Cornuéjols and Guenin [2002c], and they asked if, together with the example of Figure 56.1, these form all minimal counterexamples to the Edmonds-Giles conjecture.

56.2. Source-sink connected digraphs

Feofiloff and Younger [1987] and Schrijver [1982] showed that for source-sink connected digraphs, the min-max relation for packing directed cut covers does hold. Here a digraph is called *source-sink connected* if each strong component not left by any arc is reachable by a directed path from each strong component not entered by any arc. So an acyclic digraph is source-sink connected if each sink is reachable by a directed path from each source. We follow the proof of Schrijver [1982].

Theorem 56.1. *Let $D = (V, A)$ be a source-sink connected digraph and let $k \in \mathbb{Z}_+$. Then any directed cut k-cover C can be partitioned into k directed cut covers.*

Proof. Choose a counterexample with $|V| + |C|$ as small as possible. Then D is acyclic, since any strong component can be contracted to one vertex.

We may assume that if v is reachable in D from u and $v \neq u$, then $(u, v) \in A$. We first show:

(56.2)   for any nonempty proper subset U of V with $\delta^{\text{out}}(U) = \emptyset$ and $|\delta_C^{\text{in}}(U)| = k$, one has $|U| = 1$ or $|U| = |V| - 1$.

Suppose not. Let $D' := D/U$ and $D'' := D/\overline{U}$ be the digraphs obtained from D by contracting U and $\overline{U} := V \setminus U$, respectively. Note that D' and D'' are source-sink connected again. Let C' be the set of arcs in C with tail in $\overline{U}$, and let C'' be the set of arcs in C with head in U.

Now each directed cut in D' intersects C' in at least k arcs, as it is a directed cut in D and hence intersects C in at least k arcs. So by the minimality of $|V| + |C|$, C' can be split into k directed cut covers $B'_1, \ldots, B'_k$ for D'. As $|\delta_C^{\text{in}}(U)| = k$, each B'_i has exactly one arc entering U. Similarly, C'' can be split into k directed cut covers $B''_1, \ldots, B''_k$ for D'', such that each B''_i has exactly one arc entering U. By choosing indices appropriately, we can assume that B'_i and B''_i have an arc entering U in common, for each $i = 1, \ldots, k$ (as $|\delta_C^{\text{in}}(U)| = k$).

Then each $B'_i \cup B''_i$ is a directed cut cover for D. For suppose that there is a nonempty proper subset W of V with $\delta^{\text{out}}(W) = \emptyset$ and $\delta^{\text{in}}(W)$ disjoint from $B'_i \cup B''_i$. Then $U \cap W \neq \emptyset$ and $U \not\subseteq W$, since otherwise $\delta^{\text{in}}(W)$ is a directed cut of D', and hence some arc in B'_i enters W. So some arc in B''_i enters $U \cap W$. Similarly, some arc in B'_i enters $U \cup W$. Since exactly one arc in $B_i' \cup B_i''$ enters U, it follows that at least one arc in $B_i' \cup B_i''$ enters W, contradicting our assumption.

So each $B_i' \cup B_i''$ is a directed cut cover for D. As they are disjoint, this contradicts our assumption, thus proving (56.2).

We next show the following. Let X be the set of sources of D and let Y be the set of sinks of D. Then:

(56.3)   for each $a = (u,v) \in C$ we have $u \in X$ or $v \in Y$.

For suppose not. Then by (56.2), each directed cut of D intersects $C \setminus \{a\}$ in at least k arcs (as any directed cut intersecting C in exactly k arcs and containing a is equal to $\delta^{\text{in}}(\{v\})$ or $\delta^{\text{in}}(V \setminus \{u\})$, implying that v is a sink or u is a source). So by the minimality of $|V| + |C|$, we can split $C \setminus \{a\}$ into k directed cut covers. This implies that also C can be split into k directed cut covers, contradicting our assumption. This proves (56.3).

Next:

(56.4)   if $a = (u,v) \in C$, $a' = (u',v') \in C$, and v is reachable from u', then $u' \in X$ or $v \in Y$.

For suppose not. By (56.3), $u \in X$ and $v' \in Y$, and hence (since D is source-sink connected), $a'' = (u,v') \in A$. Now $a \neq a'$, as $u \in X$ and $u' \notin X$. So $C' := (C \setminus \{a, a'\}) \cup \{a''\}$ is smaller than C. Moreover, C' is a directed cut k-cover. Indeed, let U be a nonempty proper subset of V with $\delta^{\text{out}}(U) = \emptyset$. If $|U| = 1$ or $|U| = |V| - 1$, then $\delta^{\text{in}}_{C'}(U) = \delta^{\text{in}}_C(U)$, since then $U = \{r\}$ for some sink r or $U = V \setminus \{s\}$ for some source s. If $1 < |U| < |V| - 1$, then

(56.5) $|\delta^{\text{in}}_{C'}(U)| \geq |\delta^{\text{in}}_C(U)| - 1 \geq k$,

since if both a and a' enter U, then $u \notin U$ and $v' \in U$, and hence a'' enters U.

So C' is a directed cut k-cover, and hence, by the minimality of $|V| + |C|$, C' can be split into k directed cut covers. Let B be the directed cut cover containing a''. Then $B' := (B \setminus \{a''\}) \cup \{a, a'\}$ is a directed cut cover, since any directed cut $\delta^{\text{in}}(W)$ containing a'', contains at least one of a, a'. Indeed, otherwise $u, v \notin W$, $u', v' \in W$, but then $u' \neq v$ and arc (u', v) leaves W, contradicting the fact that W determines a directed cut.

So by replacing B by B' we obtain a splitting of C into k directed cut covers, contradicting our assumption. This proves (56.4).

This implies:

(56.6) V can be partitioned into sets R and S such that $\delta^{\text{in}}(R) = \emptyset$, $X \subseteq R$, $Y \subseteq S$, and if any $(u,v) \in C$ leaves R, then $u \in X$ and $v \in Y$.

For define

(56.7) $C' := \{(u,v) \in C \mid u \notin X \text{ or } v \notin Y\}$,
$R := \{v \in V \mid D' = (V, A \cup C'^{-1}) \text{ has a directed } v - X \text{ path}\}$,
$S := V \setminus R$.

Then $X \subseteq R$, $\delta_A^{\text{in}}(R) = \emptyset$, and any $(u,v) \in C$ leaving R satisfies $u \in X$ and $v \in Y$. To see that $Y \subseteq S$, suppose to the contrary that D' has a directed $Y - X$ path P. Choose P shortest. Then by (56.3), P has of at most three arcs. Let (v', u') and (v, u) be the first and last arc of P. So $v' \in Y$ and $u \in X$. These arcs belong to C'^{-1}, and v is reachable from u' in D. So by (56.4), $u' \in X$ or $v \in Y$. This contradicts the definition of C'. This shows (56.6).

Fix R, S as in (56.6). Let $D' = (V, A')$ be the digraph arising from D by replacing any arc (u, v) of D by k parallel arcs from v to u. Then

(56.8) $\qquad |\delta_{A' \cup C}^{\text{in}}(U)| \geq k$

for each nonempty proper subset U of V. So by Theorem 54.11, $A' \cup C$ can be split into k $R - S$ bibranchings. Let $B_1, \ldots, B_k$ be the intersections of these bibranchings with C. We show that each B_i is a directed cut cover, contradicting our assumption, and therefore finishing the proof.

Suppose that say B_1 is not a directed cut cover. Let U be a nonempty proper subset of V with $\delta^{\text{out}}(U) = \emptyset$, and suppose that no arc in B_1 enters U. Note that if U contains any source, it contains all sinks, since no arc of D leaves U. So U contains no sources or contains all sinks.

First assume that U contains no sources of D. As U contains at least one sink (as $U \neq \emptyset$ and $\delta^{\text{out}}(U) = \emptyset$), we know $U \not\subseteq R$. As $A' \cup B_1$ is an $R - S$ bibranching, we know that

(56.9) $\qquad \delta_{A' \cup B_1}^{\text{in}}(U \cap S) \neq \emptyset$.

As $\delta_A^{\text{out}}(U \cap S) = \emptyset$ (since $\delta_A^{\text{out}}(U) = \emptyset$ and $\delta_A^{\text{in}}(R) = \emptyset$), we have $\delta_{A'}^{\text{in}}(U \cap S) = \emptyset$. Hence some arc (u,v) in B_1 enters $U \cap S$. As by assumption (u,v) does not enter U, (u,v) enters S, and $u \in U$. However, by (56.6), u belongs to X. This contradicts our assumption that U contains no sources of D.

The case that U contains all sinks is symmetric, and leads again to a contradiction. ∎

A special case of Theorem 56.1 is Woodall's conjecture for source-sink connected digraphs:

Corollary 56.1a. *Let $D = (V, A)$ be a source-sink connected digraph. Then the minimum size of a directed cut is equal to the maximum number of disjoint directed cut covers.*

Proof. This is the case $C = A$ of Theorem 56.1. ∎

Also, a capacitated version can be derived from the theorem:

Corollary 56.1b. *Let $D = (V, A)$ be a source-sink connected digraph and let $c : A \to \mathbb{Z}_+$ be a capacity function. Then the minimum capacity of a directed cut is equal to the maximum number of directed cut covers such that no arc a is in more than $c(a)$ of these directed cut covers.*

Proof. Directly from Theorem 56.1, by adding, for any arc a of D, $c(a)$ arcs parallel to a, and by taking for C the set of newly added arcs. ∎

Equivalently, in TDI terms:

Corollary 56.1c. *If $D = (V, A)$ is a source-sink connected digraph, then system (56.1) is totally dual integral.*

Proof. This is a reformulation of Corollary 56.1b. ∎

Feofiloff [1983] gave a polynomial-time algorithm to find a maximum number of disjoint directed cut covers in a source-sink connected digraph. Also the proof above implies a polynomial-time algorithm.

A polynomial-time algorithm for the capacitated case can be derived from the ellipsoid method (cf. Grötschel, Lovász, and Schrijver [1988]). A semi-strongly polynomial-time algorithm also follows from Section 57.5 below.

Notes. Frank [1979b] showed the special case of Woodall's conjecture for digraphs having an arborescence. (Such digraphs are source-sink connected.) J. Edmonds observed that this can be derived from Edmonds' disjoint arborescences theorem (Corollary 53.1b): Let $D = (V, A)$ have an r-arborescence. Let k be the minimum size of a directed cut in D. Add to D, for each arc (u, v) of D, k parallel arcs from v to u. This makes the digraph $D' = (V, A')$ with $|\delta_{A'}^{\text{in}}(U)| \geq k$ for each nonempty $U \subseteq V \setminus \{r\}$. Hence D' has k disjoint r-arborescences (by Edmonds' disjoint arborescences theorem). Now for any r-arborescence B in D', the set $B \cap A$ is a directed cut cover in D, since if U is a nonempty proper subset of V with $\delta_A^{\text{out}}(U) = \emptyset$, then $\delta_{A'}^{\text{in}}(U) = \delta_A^{\text{in}}(U)$, and hence $\delta_{B \cap A}^{\text{in}}(U) = \delta_B^{\text{in}}(U) \neq \emptyset$. So we obtain k disjoint directed cut covers in D.

56.3. Other cases where Woodall's conjecture is true

Another case where Woodall's conjecture holds is given in:

Theorem 56.2. *Let $D = (V, A)$ be a digraph arising from a directed tree $T = (V, A')$ such that $(u, v) \in A$ if and only if v is reachable in T from u. Let $c : A \to \mathbb{Z}_+$ be a capacity function. Then the minimum capacity of a directed cut is equal to the maximum number of directed cut covers such that each arc a is in at most $c(a)$ of them.*

Proof. The proof is by induction on the minimum capacity k of a directed cut. Then it suffices to show that there exists a directed cut cover B with $\chi^B \leq c$ and with $(c - \chi^B)(C) \geq k - 1$ for each directed cut C.

Let M be the $A' \times A$ network matrix generated by T and D (cf. Section 13.3). Then the rows of M are precisely the incidence vectors of inclusionwise minimal directed cuts. So it suffices to show that there exists an integer solution x of

(56.10) $\quad 0 \leq x \leq c, Mx \geq 1, M(c-x) \geq (k-1)1,$

since for any such x there is a directed cut cover B satisfying $\chi^B \leq x$.

Since M is totally unimodular, it suffices to show that (56.10) has any solution. Define $x := \frac{1}{k}c$. Then x satisfies (56.10), since $Mc \geq k\mathbf{1}$ and hence $Mx \geq 1$ and $M(c-x) = \frac{k-1}{k}Mc \geq (k-1)\mathbf{1}$. ∎

The theorem can also be formulated in terms of partitioning directed cut k-covers:

Corollary 56.2a. *Let $D = (V, A)$ be a digraph such that A contains a directed spanning tree T with the property that for each arc (u, v) in A there exists a directed $u-v$ path in T. Then any directed cut k-cover in D can be partitioned into k directed cut covers.*

Proof. This follows from Theorem 56.2 by taking $c(u, v)$ equal to the number of times there is an arc from u to v in the directed cut k-cover. ∎

A. Frank also noted that Woodall's conjecture is true if the minimum size of a directed cut is at most 2:

Theorem 56.3. *Let $D = (V, A)$ be a digraph such that each directed cut has size at least two. Then there are two disjoint directed cut covers.*

Proof. As the underlying undirected graph is 2-edge-connected, it has a strongly connected orientation $D' = (V, A')$ (see Corollary 61.3a). Let B_1 be the set of arcs of D that have the same orientation in D' and let $B_2 := A \setminus B_1$. Then B_1 and B_2 are disjoint directed cut covers. ∎

Figure 56.1 shows that this cannot be generalized to each directed cut 2-cover being partitionable into two directed cut covers.

56.3a. Further notes

Karzanov [1985c] gave a strongly polynomial-time algorithm to find a minimum-*mean* capacity directed cut (cf. McCormick and Ervolina [1994]).

Chapter 57
Strong connectors

A *strong connector* is a set of new arcs whose addition to a given digraph D makes it strongly connected. If each potential new arc has been given a length, then finding a shortest strong connector is NP-complete, even if D has no arcs at all: finding a directed Hamiltonian circuit is a special case. (So even if each length is 0 or 1, the problem is NP-complete.)

However, there are a few cases where finding a shortest strong connector is tractable and where min-max relations and polyhedral characterizations hold — for instance, if D is source-sink connected. For these digraphs, packing strong connectors is similarly tractable. These results follow by reduction to directed cut covers discussed in the previous two chapters.

57.1. Making a directed graph strongly connected

Let (V, A) and (V, B) be digraphs. The set B is called a *strong connector for* D if the digraph $(V, A \cup B)$ is strongly connected.

Consider the following *strong connectivity augmentation problem*:

(57.1)   Given a digraph $D = (V, A)$ and a cost function $k : V \times V \to \mathbb{Q}$, find a minimum-cost strong connector for D.

Theorem 57.1. *The strong connectivity augmentation problem is NP-complete, even if $A = \emptyset$.*

Proof. The problem of finding a Hamiltonian circuit in a digraph $D' = (V, A')$ is equivalent to the existence of a strong connector for $(V, \emptyset)$ of cost $|V|$, where $k(u, v) := 1$ if $(u, v) \in A'$, and $k(u, v) := 2$ otherwise. ∎

Eswaran and Tarjan [1976] showed that if the cost of each new arc equals 1, then there is an easy solution:

Theorem 57.2. *If $D = (V, A)$ is an acyclic digraph with at least 2 vertices, and with ρ sources and σ sinks, then the minimum size of a strong connector for D equals $\max\{\rho, \sigma\}$.*

Proof. To see that the minimum is at least $\max\{\rho,\sigma\}$, note that for each source r one should add at least one arc entering r; similarly, for each sink s one should add at least one arc leaving s.

That the bound can be attained is shown by induction on $\max\{\rho,\sigma\}$. If there is a pair of a source r and a sink s such that s is not reachable from r, add an arc (s,r). This reduces both ρ and σ by 1, while maintaining acyclicity, and induction gives the result.

So we can assume that each sink is reachable from each source. We can also assume that $\rho \geq \sigma$ (otherwise, reverse all orientations). Let $r_1,\ldots,r_\rho$ be the sources and let $s_1,\ldots,s_\sigma$ be the sinks. Then adding arcs (s_i,r_i) for $i = 1,\ldots,\sigma$, and arcs (s_i,r_1) for $i = \sigma+1,\ldots,\rho$ makes D strongly connected, proving the theorem. ∎

This implies for not necessarily acyclic digraphs:

Corollary 57.2a. *Let $D = (V, A)$ be a digraph which is not strongly connected, let ρ be the number of strong components K of D with $d^{\text{in}}(K) = 0$ and let σ be the number of strong components K of D with $d^{\text{out}}(K) = 0$. Then the minimum size of a strong connector for D equals $\max\{\rho,\sigma\}$.*

Proof. Apply Theorem 57.2 to the digraph obtained from D by contracting each strong component of D to one vertex. ∎

These proofs also give a polynomial-time algorithm to find a minimum-size strong connector. Eswaran and Tarjan [1976] describe a linear-time implementation.

57.2. Shortest strong connectors

Let $D_0 = (V, A_0)$ and $D = (V, A)$ be digraphs. Call a subset A' of A a D_0-cut (in D) if $A' = \delta_A^{\text{in}}(U)$ for some nonempty proper subset U of V with $\delta_{A_0}^{\text{in}}(U) = \emptyset$.

It is easy to see that a set B of arcs of D is a strong connector for D_0 if and only if B intersects each D_0-cut in D. The following consequence of the Lucchesi-Younger theorem was given in Schrijver [1982]. It gives a min-max relation for the minimum length of a strong connector, if the following condition holds for digraphs $D_0 = (V, A_0)$ and $D = (V, A)$:

(57.2) for each $(u,v) \in A$ there exist $u', v' \in V$ such that in D_0, u' is reachable from u and from v', and v from v'.

We mention two special cases where this condition is satisfied:

- A is a subset of A_0^{-1},
- D_0 is source-sink connected.

We derive from the Lucchesi-Younger theorem (Schrijver [1982]):

Theorem 57.3. *Let $D_0 = (V, A_0)$ and $D = (V, A)$ be digraphs satisfying (57.2) and let $l : A \to \mathbb{Z}_+$ be a length function. Then the minimum length of a strong connector in D for D_0 is equal to the maximum number of D_0-cuts in D such that no arc a of D is in more than $l(a)$ of these cuts.*

Proof. We can assume that D_0 is acyclic, and that for any $u, v \in V$, if v is reachable in D_0 from u, then $(u, v) \in A_0$. (So $(v, v) \in A_0$ for each $v \in V$.)

We show the theorem by induction on the number τ of arcs $a = (u, v)$ of D for which $(v, u) \notin A_0$. If $\tau = 0$, the theorem is equivalent to Corollary 55.2a.

If $\tau > 0$, choose $(u, v) \in A$ with $(v, u) \notin A_0$. By assumption, there exist $u', v' \in V$ with $(u, u'), (v', u'), (v', v) \in A_0$. Introduce two new vertices, u'' and v'', and add arcs

(57.3) $\quad (u, u''), (u'', u'), (v'', u''), (v'', v), (v', v'')$

to A_0. Moreover, replace arc (u, v) of A by (u'', v''), with length equal to that of the original arc (u, v). Let $\widetilde{D}_0 = (\widetilde{V}, \widetilde{A}_0)$ and $\widetilde{D} = (\widetilde{V}, \widetilde{A})$ denote the modified graphs.

This transformation decreases the number τ by 1. Moreover,

(57.4) $\quad$ any subset C of A is a strong connector for D_0 if and only if the set $\widetilde{C} \subseteq \widetilde{A}$ is a strong connector for $\widetilde{D}_0$.

Here $\widetilde{C}$ arises from C by replacing (u, v) by (u'', v'') if $(u, v) \in C$. Proving (57.4) suffices, since it implies that the two numbers in the theorem are invariant under the transformation.

(57.4) can be seen as follows. Choose $C \subseteq A$. First let C be a strong connector for D_0. If $(u, v) \notin C$, then $\widetilde{C} = C$ is also strong connector for $\widetilde{D}_0$ (since in $\widetilde{D}_0$ the new vertex u'' is on a $u - u'$ path, and the new vertex v'' is on a $v' - v$ path). If $(u, v) \in C$, then $\widetilde{C} = (C \setminus \{(u, v)\}) \cup \{(u'', v'')\}$ is a strong connector for $\widetilde{D}_0$, since $A_0 \cup C$ contains the $u - v$ path $(u, u''), (u'', v''), (v'', v)$.

Conversely, let $\widetilde{C}$ be a strong connector for $\widetilde{D}_0$. If $(u'', v'') \notin \widetilde{C}$, then $C = \widetilde{C}$ is also a strong connector for D_0, since any directed path in $\widetilde{A}_0 \cup \widetilde{C}$ connecting two vertices in V and traversing any of the new vertices u'', v'', can be shortcut to a path avoiding u'' and v''.

If $(u'', v'') \in \widetilde{C}$, then $C = (\widetilde{C} \setminus \{(u'', v'')\}) \cup \{(u, v)\}$ is a strong connector for D_0, since any directed path in $\widetilde{A}_0 \cup \widetilde{C}$ connecting two vertices in V and traversing arc (u'', v''), must traverse $(u, u''), (u'', v'')$, and (v'', v), and hence gives a path in $A_0 \cup C$, by replacing this by (u, v). ∎

The proof gives also an algorithmic reduction to the problem of finding a minimum-length directed cut cover, and hence (by Theorem 55.5) a minimum-length strong connector for D_0 can be found in strongly polynomial time.

Theorem 57.3 includes several theorems considered earlier:

- $s, t \in V$ and $A_0 := \{(u, v) \mid u = t \text{ or } v = s\}$: max-potential min-work theorem (Theorem 8.3);
- V is the disjoint union of U and W, $A_0 := \{(u, w) \mid u \in U, w \in W\}$ and $A \subseteq \{(w, u) \mid w \in W, u \in U\}$: weighted version of the Kőnig-Rado edge cover theorem(Theorem 19.4);
- $A_0 = \{(v, r) \mid v \in V\}$ for some $r \in V$: optimum arborescence theorem(Theorem 52.3);
- V is the disjoint union of U and W and $A_0 := \{(u, w) \mid u \in U, w \in W\}$:optimum bibranching theorem(Corollary 54.8b);
- $A \subseteq \{(u, v) \mid (v, u) \in A_0\}$: Lucchesi-Younger theorem (Theorem 55.2).

A cardinality version of the previous theorem is:

Corollary 57.3a. *Let $D_0 = (V, A_0)$ and $D = (V, A)$ be digraphs satisfying (57.2). Then the minimum size of a strong connector in D for D_0 is equal to the maximum number of disjoint D_0-cuts in D.*

Proof. This is the case $l = 1$ of Theorem 57.3. ∎

We formulate this for the special case of source-sink connected digraphs. Recall that a digraph $D = (V, A)$ is called *source-sink connected* if each strong component not left by any arc is reachable by a directed path from each strong component not entered by any arc.

Corollary 57.3b. *Let $D_0 = (V, A_0)$ and $D = (V, A)$ be digraphs, with D_0 source-sink connected. Let $l : A \to \mathbb{Z}_+$ be a length function. Then the minimum length of a strong connector in D for D_0 is equal to the maximum number of D_0-cuts in D such that no arc a of D is in more than $l(a)$ of these cuts.*

Proof. Directly from Theorem 57.3, since condition (57.2) is implied by the fact that D_0 is source-sink connected. ∎

The cardinality version is:

Corollary 57.3c. *Let $D_0 = (V, A_0)$ and $D = (V, A)$ be digraphs, with D_0 source-sink connected. Then the minimum size of a strong connector in D for D_0 is equal to the maximum number of disjoint D_0-cuts in D.*

Proof. This is the case $l = 1$ in Corollary 57.3b. ∎

57.3. Polyhedrally

Theorem 57.3 can be equivalently formulated in TDI terms:

Corollary 57.3d. *Let $D_0 = (V, A_0)$ and $D = (V, A)$ be digraphs satisfying (57.2). Then the system*

(57.5) (i) $x_a \geq 0$ $a \in A$,
         (ii) $x(\delta_A^{in}(U)) \geq 1$ $U \subset V, U \neq \emptyset, \delta_{A_0}^{in}(U) = \emptyset$,

is TDI and determines the convex hull of the strong connectors of D_0.

Proof. This is a reformulation of Theorem 57.3. ∎

In fact, system (57.5) is box-TDI, as will follow from Theorem 60.3.
By the theory of blocking polyhedra, Corollary 57.3d implies:

Corollary 57.3e. *Let $D_0 = (V, A_0)$ and $D = (V, A)$ satisfy (57.2). Then the system*

(57.6) (i) $0 \leq x_a \leq 1$ $a \in A$,
         (ii) $x(B) \geq 1$ B strong connector for D_0

determines the convex hull of subsets of A containing a D_0-cut.

Proof. System (57.6) determines the blocking polyhedron of the polyhedron determined by (57.5), and hence its vertices are the incidence vectors of subsets of A that intersect all strong connectors for D_0. These are precisely the sets of arcs in A containing a D_0-cut. ∎

System (57.6) generally is not TDI, by Figure 56.1. But if D_0 is source-sink connected, system (57.6) is totally dual integral, as is shown in the following section.

57.4. Disjoint strong connectors

Similarly to the derivation of Theorem 57.3 from the Lucchesi-Younger theorem, the following generalization can be derived as a consequence of Theorem 56.1 (Schrijver [1982]):

Theorem 57.4. *Let $D_0 = (V, A_0)$ and $D = (V, A)$ be digraphs, with D_0 source-sink connected. Then the minimum size of a D_0-cut in D is equal to the maximum number of disjoint strong connectors in D for D_0.*

Proof. The proof is similar to the derivation of Theorem 57.3 from the Lucchesi-Younger theorem. We can assume that for any $u, v \in V$, if v is reachable in D_0 from u, then $(u, v) \in A_0$.

We show the theorem by induction on the number τ of arcs (u,v) of D for which $(v,u) \notin A_0$. If $\tau = 0$, the theorem is equivalent to Theorem 56.1 (by taking $C := \{(u,v) \mid (v,u) \in A\}$).

If $\tau > 0$, choose $(u,v) \in A$ with $(v,u) \notin A_0$. Let u' be a sink of D_0 with $(u,u') \in A_0$ and let v' be a source of D_0 with $(v',v) \in A_0$. As D_0 is source-sink connected we know that $(v',u') \in A_0$. Now introduce two new vertices, u'' and v'', and add arcs

(57.7) $(u,u''),(u'',u'),(v'',u''),(v'',v),(v',v'')$

to A_0. Moreover, replace arc (u,v) of A by (u'',v''). Let $\widetilde{D}_0 = (\widetilde{V}, \widetilde{A}_0)$ and $\widetilde{D} = (\widetilde{V}, \widetilde{A})$ denote the modified graphs.

This transformation decreases τ by 1. Again (57.4) holds. This implies that the two numbers in the theorem are invariant under the transformation. Hence the theorem follows by induction. ∎

The condition given in this theorem cannot be relaxed to condition (57.2), as Figure 56.1 shows.

Theorem 57.4 has the following special cases:

- $s,t \in V$ and $A_0 := \{(u,v) \mid u = t$ or $v = s\}$: Menger's theorem (Corollary 9.1b);
- V is the disjoint union of U and W, $A_0 = \{(u,w) \mid u \in U, w \in W\}$ and $A \subseteq \{(w,u) \mid w \in W, u \in U\}$: Gupta's edge-colouring theorem (Theorem 20.5);
- $r \in V$ and $A_0 = \{(v,r) \mid v \in V\}$: Edmonds' disjoint arborescences theorem (Corollary 53.1b);
- V is the disjoint union of U and W and $A_0 = \{(u,w) \mid u \in U, w \in W\}$: disjoint bibranchings theorem (Theorem 54.11);
- $D_0 = (V, A_0)$ is source-sink connected and $A \subseteq \{(u,v) \mid (v,u) \in A_0\}$: Corollary 56.1b.

An equivalent capacitated version of Theorem 57.4 reads:

Corollary 57.4a. *Let $D_0 = (V, A_0)$ and $D = (V, A)$ be digraphs, with D_0 source-sink connected, and let $c \in \mathbb{Z}_+^A$ be a capacity function. Then the minimum capacity of a D_0-cut in D is equal to the maximum number of strong connectors in D for D_0 such that any arc a of D is in at most $c(a)$ of them.*

Proof. Directly from Theorem 57.4 by replacing any arc a of D by $c(a)$ parallel arcs. ∎

Equivalently, in TDI terms:

Corollary 57.4b. *Let $D_0 = (V, A_0)$ and $D = (V, A)$ be digraphs, with D_0 source-sink connected. Then system (57.6) is totally dual integral.*

Proof. This is a reformulation of Corollary 57.4a. ∎

57.5. Complexity

As for the complexity of finding disjoint strong connectors for a source-sink connected digraph, the proof of Theorem 57.4 gives a polynomial-time reduction to finding a maximum number of disjoint directed cut covers in a subset of the arcs of a source-sink connected graph. The latter problem is solvable in polynomial time by the methods of Section 56.2.

The capacitated case can be solved in semi-strongly polynomial time (that is, where rounding is taken as one arithmetic operation) with the ellipsoid method (cf. Grötschel, Lovász, and Schrijver [1988]). A combinatorial semi-strongly polynomial-time algorithm is as follows.

Let be given a source-sink connected digraph $D_0 = (V, A_0)$, a digraph $D = (V, A)$, and a capacity function $c : A \to \mathbb{Z}_+$. We show that an optimum *fractional* packing of strong connectors subject to c can be found in strongly polynomial time. Then an *integer* packing can be found by rounding (like in Section 51.4), thus yielding a semi-strongly polynomial-time algorithm.

Define $\mathcal{C} := \{U \mid \emptyset \neq U \subset V, d^{\text{in}}_{A_0}(U) = 0\}$. To find an optimum fractional packing, let κ be the minimum of $c(\delta^{\text{in}}_A(U))$ taken over sets $U \in \mathcal{C}$. (κ can be computed with a maximum flow algorithm.) We keep a subcollection $\mathcal{U}$ of $\mathcal{C}$ with $c(\delta^{\text{in}}_A(U)) = \kappa$ for each $U \in \mathcal{U}$.

Choose a strong connector $B \subseteq A$ for A_0 with $d^{\text{in}}_B(U) = 1$ for each $U \in \mathcal{U}$. (This can be found in strongly polynomial time, by finding a minimum-length strong connector for length function $l := \sum_{U \in \mathcal{U}} \chi^{\delta^{\text{in}}_B(U)}$. It exists by Theorem 57.4.)

If $c = \mathbf{0}$, we are done. If $c \neq \mathbf{0}$, determine a rational λ as follows. First set $\lambda := \min\{c(a) \mid a \in B\}$. Next, iteratively, find a $U \in \mathcal{C}$ minimizing

$$(57.8) \qquad (c - \lambda \cdot \chi^B)(\delta^{\text{in}}_A(U)).$$

If this minimum value is less than $\kappa - \lambda$, reset

$$(57.9) \qquad \lambda := \frac{c(\delta^{\text{in}}_A(U)) - \kappa}{d^{\text{in}}_B(U) - 1},$$

and iterate. If the minimum is equal to $\kappa - \lambda$, this ends the inner iterations. Then we reset $c := c - \lambda \cdot \chi^B$, $\kappa := \kappa - \lambda$, and $\mathcal{U} := \mathcal{U} \cup \{U\}$, and (outer) iterate.

In each outer iteration, the number of arcs a with $c(a) > 0$ decreases or the intersecting family generated by $\mathcal{U}$ increases (since for the U added we have $d^{\text{in}}_B(U) > 1$). Hence the number of outer iterations is bounded by $|A| + |V|^3$ (see the argument given in the proof of Theorem 53.9).

In each outer iteration, the number of inner iterations is at most $|B|$. To see this, consider any inner iteration, and denote by λ' and U' the objects λ and U in the next inner iteration. As U minimizes (57.8), we know

(57.10) $\quad (c - \lambda \cdot \chi^B)(\delta_A^{\text{in}}(U)) \leq (c - \lambda \cdot \chi^B)(\delta_A^{\text{in}}(U')).$

Moreover, if the next iteration is not the last iteration, then

(57.11) $\quad (c - \lambda' \cdot \chi^B)(\delta_A^{\text{in}}(U')) < \kappa - \lambda' = (c - \lambda' \cdot \chi^B)(\delta_A^{\text{in}}(U))$

(the equality follows from definition (57.9), replacing λ by λ'). Now (57.10) and (57.11) imply

(57.12) $\quad \lambda'(d_B^{\text{in}}(U) - d_B^{\text{in}}(U')) < c(\delta_A^{\text{in}}(U)) - c(\delta_A^{\text{in}}(U'))$
$\qquad \leq \lambda(d_B^{\text{in}}(U) - d_B^{\text{in}}(U')).$

Since $\lambda' < \lambda$ (as (57.8) is less than $\kappa - \lambda$), we have $d_B^{\text{in}}(U') < d_B^{\text{in}}(U)$. Hence the number of inner iterations is at most $|B|$.

57.5a. Crossing families

Theorem 57.4 and part of Theorem 57.3 were generalized by Schrijver [1983b]. Let $\mathcal{C}$ be a *crossing family* of subsets of a set V; that is:

(57.13) $\quad$ if $U, W \in \mathcal{C}$ and $U \cap W \neq \emptyset$ and $U \cup W \neq V$, then $U \cap W \in \mathcal{C}$ and $U \cup W \in \mathcal{C}$.

Let $D = (V, A)$ be a digraph. Call $B \subseteq A$ a $\mathcal{C}$-*cut* if $B = \delta^{\text{in}}(U)$ for some $U \in \mathcal{C}$. Call $B \subseteq A$ a $\mathcal{C}$-*cover* if B intersects each $\mathcal{C}$-cut.

Let $\mathcal{C}$ be a crossing family of nonempty proper subsets of a set V. In Schrijver [1983b] it is shown that the following are equivalent:

(57.14) $\quad$ (i) for each digraph $D = (V, A)$, the minimum size of a $\mathcal{C}$-cut is equal to the maximum number of disjoint $\mathcal{C}$-covers;
(ii) for each digraph $D = (V, A)$ and each length function $l : A \to \mathbb{Z}_+$, the minimum length of a $\mathcal{C}$-cover is equal to the maximum number of $\mathcal{C}$-cuts such that no arc a is in more than $l(a)$ of these cuts;
(iii) there are no V_1, V_2, V_3, V_4, V_5 in $\mathcal{C}$ with $V_1 \subseteq V_3 \subseteq V_5$, $V_1 \subseteq V_2$, $V_2 \cup V_3 = V$, $V_3 \cap V_4 = \emptyset$, and $V_4 \subseteq V_5$.

The configuration described in (iii) is depicted in Figure 57.1. As directed graphs may have parallel arcs, property (57.14)(i) is equivalent to its capacitated version. So condition (57.14)(i) is equivalent to the total dual integrality of

(57.15) $\quad x_a \geq 0 \quad$ for $a \in A$,
$\qquad x(B) \geq 1 \quad$ for each $\mathcal{C}$-cover $B \subseteq A$,

for each digraph (V, A). Similarly, condition (57.14)(ii) is equivalent to the total dual integrality of

(57.16) $\quad x_a \geq 0 \quad$ for $a \in A$,
$\qquad x(B) \geq 1 \quad$ for each $\mathcal{C}$-cut $B \subseteq A$,

for each digraph (V, A).

Frank [1979b] showed that (57.14)(i) holds if $\mathcal{C}$ is an intersecting family. (For any intersecting family $\mathcal{C}$, (iii) holds if $V \notin \mathcal{C}$, which we may assume without loss of generality.)

Section 57.5a. Crossing families 977

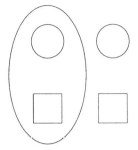

Figure 57.1
The configuration excluded in (57.14)(iii). In this Venn-diagram, the collection is represented by the *interiors* of the ellipses and by the *exteriors* of the rectangles.

In (57.14)(i) and (ii) we require the min-max relation for cuts and covers to hold for *all* directed graphs on V. It is a more general problem to characterize pairs $(\mathcal{C}, D)$ of a crossing family $\mathcal{C}$ on V and a directed graph $D = (V, A)$ having the properties described in (57.14)(i) and (ii), respectively. For example, the Lucchesi-Younger theorem (Theorem 55.2), and its extension by Edmonds and Giles [1977], assert that if $\mathcal{C}$ is a crossing family on V and no arc of D leaves any set $U \in \mathcal{C}$, then $(\mathcal{C}, D)$ has the properties described in (ii). However, the example in Figure 56.1 shows that it generally does not have the property described in (57.14)(i). So for *fixed* graphs D, (57.14)(i) and (ii) are not equivalent.

Theorem 60.3 implies that a pair $(\mathcal{C}, D)$ has property (ii) if $\mathcal{C}$ is a crossing family and D a directed graph such that if $U_1, U_2, U_3 \in \mathcal{C}$ with $U_1 \subseteq V \setminus U_2 \subseteq U_3$, then no arc enters both U_1 and U_3. This generalizes the Lucchesi-Younger theorem.

We show the equivalence of (57.14)(ii) and (iii), for which we show a lemma indicating that condition (57.14)(iii) has a natural characterization in terms of total unimodularity.

For any collection $\mathcal{C}$ of subsets of a set V, let A be the collection of all ordered pairs of elements of V (making the complete directed graph $D = (V, A)$), and let $M_\mathcal{C}$ be the $\mathcal{C} \times A$ matrix with

(57.17) $\quad (M_\mathcal{C})_{U,a} := \begin{cases} 1 & \text{if } a \text{ enters } U, \\ 0 & \text{otherwise.} \end{cases}$

Lemma 57.5α. *Let $\mathcal{C}$ be a cross-free collection of nonempty proper subsets of a set V. Then $M_\mathcal{C}$ is totally unimodular if and only if $\mathcal{C}$ satisfies* (57.14)(iii).

Proof. To see necessity, let $M_\mathcal{C}$ be totally unimodular. Suppose that condition (57.14)(iii) is violated. So there exist V_1, V_2, V_3, V_4, V_5 in $\mathcal{C}$ with $V_1 \subseteq V_3 \subseteq V_5$, $V_1 \subseteq V_2$, $V_2 \cup V_3 = V$, $V_3 \cap V_4 = \emptyset$, and $V_4 \subseteq V_5$. Choose $v_1 \in V_1$, $v_2 \in V \setminus V_2$, $v_4 \in V_4$, and $v_5 \in V \setminus V_5$. Define

(57.18) $\quad A_0 := \{(v_2, v_1), (v_4, v_1), (v_2, v_4), (v_5, v_4), (v_5, v_2)\}$

(cf. Figure 57.2). Consider the submatrix of $M_\mathcal{C}$ with rows indexed by $V_1, \ldots, V_5$, and columns indexed by the arcs in A_0. Then, as one easily checks:

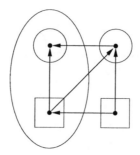

Figure 57.2

(57.19) each set in C_0 is entered by exactly two arcs from A_0, and each arc in A_0 enters exactly two sets in C_0.

So this submatrix has exactly two 1's in each row and each column, and hence is not totally unimodular.

To see sufficiency, let $\mathcal{C}$ satisfy (57.14)(iii). To prove that $M_\mathcal{C}$ is totally unimodular, we use the following characterization of Ghouila-Houri [1962b] (cf. Theorem 19.3 in Schrijver [1986b]): a matrix M is totally unimodular if and only if each collection R of rows of M can be partitioned into classes R_1 and R_2 such that the sum of the rows in R_1, minus the sum of the rows in R_2, is a vector with entries $0, \pm 1$ only.

To check this condition, we can assume that we have chosen all rows of $M_\mathcal{C}$ (as any subset of the rows gives a matrix of the same type as $M_\mathcal{C}$). Make a digraph $D = (\mathcal{C}, A')$, where A' consists of all pairs (T, U) from $\mathcal{C}$ such that

(57.20) $T \subset U$, and there is no $W \in \mathcal{C}$ with $T \subset W \subset U$.

We show that the undirected graph underlying D' is bipartite, which will verify Ghouila-Houri's criterion: let $\mathcal{C}_1$ and $\mathcal{C}_2$ be the two colour classes; then any arc $a = (u, v)$ of D enters a chain of subsets in $\mathcal{C}$ (as $\mathcal{C}$ is cross-free), which subsets are alternatingly in $\mathcal{C}_1$ and $\mathcal{C}_2$. Hence the sum of the rows with index in $\mathcal{C}_1$ minus the sum of the rows with index in $\mathcal{C}_2$, has an entry 0 or ± 1 in position a.

To show that D' is bipartite, suppose that it has an (undirected) circuit of odd length. Since this circuit is odd, and since D' is acyclic, it follows that there are distinct $U_0, U_1, \ldots, U_k, U_{k+1}$ in $\mathcal{C}$, with $k \geq 3$, such that

(57.21) $(U_1, U_0), (U_1, U_2), (U_2, U_3), \ldots, (U_{k-1}, U_k), (U_{k+1}, U_k)$

belong to A'. So U_0 and U_2 are distinct minimal sets in $\mathcal{C}$ containing U_1 as a subset. As $\mathcal{C}$ is cross-free, $U_0 \cup U_2 = V$. Similarly, U_{k-1} and U_{k+1} are distinct maximal subsets of U_k, and hence $U_{k-1} \cap U_{k+1} = \emptyset$. As $U_2 \subseteq U_{k-1}$, it follows that $U_1 \subseteq U_0 \cap U_2$, $U_0 \cup U_2 = V$, $U_2 \cup U_{k+1} \subseteq U_k$, and $U_2 \cap U_{k+1} = \emptyset$. This contradicts (57.14)(iii). ∎

This gives the box-TDI result:

Theorem 57.5. *Let $\mathcal{C}$ be a crossing family of nonempty proper subsets of a set V satisfying (57.14)(iii) and let $D = (V, A)$ be a digraph. Then the system*

Section 57.5a. Crossing families

(57.22) $\quad x_a \geq 0 \quad\quad$ for $a \in A$,
$\quad\quad\quad\quad x(\delta^{\text{in}}(U)) \geq 1 \quad$ for $U \in \mathcal{C}$,

is box-TDI.

Proof. Let $w : A \to \mathbb{R}_+$. Consider the maximum value of

(57.23) $\quad \displaystyle\sum_{U \in \mathcal{C}} y_U$

where $y : \mathcal{C} \to \mathbb{R}_+$ satisfies

(57.24) $\quad \displaystyle\sum_{U \in \mathcal{C}} y_U \chi^{\delta^{\text{in}}(U)} \leq w.$

Choose $y : \mathcal{C} \to \mathbb{R}_+$ attaining the maximum, such that

(57.25) $\quad \displaystyle\sum_{U \in \mathcal{C}} y_U |U||V \setminus U|$

is minimized. We show that the collection $\mathcal{F} := \{U \in \mathcal{C} \mid y_U > 0\}$ is cross-free; that is, for all $T, U \in \mathcal{F}$ one has

(57.26) $\quad T \subseteq U$ or $U \subseteq T$ or $T \cap U = \emptyset$ or $T \cup U = V$.

Suppose that this is not true. Let $\alpha := \min\{y_T, y_U\}$. Decrease y_T and y_U by α, and increase $y_{T \cap U}$ and $y_{T \cup U}$ by α. Now (57.24) is maintained, and (57.23) did not change. However, (57.25) decreases (Theorem 2.1), contradicting our minimality assumption.

So $\mathcal{F}$ is cross-free. As $M_{\mathcal{F}}$ is totally unimodular by Lemma 57.5α, this gives the box-total dual integrality of (57.22) by Theorem 5.35. ∎

Condition (57.14)(iii) is necessary and sufficient for integrality of the polyhedron:

Corollary 57.5a. *For any crossing family $\mathcal{C}$ of nonempty proper subsets of a set V, (57.22) defines an integer polyhedron for each digraph $D = (V, A)$ if and only if condition (57.14)(iii) holds.*

Proof. Sufficiency follows from Theorem 57.5. To see necessity, suppose that (57.22)(iii) does not hold. Let $V_1, \ldots, V_5$ in $\mathcal{C}$ with $V_1 \subseteq V_3 \subseteq V_5$, $V_1 \subseteq V_2$, $V_2 \cup V_3 = V$, $V_3 \cap V_4 = \emptyset$, and $V_4 \subseteq V_5$. Let $\mathcal{C}_0 := \{V_1, \ldots, V_5\}$ and $\mathcal{C}_1 := \mathcal{C} \setminus \mathcal{C}_0$. Choose $v_1 \in V_1$, $v_2 \in V \setminus V_2$, $v_4 \in V_4$, $v_5 \in V \setminus V_5$. Let $D = (V, A)$ be a digraph, with $A = A_0 \cup A_1$, where A_0 is as defined in (57.18) and where

(57.27) $\quad A_1 := \{(u, v) \mid u, v \in V \text{ such that } (u, v) \text{ enters no } V_i \ (i = 1, \ldots, 5)\}.$

Then

(57.28) $\quad$ each set in $\mathcal{C}_1$ is either entered by at least one arc in A_1 or by at least two arcs in A_0.

To see this, by definition of A_1, a subset U of V is entered by no arc in A_1 if and only if U belongs to the lattice generated by $\mathcal{C}_0$ (with respect to inclusion). This lattice consists of the sets

(57.29) $\emptyset, V, V_1, \ldots, V_5, V_1 \cup V_4, V_2 \cap V_3, (V_2 \cap V_3) \cup V_4, V_3 \cup V_4, V_2 \cap V_5,$

as (57.29) is closed under taking unions and intersections, and as each set in (57.29) is generated by taking unions and intersections from $\mathcal{C}_0$. Since each of the sets in (57.29), except $\emptyset$ and V, is entered by at least two arcs in A_0, we have (57.28).

(57.19) and (57.28) give:

(57.30)　　any $\mathcal{C}$-cover in A contains at least three arcs in A_0, and any $\mathcal{C}$-cut contains at least one arc in A_1 or at least two arcs in A_0.

Define $x : A \to \mathbb{Q}$ be $x := \chi^{A_1} + \frac{1}{2}\chi^{A_0}$ and a length function $l : A \to \mathbb{Z}$ by $l := \chi^{A_0}$. Then x satisfies (57.22) and $l^T x = \frac{5}{2}$. However, $l(C) \geq 3$ for each $\mathcal{C}$-cover C. So (57.22) determines no integer polyhedron. ∎

Theorem 57.5 and Corollary 57.5a imply the equivalence of (57.14)(ii) and (iii). For the proof of the equivalence of (57.14)(i) and (iii), we refer to Schrijver [1983b].

Chapter 58

The traveling salesman problem

The traveling salesman problem (TSP) asks for a shortest Hamiltonian circuit in a graph. It belongs to the most seductive problems in combinatorial optimization, thanks to a blend of complexity, applicability, and appeal to imagination.

The problem shows up in practice not only in routing but also in various other applications like machine scheduling (ordering jobs), clustering, computer wiring, and curve reconstruction.

The traveling salesman problem is an NP-complete problem, and no polynomial-time algorithm is known. As such, the problem would not fit in the scope of the present book. However, the TSP is closely related to several of the problem areas discussed before, like 2-matching, spanning tree, and cutting planes, which areas actually were stimulated by questions prompted by the TSP, and often provide subroutines in solving the TSP.

Being NP-complete, the TSP has served as prototype for the development and improvement of advanced computational methods, to a large extent utilizing polyhedral techniques. The basis of the solution techniques for the TSP is branch-and-bound, for which good bounding techniques are essential. Here 'good' is determined by two, often conflicting, criteria: the bound should be *tight* and *fast* to compute. Polyhedral bounds turn out to be good candidates for such bounds.

58.1. The traveling salesman problem

Given a graph $G = (V, E)$, a *Hamiltonian circuit* in G is a circuit C with $VC = V$. The *symmetric traveling salesman problem* (TSP) is: given a graph $G = (V, E)$ and a length function $l : E \to \mathbb{R}$, find a Hamiltonian circuit C of minimum length.

The directed version is as follows. Given a digraph $D = (V, A)$, a *directed Hamiltonian circuit*, or just a *Hamiltonian circuit*, in D is a directed circuit C with $VC = V$. The *asymmetric traveling salesman problem* (TSP or ATSP) is: given a digraph $D = (V, A)$ and a length function $l : A \to \mathbb{R}$, find a Hamiltonian circuit C of minimum length.

In the context of the traveling salesman problem, vertices are sometimes called *cities*, and a Hamiltonian circuit a *traveling salesman tour*. If the vertices are represented by points in the plane and each pair of vertices is connected by an edge of length equal to the Euclidean distance between the two points, one speaks of the *Euclidean traveling salesman problem*.

58.2. NP-completeness of the TSP

The problem of finding a Hamiltonian circuit and (hence) the traveling salesman problem are NP-complete. Indeed, in Theorem 8.11 and Corollary 8.11b we proved the NP-completeness of the directed and undirected Hamiltonian circuit problem. This implies the NP-completeness of the TSP, both in the undirected and the directed case:

Theorem 58.1. *The symmetric TSP and the asymmetric TSP are NP-complete.*

Proof. Given an undirected graph $G = (V, E)$, define $l(e) := 0$ for each edge e. Then G has a Hamiltonian circuit if and only if G has a Hamiltonian circuit of length ≤ 0. This reduces the undirected Hamiltonian circuit problem to the symmetric TSP.

One similarly shows the NP-completeness of the asymmetric TSP. ∎

This method also gives that the symmetric TSP remains NP-complete if the graph is complete and the length function satisfies the *triangle inequality*:

(58.1) $\quad l(uw) \leq l(uv) + l(vw)$ for all $u, v, w \in V$.

Indeed, to test if a graph $G = (V, E)$ has a Hamiltonian circuit, define $l(uv) := 1$ if u and v are adjacent and $l(uv) := 2$ otherwise (for $u \neq v$). Then G has a Hamiltonian circuit if and only if there exists a traveling salesman tour of length $\leq |V|$.

Garey, Graham, and Johnson [1976] and Papadimitriou [1977a] showed that even the Euclidean traveling salesman problem is NP-complete. (Similarly for several other metrics, like l_1.) More on complexity can be found in Section 58.8b below.

58.3. Branch-and-bound techniques

The traveling salesman problem is NP-complete, and no polynomial-time algorithm is known. Most exact methods known are essentially enumerative, aiming at minimizing the enumeration. A general framework is that of *branch-and-bound*. The idea of branch-and-bound applied to the traveling salesman problem roots in papers of Tompkins [1956], Rossman and Twery [1958], and Eastman [1959]. The term 'branch and bound' was introduced by Little, Murty, Sweeney, and Karel [1963].

A rough, elementary description is as follows. Let $G = (V, E)$ be a graph and let $l : E \to \mathbb{R}$ be a length function. For any set $\mathcal{C}$ of Hamiltonian circuits, let $\mu(\mathcal{C})$ denote the minimum length of the Hamiltonian circuits in $\mathcal{C}$.

Keep a collection Γ of sets of Hamiltonian circuits and a function $\lambda : \Gamma \to \mathbb{R}$ satisfying:

(58.2)   (i) $\bigcup \Gamma$ contains a shortest Hamiltonian circuit;
           (ii) $\lambda(\mathcal{C}) \leq \mu(\mathcal{C})$ for each $\mathcal{C} \in \Gamma$.

A typical iteration is:

(58.3)   Select a collection $\mathcal{C} \in \Gamma$ with $\lambda(\mathcal{C})$ minimal. Either find a circuit $C \in \mathcal{C}$ with $l(C) = \lambda(\mathcal{C})$ or replace $\mathcal{C}$ by (zero or more) smaller sets such that (58.2) is maintained.

Obviously, if we find $C \in \mathcal{C}$ with $l(C) = \lambda(\mathcal{C})$, then C is a shortest Hamiltonian circuit.

This method always terminates, but the method and its efficiency heavily depend on how the details in this framework are filled in: how to bound (that is, how to define and calculate $\lambda(\mathcal{C})$), how to branch (that is, which smaller sets replace $\mathcal{C}$), and how to find the circuit C.

As for branching, the classes $\mathcal{C}$ in Γ can be stored implicitly: for example, by prescribing sets B and F of edges such that $\mathcal{C}$ consists of all Hamiltonian circuits whose edge set contains B and is disjoint from F. Then we can split $\mathcal{C}$ by selecting an edge $e \in E \setminus (B \cup F)$ and replacing $\mathcal{C}$ by the classes determined by $B \cup \{e\}, F$ and by $B, F \cup \{e\}$ respectively.

As for bounding, one should choose $\lambda(\mathcal{C})$ that is fast to compute and close to $\mu(\mathcal{C})$. For this, polyhedral bounds seem good candidates, and in the coming sections we consider a number of them.

For finding the circuit $C \in \mathcal{C}$, a heuristic or exact method can be used. If it returns a circuit C with $l(C) > \lambda(\mathcal{C})$, we can delete all sets $\mathcal{C}'$ from Γ with $\lambda(\mathcal{C}') \geq l(C)$, thus saving computer space.

58.4. The symmetric traveling salesman polytope

The *(symmetric) traveling salesman polytope* of an undirected graph $G = (V, E)$ is the convex hull of the incidence vectors (in $\mathbb{R}^E$) of the Hamiltonian circuits. The TSP is equivalent to minimizing a function $l^\mathsf{T} x$ over the traveling salesman polytope. Hence this is NP-complete.

The NP-completeness of the TSP also implies that, unless NP=co-NP, no description in terms of inequalities of the traveling salesman polytope may be expected (Corollary 5.16a). In fact, as deciding if a Hamiltonian circuit exists is NP-complete, it is NP-complete to decide if the traveling salesman polytope is nonempty. Hence, if NP$\neq$co-NP, there exist no inequalities satisfied by the traveling salesman polytope such that their validity can be certified in polynomial time and such that they have no common solution.

58.5. The subtour elimination constraints

Polynomial-time computable lower bounds on the minimum length of a Hamiltonian circuit can be obtained by including the traveling salesman polytope in a larger polytope (a *relaxation*) over which $l^T x$ can be minimized in polynomial time.

Dantzig, Fulkerson, and Johnson [1954a,1954b] proposed the following relaxation:

(58.4)   (i) $0 \leq x_e \leq 1$ for each edge e,
           (ii) $x(\delta(v)) = 2$ for each vertex v,
           (iii) $x(\delta(U)) \geq 2$ for each $U \subseteq V$ with $\emptyset \neq U \neq V$.

The integer solutions of (58.4) are precisely the incidence vectors of the Hamiltonian circuits. If (ii) holds, then (iii) is equivalent to:

(58.5)   (iii') $x(E[U]) \leq |U| - 1$ for each $U \subseteq V$ with $\emptyset \neq U \neq V$.

These conditions are called the *subtour elimination constraints*.

It can be shown with the ellipsoid method that the minimum of $l^T x$ over (58.4) can be found in strongly polynomial time (cf. Theorem 5.10). For this it suffices to show that the conditions (58.4) can be tested in polynomial time. This is easy for (i) and (ii). If (i) and (ii) are satisfied, we can test (iii) by taking x as capacity function, and test if there is a cut $\delta(U)$ of capacity less than 2, with $\emptyset \neq U \neq V$.

No combinatorial polynomial-time algorithm is known to minimize $l^T x$ over (58.4). In practice, one can apply the simplex method to minimize $l^T x$ over the constraints (i) and (ii), test if the solution satisfies (iii) by finding a cut $\delta(U)$ minimizing $x(\delta(U))$. If this cut has capacity at least 2, then x minimizes $l^T x$ over (58.4). Otherwise, we can add the constraint $x(\delta(U)) \geq 2$ to the simplex tableau (a *cutting plane*), and iterate. (This method is implicit in Dantzig, Fulkerson, and Johnson [1954b].)

Branch-and-bound methods that incorporate such a cutting plane method to obtain bounds and that extend the cutting plane found to all other nodes of the branching tree to improve their bounds, are called *branch-and-cut*.

System (58.4) generally is not enough to determine the traveling salesman polytope: for the Petersen graph $G = (V, E)$, the vector x with $x_e = \frac{2}{3}$ for each $e \in E$ satisfies (58.4) but is not in the traveling salesman polytope of G (as it is empty).

Wolsey [1980] (also Shmoys and Williamson [1990]) showed that if G is complete and the length function l satisfies the triangle inequality, then the minimum of $l^T x$ over (58.4) is at least $\frac{2}{3}$ times the minimum length of a Hamiltonian circuit. It is conjectured (cf. Carr and Vempala [2000]) that for any length function, a lower bound of $\frac{3}{4}$ holds (which is best possible). Related results are given by Papadimitriou and Vempala [2000] and Boyd and Labonté [2002] (who verified the conjecture for $n \leq 10$).

Maurras [1975] and Grötschel and Padberg [1979b] showed that, if G is the complete graph on V and $2 \leq |U| \leq |V| - 2$, then the subtour elimination constraint (58.4)(iii) determines a facet of the traveling salesman polytope.

Chvátal [1989] showed the NP-completeness of recognizing if the bound given by the subtour elimination constraints is equal to the length of a shortest tour. He also showed that there is no nontrivial upper bound on the relative error of this bound.

58.6. 1-trees and Lagrangean relaxation

Held and Karp [1971] gave a method to find the minimum value of $l^\mathsf{T} x$ over (58.4), with the help of 1-*trees* and *Lagrangean relaxation*.

Let $G = (V, E)$ be a graph and fix a vertex, say 1, of G. A 1-*tree* is a subset F of E such that $|F \cap \delta(1)| = 2$ and such that $F \setminus \delta(1)$ forms a spanning tree on $V \setminus \{1\}$. So each Hamiltonian circuit is a 1-tree with all degrees equal to 2.

It is easy to find a shortest 1-tree F, as it consists of a shortest spanning tree of the graph $G - 1$, joined with the two shortest edges incident with vertex 1. Corollary 50.7c implies that the convex hull of the incidence vectors of 1-trees is given by:

(58.6)   (i) $0 \leq x_e \leq 1$ for each $e \in E$,
             (ii) $x(\delta(1)) = 2$,
             (iii) $x(E[U]) \leq |U| - 1$ for each nonempty $U \subseteq V \setminus \{1\}$,
             (iv) $x(E) = |V|$.

Then (58.4) is equivalent to (58.6) added with (58.4)(ii).

The Lagrangean relaxation approach to find the minimum of $l^\mathsf{T} x$ over (58.4) is based on the following result. For any $y \in \mathbb{R}^V$ define

(58.7) $l_y(e) := l(e) - y_u - y_v$

for $e = uv \in E$, and define

(58.8) $f(y) := 2y(V) + \min_F l_y(F)$,

where F ranges over all 1-trees. Christofides [1970] and Held and Karp [1970] observed that for each $y \in \mathbb{R}^V$:

(58.9) $f(y) \leq$ the minimum length of a Hamiltonian circuit,

since if C is a shortest Hamiltonian circuit, then $f(y) \leq 2y(V) + l_y(C) = l(C)$.

The function f is concave. Since a shortest 1-tree can be found fast, also $f(y)$ can be computed fast. Held and Karp [1970] showed:

Theorem 58.2. *The minimum value of $l^\mathsf{T} x$ over* (58.4) *is equal to the maximum value of $f(y)$ over $y \in \mathbb{R}^V$.*

Proof. This follows from general linear programming theory. Let $Ax = b$ be system (58.4)(ii) and let $Cx \geq d$ be system (58.6). As (58.4) is equivalent to $Ax = b$, $Cx \geq d$, we have, using LP-duality:

$$
\begin{aligned}
(58.10) \qquad \min_{\substack{Ax = b \\ Cx \geq d}} l^\mathsf{T} x &= \max_{\substack{y, z \\ z \geq 0 \\ y^\mathsf{T} A + z^\mathsf{T} C = l^\mathsf{T}}} y^\mathsf{T} b + z^\mathsf{T} d \\
&= \max_y (y^\mathsf{T} b + \max_{\substack{z \geq 0 \\ z^\mathsf{T} C = l^\mathsf{T} - y^\mathsf{T} A}} z^\mathsf{T} d) = \max_y \left(y^\mathsf{T} b + \min_{Cx \geq d} (l^\mathsf{T} - y^\mathsf{T} A) x \right) \\
&= \max_y f(y).
\end{aligned}
$$

The last inequality holds as $Cx \geq d$ determines the convex hull of the incidence vectors of 1-trees. ∎

This translates the problem of minimizing $l^\mathsf{T} x$ over (58.4) to finding the maximum of the concave function f. We can find this maximum with a subgradient method (cf. Chapter 24.3 of Schrijver [1986b]). The vector y (the *Lagrangean multipliers*) can be used as a correction mechanism to urge the 1-tree to have degree 2 at each vertex. That is, if we calculate $f(y)$, and see that the 1-tree F minimizing $l_y(F)$ has degree more than 2 at a vertex v, we can increase l_y on $\delta(v)$ by decreasing y_v. Similarly, if the degree is less than 2, we can increase y_v. This method was proposed by Held and Karp [1970,1971].

The advantage of this approach is that one need not implement a linear programming algorithm with a constraint generation technique, but that instead it suffices to apply the more elementary tools of finding a shortest 1-tree and updating y. More can be found in Jünger, Reinelt, and Rinaldi [1995].

58.7. The 2-factor constraints

A strengthening of relaxation (58.4) is obtained by using the facts that each Hamiltonian circuit is a 2-factor and that the convex hull of the incidence vectors of 2-factors is known (Corollary 30.8a) (this idea goes back to Robinson [1949] for the asymmetric TSP and Bellmore and Malone [1971] for the symmetric TSP, and was used for the symmetric TSP by Grötschel [1977a] and Pulleyblank [1979b]):

(58.11) (i) $0 \leq x_e \leq 1$ for each edge e,
(ii) $x(\delta(v)) = 2$ for each vertex v,
(iii) $x(\delta(U)) \geq 2$ for each $U \subseteq V$ with $\emptyset \neq U \neq V$,
(iv) $x(\delta(U) \setminus F) - x(F) \geq 1 - |F|$
   for $U \subseteq V$, $F \subseteq \delta(U)$, F matching, $|F|$ odd.

Since a minimum-length 2-factor can be found in polynomial time, the inequalities (i), (ii), and (iv) can be tested in polynomial time (cf. Theorem 32.5). Hence the minimum of $l^\mathsf{T} x$ over (58.11) can be found in strongly polynomial time.

System (58.11) generally is not enough to determine the traveling salesman polytope, as can be seen, by taking the Petersen graph $G = (V, E)$ and $x_e := \frac{2}{3}$ for each edge e.

Grötschel and Padberg [1979b] showed that, for complete graphs, each of the inequalities (58.11)(iv) determines a facet of the traveling salesman polytope (if $|F| \geq 3$). Boyd and Pulleyblank [1991] studied optimization over (58.11).

58.8. The clique tree inequalities

Grötschel and Pulleyblank [1986] found a large class of facet-inducing inequalities, the 'clique tree inequalities', that generalize the 'comb inequalities' (see below), which generalize both the subtour elimination constraints (58.4)(iii) and the 2-factor constraints (58.11)(iv). However, no polynomial-time test of clique tree inequalities is known.

A *clique tree inequality* is given by:

(58.12) $\displaystyle\sum_{i=1}^{r} x(\delta(H_i)) + \sum_{j=1}^{s} x(\delta(T_j)) \geq 2r + 3s - 1,$

where $H_1, \ldots, H_r$ are pairwise disjoint subsets of V and $T_1, \ldots, T_s$ are pairwise disjoint proper subsets of V such that

(58.13) (i) no T_j is contained in $H_1 \cup \cdots \cup H_r$,
(ii) each H_i intersects an odd number of the T_j,
(iii) the intersection graph of $H_1, \ldots, H_r, T_1, \ldots, T_s$ is a tree.

(Here, the *intersection graph* is the graph with vertices $H_1, \ldots, H_r, T_1, \ldots, T_s$, two of them being adjacent if and only if they intersect. Each H_i is called a *handle* and each T_j a *tooth*.)

Theorem 58.3. *The clique tree inequality (58.12) is valid for the traveling salesman polytope.*

Proof. It suffices to show that each Hamiltonian circuit C satisfies:

(58.14) $\displaystyle\sum_{i=1}^{r} d_C(H_i) + \sum_{j=1}^{s} d_C(T_j) \geq 2r + 3s - 1.$

We apply induction on r, the case $r = 0$ being easy (as it implies $s = 1$). For each $i = 1, \ldots, r$, let β_i be the number of T_j intersecting H_i.

If there is an i with $d_C(H_i) \geq \beta_i$, say $i = 1$, then, by parity, $d_C(H_1) \geq \beta_1 + 1$. The sets $H_2, \ldots, H_r, T_1, \ldots, T_s$ fall apart into β_1 collections of type (58.13), to which we can apply induction. Adding up the inequalities obtained, we get:

$$(58.15) \quad \sum_{i=2}^{r} d_C(H_i) + \sum_{j=1}^{s} d_C(T_j) \geq 2(r-1) + 3s - \beta_1.$$

Then (58.14) follows, as $d_C(H_1) \geq \beta_1 + 1$.

So we can assume that $d_C(H_i) \leq \beta_i - 1$ for each i. For all i, j, let $\alpha_{i,j} := 1$ if $T_j \cap H_i \neq \emptyset$ and C has no edge connecting $T_j \cap H_i$ and $T_j \setminus H_i$, and let $\alpha_{i,j} := 0$ otherwise. Then

$$(58.16) \quad d_C(T_j) \geq 2 + 2 \sum_{i=1}^{r} \alpha_{i,j},$$

since C restricted to T_j falls apart into at least $1 + \sum_{i=1}^{r} \alpha_{i,j}$ components (using (58.13)(i)).

Moreover, for each $i = 1, \ldots, r$, there exist at least $\beta_i - d_C(H_i)$ indices j with $\alpha_{i,j} = 1$. Hence

$$(58.17) \quad \sum_{j=1}^{s} d_C(T_j) \geq 2s + 2 \sum_{i=1}^{r} \sum_{j=1}^{s} \alpha_{i,j} \geq 2s + 2 \sum_{i=1}^{r} (\beta_i - d_C(H_i))$$
$$\geq 2s + r + \sum_{i=1}^{r} (\beta_i - d_C(H_i)) = 2r + 3s - 1 - \sum_{i=1}^{r} d_C(H_i),$$

since $\sum_{i=1}^{r} \beta_i = r + s - 1$, as the intersection graph of the H_i and the T_j is a tree with $r + s$ vertices, and hence with $r + s - 1$ edges.

(58.17) implies (58.14). ∎

Notes. Grötschel and Pulleyblank [1986] also showed that, if G is a complete graph, then any clique tree inequality determines a facet if and only if each H_i intersects at least three of the T_j.

The clique tree inequalities are not enough to determine the traveling salesman polytope, as is shown again by taking the Petersen graph $G = (V, E)$ and $x_e := \frac{2}{3}$ for all $e \in E$.

The special case $r = 1$ of the clique tree inequality is called a *comb inequality*, and was introduced by Grötschel and Padberg [1979a] and proved to be facet-inducing (if G is complete and $s \geq 3$) by Grötschel and Padberg [1979b].

The special case of the comb inequality with $|H_1 \cap T_j| = 1$ for all $j = 1, \ldots, s$ is called a *Chvátal comb inequality*, introduced by Chvátal [1973b]. The special case of the Chvátal comb inequalities with $|T_j| = 2$ for each $j = 1, \ldots, s$ gives the 2-factor constraints (58.11)(iv) (since $2x(F) + \sum_{f \in F} x(\delta(f)) = 4|F|$).

No polynomial-time algorithm is know to test the clique tree inequalities, or the comb inequalities, or the Chvátal comb inequalities. Carr [1995,1997] showed that for each constant K, there is a polynomial-time algorithm to test the clique tree inequalities with at most K teeth and handles. (This can be done by first fixing intersection points of the $H_i \cap T_j$ (if nonempty) and points in $T_j \setminus (H_1 \cup \cdots \cup H_r)$, and next finding minimum-capacity cuts separating the appropriate sets of these points (taking x as capacity function). We can make them disjoint where necessary by the usual uncrossing techniques. As K is fixed, the number of vertices to be chosen is also bounded by a polynomial in $|V|$.)

Letchford [2000] gave a polynomial-time algorithm for testing a superclass of the comb inequalities in planar graphs. Related results are given in Carr [1996], Fleischer and Tardos [1996,1999], Letchford and Lodi [2002], and Naddef and Thienel [2002a,2002b].

58.8a. Christofides' heuristic for the TSP

Christofides [1976] designed the following algorithm to find a short Hamiltonian circuit in a complete graph $G = (V, E)$ (generally not the shortest however). It assumes a nonnegative length function l satisfying the following *triangle inequality*:

(58.18) $\quad l(uw) \leq l(uv) + l(vw)$

for all $u, v, w \in V$.

First determine a shortest spanning tree T (with the greedy algorithm). Next, let U be the set of vertices that have odd degree in T. Find a shortest perfect matching M on U. Now $ET \cup M$ forms a set of edges such that each vertex has even degree. (If an edge occurs both in ET and in M, we take it as two parallel edges.) So we can make a closed path C such that each edge in $ET \cup M$ is traversed exactly once. Then C traverses each vertex at least once. By shortcutting we obtain a Hamiltonian circuit C' with $l(C') \leq l(C)$.

How far away is the length of C' from the minimum length μ of a Hamiltonian circuit?

Theorem 58.4. $l(C') \leq \frac{3}{2}\mu$.

Proof. Let C'' be a shortest Hamiltonian circuit. Then $l(T) \leq l(C'') = \mu$, since C'' contains a spanning tree. Also, $l(M) \leq \frac{1}{2}l(C'') = \frac{1}{2}\mu$, since we can split C'' into two collections of paths, each having U as set of end vertices. They give two perfect matchings on U, of total length at most $l(C'')$ (by the triangle inequality (58.18)). Hence one of these matchings has length at most $\frac{1}{2}l(C'')$. So $l(M) \leq \frac{1}{2}l(C'') = \frac{1}{2}\mu$.

Combining the two inequalities, we obtain

(58.19) $\quad l(C') \leq l(C) = l(T) + l(M) \leq \frac{3}{2}\mu$,

which proves the theorem. ∎

The factor $\frac{3}{2}$ seems quite large, but it is the smallest factor for which a polynomial-time method is known. Don't forget moreover that it is a *worst-case* bound, and that in practice (or on average) the algorithm might have a much better performance.

Wolsey [1980] showed more strongly that (if l satisfies the triangle inequality) the length of the tour found by Christofides' algorithm, is at most $\frac{3}{2}$ times the lower bound based on the subtour elimination constraints (58.4). If all distances are 1 or 2, Papadimitriou and Yannakakis [1993] gave a polynomial-time algorithm with worst-case factor $\frac{7}{6}$. Hoogeveen [1991] analyzed the behaviour of Christofides' heuristic when applied to finding shortest Hamiltonian paths.

58.8b. Further notes on the symmetric traveling salesman problem

Adjacency of vertices of the symmetric traveling salesman polytope of a graph $G = (V, E)$ is co-NP-complete, as was shown by Papadimitriou [1978].

Norman [1955] remarked that the symmetric traveling salesman polytope of the complete graph K_n has dimension $\frac{1}{2}n(n-3) = \binom{n}{2} - n$ (if $n \geq 3$). Proofs were given by Maurras [1975] and Grötschel and Padberg [1979a].

The symmetric traveling salesman polytopes of K_n for small n were studied by Norman [1955], Boyd and Cunningham [1991], Christof, Jünger, and Reinelt [1991] ($n = 8$), and Naddef and Rinaldi [1992,1993]. Weinberger [1974a] showed that the up hull of the symmetric traveling salesman polytope of K_6 is not determined by inequalities with $0, 1$ coefficients only.

Rispoli and Cosares [1998] showed that the diameter of the symmetric traveling salesman polytope of a complete graph is at most 4. Grötschel and Padberg [1985] conjecture that it is at most 2. (See Sierksma and Tijssen [1992] and Sierksma, Teunter, and Tijssen [1995] for supporting results.) Further work on the symmetric traveling salesman polytope includes Naddef and Rinaldi [1993], Queyranne and Wang [1993], Carr [2000], Cook and Dash [2001], and Naddef and Pochet [2001].

Rispoli [1998] showed that the monotonic diameter of the symmetric traveling salesman polytope of K_n is $\lfloor n/2 \rfloor - 1$ if $n \geq 6$. (The *monotonic diameter* of a polytope is the minimum λ such that for each linear function $l^\mathsf{T} x$ and each pair of vertices y, z such that $l^\mathsf{T} x$ is maximized over P at z, there is a $y - z$ path along vertices and edges of the polytope such that the function $l^\mathsf{T} x$ is monotonically nondecreasing and such that the number of edges in the path is at most λ.)

Sahni and Gonzalez [1976] showed that for any constant c, unless P=NP, there is no polynomial-time algorithm finding a Hamiltonian circuit of length at most c times the minimum length of a Hamiltonian circuit. Johnson and Papadimitriou [1985a] showed that unless P=NP there is no *fully polynomial approximation scheme* for the Euclidean traveling salesman problem (that is, there is no algorithm that gives for any $\varepsilon > 0$, a Hamiltonian circuit of length at most $1+\varepsilon$ times the minimum length of a Hamiltonian circuit, with running time bounded by a polynomial in the size of the problem and in $1/\varepsilon$).

However, Arora [1996,1997,1998] showed that for the Euclidean TSP there is a polynomial approximation scheme: there is an algorithm that gives, for any n vertices in the plane and any $\varepsilon > 0$, a Hamiltonian circuit of length at most $1 + \varepsilon$ times the minimum length of a Hamiltonian circuit, in $n^{O(1/\varepsilon)}$ time. The method also applies to several other metrics. Mitchell [1999] noticed that the methods of Mitchell [1996] imply similar results. Related work is reported in Trevisan [1997, 2000], Rao and Smith [1998], and Dumitrescu and Mitchell [2001]. Earlier work on plane TSP includes Karp [1977], Steele [1981], Moran [1984], Karloff [1989], and Clarkson [1991].

A polynomial-time approximation scheme for the traveling salesman problem where the length is determined by the shortest path metric in a weighted planar graph was given by Arora, Grigni, Karger, Klein, and Woloszyn [1998] (extending the unweighted case proved by Grigni, Koutsoupias, and Papadimitriou [1995]).

Yannakakis [1988,1991] showed that the traveling salesman problem on K_n cannot be expressed by a linear program of polynomial size that is invariant under the symmetric group on K_n. (A similar negative result was proved by Yannakakis for the perfect matching polytope.)

Section 58.8b. Further notes on the symmetric traveling salesman problem

More valid inequalities for the symmetric traveling salesman polytope were given by Grötschel [1980a], Papadimitriou and Yannakakis [1984], Fleischmann [1988], Boyd and Cunningham [1991], Naddef [1992], Naddef and Rinaldi [1992], and Boyd, Cunningham, Queyranne, and Wang [1995].

Jünger, Reinelt, and Rinaldi [1995] gave a comparison of the values of various relaxations for several instances of the symmetric traveling salesman problem. Johnson, McGeoch, and Rothberg [1996] report on an 'asymptotic experimental analysis' of the Held-Karp bound. A probabilistic analysis of the Held-Karp bound for the Euclidean TSP was presented by Goemans and Bertsimas [1991].

A worst-case comparison of several classes of valid inequalities for the traveling salesman polytope was given by Goemans [1995]. Several integer programming formulations for the TSP were compared by Langevin, Soumis, and Desrosiers [1990]. Althaus and Mehlhorn [2000,2001] showed that the subtour elimination constraints solve traveling salesman problems coming from curve reconstruction, under appropriate sampling conditions.

Semidefinite programming was applied to the symmetric TSP by Cvetković, Čangalović, and Kovačević-Vujčić [1999a,1999b] and Iyengar and Çezik [2001].

Let $G = (V, E)$ be an undirected graph. The symmetric traveling salesman polytope of G is a face of the convex hull of all integer solutions of

(58.20)   (i) $x_e \geq 0$ for each $e \in E$,
            (ii) $x(\delta(U)) \geq 2$ for each $U \subseteq V$ with $\emptyset \neq U \neq V$.

Fonlupt and Naddef [1992] characterized for which graphs G each vertex x of (58.20) is integer and has $x(\delta(v)) \equiv 0 \pmod{2}$ for each vertex v of G.

Grötschel [1980a] studied the *monotone traveling salesman polytope* of a graph, which is the convex hull of the incidence vectors of subsets of Hamiltonian circuits.

Cornuéjols, Fonlupt, and Naddef [1985] considered the related problem of finding a shortest tour in a graph such that each vertex is traversed *at least* once, and the related polytope (cf. Naddef and Rinaldi [1991]). Further and related studies (also on shortest k-connected spanning subgraphs, on the 'Steiner network problem', and on the (equivalent) 'survivable network design problem') include Bienstock, Brickell, and Monma [1990], Grötschel and Monma [1990], Monma, Munson, and Pulleyblank [1990], Kelsen and Ramachandran [1991,1995], Barahona and Mahjoub [1992,1995], Chopra [1992,1994], Goemans and Williamson [1992,1995a], Grötschel, Monma, and Stoer [1992], Han, Kelsen, Ramachandran, and Tarjan [1992,1995], Khuller and Vishkin [1992,1994], Nagamochi and Ibaraki [1992a], Cheriyan, Kao, and Thurimella [1993], Gabow, Goemans, and Williamson [1993,1998], Garg, Santosh, and Singla [1993], Naddef and Rinaldi [1993], Queyranne and Wang [1993], Williamson, Goemans, Mihail, and Vazirani [1993,1995], Aggarwal and Garg [1994], Goemans, Goldberg, Plotkin, Shmoys, Tardos, and Williamson [1994], Khuller, Raghavachari, and Young [1994,1995a,1996], Mahjoub [1994,1997], Agrawal, Klein, and Ravi [1995], Khuller and Raghavachari [1995], Ravi and Williamson [1995, 1997], Cheriyan and Thurimella [1996a,2000], Didi Biha and Mahjoub [1996], Fernandes [1997,1998], Carr and Ravi [1998], Cheriyan, Sebő, and Szigeti [1998,2001], Auletta, Dinitz, Nutov, and Parente [1999], Czumaj and Lingas [1998,1999], Jain [1998,2001], Fonlupt and Mahjoub [1999], Fleischer, Jain, and Williamson [2001], Cheriyan, Vempala, and Vetta [2002], and Gabow [2002]. This problem relates to connectivity augmentation — see Chapter 63.

58.9. The asymmetric traveling salesman problem

We next consider the *asymmetric* traveling salesman problem. Let $D = (V, A)$ be a directed graph. The *(asymmetric) traveling salesman polytope* of D is the convex hull of the incidence vectors (in $\mathbb{R}^A$) of Hamiltonian circuits in D. Again, since the asymmetric traveling salesman problem is NP-complete, we know that unless NP=co-NP there is no system of linear inequalities that describes the traveling salesman polytope of a digraph such that their validity can be certified in polynomial time.

Again, we can obtain lower bounds on the minimum length of a Hamiltonian circuit in D by including the traveling salesman polytope in a larger polytope (a *relaxation*) over which $l^T x$ can be minimized in polynomial time. The analogue of relaxation (58.4) for the directed case is:

(58.21) (i) $0 \leq x_a \leq 1$ for $a \in A$,
(ii) $x(\delta^{\text{in}}(v)) = 1$ for $v \in V$,
(iii) $x(\delta^{\text{out}}(v)) = 1$ for $v \in V$,
(iv) $x(\delta^{\text{in}}(U)) \geq 1$ for $U \subseteq V$ with $\emptyset \neq U \neq V$.

With the ellipsoid method, the minimum of $l^T x$ over (58.21) can be found in strongly polynomial time. However, no combinatorial polynomial-time algorithm is known. (The relaxation (i), (ii), (iii) is due to Robinson [1949].)

Grötschel and Padberg [1977] showed that each inequality (58.21)(iv) determines a facet of the traveling salesman polytope of the complete directed graph, if $2 \leq |U| \leq |U| - 2$. (This result was announced in Grötschel and Padberg [1975].)

(58.21) is not enough to determine the traveling salesman polytope, even not for digraphs on 4 vertices only. This is shown by Figure 58.1. Another example is obtained from the Petersen graph, by replacing each edge by two oppositely oriented edges and putting value $\frac{1}{3}$ on each arc.

Figure 58.1

Setting $x_a := \frac{1}{2}$ for each arc a, we have a vector x satisfying (58.21) but not belonging to the traveling salesman polytope.

58.10. Directed 1-trees

As in the undirected case, Held and Karp [1970] showed that the minimum of $l^\mathsf{T} x$ over (58.21) can be obtained as follows.

Let $D = (V, A)$ be a digraph and fix a vertex 1 of D. Call a subset F of A a *directed 1-tree* if F contains exactly one arc, a say, entering 1 and if $F \setminus \{a\}$ is a directed 1-tree such that exactly one arc leaves 1.[7] Each Hamiltonian circuit is a directed 1-tree, and a minimum-length directed 1-tree can be found in strongly polynomial time (by adapting Theorem 52.1).

From Corollary 52.3b one may derive that the convex hull of the incidence vectors of directed 1-trees is determined by:

(58.22)   (i) $0 \leq x_a \leq 1$ for $a \in A$,
           (ii) $x(\delta^{\text{in}}(v)) = 1$ for each $v \in V$,
           (iii) $x(\delta^{\text{out}}(1)) = 1$,
           (iv) $x(\delta^{\text{in}}(U)) \geq 1$ for each nonempty $U \subseteq V \setminus \{1\}$.

Again, a *Lagrangean relaxation* approach can find the minimum of $l^\mathsf{T} x$ over (58.21), for $l \in \mathbb{R}^A$. For any $y \in \mathbb{R}^V$ define

(58.23) $l_y(a) := l(a) - y(u)$

for any arc $a = (u, v) \in A$, and define

(58.24) $f(y) := \min_F l_y(F) + y(V)$,

where F ranges over directed 1-trees.

Then the minimum of $l^\mathsf{T} x$ over (58.21) is equal to the maximum of $f(y)$ over $y \in \mathbb{R}^V$. The proof is similar to that of Theorem 58.2.

58.10a. An integer programming formulation

The integer solutions of (58.21) are precisely the incidence vectors of Hamiltonian circuits, so it gives an integer programming formulation of the asymmetric traveling salesman problem. The system has exponentially many constraints. A.W. Tucker showed in 1960 (cf. Miller, Tucker, and Zemlin [1960]) that the asymmetric TSP can be formulated as the following integer programming problem, of polynomial size only. Set $n := |V|$, fix a vertex v_0 of D, and minimize $l^\mathsf{T} x$ where $x \in \mathbb{Z}^A$ and $z \in \mathbb{R}^V$ are such that

(58.25)   (i) $x_a \geq 0$ for $a \in A$,
           (ii) $x(\delta^{\text{in}}(v)) = 1$ for $v \in V$,
           (ii) $x(\delta^{\text{out}}(v)) = 1$ for $v \in V$,
           (iv) $z_u - z_v + n x_a \leq n - 1$ for $a = (u, v) \in A$ with $u, v \neq v_0$.

---

[7] Held and Karp used the term 1-*arborescence* for a directed 1-tree. To avoid confusion with r-arborescence (a slightly different notion), we have chosen for directed 1-tree.

The conditions (i), (ii), and (iii) and the integrality of x guarantee that x is the incidence vector of a set C of arcs forming directed circuits partitioning V. Then condition (iv) says the following. For any arc $a = (u,v)$ not incident with v_0, one has: if a belongs to C, then $z_u \leq z_v - 1$; if a does not belong to C, then $z_u - z_v \leq n - 1$. This implies that C contains no directed circuit disjoint from v_0. Hence C is a Hamiltonian circuit.

Conversely, for any incidence vector x of a Hamiltonian circuit, one can find $z \in \mathbb{R}^V$ satisfying (58.25).

Unfortunately, the linear programming bound one may derive from (58.25) is generally much worse than that obtained from (58.21).

58.10b. Further notes on the asymmetric traveling salesman problem

Bartels and Bartels [1989] gave a system of inequalities determining the traveling salesman polytope of the complete directed graph on 5 vertices (correcting Heller [1953a] and Kuhn [1955a]).

Padberg and Rao [1974] showed that the diameter of the asymmetric traveling salesman polytope of the complete directed graph on n vertices is equal to 1 if $3 \leq n \leq 5$, and to 2 if $n \geq 6$. Rispoli [1998] showed that the monotonic diameter of the asymmetric traveling salesman polytope of the complete directed graph on n vertices equals $\lfloor n/3 \rfloor$ if $n \geq 3$. (For the definition of monotonic diameter, see Section 58.8b.)

Adjacency of vertices of the asymmetric traveling salesman polytope of a graph $G = (V, E)$ is co-NP-complete, as was shown by Papadimitriou [1978][8]. The number of edges of the asymmetric traveling salesman polytope was estimated by Sarangarajan [1997].

H.W. Kuhn (cf. Heller [1953a], Kuhn [1955a]) claimed that the dimension of the asymmetric traveling salesman polytope of the complete directed graph on n vertices is equal to $n^2 - 3n + 1$ (if $n \geq 3$). A proof of this was supplied by Grötschel and Padberg [1977]. Further work on this polytope is reported in Kuhn [1991].

More valid inequalities for the asymmetric traveling salesman polytope were given by Grötschel and Padberg [1977], Grötschel and Wakabayashi [1981a,1981b], Balas [1989], Fischetti [1991,1992,1995], Balas and Fischetti [1993,1999], and Queyranne and Wang [1995].

A polytope generalizing the directed 1-tree polytope and the asymmetric traveling salesman polytope, the 'fixed-outdegree 1-arborescence polytope', was studied by Balas and Fischetti [1992]. Another polyhedron related to the asymmetric traveling salesman polytope was studied by Chopra and Rinaldi [1996].

Billera and Sarangarajan [1996] showed that each 0,1 polytope is affinely equivalent to the traveling salesman polytope of some directed graph.

Frieze, Karp, and Reed [1992,1995] investigated the tightness of the assignment bound (determined by (58.21)(i)-(iii)). Williamson [1992] compared the Held-Karp lower bound for the asymmetric TSP with the assignment bound.

Carr and Vempala [2000] related the relative error of the asymmetric TSP bound obtained from (58.21) to that of the symmetric TSP bound obtained from (58.4).

---

[8] Murty [1969] gave a characterization of adjacency that was shown to be false by Rao [1976].

Padberg and Sung [1991] compared different formulations of the asymmetric traveling salesman problem.

An analogue of Christofides' algorithm (Section 58.8a) for the asymmetric case is not known: no factor c and polynomial-time algorithm are known that give a Hamiltonian circuit in a digraph of length at most c times the length of a shortest Hamiltonian circuit, even not if the lengths satisfy the triangle inequality.

58.11. Further notes on the traveling salesman problem

58.11a. Further notes

There is an abundance of papers presenting algorithms, heuristics, and computational results for the traveling salesman problem. We give a short selection of it.

Milestones in solving large-scale symmetric traveling salesman problems were achieved by Dantzig, Fulkerson, and Johnson [1954b] (42 cities), Held and Karp [1962] (48 cities), Karg and Thompson [1964] (57 cities), Held and Karp [1971] (64 cities), Helbig Hansen and Krarup [1974] (80 cities), Camerini, Fratta, and Maffioli [1975] (100 cities), Grötschel [1980b] (120 cities), Crowder and Padberg [1980] and Padberg and Hong [1980] (318 cities), Padberg and Rinaldi [1987] (532 cities), Grötschel and Holland [1991] (666 cities), Padberg and Rinaldi [1990b,1991] (2392 cities), Applegate, Bixby, Chvátal, and Cook [1995] (7397 cities), and Applegate, Bixby, Chvátal, and Cook [1998] (13,509 cities). Although the complexity of a TSP instance is not simply a function of the number of cities, these papers represent substantial steps forward in developing computational techniques for the traveling salesman problem.

Dynamic programming approaches were proposed by Bellman [1962] and Held and Karp [1962]. Several methods were compared by computer experiments by Lin [1965]. The Lagrangean relaxation technique was introduced by Christofides [1970] and Held and Karp [1970,1971]. The Held-Karp method was implemented and extended by Helbig Hansen and Krarup [1974], Smith and Thompson [1977], and Volgenant and Jonker [1982,1983]. Related work includes Bazaraa and Goode [1977].

Miliotis [1976,1978] described a constraint generation approach, mixing subtour elimination constraints with Gomory cuts or with branching. Focusing on the asymmetric TSP are Little, Murty, Sweeney, and Karel [1963] (first reports on a branch-and-bound method), Bellmore and Malone [1971] (on the effect of the subtour elimination constraints), (cf. Garfinkel [1973], Smith, Srinivasan, and Thompson [1977], Lenstra and Rinnooy Kan [1978], Carpaneto and Toth [1980b], Zhang [1997a]), Balas and Christofides [1981] (a Lagrangean approach based on the assignment problem, solving randomly generated asymmetric TSP's with up to 325 cities), Miller and Pekny [1989,1991], Pekny and Miller [1992], and Carpaneto, Dell'Amico, and Toth [1995].

Further bounds for the symmetric and asymmetric TSP were given by Christofides [1972], Carpaneto, Fischetti, and Toth [1989] and Fischetti and Toth [1992].

Important heuristics (algorithms that yield a tour that is expected to be short, but not necessarily shortest) and local search techniques include the *nearest neighbour heuristic*: always go to the closest city not yet visited (Menger [1932a], Gavett

[1965], Bellmore and Nemhauser [1968]), the *Lin-Kernighan heuristic*: start with a Hamiltonian circuit and iteratively replace a limited number of edges by other edges as long as it makes the circuit shorter (Lin and Kernighan [1973]), and Christofides' heuristic discussed in Section 58.8a. From the further work on, and analyses of, heuristics and local search techniques we mention Christofides and Eilon [1972], Rosenkrantz, Stearns, and Lewis [1977], Cornuéjols and Nemhauser [1978], Frieze [1979], d'Atri [1980], Bentley and Saxe [1980], Ong and Moore [1984], Golden and Stewart [1985] (survey), Johnson and Papadimitriou [1985b] (survey), Karp and Steele [1985] (survey), Johnson, Papadimitriou, and Yannakakis [1988], Kern [1989], Bentley [1990,1992], Papadimitriou [1992] (showing that unless P=NP, any local search method taking polynomial time per iteration, can lead to a locally optimum tour that is arbitrarily far from the optimum), Fredman, Johnson, McGeoch, and Ostheimer [1993,1995], Chandra, Karloff, and Tovey [1994,1999], Tassiulas [1997], and Frieze and Sorkin [2001]. A survey and comparison of heuristics and local search techniques for the traveling salesman problem was given by Johnson and McGeoch [1997].

Polynomial-time solvable special cases of the traveling salesman problem were given by Gilmore and Gomory [1964a,1964b], Gilmore [1966], Lawler [1971a], Sysło [1973], Cornuéjols, Naddef, and Pulleyblank [1983], and Hartvigsen and Pulleyblank [1994]. Surveys of such problems were given by Gilmore, Lawler, and Shmoys [1985] and Burkard, Deĭneko, van Dal, van der Veen, and Woeginger [1998].

The standard reference book on the traveling salesman problem, covering a wide variety of aspects, was edited by Lawler, Lenstra, Rinnooy Kan, and Shmoys [1985]. In this book, Grötschel and Padberg [1985] considered the traveling salesman polytope, Padberg and Grötschel [1985] computation with the help of polyhedra, Johnson and Papadimitriou [1985a] the computational complexity of the TSP, and Balas and Toth [1985] branch-and-bound method methods. Computational methods and results are surveyed in the book by Reinelt [1994].

Survey articles on the traveling salesman problem were given by Gomory [1966], Bellmore and Nemhauser [1968], Gupta [1968], Tyagi [1968], Burkard [1979], Christofides [1979], Grötschel [1982] (also on other NP-complete problems), and Johnson and McGeoch [1997] (local search techniques). Introductions are given in the books by Minieka [1978], Sysło, Deo, and Kowalik [1983], Cook, Cunningham, Pulleyblank, and Schrijver [1998], and Korte and Vygen [2000]. An insightful survey of the computational methods for the symmetric TSP was given by Jünger, Reinelt, and Rinaldi [1995]. A framework for guaranteeing quality of TSP solutions was presented by Jünger, Thienel, and Reinelt [1994]. An early survey on branch-and-bound method techniques was given by Lawler and Wood [1966].

Barvinok, Johnson, Woeginger, and Woodroofe [1998] showed that there is a polynomial-time algorithm to find a *longest* Hamiltonian circuit in a complete graph with length determined by a polyhedral norm. Related work was done by Barvinok [1996]. More on the longest Hamiltonian circuit can be found in Fisher, Nemhauser, and Wolsey [1979], Serdyukov [1984], Kostochka and Serdyukov [1985], Kosaraju, Park, and Stein [1994], Hassin and Rubinstein [2000,2001], and Bläser [2002].

58.11b. Historical notes on the traveling salesman problem

Mathematically, the traveling salesman problem is related to, in fact generalizes, the question for a Hamiltonian circuit in a graph. This question goes back to Kirkman

Section 58.11b. Historical notes on the traveling salesman problem   997

[1856] and Hamilton [1856,1858] and was also studied by Kowalewski [1917b,1917a] — see Biggs, Lloyd, and Wilson [1976]. We restrict our survey to the traveling salesman problem in its general form.

The mathematical roots of the traveling salesman problem are obscure. Dantzig, Fulkerson, and Johnson [1954a] say:

> It appears to have been discussed informally among mathematicians at mathematics meetings for many years.

A 1832 manual

The traveling salesman problem has a natural interpretation, and Müller-Merbach [1983] detected that the problem was formulated in a 1832 manual for the successful traveling salesman, *Der Handlungsreisende — wie er sein soll und was er zu thun hat, um Aufträge zu erhalten und eines glücklichen Erfolgs in seinen Geschäften gewiß zu sein — Von einem alten Commis-Voyageur*[9] ('ein alter Commis-Voyageur' [1832]). (Whereas the politically correct nowadays prefer to speak of the traveling sales*person* problem, the manual presumes that the 'Handlungsreisende' is male, and it warns about the risks of women in or out of business.)

The booklet contains no mathematics, and formulates the problem as follows:

> Die Geschäfte führen die Handlungsreisenden bald hier, bald dort hin, und es lassen sich nicht füglich Reisetouren angeben, die für alle vorkommende Fälle passend sind; aber es kann durch eine zweckmäßige Wahl und Eintheilung der Tour, manchmal so viel Zeit gewonnen werden, daß wir es nicht glauben umgehen zu dürfen, auch hierüber einige Vorschriften zu geben. Ein Jeder möge so viel davon benutzen, als er es seinem Zwecke für dienlich hält; so viel glauben wir aber davon versichern zu dürfen, daß es nicht wohl thunlich sein wird, die Touren durch Deutschland in Absicht der Entfernungen und, worauf der Reisende hauptsächlich zu sehen hat, des Hin- und Herreisens, mit mehr Oekonomie einzurichten. Die Hauptsache besteht immer darin: so viele Orte wie möglich mitzunehmen, ohne den nämlichen Ort zweimal berühren zu müssen.[10]

The manual suggests five tours through Germany (one of them partly through Switzerland). In Figure 58.2 we compare one of the tours with a shortest tour, found with 'modern' methods. (Most other tours given in the manual do not qualify for 'die Hauptsache' as they contain subtours, so that some places are visited twice.)

Menger's Botenproblem 1930

K. Menger seems to be the first mathematician to have written about the traveling salesman problem. The root of his interest is given in his paper Menger [1928c]. In

---

[9] 'The traveling salesman — how he should be and what he has to do, to obtain orders and to be sure of a happy success in his business — by an old traveling salesman'

[10] Business brings the traveling salesman now here, then there, and no travel routes can be properly indicated that are suitable for all cases occurring; but sometimes, by an appropriate choice and arrangement of the tour, so much time can be gained, that we don't think we may avoid giving some rules also on this. Everybody may use that much of it, as he takes it for useful for his goal; so much of it however we think we may assure, that it will not be well feasible to arrange the tours through Germany with more economy in view of the distances and, which the traveler mainly has to consider, of the trip back and forth. The main point always consists of visiting as many places as possible, without having to touch the same place twice.

998   Chapter 58. The traveling salesman problem

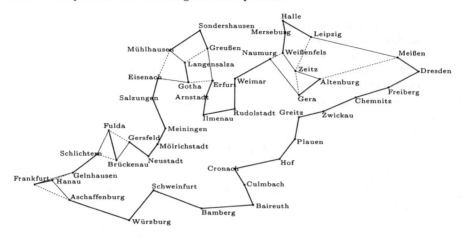

Figure 58.2
A tour along 45 German cities, as described in the 1832 traveling salesman manual, is given by the unbroken (bold and thin) lines (1285 km). A shortest tour is given by the unbroken bold and by the dashed lines (1248 km). We have taken geodesic distances — taking local conditions into account, the 1832 tour might be optimum.

this, he studies the *length* $l(C)$ of a simple curve C in a metric space S, which is, by definition,

$$(58.26) \qquad l(C) := \sup \sum_{i=1}^{n-1} \mathrm{dist}(x_i, x_{i+1}),$$

where the supremum ranges over all choices of $x_1, \ldots, x_n$ on C *in the order determined by* C. What Menger showed is that we may relax this to finite subsets X of C and minimize over all possible orderings of X. To this end he defined, for any finite subset X of a metric space, $\lambda(X)$ to be the shortest length of a path through X (in graph terminology: a *Hamiltonian path*), and he showed that

$$(58.27) \qquad l(C) = \sup_X \lambda(X),$$

where the supremum ranges over all finite subsets X of C. It amounts to showing that for each $\varepsilon > 0$ there is a finite subset X of C such that $\lambda(X) \geq l(C) - \varepsilon$.

Menger [1929a] sharpened this to:

$$(58.28) \qquad l(C) = \sup_X \kappa(X),$$

where again the supremum ranges over all finite subsets X of C, and where $\kappa(X)$ denotes the minimum length of a *spanning tree* on X.

These results were reported also in Menger [1930]. In a number of other papers, Menger [1928b,1929b,1929a] gave related results on these new characterizations of the length function.

The parameter $\lambda(X)$ clearly is close to the practical interpretation of the traveling salesman problem. This relation was made explicit by Menger in the session

Section 58.11b. Historical notes on the traveling salesman problem

of 5 February 1930 of his *mathematisches Kolloquium* in Vienna. Menger [1931a, 1932a] reported that he first asked if a further relaxation is possible by replacing $\kappa(X)$ by the minimum length of an (in current terminology) *Steiner tree* connecting X — a spanning tree on a superset of X in S. (So Menger toured along some basic combinatorial optimization problems.) This problem was solved for Euclidean spaces by Mimura [1933].

Next Menger posed the traveling salesman problem, as follows:

> Wir bezeichnen als *Botenproblem* (weil diese Frage in der Praxis von jedem Postboten, übrigens auch von vielen Reisenden zu lösen ist) die Aufgabe, für endlichviele Punkte, deren paarweise Abstände bekannt sind, den kürzesten die Punkte verbindenden Weg zu finden. Dieses Problem ist natürlich stets durch endlichviele Versuche lösbar. Regeln, welche die Anzahl der Versuche unter die Anzahl der Permutationen der gegebenen Punkte herunterdrücken würden, sind nicht bekannt. Die Regel, man solle vom Ausgangspunkt erst zum nächstgelegenen Punkt, dann zu dem diesem nächstgelegenen Punkt gehen usw., liefert im allgemeinen nicht den kürzesten Weg.[11]

So Menger asked for a shortest Hamiltonian path through the given points. He was aware of the complexity issue in the traveling salesman problem, and he realized that the now well-known nearest neighbour heuristic might not give an optimum solution.

Harvard, Princeton 1930-1934

Menger spent the period September 1930-February 1931 as visiting lecturer at Harvard University. In one of his seminar talks at Harvard, Menger presented his results (quoted above) on lengths of arcs and shortest paths through finite sets of points. According to Menger [1931b], a suggestion related to this was given by Hassler Whitney, who at that time did his Ph.D. research in graph theory at Harvard. This paper of Menger however does not mention if the practical interpretation was given in the seminar talk.

The year after, 1931-1932, Whitney was a National Research Council Fellow at Princeton University, where he gave a number of seminar talks. In a seminar talk, he mentioned the problem of finding the shortest route along the 48 States of America.

There are some uncertainties in this story. It is not sure if Whitney spoke about the 48 States problem during his 1931-1932 seminar talks (which talks he did give), or later, in 1934, as is said by Flood [1956] in his article on the traveling salesman problem:

> This problem was posed, in 1934, by Hassler Whitney in a seminar talk at Princeton University.

That memory can be shaky might be indicated by the following two quotes. Dantzig, Fulkerson, and Johnson [1954a] remark:

---

[11] We denote by *messenger problem* (since in practice this question should be solved by each postman, anyway also by many travelers) the task to find, for finitely many points whose pairwise distances are known, the shortest route connecting the points. Of course, this problem is solvable by finitely many trials. Rules which would push the number of trials below the number of permutations of the given points, are not known. The rule that one first should go from the starting point to the closest point, then to the point closest to this, etc., in general does not yield the shortest route.

Both Flood and A.W. Tucker (Princeton University) recall that they heard about the problem first in a seminar talk by Hassler Whitney at Princeton in 1934 (although Whitney, recently queried, does not seem to recall the problem).

However, when asked by David Shmoys, Tucker replied in a letter of 17 February 1983 (see Hoffman and Wolfe [1985]):

> I cannot confirm or deny the story that I heard of the TSP from Hassler Whitney. If I did (as Flood says), it would have occurred in 1931-32, the first year of the old Fine Hall (now Jones Hall). That year Whitney was a postdoctoral fellow at Fine Hall working on Graph Theory, especially planarity and other offshoots of the 4-color problem. ... I was finishing my thesis with Lefschetz on n-manifolds and Merrill Flood was a first year graduate student. The Fine Hall Common Room was a very lively place — 24 hours a day.

(Whitney finished his Ph.D. at Harvard University in 1932.)

Another uncertainty is in which form Whitney has posed the problem. That he might have focused on finding a shortest route along the 48 states in the U.S.A., is suggested by the reference by Flood, in an interview on 14 May 1984 with Tucker [1984a], to the problem as the '48 States Problem of Hassler Whitney'. In this respect Flood also remarked:

> I don't know who coined the peppier name 'Traveling Salesman Problem' for Whitney's problem, but that name certainly has caught on, and the problem has turned out to be of very fundamental importance.

TSP, Hamiltonian paths, and school bus routing

Flood [1956] remembered that in 1937, A.W. Tucker pointed out to him the connections of the TSP with Hamiltonian games and Hamiltonian paths in graphs:

> I am indebted to A.W. Tucker for calling these connections to my attention, in 1937, when I was struggling with the problem in connection with a schoolbus routing study in New Jersey.

In the following quote from the interview by Tucker [1984a], Flood referred to school bus routing in a different state (West Virginia), and he mentioned the involvement in the TSP of Koopmans, who spent 1940-1941 at the Local Government Surveys Section of Princeton University ('the Princeton Surveys'):

> Koopmans first became interested in the "48 States Problem" of Hassler Whitney when he was with me in the Princeton Surveys, as I tried to solve the problem in connection with the work by Bob Singleton and me on school bus routing for the State of West Virginia.

1940

In 1940, some papers appeared that study the traveling salesman problem, in a different context. They seem to be the first containing mathematical results on the problem.

In the American continuation of Menger's *mathematisches Kolloquium*, Menger [1940] returned to the question of the shortest path through a given set of points in a metric space, followed by investigations of Milgram [1940] on the shortest Jordan

Section 58.11b. Historical notes on the traveling salesman problem

curve that covers a given, not necessarily finite, set of points in a metric space. As the set may be infinite, a shortest curve need not exist.

Fejes [1940] investigated the problem of a shortest curve through n points in the unit square. In consequence of this, Verblunsky [1951] showed that its length is less than $2 + \sqrt{2.8n}$. Later work in this direction includes Few [1955], Beardwood, Halton, and Hammersley [1959], Steele [1981], Moran [1984], Karloff [1989], and Goddyn [1990].

Lower bounds on the expected value of a shortest path through n random points in the plane were studied by Mahalanobis [1940] in order to estimate the cost of a sample survey of the acreage under jute in Bengal. This survey took place in 1938 and one of the major costs in carrying out the survey was the transportation of men and equipment from one survey point to the next. He estimated (without proof) the minimum length of a tour along n random points in the plane, for Euclidean distance:

> It is also easy to see in a general way how the journey time is likely to behave. Let us suppose that n sampling units are scattered at random within any given area ; and let us assume that we may treat each such sample unit as a geometrical point. We may also assume that arrangements will usually be made to move from one sample point to another in such a way as to keep the total distance travelled as small as possible ; that is, we may assume that the path traversed in going from one sample point to another will follow a straight line. In this case it is easy to see that the mathematical expectation of the total length of the path travelled in moving from one sample point to another will be $(\sqrt{n} - 1/\sqrt{n})$. The cost of the journey from sample to sample will therefore be roughly proportional to $(\sqrt{n} - 1/\sqrt{n})$. When n is large, that is, when we consider a sufficiently large area, we may expect that the time required for moving from sample to sample will be roughly proportional to $\sqrt{n}$, where n is the total number of samples in the given area. If we consider the journey time per sq. mile, it will be roughly proportional to $\sqrt{y}$, where y is the density of number of sample units per sq. mile.

This research was continued by Jessen [1942], who estimated empirically a similar result for l_1-distance (Manhattan distance), in a statistical investigation of a sample survey for obtaining farm facts in Iowa:

> If a route connecting y points located at random in a fixed area is minimized, the total distance, D, of that route is[12]
>
> $$D = d\left(\frac{y-1}{\sqrt{y}}\right)$$
>
> where d is a constant.
> This relationship is based upon the assumption that points are connected by direct routes. In Iowa the road system is a quite regular network of mile square mesh. There are very few diagonal roads, therefore, routes between points resemble those taken on a checkerboard. A test wherein several sets of different members of points were located at random on an Iowa county road map, and the minimum distance of travel from a given point on the border of the county through all the points and to an end point (the county border nearest the last point on route), revealed that
>
> $$D = d\sqrt{y}$$
>
> works well. Here y is the number of randomized points (border points not included). This is of great aid in setting up a cost function.

---

[12] at this point, Jessen referred in a footnote to Mahalanobis [1940].

Marks [1948] gave a proof of Mahalanobis' bound. In fact he showed that $\sqrt{\frac{1}{2}A}(\sqrt{n}-1/\sqrt{n})$ is a lower bound, where A is the area of the region. Ghosh [1949] showed that this bound asymptotically is close to the expected value, by giving a heuristic for finding a tour, yielding an upper bound of $1.27\sqrt{An}$. He also observed the complexity of the problem:

> After locating the n random points in a map of the region, it is very difficult to find out *actually* the shortest path connecting the points, unless the number n is very small, which is seldom the case for a large-scale survey.

TSP, transportation, and assignment

As is the case for several other combinatorial optimization problems, the RAND Corporation in Santa Monica, California, played an important role in the research on the TSP. Hoffman and Wolfe [1985] write that

> John Williams urged Flood in 1948 to popularize the TSP at the RAND Corporation, at least partly motivated by the purpose of creating intellectual challenges for models outside the theory of games. In fact, a prize was offered for a significant theorem bearing on the TSP. There is no doubt that the reputation and authority of RAND, which quickly became the intellectual center of much of operations research theory, amplified Flood's advertising.

(John D. Williams was head of the Mathematics Division of RAND at that time.)

At RAND, researchers considered the idea of transferring the successful methods for the transportation problem to the traveling salesman problem. Flood [1956] mentioned that this idea was brought to his attention by Koopmans in 1948. In the interview with Tucker [1984a], Flood remembered:

> George Dantzig and Tjallings Koopmans met with me in 1948 in Washington, D.C., at the meeting of the International Statistical Institute, to tell me excitedly of their work on what is now known as the linear programming problem and with Tjallings speculating that there was a significant connection with the Traveling Salesman Problem.

The issue was taken up in a RAND Report by Julia Robinson [1949], who, in an 'unsuccessful attempt' to solve the traveling salesman problem, considered, as a relaxation, the assignment problem, for which she found a cycle reduction method. The relation is that the assignment problem asks for an optimum permutation, and the TSP for an optimum *cyclic* permutation.

Robinson's RAND report might be the earliest mathematical reference using the term 'traveling salesman problem':

> The purpose of this note is to give a method for solving a problem related to the traveling salesman problem. One formulation is to find the shortest route for a salesman starting from Washington, visiting all the state capitals and then returning to Washington. More generally, to find the shortest closed curve containing n given points in the plane.

Flood wrote (in a letter of 17 May 1983 to E.L. Lawler) that Robinson's report stimulated several discussions on the TSP of him with his research assistant at RAND, D.R. Fulkerson, during 1950-1952[13].

It was noted by Beckmann and Koopmans [1952] that the TSP can be formulated as a quadratic assignment problem, for which however no fast methods are known.

---

[13] Fulkerson started at RAND only in March 1951.

Dantzig, Fulkerson, Johnson 1954

Fundamental progress on the traveling salesman was made in a seminal paper by the RAND researchers Dantzig, Fulkerson, and Johnson [1954a] — according to Hoffman and Wolfe [1985] 'one of the principal events in the history of combinatorial optimization'. The paper introduced several new methods for solving the traveling salesman problem that are now basic in combinatorial optimization. In particular, it shows the importance of *cutting planes* for combinatorial optimization.

While the subtour elimination constraints (58.4)(iii) are enough to cut off the noncyclic permutation matrices from the polytope of doubly stochastic matrices (determined by (58.4)(i) and (ii)), they generally do not yield all facets of the traveling salesman polytope, as was observed by Heller [1953a]: there exist doubly stochastic matrices, of any order $n \geq 5$, that satisfy (58.4) but are not a convex combination of cyclic permutation matrices.

The subtour elimination constraints can nevertheless be useful for the TSP, since it gives a lower bound for the optimum tour length if we minimize over the constraints (58.4). This lower bound can be calculated with the simplex method, taking the (exponentially many) constraints (58.4)(iii) as *cutting planes* that can be added during the process when needed. In this way, Dantzig, Fulkerson, and Johnson were able to find the shortest tour along cities chosen in the 48 U.S. states and Washington, D.C. Incidentally, this is close to the problem mentioned by Julia Robinson in 1949 (and maybe also by Whitney in the 1930s).

The Dantzig-Fulkerson-Johnson paper gives no algorithm, but rather gives a tour and proves its optimality with the help of the subtour elimination constraints. This work forms the basis for most of the later work on large-scale traveling salesman problems.

Early studies of the traveling salesman polytope were reported by Heller [1953a,1953b,1955a,1955b,1956a,1956b], Kuhn [1955a], Norman [1955], and Robacker [1955b], who also made computational studies of the probability that a random instance of the traveling salesman problem needs the subtour elimination constraints (58.4)(iii) (cf. Kuhn [1991]). This made Flood [1956] remark on the intrinsic complexity of the traveling salesman problem:

> Very recent mathematical work on the traveling-salesman problem by I. Heller, H.W. Kuhn, and others indicates that the problem is fundamentally complex. It seems very likely that quite a different approach from any yet used may be required for succesful treatment of the problem. In fact, there may well be no general method for treating the problem and impossibility results would also be valuable.

Flood mentioned a number of other applications of the traveling salesman problem, in particular in machine scheduling, brought to his attention in a seminar talk at Columbia University in 1954 by George Feeney.

Other work on the traveling salesman problem in the 1950s was done by Morton and Land [1955] (a linear programming approach with a 3-exchange heuristic), Barachet [1957] (a graphic solution method), Bock [1958], Croes [1958] (a heuristic), and Rossman and Twery [1958]. In a reaction to Barachet's paper, Dantzig, Fulkerson, and Johnson [1959] showed that their method yields the optimality of Barachet's (heuristically found) solution.

Chapter 58. The traveling salesman problem

In 1962, the soap company Proctor and Gamble run a contest, requiring to solve a traveling salesman problem along 33 U.S. cities. Little, Murty, Sweeney, and Karel [1963] report:

> The traveling salesman problem recently achieved national prominence when a soap company used it as the basis of a promotional contest. Prizes up to $10,000 were offered for identifying the most correct links in a particular 33-city problem. Quite a few people found the best tour. (The tie-breaking contest for these successful mathematicians was to complete a statement of 25 words or less on "I like...because...".) A number of people, perhaps a little over-educated, wrote the company that the problem was impossible—an interesting misinterpretation of the state of the art.

Chapter 59

Matching forests

Giles [1982a,1982b,1982c] introduced the concept of a *matching forest* in a mixed graph (V, E, A), which is a subset F of $E \cup A$ such that $F \cap A$ is a branching and $F \cap E$ is a matching only covering roots of the branching $F \cap A$. Equivalently, F contains no circuit (in the underlying undirected graph) and each $v \in V$ is head of at most one $e \in F$. (Here, for an undirected edge e, both ends of e are called head of e.)

Matching forests generalize both matchings in undirected graphs and branchings in directed graphs. Giles gave a polynomial-time algorithm to find a maximum-weight matching forest, yielding as a by-product a characterization of the matching forest polytope (the convex hull of the incidence vectors of matching forests).

Giles' results generalize the polynomial-time solvability and the polyhedral characterizations for matchings (Chapters 24–26) and for branchings (Chapter 52).

59.1. Introduction

A *mixed graph* is a triple (V, E, A), where (V, E) is an undirected graph and (V, A) is a directed graph. In this chapter, a graph can have multiple edges, but no loops. The *underlying* undirected graph of a mixed graph is the undirected graph obtained from the mixed graph by forgetting the orientations of the directed edges.

As usual, if an edge e is directed from u to v, then u is called the *tail* and v the *head* of e. In this chapter, if e is undirected and connects u and v, then both u and v will be called *head* of e.

A subset F of $E \cup A$ is called a *matching forest* if F contains no circuits (in the underlying undirected graph) and any vertex v is head of at most one edge in F. We call a vertex v a *root* of F if v is head of no edge in F. We denote the set of roots of F by $R(F)$.

It is convenient to consider the relations of matching forests with matchings in undirected graphs and branchings in directed graphs: M is a matching in an undirected graph (V, E) if and only if M is a matching forest in the mixed graph $(V, E, \emptyset)$. In this case, the roots of M are the vertices not covered by M. Similarly, B is a branching in a directed graph (V, A) if and only if B

is a matching forest in the mixed graph $(V,\emptyset,A)$. In this case, the concept of root of a branching and root of a matching forest coincide.

In turn, we can characterize matching forests in terms of matchings and branchings: for any mixed graph (V,E,A), a subset F of $E\cup A$ is a matching forest if and only if $F\cap A$ is a branching in (V,A) and $F\cap E$ is a matching in (V,E) such that $F\cap E$ only covers roots of $F\cap A$.

It will be useful to observe the following formulas, for any matching forest F in a mixed graph (V,E,A), setting $M:=F\cap E$ and $B:=F\cap A$:

(59.1) $R(F)=R(M)\cap R(B)$ and $V=R(M)\cup R(B)$.

In fact, for any matching M in (V,E) and any branching B in (V,A), the set $M\cup B$ is a matching forest if and only if $R(M)\cup R(B)=V$.

59.2. The maximum size of a matching forest

Giles [1982a] described a min-max formula for the maximum size of a matching forest. It can be derived from the Tutte-Berge formula with the following direct formula:

Theorem 59.1. *Let (V,E,A) be a mixed graph and let $\mathcal{K}$ be the collection of those strong components K of the directed graph (V,A) that satisfy $d_A^{\text{in}}(K)=0$. Consider the undirected graph H with vertex set $\mathcal{K}$, where two distinct $K,L\in\mathcal{K}$ are adjacent if and only if there is an edge in E connecting K and L. Then the maximum size of a matching forest in (V,E,A) is equal to*

(59.2) $\nu(H)+|V|-|\mathcal{K}|$.

Here $\nu(H)$ denotes the maximum size of a matching in H.

Proof. Let M' be a matching in H of size $\nu(H)$. Then M' yields a matching M of size $\nu(H)$ in (V,E), where each edge in M connects two components in $\mathcal{K}$. Now there exists a branching B in (V,A) such that B has exactly $|\mathcal{K}|$ roots, such that each $K\in\mathcal{K}$ contains exactly one root, and such that each vertex covered by M is a root of B. (To see that such a branching B exists, choose, for any $K\in\mathcal{K}$ not intersecting M, an arbitrary vertex in K. Let X be the set of chosen vertices together with the vertices covered by M. As X intersects each $K\in\mathcal{K}$, each vertex in V is reachable in (V,A) by a directed path from X. Hence there exists a branching B with root set X. This B has the required properties.)

Then $M\cup B$ is a matching forest, of size $\nu(H)+|V|-|\mathcal{K}|$ (as B has size $|V|-|\mathcal{K}|$).

To see that there is no larger matching forest, let F be any matching forest. Let $U:=\bigcup\mathcal{K}$. Then F has at most $|V\setminus U|$ edges with at least one head in $V\setminus U$. Since no directed edge enters U, all other edges are contained in

U. So it suffices to show that F has at most $\nu(H)+|U|-|\mathcal{K}|$ edges contained in U.

Let N be the set of (necessarily undirected) edges in F connecting two different components in $\mathcal{K}$. For each $K \in \mathcal{K}$, let α_K be the number of edges in N incident with K. Then

(59.3) $\qquad |N| - \sum_{K \in \mathcal{K}} \max\{0, \alpha_K - 1\} \leq \nu(H),$

since by deleting, for each $K \in \mathcal{K}$, at most $\max\{0, \alpha_K - 1\}$ edges from N incident with K, we obtain a matching in the graph H defined above.

We have moreover that any $K \in \mathcal{K}$ spans at most $|K| - \max\{1, \alpha_K\}$ edges of F. With (59.3) this implies that the number of edges in F contained in U is at most

(59.4) $\qquad |N| + \sum_{K \in \mathcal{K}}(|K| - \max\{1, \alpha_K\}) \leq \nu(H) + \sum_{K \in \mathcal{K}}(|K| - 1)$
$\qquad \quad = \nu(H) + |U| - |\mathcal{K}|,$

as required. ∎

The method described in this proof also directly implies that a maximum-size matching forest can be found in polynomial time (Giles [1982a]).

59.3. Perfect matching forests

Figure 59.1
$\{e, f\}$ and $\{e, a, b\}$ are perfect matching forests.

A matching forest F is called *perfect* if each vertex is head of exactly one edge in F. (So a perfect matching forest need not be a maximum-size matching forest — cf. Figure 59.1.) The following is easy to see:

(59.5) $\qquad$ A mixed graph (V, E, A) contains a perfect matching forest F if and only if the graph (V, E) contains a matching M such that each strong component K of (V, A) with $d^{\text{in}}(K) = 0$ is intersected by at least one edge in M.

Indeed, if a perfect matching forest F exists, then $M := F \cap E$ is such a matching. Conversely, if such a matching M exists, any vertex is reachable by a directed path from at least one vertex covered by M; hence M can be augmented with directed arcs to a perfect matching forest.

This shows (59.5), which implies the following characterization for perfect matching forests of Giles [1982b]:

Theorem 59.2. *Let (V, E, A) be a mixed graph and let $\mathcal{K}$ be the collection of strong components K of (V, A) with $d_A^{\text{in}}(K) = 0$. Then (V, E, A) has a perfect matching forest if and only if for each $U \subseteq V$ and $\mathcal{L} \subseteq \mathcal{K}$ the graph $(V, E) - U$ has at most $|U| + |\bigcup \mathcal{L}| - |\mathcal{L}|$ odd components that are contained in $\bigcup \mathcal{L}$.*

Proof. Extend $G = (V, E)$ by, for each $K \in \mathcal{K}$, a clique C_K of size $|K| - 1$, such that each vertex in C_K is adjacent to each vertex in K. This makes the undirected graph H. Then (V, E, A) has a perfect matching forest if and only if graph H has a matching covering $\bigcup \mathcal{K}$. So we can apply Corollary 24.6a. ∎

This method also gives a polynomial-time algorithm to find a perfect matching forest.

59.4. An exchange property of matching forests

As a preparation for characterizing the matching forest polytope, we show an exchange property of matching forests. It generalizes the well-known and trivial exchange property of matchings in an undirected graph, based on considering the union of two matchings.

Lemma 59.3α. *Let F_1 and F_2 be matching forests in a mixed graph (V, E, A). Let $s \in R(F_2) \setminus R(F_1)$. Then there exist matching forests F_1' and F_2' such that $F_1' \cap F_2' = F_1 \cap F_2$, $F_1' \cup F_2' = F_1 \cup F_2$, $s \in R(F_1')$, and*

(59.6)   (i) $|F_1'| < |F_1|$,
        or (ii) $|F_1'| = |F_1|$ and $|R(F_1')| > |R(F_1)|$,
        or (iii) $|F_1'| = |F_1|$, $R(F_1') = (R(F_1) \setminus \{t\}) \cup \{s\}$ for some
          $t \in R(F_1)$, and $|R(F_1' \cap A) \cap K| = |R(F_1 \cap A) \cap K|$ for
          each strong component K of the directed graph (V, A).

Proof. We may assume that F_1 and F_2 partition $E \cup A$, as we can delete edges that are not in $F_1 \cup F_2$, and add parallel edges to those in $F_1 \cap F_2$.

Define $M_i := F_i \cap E$ and $B_i := F_i \cap A$ for $i = 1, 2$. Let $\mathcal{K}$ be the collection of strong components K of the directed graph (V, A) with $\delta_A^{\text{in}}(K) = \emptyset$. Then each set in $\mathcal{K}$ intersects both $R(B_1)$ and $R(B_2)$, and $\{v\} \in \mathcal{K}$ for each $v \in R(B_1) \cap R(B_2)$.

So each $K \in \mathcal{K}$ with $|K| \geq 2$ intersects $R(B_1)$ and $R(B_2)$ in disjoint subsets. Hence we can choose for each such K

Section 59.4. An exchange property of matching forests

(59.7)   a pair $e_K \subseteq K$ consisting of a vertex in $R(B_1) \setminus R(B_2)$ and a vertex in $R(B_2) \setminus R(B_1)$.

Let N be the set of pairs e_K for $K \in \mathcal{K}$ with $|K| \geq 2$. So N consists of disjoint pairs.

Then the undirected graph H on V with edge set

(59.8) $M_1 \cup M_2 \cup N$

consists of a number of vertex-disjoint paths and circuits, since no vertex in $R(B_1) \setminus R(B_2)$ is covered by M_2, and no vertex in $R(B_2) \setminus R(B_1)$ is covered by M_1.

Moreover, s has degree at most one in H. Indeed, s is not covered by M_2, as $s \in R(F_2) = R(M_2) \cap R(B_2)$. If s is covered by M_1, then $s \in R(B_1)$, and so $s \in R(B_1) \cap R(B_2)$, implying that s is not covered by N.

So s is the starting vertex of a path component P of H (possibly only consisting of s). Let Y be the set of edges in $M_1 \cup M_2$ occurring in P, and set

(59.9) $M'_1 := M_1 \triangle Y$ and $M'_2 := M_2 \triangle Y$

(where $\triangle$ denotes symmetric difference). Since Y is the union of the edge sets of some path components of the graph $(V, M_1 \cup M_2)$, we know that M'_1 and M'_2 are matchings again.

Then, obviously, $R(M'_1)$ and $R(M'_2)$ arise from $R(M_1)$ and $R(M_2)$ by exchanging these sets on VP; that is:

(59.10) $R(M'_1) = (R(M_1) \setminus VP) \cup (R(M_2) \cap VP)$ and
$R(M'_2) = (R(M_2) \setminus VP) \cup (R(M_1) \cap VP)$.

We show that a similar operation can be performed with respect to B_1 and B_2; that is, we show that there exist disjoint branchings B'_1 and B'_2 in (V, A) satisfying

(59.11) $R(B'_1) = (R(B_1) \setminus VP) \cup (R(B_2) \cap VP)$ and
$R(B'_2) = (R(B_2) \setminus VP) \cup (R(B_1) \cap VP)$.

By Lemma 53.2α, it suffices to show that each $K \in \mathcal{K}$ intersects both sets in (59.11). If $|K| = 1$, then K is contained in both $R(B_1)$ and $R(B_2)$, and hence in both sets in (59.11). If $|K| \geq 2$, then e_K intersects both $R(B_1)$ and $R(B_2)$. Since e_K is either contained in VP or disjoint from VP, e_K intersects both sets in (59.11). Hence, as $e_K \subseteq K$, also K intersects both sets in (59.11). Therefore, branchings B'_1 and B'_2 satisfying (59.11) exist.

(59.10) and (59.11) imply:

(59.12) $F'_1 := M'_1 \cup B'_1$ and $F'_2 := M'_2 \cup B'_2$ are matching forests.

To see this, we must show that $R(M'_1) \cup R(B'_1) = V$ and $R(M'_2) \cup R(B'_2) = V$. Since $R(M_1) \cup R(B_1) = V$ and $R(M_2) \cup R(B_2) = V$, this follows directly from (59.10) and (59.11). This shows (59.12).

Since $R(F) = R(M) \cap R(B)$ for any matching forest F (with $M := F \cap E$ and $B := F \cap A$), (59.10) and (59.11) imply that also $R(F_1')$ and $R(F_2')$ arise from $R(F_1)$ and $R(F_2)$ by swapping on P; that is:

(59.13) $\quad R(F_1') = (R(F_1) \setminus VP) \cup (R(F_2) \cap VP)$ and
$\quad\quad\quad\;\; R(F_2') = (R(F_2) \setminus VP) \cup (R(F_1) \cap VP)$.

This implies:

(59.14) $\quad s \in R(F_1') \setminus R(F_2')$,

since $s \in VP$ and $s \in R(F_2) \setminus R(F_1)$.

We study the effects of the exchanges (59.10) and (59.11), to show that one of the alternatives (59.6) holds. It is based on the following observations on the sizes of M_1' and B_1'. Let t be the last vertex of P (possible $t = s$).

Suppose that none of the alternatives (59.6) hold. If $s = t$, then s is not covered by M_1, and so $M_1' = M_1$ and $R(B_1') = R(B_1) \cup \{s\}$, implying $|F_1'| < |F_1|$, which is alternative (59.6)(i). So $s \neq t$.

By the exchanges made, $|M_1| - |M_1'| = |M_1 \cap EP| - |M_2 \cap EP|$ and $|R(F_1)| - |R(F_1')| = |R(F_1) \cap VP| - |R(F_2) \cap VP|$. This gives, as $|F_1'| \geq |F_1|$, since alternative (59.6)(i) does not hold:

(59.15) $\quad |M_1 \cap EP| - |M_2 \cap EP| + |R(F_1) \cap VP| - |R(F_2) \cap VP|$
$\quad\quad\quad\;\; = |M_1| + |R(F_1)| - |M_1'| - |R(F_1')| = |F_1'| - |F_1| \geq 0$.

(The last equality holds as $|F_i'| = |V| - |M_i'| - |R(F_i')|$ for $i = 1, 2$, since $|F_i'| + |M_i'|$ is the number of heads of edges in F_i'.)

We next note:

(59.16) $\quad$ no intermediate vertex v of P belongs to $R(F_1) \cup R(F_2)$.

For suppose that $v \in R(F_1)$. Then (as v is an intermediate vertex of P) v is covered by M_2 and some $e_K \in N$. Hence $v \in R(B_2)$, and therefore $v \notin R(B_1)$ (by (59.7)), contradicting the fact that $v \in R(F_1)$. One similarly shows that $v \notin R(F_2)$, proving (59.16).

As $s \in R(F_2) \setminus R(F_1)$, (59.16) implies that

(59.17) $\quad |R(F_1) \cap VP| \leq |R(F_2) \cap VP|$, with equality if and only if $t \in R(F_1) \setminus R(F_2)$.

With (59.15) this gives that $|M_1 \cap EP| \geq |M_2 \cap EP|$.

Let k be the number of edges in $M_1 \cup M_2$ on P. Note that the edges in $M_1 \cup M_2$ occur along P alternatingly in M_1 and M_2, as any intermediate $e_K \in N$ on P connects an edge in M_1 and an edge in M_2 (as by (59.7), $e_K \in N$ consists of a vertex not in $R(B_2)$ and a vertex not in $R(B_1)$).

Suppose that k is odd. Then $|M_1 \cap EP| = |M_2 \cap EP| + 1$. So the last edge in $M_1 \cup M_2$ along P (seen from s) belongs to M_1. Moreover, one has that $t \notin R(F_1)$. For if $t \in R(F_1)$, then t is not covered by M_1, and hence t belongs to some $e_K = \{v, t\} \in N$ with v covered by M_1. Hence $v \in R(B_1)$, and hence $t \notin R(B_1)$ (by (59.7)), contradicting the fact that $t \in R(F_1)$. So $t \notin R(F_1)$.

Then (59.17) implies that $|R(F_2) \cap VP| > |R(F_1) \cap VP|$. This implies with (59.15) that $|F_1'| = |F_1|$ (as $|M_1 \cap EP| = |M_2 \cap EP| + 1$), and with (59.13) that $|R(F_1')| > |R(F_1)|$. So (59.6)(ii) holds, a contradiction.

So k is even, and hence $|M_1 \cap EP| = |M_2 \cap EP|$, which implies with (59.13), (59.15), and (59.17) that $|R(F_1)| = |R(F_2)|$ and $t \in R(F_1) \setminus R(F_2)$. Therefore, $|F_1'| = |F_1|$ (by (59.16)) and $R(F_1') = (R(F_1) \setminus \{t\}) \cup \{s\}$.

Finally, $|R(B_1') \cap K| = |R(B_1) \cap K|$ for each strong component K of D. This follows directly (with (59.11)) from the fact that for any $v \in K \cap VP$ one has either $K = \{v\}$ (if $|K| = 1$) or $v \in e_K$ (if $|K| \geq 2$). For suppose that $v \in VP$ is incident with no $e_K \in N$. We show that $v \in R(B_1) \cap R(B_2)$, implying $\{v\} \in \mathcal{K}$. If v is an intermediate vertex of P, then v is covered by M_1 and M_2 and hence v belongs to $R(B_1)$ and $R(B_2)$. If $v = s$, then $v \in R(F_2)$ (so $v \in R(B_2)$) and v is covered by M_1, so $v \in R(B_1)$. If $v = t$, then $v \in R(F_1)$ (so $v \in R(B_1)$) and v is covered by M_2, so $v \in R(B_2)$. ∎

59.5. The matching forest polytope

The *matching forest polytope* of a mixed graph (V, E, A) is the convex hull of the incidence vectors of the matching forests. So the matching forest polytope is a polytope in $\mathbb{R}^{E \cup A}$.

Giles [1982b] showed that the matching forest polytope is determined by the following inequalities:

(59.18) (i) $x_e \geq 0$ for each $e \in E \cup A$,
         (ii) $x(\delta^{\text{head}}(v)) \leq 1$ for each $v \in V$,
         (iii) $x(\gamma(\mathcal{L})) \leq \lfloor |\bigcup \mathcal{L}| - \frac{1}{2}|\mathcal{L}| \rfloor$ for each subpartition $\mathcal{L}$ of V with $|\mathcal{L}|$ odd and all classes nonempty.

Here we use the following notation and terminology. $\delta^{\text{head}}(v)$ denotes the set of edges with head v. A *subpartition* of V is a collection of disjoint subsets of V. As usual, $\bigcup \mathcal{L}$ denotes the union of the sets in $\mathcal{L}$. For each subpartition $\mathcal{L}$, we define:

(59.19) $\gamma(\mathcal{L}) :=$ the set of undirected edges spanned by $\bigcup \mathcal{L}$ and directed edges spanned by any set in $\mathcal{L}$.

The inequalities (i) and (ii) in (59.18) are trivially valid for the incidence vector of any matching forest F. To see that (iii) is valid, we can assume that $F \subseteq \gamma(\mathcal{L})$ and that $V = \bigcup \mathcal{L}$. Then $|R(F \cap A)| \geq |\mathcal{L}|$, since each set in $\mathcal{L}$ contains at least one root of $F \cap A$ (since no directed edge enters any set in $\mathcal{L}$). Moreover, $|F \cap E| \leq \lfloor \frac{1}{2}|R(F \cap A)| \rfloor$, since $F \cap E$ is a matching on a subset of $R(F \cap A)$. As $|F \cap A| = |V| - |R(F \cap A)|$, this gives:

(59.20) $|F| = |F \cap E| + |F \cap A| \leq \lfloor \frac{1}{2}|R(F \cap A)| \rfloor + (|V| - |R(F \cap A)|)$
              $= \lfloor |V| - \frac{1}{2}|R(F \cap A)| \rfloor \leq \lfloor |\bigcup \mathcal{L}| - \frac{1}{2}|\mathcal{L}| \rfloor,$

as required.

Each integer solution x of (59.18) is the incidence vector of a matching forest. Indeed, as x is a 0,1 vector by (i) and (ii), we know that $x = \chi^F$ for some $F \subseteq E \cup A$. By (ii), each vertex is head of at most one edge in F. Hence, if F would contain a circuit (in the underlying undirected graph), it is a directed circuit C. But then for $\mathcal{L} := \{VC\}$, condition (iii) is violated. So F is a matching forest.

We show that system (59.18) is totally dual integral. This implies that it determines an integer polytope, which therefore is the matching forest polytope.

The proof method is a generalization of the method in Section 25.3a for proving the Cunningham-Marsh formula, stating that the matching constraints are totally dual integral.

The total dual integrality of (59.18) is equivalent to the following. For any weight function $w : E \cup A \to \mathbb{Z}$, let ν_w denote the maximum weight of a matching forest. Call a matching forest F w-$maximal$ if $w(F) = \nu_w$. Let Λ be the set of subpartitions $\mathcal{L}$ of V with $|\mathcal{L}|$ odd and with all classes nonempty.

Then the total dual integrality of (59.18) is equivalent to: for each weight function $w : E \cup A \to \mathbb{Z}$, there exist $y : V \to \mathbb{Z}_+$ and $z : \Lambda \to \mathbb{Z}_+$ satisfying

(59.21) $$\sum_{v \in V} y_v + \sum_{\mathcal{L} \in \Lambda} z(\mathcal{L})\lfloor |\bigcup \mathcal{L}| - \tfrac{1}{2}|\mathcal{L}|\rfloor \leq \nu_w$$

and

(59.22) $$\sum_{v \in V} y_v \chi^{\delta^{\text{head}}(v)} + \sum_{\mathcal{L} \in \Lambda} z(\mathcal{L})\chi^{\gamma(\mathcal{L})} \geq w.$$

Now we can derive (Schrijver [2000b]):

Theorem 59.3. *For each mixed graph (V, E, A), system (59.18) is totally dual integral.*

Proof. We must prove that for each mixed graph (V, E, A) and each function $w : E \cup A \to \mathbb{Z}$, there exist y, z satisfying (59.21) and (59.22).

In proving this, we can assume that w is nonnegative. For suppose that w has negative entries, and let w' be obtained from w by setting all negative entries to 0. As $\nu_{w'} = \nu_w$ and $w' \geq w$, any y, z satisfying (59.21) and (59.22) with respect to w', also satisfy (59.21) and (59.22) with respect to w.

Suppose that the theorem is not true. Choose a counterexample (V, E, A) and $w : E \cup A \to \mathbb{Z}_+$ with $|V| + |E \cup A| + \sum_{e \in E \cup A} w(e)$ as small as possible.

Then the underlying undirected graph of (V, E, A) is connected, since otherwise one of the components will form a smaller counterexample. Moreover, $w(e) \geq 1$ for each edge e, since otherwise we can delete e to obtain a smaller counterexample.

Next:

(59.23)   for each $v \in V$, there exists a w-maximal matching forest F with $v \in R(F)$.

For suppose that no such matching forest exists. For any edge e, let $w'(e) := w(e) - 1$ if v is head of e and $w'(e) := w(e)$ otherwise. Then $\nu_{w'} = \nu_w - 1$. By the minimality of w, there exist y, z satisfying (59.21) and (59.22) with respect to w'. Replacing y_v by $y_v + 1$ we obtain y, z satisfying (59.21) and (59.22) with respect to w, contradicting our assumption. This proves (59.23).

This implies:

(59.24)   each weak component of the directed graph (V, A) is strongly connected.

To see this, it suffices to show that each directed edge $e = (u, v)$ is contained in some directed circuit. By (59.23) there exists a w-maximal matching forest F with $v \in R(F)$. Then the weak component of F containing v is an arborescence rooted at v. As F has maximum weight, $F \cup \{e\}$ is not a matching forest, and hence $F \cap A$ contains a directed $v - u$ path. This makes a directed circuit containing e, and proves (59.24).

Let $\mathcal{K}$ denote the collection of strong components of (V, A). Define $w'(e) := w(e) - 1$ for each edge e. The remainder of this proof consists of showing that $|\mathcal{K}|$ is odd (so $\mathcal{K} \in \Lambda$), and that

(59.25) $\nu_w \geq \nu_{w'} + \lfloor |V| - \frac{1}{2}|\mathcal{K}| \rfloor$.

This is enough, since, by the minimality of w, there exist y, z satisfying (59.21) and (59.22) with respect to w'. Replacing $z(\mathcal{K})$ by $z(\mathcal{K}) + 1$ we obtain y, z satisfying (59.21) and (59.22) with respect to w (note that $\gamma(\mathcal{K}) = E \cup A$), contradicting our assumption.

To show (59.25), choose a w'-maximal matching forest F of maximum size $|F|$. Under this condition, choose F such that it maximizes $|R(F)|$.

We show that for each $s \in V$ the following holds, where r is the root of the arborescence[14] in $F \cap A$ containing s:

(59.26)   there exist a $t \in R(F)$ and a w'-maximal matching forest F' satisfying $|F'| = |F|$, $R(F') = (R(F) \setminus \{t\}) \cup \{s\}$, and $|R(F' \cap A) \cap K| = R(F \cap A) \cap K|$ for each strong component K of (V, A); if $r \in R(F)$, then moreover $t = r$ and $R(F' \cap A) = (R(F \cap A) \setminus \{r\}) \cup \{s\}$.

Let $F_1 := F$ and let F_2 be a w-maximal forest with $s \in R(F_2)$ (which exists by (59.23)). We first find F'_1 and F'_2 as follows.

If $r \notin R(F)$, then $s \notin R(F) = R(F_1)$ (since otherwise s is a root of $F \cap A$, and hence $r = s \in R(F)$). Applying Lemma 59.3α to F_1 and F_2 yields the matching forests F'_1 and F'_2.

If $r \in R(F)$, then $s \notin R(F \cap A)$. Apply Theorem 53.2 to $B_1 := F_1 \cap A$ and $B_2 := F_2 \cap A$. It yields branchings B'_1 and B'_2 in (V, A) satisfying $B'_1 \cap B'_2 = $

---

[14] An *arborescence in* a branching B is a weak component of (V, B), or just the arc set of it.

$B_1 \cap B_2$, $B_1' \cup B_2' = B_1 \cup B_2$, and $R(B_1') = R(B_1) \cup \{s\}$ or $R(B_1') = (R(B_1) \setminus \{r\}) \cup \{s\}$. This implies $R(B_2') = R(B_2) \setminus \{s\}$ or $R(B_2') = (R(B_2) \setminus \{s\}) \cup \{r\}$. Now define $F_i' := (F_i \cap E) \cup B_i'$ for $i = 1, 2$. Then the F_i' are matching forests, since $r \in R(F_1 \cap E)$ and $s \in R(F_2 \cap E)$.

In both constructions, $|F_1'| \leq |F_1|$, and if $|F_1'| = |F_1|$, then $|R(F_1')| \geq |R(F_1)|$. Moreover,

(59.27) $\quad \chi^{F_1'} + \chi^{F_2'} = \chi^{F_1} + \chi^{F_2},$

which implies that $w(F_1') + w(F_2') = w(F_1) + w(F_2)$. Hence

(59.28) $\quad \begin{aligned} w'(F_1') + w(F_2') &= w(F_1') + w(F_2') - |F_1'| \geq w(F_1) + w(F_2) - |F_1| \\ &= w'(F_1) + w(F_2). \end{aligned}$

Therefore, since F_1 is a w'-maximal matching forest and F_2 is a w-maximal matching forest, we have equality throughout in (59.28). So F_1' is w'-maximal and $|F_1'| = |F_1|$. Hence $|R(F_1')| \geq |R(F_1)|$. Then, by the maximality of $|R(F)|$, we know that $|R(F_1')| = |R(F_1)|$.

Set $F' := F_1'$. If $r \notin R(F)$, we know that (59.6)(iii) holds, which gives (59.26). If $r \in R(F)$, then (59.26) holds for $t := r$, and $R(B_1') = (R(B_1) \setminus \{t\}) \cup \{s\}$ or $R(B_2') = (R(B_2) \setminus \{s\}) \cup \{t\}$ (since $|F_1'| = |F_1|$, $F_1' \cap E = F_1 \cap E$, $|F_2'| = |F_2|$, $F_2' \cap E = F_2 \cap E$). Moreover, s and t belong to the same strong component of (V, A): as $r = t$ is the root of the arborescence in $F_1 \cap A$ containing s, there exists a $t - s$ path in (V, A); since each weak component of (V, A) is a strong component (by (59.24)), there is a directed $s - t$ path in (V, A). This implies (59.26).

Note that (59.26) implies in particular that $R(F) \neq \emptyset$. Suppose $|R(F)| \geq 2$. Choose F under the additional condition that the minimum distance in (V, E, A) between distinct vertices $u, v \in R(F)$ is as small as possible. Here, the distance in (V, E, A) is the length of a shortest $u-v$ path in the underlying undirected graph.

Necessarily, this distance is at least two, since otherwise we can extend F by an edge connecting u and v, thereby maintaining w'-maximality but increasing the size. This contradicts the maximality of $|F|$.

So we can choose an intermediate vertex s on a shortest $u - v$ path. Let F' be the matching forest described in (59.26), with $t \in R(F)$. By symmetry of u and v we can assume that $t \neq u$. So $u, s \in R(F')$, contradicting the choice of F, as the distance of u and s is smaller than that of u and v.

This implies that $|R(F)| = 1$. Let $R(F) = \{r\}$ and let K be the strong component of (V, A) containing r. We choose F (and r) under the additional constraint that $|R(F \cap A) \cap K|$ is as large as possible.

Suppose $|R(F \cap A) \cap K| \geq 2$. Choose F under the additional constraint that r has minimal distance in (V, A) from some root u of $F \cap A$ in $K \setminus \{r\}$. In this case, the distance in (V, A) from u to r is the length of a shortest directed $u - r$ path. (Such a path exists, since K is strongly connected.)

Let T be the arborescence in $F \cap A$ containing r. Let s be the first vertex on a shortest directed $u - r$ path Q in (V, A) that belongs to T. Necessarily

$s \neq r$, since otherwise we can extend F by the last edge of Q, contradicting the maximality of $|F|$.

Let F' be the matching forest described in (59.26). Then $s \in R(F')$ and $R(F' \cap A) = (R(F \cap A) \setminus \{r\}) \cup \{s\}$. Hence u remains a root of $F' \cap A$, while the distance in (V, A) from u to s is shorter than that from u to r. This contradicts our choice of F (replacing K, r by L, s).

So $|R(F \cap A) \cap K| = 1$. Suppose that there exists a component L of (V, A) with $|R(F \cap A) \cap L| \geq 2$. Choose s in L arbitrarily. Let F' be the matching forest described in (59.26). Then $s \in R(F')$ while $|R(F' \cap A) \cap L| \geq 2$, contradicting the choice of F.

So no such component L exists; that is, each $L \in \mathcal{K}$ contains exactly one root of $F \cap A$. So $|F \cap A| = |V| - |\mathcal{K}|$. Moreover, as $|R(F)| = 1$, $|\mathcal{K}|$ is odd and $|F \cap E| = \lfloor \frac{1}{2}|\mathcal{K}| \rfloor$. So $|F| = |F \cap A| + |F \cap E| = \lfloor |V| - \frac{1}{2}|\mathcal{K}| \rfloor$. Hence

(59.29) $\quad \nu_w \geq w(F) = w'(F) + |F| = \nu_{w'} + |F| = \nu_{w'} + \lfloor |V| - \frac{1}{2}|\mathcal{K}| \rfloor,$

thus proving (59.25). ∎

We remark that the optimum dual solution y, z constructed in this proof has the following additional property: if $\mathcal{K}, \mathcal{L} \in \Lambda$ and $z(\mathcal{K}), z(\mathcal{L}) > 0$, then $\mathcal{K}$ and $\mathcal{L}$ are 'laminar' in the following sense:

(59.30) $\quad \forall K \in \mathcal{K} \ \exists L \in \mathcal{L} : K \subseteq L,$
$\quad$ or $\forall L \in \mathcal{L} \ \exists K \in \mathcal{K} : L \subseteq K,$
$\quad$ or $\forall K \in \mathcal{K} \ \forall L \in \mathcal{L} : K \cap L = \emptyset.$

Theorem 59.3 implies the characterization of the matching forest polytope of Giles [1982b]:

Corollary 59.3a. *For each mixed graph (V, E, A), the matching forest polytope is determined by (59.18).*

Proof. By Theorems 59.3 and 5.22, the vertices of the polytope determined by (59.18) are integer. Since the integer solutions of (59.18) are the incidence vectors of matching forests, this proves the corollary. ∎

59.6. Further results and notes

59.6a. Matching forests in partitionable mixed graphs

Call a mixed graph $G = (V, E, A)$ *partitionable (into R and S)* if V can be partitioned into classes R and S such that each undirected edge connects R and S, while each directed arc is spanned by R or by S.

Trivially, a mixed graph is partitionable if and only if each circuit has an even number of undirected edges. That is, by contracting all directed arcs we obtain a bipartite graph. (Another characterization is: the incidence matrix is totally unimodular.)

In a different form, we have studied matching forests in partitionable mixed graphs before. Let $G = (V, E, A)$ be a mixed graph partitionable into R and S. Orient the edges in E from R to S, and turn the orientation of any arc in A spanned by R. We obtain a directed graph $D' = (V, A')$. Then it is easy to see that:

(59.31) a set of edges and arcs of G is a matching forest $\iff$ the corresponding arcs in D' form an $R - S$ bifurcation.

This implies that a number of theorems on matching forests in a partitionable mixed graph can be obtained from those on $R - S$ bifurcations. First we have:

Theorem 59.4. *Let $G = (V, E, A)$ be a partitionable mixed graph. Then the maximum size of a matching forest in G is equal to the minimum size of $|V| - |\mathcal{L}|$, where $\mathcal{L}$ is a collection of strong components K of the directed graph $D = (V, A)$ with $d_D^{\text{in}}(K) = 0$ such that no edge in E connects two components in $\mathcal{L}$.*

Proof. This is equivalent to Theorem 54.9. ∎

We similarly obtain a min-max relation for the maximum weight of a matching forest in a partitionable mixed graph, by the total dual integrality of the following system:

(59.32) (i) $x_e \geq 0$ for each $e \in E \cup A$,
(ii) $x(\delta^{\text{head}}(v)) \leq 1$ for each $v \in V$,
(iii) $x(A[U]) \leq |U| - 1$ for each nonempty U with $U \subseteq R$ or $U \subseteq S$.

Here $\delta^{\text{head}}(v)$ is the set of edges and arcs having v as head.

Theorem 59.5. *If G is a mixed graph partitionable into R and S, then (59.32) is TDI and determines the matching forest polytope.*

Proof. This is equivalent to Corollary 54.10a. ∎

For covering by matching forests in partitionable mixed graphs we have:

Theorem 59.6. *Let $G = (V, E, A)$ be a mixed graph partitionable into R and S. Then $E \cup A$ can be covered by k matching forests if and only if*

(59.33) (i) $|\delta^{\text{head}}(v)| \leq k$ for each $v \in V$;
(ii) $|A[U]| \leq k(|U| - 1)$ for each nonempty subset U of R or S.

Proof. This is equivalent to Corollary 54.11c. ∎

The case $A = \emptyset$ is Kőnig's edge-colouring theorem (Theorem 20.1).

An equivalent, polyhedral way of formulating Theorem 59.6 is:

Corollary 59.6a. *If G is a partitionable mixed graph, then the matching forest polytope has the integer decomposition property.*

Proof. Directly from Theorem 59.6. ∎

59.6b. Further notes

The facets of the matching forest polytope are characterized in Giles [1982c].

Matching forests form a special case of matroid matching. Let $G = (V, E, A)$ be a mixed graph. Consider the space $\mathbb{R}^V \times \mathbb{R}^V$. Associate with any undirected edge $e = uv \in E$, the pair $(\chi^u, 0)$, $(\chi^v, 0)$ of vectors in $\mathbb{R}^V \times \mathbb{R}^V$. Associate with any directed arc $a = (u, v) \in A$, the pair $(\chi^v, 0)$, $(0, \chi^u - \chi^v)$ of vectors in $\mathbb{R}^V \times \mathbb{R}^V$. One easily checks that $M \subseteq E \cup A$ is a matching forest if and only if its associated pairs form a matroid matching. Thus matroid matching theory implies a min-max relation and a polynomial-time algorithm for the maximum size of a matching forest. However, as we saw in Section 59.2, there is an easy direct method for this.

Chapter 60

Submodular functions on directed graphs

At two structures we came across the proof technique of making a collection of subsets cross-free: at submodular functions (like in polymatroid intersection) and at directed graphs (like in the proof of the Lucchesi-Younger theorem).

Edmonds and Giles [1977] combined the two structures into one general framework, consisting of a submodular function defined on the vertex set of a directed graph. Johnson [1975a] and Frank [1979b] designed a variant of Edmonds and Giles' framework, containing the polymatroid intersection theorem and the optimum arborescence theorem as special cases.

We first describe the results of Edmonds and Giles, and after that we present a variant, from which the results of Frank can be derived. At the base is the method of Edmonds and Giles to represent any cross-free family by a directed tree (the *tree-representation*) and to derive a network matrix if the family consists of subsets of the vertex set of a directed graph — see Section 13.4.

60.1. The Edmonds-Giles theorem

Let $D = (V, A)$ be a digraph and let $\mathcal{C}$ be a crossing family of subsets of V (that is, if $T, U \in \mathcal{C}$ with $T \cap U \neq \emptyset$ and $T \cup U \neq V$, then $T \cap U, T \cup U \in \mathcal{C}$). A function $f : \mathcal{C} \to \mathbb{R}$ is called *submodular on crossing pairs*, or *crossing submodular*, if for all $T, U \in \mathcal{C}$ with $T \cap U \neq \emptyset$ and $T \cup U \neq V$ one has

(60.1) $\quad f(T) + f(U) \geq f(T \cap U) + f(T \cup U).$

Given such $D, \mathcal{C}, f$, a *submodular flow* is a function $x \in \mathbb{R}^A$ satisfying:

(60.2) $\quad x(\delta^{\text{in}}(U)) - x(\delta^{\text{out}}(U)) \leq f(U)$ for each $U \in \mathcal{C}$.

The set P of all submodular flows is called the *submodular flow polyhedron*.

Equivalently, P is equal to the set of all vectors x in $\mathbb{R}^A$ with the property that the 'gain' vector of x is in the extended polymatroid EP_f. (The *excess function* of x equals Mx where M is the $V \times A$ incidence vector of D.)

Then Edmonds and Giles [1977] showed:

Theorem 60.1 (Edmonds-Giles theorem). *If f is crossing submodular, then (60.2) is box-TDI.*

Proof. Choose $w \in \mathbb{R}^A$, and let y be an optimum solution to the dual of maximizing $w^\mathsf{T} x$ over (60.2):

(60.3) $\quad \min\{\sum_{U \in \mathcal{C}} y(U) f(U) \mid y \in \mathbb{R}_+^\mathcal{C}, \sum_{U \in \mathcal{C}} y(U)(\chi^{\delta^{\text{in}}(U)} - \chi^{\delta^{\text{out}}(U)}) = w\}.$

Choose y such that

(60.4) $\quad \sum_{U \in \mathcal{C}} y(U)|U||V \setminus U|$

is as small as possible. Let $\mathcal{C}_0 := \{U \in \mathcal{C} \mid y(U) > 0\}$. We first prove that $\mathcal{C}_0$ is cross-free.

Suppose to the contrary that $T, U \in \mathcal{C}_0$ with $T \not\subseteq U \not\subseteq T$, $T \cap U \neq \emptyset$, $T \cup U \neq V$. Let $\alpha := \min\{y(T), y(U)\} > 0$. Then decreasing $y(T)$ and $y(U)$ by α, and increasing $y(T \cap U)$ and $y(T \cup U)$ by α, maintains feasibility of z, u, y, while its value is not increased (hence it remains optimum). However, sum (60.4) decreases (by Theorem 2.1). This contradicts the minimality of (60.4).

As $\mathcal{C}_0$ is cross-free, the submatrix formed by the constraints corresponding to $\mathcal{C}_0$ is totally unimodular (by Corollary 13.21a). Hence, by Theorem 5.35, (60.2) is box-TDI. ∎

Note that the proof also yields that the solution y in (60.3) can be taken such that the collection $\{U \in \mathcal{C} \mid y(U) > 0\}$ is cross free.

Box-TDI implies primal integrality (a polyhedron P is *box-integer* if $P \cap \{x \mid d \leq x \leq c\}$ is integer for all integer vectors d, c):

Corollary 60.1a. *If f is integer, the polyhedron determined by (60.2) is box-integer.*

Proof. By Theorem 60.1, $\max\{w^\mathsf{T} x \mid x \in P\}$ is achieved by an integer solution x, for each vector w. ∎

Complexity. The algorithmic results on polymatroid intersection of Cunningham and Frank [1985] and Fujishige, Röck, and Zimmermann [1989] imply that the optimization problem associated with the Edmonds-Giles theorem can be solved in strongly polynomial time.

Indeed, let $D = (V, A)$ be a digraph, let $\mathcal{C}$ be a crossing family, let $f : \mathcal{C} \to \mathbb{Q}$ be crossing submodular, and let $c, d, l : A \to \mathbb{Q}$. If we want to find a submodular flow x with $d \leq x \leq c$ minimizing $l^\mathsf{T} x$, we can assume that all arcs in A are vertex-disjoint. Moreover, we can assume that for each arc $a = (u, v) \in A$ we have $f(\{v\}) = c(a)$ and $f(\{u\}) = -d(a)$. Hence we can ignore d and c, and assume that we want to find a submodular flow x minimizing $l^\mathsf{T} x$.

Now define $C_2 := \{\{u,v\} \mid (u,v) \in A\}$ and $f_2(\{u,v\}) := 0$, $w(v) := l(u,v)$, and $w(u) := 0$, for each $(u,v) \in A$. Then the problem is equivalent to finding a vector x in $EP_f \cap EP_{f_2}$ with $x(V) = 0$ and minimizing $w^\mathsf{T} x$. This can be solved in strongly polynomial time by Theorem 49.9.

(Frank [1982b] gave a strongly polynomial-time algorithm for the special case if f is integer, $c = \mathbf{1}$, and $d = \mathbf{0}$.)

A similar reduction of submodular flows to polymatroid intersection was given by Kovalev and Pisaruk [1984].

60.1a. Applications

Network flows. If we take $\mathcal{C} := \{\{v\} \mid v \in V\}$ and $f = \mathbf{0}$, then (60.2) determines circulations, and Theorem 60.1 passes into a theorem on minimum-cost circulations. It may be specialized easily to several other results on flows in networks, e.g., to the max-flow min-cut theorem (Theorem 10.3; take $d = \mathbf{0}, c \geq \mathbf{0}$, and $w(a) = 0$ for $a \neq (s,r)$ and $w((s,r)) = 1$) and to Hoffman's circulation theorem (Theorem 11.2).

Lucchesi-Younger theorem. Let $D = (V,A)$ be a digraph and define

(60.5) $\qquad \mathcal{C} := \{U \subseteq V \mid \emptyset \neq U \neq V \text{ and } d_A^{\text{out}}(U) = 0\}$.

So $\mathcal{C}$ consists of all sets U such that the collection of arcs entering U forms a directed cut. Taking $f := -\mathbf{1}$, $c := \mathbf{0}$, $d := -\infty$, and $w := \mathbf{1}$, Theorem 60.1 passes into the Lucchesi-Younger theorem (Theorem 55.2, cf. Corollary 55.2b): the minimum size of a directed cut cover is equal to the maximum number of disjoint directed cuts. For arbitrary w we obtain a weighted version.

Polymatroid intersection. Let f_1 and f_2 be nonnegative submodular set function on S. Let S' and S'' be two disjoint copies of S, let $V = S' \cup S''$, and define $\mathcal{C}$ by

(60.6) $\qquad \mathcal{C} := \{U' \mid U \subseteq S\} \cup \{S' \cup U'' \mid U \subseteq S\}$

where U' and U'' denote the sets of copies of elements of U in S' and S''. Define $f : \mathcal{C} \to \mathbb{R}_+$ by

(60.7) $\qquad f(U') := f_1(U) \qquad\qquad \text{for } U \subseteq S,$
$\qquad\qquad f(V \setminus U'') := f_2(U) \qquad \text{for } U \subseteq S,$
$\qquad\qquad f(S') := \min\{f_1(S), f_2(S)\}.$

Then $\mathcal{C}$ and f satisfy (60.1). If we take $d = \mathbf{0}$ and $c = \infty$, Theorem 60.1 passes into the polymatroid intersection theorem (Corollary 46.1a, cf. Theorem 46.1).

Frank and Tardos [1989] showed that also Theorem 44.7 (a generalization of Lovász [1970a] of Kőnig's matching theorem) fits into the Edmonds-Giles model. For applications of the Edmonds-Giles theorem to graph orientation, see Chapter 61.

60.1b. Generalized polymatroids and the Edmonds-Giles theorem

The Edmonds-Giles theorem (Theorem 60.1) also comprises the total dual integrality of the system defining the intersection of two generalized polymatroids (Section 49.11b). Indeed, let S be a finite set, let, for $i = 1, 2$, $\mathcal{C}_i$ and $\mathcal{D}_i$ be collections of subsets of S, and let $f_i : \mathcal{C}_i \to \mathbb{R}$ and $g_i : \mathcal{D}_i \to \mathbb{R}$ form a paramodular pair (f_i, g_i). Then the system

(60.8) $\quad x(U) \leq f_1(U) \quad$ for $U \in \mathcal{C}_1,$
$\quad\quad\quad\;\; x(U) \geq g_1(U) \quad$ for $U \in \mathcal{D}_1,$
$\quad\quad\quad\;\; x(U) \leq f_2(U) \quad$ for $U \in \mathcal{C}_2,$
$\quad\quad\quad\;\; x(U) \geq g_2(U) \quad$ for $U \in \mathcal{D}_2,$

is box-totally dual integral, which is Corollary 49.12b.

To see this as a special case of the Edmonds-Giles theorem, let S_1 and S_2 be disjoint copies of S, and let $V := S_1 \cup S_2$. For each $s \in S$, let a_s be the arc (s_2, s_1), where s_1 and s_2 are the copies of s in S_1 and S_2 respectively. Let $A := \{a_s \mid s \in S\}$.

Let

(60.9) $\quad \mathcal{C} := \{U_1 \mid U \in \mathcal{C}_1\} \cup \{V \setminus U_1 \mid U \in \mathcal{D}_1\} \cup \{V \setminus U_2 \mid U \in \mathcal{C}_2\} \cup \{U_2 \mid U \in \mathcal{D}_2\},$

where U_i denotes the set of copies of the elements in U in S_i ($i = 1, 2$). It is easy to see that $\mathcal{C}$ is a crossing family.

Define $f : \mathcal{C} \to \mathbb{R}$ by:

(60.10) $\quad f(U_1) := f_1(U) \quad\quad\quad$ for $U \in \mathcal{C}_1,$
$\quad\quad\quad\;\; f(V \setminus U_1) := -g_1(U) \quad$ for $U \in \mathcal{D}_1,$
$\quad\quad\quad\;\; f(V \setminus U_2) := f_2(U) \quad$ for $U \in \mathcal{C}_2,$
$\quad\quad\quad\;\; f(U_2) := -g_2(U) \quad\quad$ for $U \in \mathcal{D}_2.$

(In case that $f(S_1)$ or $f(S_2)$ would be defined more than once, we take the smallest of the values.) Then f is submodular on crossing pairs. Now the system (in $x \in \mathbb{R}^A$)

(60.11) $\quad x(\delta^{\text{in}}(U)) - x(\delta^{\text{out}}(U)) \leq f(U)$ for $U \in \mathcal{C}$

is the same as (60.8) (after renaming each variable $x(s)$ to $x(a_s)$). So the box-total dual integrality of (60.8) follows from the Edmonds-Giles theorem.

Frank [1984b] showed that, conversely, the solution set of the 'Edmonds-Giles' system (60.2) is the projection of the intersection of two generalized polymatroids.

60.2. A variant

We now give a theorem similar to Theorem 60.1, which includes as special cases again the Lucchesi-Younger theorem and the polymatroid intersection theorem, and moreover theorems on optimum arborescences, bibranchings, and strong connectors.

For any digraph $D = (V, A)$ and any family $\mathcal{C}$ of subsets of V, define the $\mathcal{C} \times A$ matrix M by

(60.12) $\quad M_{U,a} := \begin{cases} 1 & \text{if } a \text{ enters } U, \\ 0 & \text{otherwise,} \end{cases}$

for $U \in \mathcal{C}$ and $a \in A$.

This matrix is totally unimodular if $\mathcal{C}$ is cross-free and the following condition holds:

(60.13) $\quad$ if $X, Y, Z \in \mathcal{C}$ with $X \subseteq V \setminus Y \subseteq Z$, then no arc of D enters both X and Z.

Theorem 60.2. *If C is cross-free and (60.13) holds, then M is totally unimodular.*

Proof. Let $T = (W, B)$ and $\pi : V \to W$ form a tree-representation for C. For any arc $a = (u, v)$ of D, the set of forward arcs in the undirected $\pi(u) - \pi(v)$ path in T is contiguous, that is, forms a directed path, say from u' to v'. This follows from the fact that there exist no arcs b, c, d in this order on the path with b and d forward and c backward, by (60.13).

Define $a' := (u', v')$, and let $D' = (W, A')$ be the digraph with $A' := \{a' \mid a \in A\}$. Then M is equal to the network matrix generated by T and D' (identifying $b \in B$ with the set X_b in C determined by b). Hence by Theorem 13.20, M is totally unimodular. ∎

Recall that a function g on a crossing family C is called *supermodular on crossing pairs*, or *crossing supermodular*, if for all $T, U \in C$:

(60.14) if $T \cap U \neq \emptyset$ and $T \cup U \neq V$, then $g(T) + g(U) \leq g(T \cap U) + g(T \cup U)$.

Consider the polyhedron P determined by:

(60.15) $\quad x_a \geq 0 \qquad\qquad$ for $a \in A$,
$\qquad\qquad x(\delta^{in}(U)) \geq g(U) \quad$ for $U \in C$.

Theorem 60.3. *If g is crossing supermodular and (60.13) holds, then system (60.15) is box-TDI.*

Proof. Let $w \in \mathbb{R}^A$ and let y achieve the maximum in the dual of minimizing $w^\top x$ over (60.15):

(60.16) $\quad \max\{\sum_{U \in C} y(U) g(U) \mid y \in \mathbb{R}_+^C, \sum_{U \in C} y(U) \chi^{\delta^{in}(U)} \geq w\},$

in such a way that

(60.17) $\quad \sum_{U \in C} y(U) |U| |V \setminus U|$

is as small as possible. Define

(60.18) $\quad C_0 := \{U \in C \mid y(U) > 0\}.$

We first show that C_0 is cross-free. Suppose to the contrary that there are T, U in C with $T \not\subseteq U \not\subseteq T$, $T \cap U \neq \emptyset$, and $T \cup U \neq V$. Let $\alpha := \min\{y(T), y(U)\}$. Now decrease $y(T)$ and $y(U)$ by α, and increase $y(T \cap U)$ and $y(T \cup U)$ by α. Then y remains feasible and optimum, while sum (60.17) decreases (Theorem 2.1), a contradiction.

Since C_0 determines a totally unimodular submatrix by Theorem 60.2, by Corollary 5.20b system (60.15) is box-TDI. ∎

Note that the proof yields that (60.16) has a solution y with $\{U \in \mathcal{C} \mid y(U) > 0\}$ cross-free. Condition (60.13) cannot be deleted, as is shown by Figure 60.1.

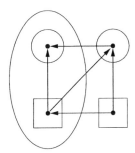

Figure 60.1
A collection and a digraph showing that condition (60.13) cannot be deleted in Theorem 60.3. In this Venn-diagram, the collection is represented by the *interiors* of the ellipses and by the *exteriors* of the rectangles.

Again, there is the following standard corollary for primal integrality:

Corollary 60.3a. *If g is integer, the polyhedron determined by (60.15) is box-integer.*

Proof. As before. ∎

Notes. Johnson [1975a] proved Theorem 60.3 for the special case that $\mathcal{C}$ is the collection of all nonempty subsets of $V \setminus \{r\}$ (where r is a fixed element of V), and Frank [1979b] extended this result to the case where $\mathcal{C}$ is any intersecting family of subsets of $V \setminus \{r\}$. Note that in this case condition (60.13) is trivially satisfied.

60.2a. Applications

We list some applications of Theorem 60.3, which may be compared with the applications of the Edmonds-Giles theorem (Section 60.1a).

Kőnig-Rado edge cover theorem. Let $G = (V, E)$ be a bipartite graph, with colour classes V_1 and V_2. Let $D = (V, A)$ be the digraph arising from G by orienting all edges from V_2 to V_1. Define $\mathcal{C} := \{\{v\} \mid v \in V_1\} \cup \{V \setminus \{v\} \mid v \in V_2\}$ and let $d := 0$, $c := \infty$, $g := 1$, $w := 1$. Then Theorem 60.3 gives the Kőnig-Rado edge cover theorem (Theorem 19.4): the minimum size of an edge cover in a bipartite graph is equal to the maximum size of a stable set. Taking w arbitrary gives a weighted version.

Optimum arborescence theorem. Let $D = (V, A)$ be a digraph, let $r \in V$, and let $\mathcal{C}$ be the collection of all nonempty subsets of $V \setminus \{r\}$. Let $g := 1$, $d := 0$, $c := \infty$, and let $w : A \to \mathbb{Z}_+$. Theorem 60.3 now gives the optimum arborescence theorem (Theorem 52.3): the minimum weight of an r-arborescence is equal to the maximum number of r-cuts such that no arc a is in more than $w(a)$ of these r-cuts.

Optimum bibranching theorem. Let $D = (V, A)$ be a digraph and let V be split into sets R and S. Define $\mathcal{C} := \{U \subseteq V \mid \emptyset \neq U \subseteq S \text{ or } S \subseteq U \subset V\}$ and $d := 0$, $c := \infty$, $g := 1$, and let $w : A \to \mathbb{Z}_+$. Then Theorem 60.3 gives Corollary 54.8b: the minimum weight of a bibranching is equal to the maximum number of subsets in $\mathcal{C}$ such that no arc a enters more than $w(a)$ of these subsets.

Lucchesi-Younger theorem. Let $D = (V, A)$ be a digraph and let $\mathcal{C}$ be the collection of all nonempty proper subsets U of V with $\delta_A^{\text{out}}(U) = \emptyset$. Let $g := 1$, $d := 0$, $c := \infty$, and $w := 1$. Then Theorem 60.3 gives the Lucchesi-Younger theorem (Theorem 55.2): the minimum size of a directed cut cover is equal to the maximum number of disjoint directed cuts. Taking w arbitrary, gives a weighted version.

Strong connectors. Suppose that $g = 1$, $d = 0$, and $c = \infty$, and that for all $V_1, V_2 \in \mathcal{C}$ we have: if $V_1 \cap V_2 \neq \emptyset$, then $V_1 \cap V_2 \in \mathcal{C}$, and if $V_1 \cup V_2 \neq V$, then $V_1 \cup V_2 \in \mathcal{C}$. Then Theorem 60.3 is equivalent to Theorem 57.3.

Indeed, let $D = (V, A)$ and $D_0 = (V, A_0)$ be digraphs such that for each arc $a = (u, v)$ of D there are vertices u' and v' such that D_0 contains directed paths from u to u', from v' to v, and from v' to u'. Let $w : A \to \mathbb{Z}_+$. Then the minimum weight of a strong connector in D for D_0 is equal to the maximum number of D_0-cuts in D such that no arc a of D is in more than $w(a)$ of these D_0-cuts.

This can be derived from Theorem 60.3 by taking $\mathcal{C} := \{U \subseteq V \mid \emptyset \neq U \neq V, \delta_{A_0}^{\text{in}}(U) = \emptyset\}$. Conversely, if $\mathcal{C}$ satisfies the condition given above, we can take $A_0 := \{(u, v) \mid u, v \in V, (u, v) \text{ enters no } U \in \mathcal{C}\}$.

Polymatroid intersection. Let g_1 and g_2 be integer supermodular nondecreasing set functions on S with $g_1(\emptyset) = g_2(\emptyset) = 0$. Then

(60.19) $\min\{x(S) \mid x \in \mathbb{Z}_+^S, x(U) \geq g_i(U) \text{ for } U \subseteq S, i = 1, 2\}$
$= \max_{U \subseteq S}(g_1(U) + g_2(S \setminus U))$.

This follows by taking disjoint copies S' and S'' of S, and setting $V := S' \cup S''$, $\mathcal{C} := \{T \subseteq V \mid T \subseteq S' \text{ or } S' \subseteq T\}$, $A := \{(s'', s') \mid s \in S\}$, $g(U') := g_1(U)$ and $g(V \setminus U'') := g_2(U)$ for $U \subseteq S$ (without loss of generality, $g_1(S) = g_2(S)$), $d := 0$, $c := \infty$, $w := 1$.

By taking d, c, w arbitrary, several other (contra)polymatroid intersection theorems follow.

60.3. Further results and notes

60.3a. Lattice polyhedra

In a series of papers, Hoffman [1976a,1978] and Hoffman and Schwartz [1978] developed a theory of 'lattice polyhedra', which extends results of Johnson [1975a]. This theory has much in common with the theories described above.

Let $(L, \leq)$ be a partially ordered set and let $\wedge : L \times L \to L$ be a function such that

(60.20) for all $a, b \in L$: $a \wedge b \leq a$ and $a \wedge b \leq b$.

Let S be a finite set and let $\phi : L \to \mathcal{P}(S)$ be such that

(60.21) if $a < b < c$, then $\phi(a) \cap \phi(c) \subseteq \phi(b)$

for a, b, c in L. Let $\vee : L \times L \to L$ and let $f : L \to \mathbb{R}_+$ satisfy:

(60.22) $f(a \wedge b) + f(a \vee b) \leq f(a) + f(b)$

for all a, b in L. So f is, in a sense, *submodular*.
Define

(60.23) $\quad S' := \{u \in S \mid \forall a, b \in L : \chi^{\phi(a \wedge b)}(u) + \chi^{\phi(a \vee b)}(u) \leq \chi^{\phi(a)}(u) + \chi^{\phi(b)}(u)\}$ and
$S'' := \{u \in S \mid \forall a, b \in L : \chi^{\phi(a \wedge b)}(u) + \chi^{\phi(a \vee b)}(u) \geq \chi^{\phi(a)}(u) + \chi^{\phi(b)}(u)\}$.

The polyhedron determined by:

(60.24) $\quad\begin{aligned} x_u &\geq 0 & (u \in S \setminus S'), \\ x_u &\leq 0 & (u \in S \setminus S''), \\ x(\phi(a)) &\leq f(a) & (a \in L). \end{aligned}$

is called a *lattice polyhedron*. Hoffman and Schwartz [1978] showed that system (60.24) is box-totally dual integral.

Theorem 60.4. *System* (60.24) *is box-TDI.*

Proof. Choose $w \in \mathbb{R}_+^S$. Consider the dual of maximizing $w^\top x$ over (60.24):

(60.25) $\quad \min\{y^\top f \mid y \in \mathbb{R}_+^L, \sum_{a \in L} y_a \chi^{\phi(a)} \leq w(u)$ if $u \in S'$ and $\sum_{a \in L} y_a \chi^{\phi(a)} \geq w(u)$ if $u \in S''\}$.

Order the elements of L as $a_1, \ldots, a_n$ such that if $a_i \leq a_j$, then $i \leq j$. Let y attain (60.25), such that $y(L)$ is minimal, and, under this condition, such that

(60.26) $(y(a_1), \ldots, y(a_n))$

is lexicographically maximal.
Then the collection $C := \{a \in L \mid y_a > 0\}$ is a chain in L. For suppose to the contrary that $a, b \in C$ with $a \not\leq b \not\leq a$. Let $\alpha := \min\{y_a, y_b\}$. Reset y by decreasing y_a and y_b by α, and increasing $y(a \wedge b)$ and $y(a \vee b)$ by α. One easily checks, using (60.22) and (60.23), that the new y again attains the minimum (60.25),

and moreover that $(y(a_1),\ldots,y(a_n))$ lexicographically increases, contradicting our assumption.

By (60.21) for each u in S, the set of a in C with $u \in \phi(a)$ forms an interval in C. So the linear inequalities corresponding to C make up a totally unimodular matrix (as it is a network matrix generated by a directed path and a directed graph (Theorem 13.20)). Therefore, by Theorem 5.35, system (60.24) is box-TDI. ∎

We give some applications of Theorem 60.4 (more applications are in Hoffman [1976a], Hoffman and Schwartz [1978], and Gröflin [1984,1987]).

Shortest paths (Johnson [1975a]). Let $D = (V, A)$ be a digraph and let $s, t \in V$. Let $L := \{U \subseteq V \mid s \in U, t \notin U\}$ and let $\preceq := \subseteq$, $\wedge := \cap$, $\vee := \cup$. Let $S := A$ and let for each $U \in L$, $\phi(U) := \delta^{\text{out}}(U)$. These data satisfy (60.20) and (60.21), where $S' := S$. If $f = -1$, Theorem 60.4 gives: the minimum length of an $s - t$ path is equal to the maximum number of $s - t$ cuts such that no arc a is in more than $c(a)$ of these $s - t$ cuts — the max-potential min-work theorem (Theorem 7.1).

Matroid intersection (Hoffman [1976a]). Let $(S, \mathcal{I})$ and $(S, \mathcal{I}_2)$ be matroids, with rank functions r_1 and r_2 and assume $r_1(S) = r_2(S)$. Let S' and S'' be two disjoint copies of S and let $V := S' \cup S''$. Let $L := \{U \subseteq V \mid U \subseteq S' \text{ or } S' \subseteq U\}$. Let $\preceq := \subseteq$, $\wedge := \cap$, $\vee := \cup$. Define for $T \subseteq S$:

(60.27) $f(T') := r_1(T), \quad \phi(T') := T,$
$f(V \setminus T'') := r_2(T), \quad \phi(V \setminus T'') := T.$

As these data satisfy (60.20), (60.21), and (60.22), Theorem 60.4 yields the matroid intersection theorem. Polymatroid intersection can be included similarly.

Chains and antichains in partially ordered sets (Hoffman and Schwartz [1978]). Let $(V, \preceq)$ be a partially ordered set and let L be the collection of lower ideals of V (a subset Y of V is a *lower ideal* if $y \preceq x \in V$ implies $y \in V$). Define $\preceq := \subseteq$, $\wedge := \cap$, $\vee := \cup$.

First, let $S := V$. For $Y \in L$, let $\phi(Y)$ be the collection of maximal elements of Y. These data satisfy (60.20) and (60.21), and $S' = S'' = S$.

Theorem 60.4 with $f(Y) := k$ for each $Y \in L$ then gives the theorem of Greene [1976] (Corollary 14.10b) that the maximum size of the union of k chains is equal to the minimum value of $|V \setminus Y| + k \cdot c_1(Y)$, where Y ranges over all subsets of V and where $c_1(Y)$ denotes the maximum size of a chain contained in Y.

Indeed, Theorem 60.4 gives the total dual integrality of

(60.28) $0 \leq x_v \leq 1$ for $v \in V$,
$x(A) \leq k$ for each antichain A.

Hence the maximum size of the union of k chains is, by Dilworth's decomposition theorem, equal to (where $\mathcal{A}$ denotes the collection of antichains in V)

(60.29) $\max\{\mathbf{1}^\mathsf{T} x \mid x \in \{0,1\}^V, x(A) \leq k \text{ for } A \in \mathcal{A}\}$
$= \min\{k \sum_{A \in \mathcal{A}} y_A + z(V) \mid y \in \mathbb{Z}_+^\mathcal{A}, z \in \mathbb{Z}_+^V, \sum_{A \in \mathcal{A}} y_A \chi^A + z \geq \mathbf{1}\}$
$= \min_{Y \subseteq V}(|V \setminus Y| + k \cdot (\text{minimum number of antichains covering } Y))$
$= \min_{Y \subseteq V}(|V \setminus Y| + k \cdot c_1(Y)).$

Also the dual result (exchanging 'chain' and 'antichain') due to Greene and Kleitman [1976] (Corollary 14.8b) can be derived. Let $L, \leq, \wedge, \vee$ be as above and let $S := V \cup \{w\}$, where w is some new element. For $Y \in L$, let $\phi(Y)$ be the collection of maximal elements of Y together with w and let $f(Y) := -|\phi(Y)|$. These data again satisfy (60.20) and (60.21), and $S' = S'' = S$.

Then Theorem 60.4 gives the box-total dual integrality of the system

(60.30) $x(A) + \lambda \leq -|A|$ for each antichain A,

and hence of the system

(60.31) $x(A) + \lambda \geq |A|$ for each antichain A.

Then the maximum union of k antichains is equal to

(60.32) $\max\{\sum_{A \in \mathcal{A}} y_A |A| \mid y \in \mathbb{Z}_+^{\mathcal{A}}, \sum_{A \in \mathcal{A}} y_A \chi^A \leq 1, \sum_{A \in \mathcal{A}} y_A = k\}$
$= \min\{x(V) + k \cdot \lambda \mid x \in \mathbb{Z}_+^V, \lambda \in \mathbb{Z}, x(A) + \lambda \geq |A| \text{ for each } A \in \mathcal{A}\}$
$\geq \min_{Y \subseteq V}(|V \setminus Y| + k \cdot (\text{maximum size of an antichain contained in } Y)).$

The equality follows from the box-total dual integrality of (60.31). The inequality follows by taking $Y := \{v \in V \mid x_v = 0\}$. Then λ is at least the maximum size of an antichain contained in Y, since for any antichain $A \subseteq Y$: $\lambda = x(A) + \lambda \geq |A|$.

Common base vectors in two polymatroids (Gröflin and Hoffman [1981]). Let f_1 and f_2 be submodular set functions on S. The polymatroid intersection theorem gives:

(60.33) $f(T) := \max\{x(T) \mid x(U) \leq f_i(U) \text{ for } U \subseteq S \text{ and } i = 1, 2\}$
$= \min_{U \subseteq T}(f_1(U) + f_2(T \setminus U)),$

for $T \subseteq S$. Gröflin and Hoffman [1981] showed that Theorem 46.4 follows from Theorem 60.4 above as follows. (The proof of Theorem 46.4 was modelled after the proof of Theorem 60.4.)

Let L be the set of all pairs (T, U) of subsets of S with $T \cap U = \emptyset$, partially ordered by $\leq$ as follows:

(60.34) $(T, U) \leq (T', U')$ and only if $T \subseteq T'$ and $U \supseteq U'$.

Then $(L, \leq)$ is a lattice with lattice operations $\wedge$ and $\vee$ (say). Define $\phi(T, U) := |S \setminus (T \cup U)|$ and $f(T, U) := f_1(T) + f_2(U) - f(S)$. As these data satisfy (60.20), (60.21), and (60.22), Theorem 60.4 applies. We have $S' = S'' = S$. Hence the system

(60.35) $x(S \setminus (T \cup U)) \geq f_1(T) + f_2(U) - f(S)$ for $(T, U) \in L$

is box-TDI. With the definition of f, this implies the box-total dual integrality of

(60.36) $x(T) \leq f(S \setminus T) - f(S)$ for $T \subseteq S$,

and (equivalently) of

(60.37) $x(T) \geq f(S) - f(S \setminus T)$ for $T \subseteq S$.

That is, we have Theorem 46.4.

Convex sets in partially ordered sets (Gröflin [1984]). Let $(S, \leq)$ be a partially ordered set. A subset C of S is called *convex* if $a, b \in C$ and $a \leq x \leq b$ imply $x \in C$. Then the system

(60.38) $\quad x(C) \leq 1$ for each convex subset C of S,

is box-TDI. Note that this system describes the polar of the convex hull of the incidence vectors of convex sets.

To see the box-total dual integrality of (60.38), define

(60.39) $\quad L := \{(A, B) \mid A \text{ lower ideal and } B \text{ upper ideal in } S \text{ with } A \cup B = S\}$.

(An *upper ideal* is a subset B such that if $b \in B$ and $x \geq b$, then $x \in B$. Similarly, a *lower ideal* is a subset B such that if $b \in B$ and $x \leq b$, then $x \in B$.) Make L to a lattice by defining a partial order $\preceq$ on L by:

(60.40) $\quad (A, B) \preceq (A', B') \iff A \subseteq A', B \supseteq B'$.

Define $f : L \to \mathbb{R}$ and $\phi : L \to \mathcal{P}(S)$ by: $f(A, B) := 1$ and $\phi(A, B) := A \cap B$, for $(A, B) \in L$. Applied to this structure, Theorem 60.4 gives the box-total dual integrality of (60.38).

('Greedy' algorithms for some lattice polyhedra problems were investigated by Kornblum [1978].)

An extension of lattice polyhedra, to handle rooted-connectivity augmentation of a digraph, was given by Frank [1999b].

60.3b. Polymatroidal network flows

Hassin [1978,1982] and Lawler and Martel [1982a,1982b] gave the following 'polymatroidal network flow' model equivalent to that of Edmonds and Giles. Let $D = (V, A)$ be a digraph. For each $v \in V$, let $\mathcal{C}_v^{\text{out}}$ and $\mathcal{C}_v^{\text{in}}$ be intersecting families of subsets of $\delta^{\text{out}}(v)$ and $\delta^{\text{in}}(v)$, respectively, and let $f_v^{\text{out}} : \mathcal{C}_v^{\text{out}} \to \mathbb{R}$ and $f_v^{\text{in}} : \mathcal{C}_v^{\text{in}} \to \mathbb{R}$ be submodular on intersecting pairs. Then the system

(60.41) $\quad\begin{aligned} & x(\delta^{\text{out}}(v)) = x(\delta^{\text{in}}(v)) && \text{for } v \in V, \\ & x(B) \leq f_v^{\text{in}}(B) && \text{for each } v \in V \text{ and } B \in \mathcal{C}_v^{\text{in}}, \\ & x(B) \leq f_v^{\text{out}}(B) && \text{for each } v \in V \text{ and } B \in \mathcal{C}_v^{\text{out}}, \end{aligned}$

is box-TDI. Frank [1982b] showed that this can be derived from the Edmonds-Giles theorem (Theorem 60.1) as follows. Make a digraph $D' = (V', A')$, where A' consists of disjoint arcs $a' := (u_a, v_a)$ for each $a \in A$. Let $\mathcal{C}$ consist of all subsets U of V' such that there exists a $v \in V$ satisfying:

(60.42) $\quad\begin{aligned} & U = \{v_a \mid a \in \delta^{\text{in}}(v)\} \cup \{u_a \mid a \in \delta^{\text{out}}(v)\}, \\ & \text{or } \exists B \in \mathcal{C}_v^{\text{in}} : U = \{v_a \mid a \in B\}, \\ & \text{or } \exists B \in \mathcal{C}_v^{\text{out}} : U = V' \setminus \{u_a \mid a \in B\}. \end{aligned}$

Define $f(U) := 0$, $f(U) := f_v^{\text{in}}(B)$, and $f(U) := f_v^{\text{in}}(B)$, respectively. Then the box-total dual integrality of (60.41) is equivalent to that of

(60.43) $\quad x(\delta_{A'}^{\text{in}}(U)) - x(\delta_{A'}^{\text{out}}(U)) \leq f(U)$ for $U \in \mathcal{C}$, which follows from Theorem 60.1.

Lawler [1982] showed that, conversely, the Edmonds-Giles model is a special case of the polymatroidal network flow model. To see this, let $D = (V, A)$ be a digraph, let $\mathcal{C}$ be a crossing family of subsets of V, and let $f : \mathcal{C} \to \mathbb{R}$ be crossing submodular. Let $\hat{\mathcal{C}}$ be the collection of all sets $U = U_1 \cap \cdots \cap U_t$ with $U_1, \ldots, U_t \in \mathcal{C} \setminus \{V\}$ such that $U_i \cup U_j = V$ for all i, j with $1 \leq i < j \leq t$. Define $\hat{f} : \hat{\mathcal{C}} \to \mathbb{R}$ by

(60.44) $\quad \hat{f}(U) := \min(f(U_1) + \cdots + f(U_t))$,

where the minimum ranges over sets $U_1, \ldots, U_t$ as above. Then $\hat{\mathcal{C}}$ is an intersecting family and $\hat{f}$ is intersecting submodular (Theorem 49.6).

Now extend D by a new vertex r, and arcs (v, r) for $v \in V$, thus making the digraph $D' = (V \cup \{r\}, A')$. Let $\mathcal{C}_r^{\text{in}}$ consist of all subsets B of $\delta_{A'}^{\text{in}}(r)$ for which there is a $U \in \hat{\mathcal{C}}$ satisfying

(60.45) $\quad B = \{(v, r) \mid v \in U\}$.

Define $f_r^{\text{in}}(B) := \hat{f}(U)$. Then

(60.46) $\quad \begin{aligned} & x(\delta_{A'}^{\text{out}}(v)) = x(\delta_{A'}^{\text{in}}(v)) \quad \text{for } v \in V, \\ & x(\delta_{A'}^{\text{in}}(r)) = 0, \\ & x(B) \leq f_r^{\text{in}}(B) \quad \text{for } B \in \mathcal{C}_r^{\text{in}}, \end{aligned}$

is a special case of (60.41). Moreover, the box-total dual integrality of (60.46) implies the box-total dual integrality of

(60.47) $\quad x(\delta_A^{\text{in}}(U)) - x(\delta^{\text{out}}(U)) \leq f(U)$ for $U \in \mathcal{C}$,

since in (60.46) we can restrict B to those B for which there exists a $U \in \mathcal{C}$ with $B = \{(v, r) \mid v \in U\}$ (since $x(\delta_{A'}^{\text{in}}(r)) = 0$). Then

(60.48) $\quad x(B) = \sum_{v \in U} \left(x(\delta_A^{\text{in}}(v)) - x(\delta_A^{\text{out}}(v)) \right) = x(\delta_A^{\text{in}}(U)) - x(\delta_A^{\text{out}}(U))$.

So it implies the Edmonds-Giles theorem (Theorem 60.1).

60.3c. A general model

In Schrijver [1984a] the following general framework was given. Let S be a finite set, let $n \in \mathbb{Z}_+$, let $\mathcal{C}$ be a collection of subsets of S, let $b, c \in (\mathbb{R} \cup \{\pm\infty\})^n$, and let $f : \mathcal{C} \to \mathbb{R}$ and $h : \mathcal{C} \to \{0, \pm 1\}^n$ satisfy:

(60.49) (i) if $\{T_1, T_2, T_3\}$ is a cross-free subcollection of $\mathcal{C}$, then for each $j = 1, \ldots, n$, there exist $u, v \in S$ such that for $i = 1, 2, 3$: $h(T_i)_j = +1$ if and only if (u, v) enters T_i, and $h(T_i)_j = -1$ if and only if (u, v) leaves T_i;

(ii) if T and U are crossing sets in $\mathcal{C}$, then there exist $T', U' \in \mathcal{C}$ such that $T' \subset T$ and

$$f(T) + f(U) - f(T') - f(U') \geq (h(T) + h(U) - h(T') - h(U'))x$$

for each x with $b \leq x \leq c$.

Then the system (in $x \in \mathbb{R}^n$)

1030   Chapter 60. Submodular functions on directed graphs

(60.50) $\quad b \leq x \leq c,$
$\qquad h(T)x \leq f(T) \quad$ for $T \in \mathcal{C},$

is box-TDI. This contains the Edmonds-Giles theorem (Theorem 60.1) and Theorems 60.3 and 60.4 as special cases.

A proof of the box-total dual integrality of (60.50) can be sketched as follows. If we maximize a linear functional $w^\mathsf{T} x$ over (60.50), condition (60.49)(ii) implies that there exists an optimum dual solution whose active constraints correspond to a cross-free subfamily of $\mathcal{C}$. Next, condition (60.49)(i) implies that these constraints form a network matrix, hence a totally unimodular matrix, proving the box-total dual integrality of (60.50) with Theorem 5.35.

60.3d. Packing cuts and Győri's theorem

Let $D = (V, A)$ be a digraph and let $g : \mathcal{P}(V) \to \mathbb{Z}_+$ satisfy the supermodular inequality

(60.51) $\quad g(U) + g(W) \leq g(U \cap W) + g(U \cup W)$

for all $U, W \subseteq V$ such that $\delta^{\text{in}}(U) \cap \delta^{\text{in}}(W) \neq \emptyset$ and $g(U) > 0$, $g(W) > 0$.

The following was shown by Frank and Jordán [1995b] (in the terminology of bisupermodular functions — see Corollary 60.5a):

Theorem 60.5. *Let $D = (V, A)$ be a digraph satisfying:*

(60.52) $\quad$ *V can be partitioned into two sets S and T such that A consists of all arcs from S to T.*

Let g be as above, with $g(U) = 0$ if $\delta^{\text{in}}(U) = \emptyset$. Then the minimum of $x(A)$ taken over all $x : A \to \mathbb{Z}_+$ satisfying

(60.53) $\quad x(\delta^{\text{in}}(U)) \geq g(U)$ *for each $U \subseteq V$,*

is equal to the maximum value of $\sum_{U \in \mathcal{B}} g(U)$, where $\mathcal{B}$ is a collection of subsets U such that the $\delta^{\text{in}}(U)$ for $U \in \mathcal{B}$ are disjoint.

Proof. Let $\tau(g)$ and $\nu(g)$ denote the minimum and maximum value, respectively. Then $\nu(g) \leq \tau(g)$, since, if $x : A \to \mathbb{Z}_+$ satisfies (60.53) and $\mathcal{B}$ is as described, then

(60.54) $\quad x(A) \geq \sum_{U \in \mathcal{B}} x(\delta^{\text{in}}(U)) \geq \sum_{U \in \mathcal{B}} g(U).$

The reverse inequality $\tau(g) \leq \nu(g)$ is shown by induction on $\nu(g)$. If $\nu(g) = 0$, then $g(U) = 0$ for all $U \subseteq V$, and hence $\tau(g) = 0$. Now let $\nu(g) \geq 1$.

For each $a \in A$, define a function g^a by

(60.55) $\quad g^a(U) := \begin{cases} g(U) - 1 & \text{if } a \in \delta^{\text{in}}(U) \text{ and } g(U) \geq 1, \\ g(U) & \text{otherwise,} \end{cases}$

for $U \subseteq V$. In other words:

(60.56) $\quad g^a(U) = \max\{g(U) - d^{\text{in}}_{\{a\}}(U), 0\}.$

Then

Section 60.3d. Packing cuts and Győri's theorem

(60.57) g^a again satisfies (60.51).

Indeed, if $\delta^{\text{in}}(U) \cap \delta^{\text{in}}(W) \neq \emptyset$, and $g^a(U) > 0$ and $g^a(W) > 0$, then $g(U) > 0$ and $g(W) > 0$, and $g^a(U) = g(U) - d^{\text{in}}_{\{a\}}(U)$ and $g^a(W) = g(W) - d^{\text{in}}_{\{a\}}(W)$. Hence

(60.58) $\quad g^a(U) + g^a(W) = g(U) + g(W) - d^{\text{in}}_{\{a\}}(U) - d^{\text{in}}_{\{a\}}(W)$
$\leq g(U \cap W) + g(U \cup W) - d^{\text{in}}_{\{a\}}(U \cap W) - d^{\text{in}}_{\{a\}}(U \cup W)$
$\leq g^a(U \cap W) + g^a(U \cup W).$

So g^a satisfies (60.51).

The following is the key of the proof:

(60.60) there exists an arc a with $\nu(g^a) \leq \nu(g) - 1$.

Suppose to the contrary that $\nu(g^a) = \nu(g)$ for all $a \in A$. As $\nu(g) \geq 1$, there exists a $W \subseteq V$ with $g(W) \geq 1$. For each $a \in \delta^{\text{in}}(W)$, as $\nu(g^a) = \nu(g)$, there exists a collection $\mathcal{B}^a$ such that any arc of D enters at most one $U \in \mathcal{B}^a$, such that $g^a(\mathcal{B}^a) = \nu_g$, and such that $g(U) > 0$ for each $U \in \mathcal{B}^a$. As $g(\mathcal{B}^a) \leq g^a(\mathcal{B}^a)$, a enters no $U \in \mathcal{B}^a$.

Now for each $U \subseteq V$, let $w(U)$ be the number of times U occurs among the $\mathcal{B}^a$ (over all $a \in \delta^{\text{in}}(W)$). Reset $w(W) := w(W) + 1$. Then w has the following properties:

(60.60) (i) $\sum_{U \subseteq V} w(U) \chi^{\delta^{\text{out}}(U)} \leq |\delta^{\text{in}}(W)| \cdot \mathbf{1}$ and

(ii) $\sum_{U \subseteq V} w(U) g(U) > |\delta^{\text{in}}(W)| \nu(g).$

Moreover, $g(U) \geq 1$ whenever $w(U) > 0$.

Now as long as there exist $U, U' \subseteq V$ with $w(U) > 0$ and $w(U') > 0$ and not satisfying:

(60.61) $\delta^{\text{in}}(U) \cap \delta^{\text{in}}(U') = \emptyset$ or $U \subseteq U'$ or $U' \subseteq U,$

decrease $w(U)$ and $w(U')$ by 1, and increase $w(U \cap U')$ and $w(U \cup U')$ by 1. This operation maintains (60.60) and decreases

(60.62) $\sum_{U \in \mathcal{P}(V)} w(U)|U||V \setminus U|$

(by Theorem 2.1). So after a finite number of these operations, w satisfies (60.60) and all U, U' with $w(U) > 0$ and $w(U') > 0$ satisfy (60.61).

Let $\mathcal{F}$ be the collection of $U \subseteq V$ with $w(U) > 0$. We apply the length-width inequality for partially ordered sets (Theorem 14.5) to $(\mathcal{F}, \subseteq)$. By (60.60)(i), the maximum of $w(\mathcal{C})$ taken over chains in $\mathcal{F}$ is at most $|\delta^{\text{in}}(W)|$, since by (60.52), there is an arc $a \in A$ entering all $U \in \mathcal{C}$ (as $\delta^{\text{in}}(U) \neq \emptyset$, since $g(U) \geq 1$, for each $U \in \mathcal{F}$). Moreover, the maximum of $g(\mathcal{B})$ taken over antichains $\mathcal{B}$ in $\mathcal{F}$ is at most $\nu(g)$, since the elements in $\mathcal{F}$ satisfy (60.61), and therefore $\mathcal{B}$ gives a collection of disjoint cuts. But then (60.60)(ii) contradicts the length-width inequality. This proves (60.59).

We now can apply induction, since trivially $\tau(g) \leq \tau(g^a) + 1$, as increasing x_a by 1 for any x satisfying (60.53) with respect to g^a, gives an x satisfying (60.53) with respect to g. So $\tau(g) \leq \tau(g^a) + 1 = \nu(g^a) + 1 \leq \nu(g)$. ∎

This theorem can be equivalently formulated as follows. Let S and T be finite sets. Consider functions $h : \mathcal{P}(S) \times \mathcal{P}(T) \to \mathbb{R}$ satisfying

1032   Chapter 60. Submodular functions on directed graphs

(60.63) $h(X_1 \cap X_2, Y_1 \cup Y_2) + h(X_1 \cup X_2, Y_1 \cap Y_2) \geq h(X_1, Y_1) + h(X_2, Y_2)$
for all $X_1, X_2 \subseteq S$ and $Y_1, Y_2 \subseteq T$ with $X_1 \cap X_2 \neq \emptyset$, $Y_1 \cap Y_2 \neq \emptyset$, $h(X_1, Y_1) > 0$, and $h(X_2, Y_2) > 0$.

Call a collection $\mathcal{F} \subseteq \mathcal{P}(S) \times \mathcal{P}(T)$ *independent* if $X_1 \cap X_2 = \emptyset$ or $Y_1 \cap Y_2 = \emptyset$ for all distinct $(X_1, Y_1), (X_2, Y_2)$ in $\mathcal{F}$. So $\mathcal{F}$ is independent if the sets $X \times Y$ for $(X, Y) \in \mathcal{F}$ are disjoint.

As usual,

(60.64) $$h(\mathcal{F}) := \sum_{(X,Y) \in \mathcal{F}} h(X, Y).$$

Moreover, if $z : S \times T \to \mathbb{R}$, denote

(60.65) $$z(X \times Y) := \sum_{(x,y) \in X \times Y} z(x, y)$$

for $X \subseteq S$ and $Y \subseteq T$.

Corollary 60.5a. *Let $h : \mathcal{P}(S) \times \mathcal{P}(T) \to \mathbb{Z}_+$ satisfy (60.63), such that $h(X, Y) = 0$ if $X = \emptyset$ or $Y = \emptyset$. Then the minimum value of $z(S \times T)$ where $z : S \times T \to \mathbb{Z}_+$ satisfies*

(60.66) $z(X \times Y) \geq h(X, Y)$ for all $X \subseteq S$, $Y \subseteq T$,

is equal to the maximum value of $h(\mathcal{F})$ where $\mathcal{F}$ is independent.

Proof. We can assume that S and T are disjoint. Let $V := S \cup T$, and define a set function g on V by:

(60.67) $g(U) := h(S \setminus U, T \cap U)$

for $U \subseteq V$. Let $D = (V, A)$ be the digraph with A consisting of all arcs from S to T. Then D and g satisfy the condition of Theorem 60.5, and the min-max equality proved in Theorem 60.5 is equivalent to the min-max equality described in the present corollary. ∎

Frank and Jordán showed that this theorem implies the following 'minimax theorem for intervals' of Győri [1984]. Let $\mathcal{I}$ and $\mathcal{J}$ be collections of sets. Then $\mathcal{J}$ is said to *generate* $\mathcal{I}$ if each set in $\mathcal{I}$ is a union of sets in $\mathcal{J}$. Győri's theorem characterizes the minimum size of a collection of intervals generating a given finite collection $\mathcal{I}$ of intervals. For this, we can take an 'interval' to be a finite, contiguous set of integers.

To describe the min-max equality, consider the undirected graph $G_\mathcal{I} = (V_\mathcal{I}, E_\mathcal{I})$ with

(60.68) $V_\mathcal{I} := \{(s, I) \mid s \in I \in \mathcal{I}\}$,

where two distinct pairs (s, I) and (s', I') are adjacent if and only if $s \in I'$ and $s' \in I$. Call a subset C of $V_\mathcal{I}$ *stable* if any two elements of C are nonadjacent (in other words, C is a stable set in $G_\mathcal{I}$).

Corollary 60.5b (Győri's theorem). *Let $\mathcal{I}$ be a finite collection of intervals. Then the minimum size of a collection of intervals generating $\mathcal{I}$ is equal to the maximum size of a stable subset of $V_\mathcal{I}$.*

Proof. To see that the minimum is not less than the maximum, observe that if $\mathcal{J}$ generates $\mathcal{I}$ and C is a stable subset of $V_{\mathcal{I}}$, then for any $J \in \mathcal{J}$, there is at most one $(s, I) \in C$ with $s \in J \subseteq I$, while for any $(s, I) \in C$ there is at least one such $J \in \mathcal{J}$. So $|\mathcal{J}| \geq |C|$.

Equality is shown with Corollary 60.5a. Let S be the union of the intervals in $\mathcal{I}$. Define a function $h : \mathcal{P}(S) \times \mathcal{P}(S) \to \{0, 1\}$ by

(60.69) $h(X, Y) = 1$ if and only if X and Y are nonempty intervals such that $\max X = \min Y$ and $X \cup Y \in \mathcal{I}$, and such that there is no $I \in \mathcal{I}$ with $X \cap Y \subseteq I \subset X \cup Y$,

for $X, Y \subseteq S$. (Here $\max Z$ and $\min Z$ denote the maximum and minimum element of Z, respectively.)

Then h satisfies (60.63). To see this, note first that each (X, Y) with $h(X, Y) = 1$ is characterized by a point $s \in S$ and an inclusionwise minimal interval $I \in \mathcal{I}$ containing s (inclusionwise minimal among all intervals in $\mathcal{I}$ containing s). The relation is that $\{s\} = X \cap Y$ and $I = X \cup Y$.

To see that h satisfies (60.63), let $h(X_1, Y_1) = 1$ and $h(X_2, Y_2) = 1$ and $X_1 \cap X_2 \neq \emptyset$ and $Y_1 \cap Y_2 \neq \emptyset$. We show $h(X_1 \cap X_2, Y_1 \cup Y_2) = 1$ (then $h(X_1 \cup X_2, Y_1 \cap Y_2) = 1$ follows by symmetry).

In fact, this is straightforward case-checking. Let $X_i = [a_i, b_i]$, $Y_i = [b_i, c_i]$, and $I_i := X_i \cup Y_i$ for $i = 1, 2$. As $X_1 \cap I_2 \neq \emptyset \neq Y_1 \cap I_2$, we know that $b_1 \in [a_2, c_2]$, and similarly $b_2 \in [a_1, c_1]$. By symmetry, we can assume that $a_1 \leq a_2$. Hence, by the minimality of $X_1 \cup Y_1$ as an interval containing b_1, $c_1 \leq c_2$. Now, if $b_2 \leq b_1$, we have $X_1 \cap X_2 = [a_2, b_2] = X_2$ and $Y_1 \cup Y_2 = [b_2, c_2] = Y_2$, and therefore $h(X_1 \cap X_2, Y_1 \cup Y_2) = h(X_2, Y_2) = 1$. If $b_1 < b_2$, then $X_1 \cap X_2 = [a_2, b_1]$ and $Y_1 \cup Y_2 = [b_1, c_2]$. Suppose that there is an $I \in \mathcal{I}$ with $b_1 \in I \subset [a_2, c_2]$. By the minimality of $[a_2, c_2]$ as an interval containing b_2, we know $b_2 \notin I$. Hence $I \subset [a_1, c_1]$, contradicting the minimality of $[a_1, c_1]$ as an interval containing b_1. Therefore, no such I exists, and hence we have $h(X_1 \cap X_2, Y_1 \cup Y_2) = 1$.

So Corollary 60.5a applies (taking $T := S$). Let z and $\mathcal{F}$ attain the minimum and maximum respectively. Let $\mathcal{J}$ be the collection of intervals $[s, t]$ with $z(s, t) \geq 1$ and $s \leq t$. Let C be the collection of pairs (s, I) with $s \in S$ and $I \in \mathcal{I}$ such that there is an $(X, Y) \in \mathcal{F}$ with $h(X, Y) = 1$, $X \cap Y = \{s\}$, and $X \cup Y = I$. Then $\mathcal{J}$ generates $\mathcal{I}$, since $z(X \times Y) \geq h(X, Y)$ for all X, Y. Moreover, C is stable as $\mathcal{F}$ is independent. Finally, $|\mathcal{J}| \leq z(S \times S) = h(\mathcal{F}) = |C|$. ∎

(Frank [1999a] gave an alternative, algorithmic proof.)

Győri's theorem in fact states that the colouring number of the complementary graph of $G_{\mathcal{I}}$ is equal to its clique number. It has the following consequence, proved by Chaiken, Kleitman, Saks, and Shearer [1981] and conjectured by V. Chvátal. Let P be a rectilinear polygon in $\mathbb{R}^2$ (where each side horizontal or vertical), such that the intersection of P with each horizontal or vertical line is convex. Then the minimum number of rectangles contained in P needed to cover P, is equal to the maximum number of points in P no two of which are contained in any rectangle contained in P.

Franzblau and Kleitman [1984] gave an $O(|\mathcal{I}|^2)$-time algorithm to find the optima in Győri's theorem, with a proof of equality as by-product.

Győri's theorem was extended by Lubiw [1991a] to a weighted version. She noted that a fully weighted version of the theorem does not hold; that is, taking integer weights $w(s, I)$ on any pair (s, I), the maximum weight of a stable set need not be equal to the minimum size of a family $\mathcal{J}$ of intervals such that for any (s, I) there are at least $w(s, I)$ intervals J in $\mathcal{J}$ satisfying $s \in J \subseteq I$. (In other words, $G_\mathcal{I}$ need not be perfect.)

However, as Lubiw showed, these two optima are equal if $w(s, I)$ only depends on s; that is, if $w(s, I) = w(s)$ for some $w : S \to \mathbb{Z}_+$. Also this can be derived from Frank and Jordán's theorem: instead of defining $h(X, Y) := 1$ in (60.69), define $h(X, Y) := w(s)$ where $\{s\} = X \cap Y$.

As Frank and Jordán observed, their method extends Győri's theorem to the case where we take 'interval' to mean: interval on the circle (instead of just the real line).

Other applications of Theorem 60.5 are to vertex-connectivity augmentation — see Theorem 63.11.

60.3e. Further notes

For another model equivalent to that of Edmonds and Giles, based on distributive lattices, see Gröflin and Hoffman [1982] — cf. Schrijver [1984b]. Grishuhin [1981] gave a lattice model requiring total unimodularity as a condition.

Further algorithms for minimum-cost submodular flow were given by Fujishige [1978a,1987], Zimmermann [1982b,1992], Barahona and Cunningham [1984], Cui and Fujishige [1988], Chung and Tcha [1991], Gabow [1993a], McCormick and Ervolina [1993], Iwata, McCormick, and Shigeno [1998,1999,2000,2002], Wallacher and Zimmermann [1999], Fleischer and Iwata [2000], and Fleischer, Iwata, and McCormick [2002]. A survey on algorithms for submodular flows was presented by Fujishige and Iwata [2000].

Zimmermann [1982a,1982b,1985] considered group-valued extensions of some of the models. Federgruen and Groenevelt [1988] extended some models to more general objective functions. Zimmermann [1986] considered duality for balanced submodular flows. Qi [1988a] and Murota [1999] gave generalizations of submodular flows. Convex-cost submodular flows were considered by Iwata [1996,1997].

An algorithm for a model comprising the Edmonds-Giles and the lattice polyhedron model (Section 60.3a) was given by Karzanov [1983]. For 0,1 problems it is polynomial-time.

The effectivity of uncrossing techniques is studied by Hurkens, Lovász, Schrijver, and Tardos [1988] and Karzanov [1996].

The facets of submodular flow polyhedra were studied by Giles [1975].

For a comparison of models, see Schrijver [1984b], and for a survey, including results on the dimension of faces of submodular flow polyhedra, see Frank and Tardos [1988]. A survey of submodular functions and flows is given by Murota [2002].

Chapter 61

Graph orientation

Orienting an undirected graph so as to obtain a k-arc-connected directed graph is the object of study in this chapter. Recall that a directed graph D is called an *orientation* of an undirected graph G if G is the underlying undirected graph of D.

Central result is a deep theorem of Nash-Williams showing that each undirected graph has an orientation that keeps at least half of the connectivity (rounded down) between any two vertices.

It implies that a $2k$-edge-connected undirected graph has a k-arc-connected orientation. This can be proved alternatively and easier with the help of submodular functions (cf. Section 61.4).

61.1. Orientations with bounds on in- and outdegrees

We first consider orientations obeying bounds on the indegrees and/or outdegrees. The results follow quite directly from bipartite matching or (equivalently) flow theory.

Hakimi [1965] considered lower bounds on the indegrees:

Theorem 61.1. *Let $G = (V, E)$ be an undirected graph and let $l : V \to \mathbb{Z}_+$. Then G has an orientation $D = (V, A)$ with $\deg_A^{in}(v) \geq l(v)$ for each $v \in V$ if and only if each $U \subseteq V$ is incident with at least $l(U)$ edges.*

Proof. Let $\mathcal{A}$ be the family of subsets of E obtained by taking set $\delta(v)$ with multiplicity $l(v)$, for each $v \in V$. Then the existence of an orientation as required is equivalent to the existence of a transversal of $\mathcal{A}$. By Hall's marriage theorem (Theorem 22.1), this is equivalent to the condition in the theorem. ∎

A direct consequence is:

Corollary 61.1a. *Let $G = (V, E)$ be an undirected graph and let $l : V \to \mathbb{Z}_+$. Then G has an orientation $D = (V, A)$ with $\deg_A^{in}(v) = l(v)$ for each $v \in V$ if and only if $l(V) = |E|$ and each $U \subseteq V$ is incident with at least $l(U)$ edges.*

Proof. Directly from Theorem 61.1. ∎

Another consequence concerns upper bounds:

Corollary 61.1b. *Let $G = (V, E)$ be an undirected graph and let $u : V \to \mathbb{Z}_+$. Then G has an orientation $D = (V, A)$ with $\deg_A^{in}(v) \leq u(v)$ for each $v \in V$ if and only if each $U \subseteq V$ spans at most $u(U)$ edges.*

Proof. For each $v \in V$, define $l(v) := \deg(v) - u(v)$. (We may assume that, for each $v \in V$, $u(v) \leq \deg(v)$, since otherwise resetting $u(v) := \deg(v)$ does not change the conditions in the theorem.)

Now G has an orientation with $\deg^{in}(v) \leq u(v)$ for each v if and only if G has an orientation with $\deg^{in}(v) \geq l(v)$ for each v (just by reversing the orientation of all edges). By Theorem 61.1, the latter is equivalent to: each $U \subseteq V$ is incident with at least $l(U)$ edges; that is: $E[U] + |\delta_E(U)| \geq l(U)$. Since

$$(61.1) \qquad l(U) = \sum_{v \in U}(\deg(v) - u(v)) = 2|E[U]| + |\delta_E(U)| - u(U),$$

it is equivalent to: $|E[U]| \leq u(U)$, as required. ∎

Frank and Gyárfás [1978] gave a characterization for the case of lower bounds on both indegrees and outdegrees:

Theorem 61.2. *Let $G = (V, E)$ be an undirected graph and let $l, u : V \to \mathbb{Z}_+$ with $l \leq u$. Then G has an orientation $D = (V, A)$ with $l(v) \leq \deg_A^{in}(v) \leq u(v)$ for each $v \in V$ if and only if each $U \subseteq V$ is incident with at least $l(U)$ edges and spans at most $u(U)$ edges.*

Proof. The condition trivially being necessary, we prove sufficiency. Let $D = (V, A)$ be an arbitrary orientation of G. It suffices to show that there exists a function $x : A \to \{0, 1\}$ such that for each $v \in V$:

$$(61.2) \qquad l(v) \leq \deg_A^{in}(v) - x(\delta_A^{in}(v)) + x(\delta_A^{out}(v)) \leq u(v),$$

since reversing the orientation of the arcs a with $x(a) = 1$ then gives an orientation as required. Condition (61.2) is equivalent to:

$$(61.3) \qquad \deg_A^{in}(v) - u(v) \leq x(\delta_A^{in}(v)) - x(\delta_A^{out}(v)) \leq \deg_A^{in}(v) - l(v).$$

By Corollary 11.2i, such an x exists if and only if

$$(61.4) \qquad |\delta_A^{in}(U)| \geq \max\{\sum_{v \in U}(\deg_A^{in}(v) - u(v)), \sum_{v \in V \setminus U}(l(v) - \deg_A^{in}(v))\}$$

for each $U \subseteq V$. Since $|\delta_A^{in}(U)| + \sum_{v \in V \setminus U} \deg_A^{in}(v)$ is equal to the number of edges incident with $V \setminus U$ and since $\sum_{v \in U} \deg_A^{in}(v) - |\delta_A^{in}(U)|$ is equal to the number of edges spanned by U, this is equivalent to the condition given in the theorem. ∎

Ford and Fulkerson [1962] observed that the undirected edges of a mixed graph (V, E, A) can be oriented so as to obtain an Eulerian directed graph if and only if

(61.5) (i) $\deg_E(v) + \deg_A^{\text{in}}(v) + \deg_A^{\text{out}}(v)$ is even for each $v \in V$,

(ii) $d_A^{\text{out}}(U) - d_A^{\text{in}}(U) \leq d_E(U)$ for each $U \subseteq V$.

This can be proved similarly.

61.2. 2-edge-connectivity and strongly connected orientations

Each $2k$-edge-connected undirected graph has a k-arc-connected orientation, which will be seen in Section 61.3. In the present section we consider the special case $k = 2$, which goes back to a theorem of Robbins [1939]. Tarjan [1972] showed that depth-first search is the tool behind. We follow his approach.

Theorem 61.3. *Given an undirected graph $G = (V, E)$ we can find an orientation D of G, in linear time, such that for each $u, v \in V$, if G has two edge-disjoint $u - v$ paths, then D has a directed $u - v$ path.*

Proof. Choose $s \in V$ arbitrarily, and consider a depth-first search tree T starting at s. Orient each edge in T away from s. For each remaining edge $e = uv$, there is a directed path in T that connects u and v. Let the path run from u to v. Then orient e from v to u. This gives the orientation D of G.

Then any edge not in T belongs to a directed circuit in D. Moreover, any edge f in T that is not a cut edge, belongs to a directed circuit in D (since there is an edge $e \notin T$ connecting the two components of $T - f$). This implies that D is as required. ∎

This implies the theorem of Robbins [1939] on strongly connected orientations:

Corollary 61.3a (Robbins' theorem). *An undirected graph G has a strongly connected orientation if and only if G is 2-edge-connected.*

Proof. Necessity is easy, and sufficiency follows from Theorem 61.3. ∎

(The proof by Robbins [1939] uses the fact that each 2-edge-connected graph has an 'ear-decomposition' — cf. Section 15.5a.)

The above proof also shows that a strongly connected orientation can be found in linear time:

Corollary 61.3b. *Given a 2-edge-connected graph G, a strongly connected orientation of G can be found in linear time.*

Proof. Directly from Theorem 61.3. ∎

Robbins' theorem (Corollary 61.3a) extends to the following result of Frank [1976a] and Boesch and Tindell [1980] for mixed graphs.

Theorem 61.4. *Let $G = (V, E)$ be a graph in which part of the edges is oriented. Then the remainder of the edges can be oriented so as to obtain a strongly connected digraph if and only if G is 2-edge-connected and there is no nonempty proper subset U of V such that all edges in $\delta(U)$ are oriented from U to $V \setminus U$.*

Proof. Necessity being easy, we show sufficiency. Let G be a counterexample with a minimum number of undirected edges. Then there is at least one undirected edge, say $e = uv$. By the minimality assumption, orienting e from s to t violates the condition. That is, there exists a $U \subseteq V$ with $u \in U$, $v \in V \setminus U$, such that each edge $\neq e$ in $\delta(U)$ is oriented from U to $V \setminus U$. Similarly, there exists a $T \subseteq V$ with $v \in T$, $u \in V \setminus T$, such that each edge $\neq e$ $\delta(T)$ is oriented from T to $V \setminus T$.

Then each edge in $\delta(U \cap T)$ is oriented from $U \cap T$ to $V \setminus (U \cap T)$, and hence $U \cap T = \emptyset$. Similarly, $U \cup T = V$. Hence $\delta(U) = \{e\}$, a contradiction. ∎

The graph $K_{2,3}$ shows that a 2-edge-connected graph need not have an orientation in which each two vertices belong to a directed circuit; that is an orientation such that for each two vertices s, t there exists an arc-disjoint pair of an $s - t$ and a $t - s$ path.

Chung, Garey, and Tarjan [1985] gave a linear-time algorithm to find an orientation as described in Theorem 61.4.

61.2a. Strongly connected orientations with bounds on degrees

Robbins' theorem (Corollary 61.3a) states that an undirected graph G has a strongly connected orientation if and only if G is 2-edge-connected. Frank and Gyárfás [1978] extended this to the case where upper and lower bounds are prescribed on the indegrees of the orientation.

Let $\kappa(G)$ denote the number of its components of any graph G.

Theorem 61.5. *Let $G = (V, E)$ be a 2-edge-connected undirected graph and let $l, u \in \mathbb{Z}_+^V$ with $l \leq u$. Then G has a strongly connected orientation $D = (V, A)$ satisfying $l(v) \leq \deg_A^{in}(v) \leq u(v)$ for each $v \in V$ if and only if for each $U \subseteq V$:*

(61.6) (i) $|E[U]| + \kappa(G - U) \leq u(U)$,
(ii) $|E[U]| + |\delta(U)| - \kappa(G - U) \geq l(U)$.

Proof. It is easy to see that condition (61.6) is necessary. To see sufficiency, let (61.6) hold. Let $D = (V, A)$ be a strongly connected orientation of G with

Section 61.2a. Strongly connected orientations with bounds on degrees 1039

(61.7) $$\sum_{v \in V} \max\{0, \deg_A^{\text{in}}(v) - u(v), l(v) - \deg_A^{\text{in}}(v)\}$$

as small as possible. (A strongly connected orientation exists by Corollary 61.3a.)

If sum (61.7) is 0, we are done, so assume that it is positive. Then there exists a vertex v_0 with $\deg_A^{\text{in}}(v_0) > u(v_0)$ or $l(v_0) > \deg_A^{\text{in}}(v_0)$. Suppose that $\deg_A^{\text{in}}(v_0) > u(v_0)$. Let U be the set of vertices v for which D has two arc-disjoint $v - v_0$ paths. Then $\deg_A^{\text{in}}(v) \geq u(v)$ for each $v \in U$, since otherwise we can reverse the orientation on the arcs of one of the two arc-disjoint $v-v_0$ paths, thereby keeping the orientation strongly connected while decreasing sum (61.7).

We claim that U violates (61.6)(i). To this end, we show that

(61.8)   each component K of $G - U$ is left by exactly one arc of D.

This can be seen as follows. For each $v \in K$, there exists a $U_v \subseteq V$ with $d_A^{\text{out}}(U_v) = 1$ and $v \in U_v$, $v_0 \notin U_v$ (as there exist no two arc-disjoint $v_0 - v$ paths). We choose each U_v inclusionwise maximal.

It suffices to show that

(61.9) $U_v = K$ for each $v \in K$.

To see this, note first that, for each $v \in K$, we have $U_v \subseteq K$. Indeed, $U_v \cap U = \emptyset$, since if say $v_1 \in U \cap U_v$, then there exist no two arc-disjoint $v_1 - v_0$ paths in D, contradicting the definition of U. If U_v would intersect another component K' of $G - U$, then $d_A^{\text{out}}(U_v) = d_A^{\text{out}}(U_v \cap K) + d_A^{\text{out}}(U_v \cap K') \geq 2$ — a contradiction.

Moreover, if $v \neq v'$ and $U_v \cap U_{v'} \neq \emptyset$, then $U_v = U_{v'}$. This follows from:

(61.10) $1 \leq d_A^{\text{out}}(U_v \cup U_{v'}) \leq d_A^{\text{out}}(U_v) + d_A^{\text{out}}(U_{v'}) - d_A^{\text{out}}(U_v \cap U_{v'}) \leq 1 + 1 - 1 = 1.$

So, if $U_v \neq U_{v'}$, we would increase U_v or $U_{v'}$ by replacing it by $U_v \cup U_{v'}$ — a contradiction.

So the U_v partition K. Now if $U_v \neq K$, there exist v and v' such that $U_v \neq U_{v'}$ and such that G has an edge connecting U_v and $U_{v'}$. We can assume that it is oriented from $U_{v'}$ to U_v. So it is the unique edge leaving $U_{v'}$. Hence $d_A^{\text{out}}(U_v \cup U_{v'}) \leq d_A^{\text{out}}(U_v) = 1$. So replacing U_v by $U_v \cup U_{v'}$ would increase U_v — a contradiction. This shows (61.9), and hence (61.8).

So $d_A^{\text{out}}(K) = 1$ for each component K of $G - U$. Therefore

(61.11) $$|E[U]| + \kappa(G - U) = \sum_{v \in U} \deg_A^{\text{in}}(v) > u(U).$$

Thus U violates (61.6)(i).

One similarly shows that $\deg_A^{\text{in}}(v_0) < l(v_0)$ implies violation of (61.6)(ii). ∎

This implies an alternative characterization:

Corollary 61.5a. *Let $G = (V, E)$ be a 2-edge-connected undirected graph and let $l, u \in \mathbb{Z}_+^V$ with $l \leq u$. Then G has a strongly connected orientation $D = (V, A)$ satisfying $l(v) \leq \deg_A^{\text{in}}(v) \leq u(v)$ for each $v \in V$ if and only if G has strongly connected orientations $D' = (V, A')$ and $D'' = (V, A'')$ with $l(v) \leq \deg_{A'}^{\text{in}}(v)$ and $\deg_{A''}^{\text{in}}(v) \leq u(v)$ for each $v \in V$.*

Proof. Directly from Theorem 61.5, as (61.6)(i) is void if $u = \infty$ and as (61.6)(ii) is void if $l = 0$ (since $\kappa(G - U) \leq d_G(U)$). ∎

For further results, see Theorem 61.7.

61.3. Nash-Williams' orientation theorem

The result of Robbins [1939] was extended deeply by Nash-Williams [1960]. Before stating and proving it, we give a useful lemma of Nash-Williams [1960]. Let $\phi : V \times V \to \mathbb{R}$ be a symmetric function (that is, $\phi(u,v) = \phi(v,u)$ for all $u, v \in V$). Define a set function R on V by:

(61.12) $\quad R(U) := \max_{u \in U, v \in V \setminus U} \phi(u,v)$ if $\emptyset \subset U \subset V$, and
$R(\emptyset) := R(V) := 0.$

Lemma 61.6α. *For all $T, U \subseteq V$:*

(61.13) $\quad R(T) + R(U) \leq R(T \cap U) + R(T \cup U)$
$\text{or } R(T) + R(U) \leq R(T \setminus U) + R(U \setminus T).$

Proof. Suppose not. Then $\emptyset \neq T \neq V$ and $\emptyset \neq U \neq V$. Choose $s \in T$, $t \in V \setminus T$, $u \in U$, $v \in V \setminus U$ such that $R(T) = \phi(s,t)$ and $R(U) = \phi(u,v)$. By symmetry, we can assume that $R(T) \leq R(U)$ and $u \in T$. So $u \in T \cap U$, and hence $T \cap U$ splits[15] $\{u, v\}$. This implies that $T \cup U$ splits neither $\{s, t\}$, nor $\{u, v\}$, as otherwise the first inequality in (61.13) holds (as $\phi(s,t) \leq \phi(u,v)$).

Hence $t, v \in T \cup U$, and so $t \in U \setminus T$ and $v \in T \setminus U$. Then $T \setminus U$ splits $\{u, v\}$, and $U \setminus T$ splits $\{s, t\}$, implying the second inequality in (61.13). ∎

For any undirected graph $G = (V, E)$ and $s, t \in V$, let $\lambda_G(s, t)$ denote the maximum number of edge-disjoint $s - t$ paths in G. Similarly, for any directed graph $D = (V, A)$ and $s, t \in V$, let $\lambda_D(s, t)$ denote the maximum number of arc-disjoint $s - t$ paths in D.

Theorem 61.6 (Nash-Williams' orientation theorem). *Any undirected graph $G = (V, E)$ has an orientation $D = (V, A)$ with*

(61.14) $\quad \lambda_D(s,t) \geq \lfloor \tfrac{1}{2} \lambda_G(s,t) \rfloor$

for all $s, t \in V$.

Proof. Call a partition of a set T into pairs, a *pairing* of T. For any number k, define

(61.15) $\quad k^* := 2\lfloor \tfrac{1}{2} k \rfloor.$

For any subset U of V, define

---
[15] Set X *splits* set Y if both $Y \cap X$ and $Y \setminus X$ are nonempty.

Section 61.3. Nash-Williams' orientation theorem   1041

(61.16) $\quad r(U) := \max_{u \in U, v \notin U} \lambda_G(u,v),$

setting $r(U) := 0$ if $U = \emptyset$ or $U = V$. Let T be the set of vertices of odd degree of G.

I. It suffices to show that T has a pairing P such that

(61.17) $\quad d_G(U) - d_P(U) \geq r(U)^*$ for each $U \subseteq V$.

To see that this is sufficient, let $G' = (V, E \cup P)$. That is, G' is the graph obtained from G by adding all pairs in P as new edges (possibly in parallel). Then all degrees in G' are even, and hence G' has an Eulerian orientation $D' = (V, A')$. So

(61.18) $\quad \deg_{D'}^{\text{out}}(v) = \deg_{D'}^{\text{in}}(v) = \frac{1}{2}\deg_{G'}(v)$

for each $v \in V$. This implies that, for each $U \subseteq V$,

(61.19) $\quad d_{D'}^{\text{out}}(U) = \frac{1}{2} d_{G'}(U).$

Let A be the restriction of A' to the original edges of G and let $D = (V, A)$. We claim that D is an orientation of G as required. Indeed, by (61.17), for each $U \subseteq V$,

(61.20) $\quad d_D^{\text{out}}(U) \geq d_{D'}^{\text{out}}(U) - d_P(U) = \frac{1}{2} d_{G'}(U) - d_P(U)$
$= \frac{1}{2}(d_G(U) - d_P(U)) \geq \lfloor \frac{1}{2} r(U) \rfloor.$

Hence, for any $u, v \in V$, if $U \subseteq V$ with $u \in U$, $v \notin U$, and $\lambda_D(u,v) = d_D^{\text{out}}(U)$, then

(61.21) $\quad \lambda_D(u,v) = d_D^{\text{out}}(U) \geq \lfloor \frac{1}{2} r(U) \rfloor \geq \lfloor \frac{1}{2} \lambda_G(u,v) \rfloor.$

II. We now prove the theorem. Define for any $Y \subseteq V$ and $U \subseteq V$,

(61.22) $\quad r_Y(U) := \max_{u \in Y \cap U, v \in Y \setminus U} \lambda_G(u,v),$

setting $r_Y(U) := 0$ if $Y \cap U = \emptyset$ or $Y \subseteq U$.
By Lemma 61.6α, for any $U, W \subseteq V$,

(61.23) $\quad r_Y(U)^* + r_Y(W)^* \leq r_Y(U \cap W)^* + r_Y(U \cup W)^*$
or $r_Y(U)^* + r_Y(W)^* \leq r_Y(U \setminus W)^* + r_Y(W \setminus U)^*.$

This follows from Lemma 61.6α by taking $\phi(u,v) := \lambda_G(u,v)^*$ if $u, v \in Y$ and $\phi(u,v) := 0$ otherwise.

Suppose that there exist graphs G for which T has no pairing P satisfying (61.17). Choose G with $|V| + |E|$ minimal.

Choose $Y \subseteq V$ such that T has no pairing P satisfying

(61.24) $\quad d_G(U) - d_P(U) \geq r_Y(U)^*$ for each $U \subseteq V$,

with $|Y|$ as small as possible. Then

(61.25) $\quad$ For any subset X of V splitting Y and satisfying $d_G(X)^* = r_Y(X)^*$, one has $|X| = 1$ or $|X| = |V| - 1$.

For suppose $1 < |X| < |V| - 1$. Consider the graph $G_1 = (V_1, E_1)$ obtained from G by contracting $V \setminus X$ to one vertex, v_1 say. Let T_1 be the set of vertices of odd degree of G_1. By the minimality of $|V|+|E|$, T_1 has a pairing P_1 such that for each subset U of X:

(61.26) $\quad d_{G_1}(U) - d_{P_1}(U) \geq r(U)^*$.

(Note that $r(U) \leq \max_{u \in U, v \in V \setminus U} \lambda_{G_1}(u,v)$.)

Similarly, consider the graph $G_2 = (V_2, E_2)$ obtained from G by contracting X to one vertex, v_2 say. Let T_2 be the set of vertices of odd degree of G_2. Again by the minimality of $|V|+|E|$, T_2 has a pairing P_2 such that for each subset U of $V \setminus X$ and for each $u \in U$, $v \in V_2 \setminus U$:

(61.27) $\quad d_{G_2}(U) - d_{P_2}(U) \geq r(U)^*$.

Now define a pairing P of T as follows. (Observe that $v_1 \in T_1$ if and only if $v_2 \in T_2$.) If $v_1 \notin T_1$ and $v_2 \notin T_2$, let $P := P_1 \cup P_2$. If $v_1 \in T_1$ and $v_2 \in T_2$, let $u_1 \in X$ and $u_2 \in V \setminus X$ be such that $u_1 v_1 \in P_1$ and $u_2 v_2 \in P_2$. Then define

(61.28) $\quad P := (P_1 \setminus \{u_1 v_1\}) \cup (P_2 \setminus \{u_2 v_2\}) \cup \{u_1 u_2\}$.

We claim that P satisfies (61.24). To show this, we may assume by (61.23) that $r_Y(U \cap X)^* + r_Y(U \cup X)^* \geq r_Y(U \setminus X)^* + r_Y(X \setminus U)^*$. (Otherwise, replace U by $V \setminus U$.)

Set $U_1 := U \cap X$ and $U_2 := V \setminus (U \cup X)$. Then

(61.29) $\quad d_G(U) + d_G(X) \geq d_G(U_1) + d_G(U_2) = d_{G_1}(U_1) + d_{G_2}(U_2)$
$\geq r(U_1)^* + d_{P_1}(U_1) + r(U_2)^* + d_{P_2}(U_2) \geq r_Y(U)^* + r_Y(X)^* + d_P(U)$
$\geq r_Y(U)^* + d_G(X) + d_P(U) - 1$,

using (61.26) and (61.27). As $d_G(U) + d_P(U)$ is even, (61.24) follows. This contradicts our assumption, showing (61.25).

We next show that

(61.30) $\quad$ each edge of G intersects Y.

For assume that G has an edge $e = st$ disjoint from Y. By (61.25), there is no U splitting Y with $d_G(U)^* = r_Y(U)^*$ and $s \in U$, $t \notin U$. So deleting edge e, changes no $r_Y(U)^*$. Let G' be the graph obtained from G by deleting e. Let T' be the set of vertices of G' of odd degree. (So $T' = T \triangle \{s,t\}$.) Then, by the minimality of $|V|+|E|$, we know that T' has a pairing P' such that, for each $U \subseteq V$,

(61.31) $\quad d_{G'}(U) - d_{P'}(U) \geq r(U)^* \geq r_Y(U)^*$.

It is not difficult to transform pairing P' of T' to a pairing P of T with the property that $|P \setminus P'| \leq 1$.[16] Then (61.24) holds. Indeed, $d_G(U) \geq d_{G'}(U)$

---

[16] If $s, t \in T'$ and $st \in P'$, define $P := P' \setminus \{st\}$. If $s, t \in T'$ and $st \notin P'$, let s' and t' be such that $ss' \in P'$ and $tt' \in P'$, and define $P := (P' \setminus \{ss', tt'\}) \cup \{s't'\}$. If $s \in T'$ and $t \notin T'$, let s' be such that $ss' \in P'$, and define $P := (P' \setminus \{ss'\}) \cup \{ts'\}$. If $s \notin T'$ and $t \in T'$, let t' be such that $tt' \in P'$, and define $P := (P' \setminus \{tt'\}) \cup \{st'\}$. If $s \notin T'$ and $t \notin T'$, define $P := P' \cup \{st\}$.

and $d_P(U) \leq d_{P'}(U)+1$ (as $|P \setminus P'| \leq 1$). Hence (61.24) follows from (61.31), with parity. This contradiction proves (61.30).

Next:

(61.32) $|Y| \geq 2$.

For suppose that $|Y| \leq 1$. In G there exist $\frac{1}{2}|T|$ edge-disjoint paths such that each vertex in T occurs exactly once as an end vertex of one of these paths. (This can be seen by taking an arbitrary pairing Q of T, and considering an Eulerian tour C in the graph $G = (V, E \cup Q)$. Then removing Q from C decomposes C into paths as required.) Let P be the set of pairs of end vertices of these paths. Then $d_G(U) \geq d_P(U)$ for each $U \subseteq V$, and (61.24) follows, contradicting our assumption. So we know (61.32).

Choose a set X splitting Y with $d_G(X)$ minimal. Then $d_G(X) = r_Y(X)$. By (61.25), we may assume that $X = \{x\}$ for some $x \in Y$. So $r_Y(U) = d_G(x)$ for any U splitting Y, since for any $y \in Y \setminus U$ we have

(61.33) $d_G(x) \leq d_G(U) \leq r_Y(U) \leq \lambda_G(x,y) \leq d_G(x)$.

Define $Y' := Y \setminus \{x\}$. Then, by the minimality of $|Y|$, T has a pairing P such that for each $U \subseteq V$,

(61.34) $d_G(U) - d_P(U) \geq r_{Y'}(U)^*$.

We show that (61.24) holds, which forms a contradiction. To this end, choose $U \subseteq V$.

First assume that U splits Y'. Then $r_{Y'}(U) \geq r_Y(U)$, since $\lambda_G(x,y) = d_G(X) \leq r_{Y'}(U)$ for each $y \in Y'$ (by the minimality of $d_G(X)$, since any splitting of Y' also splits Y). This implies (61.34).

So we can assume that U splits Y but does not split Y'; that is, $U \cap Y = \{x\}$. Consider any $u \in U \setminus \{x\}$. Let α_u denote the number of edges connecting u and x and let β_u denote the number of edges connecting u and $Y \setminus \{x\}$. By (61.30), $\alpha_u + \beta_u = \deg_G(u)$. Since $X = \{x\}$ splits Y with $d_G(X)$ minimum, we have $d_G(\{x,u\}) \geq \deg_G(x)$. Hence $\beta_u \geq \alpha_u$, with strict inequality if $u \in T$ (since then $\alpha_u + \beta_u$ is odd).

Therefore, setting $U' := U \setminus \{x\}$ and $\lambda :=$ number of edges connecting x and $V \setminus U$,

(61.35) $d_G(U) = \lambda + \sum_{u \in U'} \beta_u \geq \lambda + \sum_{u \in U'} \alpha_u + |U' \cap T| = \deg_G(x) + |U' \cap T|$

$= r_Y(U) + |U' \cap T| \geq r_Y(U) + |U \cap T| - 1 \geq r_Y(U) + d_P(U) - 1$.

Hence, with parity, we have (61.24). ∎

(This is the original proof of Nash-Williams [1960]. Mader [1978a] and Frank [1993a] gave alternative proofs.)

An orientation satisfying the condition described in Theorem 61.6 is called *well-balanced*. Nash-Williams [1969] (giving an introduction to the proof above) remarks that with methods similar to those used in the proof of Theorem 61.6, one can prove that for any graph G and any subgraph H of G, there is a well-balanced orientation of G such that the restriction to H is well-balanced again.

61.4. k-arc-connected orientations of $2k$-edge-connected graphs

Nash-Williams' orientation theorem directly implies:

Corollary 61.6a. *An undirected graph G has a k-arc-connected orientation if and only if G is $2k$-edge-connected.*

Proof. Directly from Theorem 61.6. ∎

A direct proof of this corollary, based on total dual integrality, was given by Frank [1980b] and Frank and Tardos [1984b], and is as follows.

Orient the edges of G arbitrarily, yielding the directed graph $D = (V, A)$. Consider the system

(61.36)   (i) $0 \leq x_a \leq 1$ for each $a \in A$,
           (ii) $x(\delta^{\text{in}}(U)) - x(\delta^{\text{out}}(U)) \leq d^{\text{in}}(U) - k$ for each nonempty $U \subset V$.

By the Edmonds-Giles theorem (Theorem 60.1), this system is TDI, and hence determines an integer polytope P. If G is $2k$-edge-connected, then P is nonempty, since the vector $x := \frac{1}{2} \cdot \mathbf{1}$ belongs to P.

As P is nonempty and integer, (61.36) has an integer solution x. Then G has a k-arc-connected orientation D': reversing the orientation of the arcs a of D with $x_a = 1$ gives a k-arc-connected orientation D', since

(61.37) $d_{D'}^{\text{in}}(U) = d_D^{\text{in}}(U) - x(\delta_D^{\text{in}}(U)) + x(\delta_D^{\text{out}}(U)) \geq k$

for any nonempty proper subset U of V.

Notes. The total dual integrality of (61.36) implies also the following result of Frank, Tardos, and Sebő [1984] (denoting the number of (weak) components of a (di)graph G by $\kappa(G)$). Let $G = (V, E)$ be a 2-edge-connected undirected graph and let $U \subseteq V$. Then the minimum number of arcs entering U over all strongly connected orientations of G is equal to the maximum of

(61.38) $\sum_{T \in \mathcal{P}} \kappa(G - T),$

taken over partitions $\mathcal{P}$ of U into nonempty classes such that no edge connects different classes of $\mathcal{P}$.

This implies another result of Frank, Tardos, and Sebő [1984]: Let $D = (V, A)$ be a digraph and let $C = \delta^{\text{in}}(U)$ be a directed cut. Then the minimum of $|B \cap C|$ where B is a directed cut cover is equal to the maximum of

(61.39) $$\sum_{T \in \mathcal{P}} \kappa(D - T),$$

taken over partitions $\mathcal{P}$ of U into nonempty classes such that no arc of D connects distinct classes of $\mathcal{P}$.

As A. Frank (personal communication 2002) observed, the proof above yields a stronger result of Nash-Williams [1969]: let $G = (V, E)$ be a $2k$-edge-connected graph and let $F \subseteq E$ have an Eulerian orientation; then the remaining edges have an orientation so as to obtain a k-arc-connected digraph. This follows by taking for A any orientation extending the orientation of F, and by setting $x_a := 0$ for each arcs in F, and $x_a := \frac{1}{2}$ for all other arcs a. Then x satisfies (61.36), and the result follows as above.

61.4a. Complexity

By the results in Section 60.1 on the complexity of the Edmonds-Giles problem, one can find a k-arc-connected orientation of a $2k$-edge-connected undirected graph in polynomial time; more generally, one can find a minimum-length k-arc-connected orientation in strongly polynomial time, if we are given a length for each orientation of each edge.

A direct method of finding a minimum-length k-arc-connected orientation can be based on weighted matroid intersection, similarly to the method described in Section 55.5 to find a minimum-length directed k-cover in a directed graph (such that the k-arc-connected orientations form the common bases of two matroids).

61.4b. k-arc-connected orientations with bounds on degrees

Frank [1980b] extended Corollary 61.6a to the case where lower and upper bounds on the indegrees of the vertices are prescribed:

Theorem 61.7. *Let $G = (V, E)$ be a $2k$-connected undirected graph and let $l, u \in \mathbb{Z}_+^V$ with $l \leq u$. Then G has a k-arc-connected orientation D with $l(v) \leq \deg_D^{\text{in}}(v) \leq u(v)$ for each $v \in V$ if and only if*

(61.40) $$|E[W]| + |\delta(\mathcal{P})| \geq k|\mathcal{P}| + \max\{\sum_{v \in W} l(v), \sum_{v \in W} (\deg_G(v) - u(v))\}$$

for each subpartition $\mathcal{P}$ of V with nonempty classes, where $W := V \setminus \bigcup \mathcal{P}$.

Proof. It is not difficult to see that the condition is necessary. To show sufficiency, by Corollary 61.6a, G has a k-arc-connected orientation D. Choose D such that

(61.41) $$\sum_{v \in V} \max\{0, \deg_D^{\text{in}}(v) - u(v), l(v) - \deg_D^{\text{in}}(v)\}$$

is as small as possible. If sum (61.41) is 0 we are done, so assume that it is positive. By symmetry, we can assume that there is a vertex r with $\deg_D^{\text{in}}(r) < l(r)$.

Let $\mathcal{P}$ be the collection of inclusionwise maximal nonempty subsets U of $V \setminus \{r\}$ with $d^{\text{in}}(U) = k$, and let $W := V \setminus \bigcup \mathcal{P}$.

Then the sets in $\mathcal{P}$ are disjoint. For let $U, W \in \mathcal{P}$ with $U \cap W \neq \emptyset$. Then

(61.42) $$2k \leq d_D^{\text{in}}(U \cap W) + d_D^{\text{in}}(U \cup W) \leq d_D^{\text{in}}(U) + d_D^{\text{in}}(W) = 2k,$$

implying $d_D^{in}(U \cup W) = k$, and so $U = W = U \cup W$.

Suppose that there exists a vertex $s \in W$ with $\deg_D^{in}(s) > l(s)$. Then reversing the orientations of the arcs of any $r - s$ path in D gives again a k-arc-connected orientation (since there is no $U \subseteq V$ with $d_D^{in}(U) = k$ and $s \in U$, $r \notin U$), but decreases sum (61.41). This contradicts our minimality assumption.

So $\deg_D^{in}(v) \leq l(v)$ for each $v \in W$, with strict inequality for at least one $v \in W$ (namely for r). Now each edge of G that is spanned by no set in $\mathcal{P}$, either enters some $U \in \mathcal{P}$, or has its head in W. So the number of such edges is

(61.43) $\quad k|\mathcal{P}| + \sum_{v \in W} \deg_D^{in}(v)$, which is less than $k|\mathcal{P}| + \sum_{v \in W} l(v)$.

This contradicts the condition. ∎

This has an alternative characterization as consequence:

Corollary 61.7a. *Let $G = (V, E)$ be an undirected graph, let $k \in \mathbb{Z}_+$, and let $l, u \in \mathbb{Z}_+^E$ with $l \leq u$. Then G has a k-arc-connected orientation D with $l(v) \leq \deg_D^{in}(v) \leq u(v)$ for each $v \in V$ if and only if G has k-arc-connected orientations D' and D'' with $l(v) \leq \deg_{D'}^{in}(v)$ and $\deg_{D''}^{in}(v) \leq u(v)$ for each $v \in V$.*

Proof. This follows from the fact that the condition in Theorem 61.7 can be decomposed into a condition on l and one on u. ∎

61.4c. Orientations of graphs with lower bounds on indegrees of sets

Let $G = (V, E)$ be an undirected graph and let $l : \mathcal{P}(V) \to \mathbb{Z}_+$ be such that

(61.44) $\quad l(T) + l(U) - d(T, U) \leq l(T \cap U) + l(T \cup U)$, for all $T, U \subseteq V$ with $T \cap U \neq \emptyset$ and $T \cup U \neq V$,

where $d(T, U)$ denotes the number of edges connecting $T \setminus U$ and $U \setminus T$.

Frank [1980b] showed with submodularity theory:

Theorem 61.8. *Let $G = (V, E)$ be a graph and let $l : \mathcal{P}(V) \to \mathbb{Z}_+$ satisfy (61.44). Then G has an orientation $D = (V, A)$ with*

(61.45) $\quad d_A^{in}(U) \geq l(U)$

for each $U \subseteq V$ if and only if

(61.46) $\quad |\delta(\mathcal{P})| \geq \max\{\sum_{U \in \mathcal{P}} l(U), \sum_{U \in \mathcal{P}} l(V \setminus U)\}$

for each partition $\mathcal{P}$ of V into nonempty proper subsets, where $\delta(\mathcal{P})$ denotes the set of edges of G connecting different classes of $\mathcal{P}$.

Proof. The necessity of the condition is obvious. To prove sufficiency, let $D = (V, A)$ be an arbitrary orientation of G. Define for each nonempty proper subset U of V

(61.47) $\quad f(U) := d_A^{in}(U) - l(U)$.

One easily checks, using (61.44), that f is crossing submodular. Moreover, if $x : A \to \{0,1\}$ is such that

(61.48) $\quad x(\delta_A^{\text{in}}(U)) - x(\delta_A^{\text{out}}(U)) \leq f(U)$

for each nonempty proper subset U of V, then the digraph $D' = (V, A')$ obtained from $D = (V, A)$ by reversing the direction of the arcs a with $x_a = 1$, has indegrees as required by (61.45), since

(61.49) $\quad d_{A'}^{\text{in}}(U) = d_A^{\text{in}}(U) - x(\delta_A^{\text{in}}(U)) + x(\delta_A^{\text{out}}(U)) \geq d_A^{\text{in}}(U) - f(U) = l(U).$

Hence it suffices to show that (61.48) has an integer solution x with $\mathbf{0} \leq x \leq \mathbf{1}$.

Consider x as a transshipment. The 'excess function' $\text{excess}_x \in \mathbb{R}^V$ of x is given by:

(61.50) $\quad \text{excess}_v := x(\delta_A^{\text{in}}(v)) - x(\delta_A^{\text{out}}(v))$

for $v \in V$. Then (61.48) is equivalent to

(61.51) $\quad y(U) \leq f(U).$

Now y is the excess function of some $x \in \{0,1\}^A$ if and only if x is an integer y-transshipment with $\mathbf{0} \leq x \leq \mathbf{1}$. So, by Corollary 11.2f, y is the excess function of some $x \in \{0,1\}^A$ if and only if y is integer, $y(V) = 0$, and

(61.52) $\quad y(U) \leq d_A^{\text{in}}(U)$

for each $U \subseteq V$. Since $l(U) \geq 0$, (61.52) is implied by (61.51).

So it suffices to show that (61.51) has an integer solution y with $y(V) = 0$. By Theorem 49.10, y exists if and only if

(61.53) $\quad \displaystyle\sum_{U \in \mathcal{P}} f(U) \geq 0$

for each partition or copartition $\mathcal{P}$ of V, where each set in $\mathcal{P}$ is a nonempty proper subset of V. (A *copartition* of V is a collection of sets whose complements form a partition of V.) This is equivalent to the condition given in the present theorem. ∎

61.4d. Further notes

Frank [1980b] observed that Edmonds' disjoint arborescences theorem implies:

Corollary 61.8a. *Let $G = (V, E)$ be an undirected graph and $r \in V$. Then G has an orientation such that each nonempty subset U of $V \setminus \{r\}$ is entered by at least k arcs if and only if G contains k edge-disjoint spanning trees.*

Proof. Necessity follows from the fact that if the orientation $D = (V, A)$ as required exists, then by Edmonds' disjoint arborescences theorem (Corollary 53.1b), D has k disjoint r-arborescences. Hence G has k edge-disjoint spanning trees.

Sufficiency follows from the fact that we can orient each spanning tree in G so as to become an r-arborescence. Orienting the remaining edges arbitrarily, we obtain an orientation as required. ∎

Frank [1993c] gave a direct proof of the existence of this orientation from the conditions given in the Tutte-Nash-Williams disjoint trees theorem (Corollary

51.1a), yielding (with Edmonds' disjoint arborescences theorem) a proof of the Tutte-Nash-Williams disjoint trees theorem.

Frank [1982a] showed that each k-arc-connected orientation of an undirected graph can be obtained from any other by reversing iteratively directed paths and circuits, without destroying k-arc-connectivity. This can be derived from a result of L. Lovász that two k-arc-connected orientations are adjacent on the polytope determined by (61.36) if and only if they differ on a directed circuit or on a collection of vertex-disjoint directed paths. Frank [1982b] showed that a minimum-cost k-arc-connected orientation can be found in strongly polynomial time, by reduction to the Edmonds-Giles model. Accelerations were given by Gabow [1993a,1993b,1994, 1995c].

Frank, Jordán, and Szigeti [1999,2001] and Frank and Király [1999,2002] studied graph orientations that satisfy parity and connectivity conditions. Orientations preserving prescribed shortest paths are considered by Hassin and Megiddo [1989].

Chvátal and Thomassen [1978] showed that each 2-edge-connected graph of radius r has a strongly connected orientation of radius at most $r^2 + r$. This was extended to mixed graphs by Chung, Garey, and Tarjan [1985].

For surveys on applying submodularity to orientation problems, see Frank [1993a,1996b].

Chapter 62

Network synthesis

The network synthesis problem asks for a graph having prescribed connectivity properties, with a minimum number of edges. If the edges have costs, a minimum total cost is required.

The problem can be seen as the special case of the connectivity augmentation problem where the input graph is edgeless. Connectivity augmentation in general will be discussed in Chapter 63.

62.1. Minimal k-(edge-)connected graphs

We first consider the easy problem of finding a graph of given connectivity, with a minimal number of edges. First, vertex-connectivity:

Theorem 62.1. *Let k and n be positive integers with $n \geq 2$. The minimum number of edges of a k-vertex-connected graph with n vertices is $n-1$ if $k = 1$, $\lceil \frac{1}{2} kn \rceil$ if $1 < k < n$, and $\frac{1}{2} n(n-1)$ otherwise.*

Proof. Since any k-vertex-connected graph contains a spanning tree, has minimum degree at least k if $k < n$, and is a complete graph if $k \geq n$, the values given are lower bounds. Moreover, if $k = 1$ or $k \geq n$, the bound is tight. So we can assume $1 < k < n$, and it suffices to show that there exists a k-vertex-connected graph $G = (V, E)$ with $|V| = n$ and $|E| = \lceil \frac{1}{2} kn \rceil$.

Let $V := \{1, \ldots, n\}$ and let C be the circuit on V with edge set $\{\{i, i+1\} \mid i \in V\}$, taking addition mod n. Let G be the graph on V with edges all pairs of vertices at distance at most $\frac{1}{2} k$ in C.

First assume that k is even. Then G has $\frac{1}{2} kn$ edges. We show that G is k-vertex-connected. Suppose to the contrary that G has a vertex-cut U of size less than k. There are at least two components K of $C[U]$ such that the two neighbours of K in C belong to different components of $G - U$ (as $G - U$ is disconnected). In particular, the two neighbours have distance more than $\frac{1}{2} k$ in C, and so these components each have size at least $\frac{1}{2} k$. This contradicts the fact that $|U| < k$.

Next, if k is odd, we add to G $\lceil \frac{1}{2} n \rceil$ edges $\{i, j\}$, where i and j have distance $\lfloor \frac{1}{2} n \rfloor$ in C, and such that these edges cover all vertices in V. So G has $\lceil \frac{1}{2} kn \rceil$ edges. We show that G is k-vertex-connected.

Suppose that G has a vertex-cut U of size less than k. By the above, $C[U]$ consists of two components of size $l := \frac{1}{2}(k-1)$ each. We can assume that $U = [1,l] \cup [s+1, s+l]$ for some s with $l < s$ and $s+l < n$. Now n is adjacent to no vertex in $[l+1, s]$, while n is adjacent to at least one of $\lfloor\frac{1}{2}n\rfloor$ and $\lceil\frac{1}{2}n\rceil$. So $\lfloor\frac{1}{2}n\rfloor < l+1$ or $\lceil\frac{1}{2}n\rceil > s$, implying $n > 2s$ (as $k < n$). By symmetry of the two components we similarly have $n > 2(n-s)$, that is $n < 2s$, a contradiction. ∎

For edge-connectivity the answer is almost the same:

Theorem 62.2. *Let k and n be positive integers with $n \geq 2$. The minimum number of edges of a k-edge-connected graph with n vertices is $n-1$ if $k = 1$, and $\lceil\frac{1}{2}kn\rceil$ otherwise. If $k \leq n-1$ the minimum is attained by a simple graph.*

Proof. Again, the values are lower bounds, as a k-edge-connected graph contains a spanning tree and has each degree at least k. Clearly the lower bound can be attained if $k = 1$, so assume $k \geq 2$. Let C be a graph on $V := \{1, \ldots, n\}$ with edges all pairs $\{i, i+1\}$ for $i \in V$ (with addition mod n). Let G be the graph obtained from C by replacing each edge by $\lfloor\frac{1}{2}k\rfloor$ parallel edges.

If k is even, then G is k-edge-connected as required. If k is odd, add $\lceil\frac{1}{2}n\rceil$ edges $\{i,j\}$ to G, where i and j have distance $\lfloor\frac{1}{2}n\rfloor$ in C, and such that these edges cover all vertices in V. So G has $\lceil\frac{1}{2}kn\rceil$ edges. We show that G is k-edge-connected. Suppose that $d_G(U) < k$ for some nonempty proper subset U of V. By symmetry, we can assume that $|U| \geq \frac{1}{2}n$. Now $C[U]$ is connected (as otherwise $d_G[U] \geq 4\lfloor\frac{1}{2}k\rfloor \geq k$, since $k > 1$). So we can assume that $U = [1,s]$, with $s \geq \lceil\frac{1}{2}n\rceil$. However, $n \in V \setminus U$ is adjacent to at least one of $\lfloor\frac{1}{2}n\rfloor$ and $\lceil\frac{1}{2}n\rceil$. As both of these vertices belong to U, we have $d_G(U) \geq 2\lfloor\frac{1}{2}k\rfloor + 1 = k$, a contradiction.

Finally, if $k \leq n-1$, the minimum is attained by a simple graph. Indeed, by Theorem 62.1, there is a k-vertex-connected graph $G = (V, E)$ with n vertices and $\lceil\frac{1}{2}kn\rceil$ edges. Necessarily, G is simple. We show that G is k-edge-connected. Suppose that there is a nonempty $U \subset V$ with $d_G(U) < k$. Then $|U||V \setminus U| \geq n-1 \geq k$, and hence there exist $s \in U$ and $t \in V \setminus U$ that are not adjacent. Hence G has k internally vertex-disjoint $s-t$ paths, and therefore k edge-disjoint $s-t$ paths. This contradicts the fact that $d_G(U) < k$. ∎

The directed case is even simpler. For vertex-connectivity one has:

Theorem 62.3. *Let k and n be positive integers with $n \geq 2$. Then the minimum number of arcs of a k-vertex-connected directed graph with n vertices is kn if $k \leq n-1$, and $n(n-1)$ otherwise.*

Proof. Since each vertex should have at least $\min\{k, n-1\}$ outneighbours, the values are lower bounds. Trivially it is attained if $k \geq n$.

If $k \leq n-1$, let D be the directed graph on $V := \{1, \ldots, n\}$ with arcs all pairs $(i, i+l)$ with $i \in V$ and $1 \leq l \leq k$, taking addition mod n. Then D has kn arcs. To see that D is k-vertex-connected, let U be a vertex-cut. Choose $i, j \in V \setminus U$ with j not reachable from i in $D - U$. We may assume that $1 \leq i < j \leq n$ and that $j - i$ is as small as possible. Then $j - i > k$ and $i+1, \ldots, j-1$ belong to U. So $|U| \geq k$. ∎

Finally, for arc-connectivity (Fulkerson and Shapley [1971]):

Theorem 62.4. *Let k and n be positive integers with $n \geq 2$. Then the minimum number of arcs of a k-arc-connected directed graph with n vertices is kn. If $k \leq n-1$, the minimum is attained by a simple directed graph.*

Proof. Since each vertex should be left by at least k arcs, kn is a lower bound. It is attained by the directed graph obtained from a directed circuit on n vertices, by replacing any arc by k parallel arcs.

If $k \leq n-1$, the minimum is attained by a simple directed graph. Indeed, by Theorem 62.3, there is a k-vertex-connected directed graph $D = (V, A)$ with n vertices and kn arcs. Necessarily, D is simple. We show that D is k-arc-connected. Suppose that there is a nonempty $U \subset V$ with $d_D^{\text{out}}(U) < k$. Then $|U||V \setminus U| \geq n - 1 \geq k$, and hence there exist $s \in U$ and $t \in V \setminus U$ with $(s, t) \notin A$. Hence D has k internally vertex-disjoint $s - t$ paths, and therefore k arc-disjoint $s - t$ paths. This contradicts the fact that $d_D^{\text{out}}(U) < k$. ∎

Notes. Edmonds [1964] showed that for each simple graph with all degrees at least k, there exists a k-edge-connected simple graph with the same degree-sequence.

62.2. The network synthesis problem

Let V be a finite set and let $r : V \times V \to \mathbb{R}_+$. A *realization* of r is a pair of a directed graph $D = (V, A)$ and a capacity function $c : A \to \mathbb{R}_+$ such that for all $s, t \in V$, each $s - t$ cut in G has capacity at least $r(s, t)$. The pair D, c is called an *exact realization* if for all $s, t \in V$ with $s \neq t$, the minimum capacity of an $s - t$ cut in D as equal to $r(s, t)$.

Obviously, any function r has a realization. We say that r is *exactly realizable* if it has an exact realization. The *network synthesis problem* is the problem to find an exact or cheapest realization for a given r (or to decide that none exist).

The following theorem due to Gomory and Hu [1961] characterizes the exactly realizable *symmetric*[17] functions. It also shows that if $r : V \times V \to \mathbb{R}$ is exactly realizable and symmetric, then r has an *undirected* exact realization

---
[17] A function $r : V \times V \to \mathbb{R}$ is called *symmetric* if $r(u, v) = r(v, u)$ for all $u, v \in V$.

(more precisely, an exact realization D, c where for each arc $a = (u, v)$ of D, also (v, u) is an arc, with $c(u, v) = c(v, u)$).

Theorem 62.5. *A symmetric function $r : V \times V \to \mathbb{R}_+$ is exactly realizable if and only if*

(62.1) $\qquad r(u, w) \geq \min\{r(u, v), r(v, w)\}$

for all distinct $u, v, w \in V$. If r is exactly realizable, there is a tree that gives an exact realization of r.

Proof. Necessity being easy, we show sufficiency. Let $T = (V, E)$ be a tree on V maximizing

(62.2) $\qquad \sum_{uv \in E} r(u, v).$

Taking $c(uv) := r(u, v)$ for each edge $uv \in E$ gives an exact realization of r. To see this, note that for all s, t, the minimum capacity of an $s - t$ cut is equal to $\min_{uv \in EP} r(u, v)$, where P is the $s - t$ path in T. By (62.1) we know that $r(s, t)$ is not smaller than this minimum. To show equality, suppose to the contrary that $r(u, v) < r(s, t)$ for some $uv \in P$. Then replacing T by $(T - uv) \cup st$ gives a tree with larger sum (62.2). ∎

Notes. Obviously, condition (62.1) remains necessary for exact realizability of nonsymmetric functions. Resh [1965] claimed that (62.1) also remains sufficient, but a counterexample is given by the function $r : V \times V \to \mathbb{R}_+$ with $V = \{1, 2, 3, 4\}$, and $r(1, 2) = r(1, 3) = r(1, 4) = r(2, 4) = r(3, 4) = 1$, and $r(s, t) = 0$ for all other s, t (cf. Mayeda [1962]).

62.3. Minimum-capacity network design

Theorem 62.5 yields a tree as an exact realization of a given function $r : V \times V \to \mathbb{R}_+$. A tree is a most economical realization in the sense of having a minimum number of edges with nonzero capacity. It generally gives no exact realization for which the sum of the capacities is minimum. Such an exact realization has been characterized by Chien [1960] (extending Mayeda [1960]), while Gomory and Hu [1961] showed that if r is integer, there is a half-integer optimum exact realization.

As a preparation, we first show the following lemma of Gomory and Hu [1961] ($\lambda_G(s, t)$ denotes the maximum number of edge-disjoint $s - t$ paths in G):

Lemma 62.6α. *Let $r : V \times V \to \mathbb{R}_+$ be symmetric and let T be a spanning tree on V maximizing $r(T)$. Then any graph $G = (V, E)$ satisfies*

(62.3) $\qquad \lambda_G(s, t) \geq r(s, t)$

for all $s,t \in V$ if and only if (62.3) is satisfied for each edge st of T.

Proof. Necessity being trivial, we show sufficiency. Let $s,t \in V$ and let P be the $s-t$ path in T. By the maximality of $r(T)$, we know that $r(s,t) \leq r(e)$ for each edge e on P. Hence

(62.4) $\qquad \lambda_G(s,t) \geq \min_{e=uv \in EP} \lambda_G(u,v) \geq \min_{e \in EP} r(e) \geq r(s,t)$,

as required. ∎

We also use the following lemma:

Lemma 62.6β. *If $r: V \times V \to \mathbb{R}_+$ is symmetric and exactly realizable, then there exists a spanning tree T on V that maximizes $r(T)$ over all spanning trees, and that moreover is a Hamiltonian path.*

Proof. Let T maximize $r(T)$. Choose T and k such that T contains a path $v_1, \ldots, v_k$, with k as large as possible. Choose T, k moreover such that the vector $(\deg_T(v_1), \ldots, \deg_T(v_k))$ is lexicographically minimal. If T is not a path, there is a j with $1 < j < k$ and $\deg_T(v_j) \geq 3$. Let $v_j u$ be an edge of T incident with v_j, with $u \neq v_{j-1}, v_{j+1}$. If $r(v_{j+1}, u) \geq r(v_j, u)$, we can replace edge $v_j u$ of T by $v_{j+1} u$, contradicting the lexicographic minimality. So $r(v_j, u) > r(v_{j+1}, u)$, and so $r(v_j, v_{j+1}) \leq r(v_{j+1}, u)$, since $r(v_{j+1}, u) \geq \min\{r(v_j, v_{j+1}), r(v_j, u)\}$ by (62.1). Hence replacing edge $v_j v_{j+1}$ of T by $v_{j+1} u$ would give a tree with a longer path, contradicting our assumption. ∎

Now we can formulate and prove the theorem. For any $r: V \times V \to \mathbb{R}$ and $u \in V$ define

(62.5) $\qquad R(u) := \max_{v \neq u} r(u,v)$.

Theorem 62.6. *Let $r : V \times V \to \mathbb{R}_+$ be symmetric and exactly realizable. Then the minimum value of $\sum_{e \in E} c(e)$ where $G = (V, E)$ and c form an (undirected) exact realization of r, is equal to*

(62.6) $\qquad \frac{1}{2} \sum_{u \in V} R(u)$.

Moreover, if r is integer, the minimum is attained by a half-integer exact realization c.

Proof. We may assume that r is integer. (62.6) indeed is a lower bound, since for each exact realization $G = (V, E), c$ of r one has

(62.7) $\qquad \sum_{e \in E} c(e) = \frac{1}{2} \sum_{u \in V} \sum_{e \in \delta(u)} c(e) \geq \frac{1}{2} \sum_{u \in V} R(u)$.

To see that the lower bound is attained by a half-integer exact realization, let T be a spanning tree on V maximizing $r(T)$. By Lemma 62.6β, we can assume that T is a path $v_1, \ldots, v_n$.

Let $k := \max_{u,v} r(u,v)$. For $i = 0, \ldots, k$, let E_i be the set of edges e of T with $r(e) \leq i$, and for each nonsingleton component P of $T - E_i$, make a circuit consisting of edges parallel to P and one edge connecting the end vertices of P. Let $G = (V, E)$ arise by taking the edge-disjoint union of these circuits. Let $c(e) := \frac{1}{2}$ for each $e \in E$. Then

(62.8) $\qquad \lambda_G(v_j, v_i) = 2r(v_j, v_i)$

for all i, j with $1 \leq j < i \leq n$.

Indeed, in proving $\geq$, we can assume that $j = i - 1$ (by Lemma 62.6α). As v_{i-1}, v_i are contained in $r(v_{i-1}, v_i)$ edge-disjoint circuits, we have $\lambda_G(v_{i-1}, v_i) \geq 2r(v_{i-1}, v_i)$.

Conversely, the inequality $\leq$ in (62.8) follows from

(62.9) $\qquad \lambda_G(v_j, v_i) \leq \min_{j < h \leq i} 2r(v_{h-1}, v_h) \leq 2r(v_j, v_i)$.

The first inequality here follows from the fact that for each h with $j < h \leq i$, the number of edges connecting $\{v_1, \ldots, v_{h-1}\}$ and $\{v_h, \ldots, v_n\}$ is equal to $2r(v_{h-1}, v_h)$. The second inequality follows from (62.1). ∎

Notes. Note that also any nonexact realization has size at least (62.6), and therefore, the theorem also characterizes the minimum size of any realization.

Wing and Chien [1961] observed that a minimum-capacity realization can be found by linear programming, and Gomory and Hu [1962,1964] showed that also the weighted case can be solved by linear programming. Indeed, the polyhedron P of all realizations of a given function $r : V \times V \to \mathbb{Q}_+$ can be described as follows. Let E is the collection of all unordered pairs of elements of V. Then P is determined by:

(62.10) $\qquad x_e \geq 0 \qquad$ for all $e \in E$,
$\qquad\qquad x(\delta(U)) \geq R(U) \qquad$ for all nonempty $U \subset V$,

where $R(U) := \max_{u \in U, v \in V \setminus U} r(u, v)$.

This formulation was given by Gomory and Hu [1962] and applied to finding a minimum-cost realization with linear programming, by a row-generating implementation of the simplex method (thus avoiding listing the exponential number of constraints). Bland, Goldfarb, and Todd [1981] observed that description (62.10) implies polynomial-time solvability with the ellipsoid method, since the constraints (62.10) can be tested in polynomial time.

A direct, polynomial-size linear programming formulation was given by Gomory and Hu [1964], by extending the number of variables. Indeed, P consists of those $x \in \mathbb{R}_+^E$ such that for all distinct $s, t \in V$, there exists an $s - t$ flow $f_{s,t} : E \to \mathbb{R}_+^E$ with $f \leq x$ and of value $r(s, t)$.

The latter description implies that a minimum-weight realization can be determined in polynomial time, by solving an explicit linear programming problem — in fact, in strongly polynomial time, with the method of Tardos [1986].

Note that the *exact* realizations do not form a convex set; for instance, if $V = \{u, v, w\}$ and $r(s,t) = 1$ for all $s, t \in V$, then $x(uv) = x(vw) = x(uw) = \frac{2}{3}$ is a convex combination of exact realizations, but is not itself an exact realization.

62.4. Integer realizations and r-edge-connected graphs

In Section 62.3, the fractional version of the minimum-capacity network design problem was discussed. We now consider the case where all capacities are required to be integer. It relates to: given $r : V \times V \to \mathbb{Z}_+$, find an r-edge-connected undirected graph $G = (V, E)$ with a minimum number of edges. Here a graph $G = (V, E)$ is called r-*edge-connected* if $\lambda_G(s,t) \geq r(s,t)$ for all $s, t \in V$ with $s \neq t$.

Eswaran and Tarjan [1976] observed that the weighted version of the integer realization problem is NP-complete, as finding a Hamiltonian circuit in an undirected graph can be reduced to it. (So even if $r = 2$ and all weights belong to $\{0, 1\}$, it is NP-complete.)

Chou and Frank [1970] gave a polynomial-time algorithm for finding a minimum-size integer realization, implying the following characterization of the minimum number $\gamma(r)$ of edges of an r-edge-connected graph[18].

To this end we can assume that r is symmetric and exactly realizable, that is, satisfies (62.1) (since resetting $r(s,t)$ to the maximum of $\min_{e=uv \in EP} r(u, v)$ over all $s - t$ paths P, does not change the problem).

Again, define for each $u \in V$,

(62.11) $R(u) := \max_{v \neq u} r(u, v)$.

Theorem 62.7. *Let $r : V \times V \to \mathbb{Z}_+$ be symmetric and satisfy (62.1).*
(i) *If $R(u) = 1$ for some $u \in V$, then $\gamma(r) = \gamma(r') + 1$, where r' is the restriction of r to $(V \setminus \{u\}) \times (V \setminus \{u\})$.*
(ii) *If $R(u) \neq 1$ for all $u \in V$, then*

(62.12) $\gamma(r) = \lceil \frac{1}{2} \sum_{u \in V} R(u) \rceil$.

Proof. We first show (i). The inequality $\gamma(r) \leq \gamma(r') + 1$, is easy, since an r-edge-connected graph can be obtained from an r'-edge-connected graph by adding one edge connecting u with some $v \neq u$ with $r(u, v) = 1$.

To see the reverse inequality, let $G = (V, E)$ be an r-edge-connected graph with $|E| = \gamma(r)$. As $R(u) = 1$, we have $\deg_G(u) \geq 1$. Let ut be an edge incident with u. Let H be the graph obtained from G by contracting ut. Then H is r'-edge-connected and has $|E| - 1$ edges, showing $\gamma(r') \leq |E| - 1 = \gamma(r) - 1$.

---

[18] The construction of Chou and Frank [1970] is lacunary, and does not apply, e.g., to the case where $r(u, v) = 3$ for all u, v and $|V|$ is odd.

We next show (ii). Trivially, for any r-edge-connected graph $G = (V, E)$:

(62.13) $\quad |E| = \frac{1}{2} \sum_{u \in V} \deg_G(u) \geq \frac{1}{2} \sum_{u \in V} R(u).$

This proves $\geq$ in (62.12). To prove $\leq$, order the vertices as $v_1, \ldots, v_n$ such that $R(v_1) \geq R(v_2) \geq \cdots \geq R(v_n)$. Note that $R(v_1) = R(v_2)$.

Let $k := \lfloor \frac{1}{2} R(v_1) \rfloor$. Let W be the set of vertices v with $R(v)$ odd. Let M be a set of $\lceil \frac{1}{2}|W| \rceil$ edges covering W such that if $v_1, v_2 \in W$, then $v_1 v_2 \in M$.

For $i = 1, \ldots, k$, let C_i be a circuit on $\{v \in V \mid R(v) \geq 2i\}$. So C_1 is a Hamiltonian circuit. We choose C_1 in such a way that the components of $C_1 - v_1 - v_2$ span no edge in M. Let H be the (edge-disjoint) union of M and C_1. Then for any $U \subseteq V$:

(62.14) $\quad$ if $d_H(U) = 2$ and $U \cap \{v_1, v_2\} = \emptyset$, then $U \cap W = \emptyset$.

Indeed, if $d_H(U) = 2$, then U induces a path on C_1. As $U \cap \{v_1, v_2\} = \emptyset$, U is contained in a component of $C_1 - v_1 - v_2$. Hence each edge in M incident with U belongs to $d_H(U)$. As $d_H(U) = 2$, it follows that no edge in M is incident with U. So $U \cap W = \emptyset$, proving (62.14).

Let G be the (edge-disjoint) union of $M, C_1, \ldots, C_k$. Note that the number of edges of G is equal to

(62.15) $\quad |M| + \sum_{i=1}^{k} |EC_i| = |M| + \sum_{u \in V} \lfloor \frac{1}{2} R(u) \rfloor = \lceil \frac{1}{2}|W| \rceil + \sum_{u \in V} \lfloor \frac{1}{2} R(u) \rfloor$
$\quad = \lceil \frac{1}{2} \sum_{u \in V} R(u) \rceil.$

We finally show that G is r-connected, for which it suffices to show that for $i = 2, \ldots, n$:

(62.16) $\quad \lambda_G(v_{i-1}, v_i) \geq R(v_i).$

To see that this is sufficient, note that for $h < j$ one has

(62.17) $\quad \lambda_G(v_h, v_j) \geq \min_{h < i \leq j} \lambda_G(v_{i-1}, v_i) \geq \min_{h < i \leq j} R(v_i)$
$\quad = R(v_j) \geq r(v_h, v_j).$

To prove (62.16), choose the smallest $i \geq 2$ for which it is not true. Then G has a cut $\delta(U)$ with $v_i \in U$, $v_{i-1} \notin U$, and $d_G(U) < R(v_i)$. By the minimality of i, $\delta(U)$ separates no pair among $v_1, \ldots, v_{i-1}$, and hence $v_1, \ldots, v_{i-1} \notin U$. Now, setting $l := \lfloor \frac{1}{2} R(v_i) \rfloor$:

(62.18) $\quad 2l + 1 \geq R(v_i) > d_G(U) \geq d_H(U) + \sum_{j=2}^{l} d_{C_j}(U) \geq 2l$

(as C_j covers v_{i-1} and v_i for $j = 1, \ldots, l$). Hence $d_H(U) = 2$ and $R(v_i)$ is odd. So $U \cap W \neq \emptyset$. Hence, by (62.14), $i = 2$. Then $v_1 v_2 \in M$, and so $d_H(U) \geq 3$, a contradiction. ∎

Section 62.4. Integer realizations and r-edge-connected graphs

Notes. In fact, by choosing M in this proof in such a way that $\deg_M(u) = 1$ if $u \in W \setminus \{v_1\}$, and $\deg_M(u) = 0$ if $u \in (V \setminus W) \setminus \{v_1\}$, we obtain a graph G with $\deg_G(u) = R(u)$ for each $u \neq v_1$. Hence $\lambda_G(u,v) = \min\{R(u), R(v)\}$ for all distinct $u, v \in V$.

The construction can be extended to obtain a strongly polynomial-time algorithm that, for given integer function r, finds a minimum-capacity integer realization c (Sridhar and Chandrasekaran [1990,1992]).

(Chou and Frank [1970] claim to give an algorithm to find a minimum-size *exact* realization, but their construction fails when taking $r(1,2) := r(3,4) := r(4,5) := 5$, $r(2,3) := r(5,6) := 3$, and $r(i,j) := \min_{i < h \leq j} r(h-1, h)$ for $i < j$. The construction gives 15 edges, while there is an exact realization with 14 edges only.)

Frank and Chou [1970] announced a polynomial-time algorithm for the problem: given a symmetric $r : V \times V \to \mathbb{Z}_+$, find a *simple* r-edge-connected graph $G = (V, E)$ (if any) with $|E|$ minimal.

Wang and Kleitman [1973] characterized the degree-sequences of k-vertex-connected simple undirected graphs.

Chapter 63

Connectivity augmentation

This last chapter of Part V is devoted to the connectivity augmentation problem: given a graph, find the minimum number of edges to be added to make it k-connected. There is an undirected and a directed variant, and a vertex-connectivity and an edge/arc-connectivity variant. Thus we will come across:
- making a directed graph k-arc-connected (Section 63.1),
- making an undirected graph k-edge-connected (Section 63.3),
- making a directed graph k-vertex-connected (Section 63.5),
- making an undirected graph k-vertex-connected (Section 63.6).

For the first three problems, min-max relations and polynomial-time algorithms have been found. The core is formed by fundamental theorems of Frank and Jordán. As for the fourth problem, only for fixed k the polynomial-time solvability has been proved. The complexity for general k is open.

Two special cases of connectivity augmentation have been considered before: making a digraph 1-arc-connected — that is, strongly connected (Chapter 57), and making an edge- or arcless (di)graph k-vertex- or edge/arc-connected — the network synthesis problem (Chapter 62).

63.1. Making a directed graph k-arc-connected

Let (V, A) and (V, B) be directed graphs. The set B is called a *k-arc-connector* for D if the directed graph $(V, A \cup B)$ is k-arc-connected (where in $A \cup B$ arcs are taken parallel if they occur both in A and in B). So 1-arc-connectors are precisely the strong connectors, which we discussed in Chapter 57.

Frank [1990a,1992a] characterized the minimum size of a k-arc-connector for a directed graph, with the help of the following result of Mader [1982] (we follow the proof of Frank [1992a]).

Lemma 63.1α. *Let $D = (V, A)$ be a directed graph, let $k \in \mathbb{Z}_+$, and let $x, y : V \to \mathbb{Z}_+$. Then D has a k-arc-connector B with $\deg_B^{\text{in}}(v) = x_v$ and $\deg_B^{\text{out}}(v) = y_v$ for each $v \in V$ if and only if $x(V) = y(V)$ and*

(63.1) $\qquad x(U) \geq k - d_A^{\text{in}}(U)$ *and* $y(U) \geq k - d_A^{\text{out}}(U)$

Section 63.1. Making a directed graph k-arc-connected

for each nonempty proper subset U of V.

Proof. Necessity is easy, since for each nonempty $U \subset V$,

(63.2) $\quad k \leq d^{in}_{A \cup B}(U) = d^{in}_B(U) + d^{in}_A(U) \leq x(U) + d^{in}_A(U),$

and similarly for y.

To see sufficiency, choose a counterexample with $x(V)$ minimal. Trivially, $x(V) \geq 1$. Let $\mathcal{X}$ be the collection of inclusionwise maximal proper subsets U of V satisfying $x(U) + d^{in}(U) = k$, and let $\mathcal{Y}$ be the collection of inclusionwise maximal proper subsets U of V satisfying $y(U) + d^{out}(U) = k$. (We set d^{in} and d^{out} for d^{in}_A and d^{out}_A.)

Let $R := \{v \in V \mid x_v \geq 1\}$ and $S := \{v \in V \mid y_v \geq 1\}$. Then

(63.3) $\quad$ for all $r \in R$ and $s \in S$, there exists a $U \in \mathcal{X} \cup \mathcal{Y}$ with $r, s \in U$.

Otherwise, we could augment D by a new arc (s, r) and decrease both x_r and y_s by 1. Then (63.1) is maintained, and we obtain a smaller counterexample, contradicting our assumption. This shows (63.3).

Now note that for each $U \in \mathcal{X}$:

(63.4) $\quad y(V \setminus U) \geq k - d^{out}(V \setminus U) = k - d^{in}(U) = x(U).$

This implies, for each $U \in \mathcal{X}$:

(63.5) $\quad$ if $S \subseteq U$, then $U \cap R = \emptyset$; if $R \subseteq U$, then $U \cap S = \emptyset$.

Indeed, if $S \subseteq U$, then $y(V \setminus U) = 0$, implying (with (63.4)) that $x(U) = 0$, that is, $U \cap R = \emptyset$. Similarly, if $R \subseteq U$, then $x(U) = x(V)$, implying (with (63.4)) that $y(V \setminus U) = y(V)$, that is, $U \cap S = \emptyset$. This proves (63.5).

Now choose $r \in R$, $s \in S$, and let $U \in \mathcal{X} \cup \mathcal{Y}$ with $r, s \in U$. By symmetry, we may assume that $U \in \mathcal{X}$. By (63.5), $S \not\subseteq U$. Choose $t \in S \setminus U$. Let $T \in \mathcal{X} \cup \mathcal{Y}$ contain r and t.

First assume that $T \in \mathcal{X}$. Then $T \cup U = V$, by the maximality of T and U and the submodularity of the set function $x(W) + d^{in}(W)$. This implies (using (63.4)):

(63.6) $\quad y(V) \geq y(V \setminus U) + y(V \setminus T) \geq x(U) + x(T) = x(T \cup U) + x(T \cap U)$
$\quad\quad > x(V) = y(V)$

(since $V \setminus U$ and $V \setminus T$ are disjoint, and since $r \in T \cap U$), a contradiction.

So $T \in \mathcal{Y}$. But then

(63.7) $\quad 2k = x(T) + d^{in}(T) + y(U) + d^{out}(U)$
$\quad\quad \geq x(T \setminus U) + d^{in}(T \setminus U) + y(U \setminus T) + d^{out}(U \setminus T) + x(T \cap U)$
$\quad\quad + y(T \cap U) \geq 2k,$

implying equality throughout. Hence $x(T \cap U) = 0$, contradicting the fact that $r \in T \cap U$. ∎

From this, the min-max relation for minimum-size k-arc-connectors of Frank [1990a,1992a] (generalizing Corollary 57.2a) easily follows:

Theorem 63.1. *Let $D = (V, A)$ be a directed graph and let $k, \gamma \in \mathbb{Z}_+$. Then D has a k-arc-connector of size at most γ if and only if*

(63.8) $$\gamma \geq \sum_{X \in \mathcal{P}} (k - d^{\mathrm{in}}(X)) \text{ and } \gamma \geq \sum_{X \in \mathcal{P}} (k - d^{\mathrm{out}}(X))$$

for each collection $\mathcal{P}$ of disjoint nonempty proper subsets of V.

Proof. Necessity follows since for each nonempty subset X of V, at least $k - d^{\mathrm{in}}(X)$ arcs entering X must be in any k-arc-connector, and at least $k - d^{\mathrm{out}}(X)$ arcs leaving X must be in any k-arc-connector. As any new arc can enter at most one set in $\mathcal{P}$, we have (63.8).

To see sufficiency, choose $x : V \to \mathbb{Z}_+$ satisfying $x(U) \geq k - d^{\mathrm{in}}(U)$ for each nonempty $U \subset V$, with $x(V)$ as small as possible.

We show $x(V) \leq \gamma$. Let $\mathcal{P}$ be the collection of inclusionwise maximal proper subsets U of V satisfying $x(U) = k - d^{\mathrm{in}}(U)$. Any two distinct sets $T, U \in \mathcal{P}$ satisfy $T \cup U \neq V$, since otherwise $V \setminus T$ and $V \setminus U$ are disjoint, and we obtain the contradiction

(63.9) $$\gamma \geq k - d^{\mathrm{out}}(V \setminus T) + k - d^{\mathrm{out}}(V \setminus U) = 2k - d^{\mathrm{in}}(T) - d^{\mathrm{in}}(U)$$
$$= x(T) + x(U) \geq x(T \cup U) = x(V) > \gamma.$$

Moreover, any two distinct $T, U \in \mathcal{P}$ are disjoint, since otherwise we obtain the contradiction

(63.10) $$x(T) + x(U) = 2k - d^{\mathrm{in}}(T) - d^{\mathrm{in}}(U) \leq 2k - d^{\mathrm{in}}(T \cap U) - d^{\mathrm{in}}(T \cup U)$$
$$< x(T \cap U) + x(T \cup U) = x(T) + x(U),$$

by the maximality of T.

Now each $v \in V$ with $x_v \geq 1$ is contained in some $U \in \mathcal{P}$, as otherwise we could decrease x_v. This gives

(63.11) $$x(V) = \sum_{U \in \mathcal{P}} x(U) = \sum_{U \in \mathcal{P}} (k - d^{\mathrm{in}}(U)) \leq \gamma.$$

Hence $x(V) \leq \gamma$. Similarly, there exists a $y : V \to \mathbb{Z}_+$ satisfying $y(U) \geq k - d^{\mathrm{out}}(U)$ for each nonempty proper subset U of V and $y(V) \leq \gamma$. We can assume that $x(V) = y(V) = \gamma$. So we can apply Lemma 63.1α, which gives the theorem. ∎

The proof yields a polynomial-time algorithm, as the proof reduces to a polynomial-time number of tests if a given $x : V \to \mathbb{Z}_+$ satisfies

(63.12) $x(U) \geq k - d^{\mathrm{in}}(U)$ for each nonempty $U \subset V$.

(Similarly for y.) This can be done by maximum flow calculations: add a new vertex s and for each $v \in V$, add x_v (parallel) arcs from s to v and k parallel arcs from v to s. Then (63.12) is satisfied if and only if in the extended graph there exist k arc-disjoint $u - v$ paths, for all $u, v \in V$.

Thus (Frank [1990a,1992a]):

Theorem 63.2. *Given a directed graph $D = (V, A)$ and $k \in \mathbb{Z}_+$, a minimum-size k-arc-connector can be found in time bounded by a polynomial in the size of D and in k.*

Proof. See above. ∎

63.1a. k-arc-connectors with bounds on degrees

Frank [1990a,1992a] derived similarly characterizations of the existence of k-arc-connectors of given size and satisfying given lower and upper bounds on the in- and outdegrees:

Theorem 63.3. *Let $D = (V, A)$ be an undirected graph, let $k, \gamma \in \mathbb{Z}_+$, and let $l^{\text{in}}, l^{\text{out}}, u^{\text{in}}, u^{\text{out}} \in \mathbb{Z}_+^V$ with $l^{\text{in}} \leq u^{\text{in}}$ and $l^{\text{out}} \leq u^{\text{out}}$. Then D has a k-arc-connector B of size at most γ satisfying $l^{\text{in}}(v) \leq \deg_B^{\text{in}}(v) \leq u^{\text{in}}(v)$ and $l^{\text{out}}(v) \leq \deg_B^{\text{out}}(v) \leq u^{\text{out}}(v)$ for each $v \in V$ if and only if $\gamma \leq u^{\text{in}}(V)$, $\gamma \leq u^{\text{out}}(V)$,*

(63.13) $\quad k - d^{\text{in}}(U) \leq u^{\text{in}}(U)$ and $k - d^{\text{out}}(U) \leq u^{\text{out}}(U)$

for each nonempty proper subset U of V, and

(63.14) $\quad \gamma \geq l^{\text{in}}(V \setminus \bigcup \mathcal{P}) + \sum_{X \in \mathcal{P}}(k - d^{\text{in}}(X))$ and

$\quad\quad\quad \gamma \geq l^{\text{out}}(V \setminus \bigcup \mathcal{P}) + \sum_{X \in \mathcal{P}}(k - d^{\text{out}}(X))$

for each collection $\mathcal{P}$ of disjoint nonempty proper subsets of V.

Proof. Necessity is easy. To see sufficiency, choose $x : V \to \mathbb{Z}_+$ satisfying $l^{\text{in}} \leq x \leq u^{\text{in}}$, $x(V) \geq \gamma$, and $x(U) \geq k - d^{\text{in}}(U)$ for each nonempty $U \subset V$, with $x(V)$ as small as possible.

We show $x(V) \leq \gamma$. Let $\mathcal{P}$ be the collection of inclusionwise maximal proper subsets U of V satisfying $x(U) = k - d^{\text{in}}(U)$. Any two distinct sets $T, U \in \mathcal{P}$ satisfy $T \cup U \neq V$, since otherwise $V \setminus T$ and $V \setminus U$ are disjoint, and we obtain the contradiction

(63.15) $\quad \gamma \geq k - d^{\text{out}}(V \setminus T) + k - d^{\text{out}}(V \setminus U) = 2k - d^{\text{in}}(T) - d^{\text{in}}(U)$
$\quad\quad\quad = x(T) + x(U) \geq x(T \cup U) \geq x(V) > \gamma$.

Moreover, any two distinct $T, U \in \mathcal{P}$ are disjoint, since otherwise we obtain the contradiction

(63.16) $\quad x(T) + x(U) = 2k - d^{\text{in}}(T) - d^{\text{in}}(U) \leq 2k - d^{\text{in}}(T \cap U) - d^{\text{in}}(T \cup U)$
$\quad\quad\quad < x(T \cap U) + x(T \cup U) = x(T) + x(U)$,

by the maximality of T.

Now each $v \in V$ with $x_v > l^{\text{in}}(v)$ is contained in some $U \in \mathcal{P}$, as otherwise we could decrease x_v. This gives

(63.17) $\quad x(V) = l^{\text{in}}(V \setminus \bigcup \mathcal{P}) + \sum_{U \in \mathcal{P}} x(U) = l^{\text{in}}(V \setminus \bigcup \mathcal{P}) + \sum_{U \in \mathcal{P}}(k - d^{\text{in}}(U)) \leq \gamma.$

1062   Chapter 63. Connectivity augmentation

Hence $x(V) = \gamma$. Similarly, there exists a $y : V \to \mathbb{Z}_+$ satisfying $l^{\text{out}} \leq y \leq u^{\text{out}}$, $y(V) = \gamma$, and $y(U) \geq k - d^{\text{out}}(U)$ for each nonempty $U \subset V$. So we can apply Lemma 63.1α, which gives the theorem. ∎

Again, this proof yields a polynomial-time algorithm to find a minimum-size k-arc-connector satisfying prescribed bounds on the in- and outdegrees.

Notes. The following problem is NP-complete: given a directed graph $D = (V, A)$, a function $r : V \times V \to \mathbb{Z}_+$, and a cost function $k : V \times V \to \mathbb{Q}_+$, find a minimum-cost set of new arcs whose addition to D makes the graph r-arc-connected. Frank [1990a,1992a] showed that if there are functions $k', k'' : V \to \mathbb{Q}_+$ with $k(u, v) = k'(u) + k''(v)$ for all $u, v \in V$, then this problem is solvable in polynomial time.

Gusfield [1987a] gave a linear-time algorithm to find a minimum number of directed arcs to be added to a mixed graph such that it becomes strongly connected (that is, for all vertices u, v there is a $u - v$ path traversing directed edges in the right direction only).

Frank and Jordán [1995b] gave an alternative proof of Theorem 63.1 based on bisubmodular functions, and showed a number of related results.

Frank [1993c] gave some further methods for the problems discussed in these sections.

Kajitani and Ueno [1986] showed that the minimum size of a k-arc-connector for a directed tree $D = (V, A)$ is equal to the maximum of $\sum_{v \in V} \max\{0, k - \deg^{\text{in}}(v)\}$ and $\sum_{v \in V} \max\{0, k - \deg^{\text{out}}(v)\}$.

Frank [1990a,1992a] gave a polynomial-time algorithm for: given directed graph $D = (V, A)$, $r \in V$, and $k \in \mathbb{Z}_+$, find a minimum number of arcs to be added to D such that for each $s \in V$ there exist k arc-disjoint $r - s$ paths in the augmented graph. The complexity was improved by Gabow [1991b].

63.2. Making an undirected graph 2-edge-connected

Let (V, E) and (V, F) be undirected graphs. The set F is called a *k-edge-connector* for G if the graph $(V, E \cup F)$ is k-edge-connected (where in $E \cup F$ edges are taken parallel if they occur both in E and in F).

The minimum size of a 1-edge-connector of a graph G trivially is one less than the number of components of G. Eswaran and Tarjan [1976] and Plesník [1976] characterized the minimum size of a 2-edge-connector, by first showing:

Theorem 63.4. *Let $G = (V, E)$ be a forest with at least two vertices and with p vertices of degree 1 and q isolated vertices. Then the minimum size of a 2-edge-connector for G equals $\lceil \frac{1}{2}p \rceil + q$.*

Proof. Each vertex of degree 1 should be incident with at least one new edge, and each vertex of degree 0 should be incident with at least two new edges. So any 2-edge-connector has size at least $\frac{1}{2}p + q$.

To see that $\lceil\frac{1}{2}p\rceil + q$ can be attained, first assume that G is not connected. Choose vertices u and v in different components, with $\deg(u) \leq 1$ and $\deg(v) \leq 1$. Adding edge uv, reduces $\frac{1}{2}p + q$ by 1, as one easily checks.

So we can assume that G is a tree. If $p \leq 3$, the graph is a path or a subdivision of $K_{1,3}$, and the theorem is easy.

If $p \geq 4$, there is a pair of end vertices u, v such that at least two edges of G leave the $u - v$ path P in G. Let G' be the tree obtained from G by contracting P to one vertex. Then G' has $p - 2$ end vertices. Applying induction shows that G' has a 2-edge-connector F of size $\lceil\frac{1}{2}p\rceil - 1$. By adding edge uv we obtain a 2-edge-connector of G, proving the theorem. ∎

This implies, for not necessarily forests:

Corollary 63.4a. *Let $G = (V, E)$ be a non-2-edge-connected undirected graph. For $i = 0, 1$, let p_i be the number of 2-edge-connected components K with $d_E(K) = i$. Then the minimum size of a 2-edge-connector equals $\lceil\frac{1}{2}p_1\rceil + p_0$.*

Proof. Directly from Theorem 63.4, by contracting each 2-edge-connected component to one vertex. ∎

These proofs give polynomial-time algorithms to find a minimum-size 2-edge-connector for a given undirected graph. Eswaran and Tarjan [1976] gave a linear-time algorithm.

63.3. Making an undirected graph k-edge-connected

Watanabe and Nakamura [1987] gave a min-max formula and a polynomial-time algorithm for the minimum size of a k-edge-connector for any undirected graph. Cai and Sun [1989] and Frank [1992a] showed that the min-max relation can be derived from the following lemma (given, in a different, 'vertex-splitting' terminology, by Mader [1978a] and Lovász [1979a] (Problem 6.53); the proof below follows Frank [1992a]):

Lemma 63.5α. *Let $G = (V, E)$ be a graph, let $k \in \mathbb{Z}_+$, with $k \geq 2$, and let $x : V \to \mathbb{Z}_+$. Then G has a k-edge-connector F with $\deg_F(v) = x_v$ for each $v \in V$ if and only if $x(V)$ is even and*

(63.18) $\qquad x(U) \geq k - d_E(U)$

for each nonempty proper subset U of V.

Proof. Necessity is easy, since for each nonempty $U \subset V$,

(63.19) $\qquad k \leq d_{E \cup F}(U) = d_F(U) + d_E(U) \leq x(U) + d_E(U).$

To see sufficiency, choose a counterexample with $x(V)$ minimal. Trivially, $x(V) \geq 2$.

Let $S := \{v \in V \mid x_v \geq 1\}$, and fix $s \in S$. Let $\mathcal{U}$ be the collection of inclusionwise maximal sets $U \subset V$ containing s and satisfying $x(U) + d_E(U) \leq k + 1$. Note that

(63.20) $\quad x(U) \leq \frac{1}{2}x(V)$ for each $U \in \mathcal{U}$,

since otherwise $x(V \setminus U) \leq x(U) - 2$, implying the contradiction $k \leq x(V \setminus U) + d_E(V \setminus U) \leq x(U) - 2 + d_E(U) \leq k - 1$.

Moreover,

(63.21) $\quad$ for all $t \in S \setminus \{s\}$, there exists a $U \in \mathcal{U}$ containing t.

Otherwise, we could augment G by a new edge st and decrease both x_s and x_t by 1. Then (63.18) is maintained, and we obtain a smaller counterexample, contradicting our assumption. This shows (63.21).

Next:

(63.22) $\quad$ for any two distinct $T, U \in \mathcal{U}$, G has an edge leaving $T \cap U$, and no edge connecting $T \cap U$ and $V \setminus (T \cup U)$.

Consider:

(63.23) $\quad 2(k+1) \geq x(T) + d_E(T) + x(U) + d_E(U)$
$= x(T \setminus U) + d_E(T \setminus U) + x(U \setminus T) + d_E(U \setminus T)$
$+ 2|E[T \cap U, V \setminus (T \cup U)]| + 2x(T \cap U) \geq 2k + 2$,

implying equality throughout. So $x(T \cap U) = 1$ and $|E[T \cap U, V \setminus (T \cup U)]| = \emptyset$. Since $d_E(T \cap U) \geq k - x(T \cap U) = k - 1 \geq 1$, this proves (63.22).

Now by (63.21) and (63.20) we can choose three sets $T, U, W \in \mathcal{U}$. Then

(63.24) $\quad T \cap U = T \cap W = U \cap W$.

Indeed, by symmetry it suffices to prove that $T \cap U \cap W = U \cap W$. Let $M := U \cap W$. Suppose $T \cap M \neq M$; so $M \not\subseteq T$, and hence $T \cup M \neq T$. Defining $\phi(X) := x(X) + d_E(X)$ for $X \subseteq V$, we obtain the contradiction

(63.25) $\quad k \leq \phi(T \cap M) \leq \phi(T) + \phi(M) - \phi(T \cup M)$
$\leq \phi(T) + \phi(U) + \phi(W) - \phi(U \cup W) - \phi(T \cup M)$
$\leq 3(k+1) - 2(k+2) = k - 1$,

since the maximality of T, U, and W gives $\phi(U \cup W) \geq k + 2$ and $\phi(T \cup M) \geq k + 2$. This shows (63.24).

Now by (63.22), G has an edge leaving $T \cap U$, while the other end should be in each of $T \cup U$, $T \cup W$, and $U \cup W$, and hence in $T \cap U$, a contradiction. ∎

From this, the min-max result for minimum-size k-edge-connectors of Watanabe and Nakamura [1987] follows:

Section 63.3. Making an undirected graph k-edge-connected

Theorem 63.5. *Let $G = (V, E)$ be an undirected graph and let $k, \gamma \in \mathbb{Z}_+$, with $k \geq 2$. Then G has a k-edge-connector of size at most γ if and only if*

(63.26) $\qquad 2\gamma \geq \sum_{U \in \mathcal{P}} (k - d(U))$

for each collection $\mathcal{P}$ of disjoint nonempty proper subsets of V.

Proof. Necessity follows since for each nonempty proper subset U of V, at least $k - d(U)$ edges entering U must be in any k-edge-connector. As any new edge can enter at most two sets in $\mathcal{P}$, we have (63.26).

To see sufficiency, choose $x : V \to \mathbb{Z}_+$ satisfying (63.18), with $x(V)$ as small as possible. By Lemma 63.5α it suffices to show that $x(V) \leq 2\gamma$.

Let $\mathcal{P}$ be the collection of inclusionwise maximal subsets U of V satisfying $x(U) = k - d(U)$. Any two distinct sets $T, U \in \mathcal{P}$ satisfy $T \cup U \neq V$, since otherwise we obtain the contradiction

(63.27) $\qquad 2\gamma < x(V) = x(T \cup U) \leq x(T) + x(U) = k - d(T) + k - d(U)$
$\qquad\qquad = k - d(V \setminus T) + k - d(V \setminus U) \leq 2\gamma,$

using (63.26). Moreover, any two distinct $T, U \in \mathcal{P}$ are disjoint, since otherwise we obtain the contradiction

(63.28) $\qquad x(T) + x(U) = 2k - d(T) - d(U) \leq 2k - d(T \cap U) - d(T \cup U)$
$\qquad\qquad < x(T \cap U) + x(T \cup U) = x(T) + x(U),$

by the maximality of T. Now each $v \in V$ with $x_v \geq 1$ is contained in some $U \in \mathcal{P}$, as otherwise we could decrease x_v. This gives

(63.29) $\qquad x(V) = \sum_{U \in \mathcal{P}} x(U) = \sum_{U \in \mathcal{P}} (k - d(U)) \leq 2\gamma,$

which proves the theorem. ∎

Similarly to the directed case, the proof implies that a minimum-size k-edge-connector can be found in polynomial time (Watanabe and Nakamura [1987]):

Theorem 63.6. *Given an undirected graph G and $k \in \mathbb{Z}_+$, a minimum-size k-edge-connector can be found in strongly polynomial time.*

Proof. The proof method reduces to a polynomially bounded number of tests of (63.18), which can be performed in strongly polynomial time by reducing it to maximum flow computations. ∎

Notes. Also Naor, Gusfield, and Martel [1990,1997], and Gabow [1991b] gave polynomial-time algorithms to find a minimum-size k-edge-connector. Frank [1990a, 1992a] and Benczúr [1994,1999] gave strongly polynomial-time algorithms for the (integer) capacitated version of the problem. More and related results can be found in Gabow [1994], Benczúr [1995], Nagamochi and Ibaraki [1995,1996,1997,1999c],

Nagamochi, Shiraki, and Ibaraki [1997], Benczúr and Karger [1998,2000], Nagamochi and Eades [1998], Bang-Jensen and Jordán [2000], and Nagamochi, Nakamura, and Ibaraki [2000].

63.3a. k-edge-connectors with bounds on degrees

Frank [1990a,1992a] derived similarly characterizations of the existence of k-edge-connectors of given size and satisfying given lower and upper bounds on the degrees:

Theorem 63.7. *Let $G = (V, E)$ be an undirected graph, let $k, \gamma \in \mathbb{Z}_+$, with $k \geq 2$, and let $l, u \in \mathbb{Z}_+^V$ with $l \leq u$. Then G has a k-edge-connector F of size at most γ satisfying $l(v) \leq \deg_F(v) \leq u(v)$ for each $v \in V$ if and only if $2\gamma \leq u(V)$,*

(63.30) $\quad k - d_E(U) \leq u(U)$

for each nonempty proper subset U of V, and

(63.31) $\quad 2\gamma \geq l(V \setminus \bigcup \mathcal{P}) + \sum_{U \in \mathcal{P}} (k - d_E(U)).$

for each collection $\mathcal{P}$ of disjoint nonempty proper subsets of V.

Proof. The conditions can be easily seen to be necessary.

To see sufficiency, let $x : V \to \mathbb{Z}_+$ satisfy $l \leq x \leq u$, $x(V) \geq 2\gamma$, $x(U) \geq k - d_E(U)$ for each nonempty $U \subset V$, and with $x(V)$ as small as possible. Such an x exists, by (63.30). By Lemma 63.5α, it suffices to show that $x(V) \leq 2\gamma$.

Let $\mathcal{P}$ be the collection of inclusionwise maximal proper subsets U of V satisfying $x(U) = k - d(U)$. Any two distinct sets $T, U \in \mathcal{P}$ satisfy $T \cup U \neq V$, since otherwise we obtain the contradiction

(63.32) $\quad 2\gamma < x(V) = x(T \cup U) \leq x(T) + x(U) = k - d(T) + k - d(U)$
$\quad\quad\quad = k - d(V \setminus T) + k - d(V \setminus U) \leq 2\gamma,$

by (63.31). Moreover, any two distinct $T, U \in \mathcal{P}$ are disjoint, since otherwise we obtain the contradiction

(63.33) $\quad x(T) + x(U) = 2k - d(T) - d(U) \leq 2k - d(T \cap U) - d(T \cup U)$
$\quad\quad\quad < x(T \cap U) + x(T \cup U) = x(T) + x(U),$

by the maximality of T. Now each $v \in V$ with $x_v > l(v)$ is contained in some $U \in \mathcal{P}$, as otherwise we could decrease x_v. This gives

(63.34) $\quad x(V) = l(V \setminus \bigcup \mathcal{P}) + \sum_{U \in \mathcal{P}} x(U) = l(V \setminus \bigcup \mathcal{P}) + \sum_{U \in \mathcal{P}} (k - d(U)) \leq 2\gamma,$

as required. ∎

Notes. T. Jordán (cf. Bang-Jensen and Jordán [1997]) showed that finding a minimum number of edges that makes a given simple graph k-edge-connected and keeps it simple, is NP-complete. On the other hand, Bang-Jensen and Jordán [1997,1998] gave, for any fixed k, an $O(n^4)$-time algorithm for this problem. Taoka, Watanabe, and Takafuji [1994] gave an $O(m + n \log n)$-time algorithm for $k = 4$ and an $O(n^2 + m)$-time algorithm for $k = 5$ (assuming the input graph is $k - 1$-edge-connected). Other fast algorithms for undirected edge-connectivity augmentation were given by Benczúr [1999].

Ueno, Kajitani, and Wada [1988] gave a polynomial-time algorithm for finding a minimum-size k-edge-connector for a tree.

63.4. r-edge-connectivity and r-edge-connectors

Let $G = (V, E)$ be an undirected graph and let $r : V \times V \to \mathbb{Z}_+$. G is called r-edge-connected if for all $u, v \in V$ there exist $r(u, v)$ edge-disjoint paths connecting u and v. So, by Menger's theorem, G is r-edge-connected if and only if $d_E(U) \geq r(u, v)$ for all $U \subseteq V$ and $u \in U$, $v \in V \setminus U$.

An r-edge-connector for G is a set F of edges on V such that the graph $G' := (V, E \cup F)$ satisfies $\lambda_{G'}(u, v) \geq r(u, v)$ for all $u, v \in V$. (Again, $E \cup F$ is the disjoint union, allowing parallel edges.) Define $\gamma(G, r)$ as the minimum size of an r-edge-connector for G.

Given an undirected graph $G = (V, E)$, a function $r : V \times V \to \mathbb{Z}_+$, and a cost function $k : V \times V \to \mathbb{Q}_+$, it is NP-complete to find a minimum-cost r-edge-connector (since for $E = \emptyset$, $r = 2$, it is the traveling salesman problem).

Frank [1990a,1992a] gave a polynomial-time algorithm and a min-max formula for the cardinality case: given a graph $G = (V, E)$ and $r : V \times V \to \mathbb{Z}_+$, find the minimum number of edges to be added to make G r-edge-connected. We describe the method in this section.

It is based on the following theorem of Mader [1978a] (conjectured by Lovász [1976a]; we follow the proof of Frank [1992b]):

Lemma 63.8α. *Let $G = (V \cup \{s\}, E)$ be an undirected graph, where s has even and positive degree, and s is not incident with a bridge of G. Then s has two neighbours u and v such that the graph G' obtained from G by replacing su and sv by one new edge uv satisfies*

(63.35) $\quad \lambda_{G'}(x, y) = \lambda_G(x, y)$

for all $x, y \in V$.

Proof. By induction on $|V| + \deg(s)$. For any $U \subseteq V$ with $\emptyset \neq U \neq V$, define

(63.36) $\quad R(U) := \max_{u \in U, v \in V \setminus U} \lambda_G(u, v),$

and set $R(\emptyset) := R(V) := 0$. So $R(U) \leq d(U)$ for each $U \subseteq V$.

Let $\mathcal{P}$ be the collection of nonempty proper subsets U of V with $d(U) = R(U)$, and let $\mathcal{U}$ be the collection of nonempty proper subsets U of V with $d(U) \leq R(U) + 1$ (hence $\mathcal{P} \subseteq \mathcal{U}$).

Note that $u, v \in N(s)$ are as required in the lemma if and only if there is no $U \in \mathcal{U}$ containing both u and v. So we can assume that

(63.37) $\quad$ for each pair $u, v \in N(s)$ there is a $U \in \mathcal{U}$ containing u and v.

We first show:

(63.38) $\quad |T| = 1$ for each $T \in \mathcal{P}$.

Suppose not. Consider the graph G/T obtained from G by contracting T (where T also denotes the vertex obtained by contracting T). By induction,

G/T has two edges su' and sv such that for the graph H obtained from G/T by replacing su' and sv by a new edge $u'v$, one has

(63.39) $\qquad \lambda_H(x,y) = \lambda_{G/T}(x,y)$

for all $x, y \in V(G/T) \setminus \{s\}$. By symmetry of u' and v, we may assume that $v \neq T$, that is, $v \in V \setminus T$. Let $u := u'$ if $u' \neq T$ and choose $u \in T \cap N(s)$ if $u' = T$.

Then for all $Z \in \mathcal{U}$:

(63.40) $\qquad$ if $T \subseteq Z$ or $T \cap Z = \emptyset$ then $u \notin Z$ or $v \notin Z$.

Indeed, as $R(Z) \geq d(Z) - 1$, there exist $x, y \in V$ such that Z splits x, y and $\lambda_G(x,y) \geq d(Z) - 1$. Since $T \subseteq Z$ or $T \cap Z = \emptyset$, we may assume that $x \notin T$. Define $y' := y$ if $y \notin T$, and $y' := T$ if $y \in T$. Define $Z' := Z$ if $T \cap Z = \emptyset$, and $Z' := (Z \setminus T) \cup \{T\}$ if $T \subseteq Z$. Suppose now $u, v \in Z$. Then $d_H(Z') = d_G(Z) - 2$. This gives the contradiction

(63.41) $\qquad \lambda_{G/T}(x, y') \geq \lambda_G(x, y) \geq d_G(Z) - 1 > d_H(Z') \geq \lambda_H(x, y')$
$\qquad\qquad = \lambda_{G/T}(x, y'),$

proving (63.40).

Now let $U \in \mathcal{U}$ contain u and v. By Lemma 61.6α, $R(T) + R(U)$ is at most $R(T \cap U) + R(T \cup U)$ or at most $R(T \setminus U) + R(U \setminus T)$.

If $R(T) + R(U) \leq R(T \cap U) + R(T \cup U)$, then

(63.42) $\qquad d(T) + d(U) \geq d(T \cap U) + d(T \cup U) \geq R(T \cap U) + R(T \cup U)$
$\qquad\qquad \geq R(T) + R(U) \geq d(T) + d(U) - 1,$

implying $R(T \cup U) \geq d(T \cup U) - 1$. So $T \cup U \in \mathcal{U}$ and $u, v \in T \cup U$, contradicting (63.40).

So $R(T) + R(U) \leq R(T \setminus U) + R(U \setminus T)$. Hence

(63.43) $\qquad d(T) + d(U) \geq d(T \setminus U) + d(U \setminus T) \geq R(T \setminus U) + R(U \setminus T)$
$\qquad\qquad \geq R(T) + R(U) \geq d(T) + d(U) - 1.$

So $d(T) + d(U) = d(T \setminus U) + d(U \setminus T)$, and hence $T \cap U$ contains no neighbours of s. So $u' \neq T$ (otherwise $u \in T \cap U \cap N(s)$). Hence $u' = u \in U \setminus T$. By (63.43) we also know $R(U \setminus T) \geq d(U \setminus T) - 1$. So $U \setminus T \in \mathcal{U}$ and $u, v \in U \setminus T$, contradicting (63.40). This proves (63.38).

Note that (63.38) implies

(63.44) $\qquad \lambda_G(u,v) = \min\{\deg(u), \deg(v)\}$ for all $u, v \in V$,

since $\lambda_G(u,v) = d_E(U)$ for some $U \subseteq V$ splitting $\{u,v\}$. So $U \in \mathcal{P}$, and hence $|U| = 1$, implying (63.44).

Choose a vertex $t \in N(s)$ of minimum degree. Let $\mathcal{U}'$ be a minimal collection of inclusionwise maximal sets in $\mathcal{U}$ containing t such that $\bigcup \mathcal{U}' \supseteq N(s)$ (this exists by (63.37)). Note that for each $U \in \mathcal{U}$ one has $|E[U, s]| \leq \frac{1}{2}\deg(s)$, since otherwise $d(V \setminus U) \leq d(U) - 2$ (since $\deg(s)$ is even), and hence

(63.45) $\qquad R(V \setminus U) \leq d(V \setminus U) \leq d(U) - 2 \leq R(U) - 1 = R(V \setminus U) - 1,$

a contradiction. Hence $|\mathcal{U}'| \geq 3$ (as $t \in U$ for each $U \in \mathcal{U}'$). Moreover,

(63.46) for each $U \in \mathcal{U}'$, there is a $v \in N(s)$ such that U is the only set in $\mathcal{U}'$ containing v.

(Otherwise we could delete U from $\mathcal{U}'$.)
Also,

(63.47) $R(U \setminus \{t\}) \geq R(U)$ for each $U \in \mathcal{U}'$.

Indeed, choose $x \in U$ and $y \in V \setminus U$ with $R(U) = \lambda_G(x, y)$. If $x \neq t$, then $R(U \setminus \{t\}) \geq \lambda_G(x,y) = R(U)$, as required. If $x = t$, then for any $u \in N(s) \cap (U \setminus \{t\})$,

(63.48) $R(U) = \lambda_G(t, y) = \min\{\deg(t), \deg(y)\} \leq \min\{\deg(u), \deg(y)\}$
$= \lambda_G(u, y) \leq R(U \setminus \{t\})$.

This shows (63.47).

Moreover, for any distinct $X, Y \in \mathcal{U}'$ one has

(63.49) $R(X) + R(Y) \leq R(X \setminus Y) + R(Y \setminus X)$.

Suppose not. Then, by Lemma 61.6α we know $R(X) + R(Y) \leq R(X \cap Y) + R(X \cup Y)$, and by symmetry we can assume that $R(X) > R(X \setminus Y)$. Hence by (63.47), $X \cap Y \neq \{t\}$, and hence by (63.38), $X \cap Y \notin \mathcal{P}$. By the maximality of X and Y, $R(X \cup Y) \leq d(X \cup Y) - 2$. This gives the contradiction

(63.50) $R(X \cap Y) + R(X \cup Y) \leq (d(X \cap Y) - 1) + (d(X \cup Y) - 2)$
$\leq d(X) + d(Y) - 3 \leq R(X) + R(Y) - 1$
$\leq R(X \cap Y) + R(X \cup Y) - 1$,

proving (63.49).

This implies that for any distinct $X, Y \in \mathcal{U}'$ one has

(63.51) $|X \setminus Y| = |Y \setminus X| = 1$, and st is the only edge connecting $X \cap Y$ and $(V \cup \{s\}) \setminus (X \cup Y)$.

Indeed, by (63.49) (as st connects $X \cap Y$ and $X \cup Y$),

(63.52) $d(X) + d(Y)$
$= d(X \setminus Y) + d(Y \setminus X) + 2|E[X \cap Y, (V \cup \{s\}) \setminus (X \cup Y)]|$
$\geq d(X \setminus Y) + d(Y \setminus X) + 2 \geq R(X \setminus Y) + R(Y \setminus X) + 2$
$\geq R(X) + R(Y) + 2 \geq d(X) + d(Y)$.

So we have equality throughout. Hence $X \setminus Y, Y \setminus X \in \mathcal{P}$, and therefore, by (63.38), $|X \setminus Y| = |Y \setminus X| = 1$. Moreover, $d(X \cap Y, V \setminus (X \cup Y)) = 1$, proving (63.51).

Now choose $X, Y, Z \in \mathcal{U}'$. Then (63.51) and (63.46) imply that $X \cap Y = X \cap Z = Y \cap Z$. So st is the only edge leaving $X \cap Y$, and hence st is a bridge. This contradicts the condition given in this lemma. ∎

Note that

(63.53)   the graph G' arising in Lemma 63.8α again has no bridge incident with s,

as for any two neighbours x, y of s in G' one has $\lambda_{G'}(x,y) = \lambda_G(x,y) \geq 2$. The lemma therefore can be applied iteratively to yield:

Theorem 63.8. *Let $G = (V, E)$ be an undirected graph and let $r : V \times V \to \mathbb{Z}_+$ be symmetric. Let $x : V \to \mathbb{Z}_+$ be such that $x(K) \neq 1$ for each component K of G. Then G has an r-edge-connector F satisfying $\deg_F(v) = x_v$ for each $v \in V$ if and only if $x(V)$ is even and*

(63.54) $x(U) + d_E(U) \geq r(u,v)$

for all $U \subseteq V$ and all $u \in U$, $v \in V \setminus U$.

Proof. Necessity of (63.54) follows from the fact that $x(U) + d_E(U) \geq d_F(U) + d_E(U) \geq r(u,v)$. To see sufficiency, extend V by a new vertex s and, for each $v \in V$, x_v edges connecting s and v (parallel if $x_v \geq 2$). Let H be the extended graph. Then (63.54) implies

(63.55) $d_H(U) \geq r(u,v)$

for all $U \subseteq V$ and all $u \in U$, $v \in V \setminus U$. Hence, for all $u, v \in V$,

(63.56) $\lambda_H(u,v) \geq r(u,v)$.

Now by iteratively splitting s as in Lemma 63.8α (cf. (63.53)), we obtain a set F of new edges such that adding F to G, the new graph G' satisfies

(63.57) $\lambda_{G'}(u,v) = \lambda_H(u,v) \geq r(u,v)$

for all $u, v \in V$. As moreover $\deg_F(v) = x_v$ for each $v \in V$, F is an r-edge-connector as required. ∎

The condition that $x(K) \neq 1$ for each component K cannot be deleted, as can be seen by taking $G = (V, \emptyset)$, $r := 1$, $x := 1$, with $|V| \geq 4$.

We next give the theorem of Frank [1990a,1992a] characterizing the minimum size $\gamma(G, r)$ of an r-edge-connector. To this end we can assume that r satisfies:

(63.58)   (i) $r(u,v) = r(v,u) \geq \lambda_G(u,v)$ for all $u, v \in V$;
(ii) $r(u,w) \geq \min\{r(u,v), r(v,w)\}$ for all $u, v, w \in V$.

Define

(63.59) $R(U) := \max_{u \in U, v \in V \setminus U} r(u,v)$ if $\emptyset \subset U \subset V$, and
$R(\emptyset) := R(V) := 0$.

Call a component K of G *marginal* if $K \neq V$, $r(u,v) = \lambda_G(u,v)$ for all $u, v \in K$, and $r(u,v) \leq 1$ for all $u \in K$ and $v \in V \setminus K$.

Theorem 63.9. Let $G = (V, E)$ be an undirected graph and let $r : V \times V \to \mathbb{Z}_+$ satisfy (63.58).

(i) If K is a marginal component of G, then

(63.60) $\quad \gamma(G, r) = \gamma(G - K, r') + R(K),$

where r' is the restriction of r to $(V \setminus K) \times (V \setminus K)$.

(ii) If G has no marginal components, then $\gamma(G, r)$ is equal to the maximum value of

(63.61) $\quad \lceil \frac{1}{2} \sum_{U \in \mathcal{P}} (R(U) - d_E(U)) \rceil$

taken over all collections $\mathcal{P}$ of disjoint nonempty proper subsets of V.

Proof. We first show (i). Let K be a marginal component of G and define $\alpha := R(K)$. As K is marginal, $\alpha \leq 1$. The inequality

(63.62) $\quad \gamma(G, r) \leq \gamma(G - K, r') + \alpha$

is easy, since an r-edge-connector for G can be obtained from an r'-edge-connector F for $G - K$: if $\alpha = 0$, then F is an r-edge-connector, and if $\alpha = 1$, we obtain an r-edge-connector by adding to F one edge connecting some pair $u \in K$, $v \in V \setminus K$ with $r(u, v) = 1$.

To see the reverse inequality, let F be a minimum-size r-edge-connector for G. Let $G' := (V, E \cup F)$. So G' is r-edge-connected.

If F contains no edges connecting K and $V \setminus K$, then $\alpha = 0$ and F contains an r'-edge-connector for $G - K$. Hence $\gamma(G, r) = |F| \geq \gamma(G - K, r') = \gamma(G - K, r') + \alpha$.

If F contains an edge uv with $u \in K$, $v \in V \setminus K$, then the graph H obtained from G' by contracting $(K \cup \{v\})$ to one vertex, is r'-edge-connected. Since edge $uv \in F$ is contracted, it implies that $G - K$ has an r'-edge-connector of size at most $|F| - 1$. So $\gamma(G - K, r') \leq |F| - 1 \leq \gamma(G, r) - \alpha$.

We next show (ii). Let G have no marginal components. Choose $x : V \to \mathbb{Z}_+$ such that $x(U) + d_E(U) \geq R(U)$ for each $U \subseteq V$, with $x(V)$ as small as possible. Let μ be the maximum value of (63.61). It suffices to show that $x(V) \leq 2\mu$, since then we can apply Theorem 63.8 (after increasing $x(v)$ by 1 for some $v \in V$ if $x(V)$ is odd). So assume $x(V) > 2\mu$. As $\mu > 0$ (otherwise $x = \mathbf{0}$), we know $x(V) > 2$.

Then

(63.63) $\quad x(K) \neq 1$ for each component K of G.

For suppose $x(K) = 1$. We show that K is marginal, which is a contradiction. First, $K \neq V$, since $x(V) > 2$. Second, for each $u \in K$, $v \in V \setminus K$, we have $r(u, v) \leq x(K) + d_E(K) = x(K) \leq 1$. Third, to prove that $r(u, v) = \lambda_G(u, v)$ for $u, v \in K$, there is a subset U of K with $|U \cap \{u, v\}| = 1$, $\lambda_G(u, v) = d_E(U)$, and $x(U) = 0$. Then $r(u, v) \leq x(U) + d_E(U) = d_E(U) = \lambda_G(u, v)$. So K is marginal, contradicting our assumption. This proves (63.63).

By the minimality of x, there exists a collection $\mathcal{P}$ of nonempty proper subsets U of V satisfying $x(U) = R(U) - d_E(U)$, such that $\mathcal{P}$ covers $\{v \mid x_v \geq 1\}$. Choose $\mathcal{P}$ such that

(63.64) $$\sum_{U \in \mathcal{P}} |U|$$

is as small as possible. Then

(63.65) $\quad T \cap U = \emptyset$ for distinct $T, U \in \mathcal{P}$.

For suppose $T \cap U \neq \emptyset$. Note that $T \not\subseteq U \not\subseteq T$, by the minimality of (63.64). Observe also that $T \cup U \neq V$, since otherwise we obtain the contradiction

(63.66) $\quad 2\mu < x(V) = x(T \cup U) \leq x(T) + x(U)$
$= R(T) - d(T) + R(U) - d(U)$
$= R(V \setminus T) - d(V \setminus T) + R(V \setminus U) - d(V \setminus U) \leq 2\mu.$

(by definition of μ, since $V \setminus T$ and $V \setminus U$ are disjoint).

By Lemma 61.6α, $R(T) + R(U)$ is at most $R(T \cap U) + R(T \cup U)$ or at most $R(T \setminus U) + R(U \setminus T)$.

If $R(T) + R(U) \leq R(T \cap U) + R(T \cup U)$, then

(63.67) $\quad x(T) + x(U) = R(T) - d(T) + R(U) - d(U)$
$\leq R(T \cap U) - d(T \cap U) + R(T \cup U) - d(T \cup U)$
$\leq x(T \cap U) + x(T \cup U) = x(T) + x(U),$

and hence we have equality throughout. This implies that $x(T \cup U) = R(T \cup U) - d_E(T \cup U)$, and hence replacing T and U by $T \cup U$ would decrease (63.64), a contradiction.

If $R(T) + R(U) \leq R(T \setminus U) + R(U \setminus T)$, then

(63.68) $\quad x(T) + x(U) = R(T) - d(T) + R(U) - d(U)$
$\leq R(T \setminus U) - d(T \setminus U) + R(U \setminus T) - d(U \setminus T)$
$\leq x(T \setminus U) + x(U \setminus T) \leq x(T) + x(U),$

implying equality throughout. This implies that $x(T \setminus U) = R(T \setminus U) - d_E(T \setminus U)$, and hence replacing T by $T \setminus U$ would decrease (63.64), a contradiction.

This proves (63.65), yielding the contradiction

(63.69) $\quad 2\mu < x(V) = \sum_{U \in \mathcal{P}} x(U) = \sum_{U \in \mathcal{P}} (R(U) - d(U)) \leq 2\mu,$

which proves the theorem. ∎

Frank [1990a,1992a] also gave a polynomial-time algorithm to find a minimum-cost r-edge-connector if the cost of any new edge uv is given by $k(u) + k(v)$, for some function $k : V \to \mathbb{Q}_+$. This is done with the help of the following auxiliary result:

Theorem 63.10. *Let $G = (V, E)$ be an undirected graph and let $r : V \times V \to \mathbb{Z}_+$ be symmetric. Define $R(U)$ as in (63.59). Then*

(63.70) $$Q := \{x \in \mathbb{R}_+^V \mid x(U) \geq R(U) - d_E(U) \text{ for all } U \subseteq V\}$$

is a contrapolymatroid, with associated supermodular function given by, for $X \subseteq V$:

(63.71) $$g(X) := \max_{\mathcal{U}} \sum_{U \in \mathcal{U}} (R(U) - d_E(U)),$$

where the maximum ranges over collections $\mathcal{U}$ of disjoint nonempty subsets of X.

Proof. Clearly, for any $x \in \mathbb{R}_+^V$ one has $x \in Q$ if and only if $x(U) \geq g(X)$ for each $X \subseteq V$.

To see that g is supermodular, choose $X, Y \subseteq V$. Let

(63.72) $$g(X) = \sum_{U \in \mathcal{U}} (R(U) - d(U)) \text{ and } g(Y) = \sum_{T \in \mathcal{T}} (R(T) - d(T)),$$

where $\mathcal{U}$ and $\mathcal{T}$ are collections of disjoint nonempty subsets of X and of Y, respectively. The collections $\mathcal{U}$ and $\mathcal{T}$ together form a family $\mathcal{S}$ of nonempty subsets of V satisfying

(63.73) $$\sum_{S \in \mathcal{S}} \chi^S \leq \chi^{X \cap Y} + \chi^{X \cup Y} \text{ and } g(X) + g(Y) \leq \sum_{S \in \mathcal{S}} (R(S) - d(S)).$$

We now choose $\mathcal{S}$ such that (63.73) is satisfied and such that

(63.74) $$\sum_{S \in \mathcal{S}} |S|(|V \setminus S| + 1)$$

is as small as possible.

We claim that $\mathcal{S}$ is laminar; that is,

(63.75) if $T, U \in \mathcal{S}$, then $T \subseteq U$ or $U \subseteq T$ or $T \cap U = \emptyset$.

Suppose not. By Lemma 61.6α, $R(T) + R(U)$ is at most $R(T \cap U) + R(T \cup U)$ or at most $R(T \setminus U) + R(U \setminus T)$. If $R(T) + R(U) \leq R(T \cap U) + R(T \cup U)$, then replacing T and U by $T \cap U$ and $T \cup U$ maintains (63.73) but decreases (63.74) (by Theorem 2.1), contradicting the minimality assumption. If $R(T) + R(U) \leq R(T \setminus U) + R(U \setminus T)$, then replacing T and U by $T \setminus U$ and $U \setminus T$ maintains (63.73) but decreases (63.74) (again by Theorem 2.1), again contradicting the minimality condition. This proves (63.75).

Now let $\mathcal{P}$ be the collection of inclusionwise maximal elements in $\mathcal{S}$ and let $\mathcal{Q}$ be the collection of remaining sets in $\mathcal{S}$. (If a set occurs twice in $\mathcal{S}$, it is both in $\mathcal{P}$ and in $\mathcal{Q}$.) Then each set in $\mathcal{P}$ is contained in $X \cup Y$, and each set in $\mathcal{Q}$ is contained in $X \cap Y$. Moreover, both $\mathcal{P}$ and $\mathcal{Q}$ are collections of disjoint sets. Hence

(63.76) $$g(X \cup Y) + g(X \cap Y) \geq \sum_{P \in \mathcal{P}} (R(P) - d(P)) + \sum_{Q \in \mathcal{Q}} (R(Q) - d(Q))$$
$$= \sum_{S \in \mathcal{S}} (R(S) - d(S)) \geq g(X) + g(Y);$$

that is, g is supermodular. ∎

With this theorem, also good characterizations and polynomial-time algorithms can be obtained for the minimum size of an r-edge-connector satisfying prescribed lower and upper bounds on its degrees — see Frank [1990a,1992a].

Bang-Jensen, Frank, and Jackson [1995] extended these results to mixed graphs.

63.5. Making a directed graph k-vertex-connected

Let (V, A) and (V, B) be directed graphs. The set B is called a k-vertex-connector for D if the directed graph $(V, A \cup B)$ is k-vertex-connected. (Note that parallel edges will not help the vertex-connectivity.)

Since, for directed graphs, 1-vertex-connectors and 1-arc-connectors coincide, the problem of finding a minimum-size 1-vertex-connector for a given directed graph is addressed in Section 57.1.

Frank and Jordán [1995b] showed the following min-max relation for minimum-size k-vertex connector in directed graphs (which is a special case of Frank and Jordán's Theorem 60.5 above).

Call a pair (X, Y) of subsets of V a *good pair* if $X \neq \emptyset$, $Y \neq \emptyset$, $X \cap Y = \emptyset$, and D has no arc from X to Y. Call a collection $\mathcal{F}$ of good pairs a *good collection* if $X \cap X' = \emptyset$ or $Y \cap Y' = \emptyset$ for all distinct $(X, Y), (X', Y') \in \mathcal{F}$.

Theorem 63.11. *Let $D = (V, A)$ be a directed graph and let $k \in \mathbb{Z}_+$. Then the minimum size of a k-vertex-connector for D is equal to the maximum value of*

(63.77) $$\sum_{(X,Y) \in \mathcal{F}} (k - |V \setminus (X \cup Y)|),$$

where $\mathcal{F}$ ranges over good collections of good pairs.

Proof. Let γ be the maximum value. The minimum is not less than γ, since for any $(X, Y) \in \mathcal{F}$, at least $k - |V \setminus (X \cup Y)|$ arcs from X to Y should be added, while such arcs do not run from X' to Y' for any other pair (X', Y') in $\mathcal{F}$ (as $X \cap X' = \emptyset$ or $Y \cap Y' = \emptyset$).

To see equality, we can assume that D is not k-vertex-connected. Then there exist disjoint nonempty subsets T and U of V such that D has no arc from T to U and such that $|V \setminus (T \cup U)| < k$.

If there exist $t \in T$ and $u \in U$ such that augmenting D with the arc (t, u), the maximum decreases, we are done by induction. So we can assume that no such pair t, u exists. Hence for each $t \in T$ and $u \in U$, there exists a good collection $\mathcal{F}_{t,u}$ of good pairs, with

(63.78) $$\sum_{(X,Y) \in \mathcal{F}_{t,u}} (k - |V \setminus (X \cup Y)|) = \gamma,$$

and with $t \notin X$ or $u \notin Y$ for all $(X,Y) \in \mathcal{F}_{t,u}$.

Concatenating these collections $\mathcal{F}_{t,u}$ for all $t \in T$, $u \in U$, and adding the pair (T,U), we obtain a family $\mathcal{G}$ of good pairs satisfying:

(63.79)   (i) for each $x \in T$, $y \in U$, there are at most $|T||U|$ pairs (X,Y) in $\mathcal{G}$ with $x \in X$ and $y \in Y$;

(ii) $\sum_{(X,Y) \in \mathcal{G}} (k - |V \setminus (X \cup Y)|) > \gamma |T||U|$.

Among all families $\mathcal{G}$ satisfying (63.79), we choose one minimizing

(63.80) $\sum_{(X,Y) \in \mathcal{G}} (|X| + |V \setminus Y|)(|Y| + |V \setminus X|)$.

Then

(63.81)   for all $(X,Y), (X',Y') \in \mathcal{G}$ one has $X \cap X' = \emptyset$ or $Y \cap Y' = \emptyset$ or $X \subseteq X', Y' \subseteq Y$ or $X' \subseteq X, Y \subseteq Y'$.

Suppose not. Replace (X,Y) and (X',Y') by $(X \cap X', Y \cup Y')$ and $(X \cup X', Y \cap Y')$. This maintains (63.79), while (63.80) decreases[19], contradicting our assumption. This proves (63.81).

Now consider the partial order $\leq$ on pairs (X,Y) of subsets of V, defined by $(X,Y) \leq (X',Y')$ if $X \subseteq X', Y' \subseteq Y$. For each pair (X,Y), let its 'weight' $w(X,Y)$ be the number of times (X,Y) occurs in $\mathcal{G}$, and let its 'length' $l(X,Y)$ be equal to $k - |V \setminus (X \cup Y)|$. Then by (63.79)(i), any chain has weight at most $|T||U|$. By (63.79)(ii), the sum of $l(X,Y)w(X,Y)$ over $(X,Y) \in \mathcal{G}$ is more than $\gamma |T||U|$. Hence, by the length-width inequality for partially ordered sets (Theorem 14.5), $\mathcal{G}$ contains an antichain $\mathcal{F}$ of length more than γ. Then $\mathcal{F}$ is a good collection by (63.81). This contradicts the definition of γ. ∎

The theorem implies that the minimum size of a k-vertex-connector for a given directed graph $D = (V,A)$ is equal to the minimum value of

(63.82) $\sum_{u,v \in V} x_{u,v}$

subject to

(63.83)   (i) $x_{u,v} \geq 0$ for all $u,v \in V$,

(ii) $\sum_{u \in X} \sum_{v \in Y} x_{u,v} \geq k - |V \setminus (X \cup Y)|$

for all disjoint nonempty $X, Y \subseteq V$ with no arc from X to Y.

---

[19] This can be seen with Theorem 2.1: Make a copy $\tilde{V}$ of V, and let $\tilde{Y}$ be the set of copies of elements of Y. Define $Z_{X,Y} := X \cup (\tilde{V} \setminus \tilde{Y})$. Then $|X| + |V \setminus Y| = |Z_{X,Y}|$ and $|Y| + |V \setminus X| = |(V \cup \tilde{V}) \setminus Z_{X,Y}|$. Moreover, for (X,Y) and (X',Y') we have $Z_{X \cap X', Y \cup Y'} = Z_{X,Y} \cap Z_{X',Y'}$ and $Z_{X \cup X', Y \cap Y'} = Z_{X,Y} \cup Z_{X',Y'}$. So the replacements decrease (63.80) by Theorem 2.1.

This can be seen by observing that Theorem 63.11 implies that this LP-problem has integer primal and dual solutions of equal value.

As Frank and Jordán [1995b] pointed out, this implies that a minimum-size k-vertex-connector can be found in polynomial time with the ellipsoid method, as follows.

Since the conditions (63.83) can be checked in polynomial time, the ellipsoid method (as we discuss below) implies that the minimum size of a k-arc-connector can be determined in polynomial time. Then an explicit minimum-size k-vertex-connector can be found by testing, for each pair $u, v \in V$, whether augmenting D by the new arc (u, v) decreases the minimum size of a k-arc-connector. If so, we add (u, v) to D and iterate.

The ellipsoid method applies, since given $x_{u,v} \geq 0$ $(u, v \in V)$, we can test if (63.83)(ii) holds. Indeed, let B be a set of new arcs forming a complete directed graph on V. Define a capacity function c on $A \cup B$ by: $c(a) := \infty$ for each $a \in A$ and $c(b) = x_{u,v}$ for each arc $b \in B$ from u to v. Then (63.83)(ii) is equivalent to: for each $s, t \in V$ there is an $s - t$ flow $f_{s,t}$ in $(V, A \cup B)$ subject to c of value k, such that for any vertex $v \neq s, t$, the amount of flow traversing v is at most 1 (since the set of arcs in B from X to Y, together with the vertices in $V \setminus (X \cup Y)$, form a mixed arc/vertex-cut separating s and t). As this can be tested in polynomial time, we have a polynomial-time test for (63.83).

In fact, we can transform the problem into a linear programming problem of polynomial size, by including the flow variables $f_{s,t}(a)$ (for $s, t \in V$ and $a \in A \cup B$), into the LP-problem. Thus the minimum size of a k-arc-connector can be described as the solution of a linear programming problem of polynomial size.

There is no combinatorial polynomial-time algorithm known to find a minimum-size k-vertex-connector for a given directed graph. (Frank and Jordán [1995a] describe a combinatorial polynomial-time algorithm for finding a minimum-size 2-vertex-connector for a strongly connected directed graph. Frank and Jordán [1999] extended it to a polynomial-time algorithm (for any fixed k) to find a minimum-size k-vertex-connector.)

Notes. Frank and Jordán [1995b] also showed that a directed graph $D = (V, A)$ has a k-vertex-connector B with all in- and outdegrees at most $k - \kappa(D)$ (where $\kappa(D)$ denotes the vertex-connectivity of D).

Frank [1994a] gave the following conjecture:

(63.84)   (?) Let $D = (V, A)$ be a simple acyclic directed graph. Then the minimum size of a k-vertex-connector for D is equal to the maximum of $\sum_{v \in V} \max\{0, k - \deg^{\text{in}}(v)\}$ and $\sum_{v \in V} \max\{0, k - \deg^{\text{out}}(v)\}$. (?)

An $O(kn)$-time algorithm finding a minimum-size k-vertex-connector for a rooted tree was given by Masuzawa, Hagihara, and Tokura [1987]. Frank [1994a] observed that this result easily extends to branchings.

Approximation algorithms for the minimum size of a k-vertex-connector for a directed graph were given by Jordán [1993a].

63.6. Making an undirected graph k-vertex-connected

Let (V, E) and (V, F) be undirected graphs. The set F is called a *k-vertex-connector* for G if the graph $(V, E \cup F)$ is k-vertex-connected.

Trivially, the minimum size of a 1-vertex-connector for an undirected graph G is equal to one less than the number of components of G.

The minimum size of a 2-vertex-connector for undirected graphs was given by Eswaran and Tarjan [1976] and Plesník [1976]. To this end, call a block *pendant* if it contains exactly one cut vertex of G. Moreover, call a block *isolated* if it contains no cut vertex of G. So an isolated block is a component of G.

Theorem 63.12. *Let $G = (V, E)$ be a non-2-vertex-connected undirected graph, with p pendant blocks and q isolated blocks. Let d be the maximum number of components of $G - v$, maximized over $v \in V$. Then the minimum size of a 2-vertex-connector for G is equal to*

(63.85) $\quad k := \max\{d - 1, \lceil \frac{1}{2}p \rceil + q\}.$

Proof. One needs at least $d-1$ edges, since for any $v \in V$, after deleting v the augmented graph should be connected. Any block containing no cut vertex should be incident with at least two new edges, and any block containing one cut vertex should be incident with at least one new edges. Hence the number of new edges is at least $\frac{1}{2}p + q$, and hence at least k.

To show that k can be attained, choose a counterexample G with k minimal. Then G is connected. Otherwise, we can choose two blocks B, B' from different components of G such that each of B, B' is pendant or isolated. We can choose a non-cut vertex from each of B, B', and connect them by a new edge to obtain graph G'. After that, k has decreased by exactly 1, and we can apply induction to G', implying the theorem.

So G is connected, and hence $q = 0$ (as G is not 2-vertex-connected). Moreover, $k \geq 2$, since otherwise $p \leq 2$, and we can add one edge to make G 2-vertex-connected.

Let U be the set of vertices v for which $G - v$ has at least three components and let W be the set of vertices v for which $G - v$ has $k + 1$ components. So $W \subseteq U$. Moreover, $|U| \geq 2$, since otherwise we can add $d - 1$ edges to make G connected.

We show:

(63.86) there exist two distinct pendant blocks B, B' such that each $B - B'$ path traverses all vertices in W and at least two vertices in U.

If $|W| \leq 1$, this is trivial. So we may assume that $|W| \geq 2$. Then, as $W \subseteq U$, it suffices to show that there exists a path traversing all vertices in W. If such a path would not exist, there exists a subset X of W with $|X| = 3$ that is not on a path. Then for each $v \in X$, one component K of $G - v$ contains $X \setminus \{v\}$. So for each $v \in X$, $G - v$ has k components disjoint from

X. Moreover, for distinct $v, v' \in X$, if K and K' are components of $G - v$ and $G - v'$ (respectively) each disjoint from X, then $K \cap K' = \emptyset$. Since for each $v \in X$ and each component K of $G - v$, $K \cup \{v\}$ contains at least one pendant block, we know $p \geq 3k \geq 3\lceil \frac{1}{2}p \rceil$, contradicting the fact that $p > 0$.

This shows (63.86). Now augment G by an edge connecting non-cut vertices in B and B', giving graph G'. As this augmentation decreases k (by the conditions given in (63.86)), we would obtain a counterexample with k smaller. ∎

This proof directly gives a polynomial-time algorithm to find a minimum-size 2-vertex-connector for G. Eswaran and Tarjan [1976] mention that a linear-time implementation of this algorithm was communicated to them in 1973 by R. Pecherer and A. Rosenthal — see Rosenthal and Goldner [1977]. (See also Hsu and Ramachandran [1991,1993].)

An equivalent form of Theorem 63.12 is:

Corollary 63.12a. *Let $G = (V, E)$ be a non-2-vertex-connected graph. Then G has a 2-vertex-connector of size at most γ if and only if for each vertex v, $G - v$ has at most $\gamma + 1$ components and*

$$(63.87) \qquad \sum_{U \in \mathcal{P}} (2 - |N(U)|) \leq 2\gamma$$

for each collection $\mathcal{P}$ of disjoint nonempty subsets U of V with $|U| \leq |V| - 3$.

Proof. Directly from Theorem 63.12. ∎

Jackson and Jordán [2001] showed that for each fixed k, a minimum-size k-vertex-connector for an undirected graph can be found in polynomial time.

Notes. Watanabe and Nakamura [1988,1993] give a characterization of the minimum size of a 3-vertex-connector, and Watanabe and Nakamura [1993] describe an $O(n(n+m)^2)$-time algorithm (for a sketch, see Watanabe and Nakamura [1988, 1990]). Hsu and Ramachandran [1991] gave a linear-time algorithm for this problem. Hsu [1992,2000] gave an almost-linear-time algorithm to find a minimum-size 4-vertex-connector for a 3-connected undirected graph.

Note that the natural extensions of Corollary 63.12a does not hold for k-vertex-connectors with $k \geq 4$, as is shown by the complete bipartite graph $K_{3,3}$.

For approximation algorithms, see Jordán [1993b,1995,1997a], Khuller and Thurimella [1993], Cheriyan and Thurimella [1996b,1999], Nutov and Penn [1997], Penn and Shasha-Krupnik [1997], and Jackson and Jordán [2000].

63.6a. Further notes

Corollary 53.6b implies the following characterization for connectivity augmentation, due to Frank [1979b].

Theorem 63.13. *Let $D = (V, A)$ be a digraph, let $r \in V$, and let $k \in \mathbb{Z}_+$ be such that D contains k disjoint r-arborescences. Moreover, let $D' = (V, A')$ and $l \in \mathbb{Z}_+^{A'}$. Then the minimum of $l(C)$ where $C \subseteq A'$ such that the digraph $(V, A \cup C)$ (taking arcs multiple) has $k + 1$ disjoint r-arborescences is equal to the maximum size t of a family of nonempty subsets $U_1, \ldots, U_t$ of $V \setminus \{r\}$ such that $d_D^{\text{in}}(U_j) = k$ for $j = 1, \ldots, t$ and such that each arc a of D' enters at most $l(a)$ of the U_j.*

Proof. Consider the digraph $D'' = (V, A'')$ with $A'' := A \cup A'$ (taking multiple arcs for arcs occurring both in A and in A'). Now the minimum in this corollary is equal to the minimum of $\sum_{a \in A'} l(a) x_a$ where $x \in \mathbb{Z}^{A''}$ satisfies

(63.88) $\quad 0 \leq x_a \leq 1$ if $a \in A'$,
$\quad\quad\quad\quad x(\delta_{D'}^{\text{in}}(U)) \geq k + 1 - d_D^{\text{in}}(U)$ for each nonempty $U \subseteq V \setminus \{r\}$.

Since (63.88) is TDI by Corollary 53.6b, this minimum is equal to the maximum described in the present corollary. ∎

The problem of making a bipartite directed graph strongly connected while preserving bipartiteness is considered by Gabow and Jordán [1999,2000a]. Augmenting the arc-connectivity while preserving bipartiteness is studied by Gabow and Jordán [2000b]. Making a bipartite undirected graph k-edge-connected while preserving bipartiteness, and, more generally, edge-connectivity augmentation with partition constraints, is studied by Bang-Jensen, Gabow, Jordan, and Szigeti [1998, 1999].

For the 'successive augmentation problem', see Cheng and Jordán [1999]. For NP-completeness and approximation results for connectivity augmentation, see Frederickson and Ja'Ja' [1981,1982]. Frank and Király [2001] studied problems that combine graph orientation and connectivity augmentation.

Planar graph connectivity augmentation was considered by Provan and Burk [1999].

Ishii, Nagamochi, and Ibaraki [1997,1998b,1998a,1999,2000,2001] considered the problem of making an undirected graph both k-vertex- and l-edge-connected.

For surveys on connectivity augmentation, see Frank [1993a,1994a], Jordán [1994,1997b], and Nagamochi [2000].

Part VI

Cliques, Stable Sets, and Colouring

Part VI: Cliques, Stable Sets, and Colouring

We now arrive at a class of problems that are in general NP-complete: finding a maximum-size clique or stable set or a minimum vertex-colouring in an undirected graph. These problems relate to each other: a stable set in a graph is a clique in the complementary graph, a colouring is a partitioning of the vertex set into stable sets, and the maximum size of a clique is a lower bound for the minimum number of colours.

Graph colouring was motivated originally by the four-colour conjecture formulated in the 1850s, stating that each planar map can be coloured with at most four colours — since 1977 a theorem of Appel and Haken. Later, colouring turned out to have several other applications, like in school scheduling, timetabling, and warehouse planning and in bungalow, terminal, platform, and frequency assignment. Finding optimum cliques of stable sets again can be used in frequency assignment, and in set packing problems, which show up for instance in crew scheduling.

While these problems are in general NP-complete, some are polynomial-time solvable for special classes of graphs: perfect graphs, t-perfect graphs, claw-free graphs. They form the body of this part.

Perfect graphs carry one of the deepest theorems in graph theory, the strong perfect graph theorem — recently proved by Chudnovsky, Robertson, Seymour, and Thomas. The proof is highly complicated, and we cannot give it in this book.

We refer to Part III for stable sets in and colouring of *line graphs* — equivalently, matchings and edge-colouring.

Chapters:

64. Cliques, stable sets, and colouring	1083
65. Perfect graphs: general theory	1106
66. Classes of perfect graphs	1135
67. Perfect graphs: polynomial-time solvability	1152
68. T-perfect graphs	1186
69. Claw-free graphs	1208

Chapter 64

Cliques, stable sets, and colouring

This chapter studies cliques, stable sets, and colouring for general graphs: complexity, polyhedra, fractional solutions, weighted versions.
In studying later chapters of this part, one can do largely without the results of the present chapter. Only some elementary definitions and terminology will be needed. It suffices to use this chapter just for reference.
In this chapter, all graphs can be assumed to be simple.

64.1. Terminology and notation

Let $G = (V, E)$ be an undirected graph. A *clique* is a set of vertices any two of which are adjacent. The maximum size of a clique in G is the *clique number* of G, and is denoted by $\omega(G)$.

A *stable set* is a set of vertices any two of which are nonadjacent. The maximum size of a stable set in G is called the *stable set number* of G, and is denoted by $\alpha(G)$.

A *vertex cover* is a set of vertices intersecting all edges. The minimum size of a vertex cover in G is called the *vertex cover number* of G, and is denoted by $\tau(G)$.

A *(vertex-)colouring* of G is a partition of V into stable sets $S_1, \ldots, S_k$. The sets $S_1, \ldots, S_k$ are called the *colours* of the colouring. The minimum number of colours in a vertex-colouring of G is called the *(vertex-)colouring number* of G, denoted by $\chi(G)$. A graph G is called *k-(vertex-)colourable* if $\chi(G) \leq k$, and *k-chromatic* if $\chi(G) = k$. A *minimum (vertex-)colouring* is a colouring with $\chi(G)$ colours. A *k-(vertex-)colouring* is a colouring with k colours.

A *clique cover* of G is a partition of V into cliques. The minimum number of cliques in a clique cover of G is called the *clique cover number* of G, and is denoted by $\overline{\chi}(G)$. A *minimum clique cover* is a clique cover with $\overline{\chi}(G)$ cliques.

The following relations between these parameters are immediate:

(64.1) $\alpha(G) = \omega(\overline{G})$, $\overline{\chi}(G) = \chi(\overline{G})$, $\omega(G) \leq \chi(G)$, $\alpha(G) \leq \overline{\chi}(G)$,
$\tau(G) = |V| - \alpha(G)$.

64.2. NP-completeness

It is NP-complete to find a maximum-size stable set in a graph. To be more precise, the *stable set problem*: given a graph G and a natural number k, decide if $\alpha(G) \geq k$, is NP-complete (according to Karp [1972b] this is implicit in the work of Cook [1971] and was also known to R. Reiter):

Theorem 64.1. *Determining the stable set number is* NP-*complete.*

Proof. We reduce the satisfiability problem to the stable set problem. Let $C_1 \wedge \cdots \wedge C_k$ be a Boolean expression, where each C_i is of the form $y_1 \vee \cdots \vee y_m$, with $y_1, \ldots, y_m \in \{x_1, \neg x_1, \ldots, x_n, \neg x_n\}$. Call $x_1, \neg x_1, \ldots, x_n, \neg x_n$ the *literals*. Consider the graph $G = (V, E)$ with $V := \{(\sigma, i) \mid \sigma \text{ is a literal in } C_i\}$ and $E := \{\{(\sigma, i), (\tau, j)\} \mid i = j \text{ or } \sigma = \neg \tau\}$. Then the expression is satisfiable if and only if G has a stable set of size k. ∎

It can be shown that the stable set problem remains NP-complete if the graphs are restricted to 3-regular planar graphs (Garey, Johnson, and Stockmeyer [1976]) or to triangle-free graphs (Poljak [1974]).

Since a subset U of VG is a vertex cover if and only if $VG \setminus U$ is a stable set, we also have:

Corollary 64.1a. *Determining the vertex cover number is* NP-*complete.*

Proof. By Theorem 64.1, since the vertex cover number of a graph G is equal to $|VG|$ minus the stable set number. ∎

A subset C of VG is a clique in a graph G if and only if C is a stable set in the complementary graph $\overline{G}$. So finding a maximum-size clique in G is equivalent to finding a maximum-size stable set in $\overline{G}$, and $\omega(G) = \alpha(\overline{G})$. Hence, as determining $\alpha(G)$ is NP-complete, also determining $\omega(G)$ is NP-complete.

Also, it is NP-complete to decide if a graph is k-colourable (Karp [1972b]):

Theorem 64.2. *Determining the vertex-colouring number is* NP-*complete.*

Proof. We show that the stable set problem can be reduced to the vertex-colouring problem. Let $G = (V, E)$ be an undirected graph and let $k \in \mathbb{Z}_+$. We want to decide if $\alpha(G) \geq k$. To this end, let V' be a copy of V and let C be a set of size k, where V, V', and C are disjoint. Make a graph H with vertex set $V \cup V' \cup C$ as follows. A pair of vertices in V is adjacent in H if and only if it is adjacent in G. The sets V' and C are cliques in H. Each vertex in V is adjacent to each vertex in $V' \cup C$, except to its copy in V'. No vertex in V' is adjacent to any vertex in C.

This defines the graph H. Then $\alpha(G) \geq k$ if and only if $\chi(H) \leq |V| + 1$. ∎

Well-known is the *four-colour conjecture* (or *4CC*), stating that $\chi(G) \leq 4$ for each loopless planar graph G. This conjecture was proved by Appel and Haken [1977] and Appel, Haken, and Koch [1977], and is now called the *four-colour theorem*. (A shorter proof was given by Robertson, Sanders, Seymour, and Thomas [1997], leading to an $O(n^2)$-time 4-colouring algorithm for planar graphs (Robertson, Sanders, Seymour, and Thomas [1996]).)

However, it is NP-complete to decide if a planar graph is 3-colourable, even if the graph has maximum degree 4 (Garey, Johnson, and Stockmeyer [1976]). Moreover, determining the colouring number of a graph G with $\alpha(G) \leq 4$ is NP-complete (cf. Garey and Johnson [1979]). Holyer [1981] showed that deciding if a 3-regular graph is 3-edge-colourable is NP-complete (see Section 28.3). Note that one can decide in polynomial time if a graph G is 2-colourable, since bipartiteness can be checked in polynomial time.

These NP-completeness results imply that if NP$\neq$co-NP, then one may not expect a min-max relation characterizing the stable set number $\alpha(G)$, the vertex cover number $\tau(G)$, the clique number $\omega(G)$, or the colouring number $\chi(G)$ of a graph G.

64.3. Bounds on the colouring number

A lower bound on the colouring number is given by the clique number:

(64.2) $\omega(G) \leq \chi(G)$.

This is easy, since in any clique all vertices should have different colours.

There are several graphs which have strict inequality in (64.2). We mention the odd circuits C_k, with k odd and ≥ 5: then $\omega(C_k) = 2$ and $\chi(C_k) = 3$. Moreover, for the complement $\overline{C}_k$ of any such graph we have: $\omega(\overline{C}_k) = \lfloor k/2 \rfloor$ and $\chi(\overline{C}_k) = \lceil k/2 \rceil$.

It was a conjecture of Berge [1963a] that these graphs are crucial. In May 2002, M. Chudnovsky, N. Robertson, P.D. Seymour, and R. Thomas announced that they have found a proof of this conjecture.

Strong perfect graph theorem: Each graph G with $\omega(G) < \chi(G)$ has C_k or $\overline{C}_k$ as induced subgraph for some odd $k \geq 5$.

It is convenient to define a *hole* of a graph G to be an induced subgraph of G isomorphic to C_k for some $k \geq 4$. Moreover, an *antihole* is an induced subgraph isomorphic to $\overline{C}_k$ for some $k \geq 4$. A hole or antihole is *odd* if it has an odd number of vertices. Then the strong perfect graph theorem can be formulated as: each graph G with $\omega(G) < \chi(G)$ has an odd hole or odd antihole.

For more on this we refer to Chapter 65.

64.3a. Brooks' upper bound on the colouring number

There is a trivial upper bound on the colouring number:

(64.3) $\chi(G) \leq \Delta(G) + 1$,

where $\Delta(G)$ denotes the maximum degree of G. This bound follows by colouring the vertices 'greedily' one by one: at any stage, at least one colour (out of $\Delta(G)+1$ colours) is not used by the neighbours.

Brooks [1941] sharpened this inequality as follows. We follow the proof given by Lovász [1975d].

Theorem 64.3 (Brooks' theorem). *For any connected graph G one has $\chi(G) \leq \Delta(G)$, except if G is a complete graph or an odd circuit.*

Proof. We can assume that G is 2-connected, since otherwise we can apply induction. Moreover, we can assume that $\Delta(G) \geq 3$. Let $k := \Delta(G)$.

I. First assume that G has nonadjacent vertices u and w with $G - u - w$ disconnected. Let V_1 and V_2 be proper subsets of V such that $V_1 \cup V_2 = V$, $V_1 \cap V_2 = \{u, w\}$, and no edge connects $V_1 \setminus \{u, w\}$ and $V_2 \setminus \{u, w\}$. Let $G_1 := G[V_1]$ and $G_2 := G[V_2]$.

For $i = 1, 2$, we know by induction that $\chi(G_i) \leq k$, since G_i is not complete (as u and w are nonadjacent), and since $\Delta(G_i) \leq k$ and $k \geq 3$. By symmetry of G_1 and G_2, we can assume that each k-colouring of G_1 gives u and w the same colour (otherwise G_1 and G_2 have k-colourings that coincide on u and w, yielding a k-colouring of G). This implies that both u and w have degree at least $k - 1$ in G_1. Hence they have degree at most 1 in G_2. Therefore, as $k \geq 3$, G_2 has a k-colouring giving u and w the same colour. So G is k-colourable.

II. Now choose a vertex v of maximum degree. As G is not a complete graph, v has two nonadjacent neighbours, say u and w. By part I, we can assume that $G - u - w$ is connected. Hence it has a spanning tree T. Orient T so as to obtain a rooted tree, rooted at v. Hence we can order the vertices of G as $v_1, \ldots, v_n$ such that $v_1 = v$, $v_{n-1} = u$, $v_n = w$, and such that each v_i with $i > 1$ is adjacent to some v_j with $j < i$. Give u and w colour 1. Next successively for $i = n - 2, n - 1, \ldots, 1$, we can give a colour from $1, \ldots, k$ to v_i different from the colours given to the neighbours v_j of v_i with $j > i$. Such a colour exists, since if $i > 1$, there are less than k neighbours v_j of v_i with $j > i$; and if $i = 1$, there are k such neighbours, but neighbours u and w have the same colour. ∎

(A related proof was given by Ponstein [1969], and a strengthening of Brooks' theorem by Reed [1999a]. For another proof of Brooks' theorem, see Melnikov and Vizing [1969].)

64.3b. Hadwiger's conjecture

Another upper bound on the colouring number is conjectured by Hadwiger [1943]. Since there exist graphs with $\omega(G) < \chi(G)$, it is not true that if $\chi(G) \geq k$, then G contains the complete graph K_k on k vertices as a subgraph. However, Hadwiger conjectured the following, where a graph H is called a *minor* of a graph G if H arises from some subgraph of G by contracting some (possibly none) edges.

Hadwiger's conjecture: If $\chi(G) \geq k$, then G contains K_k as a minor.

In other words, for each k, the graph K_k is the only graph G with the property that G is not $(k-1)$-colourable and each proper minor of G is $(k-1)$-colourable.

Hadwiger's conjecture is trivial for $k = 1, 2, 3$, and was shown by Hadwiger [1943] for $k = 4$ (also by Dirac [1952]):

Theorem 64.4. *If G has no K_4 minor, then $\chi(G) \leq 3$.*

Proof. One may assume that $G = (V, E)$ is not a forest or a circuit. Then G has a circuit C not covering all vertices of G. Choose $v \in V \setminus VC$. If G is 3-connected, there are three paths from v to VC, disjoint except for v. This creates a K_4 minor, a contradiction.

So G is not 3-connected, that is, G has a vertex-cut of size less than 3. Then $\chi(G) \leq 3$ follows by induction: if G is disconnected or has a 1-vertex-cut, this is trivial, and if G is 2-connected and has a 2-vertex-cut $\{u, w\}$, we can apply induction to the graphs $G - K$ after adding an edge uw, for each component K of $G - u - w$. ∎

(For another proof, see Woodall [1992].)

As planar graphs contain no K_5 minor, Hadwiger's conjecture for $k = 5$ implies the four-colour theorem. In fact, Wagner [1937a] showed that his decomposition theorem (Theorem 3.3) implies that Hadwiger's conjecture for $k = 5$ is equivalent to the four-colour conjecture. (Young [1971] gave a 'quick' proof of this equivalence.) The four-colour conjecture was proved by Appel and Haken [1977] and Appel, Haken, and Koch [1977]. (Robertson, Sanders, Seymour, and Thomas [1997] gave a shorter proof.)

Robertson, Seymour, and Thomas [1993] showed that Hadwiger's conjecture is true also for $k = 6$, by reducing it again to the four-colour theorem. For $k \geq 7$, Hadwiger's conjecture is unsettled.

Halin [1964] has proved that if G has no K_k minor, then $\chi(G) \leq 2^{k-2}$ (Wagner [1964] gave a short proof). Further progress on Hadwiger's conjecture was made by Wagner [1960], Mader [1968], Jakobsen [1971], Duchet and Meyniel [1982], Kostochka [1982], Fernandez de la Vega [1983], Thomason [1984], and Reed and Seymour [1998].

Hajós' conjecture. G. Hajós[1] conjectured (more strongly than Hadwiger) that any k-chromatic graph contains a subdivision of K_k as subgraph. For $k \leq 4$, Hajós' conjecture is equivalent to Hadwiger's conjecture.

Hajós' conjecture was refuted by Catlin [1979] for $k = 8$. He showed that the line graph $L(G)$ of the graph G obtained from the 5-circuit C_5 by replacing each

---

[1] According to Toft [1996], Hajós considered the conjecture already in the 1940s in connection with the four-colour conjecture, but he never published it. (The paper Hajós [1961] commonly referred to, does not give Hajós' conjecture.) An early written record of the conjecture is in the review of Tutte [1961b], in the January 1961 issue of *Mathematical Reviews*, of the book *Färbungsprobleme auf Flächen und Graphen* (Colouring Problems on Surfaces and Graphs) by Ringel [1959]. This book itself however does not mention the conjecture.

edge by three parallel edges, has colouring number 8 (as $L(G)$ has 15 vertices and stable set number 2), but contains no subdivision of K_8.

Catlin in fact gave a counterexample to Hajós' conjecture for each $k \geq 7$. Erdős and Fajtlowicz [1981] showed that almost all graphs are counterexamples to Hajós' conjecture.

Related is the following result of Hajós [1961]: any graph G with $\chi(G) \geq k$ can be obtained from the complete graph K_k by a series of the following operations on graphs (each preserving $\chi \geq k$):

(64.4) (i) add vertices or edges;
(ii) identify two nonadjacent vertices;
(iii) take two disjoint graphs G_1 and G_2, choose edges $e_1 = u_1v_1$ of G_1 and $e_2 = u_2v_2$ of G_2, identify u_1 and u_2, delete e_1 and e_2, and add edge v_1v_2.

64.4. The stable set, clique, and vertex cover polytope

The *stable set polytope* $P_{\text{stable set}}(G)$ of a graph $G = (V, E)$ is the convex hull of the incidence vectors of the stable sets in G. Since finding a maximum-size stable set is NP-complete, one may not expect a polynomial-time checkable system of linear inequalities describing the stable set polytope (Corollary 5.16a). More precisely, if NP$\neq$co-NP, then there do not exist inequalities satisfied by the stable set polytope such that their validity can be certified in polynomial time and such that the inequality $\mathbf{1}^T x \leq \alpha(G)$ is a nonnegative linear combination of them.

The *clique polytope* $P_{\text{clique}}(G)$ of a graph $G = (V, E)$ is the convex hull of the incidence vectors of cliques. Trivially

(64.5) $\quad P_{\text{clique}}(G) = P_{\text{stable set}}(\overline{G})$.

Hence, similar observations hold for the clique polytope.

Another related polytope is the *vertex cover polytope* $P_{\text{vertex cover}}(G)$ of G, being the convex hull of the incidence vectors of vertex covers in G. Since a subset U of V is a vertex cover if and only if $V \setminus U$ is a stable set, we have

(64.6) $\quad x \in P_{\text{vertex cover}}(G) \iff 1 - x \in P_{\text{stable set}}(G)$.

This shows that problems on the two types of polytopes can be reduced to each other.

64.4a. Facets and adjacency on the stable set polytope

Padberg [1973] (for facets induced by odd circuits) and Nemhauser and Trotter [1974] observed that

(64.7) each facet of the stable set polytope of an induced subgraph $G[U]$ of G, is the restriction to U of some unique facet of $P_{\text{stable set}}(G)$.

More precisely, for each facet F of $P_{\text{stable set}}(G[U])$ there is a unique facet F' of $P_{\text{stable set}}(G)$ with the property that $F = \{x \in \mathbb{R}^U \mid x' \in F'\}$, where $x'_v = x_v$ if $v \in U$ and $x'_v = 0$ if $v \in V \setminus U$.

To prove (64.7), it suffices to prove it for $U = V \setminus \{v\}$ for some $v \in V$. Let F be a facet of $P_{\text{stable set}}(G - v)$. We can consider F as a face of codimension 2 of $P_{\text{stable set}}(G)$ (by extending F with a 0 at coordinate v). Define $H := \{x \in \mathbb{R}^V \mid x_v = 0\}$. As F is on the facet $F'' := P_{\text{stable set}}(G) \cap H$ of $P_{\text{stable set}}(G)$, there is a unique facet F' of $P_{\text{stable set}}(G)$ with $F = F'' \cap F'$. This implies $F = F' \cap H$, since

(64.8) $\qquad F = F' \cap F'' = F' \cap P_{\text{stable set}}(G) \cap H = F' \cap H.$

Suppose now that $P_{\text{stable set}}(G)$ has another facet F'''' with $F = F'''' \cap H$. Then $F \subseteq F'''' \cap F'' \subseteq F''' \cap H = F$, and hence $F = F'' \cap F'''$, contradicting the unicity of F'. This proves (64.7).

Padberg [1973] also showed the following:

Theorem 64.5. *Let $G = (V, E)$ be a graph and let $a \in \mathbb{Z}_+^V$. Then the inequality*

(64.9) $\qquad a^\mathsf{T} x \leq 1$

is valid for the stable set polytope of G if and only if a is the incidence vector of a clique C. Moreover, (64.9) determines a facet if and only if C is an inclusionwise maximal clique.

Proof. Trivially, inequality (64.9) is valid if $a = \chi^C$ for some clique C. Conversely, if (64.9) is valid, then a is a 0,1 vector, and hence the incidence vector of a subset C of V. Then C is a clique, since otherwise C contains a stable set S of size 2, implying that (64.9) is not valid for $x := \chi^S$.

In proving the second statement, we can assume that $a = \chi^C$ for some clique C. Suppose that (64.9) determines a facet, and that C is not an inclusionwise maximal clique. Then there is a clique C' properly containing C. Hence for each $x \in P_{\text{stable set}}(G)$, if $x(C) = 1$, then $x(C') = 1$. This implies that the inequality $x(C) \leq 1$ is not facet-inducing, a contradiction.

Finally suppose that C is an inclusionwise maximal clique. To see that (64.9) determines a facet, let $a^\mathsf{T} x = \beta$ be satisfied by all x in the stable set polytope with $x(C) = 1$. So $a(S) = \beta$ for each stable set S with $|S \cap C| = 1$. Then $a_v = \beta$ for each $v \in C$, as $S := \{v\}$ is stable. Also, $a_u = 0$ for each $u \in V \setminus C$, since by the maximality of C, there is a vertex $v \in C$ that is not adjacent to u. So $S := \{u, v\}$ is stable, and hence $a_u + a_v = \beta$. So $a_u = 0$. Concluding, $a^\mathsf{T} x = \beta$ is some multiple of $x(C) = 1$, and hence $x(C) \leq 1$ determines a facet. ∎

Graphs for which the nonnegativity and clique inequalities determine all facets, are precisely the *perfect* graphs — see Chapter 65.

Trivially, the vertices of the stable set polytope are precisely the incidence vectors of the stable sets. Chvátal [1975a] characterized adjacency:

Theorem 64.6. *The incidence vectors of two different stable sets R, S are adjacent vertices of the stable set polytope if and only if $R \triangle S$ induces a connected subgraph of G.*

Proof. To see necessity, if $G[R \triangle S]$ is not connected, then (as it is bipartite) it has two colour classes U and W with $\{U, W\} \neq \{R \setminus S, S \setminus R\}$. Let $U' := U \cup (R \cap S)$ and

$W' := W \cup (R \cap S)$. Then U' and W' are stable sets and $\frac{1}{2}(\chi^R + \chi^S) = \frac{1}{2}(\chi^{U'} + \chi^{W'})$, contradicting the adjacency of χ^R and χ^S.

To see sufficiency, if χ^R and χ^S are not adjacent, then there exist stable sets U and W, and $\lambda, \mu \in (0,1)$ such that $\lambda \chi^R + (1-\lambda)\chi^S = \mu \chi^U + (1-\mu)\chi^W$ and $\{U, W\} \neq \{R, S\}$. So $U \cap W = R \cap S$. Hence $U \setminus W, W \setminus U$ forms a bipartition of $G[R \triangle S]$ different from the bipartition $R \setminus S, S \setminus R$. This contradicts the connectedness of $G[R \triangle S]$. ∎

64.5. Fractional stable sets

The incidence vectors of stable sets in an undirected graph $G = (V, E)$ are precisely the integer vectors $x \in \mathbb{R}^V$ satisfying

(64.10)   (i) $0 \leq x_v \leq 1$ for $v \in V$,
          (ii) $x_u + x_v \leq 1$ for $\{u, v\} \in E$.

(The inequalities (64.10)(ii) are called the *edge inequalities*.) Any (not necessarily integer) solution x of (64.10) is called a *fractional stable set*. By definition, its *size* is equal to $x(V)$.

The maximum size of a fractional stable set is called the *fractional stable set number* and is denoted by $\alpha^*(G)$. By linear programming duality, $\alpha^*(G)$ is equal to the *fractional edge cover number* $\rho^*(G)$ (assuming that G has no isolated vertices), which is the minimum value of $y(E)$ over all $y \in \mathbb{R}^E$ satisfying

(64.11)   (i) $0 \leq y_e \leq 1$ for $e \in E$,
          (ii) $y(\delta(v)) \geq 1$ for $v \in V$.

Any solution y of (64.11) is called a *fractional edge cover*.

This was also discussed in Section 30.11, where it was shown that each vertex of the polytope determined by (64.11) (the fractional edge cover polytope) is half-integer. A similar result holds for the *fractional stable set polytope*, which is the polytope determined by (64.10) (the result is implicit in Balinski [1965]):

Theorem 64.7. *Each vertex of the fractional stable set polytope P is half-integer.*

Proof. Let x be a vertex of P. Let $U := \{v \in V \mid 0 < x_v < \frac{1}{2}\}$ and let $W := \{v \in V \mid \frac{1}{2} < x_v < 1\}$. Then there is an $\varepsilon > 0$ such that both $x + \varepsilon(\chi^U - \chi^W)$ and $x - \varepsilon(\chi^U - \chi^W)$ belong to P. As x is a vertex, it follows that $\chi^U - \chi^W = 0$. So $U = W = \emptyset$. ∎

(This proof was provided to Nemhauser and Trotter [1974] by a referee of their paper.)

The theorem also follows from the observation of Balinski [1965] that each nonsingular submatrix of the incidence matrix of a graph has a half-integer inverse.

Theorem 64.7 implies that $\alpha^*(G) = \frac{1}{2}\alpha_2(G)$, where $\alpha_2(G)$ is the maximum size of a *2-stable set*, which is an integer vector $x \in \mathbb{R}^V$ satisfying

(64.12) (i) $x_v \geq 0$ for $v \in V$,
(ii) $x_u + x_v \leq 2$ for $\{u, v\} \in E$

(cf. Section 30.9).

Moreover, it implies a characterization of the *2-stable set polyhedron*, which is the convex hull of the 2-stable sets:

Corollary 64.7a. *The 2-stable set polyhedron is determined by* (64.12).

Proof. Directly from Theorem 64.7. ∎

With the following construction, the problem of finding a maximum-weight fractional stable set (and similarly, a maximum-weight 2-stable set), can be reduced to the problem of finding a maximum-weight stable set in a bipartite graph. The latter problem is strongly polynomial-time solvable, by Theorem 21.10.

Let $G = (V, E)$ be a graph. Let V' be a copy of V. For any $v \in V$, let v' denote the copy of V in V'. Define $\widetilde{V} := V \cup V'$. Let $\widetilde{E}$ be the set of pairs $u'v$ and uv', over all edges uv of G. Then $\widetilde{G} := (\widetilde{V}, \widetilde{E})$ is a bipartite graph.

For any weight function $w : V \to \mathbb{R}_+$, define $\widetilde{w} : \widetilde{V} \to \mathbb{R}_+$ by $\widetilde{w}(v) := \widetilde{w}(v') := w(v)$ for $v \in V$. Then any stable set S in $\widetilde{G}$ maximizing $\widetilde{w}(S)$ gives a 2-stable set x in G maximizing $w^\mathsf{T} x$, by defining $x_v := |S \cap \{v, v'\}|$. Indeed, for any 2-stable set x' in G we can define a stable set S' in $\widetilde{G}$ by

(64.13) $S' := \{v \in V \mid x'_v \geq 1\} \cup \{v \in V' \mid x'_v \geq 2\}$.

Then $w^\mathsf{T} x' = \widetilde{w}(S') \leq \widetilde{w}(S) = w^\mathsf{T} x$. (Here we assume without loss of generality that G has no isolated vertices.)

64.5a. Further on the fractional stable set polytope

Nemhauser and Trotter [1974] characterized the vertices of the fractional stable set polytope:

Theorem 64.8. *A vector $x \in \mathbb{R}^V$ is a vertex of the fractional stable set polytope P of G if and only if $x = \chi^{U_2} + \frac{1}{2}\chi^{U_1}$, where U_2 is a stable set of G, where U_1 is disjoint from $U_2 \cup N(U_2)$, and where each component of $G[U_1]$ is nonbipartite.*

Proof. *Necessity.* Let x be a vertex of P, and define $U_2 := \{v \in V \mid x_v = 1\}$ and $U_1 := \{v \in V \mid x_v = \frac{1}{2}\}$. Then U_2 is a stable set and no vertex in U_1 is adjacent to any vertex in U_2. So U_1 is disjoint from $U_2 \cup N(U_2)$.

If some component of $G[U_1]$ would be bipartite, say with colour classes S and T, then $x \pm \varepsilon(\chi^S - \chi^T)$ would belong to P for some $\varepsilon \neq 0$. This contradicts the fact that x is a vertex of P.

Sufficiency. Suppose that x satisfies the condition, and that x is not a vertex of P. Then there is a nonzero vector y such that both $x + y$ and $x - y$ belong to P. Necessarily, $y_v = 0$ if $v \notin U_1$. Moreover, for each edge uw of $G[U_1]$ one has $y_u + y_w = 0$, since $x_u + x_w = 1$. As each component of $G[U_1]$ contains an odd circuit, this implies $y_v = 0$ for each $v \in U_1$. So $y = 0$, a contradiction. ∎

A useful condition was given by Nemhauser and Trotter [1975]:

Theorem 64.9. *Let $G = (V, E)$ be a graph, let $w : V \to \mathbb{R}$ be a weight function, and let $S \subseteq V$ be a stable set. If S is a maximum-weight stable set in the subgraph of G induced by $S \cup N(S)$, then S is contained in some maximum-weight stable set of G.*

Proof. Let T be a maximum-weight stable set of G. Define $U := (S \cup T) \setminus N(S)$. Trivially, U is stable. Also, $w(N(S) \cap T) \leq w(S \setminus T)$, since $w((S \cup N(S)) \cap T) \leq w(S)$, as S has maximum weight in $G[S \cup N(S)]$. Hence

(64.14) $\quad w(U) = w(T) + w(S \setminus T) - w(N(S) \cap T) \geq w(T),$

implying that U is a maximum-weight stable set in G. ∎

This implies (Nemhauser and Trotter [1975]):

Corollary 64.9a. *Let $G = (V, E)$ be a graph, let $w : V \to \mathbb{R}$ be a weight function, and let x be a maximum-weight fractional stable set. Then $S := \{v \mid x_v = 1\}$ is contained in a maximum-weight stable set.*

Proof. This follows from Theorem 64.9, since S is a maximum-weight stable set in $G[S \cup N(S)]$. For if T would be a stable set in $G[S \cup N(S)]$ with $w(T) > w(S)$, then $x + \varepsilon(\chi^T - \chi^S)$ would belong to the fractional stable set polytope for some $\varepsilon > 0$, while it has weight larger than x, a contradiction. ∎

Picard and Queyranne [1977] showed that, for any graph $G = (V, E)$ and any weight function $w : V \to \mathbb{R}$, there is a unique minimal subset of vertices that has fractional values in some optimum fractional stable set (solving a problem posed by Nemhauser and Trotter [1975]):

Theorem 64.10. *Let $G = (V, E)$ be a graph, let $w : V \to \mathbb{R}$ be a weight function, and let x and y be maximum-weight fractional stable sets. Then there is a maximum-weight fractional stable set z such that, for each vertex v, z_v is integer if x_v or y_v integer.*

Proof. We can assume that x and y are half-integer (as we can assume that x and y are vertices of the fractional stable set polytope). For $i = 0, 1, 2$, let $X_i := \{v \mid x_v = i/2\}$ and $Y_i := \{v \mid y_v = i/2\}$. Then

(64.15) $\quad w(Y_0 \cap X_2) \leq w(X_0 \cap Y_2),$

since

(64.16) $\quad y + \frac{1}{2}(\chi^{Y_0 \cap X_2} - \chi^{X_0 \cap Y_2})$

is a fractional stable set. Otherwise, since X_2 is stable, there is an edge uv with $y_u + y_v = 1$, $u \in Y_0 \cap X_2$, and $v \notin X_0 \cap Y_2$. So $y_u = 0$, and hence $y_v = 1$. Also, $x_u = 1$, and hence $x_v = 0$. So $v \in X_0 \cap Y_2$, a contradiction. This shows (64.15).

Moreover,

(64.17) $\quad w(X_0 \setminus Y_0) \leq w(X_2 \setminus Y_2),$

since

(64.18) $\quad x + \frac{1}{2}(\chi^{X_0 \setminus Y_0} - \chi^{X_2 \setminus Y_2})$

is a fractional stable set. Otherwise there is an edge uv with $x_u + x_v = 1$, $u \in X_0 \setminus Y_0$, and $v \notin X_2 \setminus Y_2$. So $x_u = 0$, and hence $x_v = 1$. Also, $y_u > 0$, and hence $y_v < 1$. So $v \in X_2 \setminus Y_2$, a contradiction. This shows (64.17).

(64.15) and (64.17) imply that

(64.19) $\quad w(Y_1 \cap X_2) = w(X_2 \setminus Y_2) - w(X_2 \cap Y_0) \geq w(X_0 \setminus Y_0) - w(Y_2 \cap X_0)$
$\quad = w(Y_1 \cap X_0).$

Hence

(64.20) $\quad z := y + \frac{1}{2}(\chi^{Y_1 \cap X_2} - \chi^{Y_1 \cap X_0})$

has weight at least that of y. Moreover, z is a fractional stable set. Otherwise, as X_2 is stable, there is an edge uv with $y_u + y_v = 1$, $u \in Y_1 \cap X_2$ and $v \notin Y_1 \cap X_0$. So $y_u = y_v = \frac{1}{2}$, $x_u = 1$, hence $x_v = 0$. So $v \in Y_1 \cap X_0$, a contradiction. Hence z is a fractional stable set as required. ∎

Nemhauser and Trotter [1975] and Picard and Queyranne [1977] gave a polynomial-time algorithms to find a half-integer maximum-weight fractional stable set attaining the minimum number of fractional values. (This can be derived from the uniqueness of the minimal set of fractional vertices: just try $x_v = 0$ and $x_v = 1$ for each $v \in V$, and see if the fractional stable set number drops.)

Pulleyblank [1979a] and Bourjolly and Pulleyblank [1989] characterized the minimal set of fractional values. Related results were given by Grimmett [1986].

64.6. Fractional vertex covers

Similar results hold for *fractional vertex covers*, which are vectors $x \in \mathbb{R}^V$ satisfying

(64.21) $\quad$ (i) $\quad 0 \leq x_v \leq 1 \quad$ for $v \in V$,
$\qquad\quad$ (ii) $\quad x_u + x_v \geq 1 \quad$ for $\{u, v\} \in E$.

Trivially, a vector x is a fractional vertex cover if and only if $1 - x$ is a fractional stable set.

The minimum size of a fractional vertex cover is called the *fractional vertex cover number*, and is denoted by $\tau^*(G)$. So

(64.22) $\tau^*(G) + \alpha^*(G) = |V|$.

Again, by linear programming duality, $\tau^*(G)$ is equal to the *fractional matching number* $\nu^*(G)$, which is the maximum value of $y(E)$ over all $y \in \mathbb{R}^E$ satisfying

(64.23)   (i) $0 \leq y_e \leq 1$ for $e \in E$,
           (ii) $y(\delta(v)) \leq 1$ for $v \in V$.

Any solution y of (64.23) is called a *fractional matching*. This was also discussed in Section 30.3, where it was shown that each vertex of the polytope determined by (64.23) (the fractional matching polytope) is half-integer. A similar result holds for the *fractional vertex cover polytope*, which is the polytope determined by (64.21):

Theorem 64.11. *Each vertex of the fractional vertex cover polytope P is half-integer.*

Proof. Directly from Theorem 64.7, since x belongs to the fractional vertex cover polytope if and only if $\mathbf{1}-x$ belongs to the fractional stable set polytope. ∎

Theorem 64.11 implies that $\tau^*(G) = \frac{1}{2}\tau_2(G)$, where $\tau_2(G)$ is the minimum size of a *2-vertex cover*, which is an integer vector $x \in \mathbb{R}^V$ satisfying

(64.24)   (i) $x_v \geq 0$ for $v \in V$,
           (ii) $x_u + x_v \geq 2$ for $\{u,v\} \in E$

(cf. Section 30.10).

It also implies a characterization of the *2-vertex cover polyhedron*, which is the convex hull of the 2-vertex covers:

Corollary 64.11a. *The 2-vertex cover polyhedron is determined by (64.24).*

Proof. Directly from Theorem 64.11. ∎

By the results on fractional stable sets and 2-stable sets given in Section 64.5, and using the reductions described above, a minimum-weight fractional vertex cover and a minimum-weight 2-vertex cover can be found in strongly polynomial time.

Notes. Corollary 64.9a and Theorem 64.10 have direct analogues for vertex covers: given a graph $G = (V, E)$ and a weight function $w : V \to \mathbb{R}$,

(64.25)   for each minimum-weight fractional vertex cover x there is a minimum-weight vertex cover contained in $\{v \mid x_v \neq 0\}$,

and

(64.26)   for any two minimum-weight fractional vertex covers x and y there is a minimum-weight fractional vertex cover z such that for each $v \in V$: $x_v \in \mathbb{Z}$ or $y_v \in \mathbb{Z} \Rightarrow z_v \in \mathbb{Z}$.

These statements can be derived from Corollary 64.9a and Theorem 64.10 by again observing that a vector x is a (fractional) stable set if and only if $1 - x$ is a (fractional) vertex cover. Similarly, Theorem 64.8 implies a characterization of the vertices of the fractional vertex cover polytope.

64.6a. A bound of Lorentzen

The fractional stable set and vertex cover numbers give upper and lower bounds on the stable set and vertex cover number, respectively. These bounds are computable in polynomial time. A better polynomial-time computable bound was given by Lorentzen [1966]:

Theorem 64.12. *For each graph $G = (V, E)$:*

(64.27) $2\nu^*(G) - \nu(G) \leq \tau(G)$.

Proof. Since $\nu^*(G) = 2\nu_2(G)$ (cf. Section 30.2), there is a half-integer fractional matching $x : E \to \mathbb{R}_+$ with $x(E) = \nu^*(G)$, such that the support of x is the disjoint union of a matching and a number t of odd circuits. We can assume that each edge of G belongs to the support of x (as deleting edges increases neither $\nu(G)$ nor $\tau(G)$). Also we can assume that G has no isolated vertices. Then $\nu(G) = \frac{1}{2}(|V|-t)$, $\tau(G) = \frac{1}{2}(|V|+t)$, and $\nu^*(G) = \frac{1}{2}|V|$.   ∎

Bound (64.27) is generally a better lower bound on $\tau(G)$ than $\tau^*(G)$ (for example, for $G = K_3$). It implies an upper bound for $\alpha(G)$, generally better than $\alpha(G) \leq \rho^*(G)$:

Corollary 64.12a. *For each graph $G = (V, E)$ without isolated vertices:*

(64.28) $\alpha(G) \leq 2\rho^*(G) - \rho(G)$.

Proof. Using Theorem 30.9, we have $\alpha(G) = |V| - \tau(G) \leq |V| - 2\nu^*(G) + \nu(G) = 2(|V| - \nu^*(G)) - (|V| - \nu(G)) = 2\rho^*(G) - \rho(G)$.   ∎

64.7. The clique inequalities

A set of constraints stronger than the edge inequalities (64.10)(ii) is obtained by the 'clique inequalities'. Let $P(G)$ be the polytope in $\mathbb{R}^V$ determined by

(64.29)   (i) $x_v \geq 0$ for each $v \in V$,
        (ii) $x(C) \leq 1$ for each clique C.

The inequalities (64.29)(ii) are called the *clique inequalities*.

Since the integer solutions of (64.29) are exactly the incidence vectors of stable sets, the stable set polytope of G is equal to the integer hull of $P(G)$ (the convex hull of the integer vectors in $P(G)$).

We call any vector x satisfying (64.29) a *strong fractional stable set*. We denote

(64.30) $\alpha^{**}(G) :=$ *strong fractional stable set number* := the maximum size of a strong fractional stable set.

Since each strong fractional stable set is a fractional stable set, we know

(64.31) $\alpha(G) \leq \alpha^{**}(G) \leq \alpha^{*}(G)$:

So $\alpha^{**}(G)$ gives a better upper bound on $\alpha(G)$ than $\alpha^{*}(G)$ gives — however, $\alpha^{**}(G)$ is generally more difficult to compute.

Note that $P(G)$ is the antiblocking polyhedron of the clique polytope of G:

(64.32) $P(G) = A(P_{\text{clique}}(G))$.

(For background on antiblocking polyhedra, see Section 5.9.)

64.8. Fractional and weighted colouring numbers

For any graph $G = (V, E)$, the *fractional colouring number* $\chi^{*}(G)$ is the minimum value of $\lambda_1 + \cdots + \lambda_k$ with $\lambda_1, \ldots, \lambda_k \in \mathbb{R}_+$ such that there exist stable sets $S_1, \ldots, S_k$ with

(64.33) $\lambda_1 \chi^{S_1} + \cdots + \lambda_k \chi^{S_k} = \mathbf{1}$.

So if the λ_i are required to be integer, we have the colouring number.

By linear programming duality, the fractional colouring number is equal to the maximum of $\mathbf{1}^\mathsf{T} x$ over the polytope $\overline{P}(G)$ in $\mathbb{R}_+^V$ determined by

(64.34) $x_v \geq 0$ for each $v \in V$,
 $x(S) \leq 1$ for each stable set S.

(So $\overline{P}(G) = P(\overline{G})$ and $\overline{P}(G) = A(P_{\text{stable set}}(G))$.) Hence we have:

(64.35) $\chi^{*}(\overline{G}) = \alpha^{**}(G)$.

We denote

(64.36) $\overline{\chi}^{*}(G) := \chi^{*}(\overline{G})$,

which is called the *fractional clique cover number* of G.

No polynomial-time algorithm is known to calculate $\chi^{*}(G)$. Note that the separation problem for $\overline{P}(G)$ is NP-complete, since the optimization problem over $P_{\text{stable set}}(G)$ is NP-complete.

Given a graph $G = (V, E)$ and a weight function $w : V \to \mathbb{Z}_+$, the *weighted colouring number* $\chi_w(G)$ is the minimum value of $\lambda_1 + \cdots + \lambda_k$ with $\lambda_1, \ldots, \lambda_k \in \mathbb{Z}_+$ such that there exist stable sets $S_1, \ldots, S_k$ with

(64.37) $\lambda_1 \chi^{S_1} + \cdots + \lambda_k \chi^{S_k} = w$.

So if $w = 1$, then $\chi_w(G)$ is equal to the colouring number $\chi(G)$ of G. Hence determining $\chi_w(G)$ is NP-complete.

For $w : V \to \mathbb{Z}_+$, let graph G^w arise from G by replacing each vertex by a clique C_v of size $w(v)$, two vertices in different cliques C_u, C_v being adjacent if and only if u and v are adjacent. Then

(64.38) $\quad \chi_w(G) = \chi(G^w)$.

We denote

(64.39) $\quad \overline{\chi}_w(G) := \chi_w(\overline{G})$,

called the *weighted clique cover number* of G.

The fractional version is the *fractional weighted colouring number* $\chi_w^*(G)$, defined as the minimum value of $\lambda_1 + \cdots + \lambda_k$ with $\lambda_1, \ldots, \lambda_k \in \mathbb{R}_+$ such that there exist stable sets $S_1, \ldots, S_k$ with

(64.40) $\quad \lambda_1 \chi^{S_1} + \cdots + \lambda_k \chi^{S_k} = w$.

This value is equal to the maximum value of $w^\mathsf{T} x$ over the antiblocking polytope $A(P_{\text{stable set}}(G))$ pf $P_{\text{stable set}}(G)$. Since the optimization problem over $P_{\text{stable set}}(G)$ is NP-complete, determining $\chi_w^*(G)$ is NP-hard.

We denote

(64.41) $\quad \overline{\chi}_w^*(G) := \chi_w^*(\overline{G})$,

called the *fractional weighted clique cover number* of G.

The complexity results above can be specialized to classes of graphs. By the results of Grötschel, Lovász, and Schrijver [1981,1984c]:

(64.42)   For any collection $\mathcal{G}$ of graphs: there is a polynomial-time algorithm to find the fractional weighted colouring number for any graph in $\mathcal{G}$ and any weight function if and only there is a polynomial-time algorithm to find a maximum-weight stable set in any graph in $\mathcal{G}$ and for any weight function.

Since the problem of determining $\alpha(G)$ is NP-complete even if we restrict ourselves to planar cubic graphs, determining $\chi_w^*(G)$ for such graphs is NP-hard. As noticed in Grötschel, Lovász, and Schrijver [1981], determining $\chi_w^*(G)$ and $\chi(G)$ seem incomparable with respect to complexity. For cubic graphs G, $\chi(G)$ can be easily found in polynomial time (using Brooks' theorem (Theorem 64.3)), while determining $\chi_w^*(G)$ is NP-hard. In contrast to this, for the line graph G of a cubic graph H, $\chi(G)$ is NP-complete to compute by Holyer's theorem that 3-edge colourability is NP-complete (see Section 28.3), whereas $\chi_w^*(G)$ can be computed in polynomial time, since the separation problem over $A(P_{\text{stable set}}(G))$ is polynomial-time solvable, as the optimization problem over $P_{\text{stable set}}(G)$ is polynomial-time solvable (as it amounts to finding a maximum-weight matching in H).

64.8a. The ratio of $\chi(G)$ and $\chi^*(G)$

For later purposes we prove the following upper bound for the colouring number in terms of the fractional colouring number, obtained by applying a greedy-type algorithm (Johnson [1974a], Lovász [1975c]):

Theorem 64.13. *For any graph $G = (V, E)$:*

(64.43) $\chi(G) \leq (1 + \ln \alpha(G))\chi^*(G).$

Proof. Set $k := \alpha(G)$. Iteratively choose a maximum-size stable set S in G and reset G to $G - S$. We stop if VG is empty.

The stable sets found form a colouring $\mathcal{C}$ of the (original) vertex set V. So $\chi(G) \leq |\mathcal{C}|$.

For each $v \in V$, define

(64.44) $\quad x_v := \dfrac{1}{|S|},$

where S is the set in $\mathcal{C}$ containing v. Then $x(V) = |\mathcal{C}|$, and hence

(64.45) $\quad \chi(G) \leq x(V).$

Consider any stable set S' of G. Let S' consist of vertices $v_1, \ldots, v_k$, in the order by which they are covered by stable sets S in the algorithm. Then for each $i = 1, \ldots, k$, we have

(64.46) $\quad x_{v_i} \leq \dfrac{1}{k - i + 1}.$

Indeed, let v_i be covered by $S \in \mathcal{C}$. When we selected S, the vertices $v_i, v_{i+1}, \ldots, v_k$ were uncovered yet. As we chose S, we know $|S| \geq |\{v_i, v_{i+1}, \ldots, v_k\}| = k - i + 1$, implying (64.46).

(64.46) implies

(64.47) $\quad x(S') \leq \displaystyle\sum_{i=1}^{k} \dfrac{1}{k - i + 1} = \sum_{i=1}^{k} \dfrac{1}{i} \leq 1 + \ln k \leq 1 + \ln \alpha(G).$

So $(1 + \ln \alpha(G))^{-1} \cdot x$ satisfies (64.34), and hence

(64.48) $\quad (1 + \ln \alpha(G))^{-1} \cdot x(V) \leq \chi^*(G).$

Together with (64.45), this implies (64.43). ∎

This theorem will be used in proving Theorem 67.17.

64.8b. The Chvátal rank

In Section 36.7a we defined the polyhedron P' for any rational polyhedron P and the notion of the Chvátal rank of a polyhedron P.

Let $P(G)$ denote the polytope of strong fractional stable sets, that is, the polytope determined by (64.29) (the nonnegativity and clique constraints). For any polyhedron P, let P_I denote the *integer hull* of P, that is, the convex hull of the integer vectors in P.

Section 64.9a. Graphs with polynomial-time stable set algorithm

Chvátal [1973a] showed that there is no fixed t such that $P(G)^{(t)} = P(G)_I$ for each graph G, even if we restrict G to graphs with $\alpha(G) = 2$. Chvátal, Cook, and Hartmann [1989] showed that t can be at least $\frac{1}{3}\log n$ for such graphs (where n is the number of vertices).

We will see in Corollary 65.2e that the class of graphs G with $P(G)_I = P(G)$ is exactly the class of perfect graphs. By Edmonds' matching polytope theorem (Corollary 25.1a) if G is the line graph of some graph H, then $P(G)' = P(G)_I$, which is the matching polytope of H.

The smallest t for which $P(G)^{(t)} = P(G)_I$ might be an indication of the computational complexity of the stable set number $\alpha(G)$. For each fixed t, the stable set problem for graphs with $P(G)^{(t)} = P(G)_I$ belongs to NP∩co-NP. Chvátal [1973a] raised the question whether it belongs to P. (A negative indication is the result of Eisenbrand [1999] that given a polytope P by linear inequalities and given x, deciding if x belongs to P' is co-NP-complete.)

Another (weaker, but easier to compute) relaxation is: $Q(G)$ is the polytope of fractional stable sets; that is, the polytope in $\mathbb{R}^V$ determined by

(64.49) (i) $x_v \geq 0$ for each $v \in V$,
         (ii) $x_v + x_w \leq 1$ for each $vw \in E$.

Again $Q(G)_I = P_{\text{stable set}}(G)$. Since $Q(G) \supseteq P(G)$, there is no fixed t with $Q(G)^{(t)} = Q(G)_I$ for each graph G. Chvátal [1973a] noticed that for $G = K_n$ the smallest t with $Q(G)^{(t)} = P_{\text{stable set}}(G)$ is about $\log n$.

It is not difficult to see that $Q(G)'$ is the polytope determined by (64.49) together with

(64.50)   (iii) $x(VC) \leq \lfloor \frac{1}{2}|VC| \rfloor$ for each odd circuit C.

Graphs G with $Q(G)' = P_{\text{stable set}}(G)$ are called *t-perfect*. More on t-perfect graphs can be found in Chapter 68.

Chvátal [1975b] conjectures that there is no polynomial $p(n)$ such that for each graph G with n vertices we can obtain the inequality $x(V) \leq \alpha(G)$ from system (64.49) by adding at most $p(n)$ cutting planes. (That is, a list of at most $p(n)$ inequalities $a_i^T x \leq \lfloor \beta_i \rfloor$ such that, for each i, a_i is an integer vector and the inequality $a_i^T x \leq \beta_i$ is a nonnegative combination of inequalities from (64.49) and inequalities occurring earlier in the list.)

Chvátal, Cook, and Hartmann [1989] showed that the Chvátal rank of the following relaxation of the stable set polytope:

(64.51) $x_v \geq 0$ for $v \in V$,
              $x(U) \leq \alpha(G[U])$ for $U \subseteq V$,

is $\Omega((n/\log n)^{\frac{1}{2}})$, where G is a graph with n vertices. This relaxation is stronger than the polytope determined by just the nonnegativity and clique constraints.

64.9. Further results and notes

64.9a. Graphs with polynomial-time stable set algorithm

In the remaining chapters of this part we will see that a maximum-weight stable set can be found in strongly polynomial time in perfect graphs and their complements, in t-perfect graphs, and in claw-free graphs. In perfect graphs and their complements, also a minimum vertex-colouring can be found in polynomial time. In this section we list some other classes of graphs where a maximum-size stable set or a minimum vertex-colouring can be found in polynomial time.

A graph is a *circular-arc graph* if it is the intersection graph of a set of intervals on a circle. Gavril [1974a] gave polynomial-time algorithms for finding a maximum-size clique, a maximum-size stable set, and a minimum clique cover in these graphs. Karapetyan [1980] showed that $\chi(G) \leq \frac{3}{2}\omega(G)$ for any circular-arc graph G (proving a conjecture of Tucker [1975]). More on circular-arc graphs can be found in Klee [1969], Tucker [1971,1974,1975,1978,1980], Trotter and Moore [1976], Garey, Johnson, Miller, and Papadimitriou [1980], Golumbic [1980], Orlin, Bonuccelli, and Bovet [1981], Gupta, Lee, and Leung [1982], Skrien [1982], Leung [1984], Hsu [1985, 1995], Teng and Tucker [1985], Apostolico and Hambrusch [1987], Golumbic and Hammer [1988], Masuda and Nakajima [1988], Spinrad [1988], Shih and Hsu [1989a, 1989b], Bertossi and Moretti [1990], Hell, Bang-Jensen, and Huang [1990], Hsu and Tsai [1991], Deng, Hell, and Huang [1992,1996], Eschen and Spinrad [1993], Hsu and Spinrad [1995], Bhattacharya, Hell, and Huang [1996], Bhattacharya and Kaller [1997], Hell and Huang [1997], Feder, Hell, and Huang [1999], and McConnell [2001]. See also Section 65.6d.

A graph is a *circle graph* if its vertex set is a set of chords of the circle, two chords being adjacent if they intersect or cross. For these graphs, Gavril [1973] gave polynomial-time algorithms for finding a maximum-size clique and a maximum-size stable set. Bouchet [1985,1987b,1994], Naji [1985], and Gabor, Supowit, and Hsu [1989] showed that circle graphs can be recognized in polynomial time; this was improved to quadratic time by Spinrad [1994]. (Related results can be found in Fournier [1978], Garey, Johnson, Miller, and Papadimitriou [1980], Golumbic [1980], Rotem and Urrutia [1981], de Fraysseix [1984], Hsu [1985], Naji [1985], Dagan, Golumbic, and Pinter [1988], Gabor, Supowit, and Hsu [1989], Masuda, Nakajima, Kashiwabara, and Fujisawa [1990], Felsner, Müller, and Wernisch [1994], Ma and Spinrad [1994], Spinrad [1994], and Elmallah and Stewart [1998]. See also Section 65.6d.)

The weighted stable set problem was shown to be polynomial-time solvable for graphs without $K_5 - e$ minor by Barahona and Mahjoub [1994b]. (The graph $K_5 - e$ is obtained from K_5 by deleting one edge.) Descriptions of the corresponding polytopes are given by Barahona and Mahjoub [1994b,1994c].

Hsu, Ikura, and Nemhauser [1981] gave, for each fixed k, a polynomial-time algorithm for the weighted stable set problem for graphs without odd circuits of length larger than $2k + 1$. A 'nice class for the vertex packing problem' (obtained from bipartite graphs and claw-free graphs by repeated substitution) was given by Bertolazzi, De Simone, and Galluccio [1997]. Another nice class was given by De Simone [1993].

In Section 60.3d (Corollary 60.5b) we gave a proof of Győri's theorem (Győri [1984]), stating that the following class of graphs G satisfies $\alpha(G) = \overline{\chi}(G)$. Let A be a $\{0,1\}$ matrix such that the 1's in each row form a contiguous interval. Then G has vertex set $\{(i,j) \mid a_{i,j} = 1\}$, where two pairs (i,j) and (i',j') are adjacent if and only if $a_{i,j'} = a_{i',j} = 1$. The method of Frank and Jordán [1995b] also yields a polynomial-time algorithm to find a maximum-size stable set and a minimum clique cover. Frank [1999a] gave an alternative algorithmic proof. This class of graphs is not closed under taking induced subgraphs, and they need not be perfect.

Hammer, Mahadev, and de Werra [1985], Balas, Chvátal, and Nešetřil [1987], Balas and Yu [1989], De Simone and Sassano [1993], Hertz and de Werra [1993], Hertz [1995,1997], Brandstädt and Hammer [1999], Mosca [1999], and Lozin [2000] gave further classes of graphs for which the maximum-size or maximum-weight clique problem is polynomial-time solvable.

64.9b. Colourings and orientations

Let $D = (V, A)$ be an orientation of an undirected graph $G = (V, E)$. The following was shown by Gallai [1968a] and Roy [1967] (referring to conjectures by P. Erdős and C. Berge, respectively):

(64.52) $\chi(G) \leq \lambda(D)$,

where

(64.53) $\lambda(D) :=$ the maximum number of vertices on a directed path in D.

To see this, consider an inclusionwise maximal subset A' of A with the property that $D' = (V, A')$ is acyclic. For any $v \in V$, let $h(v)$ be the number of vertices in a longest directed path in D' ending at v. If $h(u) = h(v)$ for distinct vertices u and v, then u and v are nonadjacent, since otherwise we can add the arc joining u and v to A'. So h defines a colouring of V, with no more colours than the number of vertices in a longest directed path in D'.

This proves (64.52). Note that (64.52) implies that each tournament ($\equiv$ orientation of a complete graph) has a Hamiltonian path (a theorem of Rédei [1934] (Corollary 14.14a)).

Roy [1967] also observed that each undirected graph $G = (V, E)$ has an acyclic orientation in which the number of vertices in a longest directed path is equal to the colouring number of G. (This follows by colouring the vertices with colours $1, \ldots, \chi(G)$, and orienting any edge from u to v if the colour of u is smaller than that of v, which gives a digraph D with $\lambda(D) \leq \chi(G)$.)

This result is equivalent to the fact that for any undirected graph $G = (V, E)$:

(64.54) $\chi(G) = \min_D \lambda(D)$,

where D ranges over all acyclic orientations of G.

These results are essentially based on the (easy) fact that the minimum number of antichains needed to cover a partially ordered set is equal to the size of a maximum chain (Theorem 14.1).

Minty [1967] showed that for each graph $G = (V, E)$:

(64.55) $\chi(G) \leq k \iff G$ has an orientation such that each undirected circuit has at least $\frac{1}{k}|VC|$ forward arcs.

Necessity follows by colouring the vertices with colours $1, \ldots, k$, and orienting any edge from u to v if colour(u) < colour(v). To see sufficiency, let D be an orientation as described. Give each arc a length $k - 1$, and add an arc in the reverse direction of length -1. Then each directed circuit in the extended digraph has nonnegative length. Hence there is a 'potential' $p : V \to \mathbb{Z}$ with $1 \leq p(v) - p(u) \leq k-1$ for each arc (u,v) of D. Reducing p mod k gives a k-colouring as required.

Note that each orientation as in (64.55) is acyclic, and that any orientation D with $\lambda(D) \leq k$ is as in (64.55). The equivalence (64.55) gives a vertex-free description of the colouring number, and implies that $\chi(G)$ only depends on the cycle matroid of G.

Deming [1979a] showed that dual statements can be derived from Dilworth's decomposition theorem (Theorem 14.2), where 'chain' and 'antichain' are interchanged.

First one has, as a dual to (64.52), that for any orientation $D = (V, A)$ of an undirected graph $G = (V, E)$:

(64.56) $\alpha(G) \geq \xi(D)$,

where

(64.57) $\xi(D) :=$ the minimum number of directed paths in D needed to cover V.

To see (64.56), again consider an inclusionwise maximal subset A' of A with $D' = (V, A')$ acyclic. By Dilworth's decomposition theorem, V has a subset U of size $\xi(D)$ such that no two vertices in U are connected by a directed path in D. Then U is a stable set in G, since if two distinct $u, v \in U$ are adjacent in G, say $(u, v) \in A$, then $(u, v) \notin A'$, and hence $A' \cup \{(u, v)\}$ is not acyclic. But then A' contains a directed path from v to u, a contradiction.

This shows (64.56). Deming [1979a] showed also a dual form of (64.54):

(64.58) $\alpha(G) = \max_D \xi(D)$,

where D ranges over all acyclic orientations of G. Indeed, $\geq$ in (64.58) follows from (64.56). To see $\leq$, let U be a maximum-size stable set in G. Let D be any acyclic orientation of G in which each vertex in U is a source. Then $\xi(D) \geq |U| = \alpha(G)$.

64.9c. Algebraic methods

Lovász [1994] gave the following relations between stable sets, cliques, and colourings, using Hilbert's Nullstellensatz (extending Li and Li [1981] and unpublished work of D.J. Kleitman and L. Lovász). For any graph $G = (V, E)$, define the polynomial p_G in the variables x_v ($v \in V$) by:

(64.59) $p_G := \prod_{uv \in E} (x_u - x_v)$

(fixing some orientation of the edges). Then $\alpha(G) \leq k$ if and only if there are graphs $H_1, \ldots, H_t$ on V satisfying

(64.60) $p_G = p_{H_1} + \cdots + p_{H_t}$,

with $\overline{\chi}(H_i) \leq k$ for $i = 1, \ldots, t$. The number t can be exponentially large — hence (64.60) gives no good characterization for the stable set number. Similarly, $\chi(G) \geq k$ if and only if there are graphs satisfying (64.60) with $\omega(H_i) \geq k$ for $i = 1, \ldots, t$.

Let $G = (V, E)$ be a (simple) graph, with adjacency matrix A_G. Motzkin and Straus [1965] showed that the maximum value of $x^T A_G x$ over $x : V \to \mathbb{R}_+$ satisfying $x(V) = 1$, is equal to $1 - \omega(G)^{-1}$.

The proof of this is easy: for any two nonadjacent vertices u, v with $x_u > 0$ and $x_v > 0$, we can reset $x_u := x_u + \varepsilon$, $x_v := x_v - \varepsilon$ for some $\varepsilon \neq 0$ without decreasing $x^\mathsf{T} A_G x$. Hence the maximum value is attained by a vector x whose support is a clique C. As x takes the maximum value, we should have $x_v = 1/|C|$ for each $v \in C$. Then $x^\mathsf{T} A_G x$ is maximized if C is a maximum-size clique.

Motzkin and Straus' theorem implies the result of Korn [1968] that the minimum value of $x^\mathsf{T}(I + A_G)x$ over $x : V \to \mathbb{R}_+$ with $x(V) = 1$, is equal to $\alpha(G)^{-1}$. Indeed,

(64.61) $\quad \min_x x^\mathsf{T}(I + A_G)x = \min_x x^\mathsf{T}(J - A_{\overline{G}})x = 1 - \max_x x^\mathsf{T} A_{\overline{G}} x = \omega(\overline{G})^{-1}$
$= \alpha(G)^{-1}$.

More on this can be found in Gibbons, Hearn, Pardalos, and Ramana [1997].

Lovász [1982,1994] gave surveys of algebraic, topological, and other methods for the stable set and the vertex colouring problem.

64.9d. Approximation algorithms

Lund and Yannakakis [1993,1994] showed that unless NP=P, there do not exist a constant c and a polynomial-time algorithm that finds a vertex-colouring of any graph G using at most $c\chi(G)$ colours. (This was proved for $c < 2$ by Garey and Johnson [1976].)

More generally, Lund and Yannakakis [1993,1994] showed that there exists an $\varepsilon > 0$ such that, unless NP=P, there is no polynomial-time algorithm to find the colouring number of a graph up to a factor of n^ε (where n is the number of vertices).

A similar result for maximum-size stable sets was proved by Arora, Lund, Motwani, Sudan, and Szegedy [1992,1998]. Håstad [1996,1999] showed that, if NP$\neq$P, then there is no $\varepsilon > 0$ and a polynomial-time algorithm that finds a clique that is maximum-size up to a factor $n^{1/2-\varepsilon}$. Under a slightly stronger complexity assumption (NP$\neq$ZPP), Håstad proved a factor of $n^{1-\varepsilon}$.

For background, see Johnson [1992] and Papadimitriou [1994]. Related results can be found in Hochbaum [1983a], Wigderson [1983], Berger and Rompel [1990], Feige, Goldwasser, Lovász, Safra, and Szegedy [1991,1996], Berman and Schnitger [1992], Boppana and Halldórsson [1992], Bellare, Goldwasser, Lund, and Russell [1993], Khanna, Linial, and Safra [1993,2000], Bellare and Sudan [1994], Feige and Kilian [1994,1996,1998a,1998b,2000], Karger, Motwani, and Sudan [1994,1998], Bellare, Goldreich, and Sudan [1995,1998], Feige [1995,1997], Fürer [1995], Håstad [1996,1999], Alon and Kahale [1998], Arora and Safra [1998], Engebretsen and Holmerin [2000], Srinivasan [2000], and Khot [2001].

In contrast, there is an easy algorithm to obtain a vertex cover in a graph $G = (V, E)$ of size at most $2\tau(G)$ (F. Gavril 1974 (cf. Garey and Johnson [1979])): choose any inclusionwise maximal matching M (greedily); then the set of vertices covered by M is a vertex cover of size $2|M|$. Since $\tau(G) \geq |M|$, this is a vertex cover as described.

No polynomial-time algorithm yielding a factor better than 2 is known. Håstad [1997,2001] showed that, if NP$\neq$P, no factor better than $\frac{7}{6}$ is achievable in polynomial time.

See also Section 67.4f below.

64.9e. Further notes

Yannakakis [1988,1991] showed that the stable set polytope of the line graph $L(K_n)$ of a complete graph K_n cannot be represented as the projection of a polytope in higher dimensions that is symmetric under the automorphism group of $L(K_n)$. Cao and Nemhauser [1998] characterized line graphs as those graphs whose stable set polytope is determined by the inequalities corresponding to the matching polytope constraints.

Euler, Jünger, and Reinelt [1987] extended results of Padberg [1973] on facets of the stable set polytope, to more general 'independence' polytopes.

More on the stable set polytope can be found in Fulkerson [1971a], Chvátal [1973a,1975a,1985a], Padberg [1973,1974b,1977,1979,1980,1984], Nemhauser and Trotter [1974], Trotter [1975], Wolsey [1976], Balas and Zemel [1977], Naddef and Pulleyblank [1981a], Sekiguchi [1983], Ikura and Nemhauser [1985], Grötschel, Lovász, and Schrijver [1986], Lovász and Schrijver [1989,1991], Cheng and Cunningham [1995,1997], Cánovas, Landete, and Marín [2000], Lipták and Lovász [2000, 2001], and Cheng and de Vries [2002a,2002b].

The convex hull of the incidence vectors of the stable sets of size at most a given k is studied by Janssen and Kilakos [1999]. Generalizations of the stable set polytope to more general $0, \pm 1$ programming and satisfiability problems were studied by Johnson and Padberg [1982], Hooker [1996], and Sewell [1996].

Methods for and computational results on the stable set problem (or the equivalent clique, vertex cover, and set packing problems) are given by Balas and Samuelsson [1977], Chvátal [1977], Houck and Vemuganti [1977], Tarjan and Trojanowski [1977], Geoffroy and Sumner [1978], Gerhards and Lindenberg [1979], Hansen [1980b], Bar-Yehuda and Even [1981,1982,1985], Billionnet [1981], Chiba, Nishizeki, and Saito [1982] (planar graphs), Hochbaum [1982,1983a], Loukakis and Tsouros [1982], Baker [1983,1994], Clarkson [1983], Monien and Speckenmeyer [1983,1985], Balas and Yu [1986], Jian [1986], Robson [1986], Shindo and Tomita [1988], Hurkens and Schrijver [1989], Carraghan and Pardalos [1990], Nemhauser and Sigismondi [1992], Balas and Xue [1991,1996], Boppana and Halldórsson [1992], Pardalos and Rodgers [1992], Paschos [1992], Khuller, Vishkin, and Young [1993,1994], Berman and Fürer [1994], Mannino and Sassano [1994], Halldórsson [1995], Balas, Ceria, Cornuéjols, and Pataki [1996], Bourjolly, Laporte, and Mercure [1997], Halldórsson and Radhakrishnan [1997], Alon and Kahale [1998], Arkin and Hassin [1998], Feige and Kilian [1998a], Kleinberg and Goemans [1998], Chandra and Halldórsson [1999, 2001], Nagamochi and Ibaraki [1999b], Bar-Yehuda [2000], Halperin [2000,2002], Krivelevich and Vu [2000], Chen, Kanj, and Jia [2001], Krivelevich, Nathaniel, and Sudakov [2001a,2001b], and Guha, Hassin, Khuller, and Or [2002].

Methods for graph colouring are proposed and investigated by Christofides [1971], Brown [1972], Matula, Marble, and Isaacson [1972], Corneil and Graham [1973], Johnson [1974b], Wang [1974], Lawler [1976a], McDiarmid [1979], Matula and Beck [1983], Sysło, Deo, and Kowalik [1983], Wigderson [1983], Edwards [1986], Berger and Rompel [1990], Hertz [1991], Halldórsson [1993], Blum [1994], Demange, Grisoni, and Paschos [1994], Karger, Motwani, and Sudan [1994,1998], Schiermeyer [1994], Beigel and Eppstein [1995], Blum and Karger [1997], Krivelevich and Vu [2000], Eppstein [2001], Halperin, Nathaniel, and Zwick [2001], Molloy and Reed [2001], Stacho [2001], Alon and Krivelevich [2002], and Charikar [2002].

For computational results on clique, stable set, and colouring problems, consult also Johnson and Trick [1996].

A survey of graph colouring algorithms was given by Matula, Marble, and Isaacson [1972]. Chiba, Nishizeki, and Saito [1981], Thomassen [1994], and Robertson, Sanders, Seymour, and Thomas [1996] gave linear-time 5-colouring algorithms for planar graphs. The worst-case behaviour of graph colouring algorithms was investigated by Johnson [1974b].

Mycielski [1955] showed that triangle-free graphs can have arbitrarily large colouring number. King and Nemhauser [1974] and Gyárfás [1987] and Fouquet, Giakoumakis, Maire, and Thuillier [1995] studied classes of graphs for which the colouring number can be bounded by a function of the clique number.

Gyárfás [1987] conjectures that there exists a function $g : \mathbb{Z}_+ \to \mathbb{Z}_+$ such that $\chi(G) \leq g(\omega(G))$ for each graph G without odd holes. Equivalently, for $\omega \in \mathbb{Z}_+$, let $g(\omega)$ be the maximum colouring number of a graph without odd holes and cliques of size $> \omega$. Then Gyárfás' conjectures that g is finite. It is easy to see that $g(2) = 2$. N. Robertson, P.D. Seymour, and R. Thomas announced that they proved $g(3) = 4$ (this was conjectured by G. Ding).

Upper bounds for the stable set number of a graph in terms of the degrees were presented by Hansen [1979,1980b]. Relations between the colouring number and the fractional colouring number are investigated by Kilakos and Marcotte [1997]. Reed [1998] discussed bounding the chromatic number of a graph by a convex combination of its clique number and its maximum degree plus 1. Gerke and McDiarmid [2001a,2001b] investigated the ratio of the weighted colouring and the weighted clique number.

A theorem of Turán [1941] implies that any simple graph G with n vertices and m edges satisfies:

$$(64.62) \quad \alpha(G) \geq \frac{n^2}{n + 2m}.$$

Bondy [1978] showed that $m \geq 2\tau(G) - 1$ if G is connected. A study of the relations between several parameters derived from stability and colouring was given by Hell and Roberts [1982].

A survey on the stable set problem is given by Padberg [1979], on approximation methods for the stable set problem by Halldórsson [1998], and on colourings by Jensen and Toft [1995] and Toft [1995]. Colouring is also discussed in most graph theory books mentioned in Chapter 1.

Chapter 65

Perfect graphs: general theory

In this and the next two chapters, we consider the 'perfect' graphs, introduced by C. Berge in the 1960s. They turn out to unify several results in combinatorial optimization, in particular, min-max relations and polyhedral characterizations.

Berge proposed two conjectures, the *weak* and the *strong* perfect graph conjecture. The second implies the first.

The weak perfect graph conjecture says that perfection is maintained under taking the complementary graph. This was proved by Lovász [1972c]: the perfect graph theorem.

The strong perfect graph conjecture characterizes perfect graphs by excluding odd holes and odd antiholes. A proof of this was announced in May 2002 by Chudnovsky, Robertson, Seymour, and Thomas, resulting in the strong perfect graph theorem. The announced proof is highly complicated, and we cannot give it here.

Many of the results described in this and the next chapter follow directly as a consequence of the strong perfect graph theorem (while some of them are used in the proof). Where possible and appropriate, we give direct proofs of these consequences.

In this chapter, we give general theory, in Chapter 66 we discuss classes of perfect graphs, and in Chapter 67 we show the polynomial-time solvability of the maximum-weight clique and minimum colouring problems for perfect graphs.

65.1. Introduction to perfect graphs

As we saw before, the clique number $\omega(G)$ and the colouring number $\chi(G)$ of a graph $G = (V, E)$ are related by the inequality

(65.1) $\quad \omega(G) \leq \chi(G)$.

Strict inequality can occur, for instance, for any odd circuit of length at least five, and its complement.

Having equality in (65.1) does not say that much about the internal structure of a graph: any graph $G = (V, E)$ can be extended to a graph $G' = (V', E')$ satisfying $\omega(G') = \chi(G')$, simply by adding to G a clique of size $\chi(G)$, disjoint from V.

However, the condition becomes much more powerful if we require that equality in (65.1) holds for each induced subgraph of G. The idea for this was formulated by Berge [1963a]. He defined a graph $G = (V, E)$ te be *perfect* if $\omega(G') = \chi(G')$ holds for each induced subgraph G' of G.

Various classes of graphs could be shown to be perfect, like the bipartite graphs (trivially) and the line graphs of bipartite graphs (by Kőnig's edge-colouring theorem).

Berge [1960a,1963a] observed the important phenomenon that for several of these classes, also the complementary graphs are perfect. Berge therefore conjectured that the complement of any perfect graph is perfect again — the *weak perfect graph conjecture*. This conjecture was proved by Lovász [1972c], by proving an equivalent form of the conjecture given by Fulkerson [1972a] (the replication lemma — see Corollary 65.2c below).

As mentioned, obvious examples of imperfect graphs are the odd circuits of length at least five, and their complements. Berge [1963a] and P.C. Gilmore (cf. Berge [1966]) made the conjecture that this characterizes perfect graphs, which is the strong perfect graph conjecture. A proof was announced in May 2002 by Chudnovsky, Robertson, Seymour, and Thomas.

To simplify formulation, it is convenient to introduce the notions of 'hole' and 'antihole'. A *hole* in a graph G is an induced subgraph of G isomorphic to a circuit with at least four vertices. An *antihole* in G is an induced subgraph of G isomorphic to the complement of a circuit with at least four vertices. A hole or antihole is *odd* if it has an odd number of vertices.

Theorem 65.1 (Strong perfect graph theorem). *A graph G is perfect if and only if G contains no odd hole and no odd antihole.*

A graph containing no odd hole or odd antihole is called a *Berge graph*[2]. So the strong perfect graph theorem says that Berge graphs are precisely the perfect graphs.

An alternative formulation is in terms of minimally imperfect graphs. A *minimally imperfect* (or *critically imperfect*) graph is an imperfect graph such that each proper induced subgraph is perfect. Then the strong perfect graph theorem says that the only minimally imperfect graphs are the odd circuits of length at least five, and their complements.

It is (as yet) unknown if perfection of a graph can be tested in polynomial time. (Lovász [1986] 'would guess' that such an algorithm exists.) The clique number of a perfect graph can be determined in polynomial time, with the help of the ellipsoid method — see Chapter 67. However, no combinatorial polynomial-time algorithm is known.

We will next discuss perfect graph theory in greater detail (although we cannot give a proof of the strong perfect graph theorem). Let us make a useful observation:

---

[2] This term was introduced by Chvátal and Sbihi [1987].

(65.2)   any minimally imperfect graph $G = (V, E)$ has no stable set S with $\omega(G - S) < \omega(G)$.

Otherwise, $\omega(G) \geq \omega(G - S) + 1 = \chi(G - S) + 1 \geq \chi(G)$, since we can use S as colour.

Similarly, for any class $\mathcal{G}$ of graphs closed under taking induced subgraphs:

(65.3)   each graph $G \in \mathcal{G}$ is perfect $\iff$ each graph $G \in \mathcal{G}$ with $VG \neq \emptyset$ has a stable set S with $\omega(G - S) < \omega(G)$.

Here necessity follows from the fact that we can take for S any of the colours in a minimum colouring of G. Sufficiency follows by induction on $|VG|$: $\chi(G) \leq \chi(G - S) + 1 = \omega(G - S) + 1 \leq \omega(G)$.

65.2. The perfect graph theorem

Lovász [1972a] proved the weak perfect graph conjecture in the following stronger form (suggested by A. Hajnal), which we show with the elegant linear-algebraic proof found by Gasparian [1996].

Theorem 65.2. *A graph G is perfect if and only if $\omega(G')\alpha(G') \geq |VG'|$ for each induced subgraph G' of G.*

Proof. Necessity is easy, since if G is perfect, then $\omega(G') = \chi(G')$ for each induced subgraph G' of G, and since $\chi(G')\alpha(G') \geq |VG'|$ for any graph G'.

To see sufficiency, it suffices to show that each minimally imperfect graph G satisfies $|VG| \geq \alpha(G)\omega(G) + 1$. We can assume that $VG = \{1, \ldots, n\}$. Define $\omega := \omega(G)$ and $\alpha := \alpha(G)$.

We first construct

(65.4)   stable sets $S_0, \ldots, S_{\alpha\omega}$ such that each vertex is covered by exactly α of the S_i.

Let S_0 be a stable set in G of size α. By the minimality of G, we know that for each $v \in S_0$, the graph $G - v$ is perfect, and that hence $\chi(G - v) = \omega(G - v) \leq \omega$. Therefore, $V \setminus \{v\}$ can be partitioned into ω stable sets. Doing this for each $v \in S_0$, we obtain stable sets as in (65.4).

Now for each $i = 0, \ldots, \alpha\omega$, there exists a clique C_i of size ω with $C_i \cap S_i = \emptyset$ (by (65.2)). Then, for distinct i, j with $0 \leq i, j \leq \alpha\omega$, we have $|C_i \cap S_j| = 1$. This follows from the fact that C_i has size ω and intersects each S_j in at most one vertex, and hence, by (65.4), it intersects $\alpha\omega$ of the S_j. As $C_i \cap S_i = \emptyset$, we have that $|C_i \cap S_j| = 1$ if $i \neq j$.

Now consider the $(\alpha\omega + 1) \times n$ incidence matrices M and N of $S_0, \ldots, S_{\alpha\omega}$ and $C_0, \ldots, C_{\alpha\omega}$ respectively. So M and N are $\{0, 1\}$ matrices, with $M_{i,j} = 1 \iff j \in S_i$, and $N_{i,j} = 1 \iff j \in C_i$, for $i = 0, \ldots, \alpha\omega$ and $j = 1, \ldots, n$. By the above, $MN^\mathsf{T} = J - I$, where J is the $(\alpha\omega + 1) \times (\alpha\omega + 1)$ all-1 matrix, and I the $(\alpha\omega+1)\times(\alpha\omega+1)$ identity matrix. As $J-I$ has rank $\alpha\omega+1$, we have $n \geq \alpha\omega+1$. ∎

Theorem 65.2 implies (Lovász [1972c]):

Corollary 65.2a (perfect graph theorem). *The complement of a perfect graph is perfect again.*

Proof. Directly from Theorem 65.2, as the condition given in it is invariant under taking the complementary graph. ∎

As was observed by Cameron [1982], Theorem 65.2 implies that the question 'Given a graph, is it perfect?' belongs to co-NP. Indeed, to certify imperfection of a graph, it is sufficient, and possible, to specify:

(65.5) (i) an induced subgraph $G = (V, E)$,
(ii) integers $\alpha, \omega \geq 2$ with $|V| = \alpha\omega + 1$, and
(iii) for each $v \in V$, an ω-colouring of $G - v$ and an α-colouring of $\overline{G} - v$.

This is possible, since, by Theorem 65.2, a minimally imperfect subgraph G has these properties for $\omega := \omega(G)$ and $\alpha := \alpha(G)$. It is also sufficient to certify imperfection, since (65.5)(iii) implies that $\omega(G) \leq \omega$ and $\alpha(G) \leq \alpha$, and hence by (65.5)(ii), that G is not perfect.

Theorem 65.2 implies:

Corollary 65.2b. *Each minimally imperfect graph G satisfies*

(65.6) $\quad |VG| = \alpha(G)\omega(G) + 1.$

Proof. We have $|VG| \leq \alpha(G)\omega(G) + 1$, since for any vertex v of G, the graph $G - v$ is perfect, and hence

(65.7) $\quad |VG| - 1 = |V(G-v)| \leq \alpha(G-v)\omega(G-v) \leq \alpha(G)\omega(G).$

Conversely, $|VG| \geq \alpha(G)\omega(G) + 1$, since if $|VG| \leq \alpha(G)\omega(G)$, then $|VG'| \leq \alpha(G')\omega(G')$ for each induced subgraph G' of G (by the minimal imperfection of G). This implies with Theorem 65.2 that G is perfect, a contradiction. ∎

65.3. Replication

Let $G = (V, E)$ be a graph and let $v \in V$. Extend G with some new vertex, v' say, which is adjacent to v and to all vertices adjacent to v in G. In this way we obtain a new graph H, which we say is obtained from G by *duplicating* v. Repeated duplicating a vertex is called *replicating*. *Replicating* a vertex v by *a factor* k means duplicating v $k-1$ times if $k \geq 1$, and deleting v if $k = 0$.

Corollary 65.2c (replication lemma). *Let H arise from G by duplicating vertex v. Then if G is perfect, also H is perfect.*

Proof. By the perfect graph theorem, it suffices to show that $\overline{H}$ is perfect, and hence (as we can apply induction) that $\omega(\overline{H}) = \chi(\overline{H})$.

By the perfect graph theorem, if G is perfect, then $\overline{G}$ is perfect. Hence $\omega(\overline{H}) = \omega(\overline{G}) = \chi(\overline{G}) = \chi(\overline{H})$. ∎

Repeated application of Corollary 65.2c implies the following (the weighted colouring number is defined in Section 64.8):

Corollary 65.2d. *Let G be a perfect graph and let $w : V \to \mathbb{Z}_+$ be a 'weight' function. Then the maximum weight of a clique is equal to the weighted colouring number $\chi_w(G)$ of G.*

Proof. Let G^w be the graph arising from G by replicating any vertex v by a factor $w(v)$. By Corollary 65.2c, G^w is perfect, and so $\omega(G^w) = \chi(G^w)$. Since $\omega(G^w)$ is equal to the maximum weight of a clique in G and since $\chi(G^w) = \chi_w(G)$, the corollary follows. ∎

65.4. Perfect graphs and polyhedra

The *clique polytope* of a graph $G = (V, E)$ is the convex hull of the incidence vectors of the cliques. Clearly, any vector x in the clique polytope satisfies:

(65.8)   (i) $x_v \geq 0$ for each $v \in V$,
              (ii) $x(S) \leq 1$ for each stable set S.

Fulkerson [1972a] and Chvátal [1975a] showed that Corollary 65.2d implies a polyhedral characterization of perfect graphs:

Corollary 65.2e. *A graph G is perfect if and only if its clique polytope is determined by (65.8).*

Proof. *Necessity.* Let G be perfect. To prove that the clique polytope is determined by (65.8), it suffices to show that for each weight function $w : V \to \mathbb{Z}_+$, the maximum weight t of a clique in G is not less than the maximum of $w^\mathsf{T} x$ over (65.8). By Corollary 65.2d, there exist stable sets $S_1, \ldots, S_t$ with

(65.9) $w = \chi^{S_1} + \cdots + \chi^{S_t}$.

Hence for each x satisfying (65.8) we have

(65.10) $w^\mathsf{T} x = x(S_1) + \cdots + x(S_t) \leq t$.

Sufficiency. Let the clique polytope of G be determined by (65.8). Suppose that G is not perfect. Choose a minimal set U with $\omega(G[U]) < \chi(G[U])$. Let

$w := \chi^U$. The function $w^T x$ is maximized over $P_{\text{clique}}(G)$ by the incidence vector of each maximum-size clique of $G[U]$. Moreover, by linear programming duality, there exists a stable set S with $x(S) = 1$ for each optimum solution x. So S intersects each maximum-size clique of $G[U]$, and hence

(65.11) $\quad \omega(G[U \setminus S]) \leq \omega(G[U]) - 1 < \chi(G[U]) - 1 \leq \chi(G[U \setminus S])$,

contradicting the minimality of U. ∎

Corollary 65.2e is equivalent to: G is perfect if and only if $P_{\text{clique}}(G) = A(P_{\text{stable set}}(G))$. (Here $A(P)$ is the antiblocking polyhedron of P.) Hence it implies the perfect graph theorem (using the theory of antiblocking polyhedra (cf. Section 5.9)):

(65.12) $\quad G$ is perfect $\iff P_{\text{clique}}(G) = A(P_{\text{stable set}}(G))$
$\iff P_{\text{stable set}}(G) = A(P_{\text{clique}}(G))$
$\iff P_{\text{clique}}(\overline{G}) = A(P_{\text{stable set}}(\overline{G})) \iff \overline{G}$ is perfect.

Corollary 65.2d also implies that perfect graphs can be characterized by total dual integrality:

Corollary 65.2f. *A graph G is perfect if and only if system (65.8) is totally dual integral.*

Proof. Directly from Corollaries 65.2d and 65.2e. ∎

So for any graph G we have that (65.8) determines an integer polytope if *and only if* it is totally dual integral.

65.4a. Lovász's proof of the replication lemma

The proof of Lovász [1972c] of the weak perfect graph theorem is based on proving the 'replication lemma' (Corollary 65.2c above), as follows.

By (65.2), it suffices to find a stable set S in H intersecting all maximum-size cliques of H, since any induced subgraph of H is an induced subgraph of G or arises from it by replication.

Consider an $\omega(G)$-colouring of G, and let S be the colour containing v. Then S intersects each maximum-size clique C of H. Indeed, if $v' \notin C$, then C is a maximum-size clique of G, and so it intersects S. If $v' \in C$, then also $v \in C$ (as $C \cup \{v\}$ is a clique), and so C intersects S.

This proves the replication lemma, which by repeated application gives Corollary 65.2d. Since the proof of Corollary 65.2e given above only uses Corollary 65.2d, this shows (with (65.12)) that the replication lemma implies the perfect graph theorem. This is Fulkerson's proof of the equivalence of the replication lemma and the weak perfect graph conjecture ($\equiv$ perfect graph theorem).

65.5. Decomposition of Berge graphs

The proof of the strong perfect graph conjecture is based on a decomposition theorem of Berge graphs, stating that each Berge graph can be decomposed into 'basic' graphs: bipartite graphs and their complements, and line graphs of bipartite graphs and their complements. We formulate the decomposition rules.

Let $G = (V, E)$ be a graph. A *2-join* of G is a partition of V into sets V_1 and V_2 such that for $i = 1, 2$, $|V_i| \geq 3$ and V_i contains disjoint nonempty subsets A_i, B_i with the property that for all $v_1 \in V_1$ and $v_2 \in V_2$:

(65.13) $\quad v_1$ and v_2 are adjacent $\iff v_1 \in A_1$, $v_2 \in A_2$, or $v_1 \in B_1$, $v_2 \in B_2$.

A *skew partition* of G is a partition V_1, V_2 of V such that $G[V_1]$ and $\overline{G}[V_2]$ are disconnected. An *homogeneous pair* of G is a pair A, B of disjoint subsets of V such that $3 \leq |A| + |B| \leq |V| - 2$ and such that for all $x, y \in A \cup B$ and $z \in V \setminus (A \cup B)$, if $xz \in E$ and $yz \notin E$, then x and y belong to distinct sets A, B.

M. Chudnovsky, N. Robertson, P.D. Seymour, and R. Thomas announced in May 2002 that they proved the following[3]:

Theorem 65.3. *Let G be a Berge graph. Then G or $\overline{G}$ is bipartite or the line graph of a bipartite graph, or has a 2-join, a skew partition, or a homogeneous pair.*

Unfortunately, we cannot give the (highly complicated) proof of this theorem. M. Chudnovsky, N. Robertson, P.D. Seymour, and R. Thomas also showed that any minimum-size imperfect Berge graph has no skew partition[4]. Since such a graph has no 2-join (Cornuéjols and Cunningham [1985] and Kapoor [1994] — see Corollary 65.7a below) and no homogeneous pair (Chvátal and Sbihi [1987]), and since bipartite graphs and their line graphs are perfect (Kőnig [1916] — see Section 66.1), this implies:

Theorem 65.4 (strong perfect graph theorem). *A graph is perfect if and only if it is a Berge graph.*

65.5a. 0- and 1-joins

A *0-join* of a graph $G = (V, E)$ is a partition of V into nonempty sets V_1 and V_2 such that no edge connects V_1 and V_2. Let $G_1 := G[V_1]$ and $G_2 = G[V_2]$. Then G is called the *0-join* of G_1 and G_2. Trivially:

---

[3] This was conjectured by M. Conforti, G. Cornuéjols, N. Robertson, P.D. Seymour, R. Thomas, and K. Vušković (cf. Cornuéjols [2002]). It builds on work of Roussel and Rubio [2001], and it was stimulated by interaction with concurrent work of Conforti, Cornuéjols, Vušković, and Zambelli [2002] and Conforti, Cornuéjols, and Zambelli [2002b].

[4] conjectured by Chvátal [1985c].

Theorem 65.5. *G is perfect* $\iff$ G_1 *and* G_2 *are perfect.*

Proof. This follows from the facts that $\omega(G) = \max\{\omega(G_1), \omega(G_2)\}$ and $\chi(G) = \max\{\chi(G_1), \chi(G_2)\}$, and that induced subgraphs of G arise by the same construction from induced subgraphs of G_1 and G_2. ∎

Hence

(65.14) no minimally imperfect graph has a 0-join.

A 1-*join* (or *join*) of a graph $G = (V, E)$ is a partition of V into subsets V_1 and V_2 such that $|V_1| \geq 2$, $|V_2| \geq 2$, and such that there exist nonempty $A_1 \subseteq V_1$ and $A_2 \subseteq V_2$ with the property that for all $v_1 \in V_1$ and $v_2 \in V_2$:

(65.15) v_1 and v_2 are adjacent $\iff$ $v_1 \in A_1$ and $v_2 \in A_2$.

Choose $v_1 \in A_1$ and $v_2 \in A_2$, and define $G_1 := G[V_1 \cup \{v_2\}]$ and $G_2 := G[V_2 \cup \{v_1\}]$. Then G is called the 1-*join* of G_1 and G_2.

Bixby [1984] proved (generalizing a result of Lovász [1972c]):

Theorem 65.6. *G is perfect* $\iff$ G_1 *and* G_2 *are perfect.*

Proof. Necessity follows from the fact that G_1 and G_2 are induced subgraphs of G. To prove sufficiency, it suffices to show $\omega(G) = \chi(G)$, since each induced subgraph of G arises by the same construction, or by a 0-join from induced subgraphs of G_1 and G_2. Let $\omega := \omega(G)$ and $a_i := \omega(G[A_i])$ for $i = 1, 2$. It suffices to show that for each $i = 1, 2$,

(65.16) $G[V_i]$ has an ω-colouring such that A_i uses a_i colours only,

since then we can assume that we use different colours for A_1 and A_2 (as $a_1 + a_2 \leq \omega$), yielding an ω-colouring of G.

To prove (65.16), we may assume that $i = 1$. Let G'_1 be the graph obtained from G_1 by replicating v_2 by a factor $\omega - a_1$. So $\omega(G'_1) = \omega$. By the replication lemma, G'_1 is perfect. Hence $\omega(G'_1) = \chi(G'_1)$. As the clique of vertices obtained from v_2 has size $\omega - a_1$, we use only a_1 colours for A_1, as required. ∎

An alternative proof follows from Cunningham [1982b]. Cunningham [1982a] described an $O(n^3)$-time algorithm to find a 1-join (if any).

Theorem 65.6 implies:

(65.17) no minimally imperfect graph has a 1-join,

since it has no 0-join, and hence G_1 and G_2 as above are proper induced subgraphs of G, implying that they are perfect. Therefore, by Theorem 65.6, G is perfect, a contradiction.

65.5b. The 2-join

We next show that a minimally imperfect graph has no 2-join, except if it is an odd circuit. This was shown by Cornuéjols and Cunningham [1985] (for a special case) and Kapoor [1994].

The proof uses the following 'special replication lemma' (Cornuéjols and Cunningham [1985]). Let $e = uv$ be an edge of a graph G. Let G' be the graph obtained from replicating v and deleting edge uv', where v' is the new vertex.

Lemma 65.7α (special replication lemma). *If G is perfect and uv is not contained in a triangle of G, then G' is again perfect.*

Proof. It suffices to show that G' has a stable set S' such that $\omega(G' - S') < \omega(G')$. If $\omega(G') > \omega(G)$, we can take $S' = \{v'\}$. So we may assume $\omega(G') = \omega(G)$. Let S be the colour of an $\omega(G)$-colouring of G with $v \in S$. Let $S' := (S \setminus \{v\}) \cup \{v'\}$. Then S' is a stable set in G'. If $\omega(G' - S') < \omega(G')$ we are done. So assume that $\omega(G' - S') = \omega(G')$. Let C be a clique in $G' - S'$ of size $\omega(G)$. Since $\omega(G-S) < \omega(G)$ and since $G - S = G' - S' - v$, we know $v \in C$. Since $\omega(G') = \omega(G)$, we know $u \in C$. Hence, $C = \{u, v\}$ (since uv is not in a triangle). So $\omega(G') = 2$, and hence vv' is not contained in a triangle of G'. But then v' has degree 1 in G', implying $\chi(G') = \chi(G) = \omega(G) = \omega(G')$. ∎

Next we consider a *special 2-join*, namely where the sets A_i and B_i in the definition of 2-join are connected by a path in $G[V_i]$ (for $i = 1, 2$). For $i = 1, 2$, let P_i be a shortest $A_i - B_i$ path in $G[V_i]$. Define $G_1 := G[V_1 \cup VP_2]$ and $G_2 := G[V_2 \cup VP_1]$.

Theorem 65.7. *G is perfect $\iff$ G_1 and G_2 are perfect.*

Proof. Necessity follows from the fact that G_1 and G_2 are induced subgraphs of G. To prove sufficiency, it is enough to prove $\omega(G) = \chi(G)$, since each induced subgraph of G arises by the same construction, or by 1- or 0-joins, from induced subgraphs of G_1 and G_2. Define $\omega := \omega(G)$, and

(65.18) $\qquad a_i := \omega(G[A_i])$ and $b_i := \omega(G[B_i])$ for $i = 1, 2$.

Note that perfection of G_1 implies that $|EP_1| \equiv |EP_2|$ (mod 2), since $VP_1 \cup VP_2$ induces a hole in G_1.

For any colouring ϕ of a graph and any set X of vertices, let $\phi(X)$ denote the set of colours used by X. We show that, for each $i = 1, 2$, $G[V_i]$ has an ω-colouring $\phi : V \to \{1, \ldots, \omega\}$ such that

(65.19) $\qquad$ (i) $\phi(A_i) = \{1, \ldots, a_i\}$;
$\qquad\qquad$ (ii) if $|EP_i|$ is even, then $\phi(B_i) = \{1, \ldots, b_i\}$;
$\qquad\qquad$ (iii) if $|EP_i|$ is odd, then $\phi(B_i) = \{\omega - b_i + 1, \ldots, \omega\}$.

This yields an ω-colouring of G, by replacing the colour, i say, of any vertex in V_2 by $\omega - i + 1$. (The correctness follows from $a_1 + a_2 \leq \omega$ and $b_1 + b_2 \leq \omega$.)

To prove the existence of a colouring satisfying (65.19), we may assume $i = 1$. Let $v_0, v_1, \ldots, v_k$ be the vertices (in order) of the $A_2 - B_2$ path P_2.

First assume that $k > 1$ or $a_1 + b_1 \geq \omega$. Let G_1' be the graph arising from G_1 by replicating v_j by a factor

(65.20) $\qquad\begin{array}{ll} \omega - a_1 & \text{if } j < k - 1 \text{ and } j \text{ is even,} \\ a_1 & \text{if } j < k - 1 \text{ and } j \text{ is odd,} \\ \min\{\omega - a_1, b_1\} & \text{if } j = k - 1 \text{ and } j \text{ is even,} \\ \min\{a_1, b_1\} & \text{if } j = k - 1 \text{ and } j \text{ is odd,} \\ \omega - b_1 & \text{if } j = k. \end{array}$

Then $\omega(G'_1) = \omega$, and any ω-colouring of G'_1 yields a colouring satisfying (65.19). Indeed, if k is even, then (65.20) implies that the set of colours used by the copies of v_0 and the set of colours used by the copies of v_k are comparable[5]. If k is odd, then (65.20) implies that the set of colours used by the copies of v_0 and the set of colours *not* used by the copies of v_k are comparable.

Next assume that $k = 1$ and $a_1 + b_1 < \omega$. Extend G_1 by a new vertex v', adjacent to all vertices in B_1 and to v_1. By the special replication lemma (Lemma 65.7a), the new graph G'''_1 is again perfect. Let G'_1 be the graph arising from G'''_1 by replicating v_0 by a factor $\omega - a_1$, v_1 by a factor a_1, and v' by a factor $\omega - a_1 - b_1$. Again, $\omega(G'_1) = \omega$, and any ω-colouring of G'_1 yields a colouring satisfying (65.19). ∎

This implies:

Corollary 65.7a. *Any minimally imperfect graph having a 2-join is an odd circuit.*

Proof. Let G be a minimally imperfect graph, and let V_i, A_i, B_i (for $i = 1, 2$) be as in the definition of 2-join. If for some i, the graph $G[V_i]$ has no $A_i - B_i$ path, then G has a 0- or 1-join, contradicting (65.14) or (65.17). So we can assume that, for $i = 1, 2$, $G[V_i]$ has an $A_i - B_i$ path. Let P_i be a shortest such path.

By Theorem 65.7 and by symmetry, we may assume that $G[V_1 \cup VP_2]$ is not perfect. Hence, by the minimal imperfection of G, $G = G[V_1 \cup VP_2]$.

We first show $\omega(G) = 2$. Choose an internal vertex u on P_2. (This exists, since $|V_2| \geq 3$.) Choose $v \in V \setminus \{u\}$. By the minimal imperfection of G, we know $\chi(\overline{G} - v) = \alpha(G - v)$. Therefore, $VG \setminus \{v\}$ can be partitioned into $\alpha(G - v)$ cliques. Since $|VG| = \alpha(G)\omega(G) + 1$ (by (65.6)), each of these cliques has size $\omega(G)$. In particular, u is in a clique of size $\omega(G)$. Hence, since u is an internal vertex of P_2, $\omega(G) = 2$.

As $\omega(G) = 2$, $\chi(G - v) \leq 2$ for each $v \in VG$; that is, $G - v$ is bipartite for each $v \in VG$. So each odd circuit is Hamiltonian. As G is not bipartite, G has an odd circuit. This circuit has no chords, as otherwise there exists a shorter odd circuit. ∎

Cornuéjols and Cunningham [1985] gave an $O(n^2 m^2)$-time algorithm to find a 2-join in a given graph (if any).

65.6. Pre-proof work on the strong perfect graph conjecture

In this section we survey research done on the strong perfect graph conjecture before it was proved in general. Many of the results follow as a consequence of the strong perfect graph theorem. Since the proof of this theorem is very complicated, we will include proofs not based on the strong perfect graph theorem.

---

[5] Sets X and Y are called *comparable* if $X \subseteq Y$ or $Y \subseteq X$.

65.6a. Partitionable graphs

The strong perfect graph theorem implies that each minimally imperfect graph is a circuit or its complement, and hence is highly symmetric. Before the strong perfect graph theorem was proved, several regularity properties of minimally imperfect graphs were shown, initiated by the work of Padberg [1974a].

A graph $G = (V, E)$ is called *partitionable* if $|V| = \alpha(G)\omega(G)+1$ and $\chi(G-v) = \omega(G)$ and $\overline{\chi}(G - v) = \alpha(G)$ for each $v \in V$. By Corollary 65.2b, each minimally imperfect graph is partitionable. As each partitionable graph is imperfect, the strong perfect graph theorem is equivalent to: each partitionable graph has an odd hole or odd antihole.

Partitionable graphs are characterized as follows[6]. Our proof of necessity is based on Gasparian [1996] (and is similar to the proof of Theorem 65.2).

Theorem 65.8. *A graph G is partitionable if and only if $|VG| = \alpha(G)\omega(G) + 1$ and each vertex is contained in exactly $\alpha(G)$ stable sets of size $\alpha(G)$ and in exactly $\omega(G)$ cliques of size $\omega(G)$.*

Proof. Define $n := |VG|$, $\alpha := \alpha(G)$, and $\omega := \omega(G)$.

I. To see necessity, let G be partitionable. Then the proof method of Theorem 65.2 applies: We again construct

(65.21)   stable sets $S_0, \ldots, S_{\alpha\omega}$ such that each vertex is covered by exactly α of the S_i.

Indeed, let S_0 be a stable set in G of size α. For each vertex v, as G is partitionable, we know $\chi(G - v) = \omega$. Therefore, $VG \setminus \{v\}$ can be partitioned into ω stable sets. Doing this for each $v \in S_0$, we obtain stable sets as in (65.21).

Next, for each $i = 0, \ldots, \alpha\omega$, there exists a clique C_i of size ω with $C_i \cap S_i = \emptyset$. To see this, choose $v \in S_i$. As G is partitionable, $\chi(\overline{G} - v) = \alpha$, and hence $VG \setminus \{v\}$ can be partitioned into α cliques. Since $n = \alpha\omega + 1$, each clique has size ω. Since $|S_i \setminus \{v\}| \leq \alpha - 1$, at least one of these cliques is disjoint from S_i.

Then, for distinct i, j with $0 \leq i, j \leq \alpha\omega$, we have $|C_i \cap S_j| = 1$. This follows from the fact that C_i has size ω and intersects each S_j in at most one vertex, and hence, by (65.21), C_i intersects $\alpha\omega$ of the S_j. As $C_i \cap S_i = \emptyset$, we have that $|C_i \cap S_j| = 1$ if $i \neq j$.

Now consider the $(\alpha\omega + 1) \times n$ incidence matrices M and N of $S_0, \ldots, S_{\alpha\omega}$ and $C_0, \ldots, C_{\alpha\omega}$ respectively. So M and N are $\{0,1\}$ matrices, with $M_{i,j} = 1 \iff j \in S_i$, and $N_{i,j} = 1 \iff j \in C_i$, for $i = 0, \ldots, \alpha\omega$ and $j = 1, \ldots, n$. By the above, $MN^\mathsf{T} = J - I$, where J is the $(\alpha\omega + 1) \times (\alpha\omega + 1)$ all-1 matrix, and I the $(\alpha\omega + 1) \times (\alpha\omega + 1)$ identity matrix. So M and N are nonsingular.

It then suffices (by symmetry) to show that each maximum-size clique C occurs among $C_0, \ldots, C_n$. Now $(M\chi^C)_i$ is 1 if $|C \cap S_i| = 1$, and is 0 otherwise. As $|C| = \omega$ and as each $v \in V$ belongs to exactly α of the S_i, C intersects precisely $\alpha\omega$ of the S_i. That is, there is exactly one, say S_j, disjoint from C. Hence $M\chi^C = M\chi^{C_j}$, and therefore $C = C_j$, as M is nonsingular.

---

[6] Necessity of the condition for minimally imperfect graphs was shown by Padberg [1974a], and for partitionable graphs in general by Bland, Huang, and Trotter [1979]. As to sufficiency, Cameron [1982] referred to private communication with A. Lubiw in 1981, and Whitesides [1982] called it 'well known'.

II. To see sufficiency, let G satisfy the condition. As each vertex of G is in exactly α stable sets of size α, there are exactly n maximum-size stable sets. Similarly, there are exactly n maximum-size cliques.

Let M and N be the incidence matrix of the maximum-size stable sets and maximum-size cliques, respectively. We can order the rows such that $MN^\mathsf{T} = J - I$, where J is the all-one $n \times n$-matrix, and I the identity matrix of order n. To see this, each maximum-size stable set S intersects precisely $\alpha\omega$ maximum-size cliques, since $|S| = \alpha$ and each vertex $v \in S$ is in precisely ω maximum-size cliques. Hence there is a unique maximum-size clique C disjoint from S. Similarly, for each maximum-size clique C there is a unique maximum-size stable set S disjoint from C.

So $MN^\mathsf{T} = J - I$, implying

(65.22) $\quad M(J-I)N^\mathsf{T} = MJN^\mathsf{T} - MN^\mathsf{T} = \alpha JN^\mathsf{T} - (J-I) = \alpha\omega J - J + I$
$= nJ - 2J + I = (J-I)(J-I) = MN^\mathsf{T}MN^\mathsf{T}.$

Since M and N^T are nonsingular, this implies $N^\mathsf{T}M = J - I$.

Now choose $v \in V$. As $N^\mathsf{T}M = J - I$, for each $u \in V \setminus \{v\}$ there exists a unique pair of a maximum-size clique C_u and a maximum-size stable set S_u with $u \in C_u$, $v \in S_u$, and $C_u \cap S_u = \emptyset$. Then for each $w \in C_u$ we have $C_w = C_u$, since $w \in C_u$ and $v \in S_u$. So the C_u partition $V \setminus \{v\}$, and hence $\chi(\overline{G} - v) = \alpha$. Then also $\chi(G - v) = \omega$ by symmetry. ∎

A partitionable graph G with $\alpha(G) = \alpha$ and $\omega(G) = \omega$, is also called an (α, ω)-*graph*.

The proof of Theorem 65.8 also implies the following further properties of partitionable graphs (properties (i)-(iii) were proved for minimally imperfect graphs by Padberg [1974a] and for partitionable graphs by Bland, Huang, and Trotter [1979]; property (iv) was shown by Whitesides [1982]):

Theorem 65.9. *Let G be a partitionable graph with n vertices. Then:*

(65.23) (i) *G contains exactly n maximum-size cliques and exactly n maximum-size stable sets;*
   (ii) *the matrix N formed by the incidence vectors of the maximum-size cliques is nonsingular, and the matrix M formed by the incidence vectors of the maximum-size stable sets is nonsingular;*
   (iii) *each maximum-size clique intersects all but one maximum-size stable sets, and each maximum-size stable set intersects all but one maximum-size cliques;*
   (iv) *for any two distinct vertices u, v of G there is a unique pair of a maximum-size clique C and a maximum-size stable set S with $u \in C$, $v \in S$, and $C \cap S = \emptyset$.*

Proof. See the proof of Theorem 65.8. ∎

Notes. One may show that $|\det M| = \alpha(G)$ and $|\det N| = \omega(G)$ for any partitionable graph. Indeed, since $M\mathbf{1} = \alpha(G) \cdot \mathbf{1}$, we have that $M^{-1}\mathbf{1} = \alpha(G)^{-1} \cdot \mathbf{1}$. Hence $\alpha(G)$ divides $\det M$ (as $(\det M) \cdot M^{-1}$ is an integer matrix). Similarly, $\omega(G)$ divides $\det N$. Now $|\det M \cdot \det N| = |\det(MN^\mathsf{T})| = |\det(J-I)| = |VG| - 1 = \alpha(G)\omega(G)$. So $|\det M| = \alpha(G)$ and $|\det N| = \omega(G)$.

Shepherd [1994b] showed that a graph G is partitionable if and only if for some $p, q \geq 2$ with $|VG| = pq + 1$: (i) G has a family of $|VG|$ stable sets of size p such that each vertex is in precisely p of them, and (ii) G has no stable set S of size p that intersects each clique of size q. A polynomial-time recognition algorithm of partitionable graphs was given by Shepherd [2001].

65.6b. More characterizations of perfect graphs

It is not difficult to show that for any partitionable graph G one has:

(65.24) $\quad \chi^*(G) = \omega(G) + \dfrac{1}{\alpha(G)}.$

Indeed, let $n := |VG|$, $\alpha := \alpha(G)$, $\omega := \omega(G)$, $\chi^* := \chi^*(G)$. To see $\geq$, observe that the vector $\alpha^{-1} \cdot \mathbf{1}$ satisfies all stable set inequalities (64.34), and hence $\chi^* \geq n\alpha^{-1} = \omega + \alpha^{-1}$. To see $\leq$, give each stable set of size α a value α^{-1}. This gives a fractional colouring of size $\omega + \alpha^{-1}$. So $\chi^* \leq \omega + \alpha^{-1}$, proving (65.24).

Hence perfect graphs can be characterized by:

Theorem 65.10. *A graph G is perfect $\iff \chi^*(G')$ is an integer for each induced subgraph G' of G.*

Proof. See above. ∎

Berge [1973a] gave the following further characterization of perfect graphs. For any graph $G = (V, E)$, let $\chi_2(G)$ denote the *bicolouring number* of G, being the minimum number of stable sets $S_1, \ldots, S_t$ such that each vertex is in two of the S_i. Alternatively, it is the minimum number of colours such that we can assign to each vertex a pair of colours in such a way that any two adjacent vertices get two disjoint pairs of colours.

Theorem 65.11. *A graph G is perfect if and only if $\chi_2(G') = 2\chi(G')$ for each induced subgraph G' of G.*

Proof. To see necessity, we have $2\omega(G) \leq \chi_2(G) \leq 2\chi(G)$ for each graph G. Hence if G is perfect, then $\omega(G) = \chi(G)$, and hence $\chi_2(G) = 2\chi(G)$. As perfection is closed under taking induced subgraphs, necessity of the condition follows.

To see sufficiency, let G be a minimally imperfect graph. Consider two nonadjacent vertices u and v. Then $\chi_2(G) \leq \chi(G - u) + \chi(G - v) + 1$ (as we can take $\{u, v\}$ as a colour). Since, by the condition, $\chi_2(G) = 2\chi(G)$, we can assume, by symmetry, that $\chi(G) \leq \chi(G - u)$. Hence $\chi(G) \leq \chi(G - u) \leq \omega(G - u) \leq \omega(G)$, contradicting the fact that G is minimally imperfect. ∎

(This proof does not use the perfect graph theorem.)

65.6c. The stable set polytope of minimally imperfect graphs

The following theorem of Padberg [1976] is a direct consequence of the strong perfect graph conjecture, but we give a direct proof (we adapt the proof of Seymour [1990b]):

Section 65.6c. The stable set polytope of minimally imperfect graphs   1119

Theorem 65.12. *Let $G = (V, E)$ be a minimally imperfect graph. Then the polytope determined by*

(65.25)   (i) $x_v \geq 0$ for $v \in V$,
          (ii) $x(C) \leq 1$ for each clique C,

has precisely one noninteger vertex, namely $\omega^{-1} \cdot \mathbf{1}$, where $\omega := \omega(G)$.

Proof. Let $G = (V, E)$ be a minimally imperfect graph, and let x^* be a noninteger vertex of the polytope determined by (65.25). Set $n := |V|$.

First we have that

(65.26) $x_v^* > 0$ for each vertex v.

For suppose that $x_v^* = 0$. Then $x^*|V \setminus \{v\}$ is a noninteger vertex of the polytope (65.25) for $G - v$, contradicting the perfection of $G - v$ (by Corollary 65.2e). This proves (65.26).

Let $\mathcal{C}$ be a collection of cliques C such that $x^*(C) = 1$ for each $C \in \mathcal{C}$ and such that $\{\chi^C \mid C \in \mathcal{C}\}$ is a set of n linearly independent vectors. For $v \in V$, let $\mathcal{C}_v$ denote the collection of $C \in \mathcal{C}$ with $v \in C$. Then:

(65.27) $|\mathcal{C}_v| \leq \omega$ for each $v \in V$.

To see this, consider any $v \in V$ and any $C \in \mathcal{C} \setminus \mathcal{C}_v$. Since $G - v$ is perfect, the vector $x^*|V \setminus \{v\}$ is a convex combination $\sum_S \lambda_S \chi^S$ of stable sets S in $G - v$. For each $u \in C$, choose a stable set S_u with $u \in S_u$ and $\lambda_{S_u} > 0$. Then $|C' \cap S_u| = 1$ for each $C' \in \mathcal{C} \setminus \mathcal{C}_v$ (since $(x^*|V \setminus \{v\})(C') = 1$). So the incidence vectors χ^{S_u} for $u \in C$ are linearly independent. This implies that the vectors $\chi^{S_u} - x^*$ for $u \in C$ have rank at least $|C| - 1$. As each of these vectors is orthogonal to $\chi^{C'}$ for each $C' \in \mathcal{C} \setminus \mathcal{C}_v$, we have

(65.28) $|\mathcal{C} \setminus \mathcal{C}_v| \leq (n - 1) - (|C| - 1) = n - |C|$.

Let U be the set of vertices not covered by all cliques in $\mathcal{C}$. Then:

(65.29) $n = \sum_{C \in \mathcal{C}} 1 = \sum_{C \in \mathcal{C}} \sum_{v \in V \setminus C} \frac{1}{n - |C|} = \sum_{v \in U} \sum_{C \in \mathcal{C} \setminus \mathcal{C}_v} \frac{1}{n - |C|}$
$\leq \sum_{v \in U} \sum_{C \in \mathcal{C} \setminus \mathcal{C}_v} \frac{1}{|\mathcal{C} \setminus \mathcal{C}_v|} = \sum_{v \in U} 1 = |U| \leq n.$

So we have equality throughout; that is, $U = V$ and $|\mathcal{C} \setminus \mathcal{C}_v| = |V| - |C|$ for each $v \in V$ and each $C \in \mathcal{C} \setminus \mathcal{C}_v$. This gives equality in (65.28). So $|\mathcal{C}_v| = |C| \leq \omega$, proving (65.27).

Let $\mathcal{C}'$ denote the collection of maximum-size cliques in G. By Theorem 65.9, each $v \in V$ is in precisely ω sets in $\mathcal{C}'$. Hence

(65.30) $n = \sum_{C \in \mathcal{C}} 1 = \sum_{C \in \mathcal{C}} x^*(C) = \sum_{v \in V} |\mathcal{C}_v| x_v^* \leq \omega \sum_{v \in V} x_v^* = \sum_{C \in \mathcal{C}'} x^*(C)$
$\leq \sum_{C \in \mathcal{C}'} 1 = n.$

Hence we have equality throughout. Therefore, x^* satisfies $x^*(C) = 1$ for each maximum-size clique C. Hence $x^* = \omega^{-1} \cdot \mathbf{1}$.   ∎

By the antiblocking relation, Theorem 65.12 implies that the stable set polytope of a minimally imperfect graph has precisely one facet not given by the clique and nonnegative constraints:

Corollary 65.12a. *Let $G = (V, E)$ be a minimally imperfect graph. Then the stable set polytope of G is determined by:*

(65.31) $\quad x_v \geq 0 \qquad$ for $v \in V$,
$\qquad\qquad x(C) \leq 1 \qquad$ for each clique C,
$\qquad\qquad x(V) \leq \alpha(G).$

Proof. Directly from Theorem 65.12 applied to the antiblocking polytope of the polytope determined by (65.25) (for $\overline{G}$). ∎

Shepherd [1990,1994b] calls a graph *near-perfect* if its stable set polytope is determined by (65.31), and he showed that a graph G is minimally imperfect if and only if both G and its complement $\overline{G}$ are near-perfect.

65.6d. Graph classes

Before a proof of the strong perfect graph theorem in general was announced in 2002, it had been proved for several classes of graphs. Next to the classes to be discussed in Chapter 66, it was shown for (among other):

- *claw-free graphs*, that is, graphs not having $K_{1,3}$ (a *claw*) as induced subgraph (Parthasarathy and Ravindra [1976] (cf. Tucker [1979] and Maffray and Reed [1999], and Giles and Trotter [1981] for a simpler proof)).
  It follows that the line graph $L(G)$ of a graph G is perfect if and only if G contains no odd circuit with at least five vertices as (not necessarily induced) subgraph. So these graphs have edge-colouring number $\chi'(G)$ equal to the maximum degree $\Delta(G)$ (if $\Delta(G) \geq 3$); moreover, the matching number $\nu(G)$ is equal to the minimum number of stars and triangles needed to cover the edges of G; this extends Kőnig's edge-colouring and matching theorems (cf. Trotter [1977] and de Werra [1978]).
  Sbihi [1978,1980] and Minty [1980] showed that the weighted stable set problem is solvable in strongly polynomial time for claw-free graphs (see Chapter 69). A combinatorial polynomial-time algorithm for the colouring problem for claw-free perfect graphs was given by Hsu [1981], and for the weighted clique and clique cover problems by Hsu and Nemhauser [1981,1982,1984].
  Chvátal and Sbihi [1988] gave a polynomial-time algorithm to recognize claw-free perfect graphs, based on decomposition (cf. Maffray and Reed [1999]). Koch [1979] gave a polynomial-time algorithm which for any claw-free graph either finds a maximum-size stable set and a minimum-size clique cover of equal cardinalities, or else finds an odd hole or odd antihole.
  Perfection of line graphs was also studied by Cao and Nemhauser [1998]. The validity of the strong perfect graph conjecture for claw-free graphs was extended to 'pan-free' graphs by Olariu [1989b].
- K_4-*free graphs* — graphs not having K_4 as subgraph (Tucker [1977b], cf. Tucker [1979,1987a]).

- *diamond-free graphs* (Tucker [1987b][7]) — graphs not having $K_4 - e$ (a *diamond*) as induced subgraph (where $K_4 - e$ is the graph obtained from K_4 by deleting an edge) (Conforti [1989] gave an alternative proof). Tucker [1987b] gave an $O(n^2 m)$-time algorithm to colour such graphs optimally. Fonlupt and Zemirline [1987] and Conforti and Rao [1993] gave polynomial-time perfection tests for diamond-free graphs. Related results can be found in Conforti and Rao [1989, 1992a,1992b] and Fonlupt and Zemirline [1992,1993].
- *paw-free graphs* — graphs not having a *paw* (a K_4 with two incident edges deleted) as induced subgraph. This follows from the perfection of Meyniel graphs (Theorem 66.6). It also follows from a characterization of Olariu [1988e] of paw-free graphs.
- *square-free graphs* (Conforti, Cornuéjols, and Vušković [2002]) — graphs not having a C_4 (a *square*) as induced subgraph.
- *bull-free graphs* (Chvátal and Sbihi [1987]) — graphs not having a *bull* as induced subgraph, where a bull is the (self-complementary) graph on five vertices a, b, c, d, e and edges ab, ac, bc, bd, ce (see Figure 65.1). Reed and Sbihi [1995] gave a polynomial-time perfection test for bull-free graphs. More on bull-free graphs can be found in de Figueiredo [1995], de Figueiredo, Maffray, and Porto [1997,2001], and Hayward [2001].
- *chair-free graphs* (Sassano [1997]) — graphs not having a *chair* as induced subgraph, where a chair is the graph on five vertices a, b, c, d, e and edges ab, bc, cd, be (see Figure 65.1).
- *dart-free graphs* (Sun [1991]) — graphs not having a *dart* as induced subgraph (a dart is a graph with vertices a, b, c, d, e and edges ab, ac, ad, ae, bc, cd (see Figure 65.1)); Chvátal, Fonlupt, Sun, and Zemirline [2000,2002] gave a polynomial-time recognition algorithm for dart-free perfect graphs.
- graphs having neither P_5 nor K_5 as induced subgraph (Maffray and Preissmann [1994], Barré and Fouquet [1999]).
- *circular-arc graphs* (Tucker [1975]) — these are the intersection graphs of families of intervals on a circle (cf. Section 64.9a).
- *circle graphs* (Buckingham and Golumbic [1984]) — these are the intersection graphs of families of chords of a circle (cf. Section 64.9a).
- planar graphs (Tucker [1973b]). Tucker [1984b] showed that this can be derived (without appealing to the four-colour theorem) from the validity of the strong perfect graph conjecture for K_4-free graphs: a K_4 subgraph in a planar graph $G \neq K_4$ contains a triangle that is a vertex-cut of G; hence one can apply induction to find a 4-colouring of G.
Tucker and Wilson [1984] gave an $O(n^2)$ algorithm for finding a minimum colouring of a planar perfect graph, Hsu [1987b] gave an $O(n^3)$-time perfection test for planar graphs, and Hsu [1988] described strongly polynomial-time algorithms for the maximum-weight stable set, the weighted colouring, and the weighted clique cover problems for planar perfect graphs.
- graphs embeddable in the torus or in the Klein bottle (Grinstead [1980,1981]).
- *checked graphs* (Gurvich and Temkin [1992]) — graphs whose vertex set is a subset of $\mathbb{R}^2$, two vertices being adjacent if and only the line segment connecting them is horizontal or vertical.

---

[7] An partial proof was given by Parthasarathy and Ravindra [1979], cf. Tucker [1987b].

bull   chair   dart

Figure 65.1

The perfect graph theorem implies that the strong perfect graph conjecture is true also for the classes of graphs complementary to those listed above.

Since C_k and $\overline{C}_k$ (for odd $k \geq 5$) are claw-free, the result of Parthasarathy and Ravindra [1976] implies that, to show the strong perfect graph theorem, it suffices to show that each minimally imperfect graph is claw-free.

Other classes of graphs for which the strong perfect graph conjecture holds were found by Rao and Ravindra [1977], Olariu [1988d], Jamison and Olariu [1989b], Carducci [1992], Galeana-Sánchez [1993], Lê [1993b], De Simone and Galluccio [1994], Maire [1994b], Maffray and Preissmann [1995], Xue [1995,1996], Aït Haddadene and Gravier [1996], Maffray, Porto, and Preissmann [1996], Aït Haddadène and Maffray [1997], Kroon, Sen, Deng, and Roy [1997], Babaĭtsev [1998], Hoàng and Le [2000b, 2001], and Gerber and Hertz [2001].

65.6e. The P_4-structure of a graph and a semi-strong perfect graph theorem

V. Chvátal noticed that the collection of 4-sets inducing the 4-vertex path P_4 as a subgraph, provides a useful tool in studying perfection. (Note that $\overline{P_4}$ is isomorphic to P_4.) It led Chvátal [1984a] to conjecture the following 'semi-strong perfect graph theorem', which was proved by Reed [1987] (announced in Reed [1985]).

Call two graphs G and H, with $VG = VH$, P_4-*equivalent* if for each $U \subseteq V$ one has: U induces a P_4-subgraph of G if and only if U induces a P_4-subgraph of H. Then Reed's theorem states that

(65.32)   if G and H are P_4-equivalent and G is perfect, then H is perfect.

This theorem implies the perfect graph theorem, since G and $\overline{G}$ are P_4-equivalent. On the other hand, the theorem is implied by the strong perfect graph theorem, since any graph P_4-equivalent to an odd circuit of length at least 5 is equal to that circuit or to its complement.

Other relations between the P_4-structure and perfection were proved by Chvátal and Hoang [1985] and Chvátal, Lenhart, and Sbihi [1990]. Let $G = (V, E)$ be a graph and let V be partitioned into classes V_0 and V_1, with both $G[V_0]$ and $G[V_1]$ perfect. For each word $x = x_1x_2x_3x_4$ of length 4 over the alphabet $\{0,1\}$, let $\mathcal{Q}_x$ denote the set of chordless paths in G with vertices v_1, v_2, v_3, v_4 (in order) with $v_i \in V_{x_i}$ for $i = 1, 2, 3, 4$. Then G is perfect if:

(65.33)   (i) $\mathcal{Q}_{1000} = \mathcal{Q}_{0100} = \mathcal{Q}_{0111} = \mathcal{Q}_{1011} = \emptyset$, or

(ii) $\mathcal{Q}_{0000} = \mathcal{Q}_{0110} = \mathcal{Q}_{1001} = \mathcal{Q}_{1111} = \emptyset$, or
(iii) $\mathcal{Q}_{1000} = \mathcal{Q}_{0100} = \mathcal{Q}_{0110} = \mathcal{Q}_{1001} = \mathcal{Q}_{1111} = \emptyset$, or
(iv) $\mathcal{Q}_{1000} = \mathcal{Q}_{0101} = \mathcal{Q}_{0110} = \mathcal{Q}_{1001} = \mathcal{Q}_{1111} = \emptyset$, or
(v) $\mathcal{Q}_{1000} = \mathcal{Q}_{0101} = \mathcal{Q}_{0110} = \mathcal{Q}_{1001} = \mathcal{Q}_{0111} = \emptyset$, or
(vi) $\mathcal{Q}_{1000} = \mathcal{Q}_{1001} = \mathcal{Q}_{1011} = \emptyset$.

Sufficiency of (i) was shown by Chvátal and Hoang [1985], and of the other cases by Chvátal, Lenhart, and Sbihi [1990] ((ii) also by Gurvich [1993a,1993b]), who also showed that (65.33) essentially covers all cases where perfection of G follows from perfection of its constituents and the 'colouring' of the P_4-subgraphs. In fact, (65.33) and its symmetrical cases (interchanging V_0 and V_1 and/or replacing G by $\overline{G}$) describe exactly the cases excluded by G or $\overline{G}$ being an odd chordless circuit of length ≥ 5 or its complement.

A theorem of Seinsche [1974] states that each graph without an induced P_4 subgraph is perfect. (This follows from the perfection of Meyniel graphs (Theorem 66.6).[8])

Hence, case (65.33)(ii) implies the result of Hoang [1985] that any graph is perfect if there is a set U of vertices having an odd intersection with each chordless path with 4 vertices. More generally, it implies perfection of any graph if there is a set U of vertices such that each induced P_4 subgraph has exactly one of its two middle vertices in U or has exactly one of its ends in U.

Related (and more general than the results of Chvátal and Hoang quoted above) is the following theorem of Chvátal [1987a]. Let $G = (V, E)$ be a graph and let V be partitioned into two classes X and Y such that there are no $x \in X$, $y \in Y$, and $U \subseteq V$ such that both $U \cup \{x\}$ and $U \cup \{y\}$ induce a P_4 subgraph. Then G is perfect if and only if $G[X]$ and $G[Y]$ are perfect.

More work on the P_4-structure related to perfection is reported in Jung [1978], Jamison and Olariu [1989c,1992a,1992b,1995a,1995b], Hayward and Lenhart [1990], Hoàng [1990,1995,1999], Olariu [1991], Ding [1994], Rusu [1995a,1999b], Giakoumakis [1996], Hoàng, Hougardy, and Maffray [1996], Hougardy [1996b,1997,1999, 2001], Babel and Olariu [1997,1998,1999], Giakoumakis, Roussel, and Thuillier [1997], Giakoumakis and Vanherpe [1997], Hougardy, Le, and Wagler [1997], Babel [1998a,1998b], Babel, Brandstädt, and Le [1999], Brandstädt and Le [1999,2000], Roussel, Rusu, and Thuillier [1999], Brandstädt, Le, and Olariu [2000], Hoàng and Le [2000a,2001], Barré [2001], and Hayward, Hougardy, and Reed [2002].

65.6f. Further notes on the strong perfect graph conjecture

Markosyan and Karapetyan [1984] showed that the strong perfect graph conjecture is equivalent to: each minimally imperfect graph G is regular of degree $2\omega(G) - 2$.

For $k, n \in \mathbb{Z}_+$, let $C_{n,k}$ be the graph obtained from the circuit C_n (with n vertices) by adding all edges connecting two vertices at distance less than k. If $n \equiv 1 \pmod{k+1}$ and $n \geq 2k + 3$, then $C_{n,k}$ is partitionable. Chvátal [1976] showed that the strong perfect graph conjecture is equivalent to: each minimally imperfect graph G has $C_{|VG|,\omega(G)}$ as *spanning* subgraph (not necessarily induced).

---

[8] Arditti and de Werra [1976] claimed that Seinsche's result also follows from the 'fact' that any graph without induced P_4 subgraph is the comparability graph of a branching, therewith overlooking C_4.

Bland, Huang, and Trotter [1979] and Chvátal, Graham, Perold, and Whitesides [1979] gave examples of partitionable graphs G not containing $C_{|VG|,\omega(G)}$ as a spanning subgraph. Sebő [1996a] and Bacsó, Boros, Gurvich, Maffray, and Preissmann [1998] showed that these constructions give no counterexamples to the strong perfect graph conjecture. Related results are given by Chvátal [1984b]. A computer search for partitionable graphs was reported by Lam, Swiercz, Thiel, and Regener [1980].

Call a pair of vertices u, v in a graph an *even pair* if each induced $u - v$ path has even length. Meyniel [1987] showed that a minimally imperfect graph has no even pair. (This was extended to partitionable graphs by Bertschi and Reed [1987].) Hougardy [1996a] proved that the strong perfect graph conjecture is equivalent to: each properly induced subgraph of a minimally imperfect graph has an even pair or is a clique. Bienstock [1991] showed that it is NP-complete to test if a graph has no even pair. More on even pairs can be found in Hoàng and Maffray [1989], Bertschi [1990], Hougardy [1995], Rusu [1995c], Everett, de Figueiredo, Linhares-Sales, Maffray, Porto, and Reed [1997] (survey), Linhares Sales, Maffray, and Reed [1997], Linhares Sales and Maffray [1998], de Figueiredo, Gimbel, Mello, and Szwarcfiter [1999], Rusu [2000], and Everett, de Figueiredo, Linhares Sales, Maffray, Porto, and Reed [2001] (survey).

Prömel and Steger [1992] showed that 'almost all Berge graphs are perfect': the ratio of the number of n-vertex perfect graphs and the number of n-vertex Berge graphs, tends to 1 if $n \to \infty$.

The role of uniquely colourable perfect graphs for the strong perfect graph conjecture was investigated by Tucker [1983b]. Bacsó [1997] studied the conjecture that a uniquely colourable perfect graph G is either a clique or contains two maximum-size cliques intersecting each other in $\omega(G) - 1$ vertices. This is implied by the strong perfect graph theorem. Related work was given by Sakuma [2000].

Corneil [1986] investigated families of graphs 'complete' for the strong perfect graph conjecture (that is, proving the conjecture for these graphs suffices to prove the conjecture in general).

Equivalent versions of the strong perfect graph conjecture were given by Olaru [1972,1973b], Ravindra [1975], Markosyan [1981], Markosyan and Gasparyan [1987], Olariu [1990b], Huang [1991], Markosian, Gasparian, and Markosian [1992], Galeana-Sánchez [1993], De Simone and Galluccio [1994], Lonc and Zaremba [1995], and Padberg [2001].

Giles, Trotter, and Tucker [1984], Hsu [1984], Fonlupt and Sebő [1990], Croitoru and Radu [1992b], Sebő [1992], Panda and Mohanty [1997], and Rusu [1997] gave further techniques for proving the strong perfect graph conjecture.

Several other properties of minimally imperfect and partitionable graphs were derived by Olaru [1969,1972,1973a,1973b,1977,1980,1993,1998], Sachs [1970], Padberg [1974a,1974b,1975,1976], Parthasarathy and Ravindra [1976], Tucker [1977b, 1983a], Olaru and Suciu [1979], Markosyan [1981,1985], Sridharan and George [1982], Whitesides [1982], Buckingham and Golumbic [1983], Chvátal [1984c,1985c], Grinstead [1984], Olaru and Sachs [1984], Chvátal and Sbihi [1987], Meyniel [1987], Olariu [1988b,1988c,1990a,1991], Meyniel and Olariu [1989], Preissmann [1990], Sebő [1992,1996a,1996b], Cornuéjols and Reed [1993], Hougardy [1993], Maffray [1993], Olariu and Stewart [1993], Hayward [1995], Hoàng [1996c], Perz and Zaremba [1996], Fouquet, Maire, Rusu, and Thuillier [1997], Gasparyan [1998],

Barré and Fouquet [1999,2001], Croitoru [1999], de Figueiredo, Klein, Kohayakawa, and Reed [2000], Barré [2001], Roussel and Rubio [2001], and Conforti, Cornuéjols, Gasparyan, and Vušković [2002]. Surveys were given by Preissmann and Sebő [2001] and Cornuéjols [2002].

65.7. Further results and notes

65.7a. Perz and Rolewicz's proof of the perfect graph theorem

An interesting proof of the perfect graph theorem was given by Perz and Rolewicz [1990]. It does not use the replication lemma, and is based on linear algebra, in a manner different from the proof of Gasparian given in Section 65.2, namely on the value of determinants.

In fact, Perz and Rolewicz [1990] show (in a different but equivalent terminology) that a graph $G = (V, E)$ is perfect if and only if $P_{\text{stable set}}(G)$ and $P_{\text{clique}}(G)$ form an antiblocking pair of polytopes. They prove sufficiency in a way similar to the proof of Fulkerson given for sufficiency in Corollary 65.2e above.

They proved necessity as follows. Choose a counterexample with $|V|$ minimal. So G is perfect, and $P_{\text{stable set}}(G)$ and $P_{\text{clique}}(G)$ do not form an antiblocking pair. Hence there exist $x \in A(P_{\text{stable set}}(G))$ and $y \in A(P_{\text{clique}}(G))$ with $x^T y > 1$. Choose such x, y with $x^T y$ maximal. Let $\nu := x^T y$.

We first show

(65.34) $\quad \nu \leq \dfrac{n}{n-1}$,

where $n := |V|$. Indeed, for each $u \in V$, deleting the uth component of x and y, we obtain vectors in $A(P_{\text{stable set}}(G - u))$ and $A(P_{\text{clique}}(G - u))$, respectively. By the minimality of G, we have $\sum_{v \neq u} x_v y_v \leq 1$. Hence

(65.35) $\quad \nu = \sum_v x_v y_v = \dfrac{1}{n-1} \sum_u \left(\sum_{v \neq u} x_v y_v \right) \leq \dfrac{n}{n-1}$.

This proves (65.34). By the minimality of G, we also have $x_v > 0$ and $y_v > 0$ for each v.

Now $\nu^{-1} \cdot x \in P_{\text{clique}}(G)$, since otherwise there is a $z \in A(P_{\text{clique}}(G))$ with $\nu^{-1} x^T z > 1$, contradicting the maximality of $x^T y$. So there exist cliques $C_1, \ldots, C_n$ and $\lambda_1, \ldots, \lambda_n > 0$ such that

(65.36) $\quad x = \sum_{i=1}^{n} \lambda_i \chi^{C_i}$ and $\sum_{i=1}^{n} \lambda_i = \nu$.

Similarly, there exist stable sets $S_1, \ldots, S_n$ and $\mu_1, \ldots, \mu_n > 0$ such that

(65.37) $\quad y = \sum_{j=1}^{n} \mu_j \chi^{S_j}$ and $\sum_{j=1}^{n} \mu_j = \nu$.

Then $y(C_i) = 1$ for $i = 1, \ldots, n$, since $y(C_i) \leq 1$ (as $y \in A(P_{\text{clique}}(G))$), and

(65.38) $\quad \nu = x^T y = \sum_{i=1}^{n} \lambda_i y(C_i) \leq \sum_{i=1}^{n} \lambda_i = \nu$.

Similarly, $x(S_j) = 1$ for $j = 1, \ldots, n$.

Consider, for some $i = 1, \ldots, n$,

$$(65.39) \qquad 1 = y(C_i) = \sum_{j=1}^n \mu_j |C_i \cap S_j| \leq \sum_{j=1}^n \mu_j = \nu.$$

So the inequality is strict, and hence there is at least one j with $C_i \cap S_j = \emptyset$. Then

$$(65.40) \qquad 1 = x(S_j) = \sum_{i'} \lambda_{i'} |C_{i'} \cap S_j| = \sum_{i' \neq i} \lambda_{i'} |C_{i'} \cap S_j| \leq \sum_{i' \neq i} \lambda_{i'} = \nu - \lambda_i.$$

Hence with (65.34),

$$(65.41) \qquad n \leq \sum_i (\nu - \lambda_i) = n\nu - \nu \leq n,$$

implying equality in (65.40) for each i. So if $C_i \cap S_j = \emptyset$, then $C_{i'} \cap S_j \neq \emptyset$ for each $i' \neq i$. Hence for each j, there is exactly one i with $C_i \cap S_j = \emptyset$, and conversely. We can assume that $C_i \cap S_j = \emptyset$ if and only if $i = j$.

Let M and N be the incidence matrices of $S_1, \ldots, S_n$ and of $C_1, \ldots, C_n$ respectively. So $MN^\mathsf{T} = J - I$. Hence $|\det M \det N| = |\det(J - I)| = n - 1$. Since $y(S_i) = 1$ for each i, we have $My = \mathbf{1}$. So $y' := |\det M| \cdot y$ is a positive integer vector. Similarly, $x' := |\det N| \cdot x$ is a positive integer vector. Then

$$(65.42) \qquad (x')^\mathsf{T} y' = |\det M \det N| x^\mathsf{T} y = |\det(J - I)|\nu = n.$$

The kernel of the argument now is that this implies that x' and y' are the all-one vectors, and therefore x and y each are scalar multiples of the all-one vector.

As $x \in A(P_{\text{stable set}}(G))$, $x(S) \leq 1$ for any stable set S, and hence $x = \alpha'^{-1} \cdot \mathbf{1}$ for some $\alpha' \geq \alpha(G)$. Similarly, $y = \omega'^{-1} \cdot \mathbf{1}$ for some $\omega' \geq \omega(G)$. As G is perfect, $\alpha' \omega' \geq \alpha(G)\omega(G) \geq n$. Hence $\nu = x^\mathsf{T} y = (\alpha'\omega')^{-1}n \leq 1$, a contradiction.

65.7b. Kernel solvability

The following generalization of the Gale-Shapley theorem on stable matchings was conjectured by Berge and Duchet [1986,1988a][9] and proved by Boros and Gurvich [1996], using a technique from game theory due to Scarf [1967]. With the strong perfect graph theorem it characterizes perfect graph by being kernel solvable.

Call a graph $G = (V, E)$ *kernel solvable* if the following holds: if for each clique C of G we have a total order $<_C$ of C, then there exists a stable set S such that for each $v \in V$ there is an $s \in S$ and a clique C such that $v, s \in C$ and $v \leq_C s$. Berge and Duchet conjectured that kernel solvable graphs are precisely the perfect graphs. With Theorem 65.14 below, this conjecture is implied by the strong perfect graph theorem.

Kernel solvability can be formulated equivalently in terms of kernels of digraphs. A *kernel* of a directed graph $D = (V, A)$ is a subset S of V such that S spans no arc of D and such that for each $v \in V \setminus S$ there is a $u \in S$ with $(v, u) \in A$.

For any graph $G = (V, E)$, a directed graph $D = (V, A)$ is called a *superorientation* of G if $E = \{\{u, v\} \mid (u, v) \in A\}$. (So $\{u, v\}$ is an edge of $G \iff$ at least

---

[9] Berge and Duchet [1986] refer to 'Séminaire du Lundi, MSH, Paris, Janvier 1983' (Monday Seminar, MSH, Paris, January 1983). See Jensen and Toft [1995] p. 140 for further references to the history of this conjecture.

one of (u,v) and (v,u) belongs to A.) Then a graph $G = (V, E)$ is kernel solvable if and only if any superorientation D of G has a kernel if each clique C of G induces a subgraph of D with a kernel.

Kernel solvability is closed under taking induced subgraphs: if $U \subseteq V$, and each clique C of $G[U]$ has a total order $<_C$, we can choose for each clique C of G a total order which coincides with $<_{C \cap U}$ on $C \cap U$ and which has $C \cap U$ as upper ideal.

Since neither C_k nor $\overline{C}_k$ is kernel solvable for odd $k \geq 5$, the strong perfect graph theorem implies that each kernel solvable graph is perfect.

Boros and Gurvich [1996] proved that a graph G is perfect if and only if each graph H arising from G by replicating vertices is kernel solvable. It implies that the strong perfect graph theorem is equivalent to: each Berge graph is kernel solvable (since the class of Berge graphs is closed under replicating vertices).

To show that each perfect graph is kernel solvable, we follow the proof method of Aharoni and Holzman [1998]. We first prove the following results of Scarf [1967].

Let M and N be disjoint finite nonempty sets, and for each $i \in M$ let $<_i$ be a total order of N. For any U, write $y <_i U$ if $y <_i u$ for each $u \in U$. Define $K := M \cup N$.

Call a subset L of K *light* if for each $j \in N$ there is an $i \in M \setminus L$ with $j \leq_i L \setminus M$. So any subset of a light set is light again. Let $m := |M|$ and define

(65.43) $\qquad \mathcal{S} := \{M\} \cup \{L \mid L \text{ light}, |L| = m\}$.

Note that M is not light.

Now Scarf first proved:

Lemma 65.13α. *Any light set L with $|L| = m - 1$ is contained in precisely two sets in $\mathcal{S}$.*

Proof. Extend each $<_i$ to a total order on K, with $i <_i j <_i i'$ for all $j \in N$ and all $i' \in M \setminus \{i\}$. Then

(65.44) $\qquad$ any subset L of K is light if and only if for each $k \in K$ there is an $i \in M$ with $k \leq_i L$.

To see necessity in (65.44), let $L \subseteq K$ be light and let $k \in K$. If $k \in M$, then $k \leq_k L$. If $k \in N$, then there is an $i \in M \setminus L$ with $k \leq_i L \setminus M$. As $i \notin L$, we have also $k \leq_i L \cap M$ (since $k \leq_i M \setminus \{i\}$). So $k \leq_i L$.

To see sufficiency in (65.44), suppose $\forall k \in K \exists i \in M : k \leq_i L$. Let $j \in N$. Then $\exists i \in M : j \leq_i L$. Then $i \notin L$ (as otherwise $j \leq_i i$). Moreover, $j \leq_i L \setminus M$. This proves (65.44).

For any $i \in M$ and any nonempty $U \subseteq K$, let $\min_i U$ and $\max_i U$ denote the minimal and maximal element of U with respect to $<_i$.

First assume that $L \subseteq M$, say $L = M \setminus \{i\}$. Then L is contained in M, which belongs to $\mathcal{S}$. Moreover, $z := \max_i N$ is the unique element with $L \cup \{z\}$ light.[10] This proves the lemma.

So henceforth we can assume that $L \not\subseteq M$. Define $\pi : M \to L$ by $\pi(i) := \min_i L$. Then π is onto, since, as L is light, for each $r \in L$ there is an $i \in M$ with $r \leq_i L$. So $r = \min_i L = \pi(i)$.

---

[10] For let $x \in N$. Then $L \cup \{x\}$ is light $\iff \forall j \in N \exists i \in M \setminus L : j \leq_i (L \cup \{x\}) \setminus M \iff \forall j \in N : j \leq_i x \iff x = \max_i N$.

Hence, as $|L| = |M| - 1$, there exist distinct $i_1, i_2 \in M$ with $\pi(i_1) = \pi(i_2)$, while all other values of π are mutually distinct and different from $\pi(i_1)$.

For $h = 1, 2$, define

(65.45) $\quad R_h := \{k \in K \mid k \not\leq_i L \text{ for all } i \neq i_h\}$.

Then $R_h \neq \emptyset$, since $i_h \in R_h$, as there is an $r \in L \setminus M$ (as $L \not\subseteq M$), hence if $i \neq i_h$, then $i_h \not\leq_i r$. Also, $R_h \cap L = \emptyset$, since if $r \in L$, there is an $i \neq i_h$ with $\pi(i) = r$, so $\min_i L = r$, implying $r \leq_i L$, and hence $r \notin R_h$. Moreover, $R_1 \cap R_2 = \emptyset$, since otherwise there is a $k \in K$ with $k \not\leq_i L$ for all $i \in M$, contradicting the fact that L is light.

Define for $h = 1, 2$:

(65.46) $\quad r_h := \max_{i_h} R_h$.

We first show that $L \cup \{r_1\}$ and $L \cup \{r_2\}$ are light. Suppose that (say) $L \cup \{r_1\}$ is not light. So there is a $k \in K$ with $k \not\leq_i L \cup \{r_1\}$ for each $i \in M$. Since $r_1 \not\leq_i L$ for each $i \neq i_1$ (by definition of R_1), it follows that $k \not\leq_i L$ for each $i \neq i_1$. Hence $k \in R_1$. However, $r_1 <_{i_1} L$, since $r_1 \in R_1$ and $r_1 \leq_i L$ for some $i \in M$. So $k \leq_{i_1} r_1 <_{i_1} L$, and therefore $k \leq_{i_1} L \cup \{r_1\}$, a contradiction. So $L \cup \{r_1\}$ and $L \cup \{r_2\}$ are light.

Finally we show that for any $s \in K \setminus L$, if $L \cup \{s\}$ is light, then $s = r_1$ or $s = r_2$. So let $L \cup \{s\}$ be light. Then the function $\pi' : M \to L \cup \{s\}$ defined by $\pi'(i) := \min_i(L \cup \{s\})$ is onto (as $L \cup \{s\}$ is light), implying that it is one-to-one (as $|M| = |L \cup \{s\}|$). Hence π' coincides with π on all but one element of M. Necessarily the exceptional element belongs to $\{i_1, i_2\}$. Say $\pi'(i) = \pi(i)$ for each $i \neq i_1$, while $\pi'(i_1) = s$. So $\min_i L = \pi(i) = \pi'(i) <_i s$ for each $i \neq i_1$; that is, $s \in R_1$. Suppose $s \neq r_1$. So $s <_{i_1} r_1$. Then $r_1 \not\leq_i L \cup \{s\}$ for each $i \in M$, contradicting the fact that $L \cup \{s\}$ is light. ∎

From this, Scarf derived:

Theorem 65.13 (Scarf's lemma). *Let A be a nonnegative $m \times n$ matrix and let $b \in \mathbb{R}_+^m$ be such that the polytope $P := \{x \in \mathbb{R}_+^n \mid Ax \leq b\}$ is nonempty and bounded. For each $i = 1, \ldots, m$, let $<_i$ be a total order on $\{1, \ldots, n\}$. Then P has a vertex x such that*

(65.47) $\quad$ *for each $j \in \{1, \ldots, n\}$ there is an $i \in \{1, \ldots, m\}$ such that $a_i^\mathsf{T} x = b_i$ and such that $x_k = 0$ for each $k <_i j$.*

Proof. We can assume, by slightly perturbing b, that for each vertex x of P there are precisely n constraints among $x \geq \mathbf{0}$, $Ax \leq b$ satisfied with equality. Add n to each index i of $<_i$. (So $x <_{n+i} y$ in the new notation $\iff x <_i y$ in the old notation.) Let $N := \{1, \ldots, n\}$, $M := \{n+1, \ldots, n+m\}$, and $K := N \cup M$. For each face f of P define

(65.48) $\quad K_f := \{k \in K \mid \text{the } k\text{th constraint in } x \geq \mathbf{0}, Ax \leq b \text{ is not tight at some point in } f\}$.

So $|K_v| = m$ for any vertex v and $|K_e| = m + 1$ for any edge e. Call an edge e of P *good* if $1 \in K_e$ and the set $K_e \setminus \{1\}$ belongs to $\mathcal{S}$ (cf. (65.43)).

Now $\mathbf{0}$ is incident with precisely one good edge. Hence there is a vertex $v \neq \mathbf{0}$ incident with an odd number of good edges. We show that $K_v \in \mathcal{S}$, and hence K_v is light (since $K_v \neq M$, as $v \neq \mathbf{0}$), implying that v satisfies (65.47).

Let e be a good edge incident with v. Then $K_e = K_v \cup \{k\}$ for some k. As e is good, we know that $1 \in K_e$ and $K_e \setminus \{1\} \in \mathcal{S}$.

If $1 \notin K_v$, then $k = 1$ and hence $K_v \in \mathcal{S}$. So we can assume that $1 \in K_v$. Applying Lemma 65.13α to the light set $K_v \setminus \{1\}$, there is precisely one $j \notin K_v \setminus \{1\}$ with $j \neq k$ and $K_v \setminus \{1\} \cup \{j\} \in \mathcal{S}$. If $j \neq 1$, then v is incident with precisely two good edges, a contradiction. So $j = 1$, and hence $K_v \in \mathcal{S}$. ∎

This implies the theorem of Boros and Gurvich [1996]:

Corollary 65.13a. *A perfect graph is kernel solvable.*

Proof. Let $G = (V, E)$ be a perfect graph, and for each clique C, let $<_C$ be a total order on C. We must prove that

(65.49)   there exist a stable set S such that for each $v \in V$ there is a clique C and an element $s \in C \cap S$ with $v \in C$ and $v \leq_C s$.

Extend each $<_C$ to a total order on V with $w <_C v$ for each $w \in C$, $v \in V \setminus C$. Then by Theorem 65.13, the polytope in $\mathbb{R}^V$ determined by $x \geq 0$, $x(C) \leq 1$ (C clique), has a vertex x such that for each $v \in V$ there is a clique C with $x(C) = 1$ and such that $x_u = 0$ for each $u <_C v$. By Corollary 65.2e, x is the incidence vector of some stable set S. So for each $v \in V$ there is a clique C with $|C \cap S| = 1$ and with $u \notin S$ if $u <_C v$. Therefore, for the vertex s in $C \cap S$ we have $v \leq_C s$, and hence $v \in C$. This shows (65.49). ∎

It was conjectured by Berge and Duchet that conversely, each kernel solvable graph is perfect. This follows from the strong perfect graph theorem, since kernel solvability is closed under taking induced subgraphs and since odd circuits of length at least five and their complements are not kernel solvable.

It implies the following theorem found by Boros and Gurvich [1996], for which we give a direct proof. A graph H is called a *blow-up* of a graph G, if H arises from G by replicating vertices (replacing vertices by cliques).

Theorem 65.14. *A graph G is perfect if and only if each blow-up of G is kernel solvable.*

Proof. Since each blow-up of a perfect graph is perfect again (by the replication lemma (Corollary 65.2c)), necessity follows from Corollary 65.13a.

Sufficiency is shown by proving that each graph $G = (V, E)$ with $|V| \geq \alpha(G)\omega(G) + 1$ has a blow-up that is not kernel solvable. (This is sufficient by Theorem 65.2.)

Let $\mathcal{C}$ be the collection of cliques in G, and for each vertex v, let $\mathcal{C}_v$ be the collection of cliques containing v. Let $n := |V|$ and define

(65.50) $Y := \{y : \mathcal{C} \to \mathbb{Z}_+ \mid y(\mathcal{C}) \leq n|\mathcal{C}|\}$.

For each $y \in Y$, we choose a vertex v_y of G with

(65.51) $y(\mathcal{C}_{v_y}) \leq \omega(G)|\mathcal{C}|$.

This is possible since

1130   Chapter 65. Perfect graphs: general theory

$$(65.52) \qquad \sum_{v \in V} y(C_v) = \sum_{C \in \mathcal{C}} |C| y_C \leq \omega(G) \sum_{C \in \mathcal{C}} y_C \leq \omega(G) n |\mathcal{C}|.$$

Let H be the graph with vertex set Y, two distinct vertices $y, z \in Y$ being adjacent if $v_y = v_z$ or v_y and v_z are adjacent in G. So H is a blow-up of G. We show that H is not kernel solvable.

For each clique $K \subseteq Y$ of H, the set $C := \{v_y \mid y \in K\}$ is a clique of G. Then choose a total order $<_K$ on K such that for all $y, z \in K$:

$$(65.53) \qquad \text{if } y_C < z_C, \text{ then } y <_K z.$$

Assume that H is kernel solvable. Then H has a stable set $Z \subseteq Y$ such that for each $y \in Y$ there is a $z \in Z$ and a clique K of H with $y, z \in K$ and $y \leq_K z$. As Z is stable in H, the v_z for $z \in Z$ are distinct and form a stable set S in G. So for each clique C of G there is at most one $z \in Z$ with $v_z \in C$. Define $y : \mathcal{C} \to \mathbb{Z}_+$ by:

$$(65.54) \qquad y_C := \begin{cases} z_C + 1 & \text{if } v_z \in C, \text{ for } z \in Z, \\ 0 & \text{if } C \cap S = \emptyset. \end{cases}$$

Then y belongs to Y, since

$$(65.55) \qquad y(C) = \sum_{z \in Z} \sum_{C \in \mathcal{C}_{v_z}} (z_C + 1) = \sum_{z \in Z} (z(C_{v_z}) + |\mathcal{C}_{v_z}|) \leq |Z| \omega(G) |\mathcal{C}| + |\mathcal{C}|$$
$$\leq (\alpha(G)\omega(G) + 1)|\mathcal{C}| \leq n|\mathcal{C}|.$$

(The first inequality follows from (65.51).) Hence there exist a $z \in Z$ and a clique K of H with $y, z \in K$ and $y \leq_K z$. So for $C := \{v_x \mid x \in K\}$ we have, by (65.53), $y_C \leq z_C$. Since $v_z \in C$ (as $z \in K$), this contradicts (65.54). ∎

Before the strong perfect graph conjecture was settled, and hence the conjecture of Berge and Duchet, partial and related results on the latter conjecture were obtained by Blidia [1986], Maffray [1986,1992], Duchet [1987], Berge and Duchet [1988b,1990], Champetier [1989], Blidia and Engel [1992], Blidia, Duchet, and Maffray [1993,1994], Chilakamarri and Hamburger [1993], and Galeana-Sánchez [1995, 1996,1997].

65.7c. The amalgam

A composition generalizing the 1-join, the *amalgam*, was shown to preserve perfection by Burlet and Fonlupt [1984]. Let $G_1 = (V_1, E_1)$ and $G_2 = (V_2, E_2)$ be perfect graphs such that $K := V_1 \cap V_2$ is a clique in both graphs. For $i = 1, 2$, let $v_i \in V_i \setminus K$ be such that each vertex in K is adjacent to v_i and to each neighbour of v_i. Let H be the graph on $(V_1 \setminus \{v_1\}) \cup (V_2 \setminus \{v_2\})$ obtained from the union of $G_1 - v_1$ and $G_2 - v_2$ by adding all edges between $N(v_1) \setminus K$ and $N(v_2) \setminus K$.

Theorem 65.15. *If G_1 and G_2 are perfect, then H is perfect.*

Proof. It suffices to show that $\omega(H) = \chi(H)$, since each induced subgraph of H arises by the same construction.

For $i = 1, 2$, let $p_i := \omega(G_i[N(v_i)])$ and let G'_i be the graph obtained from G_i by replicating v_i by a factor $\omega(H) - p_i$. So $\omega(G'_i) = \omega(H)$. By the replication lemma, G'_i is perfect. Hence $\omega(H) = \chi(G'_i)$.

Consider colourings of G'_1 and G'_2 with colours $1, \ldots, \omega(H)$. So $N(v_i)$ uses precisely p_i colours. As $p_1 + p_2 - |K| \leq \omega(H)$, we have $(\omega(H) - p_1) + (\omega(H) - p_2) \geq \omega(H) - |K|$. Hence we can assume that in G_1 and G_2 the colourings of K are the same, and that all colours are used by the replication vertices of v_1 and v_2 and by K. Then $N(v_1) \setminus K$ and $N(v_2) \setminus K$ have no colours in common. Hence we obtain an $\omega(H)$-colouring of H. ∎

Cornuéjols and Cunningham [1985] gave an $O(n^2m)$-time algorithm to decide if a graph is the amalgam of smaller graphs.

Perfection is trivially closed under 'clique sums', that is, identifying two cliques in two graphs. Whitesides [1981] gave an $O(nm)$ algorithm to find a clique cut in a graph, that is, a vertex-cut that is a clique. Tarjan [1985] gave an $O(nm)$-time algorithm to find for any graph a decomposition by clique cuts.

Fonlupt and Uhry [1982] gave conditions under which identification of two vertices in a graph maintains perfection. Ravindra and Parthasarathy [1977], Ravindra [1978], Mândrescu [1991], and Kwaśnik and Szelecka [1997] investigated the behaviour of perfection under taking (various) products of graphs.

More on the (de)composition of perfect graphs can be found in Hsu [1986,1987a], Conforti and Rao [1992a,1992b], Corneil and Fonlupt [1993], Burlet and Fonlupt [1994], and Conforti, Cornuéjols, Kapoor, and Vušković [1995].

65.7d. Diperfect graphs

Berge [1982a] introduced a directed variant of perfect graphs. In fact, there are two symmetric variants, as no complementary phenomenon holds in the directed case.

A *stable set* or *clique* in a directed graph is a stable set of clique in the underlying undirected graph. A directed graph $D = (V, A)$ is called α-*diperfect* if for every induced subgraph $D' = (V', A')$ of D and for each maximum-size stable set S in D' there is a partition of V' into directed paths each intersecting S in exactly one vertex.

Then:

(65.56)   if the underlying undirected graph G of D is perfect, then D is α-diperfect.

Indeed, if G is perfect, there is a maximum-size stable set S and a partition of V into cliques each intersecting S. Each clique C gives a tournament on C in D, and hence, by Rédei's theorem (Corollary 14.14a), it contains a directed path spanning C.

Another class of α-diperfect digraphs is formed by the *symmetric* digraphs: directed graphs $D = (V, A)$ such that if $(u, v) \in A$, then $(v, u) \in A$:

(65.57)   each symmetric digraph is α-diperfect.

To see this, let S be a maximum-size stable set in D, and let D' arise from D by deleting all arcs entering S. By the Gallai-Milgram theorem (Theorem 14.14), V can be partitioned into $|S|$ directed paths in D'. These paths are as required.

Berge offered the following conjecture characterizing α-diperfect digraphs:

(65.58)   (?) A directed graph $D = (V, A)$ is α-diperfect if and only if D has no induced subgraph C whose underlying undirected graph is a chordless odd circuit of length ≥ 5, say with vertices $v_1, \ldots, v_{2k+1}$ (in order) such that each of $v_1, v_2, v_3, v_4, v_6, v_8, \ldots, v_{2k}$ is a source or a sink. (?)

The odd circuit described is not α-diperfect, since $\{v_1, v_4, v_6, v_8, \ldots, v_{2k}\}$ is a maximum-size stable set, but there are no directed paths as required.

A 'dual' concept is that of a χ-*diperfect* graph, which is a digraph $D = (V, A)$ such that for each induced subgraph $D' = (V', A')$ of D and for each minimum vertex-colouring (in the underlying undirected graph of D') there exists a directed path intersecting each colour exactly once.

Again one has:

(65.59)   if the underlying undirected graph G of D is perfect, then D is χ-diperfect.

Indeed, any maximum-size clique C intersects each colour in each minimum vertex-colouring, and, again by Rédei's theorem (Corollary 14.14a), there is a path spanning C.

Also:

(65.60)   any symmetric digraph is χ-diperfect.

To see this, let $S_1, \ldots, S_k$ be an optimum vertex-colouring. Let D' be the graph obtained from D by deleting all arcs from S_j to S_i for all $j > i$. By the theorem of Gallai and Roy (see (64.52)), D' has a directed path of length k. Necessarily, it intersects each S_i exactly once.

One may show that the odd undirected circuit described in (65.58) is not χ-diperfect. So conjecture (65.58) would imply that each χ-diperfect digraph is α-diperfect.

In fact, any odd undirected circuit that contains three consecutive vertices v_1, v_2, v_3 that are sources or sinks, is not χ-diperfect (since there is an optimum 3-vertex-colouring where $\{v_2\}$ is one of the colours — hence v_2 should belong to a directed path with 3 vertices). In particular, the undirected circuit with vertices $v_1, \ldots, v_7$ and arcs

(65.61) $(v_1, v_2), (v_3, v_2), (v_3, v_4), (v_4, v_5), (v_5, v_6), (v_6, v_7), (v_1, v_7)$

is α-diperfect but not χ-diperfect (cf. Figure 65.2).

Figure 65.2

65.7e. Further notes

Cameron, Edmonds, and Lovász [1986] showed that if the edges of a complete graph are coloured with three colours such that no triangle gets three different colours, and two of these colours form perfect graphs, then so does the third. (This generalizes the perfect graph theorem.) A generalization and a related characterization of perfection in terms of decomposition was given by Cameron and Edmonds [1997].

Markosyan and Karapetyan [1976] characterize perfection with the help of *critical edges* (edges e with $\alpha(G-e) > \alpha(G)$) and *essential edges* (edges e with $\chi(G-e) > \chi(G)$). More on such edges can be found in Markosyan [1975], Karapetyan [1976], Sebő [1996a], and Markossian, Gasparian, Karapetian, and Markosian [1998]. Edge-minimal perfect graphs are studied by Wagler [1999], colouring perfect 'degenerate' graphs by Aït Haddadène and Maffray [1997], 'Gallai graphs' and 'anti-Gallai graphs' by Le [1993a,1996b], and 'edge perfect graphs' by Müller [1996].

An alternative polyhedral characterization of perfection of graphs was given by Zaremba and Perz [1982]. Related is the work of Zaremba [1991] and Hujter [1999].

Chandrasekaran and Tamir [1984] and Cook, Fonlupt, and Schrijver [1986] showed that, for any perfect graph $G = (V, E)$ and any weight $w : V \to \mathbb{Z}_+$, the weighted colouring number is attained by a weighted colouring using at most $|V|$ different stable sets.

Von Rimscha [1983] showed that if $G = (V, E)$ and $H = (V, F)$ are graphs with $G - v$ isomorphic to $H - v$ for each $v \in V$, then G is perfect if and only if H is perfect.

Bienstock [1991] showed that it is NP-complete to decide if a given graph has an odd hole containing a prescribed vertex. More on the complexity of finding odd holes can be found in Reed [1990]. A survey on forbidding holes and antiholes was given by Hayward and Reed [2001].

Le [1996a] showed that if a graph G is imperfect and has no odd hole, then the intersection graph of the edge sets of chordless circuits in G has an odd hole. Akiyama and Chvátal [1990] characterized for which graphs $G = (V, E)$ the intersection graph of the triples spanning at least two edges, is perfect. Olaru and Mândrescu [1992] considered perfection of products of graphs, and de Werra and Hertz [1999] perfection of sums of graphs. Hertz [1998] characterized the graphs for which all graphs obtained by 'switching' are perfect.

Variants of the notion of perfect graph were studied by Körner [1973], Duchet [1980], Galeana-Sánchez [1982,1986,1988], Duchet and Meyniel [1983], Galeana-Sánchez and Neumann-Lara [1986,1991a,1991b,1994,1996,1998], Lehel and Tuza [1986], Conforti, Corneil, and Mahjoub [1987], Cameron [1989], Brown, Corneil, and Mahjoub [1990], Markosyan and Gasparyan [1990], Reed [1990], Scheinerman and Trenk [1990,1993], Berge [1992b,1992a,1995], Körner, Simonyi, and Tuza [1992], Lehel [1994], Trenk [1995], Cai and Corneil [1996], Markossian, Gasparian, and Reed [1996], Tamura [1997,2000], Gutin and Zverovich [1998], De Simone and Körner [1999], Huang and Guo [1999], Fachini and Körner [2000], and de Figueiredo and Vušković [2000].

Introductions to and surveys of perfect graphs are given by Berge [1973b,1975, 1986], Golumbic [1980], Lovász [1983b], Berge and Chvátal [1984] (a collection of papers on perfect graphs), Chvátal [1985b,1987b], Jensen and Toft [1995], Toft [1995], Ravindra [1997], Brandstädt, Le, and Spinrad [1999], and Ramírez Alfonsín and Reed [2001] (a collection of survey papers on perfect graphs). The latter reference includes a bibliography on perfect graphs by Chvátal [2001]. Applications of perfect graphs to graph entropy were surveyed by Simonyi [2001] (cf. Simonyi [1995]). Algorithmic aspects are discussed in Golumbic [1984].

We refer for historical remarks on perfect graphs to Section 67.4g.

Chapter 66
Classes of perfect graphs

In this chapter we consider classes of perfect graphs. The phenomenon observed by Berge that clique number and colouring number are equal for bipartite graphs, their line graphs, comparability graphs, and chordal graphs, and for their complements, formed the motivation for him to raise the conjecture that the complement of a perfect graph is perfect again ($\equiv$ perfect graph theorem).

The perfection of the graphs considered in this chapter follows directly from the strong perfect graph theorem. However, since its proof is highly complicated, we will give direct proofs of the perfection of several of these graphs.

66.1. Bipartite graphs and their line graphs

The perfect graph theorem can be used to prove several min-max relations on bipartite graphs: Kőnig's matching theorem, the Kőnig-Rado edge cover theorem, and Kőnig's edge colouring theorem.

We start from the trivial observation that:

Theorem 66.1. $\omega(G) = \chi(G)$ *for each bipartite graph* G.

Proof. Trivial. ∎

Since the class of bipartite graphs is closed under taking induced subgraphs, this gives:

Corollary 66.1a. *Each bipartite graph is perfect.*

Proof. See above. ∎

Hence, by the perfect graph theorem, also the complements of bipartite graphs are perfect. This amounts to the Kőnig-Rado edge cover theorem (Theorem 19.4):

Corollary 66.1b (Kőnig-Rado edge cover theorem). *For any bipartite graph* G, $\alpha(G) = \overline{\chi}(G)$. *Equivalently, the stable set number of any bipartite graph (without isolated vertices) is equal to its edge cover number.*

Proof. Directly from the perfect graph theorem, since by Theorem 66.1, any bipartite graph is perfect. Note that if G is a bipartite graph, then its cliques have size at most 2; hence $\overline{\chi}(G)$ is equal to the edge cover number of G if G has no isolated vertices. ∎

We saw in Section 16.2 that by Gallai's theorem (Theorem 19.1), the Kőnig-Rado edge cover theorem implies Kőnig's matching theorem (Theorem 16.2), saying that the matching number of a bipartite graph G is equal to its vertex cover number. That is, the stable set number of the line graph $L(G)$ of G is equal to the minimum number of cliques of $L(G)$ that cover all vertices of $L(G)$; in notation:

(66.1) $\quad \alpha(L(G)) = \overline{\chi}(L(G))$.

As this is true for any induced subgraph of $L(G)$ we know that the complement $\overline{L(G)}$ of the line graph $L(G)$ of any bipartite graph G is perfect.

Hence with the perfect graph theorem we know:

Corollary 66.1c. *The line graph of any bipartite graph is perfect.*

Proof. See above. ∎

This amounts to Kőnig's edge-colouring theorem (Theorem 20.1):

Corollary 66.1d (Kőnig's edge-colouring theorem). *If G is the line graph of a bipartite graph, then $\omega(G) = \chi(G)$. Equivalently, the edge-colouring number of any bipartite graph is equal to its maximum degree.*

Proof. Again directly from Kőnig's matching theorem and the perfect graph theorem. ∎

Complexity. In Part II on bipartite matching and covering, we saw that the optimization problems corresponding to the perfect graph parameters are solvable in polynomial time, and their weighted versions are solvable in strongly polynomial time, mainly by utilizing network flow techniques. We review the results.

The maximum-weight clique and the minimum colouring problem for bipartite graphs are trivially solvable in strongly polynomial time. Also the weighted colouring problem for bipartite graphs is easily solvable in strongly polynomial time.

A maximum-size stable set and a minimum clique cover in a bipartite graph can be found in polynomial time (cf. Corollary 19.3a). Note that in bipartite graphs, the minimum clique cover problem amounts to the minimum-size edge cover problem. Also the weighted versions are solvable in strongly polynomial time by max-flow techniques (cf. Corollary 21.25a). In bipartite graphs, the minimum weighted clique cover problem amounts to the minimum-size b-edge cover problem.

A bipartite graph is easily recognized, by checking if there is no odd circuit.

In *line graphs of bipartite graphs*, finding a maximum-weight clique is trivial (by checking all stars of the graph). In Sections 20.1 and 20.2 we saw that a minimum weighted colouring can be found in strongly polynomial time.

Finding a maximum clique and a minimum colouring in the complement of the line graph of a bipartite graph G amounts to finding a maximum-size matching and a minimum-size vertex cover in G, which can be found in polynomial time (cf. Theorem 16.3 and Corollary 16.6a). Their weighted versions can be found in strongly polynomial time with the methods for the assignment and the minimum-cost flow problems (cf. Theorems 17.4 and 17.6).

Van Rooij and Wilf [1965] showed that line graphs of bipartite graphs can be recognized in polynomial time, and that the corresponding bipartite graph can be reconstructed in polynomial time.

66.2. Comparability graphs

Also Dilworth's decomposition theorem (Theorem 14.2) can be derived from the perfect graph theorem. Let $(V, \leq)$ be a partially ordered set. Let $G = (V, E)$ be the graph with:

(66.2) $uv \in E$ if and only if $u < v$ or $v < u$.

Any graph G obtained in this way is called a *comparability graph*.

In Theorem 14.1 we saw the following easy 'dual' form of Dilworth's decomposition theorem:

Theorem 66.2. *In any partially ordered set $(V, \leq)$, the maximum size of a chain is equal to the minimum number of antichains needed to cover V.*

Proof. For any $v \in V$ define the *height* of v as the maximum size of a chain in V with maximum element v. Let k be the maximum height of the elements of V. For $i = 1, \ldots, k$, let A_i be the set of elements of height i. Then $A_1, \ldots, A_k$ are antichains covering V, and moreover, there is a chain of size k, since there is an element of height k. ∎

Equivalently, we have $\omega(G) = \chi(G)$ for any comparability graph. As the class of comparability graphs is closed under taking induced subgraphs we have:

Corollary 66.2a. *Each comparability graph is perfect.*

Proof. Directly from Theorem 66.2. ∎

Hence, by the perfect graph theorem, also the complement of a comparability graph is perfect. This implies:

Corollary 66.2b (Dilworth's decomposition theorem). *In any partially ordered set $(V, \leq)$, the maximum size of an antichain is equal to the minimum number of chains needed to cover V.*

Proof. Directly from Corollary 66.2a. ∎

Complexity. The optimization problems corresponding to the perfect graph parameters for comparability graphs can be solved in strongly polynomial time by path and flow techniques, as we saw in Chapter 14. With a greedy method, one can find a maximum-weight clique in a comparability graph $G = (V, E)$ with weight function $w : V \to \mathbb{Q}_+$ (if the underlying partial order $\leq$ is given): if all weights are 0, the problem is trivial; if there exist vertices of positive weights, find the set S of minimal elements of positive weight, let $\alpha := \min_{v \in S} w(v)$, reset $w(v) := w(v) - \alpha$ for $v \in S$, and find recursively a maximum-weight clique C for the new weights. Then we can assume that $C \cap S \neq \emptyset$. Hence C is also a maximum-weight clique for the original weight function.

This method also solves the weighted colouring problem in strongly polynomial time. An $O(n^2)$ algorithm for the weighted colouring problem for comparability graphs was given by Hoàng [1994]. The weighted stable set and clique cover problems can be solved in strongly polynomial time with flow techniques (see Chapter 14).

Trivially, recognizing comparability graphs belongs to NP (by giving the underlying partial order), and membership of co-NP follows from the characterizations of Ghouila-Houri [1962a,1964] and Gilmore and Hoffman [1964]. A method of Gallai [1967] implies that the problem in fact is polynomial-time solvable (cf. Pnueli, Lempel, and Even [1971], Golumbic [1977], Spinrad [1985], Muller and Spinrad [1989], and McConnell and Spinrad [1994,1997,1999] (the latter paper gives a linear-time recognition algorithm)).

Golumbic, Rotem, and Urrutia [1983] and Lovász [1983b] characterized complements of comparability graphs as those graphs that are the intersection graph of a family of continuous functions $f : (0,1) \to \mathbb{R}$. (Here f and g *intersect* if $f(x) = g(x)$ for some $x \in (0,1)$.)

Permutation graphs. A *permutation graph* is a graph on $\{1, \ldots, n\}$ for which there exists a permutation π of $\{1, \ldots, n\}$ such that $i, j \in \{1, \ldots, n\}$ are adjacent if and only if $(i-j)(\pi(i)-\pi(j)) > 0$. A graph G is (isomorphic to) a permutation graph if and only if both G and $\overline{G}$ are comparability graphs (Dushnik and Miller [1941] (also Even, Pnueli, and Lempel [1972])). McConnell and Spinrad [1997] showed that permutation graphs can be recognized in linear time (improving McConnell and Spinrad [1994]). Another characterization was given by Baker, Fishburn, and Roberts [1972].

The books by Even [1973] and Golumbic [1980] devote chapters to comparability graphs and to permutation graphs.

66.3. Chordal graphs

We next consider a further class of perfect graphs, the 'chordal graphs' (or 'rigid circuit graphs' or 'triangulated graphs'). A graph G is called *chordal* if each circuit in G of length at least 4 has a chord. (A *chord* is an edge connecting two vertices of the circuit that are nonadjacent in the circuit.) Equivalently, a graph is chordal if it has no hole.

For any set U of vertices, let $N(U)$ denote the set of vertices not in U that are adjacent to at least one vertex in U. Call a vertex v *simplicial* if $N(\{v\})$ is a clique in G.

Dirac [1961] showed the following basic property of chordal graphs:

Theorem 66.3. *Each chordal graph G contains a simplicial vertex.*

Proof. We may assume that G has at least two nonadjacent vertices a, b. Let U be a maximal nonempty subset of V with $G[U]$ connected and with $U \cup N(U) \neq V$. Such a subset U exists as $U := \{a\}$ induces a connected subgraph of G and as $\{a\} \cup N(\{a\}) \neq V$.

Let $W := V \setminus (U \cup N(U))$. Then each vertex v in $N(U)$ is adjacent to each vertex in W, since otherwise we could increase U by v. Moreover, $N(U)$ is a clique, for suppose that $u, w \in N(U)$ are nonadjacent. Choose $v \in W$. Let P be a shortest path in $U \cup N(U)$ connecting u and w. Then $P \cup \{u, v, w\}$ would form a chordless circuit of length at least 4, a contradiction.

Now inductively we know that $G[W]$ contains a vertex v that is simplicial in $G[W]$. Since $N(U)$ is a clique and since each vertex in W is adjacent to each vertex in $N(U)$, v is also simplicial in G. ∎

(The proof of Theorem 66.3 implies that, in a chordal graph, each vertex v that is nonadjacent to at least one vertex $w \neq v$, is nonadjacent to at least one simplicial vertex $w \neq v$. Hence each noncomplete chordal graph contains at least two nonadjacent simplicial vertices.)

As was observed by Fulkerson [1972a], Theorem 66.3 implies a result of Berge [1963a] (the result was announced (with partial proof) in Berge [1960a]):

Theorem 66.4. *Any chordal graph G satisfies $\omega(G) = \chi(G)$.*

Proof. By Theorem 66.3, G has a simplicial vertex v. By induction we have $\omega(G - v) = \chi(G - v)$. In particular, $G - v$ has an $\omega(G)$-vertex-colouring. As $|N(v)| \leq \omega(G) - 1$ (since $\{v\} \cup N(v)$ is a clique), we can extend this to an $\omega(G)$-vertex-colouring of G. ∎

As the class of chordal graphs is closed under taking induced subgraphs, this implies:

Corollary 66.4a. *Each chordal graph is perfect.*

Proof. Directly from Theorem 66.4. ∎

With the perfect graph theorem, this implies the following result of Hajnal and Surányi [1958] (which also can be derived directly from Theorem 66.3):

Corollary 66.4b. *For any chordal graph G, $\alpha(G) = \overline{\chi}(G)$.*

Proof. Directly from Corollary 66.4a and the perfect graph theorem (Corollary 65.2a). ∎

Complexity. Dirac's theorem (Theorem 66.3) can be used to obtain strongly polynomial-time algorithms for the basic optimization problems for chordal graphs. The proof of Theorem 66.4 yields such an algorithm to find an optimum colouring and clique, also for the weighted versions. Similarly, the strong polynomial-time solvability of the weighted stable set and clique cover problems can be derived (Gavril [1972], Frank [1976b]). $O(n^2)$ algorithms for minimum weighted colouring for chordal graphs were given by Balas and Xue [1991] and Hoàng [1994].

Dirac's theorem also directly gives a polynomial-time recognition algorithm for chordal graphs: iteratively find and delete a simplicial vertex until the graph is empty. Linear-time algorithms were given by Lueker [1974], Rose and Tarjan [1975], Rose, Tarjan, and Lueker [1976], and Tarjan and Yannakakis [1984]. (Gavril [1974b] gave another polynomial-time algorithm.)

Dirac's theorem also implies the following other characterizations of chordal graphs (Dirac [1961] (stated explicitly by Fulkerson and Gross [1965] and Rose [1970])):

(66.3)   A graph $G = (V, E)$ is chordal $\iff$ each induced subgraph has a simplicial vertex $\iff$ G has an acyclic orientation $D = (V, A)$ such that if $(u, v), (u, w) \in A$, then $\{v, w\} \in E$.

Dirac [1961] moreover showed that a graph is chordal if and only if each inclusionwise minimal vertex-cut is a clique.

Interval graphs. An *interval graph* is the intersection graph G of a family $\mathcal{C}$ of nonempty intervals on the real line[11]. Trivially, such a graph is the complement of a comparability graph: define $I < J \iff i < j$ for all $i \in I, j \in J$. This gives a partial order, and the corresponding comparability graph is equal to $\overline{G}$.

Perfection of the complements of interval graphs was observed by T. Gallai (see Hajnal and Surányi [1958]) — that is, the maximum number of disjoint intervals in $\mathcal{C}$ is equal to the minimum number of points intersecting all intervals in $\mathcal{C}$. This is not hard to prove, and can be proved similarly to the easy dual of Dilworth's decomposition theorem (Theorem 14.1). In fact, a graph is an interval graph if and only if it is chordal and its complement is a comparability graph.

The clique, stable set, colouring, and clique cover problem and their weighted versions can be solved in strongly polynomial time with the methods for comparability graphs described above. If the intervals are given in the order of their maximal elements, and we consecutively assign to each interval the smallest available colour (numbering the colours $1, 2, \ldots$), we obtain an optimum colouring. (Kierstead [1988] showed that if we get the intervals in an *arbitrary* order and we assign to any given interval the smallest possible colour ('on-line'), then we need at most $40\chi(G)$ colours.)

In fact, for any clique C in G there is a point x such that all intervals in C contain x (by Helly's theorem: a family of pairwise intersecting intervals has a nonempty intersection). So finding a maximum-weight clique is trivial. A maximum-size stable

---

[11] The *intersection graph* of a family $\mathcal{C}$ is the graph with vertex set $\mathcal{C}$, two sets in $\mathcal{C}$ being adjacent if and only if they intersect.

set can be found by a greedy method: first find an interval $I \in \mathcal{C}$ with sup I minimal. Next find recursively a maximum-size stable set S among the intervals in $\mathcal{C}$ disjoint from I. Then $S \cup \{I\}$ is a maximum-size stable set in G.

In reply to questions of Hajós [1957] and Benzer [1959], interval graphs have been characterized by Lekkerkerker and Boland [1962] (cf. Halin [1982]), Gilmore and Hoffman [1964], and Fulkerson and Gross [1965]. The latter paper gives a polynomial-time recognition algorithm. A linear-time recognition algorithm was given by Booth and Lueker [1975,1976]. This was simplified by Korte and Möhring [1987], Corneil, Olariu, and Stewart [1998], Hsu and Ma [1999], and Hsu [2002].

More on interval graphs can be found in the books by Golumbic [1980], Skrien [1982], Fishburn [1985], and Brandstädt, Le, and Spinrad [1999], and in the survey article by Golumbic [1985].

Split graphs. A *split graph* is a graph $G = (V, E)$ where V can be partitioned into a clique C and a stable set S. Trivially, a split graph is perfect, since C is contained in a maximum-size clique; hence we can assume that C is a maximum-size clique; so for each $s \in S$ there is a $c \in C$ nonadjacent to s; this yields a $|C|$-vertex-colouring of G.

A graph G is a split graph if and only if both G and $\overline{G}$ are chordal graphs (Foldes and Hammer [1977], Hammer and Simeone [1981]). The book by Golumbic [1980] devotes a chapter to split graphs.

Trivially perfect graphs. Golumbic [1978] calls a graph *trivially perfect* if for each induced subgraph, the stability number is equal to the number of inclusion-wise maximal cliques. Trivially, each trivially perfect graph is perfect. Choudom, Parthasarathy, and Ravindra [1975] and Golumbic [1978] showed that a graph is trivially perfect if and only if it has no induced subgraph equal to the path P_4 or the circuit C_4 (each with 4 vertices). This implies (by a theorem of Wolk [1962] (proof simplified in Wolk [1965])) that a graph is trivially perfect if and only if it is the comparability graph coming from a branching. Another characterization of trivially perfect graphs was given by Alexe and Olaru [1997].

Threshold graphs. A *threshold graph* is a graph on vertex set V given by a function $w : V \to \mathbb{R}$, such that two distinct vertices u, v are adjacent if and only if $w(u) + w(v) > 0$. Chvátal and Hammer [1977] showed that a graph G is a threshold graph if and only if neither G nor $\overline{G}$ has an induced subgraph equal to the path P_4 or the circuit C_4 (each with 4 vertices) — that is, both G and $\overline{G}$ are trivially perfect.

Each threshold graph is a split graph (trivially) and a permutation graph (order the vertices as $v_1, \ldots, v_n$ such that $w(v_1) \leq w(v_2) \leq \cdots \leq w(v_n)$, and let π be the permutation given by ordering $|w(v_1)|, |w(v_2)|, \ldots, |w(v_n)|$). However, the path P_4 with 4 vertices is both a split graph and a permutation graph, but no threshold graph.

The book by Mahadev and Peled [1995] focuses on threshold graphs, and the book by Golumbic [1980] devotes a chapter to threshold graphs. A related class of graphs was described by Wang [1995,1996].

'Strongly chordal' graphs have been studied by Farber [1983,1984] and Kaplan and Shamir [1994], and an analogue of chordal graphs for bipartite graphs by Golumbic and Goss [1978].

66.3a. Chordal graphs as intersection graphs of subtrees of a tree

Chordal graphs can be characterized as intersection graphs of subtrees of a tree, as was shown by L. Surányi (see Gyárfás and Lehel [1970]) and also by Walter [1972, 1978], Buneman [1974], and Gavril [1974c].

Let S be a collection of nonempty subtrees of a tree T. The *intersection graph* of S is the graph with vertex set S, where two vertices S, S' are adjacent if and only if S and S' have at least one vertex in common.

The class of graphs obtained in this way coincides with the class of chordal graphs. To see this, we first show the following elementary lemma:

Lemma 66.5α. *Let S be a collection of pairwise intersecting subtrees of a tree T. Then there is a vertex of T contained in all subtrees in S.*

Proof. By induction on $|VT|$. If $|VT| = 1$ the lemma is trivial, so assume $|VT| \geq 2$. Let t be an end vertex of T. If there exists a subtree in S consisting only of t, the lemma is trivial. Hence we may assume that each subtree in S containing t also contains the neighbour of t. So deleting t from T and from all subtrees in S gives the lemma by induction. ∎

Then we have the subtree characterization of chordal graphs:

Theorem 66.5. *A graph is chordal if and only if it is isomorphic to the intersection graph of a collection of subtrees of some tree.*

Proof. *Necessity.* Let $G = (V, E)$ be chordal. By Theorem 66.3, G contains a simplicial vertex v. By induction, the graph $G - v$ is the intersection graph of a collection S of subtrees of some tree T. Let S' be the subcollection of S corresponding to the set N of neighbours of v in G. As N is a clique, S' consists of pairwise intersecting subtrees. Hence, by Lemma 66.5α, these subtrees have a vertex t of T in common. Now we extend T and all subtrees in S' with a new vertex s and a new edge st. Moreover, we introduce a new subtree $\{s\}$ representing v. In this way we obtain a subtree representation for G.

Sufficiency. Let G be the intersection graph of some collection S of subtrees of some tree T. By (66.3) it suffices to show that G has a simplicial vertex. Let s be an end vertex of T. If S contains a subtree only consisting of s, it gives a simplicial vertex in G. If S contains no such subtree, then each subtree in S containing s also contains the neighbour t (say) of s. So deleting s from T and from all subtrees in S, does not modify the graph G. Hence we are done by induction. ∎

This theorem enables us to interpret the perfection of chordal graphs in terms of trees:

Corollary 66.5a. *Let S be a collection of nonempty subtrees of a tree T. Then the maximum number of pairwise vertex-disjoint trees in S is equal to the minimum number of vertices of T intersecting each tree in S.*

Proof. Directly from Corollary 66.4b and Theorem 66.5, using Lemma 66.5α. ∎

(This result was also stated by Cockayne, Hedetniemi, and Slater [1979].)
Similarly we have:

Corollary 66.5b. *Let S be a collection of subtrees of a tree T. Let k be the maximum number of times that any vertex of T is covered by trees in S. Then S can be partitioned into subcollections $S_1, \ldots, S_k$ such that each S_i consists of pairwise vertex-disjoint trees.*

Proof. Directly from Theorems 66.4 and 66.5, again using Lemma 66.5α. ∎

Variations of the problem of packing and covering a tree by subtrees were studied by Bárány, Edmonds, and Wolsey [1986]. More characterizations of chordal graphs were offered by Benzaken, Crama, Duchet, Hammer, and Maffray [1990]. More on chordal graphs can be found in the book of Golumbic [1980] and in Skrien [1982], Leung [1984], Seymour and Weaver [1984] (a generalization of chordal graphs), Lubiw [1987], Wallis and Wu [1995], and Nakamura and Tamura [2000] (a generalization to bidirected graphs).

66.4. Meyniel graphs

Markosyan and Karapetyan [1976] and Meyniel [1976] showed the perfection of graphs in which each odd circuit of length at least five has at least two chords (*Meyniel graphs*). This was conjectured by Olaru [1969,1972].

It implies the perfection of *Gallai graphs* — graphs in which each odd circuit of length at least five has two *noncrossing* chords (Gallai [1962], cf. Surányi [1968] for a shorter proof [12]), *parity graphs* — graphs in which each odd circuit of length at least five has two crossing chords (Olaru [1969,1972, 1977], cf. Sachs [1970]), and graphs that have no path P_4 as induced subgraph (Seinsche [1974]).

We follow the proof given by Lovász [1983b] (which is a simplification of Meyniel's original proof).

Theorem 66.6. *Each Meyniel graph is perfect.*

Proof. I. We first show that in a Meyniel graph $G = (V, E)$:

(66.4) for each odd circuit C and each vertex v on C, C has a chord disjoint from v or each vertex of $C - v$ is adjacent to v.

---

[12] Gallai [1962] published a proof that $\alpha(G) = \overline{\chi}(G)$ for graphs in which each odd circuit of length at least 5 has two noncrossing chords. Berge [1997] wrote that Gallai informed him in a letter that he knew that also $\omega(G) = \chi(G)$ holds for these graphs.

Let C have no chord disjoint from v. Then the subgraph of G induced by VC is outerplanar, with C as boundary. As each odd circuit of size at least five has a chord we know that each odd bounded face is a triangle. (A face is *odd* (*even*) if its is incident with an odd (even) number of edges.)

Moreover, as C is odd, there is at least one odd bounded face. So if v is not adjacent to all vertices of $C - v$, there is an even bounded face, neighbouring an odd bounded face. But then the union of these two faces forms an odd circuit with only one chord, contradicting the condition.

II. We now prove the theorem. It suffices to show that $\chi(G) = \omega(G)$ for any Meyniel graph $G = (V, E)$, as the class of Meyniel graphs is closed under taking induced subgraphs. We may assume that $V = \{1, \ldots, n\}$. Let $k := \chi(G)$.

For each colouring $\phi : V \to \{1, \ldots, k\}$, let the (ordered) clique $K_\phi = (v_1, \ldots, v_t)$ be obtained recursively as follows. If $v_1, \ldots, v_i$ have been determined (for $i \geq 0$), then v_{i+1} is the largest vertex of colour $i+1$ that is adjacent to each of $v_1, \ldots, v_i$. If no such vertex exists, we stop, setting $t := i$.

Let ϕ be a k-colouring with $K_\phi = (v_1, \ldots, v_t)$ lexicographically minimal. If $t = k$ we are done, so assume $t < k$. Consider the subgraph of G induced by the vertices coloured t and $t + 1$, and let H be its component containing v_t. Let ψ be the colouring obtained from ϕ by interchanging colours t and $t+1$ in H. We show that K_ψ is lexicographically less than K_ϕ, contradicting our assumption.

Trivially, $v_1, \ldots, v_{t-1}$ belong to K_ψ (since we did not change any of the colours $1, \ldots, t-1$). If no other vertex is in K_ψ we are done, so we can assume that K_ψ contains a vertex w with $\psi(w) = t$.

Then $w \neq v_t$, since $\psi(v_t) = t + 1$. If $w < v_t$ we are done, so we can assume that $w > v_t$. If $\phi(w) = t$, this contradicts the choice of $v_t \in K_\phi$. So $\phi(w) = t + 1$, and H contains a shortest path P from v_t to w. Necessarily, this path is odd, and has no chords.

Let u be the second vertex on P. So $\phi(u) = t + 1$. Since v_t is the last vertex in K_ϕ we know that there is an $i \in \{1, \ldots, t-1\}$ with v_i not adjacent to u. Let C be the circuit made by P, $v_i v_t$, and $v_i w$. As P has no chords, by (66.4) v_i is adjacent to u, a contradiction. ∎

Ravindra [1982] showed that each Meyniel graph is strongly perfect (see Section 66.5a below). This was extended by Hoàng [1987b], who showed that Meyniel graphs are precisely those graphs with the property that for each induced subgraph H and each vertex v of H, there exists a stable set in H containing v and intersecting all inclusionwise maximal cliques of H. (This was conjectured by Meyniel.)

Complexity. Burlet and Fonlupt [1984] showed that the class of Meyniel graphs is closed under amalgamation (see Section 65.7c) and that each Meyniel graph arises by amalgamation from chordal graphs and bipartite graphs added with one vertex connected to all vertices of the bipartite graph. They showed that it yields a polynomial-time recognition algorithm (speeded up by Roussel and Rusu [1999b]).

Hoàng [1987b] gave an $O(n^8)$-time algorithm to find a minimum colouring and a maximum clique. An $O(n^3)$ algorithm was given by Hertz [1990a].

(Conforti, Cornuéjols, Kapoor, and Vušković [1999] consider an extension by decomposing *cap-free* graphs (where a *cap* is a circuit with exactly one chord, connecting two vertices at distance two in the circuit) — a (not necessarily perfect) generalization of Meyniel graphs.)

Gallai graphs. As mentioned, these are graphs in which each odd circuit of length ≥ 5 has two noncrossing chords. Polynomial-time recognition algorithms were given by Burlet and Fonlupt [1984], Whitesides [1984], and Cicerone and Di Stefano [1999b] (linear-time). The latter paper also gives a linear-time maximum-weight clique algorithm. A linear-time colouring algorithm was found by Roussel and Rusu [1999a].

Parity graphs. As mentioned, these are graphs in which each odd circuit of length ≥ 5 has two crossing chords. Parity graphs can be characterized alternatively as those graphs such that for each pair u, v of vertices, all chordless $u - v$ paths have the same parity.

Combinatorial strongly polynomial-time algorithms to solve the weighted clique, stable set, colouring, and clique cover problems in parity graphs were given by Burlet and Uhry [1982], who also gave a polynomial-time recognition algorithm (by decomposition of the graph into smaller parity graphs).

The parity graphs include the *line-perfect graphs*, which are graphs whose line graph is perfect. They were characterized by Trotter [1977] — see the claw-free graphs in Section 65.6d. More on parity graphs can be found in Adhar and Peng [1990], Bandelt and Mulder [1991], Przytycka and Corneil [1991], Rusu [1995b], Jansen [1998], and Cicerone and Di Stefano [1999a].

66.5. Further results and notes

66.5a. Strongly perfect graphs

Following Berge and Duchet [1984], a graph $G = (V, E)$ is *strongly perfect* if each induced subgraph H has a stable set intersecting all inclusionwise maximal cliques of H. Each strongly perfect graph is perfect (by (65.2)). Berge and Duchet showed that comparability graphs, chordal graphs, and complements of chordal graphs are strongly perfect. Ravindra [1982] showed that Meyniel graphs are strongly perfect, and Chvátal [1984d] that perfectly orderable graphs are strongly perfect.

Berge and Duchet also showed that the recognition problem for strongly perfect graphs belongs to co-NP. No combinatorial polynomial-time algorithms are known for the optimization problems for strongly perfect graphs.

Olaru [1996] showed that the graphs that are both minimally strongly imperfect and imperfect are precisely the odd circuits of length at least five and their complements. Hence to prove the strong perfect graph theorem it suffices to show that each minimally imperfect graph is also minimally strongly imperfect.

More on strongly perfect graphs can be found in Ravindra [1981,1999], Berge [1983], Basavayya and Ravindra [1985,1987], Preissmann [1985], Preissmann and de Werra [1985], Olaru and Mîndrescu [1986a,1986b], Olaru [1987,1993], Ravindra and Basavayya [1988,1992,1994,1995], Olariu [1989a], Włoch [1995], Blidia, Duchet, and Maffray [1996], Szelecka and Włoch [1996], and Alexe and Olaru [1997].

66.5b. Perfectly orderable graphs

A graph $G = (V, E)$ is a *perfectly orderable graph* if it has an acyclic orientation $D = (V, A)$ such that if four vertices v_1, v_2, v_3, v_4 induce a chordless path in G with edges v_1v_2, v_2v_3, v_3v_4, then $(v_1, v_2) \in A$ or $(v_4, v_3) \in A$. Chvátal [1984d] showed that perfectly orderable graphs are perfect — in fact, strongly perfect:

Theorem 66.7. *Each perfectly orderable graph is strongly perfect.*

Proof. We can assume that $V = \{1, \ldots, n\}$ and that if $(i, j) \in A$, then $i < j$. Let S be the stable set with $\sum(2^{-i} \mid i \in S)$ maximal. Then each $v \notin S$ has a neighbour $u \in S$ with $u < v$, since otherwise $(S \setminus N(v)) \cup \{v\}$ is better than S.

We show that each inclusionwise maximal clique K intersects S. Suppose $K \cap S = \emptyset$. For $s \in S$, let K_s be the set of neighbours $v \in K$ with $s < v$. Choose $s \in S$ with $\sum(2^i \mid i \in K_s)$ maximal. As K is a maximal clique, s is nonadjacent to some $v \in K$. Let $u \in S$ be a neighbour of v with $u < v$. So $v \in K_u \setminus K_s$. By the choice of s, there is a vertex $i \in K_s \setminus K_u$ with $i > v$. So $u < v < i$, and hence u and i are nonadjacent (otherwise $i \in K_u$). As u and s are nonadjacent (since $u, s \in S$) and v and i are adjacent (since $v, i \in K$), u, v, i, s induce a P_4 subgraph with $(u, v), (v, i), (s, i) \in A$, a contradiction. ∎

(Another proof, and a generalization, of this was given by Duchet and Olariu [1991].)

Note that the set S in this proof can be found by a greedy method. So we can find an optimum colouring in polynomial time. Given an orientation as above, also a maximum-size clique can be found in a greedy way — see Chvátal [1984d]. Hoàng [1994] gave $O(nm)$-time algorithms, also for the weighted versions. Middendorf and Pfeiffer [1990a] showed that it is NP-complete to decide if a graph is perfectly orderable.

Comparability graphs, chordal graphs, and complements of chordal graphs are perfectly orderable.

More on perfectly orderable graphs can be found in Cochand and de Werra [1986], Preissmann, de Werra, and Mahadev [1986], Chvátal, Hoàng, Mahadev, and de Werra [1987], Lehel [1987], Hertz [1988,1990b], Hoàng and Khouzam [1988], Olariu [1988a,1993], Bielak [1989], Hoàng and Mahadev [1989], Hoàng and Reed [1989a,1989b], Jamison and Olariu [1989a], Chvátal [1990,1993], Hoàng, Maffray, and Preissmann [1991], Croitoru and Radu [1992a], Hoàng, Maffray, Olariu, and Preissmann [1992], Gavril, Toledano Laredo, and de Werra [1994], Arikati and Peled [1996], Giakoumakis [1996], Hoàng [1996a,1996b,2001], Rusu [1996], Hayward [1997a], Hoàng, Maffray, and Noy [1999], and Hoàng and Tu [2000].

More classes of graphs based on orienting or colouring edges are given by Hoàng [1987a].

66.5c. Unimodular graphs

A graph $G = (V, E)$ is *unimodular* if the following matrix M is totally unimodular: the columns are indexed by V and the rows are the incidence vectors of all inclusionwise maximal cliques of G. Any induced subgraph of a unimodular graph is unimodular again, since for each $v \in V$ and for each maximal clique C of $G - v$, either C or $C \cup \{v\}$ is a maximal clique of G.

Unimodular graphs include bipartite graphs, line graphs of bipartite graphs, and interval graphs.

Perfection of unimodular graphs and their complements was shown by Berge [1963a]. Perfection of the complements of unimodular graphs follows from the Hoffman-Kruskal theorem (Hoffman and Kruskal [1956]), since

(66.5) $\quad \alpha(G) = \max\{\mathbf{1}^\mathsf{T} x \mid x \geq \mathbf{0}, Mx \leq \mathbf{1}\} = \min\{y^\mathsf{T} \mathbf{1} \mid y \geq \mathbf{0}, y^\mathsf{T} M \geq \mathbf{1}\}$
$= \chi(\overline{G})$,

as the LP-optima are attained by integer vectors x and y.

The perfection of a unimodular graph $G = (V, E)$ can also be derived from the Hoffman-Kruskal theorem, with an idea which Berge [1963a] attributed to M.H. McAndrew. It suffices to find a stable set that intersects all maximum-size cliques. Let M' be the submatrix of M corresponding to the maximum-size cliques. The system $\mathbf{0} \leq x \leq \mathbf{1}, Mx \leq \mathbf{1}, M'x \geq \mathbf{1}$ has a solution (namely $x = \omega(G)^{-1}\mathbf{1}$). Hence, as M is total unimodular, it has an integer solution x. This is the incidence vector of a stable set as required.

By a result of Heller [1957] (cf. Theorem 21.4 in Schrijver [1986b]), a unimodular graph has at most $|V|(|V|+1)$ inclusionwise maximal cliques. As W.H. Cunningham (cf. Grötschel, Lovász, and Schrijver [1988]) observed, this gives a polynomial-time method to enumerate all maximal cliques: Choose $v \in V$. Enumerate the maximal cliques $C_1, \ldots, C_t$ of $G - v$ (recursively). Then the maximal cliques of G are among the cliques C_i ($i = 1, \ldots, t$), and $(C_i \cap N(v)) \cup \{v\}$ ($i = 1, \ldots, t$). We can select the maximal cliques among these cliques in polynomial time. Since $t \leq |V|(|V| + 1)$, this gives a polynomial-time method.

This directly gives a strongly polynomial-time method to find a maximum-weight clique. It also implies that the weighted versions of the stable set, colouring, and clique cover problems can be solved in strongly polynomial time, by solving an explicit linear programming problem (using Tardos [1986]). The colouring problem can be solved recursively by first finding (with LP-techniques) a $0, 1$ vector x satisfying $x(C) \leq 1$ for each maximal clique C and $x(C) = 1$ for each maximum-size clique C, and next colouring $G - S$ recursively (where $x = \chi^S$). The weighted version can be solved similarly.

Since by a theorem of Seymour [1980a], totally unimodular matrices can be recognized in polynomial time, this also yields a polynomial-time method to recognize a unimodular matrix.

Ghouila-Houri [1962b] showed that a graph $G = (V, E)$ is unimodular if and only if each nonempty subset U of V contains two disjoint sets U_1 and U_2 such that $U_1 \cup U_2 \neq \emptyset$ and such that each maximal clique C of G with $|C \cap U|$ even, satisfies $|C \cap U_1| = |C \cap U_2|$.

66.5d. Further classes of perfect graphs

Weakly chordal graphs. A graph $G = (V, E)$ is called *weakly chordal* (or *weakly triangulated*) if neither G nor its complement contains a chordless circuit of length at least 5. Hayward [1985] showed that weakly chordal graphs are perfect. Polynomial-time algorithms for the optimization problems related to weakly chordal graphs were given by Hayward, Hoàng, and Maffray [1989], Spinrad and Sritharan [1995], and Hayward, Spinrad, and Sritharan [2000], and polynomial-time recognition algorithms by Spinrad and Sritharan [1995] and Hayward, Spinrad, and Sritharan [2000]. The class of weakly chordal graphs contains both the chordal graphs and their complements.

More on weakly chordal graphs is given in Hoàng, Maffray, Olariu, and Preissmann [1992], Hayward [1996,1997a,1997b], and McMorris, Wang, and Zhang [1998]. Weakly chordal comparability graphs were studied by Eschen, Hayward, Spinrad, and Sritharan [1999].

Quasi-parity graphs. A graph $G = (V, E)$ is a *quasi-parity graph* if each induced subgraph H that is not a clique has two vertices that are not connected by a chordless path of odd length. Meyniel [1987] showed that these graphs are perfect, and that they include the Meyniel graphs and the perfectly orderable graphs. (A short proof of this last is given by Hertz and de Werra [1988].)

Berge [1986] showed that the class of quasi-parity graphs can be enlarged to those graphs in which for each induced subgraph H with at least two vertices, there exist two vertices such that in H or $\overline{H}$ there is no chordless odd-length path connecting them.

Edmonds-Giles graphs. Let $D = (V, A)$ be a directed graph and let $\mathcal{C}$ be a crossing collection of subsets of V with $\delta^{\text{out}}(U) = \emptyset$ for each $U \in \mathcal{C}$. Make an undirected graph G with vertex set A, two arcs a, a' being adjacent if and only if there is a $U \in \mathcal{C}$ such that both a and a' enter U. In Schrijver [1983a] such a graph is called an *Edmonds-Giles graph*. Each such graph is perfect, as can be seen as follows.

A special case of the Edmonds-Giles theorem (Theorem 60.1) is that the system (in $x \in \mathbb{R}^A$)

(66.6) (i) $0 \leq x(a) \leq 1$ for $a \in A$,
        (ii) $x(\delta^{\text{in}}(U)) \leq 1$ for $U \in \mathcal{C}$,

is totally dual integral. Hence it determines an integer polytope. Now the integer vectors x satisfying (66.6) are exactly the incidence vectors of the stable sets of G. Each inequality (66.6)(ii) is a clique inequality. The stable set polytope of G therefore is determined by the clique inequalities, and hence G is perfect (Corollary 65.2e). It in particular implies that each clique of G is contained in $\delta^{\text{in}}(U)$ for some $U \in \mathcal{C}$.

A special case of Edmonds-Giles graphs was given by Kahn [1984], where $D = (V, A)$ is a directed graph and $\mathcal{C}$ is the collection of nonempty proper subsets U of V with $\delta^{\text{out}}(U) = \emptyset$ and $|\delta^{\text{in}}(U)|$ minimal. With the perfect graph theorem this implies that the arcs of a digraph can be coloured in such a way that each minimum-size directed cut contains each colour exactly once.

p-comparability graphs. Cameron and Edmonds [1992] showed perfection of the following graphs. Let $D = (V, A)$ be a directed graph and let $U \subseteq V$ be such that each directed circuit of D has precisely one vertex in U. Let G be the undirected graph on $V \setminus U$ with any two $u, v \in V \setminus U$ adjacent if and only if there is a directed circuit containing u and v. Cameron and Edmonds [1992] call such graphs *p-comparability graphs*. Any comparability graph is a p-comparability graph, but not conversely.

Each such graph G is perfect. The proof is by reduction to minimum-cost flow, using the facts that each clique of G is contained in some directed circuit of D and that by Theorem 65.11 it suffices to show that $\overline{\chi}^*(G) = \overline{\chi}(G)$. (The class of p-comparability graphs is closed under taking induced subgraphs, since adding all arcs (u, v) for which there is a directed $u - v$ path avoiding U, maintains the above property of D.)

Now a minimum fractional clique cover of G corresponds to a minimum fractional covering of $V \setminus U$ by directed circuits. By the integer flow theorem, this last is attained by an integer covering of $V \setminus U$ by directed circuits. Hence, the minimum fractional clique cover in G is attained by an integer clique cover. This amounts to $\overline{\chi}^*(G) = \overline{\chi}(G)$.

Polyominoes. A *polyomino* is a union of unit squares in the plane. (A unit square is a square with integer coordinates and area 1.)

Given a polyomino P, make a graph G with vertices all unit squares contained in P, two of them being adjacent if and only if P contains a rectangle (with horizontal and vertical sides) containing both squares. Győri [1984] showed that if P is horizontally convex, then $\alpha(G) = \overline{\chi}(G)$ (see Section 60.3d). (P is *horizontally convex* if each horizontal line has a convex intersection with P.) This extends a result of Chaiken, Kleitman, Saks, and Shearer [1981], who proved $\alpha(G) = \overline{\chi}(G)$ if P is orthogonally convex. (P is *orthogonally convex* if each horizontal or vertical line has a convex intersection with P.) The latter paper also mentions that E. Szemerédi gave an example that one cannot delete orthogonal convexity, and it gives an example of F.R.K. Chung (1979) showing that one cannot relax it to simple connectivity.

Saks [1982] showed that if P is orthogonally convex, then the subgraph of G induced by the boundary squares is perfect. (A *boundary square* of P is a unit square having a neighbouring square not in P.) (This was proved for the subset of corner squares by Chaiken, Kleitman, Saks, and Shearer [1981]. (A *corner square* of P is a unit square having at least two neighbouring squares not in P.))

Shearer [1982] showed that also the following graph G arising from a *simply connected* polyomino P is perfect: the vertices of G are the rectangles contained in P, where two of them are adjacent if and only if they have a unit square in common.

Motwani, Raghunathan, and Saran [1989] showed that the visibility graph of a horizontally convex polyomino is perfect; in fact, a permutation graph. More on this and related problems can be found in Berge, Chen, Chvátal, and Seow [1981], Győri [1985], Motwani, Raghunathan, and Saran [1988,1990], and Maire [1994a].

66.5e. Further notes

Hayward [1990] showed that graphs containing neither C_5 nor P_6 nor $\overline{P_6}$ as induced subgraphs, are perfect. Other classes of perfect graphs were studied by Ravindra

[1976], Payan [1983], Golumbic, Monma, and Trotter [1984], Hammer and Mahadev [1985], Monma, Reed, and Trotter [1988], Hertz [1989a,1989b,1989c], Hoàng and Maffray [1989,1992], Bertschi [1990], Lubiw [1991b], Sun [1991], Croitoru and Radu [1993], Gurvich, Temkin, Udalov, and Shapovalov [1993], Thomas [1993], Maire [1994b,1996], Rusu [1995b,1999c,1999a], Cheah and Corneil [1996], Gyárfás, Kratsch, Lehel, and Maffray [1996] Giakoumakis [1997], Giakoumakis and Rusu [1997], and Maffray and Preissmann [1999]. Le [2000] gave conjectures on the perfection of certain classes of graphs. A survey of several classes of perfect graphs and their recognition and interrelations, is given in the book by Brandstädt, Le, and Spinrad [1999]. The book of Simon [1992] studies efficient algorithms for classes some of perfect graphs.

Conforti, Cornuéjols, Kapoor, and Vušković [1997] investigated 'universally signable' graphs, a generalization of chordal graphs.

Hammer and Maffray [1993] introduced 'preperfect' graphs, and showed that each preperfect graph is perfect, and that preperfect graphs include the Gallai and the parity graphs (cf. Section 66.4).

Corneil and Stewart [1990] studied the complexity of finding minimum-size dominating sets in several classes of perfect graphs. (A *dominating set* is a set U of vertices with $U \cup N(U) = V$.)

Berge and Las Vergnas [1970] showed that a graph G is perfect if for each odd circuit C and each maximal clique K, the intersection of C and K does not consist of two vertices that form an edge of C.

Vertex cuts in perfect and minimally imperfect graphs were surveyed by Rusu [2001]. A characterization of perfect total graphs was given by Rao and Ravindra [1977].

Figure 66.1

Lovász [1983b] calls a graph k-*perfect* if for each induced subgraph $G = (V, E)$ one has:

(66.7) $\qquad \omega_k(G) = \min_{U \subseteq V}(k\chi(G - U) + |U|)$

where $\omega_k(G)$ is the maximum size of a union of k cliques. By the results of Greene and Kleitman (Corollaries 14.8a and 14.10a), comparability graphs and their complements are k-perfect for each k. Also, complements of line graphs of bipartite graphs are k-perfect, by Corollary 21.4b. On the other hand, the line graph of the bipartite graph in Figure 66.1 is not 2-perfect (Greene [1976]). Related results were given by Berge [1989b,1992a,1992b] and Cameron [1989].

A.J. Hoffman and E.L. Johnson (cf. Golumbic [1980]) proposed the following sharpening of perfection. Let $G = (V, E)$ be a graph and let $w : V \to \mathbb{Z}_+$. A *k-interval colouring* is an assignment to each vertex v of an open subinterval of $[0, k]$ of length $w(v)$ such that adjacent vertices obtain disjoint intervals. Let $\chi_{\text{int}}(G, w)$ denote the minimum value of k for which G has a k-interval colouring. If $w(v) = 1$ for each vertex v, then $\chi_{\text{int}}(G, w) = \chi(G)$. Call G *superperfect* if $\chi_{\text{int}}(G, w)$ is equal to the maximum of $w(K)$ over all cliques K in G. As Hoffman observed, each comparability graph is superperfect (this can be derived from Dilworth's decomposition theorem), but none of the other known interesting classes of perfect graphs have this property.

A survey on subclasses of 'classical' perfect graphs (comparability graphs and chordal graphs) was given by Duchet [1984]. More examples and applications of perfect graphs were given by Shannon [1956], Berge [1967], and Tucker [1973a].

Chapter 67

Perfect graphs: polynomial-time solvability

In this chapter we show that a maximum-weight stable set and a minimum weighted clique cover in a perfect graph can be found in strongly polynomial time. This was shown by Grötschel, Lovász, and Schrijver [1981,1988] with the help of the ellipsoid method and of the function $\vartheta(G)$, introduced by Lovász [1979d] as upper bound on the Shannon capacity of a graph G. No combinatorial polynomial-time algorithms for these problems are known.

We should stress that the naive approach of applying the ellipsoid method to the stable set polytope of a perfect graph using the clique inequalities does not work: it reduces the problem of finding a maximum-weight stable set to deciding for any $x \in \mathbb{R}_+^V$ if there is a clique C satisfying $x(C) > 1$. This is equivalent to finding a maximum-weight clique, which is equivalent to finding a maximum-weight stable set in the complementary graph, which is perfect again. So this would give nothing but a reduction to itself.

In this chapter, all graphs can be assumed to be simple.

67.1. Optimum clique and colouring in perfect graphs algorithmically

Lovász [1979d] introduced the following real number $\vartheta(G)$ for any graph $G = (V, E)$. Let $\mathcal{M}_G$ be the collection of symmetric $V \times V$ matrices satisfying $M_{u,v} = 0$ for any two distinct adjacent vertices u and v and satisfying $\text{Tr} M = 1$. Here $\text{Tr} M$ is the trace of M (sum of diagonal elements). Define

(67.1) $\quad \vartheta(G) := \max\{\mathbf{1}^\mathsf{T} M \mathbf{1} \mid M \in \mathcal{M}_G \text{ positive semidefinite}\}.$

Here $\mathbf{1}$ denotes the all-one vector in $\mathbb{R}^V$.

$\vartheta(G)$ has two important properties: it can be calculated (at least, approximated) in polynomial time, and it gives an, often close, upper bound on the stable set number $\alpha(G)$ (Lovász [1979d]).

First we show (where $\overline{\chi}^*(G)$ denotes the fractional clique cover number — cf. Section 64.8):

Theorem 67.1. *For any graph $G = (V, E)$:*

(67.2) $\quad \alpha(G) \leq \vartheta(G) \leq \overline{\chi}^*(G)$.

Proof. To see $\alpha(G) \leq \vartheta(G)$, let S be a maximum-size stable set and let M be the matrix given by:

(67.3) $\quad M := \dfrac{1}{|S|} \chi^S (\chi^S)^\mathsf{T}$.

Here χ^S is the incidence vector of S in $\mathbb{R}^V$. Then M belongs to $\mathcal{M}_G$ and is positive semidefinite. Hence $\alpha(G) = |S| = \mathbf{1}^\mathsf{T} M \mathbf{1} \leq \vartheta(G)$.

To see $\vartheta(G) \leq \overline{\chi}^*(G)$, let M attain the maximum in (67.1). Consider cliques $C_1, \ldots, C_k$ and $\lambda_1, \ldots, \lambda_k \geq 0$ with

(67.4) $\quad \displaystyle\sum_{i=1}^{k} \lambda_i \chi^{C_i} = \mathbf{1}$ and $\displaystyle\sum_{i=1}^{k} \lambda_i = \overline{\chi}^*(G)$.

Then, setting $\gamma := \overline{\chi}^*(G)$:

(67.5) $\quad \displaystyle 0 \leq \sum_{i=1}^{k} \lambda_i (\gamma \cdot \chi^{C_i} - \mathbf{1})^\mathsf{T} M(\gamma \cdot \chi^{C_i} - \mathbf{1})$

$\displaystyle = \gamma^2 \sum_{i=1}^{k} \lambda_i (\chi^{C_i})^\mathsf{T} M \chi^{C_i} - 2\gamma \sum_{i=1}^{k} \lambda_i (\chi^{C_1})^\mathsf{T} M \mathbf{1} + \gamma \mathbf{1}^\mathsf{T} M \mathbf{1}$

$= \gamma^2 \mathrm{Tr} M - 2\gamma \mathbf{1}^\mathsf{T} M \mathbf{1} + \gamma \mathbf{1}^\mathsf{T} M \mathbf{1} = \gamma^2 - \gamma \vartheta(G)$,

since $\mathrm{Tr} M = 1$, $\mathbf{1}^\mathsf{T} M \mathbf{1} = \vartheta(G)$, and $M_{u,v} = 0$ if $u \neq v$ and $u, v \in C_i$ for some i.

(67.5) implies that $\vartheta(G) \leq \gamma = \overline{\chi}^*(G)$. ∎

Moreover, $\vartheta(G)$ can be approximated in polynomial time (Grötschel, Lovász, and Schrijver [1981]):

Theorem 67.2. *There is an algorithm that for any given graph $G = (V, E)$ and any $\varepsilon > 0$, returns a rational closer than ε to $\vartheta(G)$, in time bounded by a polynomial in $|V|$ and $\log(1/\varepsilon)$.*

Proof. This is a consequence of Corollary (4.3.12) in Grötschel, Lovász, and Schrijver [1988], stating that we can solve a convex optimization problem approximatively in polynomial time, if we know a ball contained in the feasible region and a ball containing the feasible region, and if we can test membership of the feasible region in polynomial time. These conditions are satisfied, if we restrict ourselves to the affine space $\mathcal{M}_G$. The convex body of all positive semidefinite matrices in $\mathcal{M}_G$ contains the ball with center $(1/|V|) \cdot I$ and radius $1/|V|^2$, and is contained in the ball with center the all-zero matrix and radius $|V|^2$. Membership can be tested in polynomial time, since we can test positive semidefiniteness in polynomial time. ∎

The two theorems above imply:

Corollary 67.2a. *For any graph G satisfying $\alpha(G) = \overline{\chi}(G)$, the stable set number can be found in polynomial time.*

Proof. Theorem 67.1 implies $\alpha(G) = \vartheta(G) = \overline{\chi}(G)$, and by Theorem 67.2 we can find a number closer than $\frac{1}{2}$ to $\vartheta(G)$ in time polynomial in $|V|$. Rounding to the closest integer yields $\alpha(G)$. ∎

To obtain an explicit maximum-size stable set, we need perfection of the graph:

Corollary 67.2b. *A maximum-size stable set in a perfect graph can be found in polynomial time.*

Proof. Let $G = (V, E)$ be a perfect graph. Iteratively, for each $v \in V$, replace G by $G - v$ if $\alpha(G - v) = \alpha(G)$. By the perfection of G, we can calculate these values in polynomial time, by Corollary 67.2a.

We end up with a graph that forms a maximum-size stable set in the original graph. ∎

As perfection is closed under taking complements, also a maximum-size clique in a perfect graph can be found in polynomial time.

The method described in the proof of Corollary 67.2b applies to all graphs G for which $\alpha(H) = \vartheta(H)$ holds for each induced subgraph H of G; but, as we shall see in Corollary 67.14a, these are precisely the perfect graphs.

From Corollary 67.2b one can derive that a minimum colouring of a perfect graph can also be found in polynomial time (we follow the method given in Grötschel, Lovász, and Schrijver [1988]):

Corollary 67.2c. *A minimum colouring in a perfect graph can be found in polynomial time.*

Proof. Let $G = (V, E)$ be a perfect graph. It suffices to find a stable set S intersecting each maximum-size clique in G; applying recursion to $G - S$ does the rest.

Starting with $t = 0$, we iteratively extend a list of maximum-size cliques $K_1, \ldots, K_t$ as follows. First, find a stable set S intersecting each of $K_1, \ldots, K_t$. This can be done by considering

$$(67.6) \qquad c := \chi^{K_1} + \cdots + \chi^{K_t},$$

and finding a stable set S maximizing $c(S)$. This can be found by replacing each vertex v by $c(v)$ nonadjacent vertices (adjacent to the new vertices that replace vertices adjacent to v), and finding a maximum-size stable set in the new graph. This gives a stable set S in the original graph maximizing $c(S)$.

Necessarily, $c(S) = t$, since G has a stable set intersecting each maximum-size clique (as G is perfect). So S intersects each K_i.

If $w(G - S) < w(G)$, then S intersects each maximum-size clique in G, and we are done. If $w(G - S) = w(G)$, we find a maximum-size clique K_{t+1} in $G - S$, add it to our list, and iterate.

The number of iterations is bounded by $|V|$, since in each iteration the dimension of the space L_t of vectors $x \in \mathbb{R}^V$ with $x(K_i) = 1$ for each i drops, as for the S found we have $\chi^S \in L_t$ and $\chi^S \notin L_{t+1}$. ∎

67.2. Weighted clique and colouring algorithmically

In a straightforward way, the results of the previous section can be extended to the weighted case. Let $G = (V, E)$ be a graph and let $w : V \to \mathbb{Z}_+$ be a weight function. Let G_w be the graph obtained from G by replacing each vertex v by a stable set S_v of size $w(v)$, where vertices in distinct S_u and S_v are adjacent if and only if u and v are adjacent in G. So the maximum *weight* of a stable set in G is equal to the maximum *size* of a stable set in G_w. Define:

(67.7) $\quad \alpha_w(G) := \alpha(G_w), \, \vartheta_w(G) := \vartheta(G_w), \, \overline{\chi}_w(G) := \overline{\chi}(G_w),$
$\quad \overline{\chi}_w^*(G) := \overline{\chi}^*(G_w).$

So $\alpha_w(G)$ is equal to the maximum weight of a stable set in G. The definitions of $\overline{\chi}_w(G)$ and $\overline{\chi}_w^*(G)$ agree with those in Section 64.8.

Theorem 67.1 gives the following inequalities:

Theorem 67.3. *For any graph $G = (V, E)$ and $w : V \to \mathbb{R}_+$:*

(67.8) $\quad \alpha_w(G) \leq \vartheta_w(G) \leq \overline{\chi}_w^*(G).$

Proof. Directly from Theorem 67.1 and (67.7). ∎

In order to calculate $\vartheta_w(G)$, we need not construct G_w and calculate $\vartheta(G_w)$. This would not be a polynomial-time method. We can calculate $\vartheta_w(G)$ more concisely as follows.

Define $\sqrt{w} : V \to \mathbb{R}_+$ by:

(67.9) $\quad \sqrt{w}(v) := \sqrt{w(v)}$

for $v \in V$. Then:

Theorem 67.4. *For any graph G and $w : VG \to \mathbb{Z}_+$:*

(67.10) $\quad \vartheta_w(G) = \max\{\sqrt{w}^\mathsf{T} M \sqrt{w} \mid M \in \mathcal{M}_G \text{ positive semidefinite}\}.$

Proof. We may assume that $w > 0$. Let D be the $VG_w \times VG$ matrix defined by

(67.11) $\quad D_{u,v} := \begin{cases} w(v)^{-\frac{1}{2}} & \text{if } u \in S_v, \\ 0 & \text{if } u \notin S_v, \end{cases}$

1156   Chapter 67. Perfect graphs: polynomial-time solvability

for $u \in VG_w$ and $v \in VG$.

First let M attain the maximum in (67.10). Then $M' := DMD^\mathsf{T}$ is positive semidefinite, and, moreover, belongs to $\mathcal{M}_{G_w}$. Indeed, for adjacent vertices u, u' of G_w, say $u \in S_v$ and $u' \in S_{v'}$, with v and v' adjacent vertices of G, we have $M_{v,v'} = 0$, and hence

$$(67.12) \qquad M'_{u,u'} = (DMD^\mathsf{T})_{u,u'} = \sum_{t,t' \in VG} D_{u,t} M_{t,t'} D_{u',t'}$$
$$= w(v)^{-\frac{1}{2}} w(v')^{-\frac{1}{2}} M_{v,v'} = 0.$$

Also (setting $v_u := v$ if $u \in S_v$):

$$(67.13) \qquad \mathrm{Tr}\, M' = \mathrm{Tr}(DMD^\mathsf{T}) = \sum_{u \in VG_w} \sum_{v,v' \in VG} D_{u,v} D_{u,v'} M_{v,v'}$$
$$= \sum_{u \in VG_w} w(v_u)^{-1} M_{v_u,v_u} = \sum_{v \in VG} w(v)^{-1} w(v) M_{v,v} = \mathrm{Tr}\, M = 1.$$

So $M' \in \mathcal{M}_{G_w}$. Hence

$$(67.14) \qquad \vartheta_w(G) = \vartheta(G_w) \geq \mathbf{1}^\mathsf{T} M' \mathbf{1} = \mathbf{1}^\mathsf{T} (DMD^\mathsf{T}) \mathbf{1} = \sqrt{w}^\mathsf{T} M \sqrt{w}.$$

This shows $\geq$ in (67.10).

To see the reverse inequality, let M' be a positive semidefinite matrix in $\mathcal{M}_{G_w}$ with $\mathbf{1}^\mathsf{T} M' \mathbf{1} = \vartheta(G_w)$. Then $M := D^\mathsf{T} M' D$ is positive semidefinite, and, moreover, belongs to $\mathcal{M}_G$. Indeed, for adjacent $v, v' \in VG$ we have

$$(67.15) \qquad M_{v,v'} = (D^\mathsf{T} M D)_{v,v'} = \sum_{u,u' \in VG_w} D_{u,v} D_{u',v'} M'_{u,u'}$$
$$= \sum_{u \in S_v} \sum_{u' \in S_{v'}} w(v)^{-\frac{1}{2}} w(v')^{-\frac{1}{2}} M'_{u,u'} = 0.$$

Also:

$$(67.16) \qquad \mathrm{Tr}\, M = \sum_{v \in VG} \sum_{u,u' \in VG_w} D_{u,v} D_{u',v} M'_{u,u'}$$
$$= \sum_{v \in VG} \sum_{u \in S_v} \sum_{u' \in S_v} w(v)^{-1} M'_{u,u'} \leq \sum_{v \in VG} \sum_{u \in S_v} M'_{u,u} = \mathrm{Tr}\, M' = 1.$$

The inequality holds as for any positive semidefinite matrix A one has: $\mathbf{1}^\mathsf{T} A \mathbf{1} \leq \mathbf{1}^\mathsf{T} \mathbf{1} \cdot \mathrm{Tr}\, A$, since the largest eigenvalue of A is at most $\mathrm{Tr}\, A$. This is applied to the $S_v \times S_v$ submatrix of M, for each $v \in V$.

Hence the matrix $\widetilde{M} := (\mathrm{Tr}\, M)^{-1} \cdot M$ belongs to $\mathcal{M}_G$, and so the maximum in (67.10) is at least $\sqrt{w}^\mathsf{T} \widetilde{M} \sqrt{w}$, and hence at least

$$(67.17) \qquad \sqrt{w}^\mathsf{T} M \sqrt{w} = \sqrt{w}^\mathsf{T} D^\mathsf{T} M' D \sqrt{w} = \mathbf{1}^\mathsf{T} M' \mathbf{1} = \vartheta_w(G). \qquad \blacksquare$$

This implies that $\vartheta_w(G)$ can be approximated in polynomial time:

Section 67.2. Weighted clique and colouring algorithmically

Theorem 67.5. *There is an algorithm that for any given graph $G = (V, E)$, any $w : V \to \mathbb{Z}_+$, and any $\varepsilon > 0$, returns a rational closer than ε to $\vartheta_w(G)$, in time bounded by a polynomial in $|V|$, $\log \|w\|_\infty$, and $\log(1/\varepsilon)$.*

Proof. Similar to the proof of Theorem 67.2. ∎

The two theorems above imply:

Corollary 67.5a. *For any graph G and weight function $w : V \to \mathbb{Z}_+$ satisfying $\alpha_w(G) = \overline{\chi}_w(G)$, the weighted stable set number can be found in polynomial time.*

Proof. Theorem 67.3 implies $\alpha_w(G) = \vartheta_w(G) = \overline{\chi}_w(G)$, and by Theorem 67.5 we can find a number closer than $\frac{1}{2}$ to $\vartheta_w(G)$ in time polynomial in $|V|$. Rounding to the closest integer yields $\alpha_w(G)$. ∎

To obtain a maximum-weight stable set explicitly, we again need perfection of the graph:

Corollary 67.5b. *A maximum-weight stable set in a perfect graph can be found in polynomial time.*

Proof. Let $G = (V, E)$ be a perfect graph and $w : V \to \mathbb{Z}_+$. Iteratively, for each $v \in V$, replace G by $G - v$ if $\alpha_w(G - v) = \alpha_w(G)$. By the perfection of G, we can calculate these values in polynomial time, by Corollary 67.5a.

We end up with a graph that forms a maximum-weight stable set in the original graph. ∎

As perfection is closed under taking complements, also a maximum-weight clique in a perfect graph can be found in polynomial time. So for any $w : V \to \mathbb{Z}_+$, we can determine

(67.18) $\quad \omega_w(G) :=$ maximum of $w(C)$ over cliques C of G

in polynomial time.

Moreover, a minimum weighted colouring of a perfect graph can be found in polynomial time (again, we follow the method given in Grötschel, Lovász, and Schrijver [1988]):

Corollary 67.5c. *Given a perfect graph $G = (V, E)$ and a weight function $w : V \to \mathbb{Z}_+$, a minimum weighted colouring can be found in polynomial time.*

Proof. Let $G = (V, E)$ be a perfect graph and let $w : V \to \mathbb{Z}_+$. As in the proof of Corollary 67.2c, we can find a stable set S intersecting each maximum-weight clique in G, as follows. Starting with $t = 0$, we iteratively extend a list of maximum-weight cliques $K_1, \ldots, K_t$. First find a stable set S intersecting each of $K_1, \ldots, K_t$. This can be done by considering

(67.19) $\qquad c := \chi^{K_1} + \cdots + \chi^{K_t}$,

and finding a stable set S maximizing $c(S)$. This can be found by replacing each vertex v by $c(v)$ nonadjacent vertices (adjacent to the new vertices that replace vertices adjacent to v), and finding a maximum-size stable set in the new graph. This gives a stable set S maximizing $c(S)$.

Necessarily, $c(S) = t$, since G has a stable set intersecting each maximum-weight clique (as G_w is perfect). So S intersects each K_i.

If $\omega_w(G - S) < \omega_w(G)$, then S intersects each maximum-weight clique in G, and we have the required S. If $\omega_w(G - S) = \omega_w(G)$, we find a maximum-weight clique K_{t+1} in $G - S$, add it to our list, and iterate.

The number of iterations is bounded by $|V|$, since in each iteration the dimension of the space L_t of vector $x \in \mathbb{R}^V$ with $x(K_i) = 1$ for each i drops, since for the S found we have $\chi^S \in L_t$ and $\chi^S \notin L_{t+1}$.

This describes the method to find a stable set intersecting all maximum-weight cliques. To find a minimum weighted colouring, we iteratively find stable sets $S_1, \ldots, S_i$, $\lambda_1, \ldots, \lambda_i \in \mathbb{Z}_+$, and a weight function w_i as follows. Set $w_1 := w$. Next iteratively for $i = 1, 2, \ldots$, as long as $w_i \neq \mathbf{0}$, find a stable set S_i intersecting all cliques C maximizing $w_i(C)$, calculate

(67.20) $\qquad \lambda_i := \omega_{w_i}(G) - \omega_{w_i}(G - S_i)$,

and set $w_{i+1} := w_i - \lambda_i \chi^{S_i}$.

Then the λ_i, S_i form a minimum weighted colouring, since

(67.21) $\qquad \sum_i \lambda_i \chi^{S_i} = w$ and $\sum_i \lambda_i = \omega_w(G) = \chi_w(G)$.

To prove this, we first show:

(67.22) $\qquad \omega_{w_{i+1}}(G) = \omega_{w_{i+1}}(G - S_i) = \omega_{w_i}(G - S_i) = \omega_{w_i}(G) - \lambda_i$.

Here the second equality is trivial (since w_i and w_{i+1} coincide outside S_i). The third inequality follows from definition (67.20) of λ_i. For the first equality, $\geq$ is trivial. To see $\leq$, consider a clique C intersecting S_i. Then $w_{i+1}(C) = w_i(C) - \lambda_i |C \cap S_i| \leq \omega_{w_i}(G) - \lambda_i$. This proves (67.22), which implies the second equality in (67.21).

Moreover, the number of iterations is at most $|V|$, since in each iteration the face of the clique polytope spanned by the cliques C maximizing $w_i(C)$, increases in dimension: each clique C in G maximizing $w_i(C)$ also maximizes $w_{i+1}(C)$ (since $w_{i+1}(C) \geq w_i(C) - \lambda_i = \omega_{w_i}(G) - \lambda_i = \omega_{w_{i+1}}(G)$, by (67.22)), and there is a clique C maximizing $w_{i+1}(C)$ but not $w_i(C)$ (namely any clique C of $G - S_i$ maximizing $w_i(C)$, since $w_{i+1}(C) = w_i(C) = \omega_{w_i}(G - S_i) = \omega_{w_{i+1}}(G)$). ∎

67.3. Strong polynomial-time solvability

In the previous section we showed the polynomial-time solvability of the weighted versions of the stable set and colouring problems in perfect graphs. By Theorem 5.11 of Frank and Tardos [1985,1987], this can be strengthened to *strong* polynomial-time solvability.

Theorem 67.6. *A maximum-weight clique and a minimum weighted colouring in a perfect graph can be found in strongly polynomial time.*

Proof. A maximum-weight clique can be found in strongly polynomial time by Theorem 5.11, since the class of clique polytopes of perfect graphs is polynomial-time solvable by Corollary 67.5b.

Next, a minimum weighted colouring can be found with the method described in the proof of Corollary 67.5c: it is strongly polynomial-time because we can find (by the above) a maximum-weight clique in strongly polynomial time. ∎

This implies:

Corollary 67.6a. *A maximum-weight stable set and a minimum-weight vertex cover in a perfect graph can be found in strongly polynomial time.*

Proof. Directly from Theorem 67.6, since stable sets in a perfect graph are precisely the cliques in the complementary graph, which is again perfect. Moreover, the vertex covers are precisely the complements of stable sets. ∎

67.4. Further results and notes

67.4a. Further on $\vartheta(G)$

In this section we give some further results on the function $\vartheta(G)$, and we consider the related convex body $\mathrm{TH}(G)$. We use the following notation, for vector $a, b \in \mathbb{R}_+^V$:

(67.23) b/a is the vector in $\mathbb{R}^V$ with vth entry $b(v)/a(v)$,
$\sqrt{b} = b^{\frac{1}{2}}$ is the vector in $\mathbb{R}^V$ with vth entry $b(v)^{\frac{1}{2}}$,
$b^{-\frac{1}{2}}$ is the vector in $\mathbb{R}^V$ with vth entry $b(v)^{-\frac{1}{2}}$,
Δ_b is the $V \times V$ diagonal matrix with diagonal b.

We set $(b/a)_v := 0$ if $a_v = 0$ and $(b^{-\frac{1}{2}})_v := 0$ if $b_v = 0$. (This will turn out not to harm the consistency.)

Moreover, we define, for any graph $G = (V, E)$ and any symmetric matrix M:

(67.24) $\mathcal{L}_G :=$ the set of symmetric $V \times V$ matrices A with $A_{u,v} = 0$ if $u = v$ or u and v are nonadjacent;
$\Lambda(M) :=$ the largest eigenvalue of M,
PSD $:=$ the set of symmetric positive semidefinite matrices.

We usually restrict PSD to appropriate dimensions, like $V \times V$. We define for any two matrices X, Y (of equal dimensions) the 'inner product' $X \bullet Y$ by

(67.25) $\qquad X \bullet Y := \text{Tr}(XY^\mathsf{T})$.

So if $X \in \mathcal{M}_G$ and $Y \in \mathcal{L}_G$, then $X \bullet Y = 0$.

A min-max relation for $\vartheta_w(G)$

$\vartheta_w(G)$ is defined as a maximum. Applying convex duality, we can describe $\vartheta_w(G)$ alternatively as a minimum (Lovász [1979d]):

Theorem 67.7. *For each $w \in \mathbb{R}_+^V$:*

(67.26) $\qquad \vartheta_w(G) = \min\{\Lambda(W + A) \mid A \in \mathcal{L}_G\}$,

where $W := \sqrt{w}\sqrt{w}^\mathsf{T}$.

Proof. Let M maximize $\sqrt{w}^\mathsf{T} M \sqrt{w}$ over $\text{PSD} \cap \mathcal{M}_G$. So $\vartheta_w(G) = \sqrt{w}^\mathsf{T} M \sqrt{w}$.

To prove $\leq$ in (67.26), let $A \in \mathcal{L}_G$ attain the minimum in (67.26) and let $\lambda := \Lambda(W + A)$. Then $Y := \lambda I - W - A$ is positive semidefinite, and hence

(67.27) $\qquad 0 \leq Y \bullet M = (\lambda I - W - A) \bullet M = \lambda \text{Tr} M - W \bullet M = \lambda - \sqrt{w}^\mathsf{T} M \sqrt{w}$
$\qquad\qquad = \Lambda(W + A) - \vartheta_w(G)$.

To prove $\geq$ in (67.26), we use convexity theory. Since M maximizes $W \bullet M$ over the intersection of the convex sets PSD and $\mathcal{M}_G$, there exist supporting hyperplanes $\{X \mid C \bullet X = \gamma\}$ of PSD and $\{X \mid D \bullet X = \delta\}$ of $\mathcal{M}_G$ such that

(67.28) $\qquad \text{PSD} \subseteq \{X \mid C \bullet X \geq \gamma\},\ \mathcal{M}_G \subseteq \{X \mid D \bullet X \geq \delta\},\ C \bullet M = \gamma$,
$\qquad\qquad D \bullet M = \delta$, and $W = C + D$.

Since PSD and $\mathcal{M}_G$ consist of symmetric matrices only, we can assume that C and D are symmetric (we can replace them by $\frac{1}{2}(C + C^\mathsf{T})$ and $\frac{1}{2}(D + D^\mathsf{T})$).

Since PSD is a convex cone, we have $\gamma = 0$. Then $C \in \text{PSD}$, as $xx^\mathsf{T} \in \text{PSD}$ for each $x \in \mathbb{R}^V$, hence $x^\mathsf{T} C x = C \bullet (xx^\mathsf{T}) \geq 0$.

Since $\mathcal{M}_G$ is an affine space and since $D \bullet M = \delta$, we have $\mathcal{M}_G \subseteq \{X \mid D \bullet X = \delta\}$. This implies that $D = \delta \cdot I - A$ for some $A \in \mathcal{L}_G$ (since each symmetric $0,1$ matrix containing precisely one 1 belongs to $\mathcal{M}_G$; the matrix remains to belong to $\mathcal{M}_G$ after putting a nonzero entry in any nonadjacent position and its transpose). So

(67.29) $\qquad \delta = D \bullet M = (W - C) \bullet M = W \bullet M$.

As C is positive semidefinite, $\delta \cdot I - W - A$ is positive semidefinite. Hence

(67.30) $\qquad \Lambda(W + A) \leq \delta = W \bullet M = \vartheta_w(G)$. ∎

The product $\vartheta(G)\vartheta(\overline{G})$ is at least $|V|$

For perfect graphs $G = (V, E)$, we have $\alpha(G)\omega(G) \geq |V|$, and hence $\vartheta(G)\vartheta(\overline{G}) \geq |V|$. The latter inequality holds for any graph G. To prove it, we use the following fact from matrix theory:

(67.31)   If X and Y are symmetric positive semidefinite $n \times n$ matrices, then also $X * Y$ is positive definite,

where $X * Y$ is the $n \times n$ matrix given by: $(X * Y)_{i,j} = X_{i,j}Y_{i,j}$. (67.31) follows from the fact that there exist vectors $u_1, \ldots, u_n$ and $v_1, \ldots, v_n$ with $X_{i,j} = u_i^\mathsf{T} u_j$ and $Y_{i,j} = v_i^\mathsf{T} v_j$ for all i, j. Hence $(X * Y)_{i,j} = (u_i \circ v_i)^\mathsf{T}(u_j \circ v_j)$ for all i, j, where $\circ$ denotes tensor product[13]. So $X * Y$ is positive semidefinite.

Theorem 67.8. $\vartheta(G)\vartheta(\overline{G}) \geq |V|$ *for each graph* $G = (V, E)$.

Proof. By (67.26), there exist $A \in \mathcal{L}_G$ and $B \in \mathcal{L}_{\overline{G}}$ with

(67.32) $\vartheta(G) = \Lambda(J + A)$ and $\vartheta(\overline{G}) = \Lambda(J + B)$.

So $C := \vartheta(G) \cdot I - J - A$ and $D := \vartheta(\overline{G}) \cdot I - J - B$ are positive semidefinite. Now

(67.33) $C * D + C * J + J * D = (C + J) * (D + J) - J * J$
$= (\vartheta(G) \cdot I - A) * (\vartheta(\overline{G}) \cdot I - B) - J = \vartheta(G)\vartheta(\overline{G}) \cdot I - J$

(as $A * I = I * B = A * B$ is the all-zero matrix). By (67.31), the first matrix in (67.33) is positive semidefinite, hence also the last. So

(67.34) $0 \leq \mathbf{1}^\mathsf{T}(\vartheta(G)\vartheta(\overline{G}) \cdot I - J)\mathbf{1} = \vartheta(G)\vartheta(\overline{G})|V| - |V|^2$,

implying the theorem. ∎

The convex body TH(G)

The function $\vartheta_w(G)$ is related to a convex body $\mathrm{TH}(G)$ defined in Grötschel, Lovász, and Schrijver [1986]. The following equivalent representation of $\mathrm{TH}(G)$ was given by Lovász and Schrijver [1991].

For any symmetric matrix A, define the matrix $R(A)$ by:

(67.35) $R(A) := \begin{pmatrix} 1 & a^\mathsf{T} \\ a & A \end{pmatrix}$,

where $a := \mathrm{diag}\,A$ (the diagonal vector of A; that is, $a_i = A_{i,i}$ for each coordinate i).

Given a graph $G = (V, E)$, consider the collection $\mathcal{R}_G$ of symmetric $V \times V$ matrices A with $R(A)$ positive semidefinite and with $A_{u,v} = 0$ for distinct adjacent u, v. Then define:

(67.36) $\mathrm{TH}(G) = \{\mathrm{diag}\,A \mid A \in \mathcal{R}_G\}$.

Theorem 67.9. $\mathrm{TH}(G)$ *is convex and down-monotone in* $\mathbb{R}_+^V$.

Proof. $\mathrm{TH}(G)$ is convex, as it is a projection of the convex set $\mathcal{R}_G$. Moreover, if $a \in \mathrm{TH}(G)$ and $0 \leq b \leq a$, then $b \in \mathrm{TH}(G)$. Indeed, since $a \in \mathrm{TH}(G)$, there exists a matrix $A \in \mathcal{R}_G$ with $a = \mathrm{diag}\,A$. Then the matrix

(67.37) $\begin{pmatrix} 1 & 0 \\ 0 & \Delta_{b/a} \end{pmatrix} \begin{pmatrix} 1 & a^\mathsf{T} \\ a & A \end{pmatrix} \begin{pmatrix} 1 & 0 \\ 0 & \Delta_{b/a} \end{pmatrix} = \begin{pmatrix} 1 & b^\mathsf{T} \\ b & \Delta_{b/a} A \Delta_{b/a} \end{pmatrix}$

---
[13] The *tensor product* of vectors $x \in \mathbb{R}^U$ and $y \in \mathbb{R}^V$ is the vector $x \circ y$ in $\mathbb{R}^{U \times V}$ defined by: $(x \circ y)_{(u,v)} := x_u y_v$ for $u \in U$ and $v \in V$.

is positive semidefinite. As the vth entry on the diagonal of $\Delta_{b/a}A\Delta_{b/a}$ is equal to $b(v)^2/a(v)$ (or 0 if $a(v) = 0$), which is at most $b(v)$, we have that

(67.38) $\quad \Delta_{b/a}A\Delta_{b/a} + (\Delta_{b-b^2/a})$

belongs to $\mathcal{R}_G$ and has diagonal equal to b. This proves that $b \in \mathrm{TH}(G)$, and hence $\mathrm{TH}(G)$ is down-monotone. ∎

To obtain a relation of $\mathrm{TH}(G)$ with the function $\vartheta_w(G)$, we first show the following, where for $x, y \in \mathbb{R}^V$, $x * y$ is the vector in $\mathbb{R}^V$ defined by:

(67.39) $\quad (x * y)_v := x_v y_v$ for $v \in V$.

Theorem 67.10. *Let M maximize $\sqrt{w}^T M \sqrt{w}$ over $\mathrm{PSD} \cap \mathcal{M}_G$. Then*

(67.40) $\quad M\sqrt{w} = \vartheta_w(G) \cdot b * w^{-\frac{1}{2}},$

where $b := \mathrm{diag} M$.

Proof. The maximum of

(67.41) $\quad \sqrt{w}^T \Delta_x M \Delta_x \sqrt{w}$

over $x \in \mathbb{R}^V$ satisfying $x^T \Delta_b x = 1$, is attained by $x = \mathbf{1}$. (Otherwise we can replace M by $\Delta_x M \Delta_x$ to increase $\sqrt{w}^T M \sqrt{w}$.) Now (67.41) is equal to

(67.42) $\quad x^T \Delta_{\sqrt{w}} M \Delta_{\sqrt{w}} x.$

So the maximum of (67.42) over $x \in \mathbb{R}^V$ satisfying $x^T \Delta_b x = 1$, is attained by $x = \mathbf{1}$. Hence, by Lagrange's theorem, there exists a $\mu \in \mathbb{R}$ with

(67.43) $\quad \Delta_{\sqrt{w}} M \Delta_{\sqrt{w}} \mathbf{1} = \mu \cdot \Delta_b \mathbf{1} = \mu \cdot b.$

Then

(67.44) $\quad \vartheta_w(G) = \sqrt{w}^T M \sqrt{w} = \mathbf{1}^T \Delta_{\sqrt{w}} M \Delta_{\sqrt{w}} \mathbf{1} = \mu \mathbf{1}^T b = \mu \mathrm{Tr} M = \mu.$

(67.43) and (67.44) give

(67.45) $\quad M\sqrt{w} = M \Delta_{\sqrt{w}} \mathbf{1} = \mu \cdot w^{-\frac{1}{2}} * b = \vartheta_w(G) \cdot b * w^{-\frac{1}{2}},$

which is (67.40). ∎

Now the relation of $\mathrm{TH}(G)$ with $\vartheta_w(G)$ is:

Theorem 67.11. *For each $w \in \mathbb{R}_+^V$:*

(67.46) $\quad \vartheta_w(G) = \max\{w^T x \mid x \in \mathrm{TH}(G)\}.$

Proof. I. We first show $\leq$ in (67.46). Let M be a matrix maximizing $\sqrt{w}^T M \sqrt{w}$ over the positive semidefinite matrices $M \in \mathcal{M}_G$. It suffices to show that the matrix

(67.47) $\quad A := \vartheta_w(G) \cdot \Delta_{w^{-\frac{1}{2}}} M \Delta_{w^{-\frac{1}{2}}}$

belongs to $\mathcal{R}_G$, since $w^T \mathrm{diag} A = \vartheta_w(G) \mathrm{Tr} M = \vartheta_w(G)$.

Trivially, $A_{u,v} = 0$ for distinct adjacent u, v (since $M_{u,v} = 0$ for distinct adjacent u, v). To see that $R(A)$ is positive semidefinite, write $a := \mathrm{diag} A$, $b := \mathrm{diag} M$, and $\vartheta := \vartheta_w(G)$. By (67.40) we have $M\sqrt{w} = \vartheta \cdot b * w^{-\frac{1}{2}}$. So $\Delta_{w^{-\frac{1}{2}}} M \sqrt{w} = \vartheta \cdot \Delta_{w^{-\frac{1}{2}}}(b * w^{-\frac{1}{2}}) = \vartheta \cdot (b/w)$. Hence

(67.48) $R(A) = \begin{pmatrix} 1 & a^T \\ a & A \end{pmatrix} = \begin{pmatrix} 1 & \vartheta \cdot (b/w)^T \\ \vartheta \cdot (b/w) & \vartheta \cdot \Delta_{w^{-\frac{1}{2}}} M \Delta_{w^{-\frac{1}{2}}} \end{pmatrix}$

$= \begin{pmatrix} \vartheta^{-1} \sqrt{w}^T M \sqrt{w} & \sqrt{w}^T M \Delta_{w^{-\frac{1}{2}}} \\ \Delta_{w^{-\frac{1}{2}}} M \sqrt{w} & \vartheta \cdot \Delta_{w^{-\frac{1}{2}}} M \Delta_{w^{-\frac{1}{2}}} \end{pmatrix} = \vartheta^{-1} \cdot U^T M U,$

where U is the matrix given by

(67.49) $U := (\sqrt{w} \quad \vartheta \cdot \Delta_{w^{-\frac{1}{2}}}).$

So $R(A)$ is positive semidefinite.

II. To see $\geq$ in (67.46), let $A \in \mathcal{R}_G$ maximize $w^T \text{diag} A$. Define $a := \text{diag} A$, $\eta := w^T a$, and

(67.50) $M := \eta^{-1} \cdot \Delta_{w^{\frac{1}{2}}} A \Delta_{w^{\frac{1}{2}}}.$

Trivially, M is positive semidefinite and belongs to $\mathcal{M}_G$. Also

(67.51) $0 \leq (\eta, -w^T) \begin{pmatrix} 1 & a^T \\ a & A \end{pmatrix} \begin{pmatrix} \eta \\ -w \end{pmatrix} = \eta^2 - 2\eta \cdot w^T a + w^T A w$

$= \eta \sqrt{w}^T M \sqrt{w} - \eta^2.$

Therefore $\sqrt{w}^T M \sqrt{w} \geq \eta$, which proves $\geq$ in (67.46). ∎

(67.46) implies that $\vartheta_w(G)$ is a convex function of w and that

(67.52) $\text{TH}(G) = \{x \in \mathbb{R}_+^V \mid w^T x \leq \vartheta_w(G) \text{ for each } w \in \mathbb{R}_+^V\}.$

By (67.8),

(67.53) $\alpha_w(G) \leq \vartheta_w(G) \leq \overline{\chi}_w^*(G).$

This gives:

Corollary 67.11a. *For each graph* $G = (V, E)$:

(67.54) $P_{\text{stable set}}(G) \subseteq \text{TH}(G) \subseteq A(P_{\text{clique}}(G)).$

Proof. This follows directly from Theorem 67.11 with the inequalities (67.53), since for each $w \in \mathbb{R}_+^V$:

(67.55) $\alpha_w(G) = \max\{w^T x \mid x \in P_{\text{stable set}}(G)\},$
$\vartheta_w(G) = \max\{w^T x \mid x \in \text{TH}(G)\},$
$\overline{\chi}_w^*(G) = \max\{w^T x \mid x \in A(P_{\text{clique}}(G))\}.$ ∎

The antiblocking body of TH(G)

It turns out that taking the antiblocking body $A(\text{TH}(G))$ of $\text{TH}(G)$ corresponds to replacing G by its complement (Grötschel, Lovász, and Schrijver [1986]). We first observe that

(67.56) $A(\text{TH}(G)) = \{w \in \mathbb{R}_+^V \mid \vartheta_w(G) \leq 1\},$

1164   Chapter 67. Perfect graphs: polynomial-time solvability

since for each $w: V \to \mathbb{R}_+$: $w \in A(\text{TH}(G)) \iff \max\{w^\mathsf{T} x \mid x \in \text{TH}(G)\} \leq 1$ $\iff \vartheta_w(G) \leq 1$.

Theorem 67.12. $A(\text{TH}(G)) = \text{TH}(\overline{G})$.

Proof. I. We first show $A(\text{TH}(G)) \subseteq \text{TH}(\overline{G})$. Let $w \in A(\text{TH}(G))$; that is (by (67.56)), $\vartheta_w(G) \leq 1$. To show that w belongs to $\text{TH}(\overline{G})$ we should show by (67.52) that

(67.57) $w^\mathsf{T} a \leq \vartheta_a(\overline{G})$

for each $a \in \mathbb{R}_+^V$.

By (67.26), there exist $A \in \mathcal{L}_G$ and $B \in \mathcal{L}_{\overline{G}}$ such that

(67.58) $\vartheta_w(G) = \Lambda(\sqrt{w}\sqrt{w}^\mathsf{T} + A)$ and $\vartheta_a(\overline{G}) = \Lambda(\sqrt{a}\sqrt{a}^\mathsf{T} + B)$.

So $C := \vartheta_w(G) \cdot I - \sqrt{w}\sqrt{w}^\mathsf{T} - A$ and $D := \vartheta_a(\overline{G}) \cdot I - \sqrt{a}\sqrt{a}^\mathsf{T} - B$ are positive semidefinite. Therefore, the matrix

(67.59) $\vartheta_w(G)\vartheta_a(\overline{G}) \cdot I - \sqrt{w*a}\sqrt{w*a}^\mathsf{T} = C*D + C*(\sqrt{a}\sqrt{a}^\mathsf{T}) + (\sqrt{w}\sqrt{w}^\mathsf{T})*D$

is positive semidefinite by (67.31) (note that $A*I = I*B = A*B$ is the all-zero matrix). Hence

(67.60) $\begin{aligned} 0 &\leq \sqrt{w*a}^\mathsf{T}(\vartheta_w(G)\vartheta_a(\overline{G}) \cdot I - \sqrt{w*a}\sqrt{w*a}^\mathsf{T})\sqrt{w*a} \\ &= \vartheta_w(G)\vartheta_a(\overline{G})\sqrt{w*a}^\mathsf{T}\sqrt{w*a} - \sqrt{w*a}^\mathsf{T}\sqrt{w*a}\sqrt{w*a}^\mathsf{T}\sqrt{w*a} \\ &= \vartheta_w(G)\vartheta_a(\overline{G})w^\mathsf{T} a - (w^\mathsf{T} a)^2, \end{aligned}$

implying (67.57).

II. To prove $\text{TH}(\overline{G}) \subseteq A(\text{TH}(G))$, let $w \in \text{TH}(\overline{G})$. By (67.56) we should prove $\vartheta_w(G) \leq 1$.

Let B maximize $\sqrt{w}^\mathsf{T} B\sqrt{w}$ over $\text{PSD} \cap \mathcal{M}_G$. Let $b := \text{diag}B$ and define

(67.61) $C := \Delta_{\sqrt{w/b}} B \Delta_{\sqrt{w/b}}$.

Then, with (67.40),

(67.62) $C\sqrt{b} = \Delta_{\sqrt{w/b}} B\sqrt{w} = \mu \cdot \Delta_{\sqrt{w/b}} b * w^{-\frac{1}{2}} = \mu \cdot \sqrt{b}$,

where $\mu := \vartheta_w(G)$. So C has $\sqrt{b}$ as eigenvector, with eigenvalue μ. Since C is positive semidefinite, also the matrix

(67.63) $C - \mu(\sqrt{b}\sqrt{b}^\mathsf{T})$

is positive semidefinite. Hence the matrix

(67.64) $\Delta_{w^{-\frac{1}{2}}}(C - \mu \cdot \sqrt{b}\sqrt{b}^\mathsf{T})\Delta_{w^{-\frac{1}{2}}} = \Delta_{b^{-\frac{1}{2}}} B \Delta_{b^{-\frac{1}{2}}} - \mu \cdot \sqrt{b/w}\sqrt{b/w}^\mathsf{T}$

is positive semidefinite.

Define $A := I - \Delta_{b^{-\frac{1}{2}}} B \Delta_{b^{-\frac{1}{2}}}$ and $z := b/w$. So $A \in \mathcal{L}_{\overline{G}}$ and $\mu \cdot \sqrt{z}\sqrt{z}^\mathsf{T} + A$ has largest eigenvalue at most 1. Hence $\vartheta_z(\overline{G}) \leq \mu^{-1}$, and so

(67.65) $\vartheta_w(G)\vartheta_z(\overline{G}) = \mu\vartheta_z(\overline{G}) \leq 1 = \text{Tr}B = b^\mathsf{T} \mathbf{1} = w^\mathsf{T} z \leq \vartheta_z(\overline{G})$,

where the last inequality holds as $w \in \text{TH}(\overline{G})$. Hence $\vartheta_w(G) \leq 1$. ∎

Facets of TH(G)

A subset F of TH(G) is called a *facet* of TH(G) if there is an inequality $c^\mathsf{T} x \leq \gamma$ (with $c \neq 0$) which is valid for TH(G), such that F is the set of vectors in TH(G) having equality and such that F has dimension $|V| - 1$. Then (Grötschel, Lovász, and Schrijver [1986]):

Theorem 67.13. *For each graph $G = (V, E)$, each facet F of* TH(G) *is determined by an inequality $x_v \geq 0$ for some $v \in V$ or by $x(C) \leq 1$ for some clique C of G.*

Proof. Let F be determined by the inequality $c^\mathsf{T} x \leq \gamma$. If there is a $v \in V$ with $x_v = 0$ for each $x \in F$, then F is determined by the inequality $x_v \geq 0$. So we can assume that $x > 0$ for some $x \in F$. Since TH(G) = A(TH($\overline{G}$)), there is a $w \in \mathbb{R}^V_+$ with $\vartheta_w(\overline{G}) = 1$ and F is determined by $w^\mathsf{T} x \leq 1$. So $w \in \text{TH}(\overline{G})$, and therefore there is a matrix $A \in \mathcal{R}_{\overline{G}}$ with $\text{diag} A = w$. As $A \in \mathcal{R}_{\overline{G}}$, the matrix $R(A)$ is positive semidefinite. Hence there exist linearly independent vectors $\begin{pmatrix} \alpha_i \\ a_i \end{pmatrix}$ $(i = 1, \ldots, k)$ such that

(67.66) $\quad \begin{pmatrix} 1 & w^\mathsf{T} \\ w & A \end{pmatrix} = R(A) = \sum_{i=1}^{k} \begin{pmatrix} \alpha_i \\ a_i \end{pmatrix} (\alpha_i, a_i^\mathsf{T}).$

We can assume that $\alpha_i \geq 0$ for each $i = 1, \ldots, k$. Now

(67.67) $\quad a_i^\mathsf{T} x = \alpha_i$ for each $x \in F$ and each $i = 1, \ldots, k$.

To see this, choose $x \in F$. As $x \in \text{TH}(G)$, there is a matrix $B \in \mathcal{R}_G$ with $\text{diag} B = x$. Since $R(B)$ is positive semidefinite, also the matrix

(67.68) $\quad B' := \begin{pmatrix} 1 & -x^\mathsf{T} \\ -x & B \end{pmatrix}$

is positive semidefinite. We therefore have (where again $X \bullet Y := \text{Tr}(XY^\mathsf{T})$):

(67.69) $\quad \sum_{i=1}^{k} (\alpha_i, a_i^\mathsf{T}) B' \begin{pmatrix} \alpha_i \\ a_i \end{pmatrix} = R(A) \bullet B' = 1 - 2w^\mathsf{T} x + A \bullet B = 1 - 2w^\mathsf{T} x + w^\mathsf{T} x = 0.$

(Here $A \bullet B = w^\mathsf{T} x$ follows from the fact that $A \in \mathcal{R}_G$, $B \in \mathcal{R}_{\overline{G}}$, $\text{diag} A = w$, and $\text{diag} B = x$.)

Since B' is positive semidefinite, (67.69) implies that, for each $i = 1, \ldots, k$:

(67.70) $\quad (\alpha_i, a_i^\mathsf{T}) B' \begin{pmatrix} \alpha_i \\ a_i \end{pmatrix} = 0,$

and therefore

(67.71) $\quad B' \begin{pmatrix} \alpha_i \\ a_i \end{pmatrix} = \mathbf{0}.$

In particular,

(67.72) $\quad (1, -x^\mathsf{T}) \begin{pmatrix} \alpha_i \\ a_i \end{pmatrix} = 0,$

that is, $a_i^\mathsf{T} x = \alpha_i$, proving (67.67).

Since F is a facet, and since the $\begin{pmatrix}\alpha_i \\ a_i\end{pmatrix}$ are linearly independent, we know $k = 1$. So

(67.73) $\quad \begin{pmatrix} 1 & w^\mathsf{T} \\ w & A \end{pmatrix} = \begin{pmatrix} \alpha_1 \\ a_1 \end{pmatrix}(\alpha_1, a_1^\mathsf{T}).$

Since $\alpha_1 \geq 0$, this implies $\alpha_1 = 1$ and $a_1 = w$. Since $\mathrm{diag} A = w$, we know $w(v)^2 = w(v)$ for each $v \in V$, and so $w \in \{0,1\}^V$. Hence $A = \chi^C(\chi^C)^\mathsf{T}$ for some $C \subseteq V$. As $A_{u,v} = 0$ for distinct nonadjacent u, v, we know that C is a clique. ∎

This gives as consequence:

Corollary 67.13a. $\mathrm{TH}(G)$ *is a polytope if and only if G is perfect.*

Proof. If G is perfect, we have

(67.74) $\quad P_{\text{stable set}}(G) \subseteq \mathrm{TH}(G) \subseteq A(P_{\text{clique}}(G)) = P_{\text{stable set}}(G),$

implying that $\mathrm{TH}(G) = P_{\text{stable set}}(G)$, and therefore is a polytope.

To see the reverse implication, if $\mathrm{TH}(G)$ is a polytope, by (67.54) and Theorem 67.13, $\mathrm{TH}(G)$ is fully determined by the nonnegativity and clique inequalities; that is,

(67.75) $\quad \mathrm{TH}(G) = A(P_{\text{clique}}(G)).$

Since also $A(\mathrm{TH}(G)) = \mathrm{TH}(\overline{G})$ is a polytope, we know similarly that $\mathrm{TH}(\overline{G}) = A(P_{\text{clique}}(\overline{G}))$. Hence

(67.76) $\quad \mathrm{TH}(G) = A(\mathrm{TH}(\overline{G})) = P_{\text{clique}}(\overline{G}) = P_{\text{stable set}}(G).$

(67.75) and (67.76) imply that $P_{\text{stable set}}(G) = A(P_{\text{clique}}(G))$, and therefore G is perfect by Corollary 65.2e. ∎

Characterizing perfection by $\vartheta(G)$

Lovász [1983b] showed that perfection can be characterized by the function $\vartheta(G)$. To this end, Lovász first proved:

Theorem 67.14. *If G is a partitionable graph, then*

(67.77) $\quad \alpha(G) < \vartheta(G) < \overline{\chi}^*(G).$

Proof. Let M be the incidence matrix of the maximum-size stable sets in G and let N be the incidence matrix of the maximum-size cliques of G. Define $n := |VG|$, $\alpha := \alpha(G)$, and $\omega := \omega(G)$. We first show the second inequality.

Let λ be the smallest eigenvalue of $N^\mathsf{T} N$. Since N is nonsingular (Theorem 65.9), we know $\lambda > 0$, and since $\mathrm{Tr}(N^\mathsf{T} N) = n\omega$ and $N^\mathsf{T} N \mathbf{1} = \omega^2 \cdot \mathbf{1}$, we know $\lambda < \omega$ (otherwise $\mathrm{Tr}(N^\mathsf{T} N) \geq \omega^2 + (n-1)\omega > n\omega$). So

(67.78) $\quad N^\mathsf{T} N - \lambda I - \dfrac{\omega^2 - \lambda}{n} J$

is positive semidefinite, and therefore

(67.79) $$\frac{n(\omega - \lambda)}{\omega^2 - \lambda}I - J + \frac{n}{\omega^2 - \lambda}(N^\mathsf{T}N - \omega I)$$

is positive semidefinite. So (using (67.26) and (65.24))

(67.80) $$\vartheta(G) \leq \Lambda(J - \frac{n}{\omega^2 - \lambda}(N^\mathsf{T}N - \omega I)) \leq \frac{n(\omega - \lambda)}{\omega^2 - \lambda} < \frac{n}{\omega} = \chi^*(\overline{G}).$$

So we have the second inequality in (67.77), which implies the first, since:

(67.81) $$\vartheta(G) \geq \frac{n}{\vartheta(\overline{G})} > \frac{n}{\chi^*(G)} = \alpha,$$

by (65.24) and Theorem 67.8. ∎

This implies a characterization of perfect graphs:

Corollary 67.14a. *For any graph G, the following are equivalent:*

(67.82) (i) *G is perfect,*
 (ii) *$\alpha(H) = \vartheta(H)$ for each induced subgraph H of G,*
 (iii) *$\vartheta(H) = \overline{\chi}^*(H)$ for each induced subgraph H of G,*
 (iv) *$\vartheta(H)$ is an integer for each induced subgraph H of G.*

Proof. Directly from Theorem 67.14, using (65.24). ∎

67.4b. The Shannon capacity $\Theta(G)$

Shannon [1956] introduced the following parameter $\Theta(G)$, now called the Shannon capacity of a graph G.

The *strong product* $G \cdot H$ of graphs G and H is the graph with vertex set $VG \times VH$, with two distinct vertices (u, v) and (u', v') adjacent if and only if u and u' are equal or adjacent in G and v and v' are equal or adjacent in H.

The strong product of k copies of G is denoted by G^k. Then the *Shannon capacity* $\Theta(G)$ of G is defined by:

(67.83) $$\Theta(G) = \sup_k \sqrt[k]{\alpha(G^k)}.$$

(The interpretation is that if V is an alphabet, and adjacency means 'confusable', then $\alpha(G^k)$ is the maximum number of k-letter words any two of which have unequal and inconfusable letters in at least one position. Then $\Theta(G)$ is the maximum possible 'information rate'.)

Since $\alpha(G^{k+l}) \geq \alpha(G^k)\alpha(G^l)$, we know by Fekete's lemma (Corollary 2.2a) that

(67.84) $$\Theta(G) = \lim_{k \to \infty} \sqrt[k]{\alpha(G^k)}.$$

Guo and Watanabe [1990] showed that there exist graphs G for which $\Theta(G)$ is not achieved by a finite product (that is, $\sqrt[k]{\alpha(G^k)} < \Theta(G)$ for each k).

Since $\alpha(G^k) \geq \alpha(G)^k$, we have

(67.85) $\alpha(G) \leq \Theta(G),$

while strict inequality may hold: the 5-circuit C_5 has $\alpha(C_5) = 2$ and $\alpha(C_5^2) = 5$. (If C_5 has vertices $1,\ldots,5$ and edges 12, 23, 34, 45, and 51, then $(1,1)$, $(2,3)$, $(3,5)$, $(4,2)$, $(5,4)$ is a stable set in C_5^2.) So $\Theta(C_5) \geq \sqrt{5}$, and Shannon [1956] raised the question if equality holds here. Shannon proved $\Theta(C_5) \leq \frac{5}{2}$; more generally, he proved, for any graph G:

(67.86) $\quad \Theta(G) \leq \overline{\chi}^*(G),$

where $\overline{\chi}^*(G)$ is the fractional clique cover number. This bound can be proved by showing that

(67.87) $\quad \overline{\chi}^*(G \cdot H) \leq \overline{\chi}^*(G)\overline{\chi}^*(H).$

This follows from the fact that if C and D are cliques of G and H respectively, then $C \times D$ is a clique of $G \cdot H$; hence if $\lambda : \mathcal{C} \to \mathbb{R}_+$ and $\mu : \mathcal{D} \to \mathbb{R}_+$ are minimum fractional clique covers for G and H respectively, where $\mathcal{C}$ and $\mathcal{D}$ denote the collections of cliques of G and H respectively, then (where $\circ$ denotes tensor product — see footnote on page 1161, and $\mathbf{1}_U$ denotes the all-one vector in $\mathbb{R}^U$, for any set U)

(67.88) $\displaystyle\sum_{C \in \mathcal{C}} \sum_{D \in \mathcal{D}} \lambda_C \mu_D \chi^{C \times D} = \sum_{C \in \mathcal{C}} \sum_{D \in \mathcal{D}} \lambda_C \mu_D (\chi^C \circ \chi^D)$
$= \Big(\displaystyle\sum_{C \in \mathcal{C}} \lambda_C \chi^C\Big) \circ \Big(\sum_{D \in \mathcal{D}} \mu_D \chi^D\Big) = \mathbf{1}_{VG} \circ \mathbf{1}_{VH} = \mathbf{1}_{VG \times VH}$

and hence

(67.89) $\quad \overline{\chi}^*(G \cdot H) \leq \displaystyle\sum_{C \in \mathcal{C}} \sum_{D \in \mathcal{D}} \lambda_C \mu_D = \Big(\sum_{C \in \mathcal{C}} \lambda_C\Big)\Big(\sum_{D \in \mathcal{D}} \mu_D\Big) = \overline{\chi}^*(G)\overline{\chi}^*(H).$

This proves (67.87) (in (67.112) we show equality).

(67.87) implies (67.86), since

(67.90) $\quad \sqrt[k]{\alpha(G^k)} \leq \sqrt[k]{\overline{\chi}^*(G^k)} \leq \sqrt[k]{\overline{\chi}^*(G)^k} = \overline{\chi}^*(G).$

This bound was improved by Lovász [1979d] as follows (which will imply that $\Theta(C_5) = \sqrt{5}$):

Theorem 67.15. $\Theta(G) \leq \vartheta(G)$ for each graph G.

Proof. Since $\alpha(G) \leq \vartheta(G)$, it suffices to show that for each k: $\alpha(G^k) \leq \vartheta(G)^k$. For this it suffices to show that

(67.91) $\quad \vartheta(G \cdot H) \leq \vartheta(G)\vartheta(H)$

for any graphs G and H.

By (67.26), there exist matrices $A \in \mathcal{L}_G$ and $B \in \mathcal{L}_H$ such that

(67.92) $\quad \vartheta(G) = \Lambda(J_{VG} + A)$ and $\vartheta(H) = \Lambda(J_{VH} + B),$

where J_U denotes the $U \times U$ all-one matrix, for any set U. Hence the matrices

(67.93) $\quad C := \vartheta(G) \cdot I_{VG} - J_{VG} - A$ and $D := \vartheta(H) \cdot I_{VH} - J_{VH} - B$

are positive semidefinite, where I_U denotes the $U \times U$ identity matrix, for any set U.

Therefore, also the following matrix[14] is positive semidefinite:

---

[14] The *tensor product* of a $W \times X$ matrix M and a $Y \times Z$ matrix N (where W, X, Y, Z are sets), is the $(W \times Y) \times (X \times Z)$ matrix $M \circ N$ defined by

(67.94) $\quad C \circ D + C \circ J_{VH} + J_{VG} \circ D = (C + J_{VG}) \circ (D + J_{VH}) - J_{VG} \circ J_{VH}$
$= (\vartheta(G) \cdot I_{VG} - A) \circ (\vartheta(H) \cdot I_{VH} - B) - J_{VG \times VH}$
$= \vartheta(G)\vartheta(H) \cdot I_{VG \times VH} - J_{VG \times VH} - M,$

where $M := \vartheta(G) \cdot I_{VG} \circ B + \vartheta(H) A \circ I_{VH} - A \circ B$. Since $I_{VG} \circ B$, $A \circ I_{VH}$, and $A \circ B$ belong to $\mathcal{L}_{G \cdot H}$,[15] also M belongs to $\mathcal{L}_{G \cdot H}$. Therefore,

(67.95) $\quad \vartheta(G \cdot H) \leq \Lambda(J_{VG \times VH} + M) \leq \vartheta(G)\vartheta(H),$

giving (67.91). ∎

This proof consists of showing the inequality (67.91) for any two graphs G and H. In fact, equality holds (Lovász [1979d]):

(67.96) $\quad \vartheta(G \cdot H) = \vartheta(G)\vartheta(H).$

Indeed, let M and N attain the maximum in definition (67.1) for $\vartheta(G)$ and $\vartheta(H)$ respectively. Then $M \circ N \in \mathcal{M}_{G \cdot H}$, and hence

(67.97) $\quad \vartheta(G \cdot H) \geq \mathbf{1}_{VG \times VH}^\mathsf{T} (M \circ N) \mathbf{1}_{VG \times VH} = (\mathbf{1}_{VG}^\mathsf{T} M \mathbf{1}_{VG})(\mathbf{1}_{VH}^\mathsf{T} N \mathbf{1}_{VH})$
$= \vartheta(G)\vartheta(H).$

Theorem 67.15 implies that $\Theta(C_5) = \sqrt{5}$. One may give an explicit construction to prove this, but it also follows from the following general result (Lovász [1979d]):[16]

Theorem 67.16. *For each graph $G = (V, E)$: $\vartheta(G)\vartheta(\overline{G}) \geq |V|$, with equality if G is vertex-transitive.*

Proof. The inequality is Theorem 67.8. If G is vertex-transitive, then $\mathbf{1}^\mathsf{T} x$ is maximized over $\mathrm{TH}(G)$ at a vector $x = \mu \cdot \mathbf{1}$ for some $\mu \in \mathbb{R}$, since if it is maximized at x we can replace it by

(67.98) $\quad \dfrac{1}{|\Gamma|} \displaystyle\sum_{P \in \Gamma} Px,$

where Γ is the group of permutation matrices representing automorphisms of G. (This follows from the fact that $Px \in \mathrm{TH}(G)$ and $\mathbf{1}^\mathsf{T} Px = \mathbf{1}^\mathsf{T} x$.)

As the maximum value is equal to $\vartheta := \vartheta(G)$, we know $\mathbf{1}^\mathsf{T} x = \vartheta$, and so $\mu = \vartheta/n$, where $n := |V|$. Since $x \in \mathrm{TH}(G) = A(\mathrm{TH}(\overline{G}))$ (by Theorem 67.12), we have $\vartheta_x(\overline{G}) \leq 1$; hence (as $x = \mu \cdot \mathbf{1}$) $\vartheta(\overline{G}) \leq \mu^{-1} = n/\vartheta$. This shows $\vartheta(G)\vartheta(\overline{G}) \leq n$. ∎

$(M \circ N)_{(w,y),(x,z)} := M_{w,x} N_{y,z}$

for $w \in W$, $x \in X$, $y \in Y$, $z \in Z$. If M and N are symmetric positive semidefinite matrices, then $M \circ N$ is symmetric and positive semidefinite again, since if $M = U^\mathsf{T} U$ and $N = V^\mathsf{T} V$, then $M \circ N = (U \circ V)^\mathsf{T} (U \circ V)$.

[15] To see this, let (u, v) and (u', v') be equal or nonadjacent. Then (by definition of $G \cdot H$) $u = u'$ and $v = v'$, or $u \neq u'$ and u and u' are nonadjacent, or $v \neq v'$ and v and v' are nonadjacent. Hence $(I_{VG})_{u,u'} = 0$ or $B_{v,v'} = 0$, and $A_{u,u'} = 0$ or $(I_{VH})_{v,v'} = 0$, and $A_{u,u'} = 0$ or $B_{v,v'} = 0$.

[16] An *automorphism* of a graph $G = (V, E)$ is a permutation $\pi : V \to V$ with $E = \{\{\pi(u), \pi(v)\} \mid \{u, v\} \in E\}$. The graph G is *vertex-transitive* if for all $u, v \in V$ there exists an automorphism π with $\pi(u) = v$.

Since $\overline{C}_5$ is isomorphic to C_5, Theorem 67.16 gives $\vartheta(C_5) = \sqrt{5}$. So $\Theta(G) \leq \sqrt{5}$. As $\Theta(G) \geq \sqrt{\alpha(C_5^2)} = \sqrt{5}$, one has $\Theta(G) = \sqrt{5}$.

Another consequence of Theorem 67.16 is that for any vertex-transitive graph G: $\Theta(G \cdot \overline{G}) = |VG|$, since the pairs (v,v) for $v \in VG$ form a stable set in $G \cdot \overline{G}$ (so $\Theta(G \cdot \overline{G}) \geq |VG|$), and since $\Theta(G \cdot \overline{G}) \leq \vartheta(G \cdot \overline{G}) = \vartheta(G)\vartheta(\overline{G}) = |VG|$. If moreover G is self-complementary (like C_5), then $\Theta(G) = \sqrt{|VG|}$.

For graphs that are not vertex-transitive, $\vartheta(G)\vartheta(\overline{G}) > |VG|$ may hold, even $\alpha(G)\alpha(\overline{G}) > |VG|$, for instance for $G = K_{1,2}$.

Lovász [1979d] also gave the value of $\vartheta(C_n)$ for any odd circuit C_n:

(67.99) $\quad \vartheta(C_n) = \dfrac{n\cos(\pi/n)}{1+\cos(\pi/n)}$ for odd n.

For odd $n \geq 7$, it is unknown if this is the value of $\Theta(C_n)$. Since each C_n is vertex-transitive, by Theorem 67.16 we can derive from (67.99) the value of $\vartheta(\overline{C_n})$ for odd n.

Lovász asked the question if $\Theta(G) = \vartheta(G)$ for each graph G. This was answered in the negative by Haemers [1979], by giving the following alternative upper bound on the Shannon capacity of a graph $G = (V, E)$. Let $\eta(G)$ be the minimum rank of a $V \times V$ matrix M (over any field) such that $M_{v,v} = 1$ for each $v \in V$ and $M_{u,v} = 0$ for distinct nonadjacent u and v. Then

(67.100) $\quad \Theta(G) \leq \eta(G)$.

This follows from the facts that $\alpha(G) \leq \eta(G)$ (since any stable set S in G gives an $S \times S$ identity submatrix of M), and that $\eta(G \cdot H) \leq \eta(G)\eta(H)$ (since $\mathrm{rank}(M \circ N) = \mathrm{rank}(M)\mathrm{rank}(N)$ for any two matrices (over the same field)). Moreover, one has $\eta(G) \leq \overline{\chi}(G)$ (by considering, for any clique cover of G, the $\{0,1\}$ matrix M with $M_{u,v} = 1$ if and only if u and v belong to some clique in the clique cover).

Haemers gave a graph G on 27 vertices (the complement of the 'Schläfli graph') with $\eta(G) \leq 7$ and $\vartheta(G) = 9$, implying $\Theta(G) \leq 7 < \vartheta(G)$. Since $\vartheta(\overline{G}) = 3$, this also gives an example of a graph G satisfying $\Theta(G)\Theta(\overline{G}) < |VG|$ and (hence) $\Theta(G)\Theta(\overline{G}) < \Theta(G \cdot \overline{G})$. (This disproves the conjecture of Shannon [1956] that $\Theta(G)\Theta(H) = \Theta(G \cdot H)$ for all graphs G, H, and answering to the negative the question of Lovász [1979d] whether $\Theta(G)\Theta(\overline{G}) \geq |VG|$ for all graphs G.)

It is unknown if Haemers' bound $\eta(G)$ can be computed in polynomial time. (Peeters [1996] reports results on this. More work on Haemers' bound in Haemers [1981].)

The following bound follows with a method of Rosenfeld [1967]:

(67.101) $\quad \alpha(G \cdot H) \leq \overline{\chi}^*(G)\alpha(H)$.

To see this, let $C_1, \ldots, C_k$ be cliques in G and $\lambda_1, \ldots, \lambda_k \geq 0$ be such that

(67.102) $\quad \lambda_1 \chi^{C_1} + \cdots + \lambda_k \chi^{C_k} = \mathbf{1}_{VG}$ and $\lambda_1 + \cdots + \lambda_k = \overline{\chi}^*(G)$.

Let $S \subseteq VG \times VH$ be a stable set in $G \cdot H$ of size $\alpha(G \cdot H)$. For each $u \in VG$, let $S_u := \{v \in VH \mid (u,v) \in S\}$. Then S_u is a stable set of H, and if u and u' are adjacent vertices of G, then $S_u \cap S_{u'} = \emptyset$. For each $i = 1, \ldots, k$, let

(67.103) $\quad T_i := \{v \in VH \mid \exists u \in C_i : (u,v) \in S\} = \bigcup_{u \in C_i} S_u$.

Since C_i is a clique in G, T_i is a stable set in H, and $|T_i| = \sum_{u \in C_i} |S_u|$. Hence

(67.104) $$|S| = \sum_{u \in VG} |S_u| = \sum_{i=1}^{k} \lambda_i \sum_{u \in C_i} |S_u| = \sum_{i=1}^{k} \lambda_i |T_i| \leq \sum_{i=1}^{k} \lambda_i \alpha(H)$$
$$= \overline{\chi}^*(G)\alpha(H).$$

This shows (67.101).

Rosenfeld [1967] showed that for each graph G:

(67.105) $\alpha(G \cdot H) = \alpha(G)\alpha(H)$ for each graph $H \iff \alpha(G) = \overline{\chi}^*(G)$.

Here $\Longleftarrow$ follows from (67.101). To see $\Longrightarrow$, let $x \in \mathbb{Q}_+^{VG}$ be a vector satisfying $x(C) \leq 1$ for each clique C, and $\mathbf{1}^\mathsf{T} x = \overline{\chi}^*(G)$. Let K be a positive integer such that $w := K \cdot x$ is integer. Let G^w be the graph obtained from G by replacing each vertex u by a clique C_u of size $w(u)$ (where vertices in distinct $C_u, C_{u'}$ are adjacent if and only if u and u' are adjacent). Then $\omega(G^w) \leq K$. Hence for $H := \overline{G^w}$ we have $\alpha(H) \leq K$.

Now let

(67.106) $S := \{(u,v) \mid u \in VG, v \in C_u\}$.

Then S is a stable set in $G \cdot H$, since if (u,v) and (u',v') are distinct elements in S, then, if $u = u'$, v and v' belong to C_u and hence are nonadjacent in H, and, if $u \neq u'$, u and u' are nonadjacent in G or v and v' are nonadjacent in H.

So $|S| \leq \alpha(G \cdot H) = \alpha(G)\alpha(H)$. Hence

(67.107) $\overline{\chi}^*(G) = \mathbf{1}^\mathsf{T} x = \frac{1}{K}\mathbf{1}^\mathsf{T} w = \frac{1}{K}|S| \leq \frac{1}{K}\alpha(G)\alpha(H) \leq \alpha(G).$

Hence $\overline{\chi}^*(G) = \alpha(G)$.

More results on the stable set number of products of graphs are given by Vizing [1963], Barnes and Mackey [1978], and Jha and Slutzki [1994].

The stable set number of products of circuits

The following equality was given by Baumert, McEliece, Rodemich, Rumsey, Stanley, and Taylor [1971] and Markosyan [1971]:

(67.108) $\alpha(C_{2k+1}^2) = k^2 + \lfloor \frac{1}{2}k \rfloor$.

$\leq$ directly follows from (67.101), since $\alpha(C_{2k+1}) = k$ and $\overline{\chi}^*(C_{2k+1}) = k + \frac{1}{2}$. To see $\geq$, we may assume that the vertices of C_{2k+1} are $0, 1, \ldots, 2k$, in order. Then the pairs $(2i, \lfloor 2i/k \rfloor)$, for $i = 1, \ldots, k^2 + \lfloor \frac{1}{2}k \rfloor$, where we take integers mod $2k+1$, form a stable set of size $k^2 + \lfloor \frac{1}{2}k \rfloor$ in C_{2k+1}^2.

Baumert, McEliece, Rodemich, Rumsey, Stanley, and Taylor [1971] showed moreover the following inequalities (next to several other estimates for $\alpha(C_n^k)$):

(67.109) $\alpha(C_{n+2}^k) \geq 1 + \frac{(n+2)^k - 2^k}{n^k}\alpha(C_n^k),$

$\alpha(C_n^k) \leq \frac{n^k - n^{k-1}}{2^k},$

$\alpha(C_5^3) = 10, \alpha(C_5^4) = 25, \alpha(C_7^3) = 33.$

Hales [1973] extended (67.108) to:

(67.110) $\alpha(C_{2k+1} \cdot C_{2l+1}) = kl + \lfloor \frac{1}{2}\min\{k,l\} \rfloor.$

Related results on the stable set number of products of circuits are given by Sonnemann and Krafft [1974], Stein [1977], Hell and Roberts [1982], Mead and Narkiewicz [1982], Vesel [1998], and Vesel and Žerovnik [1998].

67.4c. Clique cover numbers of products of graphs

As for the analogue of the Shannon capacity for clique cover numbers, McEliece and Posner [1971] showed that it gives no new parameter. We follow the proof of Lovász [1975c].

Theorem 67.17. *For any graph G:*
(67.111) $$\inf_k \sqrt[k]{\overline{\chi}(G^k)} = \lim_{k\to\infty} \sqrt[k]{\overline{\chi}(G^k)} = \overline{\chi}^*(G).$$

Proof. We first show that for any two graphs G, H:
(67.112) $\quad \overline{\chi}^*(G \cdot H) = \overline{\chi}^*(G)\overline{\chi}^*(H).$

Here $\leq$ follows from (67.87). To see $\geq$, choose vectors $x : VG \to \mathbb{R}_+$ with $x(C) \leq 1$ for each clique, and with $x(VG) = \overline{\chi}^*(G)$, and $z : VH \to \mathbb{R}_+$ with $z(C) \leq 1$ for each clique, and with $z(VH) = \overline{\chi}^*(H)$. Define $y : VG \times VH \to \mathbb{R}_+$ by
(67.113) $\quad y(u,v) := x(u)z(v)$

for $(u,v) \in VG \times VH$. Then $y(C) \leq 1$ for each clique C of $G \cdot H$, since there are cliques C' and C'' of G and H, respectively, such that $C \subseteq C' \times C''$; then $y(C) \leq y(C' \times C'') = x(C')z(C'') \leq 1$.

Hence
(67.114) $\quad \overline{\chi}^*(G \cdot H) \geq y(VG \times VH) = x(VG)z(VH) = \overline{\chi}^*(G)\overline{\chi}^*(H).$

This proves (67.112).

To prove (67.111), the first equality follows from Fekete's lemma (Corollary 2.2a), since $\overline{\chi}(G^{k+l}) = \overline{\chi}(G^k) \cdot \overline{\chi}(G^l)$. Also we have by (67.112):
(67.115) $\quad \inf_k \sqrt[k]{\overline{\chi}(G^k)} \geq \inf_k \sqrt[k]{\overline{\chi}^*(G^k)} = \overline{\chi}^*(G),$

So it suffices to prove the reverse inequality in (67.115). Since $\omega(G^k) = \omega(G)^k$ and since $\overline{\chi}^*(G^k) = \overline{\chi}^*(G)^k$, we have by Theorem 64.13 (applied to G^k):
(67.116) $\quad \inf_k \sqrt[k]{\overline{\chi}(G^k)} \leq \inf_k \sqrt[k]{(1 + \ln \omega(G^k))\overline{\chi}^*(G^k)}$
$\quad = \inf_k \sqrt[k]{(1 + k \ln \omega(G))}\overline{\chi}^*(G) = \overline{\chi}^*(G),$

as required. ∎

An alternative proof was given by Hell and Roberts [1982]. A related information-theoretic characterization of perfect graphs was given by Csiszár, Körner, Lovász, Marton, and Simonyi [1990] (proving a conjecture of Körner and Marton [1988]). More on the colouring number of products of graphs can be found in Borowiecki [1972], Greenwell and Lovász [1974], Vesztergombi [1980,1981], Turzík [1983], Duffus, Sands, and Woodrow [1985], El-Zahar and Sauer [1985], Puš [1988], Soukop [1988], Linial and Vazirani [1989], and Klavžar [1996] (survey).

Hales [1973] showed that for all graphs G, H:
(67.117) $\quad \overline{\chi}(G \cdot H) \geq \overline{\chi}^*(G)\overline{\chi}(H),$
and
(67.118) $\quad \overline{\chi}(C_{2k+1} \cdot C_{2l+1}) = (k+1)(l+1) - \lceil \frac{1}{2} \min\{k,l\} \rceil.$

McEliece and Taylor [1973] showed that $\overline{\chi}(C_{n,t}^2) = \lceil n/t\lceil n/t \rceil \rceil$, where $C_{n,t}$ is the graph obtained from the circuit C_n by adding all chords connecting vertices at distance less than t in C_n.

67.4d. A sharper upper bound $\vartheta'(G)$ on $\alpha(G)$

McEliece, Rodemich, and Rumsey [1978] and Schrijver [1979a] gave the following sharper bound $\vartheta'(G)$ on the stable set number $\alpha(G)$, generally sharper than $\vartheta(G)$. Again, let $\mathcal{M}_G$ be the collection of symmetric $V \times V$ matrices satisfying $M_{u,v} = 0$ for any two distinct adjacent vertices u and v, and $\text{Tr}M = 1$. (Here $\text{Tr}M$ is the trace of M (sum of diagonal elements).) Then define

(67.119) $\vartheta'(G) := \max\{\mathbf{1}^\mathsf{T} M \mathbf{1} \mid M \in \mathcal{M}_G \text{ nonnegative and positive semi-definite}\}$.

Here $\mathbf{1}$ denotes the all-one vector in $\mathbb{R}^V$. Similarly to $\vartheta(G)$, the value of $\vartheta'(G)$ can be calculated in polynomial time. Moreover

(67.120) $\alpha(G) \leq \vartheta'(G) \leq \vartheta(G)$

for each graph G. The first inequality is proved similarly to the proof of the first inequality in Theorem 67.1, while the second inequality follows from the fact that the range of the maximization problem for $\vartheta'(G)$ is contained in that for $\vartheta(G)$.

$\vartheta'(G)$ indeed can be a sharper upper bound on the stable set number than $\vartheta(G)$, as M.R. Best (cf. Schrijver [1979a]) found the following example of a graph G with $\vartheta'(G) < \vartheta(G)$. The vertex set is $\{0,1\}^6$, two vectors being adjacent if and only if their Hamming distance[17] is at most 3. Then $\vartheta'(G) = 4$ whereas $\vartheta(G) = 16/3$.

Schrijver [1979a] gave relations of $\vartheta'(G)$ with the linear programming bound for codes of Delsarte [1973]. Related work can be found in Schrijver [1981a] and Miklós [1996]. (The polynomial-time computable upper bound for $\alpha(G)$ given by Luz [1995] is at least $\vartheta'(G)$ for all graphs G.)

67.4e. An operator strengthening convex bodies

The matrix method describing $\text{TH}(G)$ given in Section 67.4a can be seen as a special case of a method of improving approximations of the stable set polytope — in fact, of any polytope with $\{0,1\}$ vertices (Lovász and Schrijver [1989,1991]).

Let K be a convex set, let $R(A)$ be defined as in (67.35), and define

(67.121) $\mathcal{N}_K :=$ the collection of symmetric $n \times n$ matrices A with $R(A)$ positive semidefinite, and with $A_i \in A_{i,i} \cdot K$ and $\text{diag}A - A_i \in (1 - A_{i,i}) \cdot K$ for each $i = 1, \ldots, n$,

where A_i denotes the ith column of A.

Define the following new convex set $N_+(K)$:

(67.122) $N_+(K) := \{\text{diag}A \mid A \in \mathcal{N}_K\}$.

Then $N_+(K) \subseteq [0,1]^n$, since $R(A)$ is positive semidefinite. The ellipsoid method gives, for any collection $\mathcal{K}$ of convex sets:

(67.123) if the optimization problem over K is polynomial-time solvable for each $K \in \mathcal{K}$, then also the optimization problem over $N_+(K)$ is polynomial-time solvable for each $K \in \mathcal{K}$.

---

[17] The *Hamming distance* of two vectors of equal dimension is equal to the number of coordinates in which they differ.

Indeed, if the optimization problem over K is polynomial-time solvable, then the membership problem over K is polynomial-time solvable. Hence the membership problem over $\mathcal{N}_K$ is polynomial-time solvable, implying that the optimization problem over $\mathcal{N}_K$ is polynomial-time solvable. Therefore, the optimization problem over $N_+(K)$ is polynomial-time solvable. (Cf. Chapter 4 of Grötschel, Lovász, and Schrijver [1988].)

Before proving further properties of the operator N_+, we note that it commutes with the following reflection. Define $r: \mathbb{R}^n \to \mathbb{R}^n$ by $r(x)_1 := 1 - x_1$ and $r(x)_i := x_i$ for $i = 2, \ldots, n$, for $x \in \mathbb{R}^n$: Then

(67.124) $\quad N_+(r(K)) = r(N_+(K))$.

To see this, let, for any $n \times n$ matrix A, the matrix A' be defined by:

(67.125) $\quad A'_{1,1} := 1 - A_{1,1};\ A'_{1,i} := A'_{i,1} := A_{i,i} - A_{i,1}$ for $i = 2, \ldots, n$;
$A'_{i,j} := A_{i,j}$ for $i, j = 2, \ldots, n$.

Then $R(A)$ is positive semidefinite if and only if $R(A')$ is positive semidefinite, since

(67.126) $\quad R(A') = \begin{pmatrix} 1 & 0 & 0 \\ 1 & -1 & 0 \\ 0 & 0 & I \end{pmatrix} R(A) \begin{pmatrix} 1 & 1 & 0 \\ 0 & -1 & 0 \\ 0 & 0 & I \end{pmatrix}$.

Moreover, $A \in \mathcal{N}_K \iff A' \in \mathcal{N}_{r(K)}$ and $\mathrm{diag}A' = r(\mathrm{diag}A)$. This gives (67.124).

From this one can derive, if K is compact and convex and intersects $[0,1]^n$:

(67.127) $\quad N_+(K) \subseteq K$.

For let $A \in \mathcal{N}_K$ and define $a := \mathrm{diag}A$. If $a \notin K$, there exists a $w \in \mathbb{R}^n$ and $\beta \in \mathbb{R}$ with $w^\mathsf{T} x \leq \beta$ for each $x \in K$ and $w^\mathsf{T} a > \beta$. Since by (67.124) we can flip signs if necessary, we can assume $w \geq \mathbf{0}$. Then, since for each i the vector A_i belongs to $A_{i,i} \cdot K$,

(67.128) $\quad w^\mathsf{T} A w = \sum_i w_i \left(\sum_j w_j A_{i,j} \right) = \sum_i w_i (w^\mathsf{T} A_i) \leq \sum_i w_i A_{i,i} \beta = \beta w^\mathsf{T} a$.

Hence

(67.129) $\quad 0 \leq (w^\mathsf{T} a, -w^\mathsf{T}) \begin{pmatrix} 1 & a^\mathsf{T} \\ a & A \end{pmatrix} \begin{pmatrix} w^\mathsf{T} a \\ -w \end{pmatrix} = (w^\mathsf{T} a)^2 - 2(w^\mathsf{T} a)^2 + w^\mathsf{T} A w$
$\leq -(w^\mathsf{T} a)^2 + \beta \cdot w^\mathsf{T} a = (\beta - w^\mathsf{T} a) w^\mathsf{T} a < 0$,

since $\beta - w^\mathsf{T} a < 0$ and $w^\mathsf{T} a > \beta \geq 0$ (since $\beta \geq w^\mathsf{T} x \geq 0$ for any $x \in K \cap [0,1]^n$). This is a contradiction, showing (67.127).

Moreover, if $K \subseteq [0,1]^n$, then $N_+(K)$ remains to contain the integer hull of K:

(67.130) $\quad (N_+(K))_\mathrm{I} = K_\mathrm{I}$.

To see this, it suffices to show that $x \in N_+(K)$ for each 0,1 vector x in K. Obviously, $A := xx^\mathsf{T}$ belongs to $\mathcal{N}_K$. Hence $x = \mathrm{diag}A$ belongs to $N_+(K)$. This proves (67.130).

Finally, if $K \subseteq [0,1]^n$, then repeated application of the N_+ operator gives the integer hull K_I of K. In fact, one has:

(67.131) $\quad N_+^n(K) = K_\mathrm{I}$.

This follows from the fact that for each $j = 1, \ldots, n$:

(67.132) $N_+(K) \subseteq \text{conv.hull}\{x \in K \mid x_j \in \{0,1\}\}$.

To see this, we may assume that $j = n$. Let $a \in N_+(K)$, with $a = \text{diag} A$ and $A \in \mathcal{N}_K$. Then $A_n \in a_n \cdot K$ and $(a - A_n) \in (1 - a_n) \cdot K$. If $a_n \in \{0,1\}$, then a belongs to the right-hand side of (67.132). So we can assume that $0 < a_n < 1$. Set

(67.133) $a' := \frac{1}{a_n} A_n$ and $a'' := \frac{1}{1-a_n}(a - A_n)$.

Then a' and a'' belong to K, and $a'_n = 1$, $a''_n = 0$. As $a = a_n \cdot a' + (1 - a_n) \cdot a''$, we have that a belongs to the right-hand side of (67.132). This proves (67.132).

(67.131) implies that, when starting with $K := \text{TH}(G)$, we can obtain better and better approximations of $P_{\text{stable set}}(G)$ by applying the N_+ operator. After any fixed number of iterations, we can optimize over the convex body in polynomial time, by (67.123).

Stephen and Tunçel [1999] showed that for the line graph $G = L(K_{2n+1})$ of the complete graph K_{2n+1}, when starting with the polytope determined by the non-negativity and edge constraints ((64.10) in Section 64.5), the number of iterations is precisely n. Related results were given by Cook and Dash [2001].

Leaving out the positive semidefiniteness condition in $\mathcal{N}_K$ yields a weaker operator $N(K)$, which however still satisfies a number of the above properties, including (67.131). The operator $N(K)$ is a special case of a more general operator introduced by Sherali and Adams [1990].

Results relating a related operator to perfection of graphs were given by Aguilera, Escalante, and Nasini [2002].

67.4f. Further notes

Juhász [1982] showed that for a random graph G on n vertices, $\vartheta(G)$ is of the order $\sqrt{n}$, while $\Theta(G)$ is 'likely' to be of the order $\log n$. Knuth [1994] asked if there is a constant c such that $\vartheta(G) \leq c\sqrt{n}\alpha(G)$ for each graph G. This was answered negatively by Feige [1995,1997], who showed that there is a constant $c > 0$ such that

(67.134) $\vartheta(G) > \alpha(G)n/2^{c\sqrt{\log n}}$

for infinitely many graphs G (where $n := |VG|$).

The results of Kashin and Konyagin [1981] and Konyagin [1981] imply that if $\alpha(G) \leq 2$, then $\vartheta(G) \leq 2^{\frac{2}{3}} n^{\frac{1}{3}}$ and (in the worst case) $\vartheta(G) = \Omega(n^{\frac{1}{3}}/\sqrt{\log n})$.

Karger, Motwani, and Sudan [1994,1998] showed the existence of a constant $c > 0$ such that

(67.135) $\overline{\chi}(G) \leq n^{1 - \frac{c}{\vartheta(G)}}$

for each graph G (where $n := |VG|$). More on approximating $\alpha(G)$ or $\overline{\chi}(G)$ by $\vartheta(G)$ can be found in Szegedy [1994] and Charikar [2002].

Kleinberg and Goemans [1998] observed that for any graph G:

(67.136) $\tau(G) \leq 2(|V| - \vartheta(G)) \leq 2\tau(G)$

(where $\tau(G)$ is the vertex cover number of G), and they showed that the factor 2 cannot be improved. Thus the factor 2 as relative error of $\nu(G)$ for approximating $\tau(G)$ is not improved by $2(|V| - \vartheta(G))$.

Fast practical algorithms to compute $\vartheta(G)$, based on interior-point methods, were developed by Alizadeh [1991,1995]. The latter paper also gives a survey on applying semidefinite programming to combinatorial optimization.

A colouring algorithm for perfect graphs based on decomposition was described by Hsu [1986]. An on-line colouring algorithm for perfect graphs (not necessarily yielding an optimum colouring) was given by Kierstead and Kolossa [1996]. An algorithm for colouring some perfect graphs was given by Aït Haddadène, Gravier, and Maffray [1998]. Kratochvíl and Sebő [1997] studied the complexity of colouring a perfect graph if some vertices are pre-coloured. Brandstädt [1987] showed the NP-completeness of several optimization problems for special classes of perfect graphs, like finding a minimum feedback vertex set or a minimum dominating set.

Introductory surveys were given by Knuth [1994] and Goemans [1997] on $\vartheta(G)$, by Grötschel, Lovász, and Schrijver [1984c] on polynomial-time algorithms for clique and colouring problems in perfect graphs, and by Reed [2001a] on semi-definite programming in relation to perfect graphs. Another characterization of perfection in terms of TH(G) was given by Shepherd [2001].

A generalization of $\vartheta(G)$ was given by Narasimhan and Manber [1990]. A generalization of the Shannon capacity to directed graphs was studied by Bidamon and Meyniel [1985]. An analogue of the Shannon capacity based on the 'independent domination number' of a graph, was investigated by Farber [1986]. The Shannon capacity of probabilistic graphs was investigated by Marton [1993].

Further investigations of eigenvalue methods to bound the Shannon capacity are reported by Haemers [1995] and Fiol [1999]. Further convex programming duality phenomena for perfect graphs were found by Wei [1988].

67.4g. Historical notes on perfect graphs

Shannon

As Berge [1997] mentioned, the perfect graph conjectures root in work of Shannon [1956] concerning the 'zero error capacity of a noisy channel'. It amounts to a study of what we now call the Shannon capacity of a graph. Shannon gave the example of C_5 where $\alpha(C_5) = 2$ and $\alpha(C_5^2) = 5$, implying $\Theta(C_5) \geq \sqrt{5} > \alpha(C_5)$. Denoting the logarithm of the Shannon capacity by C_0, Shannon remarked:

> No method has been found for determining C_0 for the general discrete channel, and this we propose as an interesting problem in coding theory.

Shannon proved the following lower and upper bounds on the Shannon capacity $\Theta(G)$ of a graph $G = (V, E)$. First:

(67.137) $\max_p \big(\sum (p_u p_v \mid u, v \in V, u = v \text{ or } uv \in E) \big)^{-1} \leq \Theta(G),$

where p ranges over all $p \in \mathbb{R}_+^V$ with $\sum_{v \in V} p(v) = 1$. It was observed by Korn [1968] that this lower bound (and also the lower bound given by Gallager [1965]) is equal to the stable set number $\alpha(G)$: if $p_u > 0$ and $p_v > 0$ for two adjacent vertices u and v, either resetting $p_u := p_u + p_v$ and $p_v := 0$, or resetting $p_v := p_u + p_v$ and $p_u := 0$, would increase the value in (67.137), a contradiction. So the set $S := \{v \mid p_v > 0\}$ is a stable set. Then the value in (67.137) is maximized by taking $p_v := 1/|S|$ for $v \in S$. (As we saw in Section 64.9c, this also follows from a theorem of Motzkin and Straus [1965].)

The upper bound given by Shannon [1956] amounts to:

Section 67.4g. Historical notes on perfect graphs   1177

(67.138) $\Theta(G) \leq \overline{\chi}^*(G)$.

Shannon formulated and proved this upper bound in terms of information theory as follows. Let V be an alphabet, let Σ be a set of 'signals', and for $v \in V$ and $\sigma \in \Sigma$, let $p_{v,\sigma}$ be the probability that when transmitting symbol v, signal σ is received. So $\sum_{\sigma \in \Sigma} p_{v,\sigma} = 1$ for each $v \in V$. Let G be the graph on V where two elements $u, v \in V$ are adjacent if and only if there is a signal σ with $p_{u,\sigma} > 0$ and $p_{v,\sigma} > 0$. For each $\sigma \in \Sigma$, define the clique $K_\sigma := \{v \in V \mid p_{v,\sigma} > 0\}$ and the real number $\lambda_\sigma := \max\{p_{v,\sigma} \mid v \in V\}$. So

(67.139) $\sum_{\sigma \in \Sigma} \lambda_\sigma \chi^{K_\sigma} \geq 1.$

Hence, by definition of $\overline{\chi}^*(G)$,

(67.140) $\overline{\chi}^*(G) \leq \sum_{\sigma \in \Sigma} \lambda_\sigma.$

Moreover, for any fixed G, the minimum of the right-hand side in (67.140) is equal to the left-hand side.

For any $v = (v_1, \ldots, v_k) \in V^k$ and $s = (s_1, \ldots, s_k) \in \Sigma^k$ define

(67.141) $p_{v,s} := \prod_{i=1}^{k} p_{v_i, s_i}$ and $\lambda_s := \prod_{i=1}^{k} \lambda_{s_i}.$

So $p_{v,s}$ is the probability that transmitted word v is received as word s.

Now consider any nonempty 'code' $C \subseteq V^k$. The 'error probability' of C is equal to

(67.142) $q(C) := \min_\phi \frac{1}{|C|} \sum_{v \in C} \sum (p_{v,s} \mid s \in \Sigma^k, \phi(s) \neq v),$

where ϕ ranges over all functions $\phi : \Sigma^k \to C$. So it is the minimum error probability taken over all possible 'decoding schemes' ϕ. Trivially, this minimum is attained by the function ϕ with $\phi(s)$ equal to any $v \in C$ maximizing $p_{v,s}$ over $v \in C$. So

(67.143) $1 - q(C) = \frac{1}{|C|} \sum_{s \in \Sigma^k} \max_{v \in C} p_{v,s} \leq \frac{1}{|C|} \sum_{s \in \Sigma^k} \lambda_s = \frac{1}{|C|} \left(\sum_{\sigma \in \Sigma} \lambda_\sigma \right)^k.$

Therefore,

(67.144) $\sqrt[k]{|C|} \leq \dfrac{\sum_{\sigma \in \Sigma} \lambda_\sigma}{\sqrt[k]{1 - q(C)}}.$

Now $q(C) = 0$ if and only if C is stable in G^k. Minimizing over all Σ and probability distributions $p_{v,\sigma}$ then yields

(67.145) $\sqrt[k]{|C|} \leq \overline{\chi}^*(G).$

So this gives (67.138).

Shannon next observed that if a graph $G = (V, E)$ has a function $f : V \to V$ such that $f(u) \neq f(v)$ for any distinct nonadjacent vertices u and v, and such that $f(V)$ is a stable set, then $\Theta(G) = \alpha(G)$. The condition clearly is equivalent to: $\alpha(G) = \overline{\chi}(G)$. Shannon noticed that this yields the value of $\Theta(G)$ for all graphs G

with at most 5 vertices, except for C_5, for which he derived $\sqrt{5} \leq \Theta(C_5) \leq \frac{5}{2}$ from (67.138). Shannon observed that on 6 vertices all but four graphs have $\alpha(G) = \overline{\chi}(G)$, and that the Shannon capacity of these four graphs can be expressed in terms of $\Theta(C_5)$. On 7 vertices, he stated that 'at least one new situation arises', namely C_7.

Shannon proved that if G and H are disjoint graphs, then $\Theta(G + H) \geq \Theta(G) + \Theta(H)$ and $\Theta(G \cdot H) \geq \Theta(G) \cdot \Theta(H)$, and that equality holds if $\alpha(G) = \overline{\chi}(G)$. Moreover, he conjectured equality for all G, H, but for the product this was disproved by Haemers [1979], and for the sum by Alon [1998].

Berge

As remarked, in developing the concept of perfect graph Berge was motivated by Shannon's problem on the capacity of graphs. We quote from the article 'Motivations and history of some of my conjectures' of Berge [1997]:

> *June 1957:* When he heard that I was writing a book on graph theory, my friend M.P. Schützenberger drew my attention on an interesting paper of Shannon [51] which was presented at a meeting for engineers and statisticians, but which could have been missed by mathematicians working in algebra or combinatorics.

(Berge's reference [51] is Shannon [1956].)

In his book 'Théorie des graphes' (Theory of Graphs), Berge [1958b] called a function $\sigma : VG \to VG$ a *preserving function* ('application préservante'), if for any two distinct nonadjacent vertices u, v, also $\sigma(u)$ and $\sigma(v)$ are distinct and nonadjacent. Then, like Shannon, he considered graphs G having a preserving function σ mapping VG to a stable subset of VG. Clearly, these are exactly the graphs with $\alpha(G) = \overline{\chi}(G)$.

Berge [1958b] also mentioned that M.P. Schützenberger conjectured that

(67.146) $\quad \Theta(G) = \lim_{k \to \infty} \sqrt[k]{\alpha(G^k)},$

which was shown by Lyubich [1964] to follow directly from Fekete's lemma (Corollary 2.2a).

According to Berge [1997], the problem of finding the minimal graphs G with $\alpha(G) < \Theta(G)$ was discussed in January 1960 at the Seminar of R. Fortet, where he asked (prompted by graphs found by A. Ghouila-Houri) if it is true that each graph G not having an odd hole or odd antihole satisfies $\alpha(G) = \Theta(G)$:

> This conjecture, somewhat weaker than the Perfect Graph Conjecture, was motivated by the remark that for the most usual channels, the graphs representing the possible confusions between a set of signals (in particular the interval graphs) have no odd holes and no odd antiholes, and are *optimal in the sense of Shannon.*

At the first international meeting on graph theory held at Dobogókő (Hungary) in October 1959, Hajnal and Surányi [1958] presented the result that $\alpha(G) = \chi(\overline{G})$ for each chordal graph G. This motivated Berge to show that the same holds for complements of chordal graphs. This result was announced, with partial proof, in the paper Berge [1960a], which moreover mentions that several known results yield other classes of graphs G with $\omega(G) = \chi(G)$. In particular, it is observed that theorems of Kőnig imply that $\omega(G) = \chi(G)$ if G or $\overline{G}$ is the line graph of a bipartite graph — 'propriétés remarquables' (remarkable properties) according to Berge.

These results were presented at the Second International Symposium on Graph Theory at the Martin-Luther-Universität in Halle an der Saale (German Democratic Republic) in April 1960. In his memoirs, Berge [1997] mentioned that[18]

> At that time, we were pretty sure that there were no other minimal obstructions; for that reason, at the end of my talk in Halle, I proposed the following open problem: *If a graph G and its complement are semi-Gallai graphs, is it ture that $\gamma(G) = \omega(G)$?*

where a graph is *semi-Gallai* if it has no odd hole, and where $\gamma(G)$ is Berge's notation for the colouring number of G.

So, according to Berge, the strong perfect graph conjecture was stated in 1960 in Halle. It seems however that Berge was hesitating in putting the conjecture in print. It is not quoted in the written abstract of the talk (Berge [1961]), which in this respect only says that

> Angesichts einer solchen Menge von Beispielen könnte man vermuten, daß für jeden semi-Gallaischen Graphen G die Beziehung $\omega(G) = \gamma(G)$ gilt. Aber das stimmt nicht, wie das folgende, von einem unserer Schüler, Herrn GHOUILA-HOURI, angegebene Gegenbeispiel zeigt:
> G ist ein Graph mit den Knoten a, b, c, d, e, f, g und den Kanten ac, ad, ae, af, bd, be, bf, bg, ce, cf, cg, df, dg, eg. Man kann leicht zeigen, daß G ein semi-Gallaischer Graph ist mit $\omega(G) = 3$, aber $\gamma(G) = 4$ (siehe Abbildung 1).[19]

(This example $\overline{C_7}$) was also given by Shannon [1956].) Incidentally, in this paper, Berge called graphs G satisfying $\alpha(G) = \overline{\chi}(G)$ *perfect graphs of Shannon* ('vollkommenen Graphen von Shannon').

About the strong perfect graph conjecture, Berge and Chvátal [1984] wrote:

> An early effort of Alain Ghouila-Houri failed to produce a counterexample to this conjecture. Despite this encouraging sign, Berge felt that the conjecture might be too ambitious. Therefore he restricted himself to a weaker conjecture in the hope that it might be easier to settle.

According to Berge and Chvátal [1984] (where a triangulated graph is a chordal graph),

> After the meeting at Halle an der Saale in 1960, the Strong Perfect Graph Conjecture received the enthusiastic support of G. Hajós and T. Gallai. In fact, Gallai provided further evidence in support of the conjecture by strengthening the results on triangulated graphs: he proved that a graph is α-perfect and γ-perfect whenever each of its odd cycles of length at least five has at least two non-crossing chords.

In Gallai [1962], only a proof of $\alpha(G) = \overline{\chi}(G)$ is given, for graphs G in which any odd circuit of length at least 5 has two noncrossing chords. Berge [1997] reported that Gallai informed him in a letter that he knew that also $\omega(G) = \chi(G)$ holds for such graphs. However, Gallai's paper does not mention this, and no reference is made to Berge's conjectures.

Berge and Chvátal [1984] continued:

---

[18] As we aim at verbatim quotations, we leave the typo unchanged.

[19] In view of such a multitude of examples one could conjecture that for each semi-Gallai graph G the relation $\omega(G) = \gamma(G)$ holds. But that does not hold, as the following counterexample, presented by one of our students, Mr GHOUILA-HOURI, shows:
G is a graph with nodes a, b, c, d, e, f, g and edges ac, ad, ae, af, bd, be, bf, bg, ce, cf, cg, df, dg, eg. One can easily show, that G is a semi-Gallai graph with $\omega(G) = 3$, but $\gamma(G) = 4$ (see Figure 1).

Nevertheless, Berge still felt that the weak conjecture was more promising. At a conference at Rand Corporation in the summer of 1961, he had fruitful discussions with Alan Hoffman, Ray Fulkerson and others. Later on, discussions between Alan Hoffman and Paul Gilmore led Gilmore to a rediscovery of the Strong Perfect Graph Conjecture and to an attempt to axiomatize the relevant properties of cliques in perfect graphs.

Berge [1997] wrote that the discussions at the RAND Corporation with Alan Hoffman encouraged him to write a paper 'in English'. This paper might have been the first version of the paper 'Some classes of perfect graphs' (Berge [1963a]), published in a booklet 'Six Papers on Graph Theory' by the Indian Statistical Institute in Calcutta, which Berge visited in March-April 1963 and where he gave a series of lectures. The booklet contains no year of publication, and the preface mentions that it is intended for private circulation, and that the papers will be given for publication by journals.

The paper contains as new results that $\omega(G) = \chi(G)$ for unimodular graphs and their complements, and also a full proof that it holds for chordal graphs (announced earlier). The paper seems to be the first written account of the concept of perfect graph, and of the perfect graph conjectures, in the last section of the paper:

V. CONJECTURES

The problem of characterizing α-perfect and γ-perfect graphs seems difficult, but the preceding results enable us to state several conjectures. For instance :

Conjecture 1. A graph is α-perfect if and only if it is γ-perfect

Conjecture 2. A graph is γ-perfect if and only if it does not contain an elementary odd cycle of one of the following types :

type 1 : the cycle is of length greater than 3 and does not possess any chord ;

type 2 : the cycle is of length greater than 3, and does not possess any triangular chord, but possesses all its non-triangular chords (a chord is triangular if it determines a triangle with the edges of the cycle)

Conjecture 3. A graph is α-perfect if and only if it does not contain an elementary odd cycle of type 1 or 2.

It is easy to show that conjecture 2 is equivalent to conjecture 3, and implies conjecture 1. It is also easy to show that if a graph is γ-perfect (or α-perfect), then it does not contain an elementary odd cycle of type 1 or 2.

At the General Assembly of the U.R.S.I. (Union Radio Scientifique Internationale) in Tokyo in September 1963, Berge developed further on the relations between perfection and optimum codes in the sense of Shannon. We quote the abstract (Berge [1963b]):

3. Claude BERGE : *Sur une conjecture relative au problème des codes optimaux de Shannon*, on considère un émetteur qui peut émettre un ensemble de signaux, par suite du bruit chaque signal peut donner plusieurs interprétations à la réception. On trace le graphe dont les sommets représentent les différents signaux, deux points étant liés par une arête si les signaux correspondants peuvent être confondus à la réception. Le problème essentiel est de caractériser les graphes que l'on peut enrichir, on aboutit ainsi à une conjecture que l'on démontre pour certaines classes particulières.[20]

---

[20] 3. Claude BERGE : *On a conjecture related to the problem of the optimal codes of Shannon*, we consider a transmitter that can transmit a set of signals, as a consequence of noise each signal can give several interpretations at the reception. We make the graph

Berge [1997] wrote that the paper Berge [1963a] was distributed to all participants of the U.R.S.I. meeting in 1963, and that a French version of it was published as Berge [1966], added with some new results and an appendix with some results proved in Berge [1967], in order to make the conjecture more plausible and more interesting.

The paper Berge [1966] is more descriptive, but gives more relations to the Shannon problem, and also mentions the strong perfect graph conjecture, attributing it jointly to P.C. Gilmore. After remarking that $\alpha(G) \neq \overline{\chi}(G)$ for odd circuits of length at least 5 and their complements, the paper states:

> Nous nous sommes proposés de voir si la réciproque était vraie, et sommes arrivés à la conjecture suivante avec P. GILMORE:
> **Conjecture.** *Soit G un graphe de signaux; il est parfait si et seulement s'il ne contient pas un cycle impair sans cordes (de longueur > 3), ni le complémentaire d'un cycle impair sans cordes (de longueur > 3).*[21]

Berge [1966] also claimed, without proof, that $\Theta(G) = \alpha(G)$ if and only if $\overline{\chi}(G) = \alpha(G)$:

> On voit aussi *que la condition nécessaire et suffisante pour que la capacité du graphe de signaux G soit égale à* $\alpha(G)$ *est que* $\alpha(G) = \theta(G)$.[22]

(Italics of Berge, who denoted the clique cover number $\overline{\chi}(G)$ of G by $\theta(G)$.) However, the line graph $L(K_6)$ of K_6 is a counterexample to this (it has $\alpha = \Theta = \overline{\chi}^* = 3$ and $\overline{\chi} = 4$).

The paper 'Some classes of perfect graphs' was published again in a book on *Graph Theory and Theoretical Physics* edited by F. Harary (Berge [1967]). According to Berge [1997], this paper is 'a final version' of the manuscript, with suggestions by Hoffman, and was handed over to Harary at the end of a NATO Advanced Study Institute on Graph Theory in Frascati, Italy in March-April 1964. Compared with Berge [1963a], the paper contains no new results, and moreover the last section with the perfect graph conjectures (quoted above) has been omitted.

This paper was published also in the Proceedings of a Conference on Combinatorial Mathematics and Its Applications at the University of North Carolina at Chapel Hill, 10–14 April 1967. It is followed by a 'Discussion on Professor Berge's Paper' by M.E. Watkins stating that 'it seems likely that G is perfect if and only if $\overline{G}$ is perfect'. Berge [1996] mentioned that this addendum

> contributed to make the perfect graph conjecture popular. Before the Chapel Hill conference, I did not get much interest for my problems from the mathematics community; the first symposium lecture about perfect graphs from other mathematicians was delivered by Horst Sachs [20] at the Calgary conference in 1969.

(Berge's reference [20] is Sachs [1970].)

---

the vertices of which represent the different signals, two points being connected by an edge if the corresponding signals can be confused at the reception. The essential problem is of characterizing the graphs that one can enrich, we arrive this way at a conjecture that we prove for certain particular classes.

[21] We have resolved to see if the reverse would be true, and have arrived at the following conjecture with P. GILMORE:
   Conjecture. *Let G be a graph of signals; it is perfect if and only if it neither contains an odd circuit without chords (of length > 3), nor the complement of an odd circuit without chords (of length > 3).*

[22] One also sees *that the necessary and sufficient condition for that the capacity of the graph of signals G is equal to* $\alpha(G)$ *is that* $\alpha(G) = \theta(G)$.

Fulkerson

The results on perfect graphs obtained until then being restricted to specific classes of graphs, the first serious dent in solving the perfect graph conjectures in general was made by Fulkerson in a RAND Report of 1970 on antiblocking polyhedra. They led Fulkerson to prove a 'pluperfect graph theorem', but also to doubt the validity of the weak perfect graph conjecture, which blocked him finishing it off.

The RAND Report (Fulkerson [1970c]) was published as Fulkerson [1972a], and the results were presented at the Second Chapel Hill Conference on Combinatorial Mathematics and Its Applications at the University of North Carolina at Chapel Hill in May 1970 (Fulkerson [1970d]), and at the 7th International Mathematical Programming Symposium in 1970 in The Hague, for which a survey paper on blocking and antiblocking pairs of polyhedra was written (Fulkerson [1970a,1971a]).

Fulkerson called a graph G γ-*pluperfect* if $\chi(H) = \omega(H)$ for each graph H obtained from G by deleting and replicating vertices. In particular, if G is γ-pluperfect, then G is γ-perfect.

What Fulkerson [1970a,1971a] proved is that:

(67.147) G is γ-pluperfect $\iff$ $\overline{G}$ is γ-pluperfect.

The proof is not hard, but is based on a series of pioneering observations and general polyhedral insights that are now fundamental in polyhedral combinatorics. It uses the linear programming duality equality

(67.148) $\max\{w^T x \mid x \geq 0, Mx \leq \mathbf{1}\} = \min\{y^T \mathbf{1} \mid y \geq 0, y^T M \geq w^T\}$,

where M is the incidence matrix of the stable sets of G and where $w : V \to \mathbb{R}_+$. Then:

(67.149) $G = (V, E)$ is γ-pluperfect

$\overset{1}{\iff}$ $\forall w : V \to \mathbb{Z}_+$, both optima in (67.148) are attained by integer solutions x and y

$\overset{2}{\iff}$ $\forall w : V \to \mathbb{Z}_+$, the maximum in (67.148) is attained by an integer solution x

$\overset{3}{\iff}$ $\forall w : V \to \mathbb{Q}_+$, the maximum in (67.148) is attained by an integer solution x

$\overset{4}{\iff}$ $\forall w : V \to \mathbb{R}_+$, the maximum in (67.148) is attained by an integer solution x

$\overset{5}{\iff}$ each vertex of the polytope $\{x \mid x \geq 0, Mx \leq \mathbf{1}\}$ is integer

$\overset{6}{\iff}$ the clique polytope of G is determined by the nonnegativity and stable set constraints.

The first equivalence in (67.149) follows by observing that a weight $w(v)$ of a vertex v corresponds to replacing v by a clique of size $w(v)$; this is equivalent to duplicating v $w(v) - 1$ times, or, if $w(v) = 0$, deleting v. The second equivalence can be derived by considering, for any $w : V \to \mathbb{Z}_+$ an inequality $x(S) \leq 1$ in $Mx \leq \mathbf{1}$ satisfied with equality by all optimum solutions; hence replacing w by $w - \chi^S$ the maximum decreases, hence by at least 1 (as it has an integer value); as the minimum decreases by at most 1, we obtain an integer optimum dual solution by induction. The third and fourth equivalences follow by scaling w and by continuity. The fifth equivalence is general polyhedral theory, and the sixth one follows by observing that the integer solutions of $x \geq 0$, $Mx \leq \mathbf{1}$ are precisely the incidence vectors of cliques.

Section 67.4g. Historical notes on perfect graphs

Now by Fulkerson's theory of antiblocking polyhedra, the last statement in (67.149) is invariant under interchanging 'clique' and 'stable set'; that is, under replacing G by the complementary graph $\overline{G}$. Hence the same holds for the first statement.

Fulkerson [1970c,1970a,1971a,1972a] gave another, symmetrical characterization of γ-pluperfect graphs:

(67.150) a graph $G = (V, E)$ is γ-pluperfect if and only if for all $l, w : V \to \mathbb{Z}_+$, the maximum of $l(S)w(C)$ over all stable sets S and cliques C is at least $\sum_v l(v)w(v)$.

For this, Fulkerson was inspired by the length-width inequality for blocking pairs of hypergraphs given in a 1965 preprint of Lehman [1965,1979].

The weak perfect graph conjecture implies that each perfect graph G is γ-pluperfect, since trivially if $\chi(\overline{H}) = \omega(\overline{H})$ for each induced subgraph H of G, then $\chi(\overline{H}) = \omega(\overline{H})$ for each H obtained from G by deleting and replicating vertices. (Note that $\chi(\overline{H}) = \chi(\overline{G})$ and $\omega(\overline{H}) = \omega(\overline{G})$ if H arises from G by duplicating a vertex.)

So, as Fulkerson [1970a,1971a] remarked ('theorem 14'), the perfect graph conjecture is equivalent to: each γ-perfect graph is γ-pluperfect; or: γ-perfection is maintained under duplicating vertices (later called the *replication lemma*):

> Thus to prove the perfect graph conjecture, it would suffice to prove that γ-perfection implies γ-pluperfection. For this it would suffice to show that if G is γ-perfect, and if we duplicate an arbitrary vertex v in G and join v to its duplicate vertex, the new graph G' is again γ-perfect.

Another way of stating it is: if for each $w : V \to \{0,1\}$ both optima in (67.148) have integer solutions, then likewise for each $w : V \to \mathbb{Z}_+$. This might seem too strong from a general polyhedral point of view, and it made Fulkerson [1970a,1971a] mistrust the conjecture:

> It is our feeling that theorem 14 casts some doubt on the validity of the perfect graph conjecture.

Lovász

The weak perfect graph conjecture was finally proved by Lovász [1972c], stating:

> Fulkerson [5] reduced the problem to the following conjecture, using the theory of antiblocking polyhedra:
> *Duplicating an arbitrary vertex of a perfect graph and joining the obtained two vertices by an edge, the arising graph is perfect.*
> In §1 we prove a theorem which contains this conjecture.

(Reference [5] is Fulkerson [1972a].) Lovász also wrote:

> It should be pointed out that thus the proof consists of two steps and the more difficult second step was done first by Fulkerson.

With respect to this, Fulkerson [1973] remarked in his comments 'On the perfect graph theorem':

Concerning this proof, Lovász states: "It should be pointed out that thus the proof consists of two steps, and the most difficult second step was done first by Fulkerson." I would be less than candid if I did not say that I agree with this remark, at least in retrospect. But the fact remains that, while part of my aim in developing the anti-blocking theory had been to settle the perfect graph conjecture, and that while I had succeeded via this theory in reducing the conjecture to a simple lemma about graphs [3,4] (the "replication lemma", a proof of which is given in this paper) and had developed other seemingly more complicated equivalent versions of the conjecture [3,4,5], I eventually began to feel that the conjecture was probably false and thus spent several fruitless months trying to construct a counterexample. It is not altogether clear to me now just why I felt the conjecture was false, but I think it was due mainly to one equivalent version I had found [4,5], a version that does not explicitly mention graphs at all.

(The references [3,4,5] correspond to Fulkerson [1972a,1971a,1970d].)

In the preprint of this article, Fulkerson [1972b] wrote moreover, after stating the replication lemma:

Actually I knew more: Namely that the truth or falsity of the perfect graph conjecture rested entirely on the truth or falsity of the replication lemma. I tried for awhile to prove this lemma, without success, and then, as was mentioned earlier, became convinced on other grounds that the perfect graph conjecture was probably false, and began to look for a graph that was perfect but not pluperfect. (I knew that it would do no good to look at known classes of perfect graphs, since I had been able to prove that all of these were pluperfect.) The fact is that such graphs don't exist, of course. After some months of sporadic effort along these lines, I quit working on the perfect graph conjecture, thinking that I would come back to it later. There were other aspects of anti-blocking pairs of polyhedra, and of blocking pairs of polyhedra, that I wanted to study, and, in any event, I felt that the pluperfect graph theorem was a beautiful result in its own right.

In the spring of 1971 I received a postcard from Berge, who was then visiting the University of Waterloo, saying that he had just heard that Lovász had a proof of the perfect graph conjecture. This immediately rekindled my interest, naturally, and so I sat down at my desk and thought again about the replication lemma. Some four or five hours later, I saw a simple proof of it.

After having given a simple proof of the replication lemma, Fulkerson [1972b] continued:

As can be seen, there is nothing deep or complicated about the proof of this lemma. Perhaps the fact that I saw a proof of it only after knowing it had to be true may say something about the psychology of invention (or, better yet, anti-invention) in mathematics, at least for me.

This is indeed an instructive illustration that believing a conjecture may help in proving it.

In a subsequent paper, Lovász [1972a] proved more strongly that a graph G is perfect if and only if $\alpha(H)\omega(H) \geq |VH|$ for each induced subgraph H of G. This generalizes the perfect graph theorem, and was suggested by A. Hajnal. It also sharpens Fulkerson's result (67.150), implying that one may restrict l and w to $\{0,1\}$-valued functions with $l = w$.

The problem of Shannon [1956] concerning the Shannon capacity of C_5 was solved by Lovász [1979d].

In May 2002, M. Chudnovsky, N. Robertson, P.D. Seymour, and R. Thomas announced that they found a proof of the strong perfect graph conjecture, by proving a number of deep results, and building on and inspired by earlier results of, among others, V. Chvátal, M. Conforti, G. Cornuéjols, W.H. Cunningham, A. Kapoor, F. Roussel, P. Rubio, N. Sbihi, K. Vušković, and G. Zambelli.

More historical notes are given by Berge and Ramírez Alfonsín [2001] and Reed [2001b].

Chapter 68

T-perfect graphs

The class of t-perfect graphs is defined polyhedrally: the stable set polytope should be determined by the nonnegativity, edge, and odd circuit constraints. It implies that a maximum-weight stable set in such graphs can be found in polynomial time. LP duality gives a min-max relation for the maximum-weight of a stable set in t-perfect graphs.

A characterization of t-perfect graphs is not known. The widest class of t-perfect graphs known consists of those not containing certain subdivisions of K_4 as subgraph.

68.1. T-perfect graphs

A graph $G = (V, E)$ is called *t-perfect*[23] if the stable set polytope of G is determined by

(68.1)   (i) $0 \leq x_v \leq 1$ for each $v \in V$,
             (ii) $x_u + x_v \leq 1$ for each edge $uv \in E$,
             (iii) $x(VC) \leq \lfloor \frac{1}{2}|VC| \rfloor$ for each odd circuit C.

A prominent non-t-perfect graph is K_4. Below we shall see that, on the other hand, if K_4 does not occur in a graph in a certain way, then the graph is t-perfect. But no exact characterization of t-perfection is known.

A motivation for studying t-perfection is algorithmic, since the definition implies:

Theorem 68.1. *A maximum-weight stable set in a t-perfect graph can be found in strongly polynomial time.*

Proof. By Theorems 5.10 and 5.11, it suffices to show that the separation problem over the stable set polytope is polynomial-time solvable. Conditions (i) and (ii) in (68.1) can be tested one by one. If they are satisfied, define a function $y : E \to \mathbb{R}_+$ by:

(68.2) $y_e := 1 - x_u - x_v$

for each $e = uv \in E$. Then condition (iii) is equivalent to:

---
[23] t stands for 'trou' (French for 'hole').

(68.3) $\quad y(EC) \geq 1$ for each odd circuit C

(since $y(EC) = |EC| - 2x(VC)$). The latter condition can be checked in polynomial time: Consider y as a length function, and for each $u \in V$, find an odd circuit C through u with $y(EC)$ minimal. This can be done by replacing each vertex v by two vertices v', v'', and each edge $e = vw$ by two edges $v'w''$ and $v''w'$, each of length y_e; then a shortest path from u' to u'' gives the required circuit.

If $y(EC) < 1$, we have a violated inequality. ∎

A combinatorial polynomial-time algorithm to find the stable set number of a t-perfect graph was given by Eisenbrand, Funke, Garg, and Könemann [2002]. It is based on finding (by a greedy method similar to that used in the proof of Theorem 64.13) an approximative fractional dual solution to the problem of maximizing $\mathbf{1}^\mathsf{T} x$ over (68.1), with relative error less than $1/|V|$. Rounding then gives the stable set number. Applying this iteratively gives an explicit maximum-size stable set.

Notes. The construction given in the proof of Theorem 68.1 shows that the maximum-weight stable set problem in a t-perfect graph can be described by a 'compact' linear programming: the stable set polytope is the projection of a polytope whose dimension and number of facets are polynomially bounded. To see this, introduce, next to the variables $y \in \mathbb{R}_+^E$, a variable $z_{u,v}$ for each $u, v \in V$. Requiring:

(68.4) $\quad z_{v,v} \geq 1 \qquad\qquad$ for each $v \in V$,
$\qquad\qquad z_{u,v} \leq y_{uv} \qquad\qquad$ for each edge $uv \in E$,
$\qquad\qquad z_{t,w} \leq z_{t,u} + y_{uv} + y_{vw} \quad$ for all $t, u, v, w \in V$ with $uv, vw \in E$,

is equivalent to the odd circuit constraints. (In fact, one can do without the variables y_e, as they can be expressed in the x_v.) So a maximum-weight stable set in a t-perfect graph can be found in polynomial time with any polynomial-time linear programming algorithm.

T-perfection can also be characterized in terms of the vertex cover polytope:

Theorem 68.2. *A graph $G = (V, E)$ is t-perfect if and only if the vertex cover polytope of G is determined by:*

(68.5) $\quad$ (i) $\quad 0 \leq x_v \leq 1 \qquad$ for each $v \in V$,
$\qquad\quad$ (ii) $\quad x_u + x_v \geq 1 \qquad$ for each edge $uv \in E$,
$\qquad\quad$ (iii) $\quad x(VC) \geq \lceil \frac{1}{2}|VC| \rceil \quad$ for each odd circuit C.

Proof. System (68.5) arises from (68.1) by the reflection $x \to \mathbf{1} - x$. So integrality of the two polytopes is equivalent. ∎

68.2. Strongly t-perfect graphs

A graph $G = (V, E)$ is called *strongly t-perfect* if system (68.1) is totally dual integral. So each strongly t-perfect graph is t-perfect (Theorem 5.22). It is unknown if the reverse implication holds:

1188   Chapter 68. T-perfect graphs

(68.6)   Is every t-perfect graph strongly t-perfect?

Strong t-perfection can be characterized by the weighted version of the stable set number and a certain weighted 'edge and circuit' cover number. Let $G = (V, E)$ be a graph and let $w : V \to \mathbb{Z}_+$. In this chapter, a *w-cover* is a family of vertices, edges, and odd circuits covering each vertex v at least $w(v)$ times. By definition, the *cost* of a vertex or edge is 1, and the *cost* of an odd circuit C is $\lfloor \frac{1}{2}|VC| \rfloor$. The *cost* of a w-cover $\mathcal{F}$ is the sum of the costs of the elements of $\mathcal{F}$. Define

(68.7) $\alpha_w(G) :=$ the maximum weight of a stable set in G,
    $\tilde{\rho}_w(G) :=$ the minimum cost of a w-cover.

Obviously, $\alpha_w(G) \leq \tilde{\rho}_w(G)$ for any graph G. Moreover:

(68.8) G is strongly t-perfect $\iff \alpha_w(G) = \tilde{\rho}_w(G)$ for each $w : V \to \mathbb{Z}_+$.

This follows directly from a combinatorial interpretation of total dual integrality.

Notes. W.R. Pulleyblank (cf. Gerards [1989a]) observed that, even for $w = 1$, determining $\tilde{\rho}_w(G)$ is NP-complete, since the vertex set of a graph G can be partitioned into triangles if and only if $\tilde{\rho}_w(G) = \frac{1}{3}|V|$ where $w = 1$. The problem of partitioning a graph into triangles is NP-complete. Since partitioning into triangles remains to be NP-complete for planar graphs (Dyer and Frieze [1986]), even determining $\tilde{\rho}_w(G)$ for planar graphs is NP-complete.

Again, strong t-perfection is equivalent to the total dual integrality of the vertex cover constraints (68.5).

68.3. Strong t-perfection of odd-K_4-free graphs

K_4 is the smallest graph that is not t-perfect. Gerards and Schrijver [1986] showed that any graph not containing an 'odd K_4-subdivision' is t-perfect — in fact, as Gerards [1989a] showed, strongly t-perfect. We will prove this in this section (with a method inspired by Geelen and Guenin [2001]).

Call a subdivision of K_4 *odd* if each triangle of K_4 has become an odd circuit — equivalently, if the evenly subdivided edges of K_4 form a cut of K_4. We say that a graph *contains no* odd K_4-subdivision if it has no subgraph which is an odd K_4-subdivision.

Theorem 68.3. *A graph containing no odd K_4-subdivision is strongly t-perfect.*

Proof. Let $G = (V, E)$ be a counterexample with $|V| + |E|$ minimum. Then G has no isolated vertices. So we can assume that any minimum-cost w-cover contains no vertices (for any w).

For any weight function $w : V \to \mathbb{Z}_+$, denote $\alpha_w := \alpha_w(G)$ and $\tilde{\rho}_w := \tilde{\rho}_w(G)$. As G is a counterexample, there exists a $w : V \to \mathbb{Z}_+$ with $\alpha_w < \tilde{\rho}_w$.
For any such w we have, for each edge $e = uv$,

(68.9)   if S maximizes $w(S)$ over stable sets S of $G - e$, then S contains u and v.

Otherwise, S is a stable set of G, implying that (by the minimality of $|V| + |E|$):

(68.10) $\alpha_w(G) \geq \alpha_w(G - e) = \tilde{\rho}_w(G - e) \geq \tilde{\rho}_w(G)$,

a contradiction.
This implies

(68.11) $w \geq 1$,

since if $w(v) = 0$ for some vertex v, then for any edge e incident with v there is a stable set S of $G - e$ maximizing $w(S)$ and not containing v (since deleting v from S does not decrease $w(S)$). This contradicts (68.9).

We next show that we can assume w to have some additional properties (for an edge $e = uv$, χ^e is the incidence vector of the set $\{u, v\}$, that is, it is the 0,1 vector in $\mathbb{R}^V$ having 1's in positions u and v):

Claim 1. *There exist $w : V \to \mathbb{Z}_+$ and $f \in E$ such that*

(68.12) $\tilde{\rho}_{w+\chi^f} = \alpha_w + 1 = \tilde{\rho}_w = \alpha_{w+\chi^f}$

and such that

(68.13) $\alpha_{w-\chi^{VC}} = \tilde{\rho}_{w-\chi^{VC}}$

for each odd circuit C.

Proof of Claim 1. As G is not bipartite (by Theorem 19.7) and not just an odd circuit (as this is trivially strongly t-perfect), we know that H has a chordless odd circuit C_0 that has at least one vertex of degree at least 3. Let v be such a vertex, and let e be an edge incident with v but which is not on C_0.

Let $B := VC_0 \setminus \{v\}$. We choose w such that $w(V \setminus B)$ is minimal. There exists a $k \in \mathbb{Z}_+$ such that for $w' := w + k \cdot \chi^B$, each stable set S of $G - e$ maximizing $w'(S)$ satisfies $|S \cap B| = \frac{1}{2}|B|$. Hence no such set S contains v, and therefore, by (68.9), $\alpha_{w'} = \tilde{\rho}_{w'}$.

Now let M be the perfect matching in $C_0 - v$. For $y : M \to \mathbb{Z}_+$ define

(68.14) $w^y := w + \sum_{f \in M} y_f \chi^f$.

As $\alpha_{w'} = \tilde{\rho}_{w'}$, there exists a $y : M \to \mathbb{Z}_+$ such that

(68.15) $\alpha_{w^y} = \tilde{\rho}_{w^y}$.

We choose such a y with $\sum_{f \in M} y_f$ minimal. Since $\alpha_w < \tilde{\rho}_w$, there exists an $f \in M$ with $y_f \geq 1$. Then, by the minimality of y, we have $\alpha_{w^y - \chi^f} < \tilde{\rho}_{w^y - \chi^f}$. So we can assume that $y_f = 1$ and $y_{f'} = 0$ for each $f' \in M \setminus \{f\}$. We show that w and f are as required.

To show (68.12), we have $\alpha_{w+\chi^f} \leq \alpha_w + 1$, since any stable set S satisfies $(w + \chi^f)(S) = w(S) + |f \cap S| \leq w(S) + 1$. This implies

(68.16) $\quad \alpha_w + 1 \leq \tilde{\rho}_w \leq \tilde{\rho}_{w+\chi^f} = \alpha_{w+\chi^f} \leq \alpha_w + 1,$

implying (68.12).

Next, consider any odd circuit C in G. Then $(w - \chi^{VC})(V \setminus B) < w(V \setminus B)$, since VC is not contained in B. Therefore, by the choice of w, we have (68.13).

End of Proof of Claim 1

As from now we fix w and f satisfying (68.12) and (68.13). Let f connect vertices u and u'. Since by the minimality of G, G has no isolated vertices, there exists a minimum-cost $w + \chi^f$-cover $\mathcal{F}$ consisting only of edges and odd circuits, say, $e_1, \ldots, e_t, C_1, \ldots, C_k$. We choose them such that

(68.17) $\quad |VC_1| + \cdots + |VC_k|$

is as small as possible. Then:

(68.18) $\quad$ at least two of the C_i traverse f.

To see this, let $G' := G - f$ (the graph obtained by deleting edge f). If $\alpha_w(G') = \alpha_w(G)$, then by the minimality of G, G' has a w-cover of cost α_w. As this is a w-cover in G as well, this would imply $\alpha_w = \tilde{\rho}_w$, a contradiction.

So $\alpha_w(G') > \alpha_w(G)$. That is, there exists a stable set S in G' with $w(S) > \alpha_w$. Necessarily, S contains both u and u'. Then for any circuit C traversing f:

(68.19) $\quad |VC \cap S| \leq \lfloor \tfrac{1}{2}|VC| \rfloor + 1.$

Also, f is not among $e_1, \ldots, e_t$, since otherwise $\mathcal{F} \setminus \{f\}$ is a w-cover of cost $\tilde{\rho}_{w+\chi^f} - 1 = \tilde{\rho}_w - 1$, contradicting the definition of $\tilde{\rho}_w$. Setting l to the number of C_i traversing f, we obtain:

(68.20) $\quad \tilde{\rho}_{w+\chi^f} \leq \alpha_w + 1 \leq w(S) = (w + \chi^f)(S) - 2$
$$\leq -2 + \sum_{j=1}^{t} |e_j \cap S| + \sum_{i=1}^{k} |VC_i \cap S| \leq -2 + t + \sum_{i=1}^{k} \lfloor \tfrac{1}{2}|VC_i| \rfloor + l$$
$$= \tilde{\rho}_{w+\chi^f} + l - 2.$$

So $l \geq 2$, which is (68.18).

By (68.18) we can assume that C_1 and C_2 traverse f. It is convenient to assume that $EC_1 \setminus \{f\}$ and $EC_2 \setminus \{f\}$ are disjoint; this can be achieved by adding parallel edges. So $EC_1 \cap EC_2 = \{f\}$.

Then:

Section 68.3. Strong t-perfection of odd-K_4-free graphs

(68.21)   if C is an odd circuit with $EC \subseteq EC_1 \cup EC_2$, then $f \in EC$ and $EC_1 \triangle EC_2 \triangle EC$ is again an odd circuit.

Indeed, as $EC_1 \triangle EC_2 \triangle EC$ is an odd cycle, it can be decomposed into circuits $C'_2, \ldots, C'_p$, with $C'_2, \ldots, C'_q$ odd and $C'_{q+1}, \ldots, C'_p$ even ($q \geq 2$). Then

(68.22)
$$\sum_{i=2}^{p} |EC'_i| = |EC_1 \triangle EC_2 \triangle EC|$$
$$= |EC_1| + |EC_2| - |EC| - 2|\{f\} \setminus EC|.$$

Choose for each $i = q+1, \ldots, p$ a perfect matching M_i in C'_i. Let $e'_1, \ldots, e'_r$ be the edges in the matchings M_i and in $\{f\} \setminus EC$. Then, defining $C'_1 := C$,

(68.23) $\chi^{VC_1} + \chi^{VC_2} = \sum_{i=1}^{q} \chi^{VC'_i} + \sum_{j=1}^{r} \chi^{e'_j}$

and (using (68.22))

(68.24) $\lfloor \frac{1}{2}|VC_1| \rfloor + \lfloor \frac{1}{2}|VC_2| \rfloor = \frac{1}{2}|EC_1| + \frac{1}{2}|EC_2| - 1$
$$= -1 + |\{f\} \setminus EC| + \frac{1}{2}\sum_{i=1}^{p}|EC'_i| = -1 + r + \frac{1}{2}\sum_{i=1}^{q}|EC'_i|$$
$$\geq r + \sum_{i=1}^{q} \lfloor \frac{1}{2}|VC'_i| \rfloor.$$

So replacing C_1, C_2 by $C'_1, \ldots, C'_q$ and adding $e'_1, \ldots, e'_r$ to $e_1, \ldots, e_t$, gives again a $w + \chi^f$-cover of cost at most $\tilde{\rho}_{w+\chi^f}$. This also implies $q = 2$, since otherwise we have strict inequality in (68.24), and we would obtain a w-cover of cost less than $\tilde{\rho}_w$.

If $f \notin EC$, then f is among $e'_1, \ldots, e'_r$. Hence deleting f gives a w-cover of cost at most $\tilde{\rho}_{w+\chi^f} - 1 \leq \alpha_w$, contradicting (68.12). So $f \in EC$. As this is true for any odd circuit in $EC_1 \cup EC_2$ we know that $f \in EC'_i$ for $i = 1, 2$.

If $p \geq 3$ or $r \geq 1$, then $|EC'_1| + |EC'_2| < |EC_1| + |EC_2|$, contradicting the minimality of (68.17). So $p = q = 2$ and $r = 0$, which proves (68.21).

First, it implies

(68.25)   a circuit in $EC_1 \cup EC_2$ is odd if and only if it traverses f.

A second consequence is as follows. Let P_i be the $u - u'$ path $C_i \setminus \{f\}$. Orient the edges occurring in the path $P_i := C_i \setminus \{f\}$ in the direction from u to u', for $i = 1, 2$. Then

(68.26)   the orientation is acyclic.

For suppose that it contains a directed circuit C. Then $(EC_1 \cup EC_2) \setminus EC$ contains a directed $u-u'$ path, and hence an odd circuit C'. Hence by (68.21), $EC_1 \triangle EC_2 \triangle EC'$ is an odd circuit, however containing the even circuit EC, a contradiction.

Define

(68.27) $\quad W := VP_1 \cup VP_2$ and $F := EP_1 \cup EP_2$.

Consider the graph (W, F). It is bipartite, as it contains no odd circuits by (68.25). Moreover, u and u' belong to the same colour class. Let A and B be the colour classes of (W, F), such that $u, u' \in A$. So

(68.28) $\quad A := \{v \in W \mid \text{there exists an even-length directed } u - v \text{ path}\}$,
$ B := \{v \in W \mid \text{there exists an odd-length directed } u - v \text{ path}\}$.

(Here and below, when speaking of a directed path, it is assumed to use only the edges in $EP_1 \cup EP_2$.) Define

(68.29) $\quad X := VP_1 \cap VP_2$ and
$$U := \{v \in V \mid w(v) = \sum_{j=1}^{t} |e_j \cap \{v\}| + \sum_{j=1}^{k} |VC_j \cap \{v\}|\}.$$

So $u, u' \notin U$, $u, u' \in X$, and $X \setminus \{u, u'\}$ is the set of vertices in W having degree 4 in the graph (W, F).

We next show the following technical, but straightforward to prove, claim:

Claim 2. *Let $z \in A$, let Q be an even-length directed $u - z$ path, and let S be a stable set in G. Then*

(68.30) $\quad (w - \chi^{VQ})(S) \geq \alpha_w - \lfloor \frac{1}{2}|VQ| \rfloor + 1$

if and only if

(68.31) $\quad$ (i) $|e_j \cap S| = 1$ for each $j = 1, \ldots, t$,
 (ii) $|VC_j \cap S| = \lfloor \frac{1}{2}|VC_j| \rfloor$ for $j = 3, \ldots, k$,
 (iii) $S \subseteq U$,
 (iv) S contains $B \setminus VQ$ and is disjoint from $A \setminus VQ$,
 (v) S contains $B \cap X$ and is disjoint from $A \cap X$.

Proof of Claim 2. By rerouting C_1 and C_2, we can assume that $EQ \subseteq EC_1$. Define $Z := VC_1 \setminus VQ$. So $|Z|$ is even. Consider the following sequence of (in)equalities:

(68.32) $\quad (w - \chi^{VQ})(S) = w(S) - |VQ \cap S|$
$$\leq \sum_{j=1}^{t} |e_j \cap S| + \sum_{j=1}^{k} |VC_j \cap S| - |VQ \cap S|$$
$$= \sum_{j=1}^{t} |e_j \cap S| + \sum_{j=2}^{k} |VC_j \cap S| + |Z \cap S| \leq t + \sum_{j=2}^{k} \lfloor \frac{1}{2}|VC_j| \rfloor + |Z \cap S|$$
$$= \tilde{p}_{w+\chi^f} - \lfloor \frac{1}{2}|VC_1| \rfloor + |Z \cap S| \leq \tilde{p}_{w+\chi^f} - \lfloor \frac{1}{2}|VC_1| \rfloor + \frac{1}{2}|Z|$$
$$= \alpha_w + 1 - \lfloor \frac{1}{2}|VQ| \rfloor.$$

Hence (68.30) holds if and only if equality holds throughout in (68.32), which is equivalent to (68.31). Note that (68.31)(iv) and (v) are equivalent to: S contains $VC_2 \cap B$ and is disjoint from $VC_2 \cap A$, and S contains $Z \cap B$ and

Section 68.3. Strong t-perfection of odd-K_4-free graphs

is disjoint from $Z \cap \Lambda$. Hence it is equivalent to (as $u, u' \notin S$ by (68.31)(iii)): $|VC_2 \cap S| = \lfloor \frac{1}{2}|VC_2|\rfloor$ and $|Z \cap S| = \frac{1}{2}|Z|$. *End of Proof of Claim 2*

By (68.26), we can order the vertices in X as $x_0 = u, x_1, \ldots, x_s = u'$ such that both P_1 and P_2 traverse them in this order. For $j = 0, \ldots, s$, let $\mathcal{P}_j$ be the collection of directed $u - x$ paths, where $x = x_j$ if $x_j \in A$, and x is an inneighbour of x_j if $x_j \in B$. So $x \in A$ and each path in each $\mathcal{P}_j$ has even length.

Let j be the largest index for which there exists a path $Q \in \mathcal{P}_j$ with

(68.33) $\qquad \alpha_{w-\chi^{VQ}} \leq \alpha_w - \lfloor \frac{1}{2}|VQ|\rfloor.$

Such a j exists, since (68.33) holds for the trivial directed $u - u$ path, as $\alpha_{w-\chi^u} \leq \alpha_w$. Also, $j < s$, since otherwise $VQ = VC$ for some odd circuit C, and hence, with (68.13) we have

(68.34) $\qquad \tilde{\rho}_w \leq \tilde{\rho}_{w-\chi^{VC}} + \lfloor \frac{1}{2}|VC|\rfloor = \alpha_{w-\chi^{VC}} + \lfloor \frac{1}{2}|VC|\rfloor \leq \alpha_w,$

contradicting (68.12).

Let Q_1 and Q_2 be the two paths in $\mathcal{P}_{j+1}$ that extend Q. By the maximality of j, we know

(68.35) $\qquad \alpha_{w-\chi^{VQ_i}} \geq \alpha_w - \lfloor \frac{1}{2}|VQ_i|\rfloor + 1.$

So there exist stable sets S_1 and S_2 with

(68.36) $\qquad (w - \chi^{VQ_i})(S_i) \geq \alpha_w - \lfloor \frac{1}{2}|VQ_i|\rfloor + 1$

for $i = 1, 2$. So for $i = 1, 2$, (68.31) holds for Q_i, S_i. By (68.31)(iv), S_1 and S_2 coincide on $W \setminus (VQ_1 \cup VQ_2)$, and they coincide on X. In other words:

(68.37) $\qquad (S_1 \triangle S_2) \cap W \subseteq (VQ_1 \cup VQ_2) \setminus X.$

Let H be the subgraph of G induced by $S_1 \triangle S_2$. So H is a bipartite graph, with colour classes $S_1 \setminus S_2$ and $S_2 \setminus S_1$. Define

(68.38) $\qquad Y_i := VQ_i \setminus VQ$

for $i = 1, 2$. Then

(68.39) $\qquad$ H contains a path connecting Y_1 and Y_2.

For suppose not. Let K be the union of the components of H that intersect Y_1. So K is disjoint from Y_2. Define $S := S_1 \triangle K$. Then $S \cap Y_1 = S_2 \cap Y_1$ and $S \cap Y_2 = S_1 \cap Y_2$. This implies that Q, S satisfy (68.31). Hence (68.30) holds, contradicting (68.33). This proves (68.39).

Let C be the (even) circuit formed by the two directed $x_j - x_{j+1}$ paths. So Y_1 and Y_2 are subsets of VC. Let R be a shortest path in H that connects Y_1 and Y_2; say it connects $y_1 \in Y_1$ and $y_2 \in Y_2$.

Since $y_1, y_2 \in S_1 \triangle S_2$, we know by (68.37) that $y_1, y_2 \notin X$. By (68.31)(iv), if $y_1 \in S_1 \setminus S_2$, then $y_1 \in A$ (since if $y_1 \in B$, then $y_1 \in B \setminus VQ_2$, and so $y_1 \in S_2$), and if $y_1 \in S_2 \setminus S_1$, then $y_1 \in B$ (since if $y_1 \in A$, then $y_1 \in A \setminus VQ_2$, and so $y_1 \notin S_2$). Similarly, if $y_2 \in S_2 \setminus S_1$, then $y_2 \in A$ and if $y_2 \in S_1 \setminus S_2$, then $y_2 \in B$.

So if R is even, then y_1 and y_2 belong to the same set among $S_1 \setminus S_2$, $S_2 \setminus S_1$, and hence they belong to different sets A, B. Similarly, if R is odd, then y_1 and y_2 belong to the same set among A, B. Hence R forms with part of C an odd circuit.

By (68.37), there exist a directed $u - x_j$ path N' and a directed $x_{j+1} - u'$ path N'' that are (vertex-)disjoint from $S_1 \triangle S_2$. Concatenating N', f, and N'' makes an $x_{j+1} - x_j$ path N. Then N, R, and C make an odd K_4-subdivision, with 3-valent vertices x_j, x_{j+1}, y_1, y_2. ∎

(The above proof of Claim 1 was given by D. Gijswijt.)

Notes. Theorem 68.3 includes the t-perfection of series-parallel graphs (conjectured by Chvátal [1975a], and proved by M.J. Clancy in 1977 and by Mahjoub [1988]), the strong t-perfection of series-parallel graphs (Boulala and Uhry [1979], who also gave a polynomial-time algorithm to find a maximum-weight stable set in series-parallel graphs), the t-perfection of almost bipartite graphs — graphs G having a vertex v with $G - v$ bipartite (Fonlupt and Uhry [1982]), the strong t-perfection of almost bipartite graphs (this is implicit in Sbihi and Uhry [1984]), and the t-perfection of odd-K_4-free graphs (Gerards and Schrijver [1986]).

68.4. On characterizing t-perfection

The problem if a given graph $G = (V, E)$ is t-perfect, belongs to co-NP: non-t-perfection can be certified by a noninteger vertex x^* of the polytope determined by (68.1), together with a nonsingular system of constraints that are tight for x^*. One must check that x^* satisfies all constraints among (68.1) — this can be done in polynomial time by the methods described in the proof of Theorem 68.1. A polynomial-time algorithm for, or a combinatorial certificate of, non-t-perfection is not known.

T-perfection and strong t-perfection are not closed under taking subgraphs, as is shown by Figure 68.1. However, t-perfection is closed under taking induced subgraphs. This is easy to check, as well as that it is closed under the following operation:

(68.40)   choose a vertex v with $N(v)$ a stable set, and contract all edges in $\delta(v)$.

So one may ask for the minimally non-t-perfect graphs — minimal with respect to taking induced subgraphs and applying operation (68.40). Known minimal graphs include the wheels[24] with an even number of vertices and the graphs consisting of a circuit of length $4k$ and all chords connecting a

---
[24] A *wheel* is a graph obtained from a circuit C by adding a new vertex, adjacent to all vertices in C.

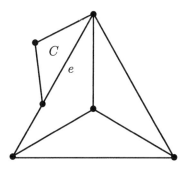

Figure 68.1
A strongly t-perfect graph G with $G - e$ not t-perfect. The strong t-perfection of G can be derived from the fact that each inclusionwise maximal stable set intersects the triangle C. Hence for any integer weight function, subtracting the incidence vector of VC, reduces the maximum weight of a stable set by 1. We therefore can assume that at least one of the vertices of C has weight 0, and hence we can delete it. We are left with a graph containing no odd K_4-subdivision — hence being strongly t-perfect (Theorem 68.3).

vertex with its opposite vertex ($k \geq 1$). Also strong t-perfection is closed under taking induced subgraphs and the operation (68.40). So one may ask a similar question for strong t-perfection.

A characterization that has been achieved is of those graphs for which each, also noninduced, subgraph is t-perfect. Here subdivisions of K_4 play a role. Call a subdivision of K_4 *bad* if it is not t-perfect.

It has been shown by Gerards and Shepherd [1998] that any graph without bad K_4-subdivision is t-perfect. Hence, each subgraph of a graph G is t-perfect if and only if G contains no bad K_4-subdivision. This was extended to: any graph without bad K_4-subdivision is strongly t-perfect (Schrijver [2002b]). So each subgraph of a graph is t-perfect if and only if each subgraph is strongly t-perfect.

The K_4-subdivisions that are bad have been characterized by Barahona and Mahjoub [1994c]. They showed that a K_4-subdivision is not t-perfect if and only if it is an odd K_4-subdivision such that the following does *not* hold: the edges of K_4 that have become an even path, form a 4-cycle in K_4, while the two other edges of K_4 are not subdivided. One may check that this is equivalent to the fact that one cannot obtain K_4 by the operations (68.40). So necessity in this characterization follows from the closedness of t-perfection under operation (68.40).

68.5. A combinatorial min-max relation

A subdivision of K_4 is called *totally odd* if it arises from K_4 be replacing each edge by an odd-length path. So a totally odd K_4-subdivision is an odd K_4-subdivision. A graph containing no totally odd K_4-subdivision need not be t-perfect (see Figure 68.2, from Chvátal [1975a]). However, Sewell and Trotter [1990,1993] showed that for weight function $w = 1$, the min-max relation is maintained for totally odd K_4-free graphs.

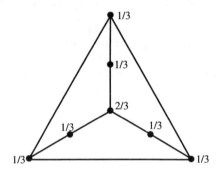

Figure 68.2
A graph containing no totally odd K_4-subdivision and not being t-perfect. The values at the vertices represent a vector satisfying (68.1) but not belonging to the stable set polytope.

This can be formulated in terms of the nonweighted version $\tilde{\rho}(G)$ of $\tilde{\rho}_w(G)$ defined in (68.7):

(68.41) $\tilde{\rho}(G) :=$ the minimum cost of a family of vertices, edges, and odd circuits covering V.

One easily checks that the minimum is attained by a vertex-disjoint family.
Obviously, for any graph G,

(68.42) $\alpha(G) \leq \tilde{\rho}(G)$.

So Sewell and Trotter [1990,1993] showed that equality holds for graphs without totally odd K_4-subdivision (generalizing a result of Gerards [1989a], who proved it for graphs without odd K_4-subdivision — a consequence of Theorem 68.3; Chvátal [1975a] proved it for series-parallel graphs).

Theorem 68.4. *For any graph G containing no totally odd K_4-subdivision, the stable set number $\alpha(G)$ is equal to $\tilde{\rho}(G)$.*

Proof. Let $G = (V, E)$ be a counterexample with $|V| + |E|$ minimal. Set $\alpha := \alpha(G)$. Then G is connected, and

Section 68.5. A combinatorial min-max relation   1197

(68.43) $\alpha(G-v) = \alpha$ for each $v \in V$ and $\alpha(G-e) > \alpha$ for each $e \in E$,

since otherwise $G-v$ or $G-e$ would be a smaller counterexample.

Hence for each vertex v, there exists a vertex-disjoint collection of vertices, edges, and odd circuits, covering $V \setminus \{v\}$ and of cost α. Let F_v be the set of edges contained in this collection or in one of the circuits in it. Let G_v be the graph $(V \setminus \{v\}, F_v)$. So

(68.44) $\alpha(G_v) = \alpha$,

and G_v has maximum degree at most 2. Moreover, the minimality of G implies:

(68.45) $F_u \cup F_v = E \setminus \{uv\}$ for each edge uv.

To see this, trivially, $uv \notin F_u \cup F_v$. Suppose that $e \neq uv$ is an edge not contained in $F_u \cup F_v$. As $\alpha(G-e) > \alpha$, $G-e$ has a stable set S of size $\alpha+1$. By symmetry, we can assume that $v \notin S$. Then S is a stable set in the graph G_v, contradicting (68.44).

This proves (68.45), which gives:

(68.46)   for each edge uv, each edge $e \neq uv$ incident with u belongs to F_v.

This follows directly from (68.45), since $e \notin F_u$ (as $u \in e$).

This is used in proving:

(68.47) G is 3-regular.

For if vertex v has degree 1, with neighbour u, then $\alpha(G-v-u) < \alpha$; since moreover $\tilde{\rho}(G) \leq \tilde{\rho}(G-u-v)+1$ (since we can add edge uv to any collection attaining the minimum for $G-u-v$), we have $\alpha(G) \geq \alpha(G-u-v)+1 = \tilde{\rho}(G-u-v)+1 \geq \tilde{\rho}(G)$. This contradicts the fact that G is a counterexample.

If v has degree 2, let G' be the graph obtained by contracting the edges incident with v. Then G' contains no totally odd K_4-subdivision. Moreover, it is straightforward to check that $\alpha(G) \geq \alpha(G')+1$ and $\tilde{\rho}(G) \leq \tilde{\rho}(G')+1$. As G' is smaller than G, we have $\tilde{\rho}(G') = \alpha(G')$. Hence $\alpha(G) \geq \alpha(G')+1 = \tilde{\rho}(G')+1 \geq \tilde{\rho}(G)$. Again, this contradicts the fact that G is a counterexample.

So v has degree at least 3. Let u be one of its neighbours. Then $\delta(v) \subseteq F_u \cup \{uv\}$ by (68.46). As G_u has maximum degree at most 2, we have $|\delta(v)| = 3$. This proves (68.47).

By (68.45) and (68.47),

(68.48)   for each edge uv of G, u is traversed by an odd circuit in F_v.

Moreover:

(68.49)   Let uv be an edge of G and let C be the odd circuit in F_v traversing u. Consider any edge $e = xy$ on C with $e \notin F_u$. Then both x and y have even distance from u along $C - e$.

Let S be a stable set of $G-e$ of size $\alpha+1$. So $x,y \in S$. Moreover, $u \in S$, since otherwise $\alpha(G_u) > \alpha$ (since $e \notin F_u$), contradicting (68.44). So $v \notin S$, and hence $S \setminus \{y\}$ is a maximum-size stable set of $G-v$. Hence, $S \setminus \{y\}$ intersects C in $\lfloor \frac{1}{2}|VC| \rfloor$ vertices. Therefore, S intersects C in $\lceil \frac{1}{2}|VC| \rceil$ vertices. As $x, y \in S$, (68.49) follows.

Now choose a vertex, r say, and its neighbours, u_1, u_2, u_3 say. For each $i \in \{1,2,3\}$, F_{u_i} contains an odd circuit C_i traversing r (by (68.48)), and hence traversing the edges ru_{i+1} and ru_{i+2} (taking indices mod 3). We will construct a totally odd K_4-subdivision from them, which contradicts the condition of the theorem.

For $i = 1, 2, 3$, let P_i be the path in C_i from u_{i+1} to u_{i+2} obtained by deleting vertex r from C_i. Since $\alpha(G - ru_i) > \alpha$, G has a stable set S_i of size α, intersecting $\{r, u_1, u_2, u_3\}$ precisely in $\{u_i\}$. Then for all distinct $i, j \in \{1, 2, 3\}$:

(68.50) S_j contains all vertices along P_i at even distance from u_j.

To see this, we may assume that $i = 1$, $j = 2$. Since S_2 is a maximum-size stable set in $G - u_1$, it intersects C_1 in $\lfloor \frac{1}{2}|VC_1| \rfloor$ vertices. Since $r, u_3 \notin S_2$, S_2 contains all vertices along P_1 at even distance from u_2, proving (68.50).

This implies, for distinct $i, j, k \in \{1, 2, 3\}$:

(68.51) $VP_i \subseteq S_j \triangle S_k$.

One similarly shows, for distinct $i, j, k \in \{1, 2, 3\}$:

(68.52) P_i contains an edge that splits P_i into two even-length paths $P_{i,j}$ (containing u_j) and $P_{i,k}$ (containing u_k), in such a way that S_i contains all vertices along $P_{i,j}$ at odd distance from u_j and all vertices along $P_{i,k}$ at odd distance from u_k.

To prove this, we may assume that $i = 1$, $j = 2$, $k = 3$. Since $S := S_1 \setminus \{u_1\} \cup \{r\}$ is a maximum-size stable set in $G - u_1$, it intersects C_1 in $\lfloor \frac{1}{2}|VC_1| \rfloor$ vertices. Since S contains r, there is precisely one edge on P_1 not intersected by S. This gives the edge as required in (68.52).

This implies, for distinct $i, j \in \{1, 2, 3\}$:

(68.53) $VP_{i,j} = VP_i \cap (S_i \triangle S_j)$,

and hence, for distinct $i, j, k \in \{1, 2, 3\}$:

(68.54) $VP_i \cap VP_j = VP_{i,k} \cap VP_{j,k}$,

since

(68.55) $VP_{i,k} \cap VP_{j,k} = VP_i \cap (S_i \triangle S_k) \cap VP_j \cap (S_j \triangle S_k) = VP_i \cap VP_j$

(using (68.51)).

For each $i = 1, 2, 3$, vertex u_{i+2} is on P_i and P_{i+1}. Hence there is a first vertex v_i on P_i (starting from u_{i+1}), that also belongs to P_{i+1}. By (68.54), v_i occurs after v_{i+2} along P_i (seen from u_{i+1}), since $v_i \in VP_i \cap VP_{i+1} \subseteq VP_{i,i+2}$ and $v_{i+2} \in VP_{i+2} \cap VP_i \subseteq VP_{i,i+1}$. Moreover,

(68.56) v_i has even distance from u_{i+2} along P_i and along $P_{i|1}$.

To prove this, we may assume that $i = 1$. Suppose that v_1 has odd distance from u_3 along P_1. Let f and e be the previous and next edge along P_1 (seen from u_2) and let g be the third edge incident with v_1. Since v_1 is the first vertex along P_1 belonging to P_2, we know that f is not on P_2. So $f \notin F_{u_2}$, and hence (by (68.45)) $f \in F_r$. Since g is not on P_1, we have $g \notin F_{u_1}$, and hence (again by (68.45)) $g \in F_r$. Therefore (as F_r has maximum degree at most degree 2), $e \notin F_r$. Then (68.49) implies that v_1 has even distance from u_3 along P_1. Hence $v_1 \in S_3$, and so v_1 has also even distance from u_3 along P_2 (by (68.50)). This proves (68.56).

For $i = 1, 2, 3$, let Q_i be the $u_{i+1} - v_i$ part of P_i. Then Q_i and Q_{i+1} intersect each other only in v_i (since v_i is the first vertex along P_i that is on P_{i+1}). This implies that Q_1, Q_2, Q_3 together with the edges ru_1, ru_2, and ru_3, form a totally odd K_4-subdivision, a contradiction. ∎

Recall that a graph is bipartite if and only if for each subgraph H, the stable set number $\alpha(H)$ is equal to the edge cover number $\rho(H)$. An extension of this is implied by the theorem above:

Corollary 68.4a. *A graph G contains no totally odd K_4-subdivision if and only $\alpha(H) = \tilde{\rho}(H)$ for each subgraph H of G.*

Proof. Necessity follows from Theorem 68.4. Sufficiency follows from the fact that if G is a totally odd K_4-subdivision, then $\alpha(G) < \tilde{\rho}(G)$. This can be seen by induction on $|VG|$. If $|VG| = 4$, then $G = K_4$, and $\alpha(G) = 1$, $\tilde{\rho}(G) > 1$. If $|VG| > 4$, G has a vertex v of degree 2. Let G' arise by contracting the two edges incident with v. Then, using the induction hypothesis, $\alpha(G) \leq \alpha(G') + 1 < \tilde{\rho}(G') + 1 \leq \tilde{\rho}(G)$. ∎

Theorem 68.4 also implies (in fact, is equivalent to) the following. A graph $G = (V, E)$ is called α-*critical* if $\alpha(G - e) > \alpha(G)$ for each $e \in E$. Then each connected α-critical graph is either K_1, or K_2, or an odd circuit, or contains a totally odd K_4-subdivision (answering a question of Chvátal [1975a]).

We note that Theorem 68.4 implies that the stable set number $\alpha(G)$ of a graph G without totally odd K_4-subdivision can be determined in polynomial time, as $\alpha(G)$ is equal to the maximum of $\mathbf{1}^T x$ over (68.1) (since the separation problem is polynomial-time solvable — see Theorem 68.1). This implies that an explicit maximum-size stable set can be found in polynomial time (just by deleting vertices as long as the stable set number does not decrease).

The vertex cover number. Another consequence of Theorem 68.4 concerns the vertex cover number $\tau(G)$ of a graph $G = (V, E)$. Trivially, $\tau(G) + \alpha(G) = |V|$. Define the *profit* of an edge to be 1, and the *profit* of a circuit C to be $\lceil \frac{1}{2}|VC| \rceil$. The *profit* of a family of edges and circuits is equal to the sum of the profits of its elements. Let $\tilde{\nu}(G)$ denote the maximum profit of a collection of pairwise vertex-disjoint edges and odd circuits in G. Then there is the following analogue to Gallai's theorem (Theorem 19.1):

Theorem 68.5. *For any graph* $G = (V, E)$: $\tilde{\nu}(G) + \tilde{\rho}(G) = |V|$.

Proof. Define the profit of any vertex to be 0. Then $\tilde{\nu}(G)$ is equal to the maximum profit of a collection of vertices, edges, and circuits partitioning V. Similarly, $\tilde{\rho}(G)$ is equal to the minimum cost of a collection of vertices, edges, and circuits partitioning V. Now for any collection $\mathcal{C}$ of vertices, edges, and circuits partitioning V we have $\text{cost}(\mathcal{C}) + \text{profit}(\mathcal{C}) = |V|$. Hence the minimum cost over all such collections equals $|V|$ minus the maximum profit over all such collections. This gives the required equality. ∎

With Theorem 68.4, this implies a min-max relation for the vertex cover number of totally-odd-K_4-free graphs:

Corollary 68.5a. *For any graph G containing no totally odd K_4-subdivision, the vertex cover number $\tau(G)$ is equal to $\tilde{\nu}(G)$.*

Proof. Directly from Theorems 68.4 and 68.5, and from the fact that $\alpha(G) + \tau(G) = |V|$ for any graph G. ∎

68.6. Further results and notes

68.6a. The w-stable set polyhedron

The t-perfection of odd-K_4-free graphs can be extended to apply to w-stable sets. Given a graph $G = (V, E)$ and a function $w : E \to \mathbb{Z}_+$, a *w-stable set* is a function $x : V \to \mathbb{Z}_+$ such that $x_u + x_v \leq w_e$ for each edge $e = uv$. So if $w = 1$ and G has no isolated vertices, w-stable sets are the incidence vectors of stable sets. The *w-stable set polyhedron* is the convex hull of the w-stable sets.

Theorem 68.3 implies a characterization of the w-stable set polyhedron of odd-K_4-free graphs. Consider the following system:

(68.57)   (i) $x_v \geq 0$ for each $v \in V$,
                  (ii) $x(e) \leq w_e$ for each $e \in E$,
                  (iii) $x(VC) \leq \lfloor \frac{1}{2} w(EC) \rfloor$ for each odd circuit C,

where $x(e) = x_u + x_v$ for $e = uv$.

Theorem 68.6. *For any graph $G = (V, E)$ containing no odd K_4-subdivision and for any $w : E \to \mathbb{Z}_+$, system (68.57) determines the w-stable set polyhedron.*

Proof. We show that (68.57) determines an integer polyhedron, and hence is equal to the w-stable set polyhedron. Let x be a noninteger vertex of P. By resetting $w_e := w_e - \lfloor x_u \rfloor - \lfloor x_v \rfloor$ for $e = uv \in E$ and $x_v := x_v - \lfloor x_v \rfloor$ for $v \in V$, x remains a noninteger vertex of the new P. So we can assume that $0 \leq x_v < 1$ for each $v \in V$.

Let E' be the set of edges e of G with $w_e = 1$. Then $G' = (V, E')$ contains no odd K_4-subdivision, and hence is t-perfect (Theorem 68.3). So x is a convex combination of incidence vectors of stable sets of G'. As each such incidence vector satisfies (i) and (ii) of (68.57) (since $x_u + x_v \leq 1 + 1 = 2 \leq w_e$ for each edge $e = uv$ in $E \setminus E'$), it also satisfies (iii) (as it is integer). Hence x is a convex combination of integer solutions of (68.57). So P is integer. ∎

It was shown by Gijswijt and Schrijver [2002] that system (68.57) is totally dual integral for each $w : E \to \mathbb{Z}_+$ if and only if G contains no bad K_4-subdivision.

68.6b. Bidirected graphs

We saw bidirected graphs before in Chapter 36. We recall some definitions and terminology. A *bidirected graph* is a triple $G = (V, E, \sigma)$, where (V, E) is an undirected graph and where σ assigns to each $e \in E$ and each $v \in e$ a 'sign' $\sigma_{e,v} \in \{+1, -1\}$. The graph (V, E) may have loops, but we will assume that the 'two' ends of the loop have the same sign. (Other loops will be meaningless in our discussion.)

The edges e for which $\sigma_{e,v} = 1$ for each $v \in e$ are called the *positive edges*, those with $\sigma_{e,v} = -1$ for each $v \in e$ are the *negative edges*, and the remaining edges are the *directed edges*.

Clearly, undirected graphs and directed graphs can be considered as special cases of bidirected graphs. Graph terminology extends in an obvious way to bidirected graphs. The undirected graph (V, E) is called the *underlying undirected graph* of G. We also will need the *underlying signed graph* $G = (V, E, \Sigma)$, where Σ is the family of positive and negative edges. We call a circuit C in (V, E) *odd* or *even*, if $|EC \cap \Sigma|$ is odd or even, respectively.

A signed graph $G = (V, E, \Sigma)$ is called an *odd K_4-subdivision* if (V, E) is a subdivision of K_4 such that each triangle has become an odd circuit (with respect to Σ). A bidirected graph is called an *odd K_4-subdivision* if its underlying signed graph is an odd K_4-subdivision.

The $E \times V$ *incidence matrix* of a bidirected graph $G = (V, E, \sigma)$ is the $E \times V$ matrix M defined by, for $e \in E$ and $v \in V$:

(68.58) $\quad M_{e,v} := \begin{cases} \sigma_{e,v} & \text{if } e \text{ is not a loop,} \\ 2\sigma_{e,v} & \text{if } e \text{ is a loop,} \end{cases}$

setting $\sigma_{e,v} := 0$ if $v \notin e$.

For $b \in \mathbb{Z}^E$, we consider integer solutions of the system $Mx \leq b$. To this end, define for any circuit C (in the undirected graph (V, E)) and any vertex v:

(68.59) $\quad a_{C,v} := \frac{1}{2} \sum_{e \in EC} M_{e,v}$ and $d_C := \lfloor \frac{1}{2} \sum_{e \in EC} b_e \rfloor$.

As C is a circuit, $a_{C,v}$ is an integer. Hence each integer solution x of $Mx \leq b$ satisfies

(68.60) $\quad \sum_{v \in V} a_{C,v} x_v = \frac{1}{2} \sum_{e \in EC} \sum_{v \in V} M_{e,v} x_v \leq \lfloor \frac{1}{2} \sum_{e \in EC} b_e \rfloor = d_C.$

Therefore, each integer solution of $Mx \leq b$ satisfies:

(68.61)   (i) $Mx \leq b$,
            (ii) $\sum_{v \in V} a_{C,v} x_v \leq d_C$ for each odd circuit C.

(Again, 'odd' is with respect to Σ.) Then Theorem 68.6 implies:

Corollary 68.6a. *If a bidirected graph G contains no odd K_4-subdivision, then system (68.61) determines an integer polyhedron.*

Proof. Make from the bidirected graph $G = (V, E, \sigma)$ the following auxiliary undirected graph $G' = (V', E')$. For each $e \in E$ which is not a positive loop, let $c_e := 1$ if e is positive, $c_e := 2$ if e is directed, and $c_e := 3$ if e is negative. Then replace e by a path P_e of length c_e connecting the two vertices in V incident with e. Let $\tilde{e}$ be the unique edge on P_e that is not incident with a vertex v of G with $\sigma_{e,v} = -1$.

If e is a positive loop at v, make a circuit P_e of length 3 starting and ending at v. Let $\tilde{e}$ be one of the two edges on P_e incident with v.

Let F be the set of edges f of G' that are on P_e for some $e \in E$ and satisfy $f \neq \tilde{e}$. As G has no odd K_4-subdivision (as a bidirected graph), G' has no odd K_4-subdivision (as an undirected graph). Hence by Theorem 68.6, the following system (in $x \in \mathbb{R}^{V'}$) determines an integer polyhedron:

(68.62)   (i) $x(\tilde{e}) \leq b_e$ for each edge $e \in E$,
               (ii) $x(f) = 0$ for each edge $f \in F$,
               (iii) $x(VC) \leq \lfloor \frac{1}{2}|VC| \rfloor$ for each odd circuit C in G'.

(Here 'odd' refers to the length of the circuit. As usual, $x(f) := x_u + x_v$ where u and v are the ends of f for $f \in E'$.) This implies that system (68.61) determines an integer polyhedron, since the conditions (68.62)(ii) allow elimination of the variables x_v for $v \in V' \setminus V$. ∎

This theorem may be used to characterize odd-K_4-free bidirected graphs. Let $G = (V, E, \sigma)$ be a bidirected graph, with $E \times V$ incidence matrix M. For $a, b \in \mathbb{Z}^E$ consider integer solutions of

(68.63) $a \leq Mx \leq b$.

As the matrix

(68.64) $\begin{pmatrix} M \\ -M \end{pmatrix}$

is again the incidence matrix of some bidirected graph, we can consider the inequalities (68.61)(ii) corresponding to matrix (68.64). They amount to:

(68.65) $\sum_{v \in V} \frac{1}{2} \left(\sum_{e \in F} M_{e,v} - \sum_{e \in EC \setminus F} M_{e,v} \right) x_v \leq \lfloor \frac{1}{2} \left(\sum_{e \in F} b_e - \sum_{e \in EC \setminus F} a_e \right) \rfloor$ for each odd circuit C and each $F \subseteq EC$.

To describe the characterization, we define 'minor' of a signed graph $G = (V, E, \Sigma)$. For $e \in E$, *deletion* of e means resetting E and Σ to $E \setminus \{e\}$ and $\Sigma \setminus \{e\}$. Deletion of a vertex v means deleting all edges incident with v, and deleting v from V. If e is not a loop, *contraction* of e means the following. Let e have ends u and v. If $e \in \Sigma$, reset $\Sigma := \Sigma \triangle \delta(u)$. Otherwise, let Σ be unchanged. Next contract e in (V, E). This definition depends on the choice of the end u of e, but for the application below this will be irrelevant. A *resigning* means choosing $U \subseteq V$ and resetting Σ to $\Sigma \triangle \delta(U)$. A signed graph H is called a *minor* of a signed graph G if

H arises from G by a series of deletions of edges and vertices, contractions of edges, and resignings.

Then we have the following characterization (Gerards and Schrijver [1986]), where odd-K_4 stands for the signed graph (VK_4, EK_4, EK_4).

Corollary 68.6b. *For any bidirected graph G the following are equivalent:*

(68.66) (i) *G contains no odd K_4-subdivision as subgraph;*
(ii) *the signed graph underlying G has no odd-K_4 minor;*
(iii) *for all integer vectors a, b, system (68.63)(68.65) determines a box-integer polyhedron.*

Proof. The implication (ii)$\Rightarrow$(i) follows from the easy fact that any odd K_4-subdivision in G would yield an odd-K_4 minor of the signed graph underlying G.

The implication (i)$\Rightarrow$(iii) can be derived from Corollary 68.6a as follows. Replace any 'box' constraint $d_v \leq x_v \leq c_v$ by $2d_v \leq 2x_v \leq 2c_v$, and incorporate it into M, by adding loops at v. Then the constraint (68.65) corresponding to such a loop C at v is $x_v \leq c_v$ or $-x_v \leq -d_v$. This gives a reduction to Corollary 68.6a.

To see the implication (iii)$\Rightarrow$(ii), note that (iii) is invariant under deleting rows of M and under multiplying rows or columns by -1. It is also closed under contractions of any edge e, as it amounts to taking $a_e = b_e = 0$ in (68.63). So, in proving (iii)$\Rightarrow$(ii), if the signed graph underlying G has an odd-K_4 minor, we may assume that it is odd-K_4. By multiplying rows and columns of M by -1, we may assume that M is nonnegative. Then we do not have an integer polytope for $a = \mathbf{0}$, $b = \mathbf{1}$, $d = \mathbf{0}$, $c = \mathbf{1}$. ∎

In other words, the bidirected graphs without odd K_4-subdivision are precisely those whose $E \times V$ incidence matrix has strong Chvátal rank at most 1 (cf. Section 36.7a, where it is shown that the transpose of *each* such matrix has strong Chvátal rank at most 1).

68.6c. Characterizing odd-K_4-free graphs by mixing stable sets and vertex covers

A similar characterization can be formulated in terms of just undirected graphs, by mixing stable sets and vertex covers. Call a graph H an *odd minor* of a graph G if H arises from G by deleting edges and vertices, and by contracting all edges in some cut $\delta(U)$ (in the graph without the deleted edges). The following is easy to show:

(68.67) A graph G contains an odd K_4-subdivision $\iff$ G contains K_4 as odd minor.

For a graph $G = (V, E)$ and $F \subseteq E$, a subset U of V is called *F-stable* if U is a stable set of the graph (V, F). U is called an *F-cover* if U is a vertex cover of (V, F). Let F_1 and F_2 be disjoint subsets of E, and consider the system:

(68.68)
$$0 \leq x_v \leq 1 \quad \text{for } v \in V,$$
$$x(e) \leq 1 \quad \text{for } e \in F_1,$$
$$x(e) \geq 1 \quad \text{for } e \in F_2,$$
$$\sum_{e \in EC \cap F_1} x(e) - \sum_{e \in EC \cap F_2} x(e) \leq |EC \cap F_1| - |EC \cap F_2| - 1$$
$$\text{for each odd circuit } C \text{ with } EC \subseteq F_1 \cup F_2$$

(where $x(e) := x_u + x_v$ for $e = uv \in E$).

Corollary 68.6c. *For any graph $G = (V, E)$ the following are equivalent:*

(68.69) (i) *G contains no odd K_4-subdivision;*
(ii) *for all disjoint $F_1, F_2 \subseteq E$, the convex hull of the incidence vectors of the F_1-stable F_2-covers is determined by (68.68).*

Proof. The implication (i)⇒(ii) follows from Corollary 68.6a. To see (ii)⇒(i), we first show that (ii) is maintained under taking odd minors. Maintenance under deletion of edges or vertices is trivial. To see that it is maintained under contraction of cuts, let $U \subseteq V$ and let $G' = (V', E')$ be the contracted graph. Let F_1' and F_2' be disjoint subsets of E', and let x' satisfy (68.68) for G', F_1', F_2'. Define $x : V \to \mathbb{R}$ as follows, where, for $v \in V$, v' denotes the vertex of G' to which v is contracted:

(68.70) $$x_v := \begin{cases} x'_{v'} & \text{if } v \in U, \\ 1 - x'_{v'} & \text{if } v \in V \setminus U. \end{cases}$$

Moreover, define F_1 and F_2 by:

(68.71) $$F_1 := (F_1' \cap E[U]) \cup (F_2' \cap E[V \setminus U]) \cup \delta(U),$$
$$F_2 := (F_2' \cap E[U]) \cup (F_1' \cap E[V \setminus U]).$$

Then x satisfies (68.68) with respect to G, F_1, F_2. Hence x is a convex combination of integer solutions of (68.68). Applying the construction in reverse to (68.70), we obtain x' as a convex combination of integer solutions of (68.68) with respect to G', F_1', F_2'.

This shows that (68.69)(ii) is maintained under taking odd minors. Moreover, K_4 violates the condition (taking $F_1 := E$, $F_2 := \emptyset$, $x_v := \frac{1}{3}$ for each $v \in V$). This shows sufficiency of the condition. ∎

68.6d. Orientations of discrepancy 1

A directed graph $D = (V, A)$ is said to have *discrepancy k* if for each (undirected) circuit, the number of forward arcs differs by at most k from the number of backward arcs.

The proof of Gerards [1989a] of the strong t-perfection of odd-K_4-free graphs (Theorem 68.3) is by showing that each such graph can be decomposed into graphs that have an orientation of discrepancy 1, using a characterization of Gerards [1994] of orientability of discrepancy 1 and a decomposition theorem of Gerards, Lovász, Schrijver, Seymour, Shih, and Truemper [1993] (cf. Gerards [1990]). As the graphs having an orientation of discrepancy 1 can be shown to be strongly t-perfect with minimum-cost flow techniques (see Theorem 68.7 below), and as the composition maintains total dual integrality of (68.1), the required result follows.

It is not difficult to show that the underlying undirected graph of any digraph of discrepancy 1, contains no odd K_4-subdivision. So, by Theorem 68.3, any undirected graph having an orientation of discrepancy 1, is strongly t-perfect. Gerards gave a direct proof of the strong t-perfection of such graphs, based on the following minimum-cost circulation argument:

Lemma 68.7α. *Let $D = (V, A)$ be a directed graph and let $b : A \to \mathbb{Z}_+$. Then the following system is totally dual integral:*

(68.72)   (i) $x_v \geq 0$ for $v \in V$,
              (ii) $x(VC) \leq b(AC)$ for each directed circuit C in D.

Proof. Choose $w : V \to \mathbb{Z}_+$. We must show that the dual of maximizing $w^\mathsf{T} x$ over (68.72) has an integer optimum solution.

Make another directed graph $\widetilde{D} = (\widetilde{V}, \widetilde{A})$ as follows. For each vertex v of D, make two vertices v', v'' and an arc (v', v''), and for each arc (u, v) of D, make an arc (u'', v'). This defines $\widetilde{D}$.

Define $g, f : \widetilde{A} \to \mathbb{Z}_+$ by:

(68.73) $g(v', v'') := w(v)$ and $f(v', v'') := 0$ for $v \in V$,
             $g(u'', v') := 0$ and $f(u'', v') := b(u, v)$ for $(u, v) \in A$.

Then the maximum of $w^\mathsf{T} x$ over (68.72) is equal to the maximum of $g^\mathsf{T} z$ where $z : \widetilde{A} \to \mathbb{R}_+$ satisfies

(68.74) $z(A\widetilde{C}) \leq f(A\widetilde{C})$ for each directed circuit $\widetilde{C}$ in $\widetilde{D}$.

So if we consider $f - z$ as length function on $\widetilde{A}$, then (68.74) says that each directed circuit in $\widetilde{D}$ has nonnegative length. Hence, by Theorem 8.2, the maximum is equal to the maximum of $g^\mathsf{T} z$ over $z : \widetilde{A} \to \mathbb{R}_+$ for which there exists a $p : \widetilde{V} \to \mathbb{R}$ such that

(68.75) $z(\tilde{a}) + p(t) - p(s) \leq f(\tilde{a})$ for each $\tilde{a} = (s, t) \in \widetilde{A}$.

The latter system has a totally unimodular constraint matrix, and hence the LP has integer optimum primal and dual solutions. The dual asks for the minimum of $y^\mathsf{T} f$ where $y : \widetilde{A} \to \mathbb{Z}_+$ satisfies

(68.76) $y(\tilde{a}) \geq g(\tilde{a})$ for each $\tilde{a} \in \widetilde{A}$,
             $y(\delta^{\text{in}}(\tilde{v})) = y(\delta^{\text{out}}(\tilde{v}))$ for each $\tilde{v} \in \widetilde{V}$.

So y is a circulation in $\widetilde{D}$. Hence y is a nonnegative integer combination of incidence vectors of directed circuits $\widetilde{C}$ in $\widetilde{D}$:

(68.77) $y = \sum_{\widetilde{C}} \lambda_{\widetilde{C}} \chi^{A\widetilde{C}}$.

For each directed circuit $\widetilde{C}$ in $\widetilde{D}$, let C denote the corresponding directed circuit in D (obtained by contracting all arcs (v', v'') occurring in $\widetilde{C}$). Then

(68.78) $y^\mathsf{T} f = \sum_{\widetilde{C}} \lambda_{\widetilde{C}} (\chi^{A\widetilde{C}})^\mathsf{T} f = \sum_{\widetilde{C}} \lambda_{\widetilde{C}} f(A\widetilde{C}) = \sum_{\widetilde{C}} \lambda_{\widetilde{C}} b(AC)$

and

(68.79) $$\sum_{\tilde{C}} \lambda_{\tilde{C}} \chi^{VC} \geq w.$$

Hence we have obtained an integer dual solution for the problem of maximizing $w^T x$ over (68.72). ∎

This lemma implies:

Theorem 68.7. *Let $G = (V, E)$ be an undirected graph having an orientation D of discrepancy 1. Then G is strongly t-perfect.*

Proof. Let $D' = (V, A')$ be the digraph obtained from D by adding a reverse arc (v, u) for each arc (u, v) of D, defining $b(u, v) := 1$ and $b(v, u) := 0$. Then the total dual integrality of (68.1) follows directly from the total dual integrality of (68.72). Note that each directed circuit C' in D' gives an undirected circuit C in D, with $b(AC')$ equal to the number of forward arcs in C. As D has discrepancy 1, $\lfloor \frac{1}{2}|VC|\rfloor$ is equal to the minimum value of $b(AC')$ and $b(AC'^{-1})$. ∎

This immediately implies the strong t-perfection of *almost bipartite* graphs — graphs having a vertex v with $G - v$ bipartite, since they have an orientation of discrepancy 1, as one easily checks.

68.6e. Colourings and odd K_4-subdivisions

Zang [1998] and Thomassen [2001] showed that any graph G without totally odd K_4-subdivision satisfies $\chi(G) \leq 3$.[25] We may interpret this in terms of the integer decomposition and rounding properties. Consider the antiblocking polytope Q of the stable set polytope of a graph $G = (V, E)$:

(68.80) $x_v \geq 0$ for each $v \in V$,
$x(S) \leq 1$ for each stable set S.

If G is t-perfect, the vertices of Q are: the origin, the unit base vectors, the incidence vectors of the edges, and the vectors $\chi^{VC}/\lfloor \frac{1}{2}|VC|\rfloor$ where C is an odd circuit. (This follows from the definition of t-perfection with antiblocking polyhedra theory.) Hence the fractional colouring number $\chi^*(G)$ of G, which is equal to the maximum of $\mathbf{1}^T x$ over (68.80) (cf. Section 64.8), is equal to

(68.81) $\max\{2, \max\{\dfrac{|VC|}{\lfloor \frac{1}{2}|VC|\rfloor} \mid C \text{ odd circuit}\}\}$

(assuming $E \neq \emptyset$). For nonbipartite graphs, this value is equal to 3. So for graphs G without totally odd K_4-subdivision, the colouring number $\chi(G)$ is equal to the round-up $\lceil \chi^*(G) \rceil$ of the fractional colouring number.

---
[25] This was conjectured by Toft [1975], and extends results of Hadwiger [1943] that a 4-chromatic graph contains a K_4-subdivision, of Catlin [1979] that it contains an odd K_4-subdivision, and of Gerards and Shepherd [1998] that it contains a bad K_4-subdivision. Zeidl [1958] showed that any vertex of a minimally 4-chromatic graph lies in a subdivided K_4 that contains an odd circuit. Other partial and related results were found by Krusensjterna-Hafstrøm and Toft [1980], Thomassen and Toft [1981], and Jensen and Shepherd [1995].

A.M.H. Gerards (personal communication 2001) showed that system (68.80) has the integer rounding property if G has no odd K_4-subdivision. It implies that the corresponding stable set polytope has the integer decomposition property. This is equivalent to:

(68.82) $\quad \chi_w(G) = \lceil \chi_w^*(G) \rceil$

for each odd-K_4-free graph G and each $w : VG \to \mathbb{Z}_+$.

This does not hold for any t-perfect graph: M. Laurent and P.D. Seymour showed in 1994 that the complement of the line graph of a prism (complement of C_6) is t-perfect, but is not 3-colourable; hence its stable set polytope does not have the integer decomposition property.

68.6f. Homomorphisms

Let G and H be simple graphs. A *homomorphism* $G \to H$ is a function $\phi : VG \to VH$ such that if $uv \in EG$, then $\phi(u)\phi(v) \in EH$ (in particular, $\phi(u) \neq \phi(v)$). Obviously, if there exists a homomorphism $G \to H$, then $\chi(G) \leq \chi(H)$.

For any k, let $K_4^{(k)}$ be the graph obtained from K_4 by replacing each edge by a path of length k. Then one may check that for odd k there is no homomorphism $K_4^{(k)} \to C_{2k+1}$.

Catlin [1985] showed that this is essentially the only counterexample: if G is a connected graph of maximum degree 3 and $k \in \mathbb{Z}_+$, such that any two vertices of G of degree 3 have distance at least k, and such that there is no homomorphism $G \to C_{2k+1}$, then k is odd and $G = K_4^{(k)}$. (This extends Brooks' theorem (Theorem 64.3) for $k = 1$.)

Gerards [1988] extended this to: if a nonbipartite graph G has no odd minor equal to K_4 or to the graph obtained from the triangle by adding for each edge a new vertex adjacent to the ends of the edge, then there is a homomorphism $G \to C_t$, where t is the shortest length of an odd circuit of G. Further results are given by Catlin [1988].

68.6g. Further notes

Sbihi and Uhry [1984] call a graph $G = (V, E)$ *h-perfect*[26] if the stable set polytope is determined by

(68.83) $\quad$ (i) $\quad x_v \geq 0$ $\quad\quad\quad\quad\quad\quad$ for $v \in V$,
$\quad\quad\quad\quad$ (ii) $\quad x(C) \leq 1$ $\quad\quad\quad\quad\quad$ for each clique C,
$\quad\quad\quad\quad$ (iii) $\quad x(VC) \leq \lfloor \frac{1}{2}|VC| \rfloor$ $\quad$ for each odd circuit C.

So perfect graphs and t-perfect graphs are h-perfect. Sbihi and Uhry showed that substituting bipartite graphs for edges of a series-parallel graph preserves h-perfection.

The t-perfection of line graphs, and classes of graphs that are h-perfect but not t-perfect, were studied by Cao and Nemhauser [1998]. Gerards [1990] gave a survey on signed graphs without odd K_4-subdivision.

---

[26] h stands for 'hole' (English for 'trou').

Chapter 69

Claw-free graphs

Claw-free graphs are graphs not having $K_{1,3}$ as induced subgraph. We show the result of Minty and Sbihi that a maximum-size stable set in a claw-free graph can be found in strongly polynomial time, and the extension of Minty to the weighted case.

69.1. Introduction

A graph $G = (V, E)$ is called *claw-free* if no induced subgraph of G is isomorphic to $K_{1,3}$. Minty [1980] and Sbihi [1980] showed that a maximum-size stable set in a claw-free graph can be found in polynomial time. Since the line graph of any graph is claw-free, this generalizes Edmonds' polynomial-time algorithm for finding a maximum-size matching in a graph.

Sbihi's algorithm is an extension of Edmonds' blossom shrinking technique, while Minty gave a reduction to the maximum-size matching problem. Minty [1980] also indicated that his algorithm can be extended to the weighted case by reduction to Edmonds' weighted matching algorithm. The final argument for this was given by Nakamura and Tamura [2001].

In Section 69.2, we describe Minty's method for finding a maximum-size stable set in claw-free graphs, and in Section 69.3 we describe the extension to the weighted case.

69.2. Maximum-size stable set in a claw-free graph

An important property of claw-free graphs is that any vertex has at most two neighbours in any stable set. This enables us to augment stable sets by S-augmenting paths, which we define now.

Let $G = (V, E)$ be a graph and let S be a stable set in G. A walk $P = (v_0, v_1, \ldots, v_k)$ (given by its vertex-sequence) is called S-*alternating* if precisely one of v_{i-1}, v_i belongs to S, for each $i = 1, \ldots, k$. It is an S-*augmenting path* if moreover P is a path, $v_0, v_k \notin S$, and $(S \setminus \{v_1, v_3, \ldots, v_{k-1}\}) \cup \{v_0, v_2, \ldots, v_k\}$ is stable. This implies that (if $k \geq 2$) each of v_0 and v_k has precisely one neighbour in S, and each of $v_2, v_4, \ldots, v_{k-2}$ precisely two.

Section 69.2. Maximum-size stable set in a claw-free graph

It is easy to see that if G is claw-free, then there is a stable set larger than S if and only if there exists an S-augmenting path. Indeed, sufficiency follows from the definition of S-augmenting path. To see necessity, let S' be a stable set larger than S. Then the subgraph of G induced by $S \triangle S'$ has a component K with more vertices in S' than in S. Since G is claw-free, this subgraph has maximum degree 2, and hence K forms an S-augmenting path.

So in order to find a maximum-size stable set, it suffices to have a polynomial-time algorithm to find for given S, an S-augmenting path, if any. For this, it suffices to describe a polynomial-time algorithm to find an S-augmenting $a - b$ path for prescribed $a, b \in V \setminus S$ (if any). Varying over all $a, b \in V \setminus S$, we find an S-augmenting path (if any).

Therefore, from now on we fix $a, b \in V \setminus S$. Then we can assume:

(69.1) $a \neq b$; a and b have degree 1, each with neighbour in S, say s_a and s_b; $s_a \neq s_b$; each $v \in V \setminus S$ with $v \neq a, b$ has precisely two neighbours in S; for each $s \in S$ with $s \neq s_a, s_b$ there are at least two vertices in S at distance two from s; G is connected.

Indeed, otherwise finding an S-augmenting path is trivial, or it does not exist; moreover, we can delete all neighbours of a or b distinct from s_a or s_b, and all vertices in $S \setminus \{s_a, s_b\}$ that have less than two vertices in S at distance two.

The assumptions (69.1) imply that any S-augmenting path connects a and b. Consider an S-alternating path

(69.2) $P = (v_0, s_1, v_1, \ldots, s_k, v_k)$

(given by its vertex-sequence), with $v_0 = a$ and $v_k = b$. So $s_1 = s_a$ and $s_k = s_b$. Then (under the assumptions (69.1)):

Lemma 69.1α. *P is S-augmenting if and only if v_{i-1} and v_i are nonadjacent for each $i = 2, \ldots, k - 1$.*

Proof. Necessity being trivial, we show sufficiency. It suffices to show that $(S \setminus \{s_1, \ldots, s_k\}) \cup \{v_0, \ldots, v_k\}$ is a stable set. Any two vertices in S are nonadjacent. All neighbours in S of any v_i are among $s_1, \ldots, s_k$. Finally, suppose that any v_i, v_j are adjacent, with $i < j$. Then $j \geq i + 2$, since v_i and v_{i+1} are nonadjacent by the condition. But then v_i is adjacent to the three pairwise nonadjacent vertices s_i, s_{i+1}, and v_j. This contradicts the claw-freeness of G. ∎

We next prove a basic lemma of Minty [1980]. Define, for $u, v \in V \setminus S$:

(69.3) $u \sim v \iff N(u) \cap S = N(v) \cap S$.

Clearly, $\sim$ is an equivalence relation. We call any equivalence class a *similarity class*, and if $u \sim v$ we say that u and v are *similar*. So for each $s \in S$, $N(s)$ is a union of similarity classes.

We call a vertex $s \in S$ *splittable* if $N(s)$ can be partitioned into two classes X, Y such that

(69.4) $\quad uv \in E \iff u, v \in X$ or $u, v \in Y$

for all $u, v \in N(s)$ with $u \not\sim v$. If s is splittable, we call X and Y the *classes* of s. Define

(69.5) $\quad S' := \{s \in S \mid s \text{ is splittable}\}$ and $S'' := S \setminus S'$.

Then $s_a, s_b \in S'$, since $N(s_a) \setminus \{a\}$ is a clique, as a has no neighbours in $N(s_a)$ (by assumption (69.1)) and as G is claw-free — similarly for s_b. Moreover:

Lemma 69.1β. *Each vertex $s \in S$ having at least three vertices in S at distance two, belongs to S'.*

Proof. Since $s_a, s_b \in S'$, we may assume that $s \neq s_a, s_b$. Let $G' = (N(s), F)$ be the subgraph of G with

(69.6) $\quad F := \{uv \in E \mid u, v \in N(s), u \not\sim v\}$.

Then

(69.7) $\quad$ each component of G' induces a clique of G.

Suppose not. Let $P = (v_0, v_1, \ldots, v_k)$ be a shortest path in G' with $v_0 v_k \notin E$. If $k = 2$, then $v_0 \not\sim v_1 \not\sim v_2$, and hence v_1 has a neighbour $t \in S$ which is not a neighbour of v_0 or v_2. But then v_1 is adjacent to the pairwise nonadjacent t, v_0, v_2, contradicting the claw-freeness of G.

If $k = 3$, then as P is shortest, $v_0 v_2, v_1 v_3 \in E \setminus F$. So $v_0 \sim v_2$ and $v_1 \sim v_3$. Choose a vertex p with $p \not\sim v_0$ and $p \not\sim v_1$. (This is possible since $N(s)$ contains at least three similarity classes.) Then p has a neighbour t in S which is not a neighbour of any of v_0, v_1, v_2, v_3. Since $N(s)$ contains no three pairwise nonadjacent vertices (as G is claw-free), we know that $v_0 p \in E$ or $v_3 p \in E$. By symmetry, we can assume that $v_0 p \in E$, and hence $v_0 p \in F$. Then, by the minimality of k, we know successively that $v_1 p \in F$, $v_2 p \in F$, and $v_3 p \in F$. But then $v_0 p$ and $p v_3$ are in F, and hence, by the minimality of k, $v_0 v_3 \in E$.

If $k \geq 4$, then $v_0 v_2, v_0 v_3 \in E$, hence (since $v_2 \not\sim v_3$) $v_0 v_2 \in F$ or $v_0 v_3 \in F$, contradicting the minimality of k. This proves (69.7).

Since G is claw-free, G' has at least one component, X say, that intersects at least two of the similarity classes. If G' has at most two components, or if X contains all but at most one similarity class, we are done, taking $Y := N(s) \setminus X$. If G' has at least three components and $N(s) \setminus X$ intersects at least two similarity classes, then G' has two other components Y, Z for which there exist $x \in X$, $y \in Y$, and $z \in Z$ with $x \not\sim y \not\sim z \not\sim x$, as one easily checks[27]. But then s is adjacent to the three pairwise nonadjacent vertices x, y, z, contradicting the claw-freeness of G. ∎

---

[27] Let $y \in N(s) \setminus X$ be such that there exist $x', x'' \in X$ with $y \not\sim x' \not\sim x'' \not\sim y$. Let Y be the component of G' containing y. Let Z be a third component, if possible containing

Section 69.2. Maximum-size stable set in a claw-free graph

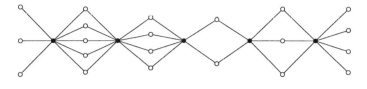

• vertex in S ○ vertex not in S

Figure 69.1
A typical bone

Now consider the subgraph

(69.8) $(V \setminus S', \delta(S''))$

of G. It is a bipartite graph, with colour classes S'' and $V \setminus S$. We call each component of this graph a *bone*. (A typical bone is depicted in Figure 69.1.) By Lemma 69.1β, each $s \in S''$ has at most two vertices in S at distance two. Hence any bone B consists of a series of vertices $s_1, \ldots, s_k$ in S'', together with disjoint nonempty sets $V_0, V_1, \ldots, V_k$ of vertices such that s_i is incident with each vertex in $V_{i-1} \cup V_i$ for each $i = 1, \ldots, k$. Moreover, B has two neighbours in S', say s and t, where s is adjacent to all vertices in V_0 and t is adjacent to all vertices in V_k. (It might be that $s = t$.) The degenerate case is that $k = 0$, where B is a singleton vertex in $V \setminus S$ with two neighbours in S'.

The relevance of bones is that if we leave out from any S-augmenting path the vertices that belong to S', we are left with a number of subpaths, each of which is an S''-augmenting path contained in some bone. So in constructing or analyzing an S-augmenting path, we can decompose it into S''-augmenting paths, glued together at vertices in S'. Here the classes of the vertices in S' come in, since the ends of the two subpaths glued together at $s \in S'$ should belong to different classes of s. This motivates the following graph H (called the *Edmonds graph* by Minty [1980])[28].

H has vertex set

(69.9) $\{(s, X) \mid s \in S', X \text{ class of } s\} \setminus \{(s_a, \{a\}), (s_b, \{b\})\}$

and the following edges:

(69.10)   (i) $\{(s, X), (s, Y)\}$ for $s \in S' \setminus \{s_a, s_b\}$ and X, Y the classes of s;

---

a vertex nonsimilar to y. Then, if Z contains a vertex $z \not\sim y$, we can take for x one of x', x''. If Z contains no such vertex, let $z \in Z$. Then Y contains a vertex $y' \not\sim z$. As $z \not\sim x'$ and $z \not\sim x''$, we are done again.

[28] We note that in constructing H we could restrict S' to those vertices in S that have at least 3 vertices in S at distance two, together with s_a and s_b. However, for the extension to the weighted case, we need S' as defined above (namely, by being splittable).

(ii) $\{(s,X),(t,Y)\}$ for vertices $(s,X),(t,Y)$ of H such that there exists an S''-augmenting $X-Y$ path P.

So the path P is contained in the bone B containing x and y for some $x \in X$ and some $y \in Y$. Its existence can be checked as follows. Let $V_0, V_1, \ldots, V_k$ be as above. Make the digraph D on $V_0 \cup V_1 \cup \ldots \cup V_k$ with an arc from $u \in V_{i-1}$ to $v \in V_i$ if $uv \notin E$ (for $i = 1, \ldots, k$). Then a directed $X-Y$ path in D gives a path P as required, and conversely.

Let M be the matching of edges in (69.10)(i). So M covers all vertices of H, except the vertices $(s_a, N(s_a) \setminus \{a\})$ and $(s_b, N(s_b) \setminus \{b\})$. Then (under the assumptions (69.1)):

Lemma 69.1γ. *G has an S-augmenting path $\iff$ H has an M-augmenting path. We can obtain one from the other in polynomial time.*

Proof. Let $P = (v_0, s_1, v_1, \ldots, s_k, v_k)$ be an S-augmenting path in G, with $v_0 = a$ and $v_k = b$. Let $s_{i_1}, \ldots, s_{i_t}$ be those vertices in P that belong to S' (in order). So $i_1 = 1$ and $i_t = k$. For $j = 1, \ldots, t$, let X_j and Y_j be the classes of s_{i_j} that contain v_{i_j-1} and v_{i_j}, respectively. Then $X_j \neq Y_j$, since v_{i_j-1} and v_{i_j} are nonsimilar and nonadjacent. Moreover, the subpath of P between any two s_{i_j} and $s_{i_{j+1}}$ forms an S''-augmenting $Y_j - X_{j+1}$ path. Hence

(69.11) $\quad ((s_{i_1}, Y_1), (s_{i_2}, X_2), (s_{i_2}, Y_2), \ldots, (s_{i_{t-1}}, X_{t-1}), (s_{i_{t-1}}, Y_{t-1}),$
$(s_{i_t}, X_t))$

is an M-augmenting path in H.

We can reverse this construction. Indeed, any M-augmenting path Q yields an S-alternating $a-b$ walk P in G, by inserting appropriate S''-augmenting paths.

In fact, P is a path. For suppose that P traverses some vertex u of G more than once. Then u belongs to two of the inserted paths. Necessarily, they belong to the same bone B. Hence B has a neighbour in S' that is traversed more than once. But then Q traverses some matching edge more than once, a contradiction.

So P is a path. Moreover, any two vertices at distance two in P are nonadjacent, by construction of P. So P is S-augmenting, by Lemma 69.1α.
∎

Concluding, we have obtained the result of Minty [1980] and Sbihi [1980]:

Theorem 69.1. *A maximum-size stable set in a claw-free graph can be found in polynomial time.*

Proof. From Lemma 69.1γ, since finding an M-augmenting path in H is equivalent to finding a perfect matching in M. The latter problem is polynomial-time solvable by Corollary 24.4a.
∎

69.3. Maximum-weight stable set in a claw-free graph

There is an obvious way of extending the above construction to the weighted case, but there is a catch in it. The idea was noted by Minty [1980], and finalized by Nakamura and Tamura [2001].

Let $G = (V, E)$ be a graph and let $w : V \to \mathbb{R}_+$ be a weight function. Call a stable set S *extreme* if it has maximum weight among all stable sets of size $|S|$. It suffices to describe an algorithm to derive from any extreme stable set S, an extreme set of size $|S|+1$, if any (since then we can start with $S := \emptyset$, enumerate extreme stable sets of all possible sizes, and choose one of maximum weight among them).

The following observations are basic:

Lemma 69.2α. *Let $G = (V, E)$ be a claw-free graph, let $w : V \to \mathbb{R}_+$, and let S be an extreme stable set. Then:*

(69.12)   (i) *each S-alternating chordless circuit satisfies $w(VC \setminus S) \leq w(VC \cap S)$;*
          (ii) *if P is an S-augmenting path maximizing $w(VP \setminus S) - w(VP \cap S)$, then $S \triangle VP$ is an extreme stable set of size $|S|+1$.*

Proof. (i) follows from the fact that $S \triangle VC$ is a stable set of size $|S|$, and hence $w(S) \geq w(S \triangle VC) = w(S) + w(VC \setminus S) - w(VC \cap S)$.

(ii) can be seen as follows. Let $\widetilde{S}$ be an extreme stable set of size $|S|+1$. The subgraph induced by $S \triangle \widetilde{S}$ has a component K with $|K \cap \widetilde{S}| > |K \cap S|$. Since G is claw-free, K has maximum degree at most 2. So K is an S-augmenting path, and hence $|K \cap \widetilde{S}| = |K \cap S| + 1$. Let $L := (S \triangle \widetilde{S}) \setminus K$. Then $S \triangle L$ and $\widetilde{S} \triangle L$ are stable sets of size $|S|$ and $|S|+1$ respectively. Since S is extreme, $w(L \cap \widetilde{S}) \leq w(L \cap S)$. Hence $w(\widetilde{S} \triangle L) \geq w(\widetilde{S})$. So $\widetilde{S} \triangle L$ is extreme again. Hence we can assume that $L = \emptyset$. Then, since K is an S-augmenting path:

(69.13) $w(S \triangle VP) = w(S) + w(VP \setminus S) - w(VP \cap S)$
            $\geq w(S) + w(K \setminus S) - w(K \cap S) = w(\widetilde{S})$.

So $S \triangle VP$ is extreme. ∎

Statement (ii) of Lemma 69.2α implies that, to find an extreme stable set of size $|S|+1$, it suffices to find an S-augmenting path P maximizing $w(VP \setminus S) - w(VP \cap S)$. By enumerating over all pairs $a, b \in V \setminus S$, it suffices to find, for each fixed $a, b \in V \setminus S$, an S-augmenting $a-b$ path P maximizing $w(VP \setminus S) - w(VP \cap S)$ (if any). Then we can make again the assumptions (69.1), and construct the graph H. Define a weight function ω on the edges of H (following the items in (69.10)) as follows:

(69.14)   (i) $\omega(\{(s, X), (s, Y)\}) := w(s)$,
          (ii) $\omega(\{(s, X), (t, Y)\}) :=$ the maximum of $w(VP \setminus S'') - w(VP \cap S'')$ over all S''-augmenting $X - Y$ paths P.

1214   Chapter 69. Claw-free graphs

The maximum in (69.14)(ii) can be found in strongly polynomial time, since it amounts to finding a longest directed $X - Y$ path in the acyclic digraph D described just after (69.10).

Now a maximum-weight perfect matching in H need not yield a maximum-weight stable set in G, as was pointed out by Nakamura and Tamura [2001], since there might exist M-alternating circuits that increase the weight of M, while they do not correspond to a chordless S-alternating circuit. However, this can be avoided by preprocessing as follows.

We can assume that for each $v \in V \setminus S$ with $v \neq a, b$:

(69.15) (i) there exist s, t, x, y such that (x, s, v, t, y) is a chordless S-alternating path and such that $N(x) \cap N(y) \cap S = \emptyset$;
(ii) there exist $s, t \in S'$ and classes X of s and Y of t such that there exists an S''-augmenting $X - Y$ path and such that each S''-augmenting $X - Y$ path P attaining the maximum in (69.14)(ii), traverses v.

Otherwise v is on no maximum-weight S-augmenting path, and hence we can delete v. The conditions (69.15) can be tested in strongly polynomial time (for (ii) using digraph D). Hence the deletions take strongly polynomial time only.

Fix for each edge e of H in (69.14)(ii), a path P_e attaining the maximum. Then we can transform any M-alternating path or circuit to an S-alternating walk or closed walk, by replacing each such edge e by P_e. We call this the *corresponding* walk or closed walk in G.

Lemma 69.2β. *Under the assumptions* (69.15), *each M-alternating circuit C in H satisfies $w(EC \setminus M) \leq w(EC \cap M)$.*

Proof. Suppose not. Choose C maximizing $w(EC \setminus M) - w(EC \cap M)$. Let Γ be the corresponding S-alternating closed walk in G. Then Γ is not a chordless circuit, since otherwise

(69.16) $w(V\Gamma \setminus S) - w(V\Gamma \cap S) = w(EC \setminus M) - w(EC \cap M) > 0$,

which contradicts (i) of Lemma 69.2α.

Since each P_e is simple and chordless, it follows that $EC \setminus M$ contains distinct edges e, f for which there exist $u \in VP_e$ and $v \in VP_f$ with $u = v$ or $uv \in E$. This implies that C has length 4, and that e and f are the only edges in $EC \setminus M$. So P_e and P_f are in the same bone B. Let s and t be the neighbours of B in S'. Let s have classes Y, Z and t have classes W, X such that P_e connects Y and W and P_f connects Z and X. Write

(69.17) $P_e = (u_0, s_1, u_1, \ldots, s_k, u_k)$ and $P_f = (v_0, s_1, v_1, \ldots, s_k, v_k)$

for some $k \geq 0$ and $s_1, \ldots, s_k \in S''$, where $u_0 \in Y$, $u_k \in W$, $v_0 \in Z$, $v_k \in X$. We define $s_0 := s$ and $s_{k+1} := t$. Now

(69.18)   for each $i = 1, \ldots, k$: $u_{i-1}v_i \in E$ or $v_{i-1}u_i \in E$.

Otherwise, we can 'switch' P_c and P_f at s_i to obtain the S''-augmenting paths

(69.19) $\quad Q := (u_0, s_1, \ldots, u_{i-1}, s_i, v_i, \ldots, s_k, v_k)$ and
$\quad\quad\quad R := (v_0, s_1, \ldots, v_{i-1}, s_i, u_i, \ldots, s_k, u_k).$

Hence H has edges $\{(s,Y),(t,X)\}$ and $\{(s,Z),(t,W)\}$, and

(69.20) $\quad w(\{(s,Y),(t,X)\}) + w(\{(s,Z),(t,W)\})$
$\quad\quad\quad \geq w(\{(s,Y),(t,W)\}) + w(\{(s,Z),(t,X)\}).$

By the choice of C, we have equality, and hence the paths Q and R attain the corresponding maxima in (69.14)(ii). It implies, by assumption (69.15)(ii), that u_{i-1}, v_{i-1}, u_i, and v_i are the only neighbours of s_i. Since none of u_{i-1}, v_{i-1} are adjacent to any of u_i, v_i, we have that s_i is splittable, that is, $s_i \in S'$, a contradiction. This proves (69.18).

Next

(69.21) $\quad u_0 v_0 \notin E$ and $u_k v_k \notin E$.

For suppose that (say) $u_0 v_0 \in E$. By (69.15)(i), there exist $x, y \in V \setminus S$ such that (x, s, u_0, s_1, y) is a chordless path and such that $N(x) \cap N(y) \cap S = \emptyset$. As x is nonadjacent to u_0, and as $u_0 \in X$, we have $x \in Y$, and so (as $v_0 \in Y$) $xv_0 \in E$.

If $k = 0$, we have similarly $yv_0 \in E$. Then v_0 is adjacent to the pairwise nonadjacent x, u_0, y, a contradiction.

So $k \geq 1$. Then $y \sim u_1$ and $N(y) \cap S = \{s_1, s_2\}$. So $xs_1, xs_2 \notin E$ (since $N(x) \cap N(y) \cap S = \emptyset$). This implies $xu_1 \notin E$, since otherwise u_1 is adjacent to the pairwise nonadjacent s_1, s_2, x. Hence $v_0 u_1 \notin E$, since otherwise v_0 is adjacent to the pairwise nonadjacent x, u_0, u_1. By symmetry, also $u_0 v_1 \notin E$. This contradicts (69.18), and hence proves (69.21).

Moreover,

(69.22) $\quad$ there is an i with $0 \leq i \leq k$ and $u_i v_i \in E$,

as otherwise each circuit $(s_i, u_i, s_{i+1}, v_i, s_i)$ is S-alternating and chordless, which implies $w(u_i) + w(v_i) - w(s_i) - w(s_{i+1}) \leq 0$ by Lemma 69.2α. This gives the contradiction

(69.23) $\quad 0 < w(EC \setminus M) - w(EC \cap M)$
$\quad\quad\quad = w(VP_e \setminus S'') - w(VP_e \cap S'') + w(VP_f \setminus S'') - w(VP_f \cap S'')$
$\quad\quad\quad - w(s) - w(t) = \sum_{i=0}^{k}(w(u_i) + w(v_i) - w(s_i) - w(s_{i+1})) \leq 0,$

proving (69.22).

Now let i be the smallest index with $u_i v_i \in E$. By (69.21), we know $1 \leq i \leq k-1$. By (69.18) and by symmetry we can assume that $v_i u_{i+1} \in E$. Since s_i is adjacent to u_{i-1}, v_{i-1}, and v_i, and since $u_{i-1} v_{i-1} \notin E$ and $v_{i-1} v_i \notin E$, we know $u_{i-1} v_i \in E$. Then v_i is adjacent to the pairwise nonadjacent u_{i-1}, u_i, and u_{i+1}, a contradiction. ∎

Now find a maximum-weight perfect matching N in H, with the maximum-weight perfect matching algorithm (Chapter 26). By Lemma 69.2β, we can assume that $N = M\triangle EQ$ for some M-augmenting path Q in H (since if $N\triangle M$ contains a circuit C, then $N\triangle EC$ again is a maximum-weight perfect matching in H). Then Q maximizes $\omega(EQ \setminus M) - \omega(EQ \cap M)$ over all M-augmenting paths. Let P be the corresponding path in G. Then P is an S-augmenting path in G maximizing $w(VP \setminus S) - w(VP \cap S)$, as required.

We conclude:

Theorem 69.2. *A maximum-weight stable set in a claw-free graph can be found in strongly polynomial time.*

Proof. See above. ∎

69.4. Further results and notes

69.4a. On the stable set polytope of a claw-free graph

The polynomial-time solvability of the maximum-weight stable set problem for claw-free graphs implies that the optimization problem over the stable set polytope $P_{\text{stable set}}(G)$ of a claw-free graph $G = (V, E)$ is polynomial-time solvable. Hence also the separation problem is polynomial-time solvable (with the ellipsoid method (Theorem 5.10)). It implies (cf. Theorem 5.11) that, given a vector $x \in \mathbb{Q}^V$, one can decide in strongly polynomial time if x belongs to $P_{\text{stable set}}(G)$, and if not, find a facet-inducing inequality violated by x.

So in this respect, the stable set polytope of a claw-free graph is under control. However, no explicit description is known of a system that determines $P_{\text{stable set}}(G)$. As we saw in Section 25.2, such a description is known for the special case where G is the line graph of some graph H — that is, for the matching polytope of H. In this special case, each facet can be described by an inequality with coefficients in $\{0, 1\}$.

The latter fact does not generalize to claw-free graphs. Giles and Trotter [1981] showed that for each $k \in \mathbb{Z}_+$ there exists a claw-free graph such that its stable set polytope has a facet that is described by a linear inequality with coefficients k and $k+1$. (This refutes a conjecture of Sbihi [1978].)

Galluccio and Sassano [1997] characterized those facets of the stable set polytope of a claw-free graph that can be described by an inequality with all coefficients in $\{0, 1\}$ (the *rank facets*).

More on facets of the stable set polytope of special classes of claw-free graphs can be found in Ben Rebea [1981] and Oriolo [2002] (for graphs such that for each vertex v, the graph induced by $N(v)$ is the complement of a bipartite graph) and Pulleyblank and Shepherd [1993] (for claw-free graphs such that no vertex has three pairwise nonadjacent vertices at distance two).

69.4b. Further notes

Minty [1980] observed that finding a maximum-size stable set in a graph without induced $K_{1,4}$ is NP-complete. This follows from the fact that the 3-dimensional assignment problem can be reduced to it (its intersection graph has no induced $K_{1,4}$).

Poljak [1974] showed that finding a maximum-size stable set in a triangle-free graph is NP-complete. It implies that finding a maximum-size clique in a claw-free graph is NP-complete.

Shepherd [1995] characterized the stable set polytope of *near-bipartite graphs*, that is, graphs with $G - N(v)$ bipartite for each $v \in VG$. They include the complements of line graphs, and the complement of any near-bipartite graph is claw-free.

Ben Rebea [1981] showed that each connected claw-free graph G with $\alpha(G) \geq 3$ not containing an induced C_5, contains no odd antihole. This was extended by Fouquet [1993] who showed that each connected claw-free graph G with $\alpha(G) \geq 4$ contains no odd antihole with at least 7 vertices.

Lovász and Plummer [1986] gave a variant of Minty's reduction of the maximum-size stable set problem in claw-free graphs to the maximum-size matching problem.

Beineke [1970] (for simple graphs), N. Robertson (unpublished), Hemminger [1971] (abstract only), and Bermond and Meyer [1973] characterized line graphs by means of forbidden induced subgraphs (six graphs next to $K_{1,3}$).

The polynomial-time solvability of the weighted stable set problem for claw-free graphs was extended to claw-free bidirected graphs by Nakamura and Tamura [1998]. A linear-time algorithm for 'triangulated' bidirected graphs was given by Nakamura and Tamura [2000].